OXFORD MATHEMATICAL MONOGRAPHS

Series Editors

J. M. BALL   E. M. FRIEDLANDER   I. G. MACDONALD
L. NIRENBERG   R. PENROSE   J. T. STUART

OXFORD MATHEMATICAL MONOGRAPHS

A. Belleni-Moranti: *Applied semigroups and evolution equations*
A.M. Arthurs: *Complementary variational principles* 2nd edition
M. Rosenblum and J. Rovnyak: *Hardy classes and operator theory*
J.W.P. Hirschfeld: *Finite projective spaces of three dimensions*
A. Pressley and G. Segal: *Loop groups*
D.E. Edmunds and W.D. Evans: *Spectral theory and differential operators*
Wang Jianhua: *The theory of games*
S. Omatu and J.H. Seinfeld: *Distributed parameter systems: theory and applications*
J. Hilgert, K.H. Hofmann, and J.D. Lawson: *Lie groups, convex cones, and semigroups*
S. Dineen: *The Schwarz lemma*
S.K. Donaldson and P.B. Kronheimer: *The geometry of four-manifolds*
D.W. Robinson: *Elliptic operators and Lie groups*
A.G. Werschulz: *The computational complexity of differential and integral equations*
L. Evens: *Cohomology of groups*
G. Effinger and D.R. Hayes: *Additive number theory of polynomials*
J.W.P. Hirschfeld and J.A. Thas: *General Galois geometries*
P.N. Hoffman and J.F. Humphreys: *Projective representations of the symmetric groups*
I. Györi and G. Ladas: *The oscillation theory of delay differential equations*
J. Heinonen, T. Kilpelainen, and O. Martio: *Non-linear potential theory*
B. Amberg, S. Franciosi, and F. de Giovanni: *Products of groups*
M.E. Gurtin: *Thermomechanics of evolving phase boundaries in the plane*
I. Ionescu and M. Sofonea: *Functional and numerical methods in viscoplasticity*
N. Woodhouse: *Geometric quantization* 2nd edition
U. Grenander: *General pattern theory*
J. Faraut and A. Koranyi: *Analysis on symmetric cones*
I.G. Macdonald: *Symmetric functions and Hall polynomials* 2nd edition
B.L.R. Shawyer and B.B. Watson: *Borel's methods of summability*
M. Holschneider: *Wavelets: an analysis tool*
Jacques Thévenaz: *G-algebras and modular representation theory*
Hans-Joachim Baues: *Homotopy type and homology*
P.D.D'Eath: *Black holes: gravitational interactions*
R. Lowen: *Approach spaces: the missing link in the topology–uniformity–metric triad*
Nguyen Dinh Cong: *Topological dynamics of random dynamical systems*
J.W.P. Hirschfeld: *Projective geometries over finite fields* 2nd edition
K. Matsuzaki and M. Taniguchi: *Hyperbolic manifolds and Kleinian groups*
David E. Evans and Yasuyuki Kawahigashi: *Quantum symmetries on operator algebras*
Norbert Klingen: *Arithmetical similarities: prime decomposition and finite group theory*
Isabelle Catto, Claude Le Bris, and Pierre-Louis Lions: *The mathematical theory of thermodynamic limits: Thomas–Fermi type models*
D. McDuff and D. Salamon: *Introduction to symplectic topology* 2nd edition
William M. Goldman: *Complex hyperbolic geometry*
Charles J. Colbourn and Alexander Rosa: *Triple systems*
V. A. Kozlov, V. G. Maz'ya and A. B. Movchan: *Asymptotic analysis of fields in multi-structures*
Gérard A. Maugin: *Nonlinear waves in elastic crystals*
George Dassios and Ralph Kleinman: *Low frequency scattering*
Gerald W. Johnson and Michel L. Lapidus: *The Feynman Integral and Feynman's Operational Calculus*
W. Lay and S. Y. Slavyanov: *Special functions: A unified theory base on singularities*
D. Joyce: *Compact manifolds with special holonomy*

Compact Manifolds with Special Holonomy

Dominic D. Joyce
Lincoln College, Oxford

This book has been printed digitally and produced in a standard specification in order to ensure its continuing availability

OXFORD
UNIVERSITY PRESS

Great Clarendon Street, Oxford OX2 6DP

Oxford University Press is a department of the University of Oxford.
It furthers the University's objective of excellence in research, scholarship,
and education by publishing worldwide in

Oxford New York

Auckland Cape Town Dar es Salaam Hong Kong Karachi
Kuala Lumpur Madrid Melbourne Mexico City Nairobi
New Delhi Shanghai Taipei Toronto
With offices in
Argentina Austria Brazil Chile Czech Republic France Greece
Guatemala Hungary Italy Japan South Korea Poland Portugal
Singapore Switzerland Thailand Turkey Ukraine Vietnam

Oxford is a registered trade mark of Oxford University Press
in the UK and in certain other countries

Published in the United States
by Oxford University Press Inc., New York

© Dominic D. Joyce 2000

The moral rights of the author have been asserted

Database right Oxford University Press (maker)

Reprinted 2008

All rights reserved. No part of this publication may be reproduced,
stored in a retrieval system, or transmitted, in any form or by any means,
without the prior permission in writing of Oxford University Press,
or as expressly permitted by law, or under terms agreed with the appropriate
reprographics rights organization. Enquiries concerning reproduction
outside the scope of the above should be sent to the Rights Department,
Oxford University Press, at the address above

You must not circulate this book in any other binding or cover
And you must impose this same condition on any acquirer

ISBN 978-0-19-850601-0

For my father, David Joyce

PREFACE

This book is, in effect, a marriage of two books: a graduate textbook on Riemannian holonomy groups, Chapters 1–7, and a research monograph on the exceptional holonomy groups G_2 and Spin(7), Chapters 8–15. I hope the marriage is a harmonious one. The broad subject of the book is compact Riemannian manifolds with special holonomy groups, in particular the Ricci-flat holonomy groups SU(m), Sp(m), G_2 and Spin(7).

The book is aimed at mathematicians—researchers and graduate students—working in differential and Riemannian geometry, and also at physicists working in String Theory, who are interested in compact manifolds with holonomy SU(3), G_2 and Spin(7) as ingredients in constructing a theory of Quantum Gravity. However, the book is written strictly from the mathematical point of view.

We place particular emphasis on constructions and existence theorems for compact manifolds with special holonomy. These results generally come down to proving the existence of solutions of nonlinear elliptic partial differential equations on compact manifolds. Because of this, a large part of the book—much of Chapters 5, 8, 9, 11, 13 and 15—is taken up with proofs using a lot of analysis.

These proofs are mostly given in full, and they are difficult, very technical mathematics. Now it seems likely that a large part of the intended readership (in particular, many physicists working in String Theory) may not have the inclination, or the necessary mathematical training, to follow them in detail.

Therefore I have tried to make the book easy to use for readers who wish to understand these analytic results, but not their proofs. The important results, discussion and examples are generally grouped together at the beginning of each chapter, and the proofs come later and can be missed out. At the same time, for the benefit of mathematicians wishing to understand the analysis, I have tried to make the book fairly self-contained in this area.

The book came to be written for two reasons. Firstly, it seemed to me that the subject of compact manifolds with special holonomy (and exceptional holonomy in particular) is now of interest to a large audience, and there is a clear need for a book about it. Secondly, I had just completed a substantial piece of research into new constructions of compact manifolds with exceptional holonomy, and was looking for a suitable form in which to publish it.

The main innovation was to use complex geometry and Calabi conjecture methods to define constructions of compact manifolds with exceptional holonomy which were more elaborate and powerful than those known at the time, and yielded many more examples. The best way to present these ideas coherently appeared to be to write a book about it.

This new research material takes up Chapters 8, 9 and 11–15, much of which is published here for the first time. I believe that these ideas are more-or-less in their final form, and are unlikely to be succeeded by more powerful constructions for some time to come, so they are a suitable subject for a book.

Acknowledgements. I would like to thank Simon Salamon for first introducing me to the exceptional holonomy groups, and for many interesting conversations over the years. I would also like to thank two people who have been rôle models for my mathematical life: Dominic Welsh, my undergraduate tutor, and Simon Donaldson, my supervisor. Finally, I would like to thank my wife, Jayne, and friends at St. Aldate's church, Oxford, for their support, patience, and good humour whilst I was writing this book.

Oxford
January 2000
D.D.J.

CONTENTS

1 Background material ... 1
   1.1 Exterior forms on manifolds ... 1
   1.2 Introduction to analysis ... 4
   1.3 Introduction to elliptic operators ... 7
   1.4 Regularity of solutions of elliptic equations ... 13
   1.5 Existence of solutions of linear elliptic equations ... 16

2 Introduction to connections, curvature and holonomy groups ... 20
   2.1 Bundles, connections and curvature ... 20
   2.2 Vector bundles, connections and holonomy groups ... 25
   2.3 Holonomy groups and principal bundles ... 29
   2.4 Holonomy groups and curvature ... 32
   2.5 Connections on the tangent bundle, and torsion ... 33
   2.6 G-structures and intrinsic torsion ... 38

3 Riemannian holonomy groups ... 42
   3.1 Introduction to Riemannian holonomy groups ... 42
   3.2 Reducible Riemannian manifolds ... 46
   3.3 Riemannian symmetric spaces ... 50
   3.4 The classification of Riemannian holonomy groups ... 55
   3.5 Holonomy groups and de Rham cohomology ... 59
   3.6 Spinors and holonomy groups ... 64
   3.7 Calibrated geometry and holonomy groups ... 68

4 Kähler manifolds ... 71
   4.1 Introduction to complex manifolds ... 72
   4.2 Tensors on complex manifolds ... 75
   4.3 Holomorphic vector bundles ... 77
   4.4 Introduction to Kähler manifolds ... 78
   4.5 Kähler potentials ... 80
   4.6 Curvature of Kähler manifolds ... 81
   4.7 Exterior forms on Kähler manifolds ... 82
   4.8 Complex algebraic varieties ... 86
   4.9 Singular varieties, resolutions, and deformations ... 90
   4.10 Line bundles and divisors ... 94

5 The Calabi conjecture — 98
- 5.1 Reformulating the Calabi conjecture — 99
- 5.2 Overview of the proof of the Calabi conjecture — 101
- 5.3 Calculations at a point — 105
- 5.4 The proof of Theorem C1 — 108
- 5.5 The proof of Theorem C2 — 115
- 5.6 The proof of Theorem C3 — 117
- 5.7 The proof of Theorem C4 — 119
- 5.8 A discussion of the proof — 119

6 Calabi–Yau manifolds — 121
- 6.1 The holonomy groups $SU(m)$ — 122
- 6.2 Compact Ricci-flat Kähler manifolds — 123
- 6.3 Crepant resolutions, small resolutions, and flops — 126
- 6.4 Crepant resolutions of quotient singularities — 128
- 6.5 Complex orbifolds — 132
- 6.6 Crepant resolutions of orbifolds — 136
- 6.7 Complete intersections — 139
- 6.8 Deformations of Calabi–Yau manifolds — 142
- 6.9 String Theory and Mirror Symmetry — 144
- 6.10 Further reading on Calabi–Yau manifolds — 146

7 Hyperkähler manifolds — 148
- 7.1 An introduction to hyperkähler geometry — 149
- 7.2 Hyperkähler ALE spaces — 153
- 7.3 $K3$ surfaces — 155
- 7.4 Higher-dimensional compact hyperkähler manifolds — 162
- 7.5 The other quaternionic geometries — 167
- 7.6 A reading list on quaternionic geometry — 169

8 Asymptotically locally Euclidean metrics with holonomy $SU(m)$ — 172
- 8.1 Introduction to ALE metrics — 173
- 8.2 Ricci-flat ALE Kähler manifolds — 175
- 8.3 Analysis on ALE manifolds — 178
- 8.4 Exterior forms and de Rham cohomology — 182
- 8.5 The Calabi conjecture for ALE manifolds — 186
- 8.6 A priori estimates of ϕ — 188
- 8.7 The proofs of Theorems A1–A4 — 196
- 8.8 The proofs of Theorems 8.2.3 and 8.2.4 — 199
- 8.9 The case $m=2$, and deformations — 201

9 QALE metrics with holonomy SU(m) and Sp(m) — 203
- 9.1 Local product resolutions — 204
- 9.2 Quasi-ALE Kähler metrics — 208
- 9.3 Ricci-flat QALE Kähler manifolds — 210
- 9.4 Kähler potentials on QALE Kähler manifolds — 215
- 9.5 Analysis on QALE Kähler manifolds — 219
- 9.6 The Calabi conjecture for QALE manifolds — 225
- 9.7 A more complicated QALE Calabi conjecture — 228
- 9.8 The proofs of Theorems 9.3.3 and 9.3.4 — 232
- 9.9 Generalized QALE manifolds — 235

10 Introduction to the exceptional holonomy groups — 242
- 10.1 The holonomy group G_2 — 242
- 10.2 The topology of compact G_2-manifolds — 245
- 10.3 Exterior forms on G_2-manifolds — 247
- 10.4 The moduli space of holonomy G_2 metrics — 251
- 10.5 The holonomy group Spin(7) — 254
- 10.6 The topology of compact Spin(7)-manifolds — 259
- 10.7 The moduli space of holonomy Spin(7) metrics — 261
- 10.8 Exceptional holonomy and calibrated geometry — 264
- 10.9 A reading list on the exceptional holonomy groups — 268

11 Construction of compact G_2-manifolds — 270
- 11.1 Resolving G_2-singularities with holonomy SU(2), SU(3) — 271
- 11.2 Quasi-ALE G_2-manifolds — 274
- 11.3 Some notation to describe the orbifolds T^7/Γ — 278
- 11.4 R-data and resolutions of T^7/Γ — 280
- 11.5 G_2-structures on M with small torsion — 284
- 11.6 An existence result for compact G_2-manifolds — 294
- 11.7 The proof of Theorem G1 — 297
- 11.8 The proof of Theorem G2 — 299
- 11.9 Other constructions of compact G_2-manifolds — 303

12 Examples of compact 7-manifolds with holonomy G_2 — 306
- 12.1 Topological invariants of resolutions of T^7/Γ — 306
- 12.2 A simple example — 309
- 12.3 A modification of the example of §12.2 — 312
- 12.4 Examples with nontrivial fundamental group — 314
- 12.5 A more complicated example — 315
- 12.6 Examples of associative and coassociative submanifolds — 326
- 12.7 An example of a resolution of T^7/Γ with Γ nonabelian — 329
- 12.8 More examples — 334
- 12.9 Conclusions — 338

13 Construction of compact Spin(7)-manifolds — 341
13.1 Resolving Spin(7)-singularities and QALE Spin(7)-manifolds — 342
13.2 Orbifolds T^8/Γ, R-data, and resolutions of T^8/Γ — 346
13.3 Construction of various exterior forms — 349
13.4 Introducing a real parameter $t \in (0, 1]$ — 352
13.5 Spin(7)-structures on M with small torsion — 355
13.6 An existence result for compact Spin(7)-manifolds — 360
13.7 The proof of Theorem S2 — 362

14 Examples of compact 8-manifolds with holonomy Spin(7) — 366
14.1 Topological invariants of resolutions of T^8/Γ — 366
14.2 A simple example — 368
14.3 Examples of Cayley submanifolds — 371
14.4 A more complicated example — 373
14.5 Another example with $\Gamma = \mathbb{Z}_2^4$ — 380
14.6 An example with Γ nonabelian — 386

15 A second construction of compact 8-manifolds with holonomy Spin(7) — 392
15.1 ALE Spin(7)-manifolds — 393
15.2 Proof of the construction — 395
15.3 How to apply the construction — 401
15.4 A simple example — 406
15.5 Examples from hypersurfaces in $\mathbb{CP}^5_{a_0,\ldots,a_5}$ — 408
15.6 A hypersurface of degree 8 in $\mathbb{CP}^5_{1,1,1,1,2,2}$ over $\mathbb{Z}_2$ — 411
15.7 Examples from complete intersections in $\mathbb{CP}^6_{a_0,\ldots,a_6}$ — 413
15.8 Discussion, and questions for future research — 416

References — 419

Index — 432

1

BACKGROUND MATERIAL

In this chapter we explain some background necessary for understanding the rest of the book. We shall assume that the reader is already familiar with the basic ideas of differential and Riemannian geometry (in particular, manifolds, tensors, and Riemannian metrics) and of algebraic topology (in particular, fundamental group, homology and cohomology).

We start in §1.1 with a short introduction to exterior forms on manifolds, de Rham cohomology, and Hodge theory. These will be essential tools later in the book, and we discuss them out of completeness, and to fix notation.

The rest of the chapter is an introduction to the analysis of elliptic operators on manifolds. This will be a major focus of the book. The author has tried to make the book fairly self-contained in this area, so that those without much background in analysis should hopefully be able to follow it.

Section 1.2 defines *Sobolev* and *Hölder spaces*, which are Banach spaces of functions and tensors on a manifold, and discusses their basic properties. Then §1.3–§1.5 define *elliptic operators*, a special class of partial differential operators, and explain how solutions of elliptic equations have good existence and regularity properties in Sobolev and Hölder spaces.

1.1 Exterior forms on manifolds

In this section we introduce exterior forms on manifolds, and summarize two theories involving exterior forms—de Rham cohomology and Hodge theory. The books by Warner [219] and Bott and Tu [38] are good references for the material in this section.

Let M be an n-manifold, with tangent bundle TM and cotangent bundle T^*M. The kth exterior power of the bundle T^*M is written $\Lambda^k T^*M$. It is a real vector bundle over M, with fibres of dimension $\binom{n}{k}$. Smooth sections of $\Lambda^k T^*M$ are called *k-forms*, and the vector space of k-forms is written $C^\infty(\Lambda^k T^*M)$.

Now $\Lambda^k T^*M$ is a subbundle of $\otimes^k T^*M$, so k-forms are tensors on M, and may be written using index notation. We shall use the common notation that a collection of tensor indices enclosed in square brackets [...] are to be antisymmetrized over. That is, if $T_{a_1 a_2 \ldots a_k}$ is a tensor with k indices, then

$$T_{[a_1 \ldots a_k]} = \frac{1}{k!} \sum_{\sigma \in S_k} \text{sign}(\sigma) T_{a_{\sigma(1)} \ldots a_{\sigma(k)}},$$

where S_k is the group of permutations of $\{1, 2, \ldots, k\}$, and $\text{sign}(\sigma)$ is 1 if σ is even, and -1 if σ is odd. Then a k-form α on M is a tensor $\alpha_{a_1 \ldots a_k}$ with k covariant indices that is antisymmetric, i.e. that satisfies $\alpha_{a_1 \ldots a_k} = \alpha_{[a_1 \ldots a_k]}$.

The *exterior product* $\wedge$ and the *exterior derivative* d are important natural operations on forms. If α is a k-form and β an l-form then $\alpha \wedge \beta$ is a $(k+l)$-form and dα a $(k+1)$-form, which are given in index notation by

$$(\alpha \wedge \beta)_{a_1\ldots a_{k+l}} = \alpha_{[a_1\ldots a_k}\beta_{a_{k+1}\ldots a_{k+l}]} \quad \text{and}$$

$$(\mathrm{d}\alpha)_{a_1\ldots a_{k+1}} = T_{[a_1\ldots a_{k+1}]}, \quad \text{where} \quad T_{a_1\ldots a_{k+1}} = \frac{\partial \alpha_{a_2\ldots a_{k+1}}}{\partial x^{a_1}}.$$

If α is a k-form and β an l-form then

$$\mathrm{d}(\mathrm{d}\alpha)=0, \quad \alpha \wedge \beta = (-1)^{kl}\beta \wedge \alpha \quad \text{and} \quad \mathrm{d}(\alpha \wedge \beta) = (\mathrm{d}\alpha) \wedge \beta + (-1)^k \alpha \wedge (\mathrm{d}\beta).$$

The first of these is written $\mathrm{d}^2 = 0$, and is a fundamental property of d. If $\mathrm{d}\alpha = 0$, then α is *closed*, and if $\alpha = \mathrm{d}\beta$ for some β then α is *exact*. As $\mathrm{d}^2 = 0$, every exact form is closed. If M is a compact, oriented n-manifold and α an $(n-1)$-form, then *Stokes' Theorem* says that $\int_M \mathrm{d}\alpha = 0$.

1.1.1 De Rham cohomology

Let M be a smooth n-manifold. As $\mathrm{d}^2 = 0$, the chain of operators

$$0 \to C^\infty(\Lambda^0 T^*M) \xrightarrow{\mathrm{d}} C^\infty(\Lambda^1 T^*M) \xrightarrow{\mathrm{d}} \cdots \xrightarrow{\mathrm{d}} C^\infty(\Lambda^n T^*M) \to 0$$

forms a *complex*, and therefore we may find its cohomology groups. For $k = 0, 1, \ldots, n$, define the *de Rham cohomology groups* $H^k_{DR}(M, \mathbb{R})$ of M by

$$H^k_{DR}(M, \mathbb{R}) = \frac{\mathrm{Ker}\bigl(\mathrm{d} : C^\infty(\Lambda^k T^*M) \to C^\infty(\Lambda^{k+1} T^*M)\bigr)}{\mathrm{Im}\bigl(\mathrm{d} : C^\infty(\Lambda^{k-1} T^*M) \to C^\infty(\Lambda^k T^*M)\bigr)}.$$

That is, $H^k_{DR}(M, \mathbb{R})$ is the quotient of the vector space of closed k-forms on M by the vector space of exact k-forms on M. If η is a closed k-form, then the *cohomology class* $[\eta]$ of η in $H^k_{DR}(M, \mathbb{R})$ is $\eta + \mathrm{Im}\,\mathrm{d}$, and η is a *representative* for $[\eta]$.

There are several different ways to define the *cohomology* of topological spaces, for example, singular, Alexander–Spanier and Čech cohomology. If the topological space is well-behaved (e.g. if it is paracompact and Hausdorff) then the corresponding cohomology groups are all isomorphic. The de Rham Theorem [219, p. 206], [38, Th. 8.9] is a result of this kind.

Theorem 1.1.1. (The de Rham Theorem) *Let M be a smooth manifold. Then the de Rham cohomology groups $H^k_{DR}(M, \mathbb{R})$ are canonically isomorphic to the singular, Alexander–Spanier and Čech cohomology groups of M over $\mathbb{R}$.*

Thus the de Rham cohomology groups are topological invariants of M. As there is usually no need to distinguish between de Rham and other sorts of cohomology, we will write $H^k(M, \mathbb{R})$ instead of $H^k_{DR}(M, \mathbb{R})$ for the de Rham cohomology groups. The kth *Betti number* b^k is $b^k = \dim H^k(M, \mathbb{R})$. The Betti numbers are important topological invariants of a manifold.

Theorem 1.1.2 (Poincaré duality) *Let M be a compact, oriented n-manifold. Then there is a canonical isomorphism $H^{n-k}(M, \mathbb{R}) \cong \left(H^k(M, \mathbb{R})\right)^*$, and the Betti numbers satisfy $b^k = b^{n-k}$.*

1.1.2 Exterior forms on Riemannian manifolds

Now let M be a compact, oriented Riemannian n-manifold, with metric g. The metric and the orientation combine to give a *volume form* dV_g on M, which can be used to integrate functions on M. We shall define two sorts of inner product on k-forms. Let α, β be k-forms on M, and define (α, β) by

$$(\alpha, \beta) = \alpha_{a_1 \ldots a_k} \beta_{b_1 \ldots b_k} g^{a_1 b_1} \ldots g^{a_k b_k},$$

in index notation. Then (α, β) is a function on M. We call (α, β) the *pointwise inner product* of α, β. Now for k-forms α, β, define $\langle \alpha, \beta \rangle = \int_M (\alpha, \beta) dV_g$. As M is compact, $\langle \alpha, \beta \rangle$ exists in $\mathbb{R}$ provided α, β are (for instance) continuous. We call $\langle \alpha, \beta \rangle$ the L^2 *inner product* of α, β. (This is because it is the inner product of the Hilbert space $L^2(\Lambda^k T^*M)$, which will be defined in §1.2.)

The *Hodge star* is an isomorphism of vector bundles $* : \Lambda^k T^*M \to \Lambda^{n-k} T^*M$, which is defined as follows. Let β be a k-form on M. Then $*\beta$ is the unique $(n-k)$-form that satisfies the equation $\alpha \wedge (*\beta) = (\alpha, \beta) dV_g$ for all k-forms α on M. The Hodge star is well-defined, and depends upon g and the orientation of M. It satisfies the identities $*1 = dV_g$ and $*(*\beta) = (-1)^{k(n-k)} \beta$, for β a k-form, so that $*^{-1} = (-1)^{k(n-k)} *$.

Define an operator $d^* : C^\infty(\Lambda^k T^*M) \to C^\infty(\Lambda^{k-1} T^*M)$ by

$$d^*\beta = (-1)^{kn+n+1} * d(*\beta). \tag{1.1}$$

Let α be a $(k-1)$-form and β a k-form on M. Then

$$\langle \alpha, d^*\beta \rangle = \int_M (\alpha, d^*\beta) dV_g = \int_M \alpha \wedge (*d^*\beta) = (-1)^k \int_M \alpha \wedge d * \beta.$$

But $d(\alpha \wedge *\beta) = (d\alpha) \wedge *\beta + (-1)^{k-1} \alpha \wedge d*\beta$, and as M is compact $\int_M d(\alpha \wedge *\beta) = 0$ by Stokes' Theorem. Therefore

$$(-1)^k \int_M \alpha \wedge d * \beta = \int_M d\alpha \wedge *\beta = \int_M (d\alpha, \beta) dV_g = \langle d\alpha, \beta \rangle.$$

Combining the two equations shows that $\langle \alpha, d^*\beta \rangle = \langle d\alpha, \beta \rangle$. This technique is called *integration by parts*. Thus d^* has the formal properties of the adjoint of d, and is sometimes called the *formal adjoint* of d.

As $d^2 = 0$ we see that $(d^*)^2 = 0$. If a k-form α satisfies $d^*\alpha = 0$, then α is *coclosed*, and if $\alpha = d^*\beta$ for some β then α is *coexact*. The *Laplacian* Δ is $\Delta = dd^* + d^*d$. Then $\Delta : C^\infty(\Lambda^k T^*M) \to C^\infty(\Lambda^k T^*M)$ is a linear elliptic partial differential operator of order 2. By convention $d^* = 0$ on functions, so $\Delta = d^*d$ on functions.

Several different operators are called Laplacians. When we need to distinguish between them we will refer to this one as the d-*Laplacian*, and write it Δ_d. If α is a k-form and $\Delta \alpha = 0$, then α is called a *harmonic form*.

1.1.3 *Hodge theory*

Let M be a compact, oriented Riemannian manifold, and define

$$\mathcal{H}^k = \text{Ker}\big(\Delta : C^\infty(\Lambda^k T^*M) \to C^\infty(\Lambda^k T^*M)\big),$$

so that $\mathcal{H}^k$ is the vector space of harmonic k-forms on M. Suppose $\alpha \in \mathcal{H}^k$. Then $\Delta\alpha = 0$, and thus $\langle \alpha, \Delta\alpha \rangle = 0$. But $\Delta = dd^* + d^*d$, so

$$0 = \langle \alpha, dd^*\alpha \rangle + \langle \alpha, d^*d\alpha \rangle = \langle d^*\alpha, d^*\alpha \rangle + \langle d\alpha, d\alpha \rangle = \|d^*\alpha\|_{L^2}^2 + \|d\alpha\|_{L^2}^2,$$

where $\|.\|_{L^2}$ is the L^2 norm defined in §1.2. Thus $\|d^*\alpha\|_{L^2} = \|d\alpha\|_{L^2} = 0$, so that $d^*\alpha = d\alpha = 0$. Conversely, if $d^*\alpha = d\alpha = 0$ then $\Delta\alpha = (dd^* + d^*d)\alpha = 0$, so that a k-form α lies in $\mathcal{H}^k$ if and only if it is closed and coclosed. Note also that if $\alpha \in \mathcal{H}^k$, then $*\alpha \in \mathcal{H}^{n-k}$.

The next result is proved in [219, Th. 6.8].

Theorem 1.1.3. (The Hodge Decomposition Theorem) *Let M be a compact, oriented Riemannian manifold, and write d_k for d acting on k-forms and d_k^* for d^* acting on k-forms. Then*

$$C^\infty(\Lambda^k T^*M) = \mathcal{H}^k \oplus \text{Im}(d_{k-1}) \oplus \text{Im}(d_{k+1}^*).$$

Moreover, $\text{Ker}(d_k) = \mathcal{H}^k \oplus \text{Im}(d_{k-1})$ and $\text{Ker}(d_k^*) = \mathcal{H}^k \oplus \text{Im}(d_{k+1}^*)$.

Now $H_{DR}^k(M, \mathbb{R}) = \text{Ker}(d_k)/\text{Im}(d_{k-1})$, and as $\text{Ker}(d_k) = \mathcal{H}^k \oplus \text{Im}(d_{k-1})$ there is a canonical isomorphism between $\mathcal{H}^k$ and $H_{DR}^k(M, \mathbb{R})$. Thus we have:

Theorem 1.1.4. (Hodge's Theorem) *Let M be a compact, oriented Riemannian manifold. Then every de Rham cohomology class on M contains a unique harmonic representative, and $\mathcal{H}^k \cong H_{DR}^k(M, \mathbb{R})$.*

1.2 Introduction to analysis

Let M be a Riemannian manifold with metric g. In problems in analysis it is often useful to consider infinite-dimensional vector spaces of functions on M, and to equip these vector spaces with norms, making them into Banach spaces. In this book we will meet four different types of Banach spaces of this sort, written $L^q(M)$, $L_k^q(M)$, $C^k(M)$ and $C^{k,\alpha}(M)$, and they are defined below.

1.2.1 *Lebesgue spaces and Sobolev spaces*

Let M be a Riemannian manifold with metric g. For $q \geq 1$, define the *Lebesgue space* $L^q(M)$ to be the set of locally integrable functions f on M for which the norm

$$\|f\|_{L^q} = \left(\int_M |f|^q dV_g\right)^{1/q}$$

is finite. Here dV_g is the volume form of the metric g. Suppose that $r, s, t \geq 1$ and that $1/r = 1/s + 1/t$. If $\phi \in L^s(M)$, $\psi \in L^t(M)$, then $\phi\psi \in L^r(M)$, and $\|\phi\psi\|_{L^r} \leq \|\phi\|_{L^s}\|\psi\|_{L^t}$; this is *Hölder's inequality*.

Let $q \geqslant 1$ and let k be a nonnegative integer. Define the *Sobolev space* $L_k^q(M)$ to be the set of $f \in L^q(M)$ such that f is k times weakly differentiable and $|\nabla^j f| \in L^q(M)$ for $j \leqslant k$. Define the *Sobolev norm* on $L_k^q(M)$ to be

$$\|f\|_{L_k^q} = \left(\sum_{j=0}^k \int_M |\nabla^j f|^q dV_g \right)^{1/q}.$$

Then $L_k^q(M)$ is a Banach space with respect to the Sobolev norm. Furthermore, $L_k^2(M)$ is a Hilbert space.

The spaces $L^q(M)$, $L_k^q(M)$ are vector spaces of *real functions* on M. It is useful to generalize this idea to vector spaces of *sections of a vector bundle* over M. So, let M be a Riemannian manifold, and let $V \to M$ be a vector bundle on M, equipped with Euclidean metrics on its fibres. Let $\hat{\nabla}$ be a connection on V preserving these metrics.

Then as above, for $q \geqslant 1$, define the *Lebesgue space* $L^q(V)$ to be the set of locally integrable sections v of V over M for which the norm

$$\|v\|_{L^q} = \left(\int_M |v|^q dV_g \right)^{1/q}$$

is finite, and the *Sobolev space* $L_k^q(V)$ to be the set of $v \in L^q(V)$ such that v is k times weakly differentiable and $|\hat{\nabla}^j v| \in L^q(M)$ for $j \leqslant k$, with the obvious Sobolev norm.

1.2.2 C^k spaces and Hölder spaces

Let M be a Riemannian manifold with metric g. For each integer $k \geqslant 0$, define $C^k(M)$ to be the space of continuous, bounded functions f on M that have k continuous, bounded derivatives, and define the norm $\|.\|_{C^k}$ on $C^k(M)$ by $\|f\|_{C^k} = \sum_{j=0}^k \sup_M |\nabla^j f|$, where ∇ is the Levi-Civita connection.

The fourth class of vector spaces are the Hölder spaces $C^{k,\alpha}(M)$ for $k \geqslant 0$ an integer and $\alpha \in (0,1)$. We begin by defining $C^{0,\alpha}(M)$. Let $d(x,y)$ be the distance between $x, y \in M$ calculated using g, and let $\alpha \in (0,1)$. Then a function f on M is said to be *Hölder continuous with exponent* α if

$$[f]_\alpha = \sup_{x \neq y \in M} \frac{|f(x) - f(y)|}{d(x,y)^\alpha}$$

is finite. Any Hölder continuous function f is continuous. The vector space $C^{0,\alpha}(M)$ is the set of continuous, bounded functions on M which are Hölder continuous with exponent α, and the norm on $C^{0,\alpha}(M)$ is $\|f\|_{C^{0,\alpha}} = \|f\|_{C^0} + [f]_\alpha$.

In the same way, we shall define Hölder norms on spaces of sections v of a vector bundle V over M, equipped with Euclidean metrics in the fibres, and a connection $\hat{\nabla}$ preserving these metrics. Let $\delta(g)$ be the injectivity radius of the metric g on M, which we suppose to be positive, and set

$$[v]_\alpha = \sup_{\substack{x \neq y \in M \\ d(x,y) < \delta(g)}} \frac{|v(x) - v(y)|}{d(x,y)^\alpha}, \qquad (1.2)$$

whenever the supremum exists. Now we have a problem in making sense of $|v(x) - v(y)|$ in this equation, since $v(x)$ and $v(y)$ lie in different vector spaces. We interpret it in the following way. When $x \neq y \in M$ and $d(x, y) < \delta(g)$, there is a unique geodesic γ of length $d(x, y)$ joining x and y in M. Parallel translation along γ using $\hat{\nabla}$ identifies the fibres of V over x and y, and the metrics on the fibres. With this understanding, the expression $|v(x) - v(y)|$ is well-defined.

So, define $C^{k,\alpha}(M)$ to be the set of f in $C^k(M)$ for which the supremum $[\nabla^k f]_\alpha$ defined by (1.2) exists, working in the vector bundle $\otimes^k T^*M$ with its natural metric and connection. The Hölder norm on $C^{k,\alpha}(M)$ is $\|f\|_{C^{k,\alpha}} = \|f\|_{C^k} + [\nabla^k f]_\alpha$. With this norm, $C^{k,\alpha}(M)$ is a Banach space, called a *Hölder space*.

The condition of Hölder continuity is analogous to a sort of fractional differentiability. To see this, observe that if $f \in C^1(M)$, then by the mean value theorem $[f]_\alpha$ exists, and

$$[f]_\alpha \leq \left(2\|f\|_{C^0}\right)^{1-\alpha} \|\nabla f\|_{C^0}^\alpha.$$

Thus $[f]_\alpha$ is a sort of interpolation between the C^0 and C^1 norms of f. It can be helpful to think of $C^{k,\alpha}(M)$ as the space of functions on M that are $(k+\alpha)$ times differentiable.

Now suppose that V is a vector bundle on M with Euclidean metrics on its fibres, and ∇^V is a connection on V preserving these metrics. As in the case of Lebesgue and Sobolev spaces, we may generalize the definitions above in an obvious way to give Banach spaces $C^k(V)$ and $C^{k,\alpha}(V)$ of sections of V, and we leave this to the reader.

1.2.3 Embedding theorems

An important tool in problems involving Sobolev spaces is the Sobolev Embedding Theorem, which includes one Sobolev space inside another. Embedding theorems are dealt with at length by Aubin in [17, §2.3–§2.9]. The following result is taken from [17, Th. 2.30].

Theorem 1.2.1. (Sobolev Embedding Theorem) *Suppose M is a compact Riemannian n-manifold, k, l are integers with $k \geq l \geq 0$, q, r are real numbers with $q, r \geq 1$, and $\alpha \in (0, 1)$. If*

$$\frac{1}{q} \leq \frac{1}{r} + \frac{k-l}{n},$$

then $L_k^q(M)$ is continuously embedded in $L_l^r(M)$ by inclusion. If

$$\frac{1}{q} \leq \frac{k-l-\alpha}{n},$$

then $L_k^q(M)$ is continuously embedded in $C^{l,\alpha}(M)$ by inclusion.

Next we define the idea of a compact linear map between Banach spaces.

Definition 1.2.2 Let U_1, U_2 be Banach spaces, and let $\psi : U_1 \to U_2$ be a continuous linear map. Let $B_1 = \{u \in U_1 : \|u\|_{U_1} \leq 1\}$ be the unit ball in U_1. We call ψ a *compact linear map* if the image $\psi(B_1)$ of B_1 is a precompact subset of U_2, that is, if its closure $\overline{\psi(B_1)}$ is a compact subset of U_2.

It turns out that some of the embeddings of Sobolev and Hölder spaces given in the Sobolev Embedding Theorem are compact linear maps in the above sense. This is called the Kondrakov Theorem, and can be found in [17, Th. 2.34].

Theorem 1.2.3. (The Kondrakov Theorem) *Suppose M is a compact Riemannian n-manifold, k, l are integers with $k \geq l \geq 0$, q, r are real numbers with $q, r \geq 1$, and $\alpha \in (0, 1)$. If*

$$\frac{1}{q} < \frac{1}{r} + \frac{k-l}{n}$$

then the embedding $L_k^q(M) \hookrightarrow L_l^r(M)$ is compact. If

$$\frac{1}{q} < \frac{k-l-\alpha}{n}$$

then $L_k^q(M) \hookrightarrow C^{l,\alpha}(M)$ is compact. Also $C^{k,\alpha}(M) \hookrightarrow C^k(M)$ is compact.

Finally, we state two related results, the Inverse Mapping Theorem and the Implicit Mapping Theorem for Banach spaces, which can be found in Lang [144, Th. 1.2, p. 128] and [144, Th. 2.1, p. 131].

Theorem 1.2.4. (Inverse Mapping Theorem) *Let X, Y be Banach spaces, and U an open neighbourhood of x in X. Suppose that the function $F : U \to Y$ is C^k for some $k \geq 1$, with $F(x) = y$, and that the first derivative $\mathrm{d}F_x : X \to Y$ of F at x is an isomorphism of X, Y as both vector spaces and topological spaces. Then there are open neighbourhoods $U' \subset U$ of x in X and V' of y in Y, such that $F : U' \to V'$ is a C^k-isomorphism.*

Theorem 1.2.5. (Implicit Mapping Theorem) *Let X, Y and Z be Banach spaces, and U, V open neighbourhoods of 0 in X and Y. Suppose that the function $F : U \times V \to Z$ is C^k for some $k \geq 1$, with $F(0, 0) = 0$, and that $\mathrm{d}F_{(0,0)}|_Y : Y \to Z$ is an isomorphism of Y, Z as vector and topological spaces. Then there exists a connected open neighbourhood $U' \subset U$ of 0 in X and a unique C^k map $G : U' \to V$ such that $G(0) = 0$ and $F(x, G(x)) = 0$ for all $x \in U'$.*

1.3 Introduction to elliptic operators

In this section we define *elliptic operators*, which are a special sort of partial differential operator on a manifold. Many of the differential operators that crop up in problems in geometry, applied mathematics and physics are elliptic. For example, consider the equation $\Delta u = f$ on a Riemannian manifold M, where Δ is the Laplacian, and u and f are real functions on M. It turns out that Δ is a linear elliptic operator.

The theory of linear elliptic operators tells us two things about the equation $\Delta u = f$. First, there is a theory about the existence of solutions u to this equation. If f is a given function, there are simple criteria to decide whether or not there exists a function u with $\Delta u = f$. Secondly, there is a theory about the regularity of solutions u, that is, how smooth u is. Roughly speaking, u is as smooth as the problem allows, so that if f is k times differentiable, then u is $k+2$ times differentiable, but this is an oversimplification. These theories of regularity and existence of solutions to elliptic equations will be explained in §1.4 and §1.5.

Here we will define elliptic operators, and give a few examples and basic facts. Although the underlying idea of ellipticity is fairly simple, there are many variations on the theme—elliptic operators can be linear, quasilinear or nonlinear, for instance, and they can operate on functions or on sections of vector bundles, and so on. Because of this, to reduce confusion we will define the idea of elliptic operator in stages, starting with elliptic operators on functions, and illustrating the ideas with the simplest case of linear elliptic operators of order two. Some useful references for the material in this section are the books by Gilbarg and Trudinger [83] and Morrey [165], and the appendix in Besse [33].

1.3.1 *Partial differential operators on functions*

Let M be a manifold, and ∇ a connection on the tangent bundle of M, for instance, the Levi-Civita connection of a Riemannian metric on M. Let u be a smooth function on M. Then the kth derivative of u using ∇ is $\nabla^k f$, or in index notation $\nabla_{a_1} \cdots \nabla_{a_k} u$. We will write $\nabla_{a_1 \ldots a_k} u$ as a shorthand for this k^{th} derivative $\nabla_{a_1} \cdots \nabla_{a_k} u$. Here is the definition of a partial differential operator on functions.

Definition 1.3.1 A *partial differential operator* or *differential operator* P on M *of order* k is an operator taking real functions u on M to real functions on M, that depends on u and its first k derivatives. Explicitly, if u is a real function on M such that the first k derivatives $\nabla u, \ldots, \nabla^k u$ of u exist (possibly in some weak sense), then $P(u)$ or Pu is a real function on M given by

$$(Pu)(x) = Q(x, u(x), \nabla u(x), \ldots, \nabla^k u(x)) \tag{1.3}$$

for $x \in M$, where Q is some real function of its arguments.

It is usual to require that this function Q is at least continuous in all its arguments. If Q is a smooth function of its arguments, then P is called a *smooth* differential operator. If Pu is linear in u (that is, $P(\alpha u + \beta v) = \alpha Pu + \beta Pv$ for u, v functions and $\alpha, \beta \in \mathbb{R}$) then P is called a *linear* differential operator. If P is not linear, it is called *nonlinear*.

Here is an example. Let P be a linear differential operator of order 2, and let $(x_1, \ldots, x_n)$ be coordinates on an open set in M. Then we may write

$$(Pu)(x) = \sum_{i,j=1}^{n} a^{ij}(x) \frac{\partial^2 u}{\partial x_i \partial x_j}(x) + \sum_{i=1}^{n} b^i(x) \frac{\partial u}{\partial x_i}(x) + c(x)u(x), \tag{1.4}$$

where for $i, j = 1, \ldots, n$, each of a^{ij}, b^i and c are real functions on this coordinate patch, and $a^{ij} = a^{ji}$. We call a^{ij}, b^i and c the *coefficients* of the operator P, so that, for instance, we say P has Hölder continuous coefficients if each of a^{ij}, b^i and c are Hölder continuous functions. Also, a^{ij} are called the *leading coefficients*, as they are the coefficients of the highest order derivative of u.

Now in §1.2 we defined various vector spaces of functions: $C^k(M)$, $C^\infty(M)$, Hölder spaces and Sobolev spaces. It is often useful to regard a differential operator as a mapping between two of these vector spaces. For instance, if P is a *smooth* differential operator of order k, and $u \in C^\infty(M)$, then $Pu \in C^\infty(M)$, so P maps $C^\infty(M) \to C^\infty(M)$. On the other hand, if $u \in C^{k+l}(M)$ then $Pu \in C^l(M)$, so that P also maps $C^{k+l}(M) \to C^l(M)$.

It is not necessary to assume P is a smooth operator. For instance, let P be a linear differential operator of order k. It is easy to see that if the coefficients of P are bounded, then $P : L^q_{k+l}(M) \to L^q_l(M)$ is a linear map, and if the coefficients of P are at least $C^{l,\alpha}$, then $P : C^{k+l,\alpha}(M) \to C^{l,\alpha}(M)$ is also a linear map, and so on. In this way we can consider an operator P to act on several different vector spaces of functions.

Definition 1.3.2 Let P be a (nonlinear) differential operator of order k, that is defined as in (1.3) by a function Q that is at least C^1 in the arguments $u, \nabla u, \ldots, \nabla^k u$. Let u be a real function with k derivatives. We define the *linearization* $L_u P$ of P at u to be the derivative of $P(v)$ with respect to v at u, that is,

$$L_u P v = \lim_{\alpha \to 0} \left(\frac{P(u + \alpha v) - P(u)}{\alpha} \right). \tag{1.5}$$

Then $L_u P$ is a *linear* differential operator of order k. If P is linear then $L_u P = P$. Note that even if P is a smooth operator, the linearization $L_u P$ need not be smooth if u is not smooth. For instance, if P is of order k and $u \in C^{k+l}(M)$, then $L_u P$ will have C^l coefficients in general, as they depend on the kth derivatives of u.

Many properties of a linear differential operator P depend only on the highest order derivatives occurring in P. The *symbol* of P is a convenient way to isolate these highest order terms.

Definition 1.3.3 Let P be a linear differential operator on functions of order k. Then in index notation, we may write

$$Pu = A^{i_1 \ldots i_k} \nabla_{i_1 \ldots i_k} u + B^{i_1 \ldots i_{k-1}} \nabla_{i_1 \ldots i_{k-1}} u + \cdots + K^{i_1} \nabla_{i_1} u + Lu,$$

where $A, B, \ldots, K$ are symmetric tensors and L a real function on M. For each point $x \in M$ and each $\xi \in T_x^* M$, define $\sigma_\xi(P; x) = A^{i_1 \ldots i_k} \xi_{i_1} \xi_{i_2} \ldots \xi_{i_k}$. Let $\sigma(P) : T^*M \to \mathbb{R}$ be the function with value $\sigma_\xi(P; x)$ at each $\xi \in T_x^* M$. Then $\sigma(P)$ is called the *symbol* or *principal symbol* of P. It is a homogeneous polynomial of degree k on each cotangent space.

1.3.2 Elliptic operators on functions

Now we can define linear elliptic operators on functions.

Definition 1.3.4 Let P be a linear differential operator of degree k on M. We say P is an *elliptic operator* if for each $x \in M$ and each nonzero $\xi \in T_x^* M$, we have $\sigma_\xi(P; x) \neq 0$, where $\sigma(P)$ is the principal symbol of P.

Thus, $\sigma(P)$ must be nonzero on each $T_x^*M \setminus \{0\}$, that is, on the complement of the zero section in T^*M. Suppose dim $M > 1$. Then $T_x^*M \setminus \{0\}$ is connected, and as $\sigma(P)$ is continuous on T_x^*M, either $\sigma_\xi(P; x) > 0$ for all $\xi \in T_x^*M \setminus \{0\}$, or $\sigma_\xi(P; x) < 0$ for all $\xi \in T_x^*M \setminus \{0\}$. However, $\sigma_{-\xi}(P; x) = (-1)^k \sigma_\xi(P; x)$. It follows that if dim $M > 1$, then the degree k of an elliptic operator P must be *even*. Also, if M is connected and P has continuous leading coefficients, then $\sigma(P)$ is continuous on a connected space, so that either $\sigma(P) > 0$ or $\sigma(P) < 0$ on the whole of the complement of the zero section in T^*M.

For example, let P be a linear differential operator of order 2, given in a coordinate system $(x_1, \ldots, x_n)$ by (1.4). At each point $x \in M$, the leading coefficients $a^{ij}(x)$ form a real symmetric $n \times n$ matrix. The condition for P to be elliptic is that $a^{ij}\xi_i\xi_j \neq 0$ whenever $\xi \neq 0$, that is, either $a^{ij}\xi_i\xi_j > 0$ for all nonzero ξ or $a^{ij}\xi_i\xi_j < 0$ for all nonzero ξ. This is equivalent to saying that the eigenvalues of the matrix $a^{ij}(x)$ must either all be positive, or all be negative.

The best known example of a linear elliptic operator is the Laplacian on a Riemannian manifold, defined by $\Delta u = -g^{ij}\nabla_{ij}u$. The symbol $\sigma(\Delta)$ is $\sigma_\xi(\Delta; x) = -g^{ij}\xi_i\xi_j = -|\xi|^2$, so that if $\xi \neq 0$ then $\sigma_\xi(\Delta; x) < 0$, and Δ is elliptic. Next we define *nonlinear* elliptic operators.

Definition 1.3.5 Let P be a (nonlinear) differential operator of degree k on M, and let u be a function with k derivatives. We say P is *elliptic at u* if the linearization L_uP of P at u is elliptic. A nonlinear operator P may be elliptic at some functions u and not at others.

1.3.3 Differential operators on vector bundles

Now let M be a manifold, and let V, W be vector bundles over M. As above, let ∇ be some connection on TM, and let ∇^V be a connection on V. Let v be a section of V. By coupling the connections ∇ and ∇^V, one can form repeated derivatives of v. We will write $\nabla^V_{a_1 a_2 \ldots a_k} v$ for the kth derivative of v defined in this way. Here is the idea of differential operator on vector bundles.

Definition 1.3.6 A *differential operator P of order k taking sections of V to sections of W* is an operator taking sections v of V to sections of W, that depends on v and its first k derivatives. Explicitly, if v is a k times differentiable section of V then Pv is given by

$$(Pv)(x) = Q\bigl(x, v(x), \nabla^V_{a_1}v(x), \ldots, \nabla^V_{a_1\ldots a_k}v(x)\bigr) \in W_x$$

for $x \in M$. If Q is a smooth function of its arguments, then P is called *smooth*, and if Pv is linear in v then P is called *linear*. If P is not linear, it is *nonlinear*. If P is a (nonlinear) differential operator defined by a function Q that is C^1 in the arguments $v, \nabla^V_{a_1}v, \ldots, \nabla^V_{a_1\ldots a_k}v$, then we define the *linearization L_uP at u* by (1.5). Although P maps sections of V to sections of W, by an abuse of notation we may also say that P is a differential operator *from V to W*.

This is a natural generalization of differential operators on functions. Since real functions are the same thing as sections of the trivial line bundle over M with fibre $\mathbb{R}$, a differential operator on functions is just the special case when $V = W = \mathbb{R}$. Here are some examples. The operators

$$\mathrm{d} : C^\infty(\Lambda^k T^*M) \to C^\infty(\Lambda^{k+1} T^*M),$$
$$\mathrm{d}^* : C^\infty(\Lambda^k T^*M) \to C^\infty(\Lambda^{k-1} T^*M),$$
$$\text{and}\quad \Delta : C^\infty(\Lambda^k T^*M) \to C^\infty(\Lambda^k T^*M)$$

introduced in §1.1 are all smooth linear differential operators acting on the vector bundle $\Lambda^k T^*M$, where d, d^* have order 1 and Δ has order 2. A connection ∇^V on a vector bundle V is a smooth linear differential operator of order 1, mapping from V to $V \otimes T^*M$, and so on.

As in the case of differential operators on functions, we can regard differential operators on vector bundles as mapping a vector space of sections of V to a vector space of sections of W. For instance, if P is a smooth, linear differential operator of order k from V to W, then P acts by $P : C^\infty(V) \to C^\infty(W)$, $P : C^{k+l,\alpha}(V) \to C^{l,\alpha}(W)$ and $P : L^q_{k+l}(V) \to L^q_l(W)$.

Let P be a linear differential operator of order k from V to W. Then in index notation, we write

$$Pv = A^{i_1\ldots i_k} \nabla_{i_1 \ldots i_k} v + B^{i_1 \ldots i_{k-1}} \nabla_{i_1 \ldots i_{k-1}} v + \cdots + K^{i_1} \nabla_{i_1} v + Lv. \tag{1.6}$$

However, here $A^{i_1\ldots i_k}, B^{i_1\ldots i_{k_1}}, \ldots$ are not ordinary tensors. Instead, they are tensors taking values in $V^* \otimes W$. For instance, if ξ_i is a 1-form at $x \in M$, then $A^{i_1\ldots i_k}(x)\xi_{i_1} \ldots \xi_{i_k}$ is not a real number, but an element of $V_x^* \otimes W_x$, or equivalently, a linear map from V_x to W_x, the fibres of V and W at x.

One can represent this in index notation by writing $A^{\alpha\, i_1\ldots i_k}_\beta$ in place of $A^{i_1\ldots i_k}$, where $i_1, \ldots, i_k$ are indices for TM, α is an index for W, and β is an index for V^*, but we prefer to suppress the indices for V and W. We call $A^{i_1\ldots i_k}, \ldots, L$ the *coefficients* of P. Next we define the *symbol* of a linear differential operator on vector bundles.

Definition 1.3.7 Let P be a linear differential operator of order k, mapping sections of V to sections of W, that is given by (1.6) in index notation. For each point $x \in M$ and each $\xi \in T^*_x M$, define $\sigma_\xi(P; x) = A^{i_1\ldots i_k} \xi_{i_1} \xi_{i_2} \ldots \xi_{i_k}$. Then $\sigma_\xi(P; x)$ is a linear map from V_x to W_x. Let $\sigma(P) : T^*M \times V \to W$ be the bundle map defined by $\sigma(P)(\xi, v) = \sigma_\xi(P; x)v \in W_x$ whenever $x \in M, \xi \in T^*_x M$ and $v \in V_x$. Then $\sigma(P)$ is called the *symbol* or *principal symbol* of P, and $\sigma(P)(\xi, v)$ is homogeneous of degree k in ξ and linear in v.

1.3.4 Elliptic operators on vector bundles

Now we define linear elliptic operators on vector bundles.

Definition 1.3.8 Let V, W be vector bundles over a manifold M, and let P be a linear differential operator of degree k from V to W. We say P is an *elliptic operator* if for each $x \in M$ and each nonzero $\xi \in T^*_x M$, the linear map $\sigma_\xi(P; x) : V_x \to W_x$ is invertible, where $\sigma(P)$ is the principal symbol of P.

Also, we say that P is an *underdetermined elliptic operator* if for each $x \in M$ and each $0 \neq \xi \in T_x^* M$, the map $\sigma_\xi(P; x) : V_x \to W_x$ is *surjective*, and that P is an *overdetermined elliptic operator* if for each $x \in M$ and each $0 \neq \xi \in T_x^* M$, the map $\sigma_\xi(P; x) : V_x \to W_x$ is *injective*. If P is a (nonlinear) differential operator of degree k from V to W, and v is a section of V with k derivatives, then we say P is *elliptic at v* if the linearization $L_v P$ of P at v is elliptic.

Suppose the vector bundles V, W have fibres $\mathbb{R}^l$ and $\mathbb{R}^m$ respectively. If $x \in M$ then $V_x \cong \mathbb{R}^l$ and $W_x \cong \mathbb{R}^m$, so that $\sigma_\xi(P; x) : \mathbb{R}^l \to \mathbb{R}^m$. Thus, $\sigma_\xi(P; x)$ can only be invertible if $l = m$, it can only be surjective if $l \geq m$, and it can only be injective if $l \geq m$. So, if P is *elliptic* then $\dim V = \dim W$, if P is *underdetermined elliptic* then $\dim V \geq \dim W$, and P is *overdetermined elliptic* if $\dim V \leq \dim W$.

Consider the equation $P(v) = w$ in a small region of M. Locally we can think of v as a collection of l real functions, and the equation $P(v) = w$ as being m simultaneous equations on the l functions of v. Now, guided by elementary linear algebra, we expect that a system of m equations in l variables is likely to have many solutions if $l > m$ (underdetermined), one solution if $l = m$, and no solutions at all if $l < m$ (overdetermined). This can help in thinking about differential operators on vector bundles.

Some authors (particularly of older texts) make a distinction between *elliptic equations*, by which they mean elliptic equations in one real function, and *elliptic systems*, by which they mean systems of l real equations in l real functions for $l > 1$, which we deal with using vector bundles. We will not make this distinction, but will refer to both cases as elliptic equations.

Papers about elliptic systems often use a more general concept than we have given, in which the operators can have mixed degree. (See Morrey [165], for instance). It seems to be a general rule that results proved for elliptic equations (in one real function), can also be proved for elliptic systems (in several real functions). However, it can be difficult to locate the proof for elliptic systems in the literature, as many papers deal only with elliptic equations in one real function.

Here are some examples. Let M be a Riemannian manifold of dimension n, and consider the operators d, d* and Δ on M defined in §1.1. Now $d : C^\infty(\Lambda^0 T^* M) \to C^\infty(\Lambda^1 T^* M)$ is a smooth linear differential operator of order 1. For $x \in M$ and $\xi \in T_x^* M$, the symbol is $\sigma_\xi(d; x) v = v \xi$, for $v \in \mathbb{R} = \Lambda^0 T_x^* M$. Thus, if $\xi \neq 0$, $\sigma_\xi(d; x)$ is injective, and d is overdetermined elliptic. But if $n > 1$ then $\sigma_\xi(d; x)$ is not surjective, so d is not elliptic. Similarly, $d^* : C^\infty(\Lambda^1 T^* M) \to C^\infty(\Lambda^0 T^* M)$ is underdetermined elliptic.

It can also be shown that the operator

$$d + d^* : C^\infty\left(\bigoplus_{j=0}^n \Lambda^j T^* M\right) \to C^\infty\left(\bigoplus_{j=0}^n \Lambda^j T^* M\right)$$

is a smooth linear elliptic operator of order 1, and the Laplacian on j-forms $\Delta : C^\infty(\Lambda^j T^* M) \to C^\infty(\Lambda^j T^* M)$ is smooth, linear and elliptic of order 2 for each j.

1.3.5 Elliptic operators over compact manifolds

Let M be a compact Riemannian manifold (without boundary). Then from §1.2, $L^2(M)$ is a Banach space of functions on M. In fact, it is a Hilbert space, with the L^2 inner product $\langle u_1, u_2 \rangle = \int_M u_1 u_2 \, dV_g$ for $u_1, u_2 \in L^2(M)$. We can also use this inner product on any vector subspace of $L^2(M)$, such as $C^\infty(M)$. In the same way, if V is a vector bundle over M equipped with Euclidean metrics on its fibres, then $L^2(V)$ is a Hilbert space of sections of V, with inner product $\langle \, , \, \rangle_V$ given by $\langle v_1, v_2 \rangle_V = \int_M (v_1, v_2) dV_g$.

Now suppose that V, W are vector bundles over M, equipped with metrics on the fibres, and let P be a linear differential operator of order k from V to W, with coefficients at least k times differentiable. It turns out that there is a unique linear differential operator P^* of order k from W to V, with continuous coefficients, such that $\langle Pv, w \rangle_W = \langle v, P^*w \rangle_V$ whenever $v \in L^2_k(V)$ and $w \in L^2_k(W)$. This operator P^* is called the *adjoint* or *formal adjoint* of P. We have already met an example of this in §1.1.2, where the adjoint d* of the exterior derivative d was explicitly constructed.

Here are some properties of adjoint operators. We have $(P^*)^* = P$ for any P. If P is smooth then P^* is smooth. If $V = W$ and $P = P^*$, then P is called *self-adjoint*; the Laplacian Δ on functions or k-forms is an example of a self-adjoint elliptic operator. If P is elliptic then P^* is elliptic, and if P is overdetermined elliptic then P^* is underdetermined elliptic, and vice versa.

One can write down an explicit formula for P^* in terms of the coefficients of P and the metric g. Because of this, adjoint operators are still well-defined when the manifold M is not compact, or has nonempty boundary. However, in these cases the equation $\langle Pv, w \rangle_W = \langle v, P^*w \rangle_V$ no longer holds and must be modified.

1.4 Regularity of solutions of elliptic equations

Let M be a compact manifold and V, W vector bundles over M, and suppose P is a smooth linear elliptic operator of order k from V to W. Consider the equation $Pv = w$. Clearly, if $v \in C^{k+l}(V)$ then $w \in C^l(W)$, as w is a function of v and its first k derivatives, all of which are l times differentiable. It is natural to ask whether the converse holds, that is, if $w \in C^l(W)$, is it necessarily true that $v \in C^{k+l}(V)$?

In fact this is false, and an example is given by Morrey [165, p. 54]. However, it is in general true that for $\alpha \in (0, 1)$, if $w \in C^{l,\alpha}(W)$ then $v \in C^{k+l,\alpha}(V)$, and for $p > 1$, if $w \in L^p_l(W)$ then $v \in L^p_{k+l}(V)$. One way to interpret this is that if v is the solution of a linear elliptic equation, then v must be as smooth as the problem allows it to be. This property is called *elliptic regularity*. The main reason that Hölder and Sobolev spaces are used a lot in analysis, instead of the simpler C^k spaces, is that they have this regularity property but the C^k spaces do not.

Let us begin by quoting a rather general elliptic regularity result, taken from [33, Th. 27, Th. 31, p. 463–4]. For a proof, see [165, Th. 6.4.8, p. 251].

Theorem 1.4.1 *Suppose M is a compact Riemannian manifold, V, W are vector bundles over M of the same dimension, and P is a smooth, linear, elliptic differential operator of order k from V to W. Let $\alpha \in (0, 1)$, $p > 1$, and $l \geqslant 0$ be an integer.*

Suppose that $P(v) = w$ holds weakly, with $v \in L^1(V)$ and $w \in L^1(W)$. If $w \in C^\infty(W)$, then $v \in C^\infty(V)$. If $w \in L^p_l(W)$ then $v \in L^p_{k+l}(V)$, and

$$\|v\|_{L^p_{k+l}} \leqslant C(\|w\|_{L^p_l} + \|v\|_{L^1}), \tag{1.7}$$

for some $C > 0$ independent of v, w. If $w \in C^{l,\alpha}(W)$, then $v \in C^{k+l,\alpha}(V)$, and

$$\|v\|_{C^{k+l,\alpha}} \leqslant C(\|w\|_{C^{l,\alpha}} + \|v\|_{C^0}), \tag{1.8}$$

for some $C > 0$ independent of v, w.

The estimates (1.7) and (1.8) are called the L^p *estimates* and *Schauder estimates* for P respectively. Theorem 1.4.1 is for smooth linear elliptic operators. However, when studying nonlinear problems in analysis, it is often necessary to deal with linear elliptic operators that are not smooth. Here are the Schauder estimates for operators with Hölder continuous coefficients, taken from the same references as the previous result.

Theorem 1.4.2 *Suppose M is a compact Riemannian manifold, V, W are vector bundles over M of the same dimension, and P is a linear, elliptic differential operator of order k from V to W. Let $\alpha \in (0, 1)$ and $l \geqslant 0$ be an integer. Suppose that the coefficients of P are in $C^{l,\alpha}$, and that $P(v) = w$ for some $v \in C^{k,\alpha}(V)$ and $w \in C^{l,\alpha}(W)$. Then $v \in C^{k+l,\alpha}(V)$, and*

$$\|v\|_{C^{k+l,\alpha}} \leqslant C(\|w\|_{C^{l,\alpha}} + \|v\|_{C^0}), \tag{1.9}$$

for some constant C independent of v, w.

1.4.1 How elliptic regularity results are proved

We shall now digress briefly to explain how the proofs of results like Theorems 1.4.1 and 1.4.2 work. For simplicity we will confine our attention to linear elliptic operators of order 2 on functions, but the proofs in the more general cases follow similar lines.

First, let $n > 2$ and consider $\mathbb{R}^n$ with coordinates $(x_1, \ldots, x_n)$, with the Euclidean metric $(dx_1)^2 + \cdots + (dx_n)^2$. The Laplacian Δ on $\mathbb{R}^n$ is given by

$$\Delta u = -\sum_{j=1}^{n} \frac{\partial^2 u}{(\partial x_j)^2}.$$

Define a function $\Gamma : \mathbb{R}^n \setminus \{0\} \to \mathbb{R}$ by

$$\Gamma(x) = \frac{1}{(n-2)\Omega_{n-1}} |x|^{2-n},$$

where Ω_{n-1} is the volume of the unit sphere $\mathcal{S}^{n-1}$ in $\mathbb{R}^n$. Then $\Delta \Gamma(x) = 0$ for $x \neq 0$ in $\mathbb{R}^n$. Now suppose that $\Delta u = f$, for u, f real functions on $\mathbb{R}^n$. It turns out that if $u(x)$ and $f(x)$ decay sufficiently fast as $x \to \infty$ in $\mathbb{R}^n$, we have

$$u(y) = \int_{x \in \mathbb{R}^n} \Gamma(x-y) f(x) dx. \tag{1.10}$$

This is called *Green's representation* for u, and can be found in [83, §2.4].

Because (1.10) gives u in terms of f, if we know something about f or its derivatives, we can deduce something about u. For instance, differentiating (1.10) with respect to x_j, we see that

$$\frac{\partial u}{\partial x_j}(y) = -\int_{x\in\mathbb{R}^n} \frac{\partial \Gamma(x-y)}{\partial x_j} f(x)dx = \int_{x\in\mathbb{R}^n} \Gamma(x-y)\frac{\partial f}{\partial x_j}(x)dx$$

by integration by parts, provided $\partial f/\partial x_j$ exists, and using this equation one can deduce bounds on ∇u. Working directly from (1.10), one can deduce L^p estimates and Schauder estimates analogous to those in Theorem 1.4.1, for the operator Δ on $\mathbb{R}^n$.

Now Δ is an operator with *constant coefficients*, that is, the coefficients are constant in coordinates. The next stage in the proof is to extend the results to operators P with *variable coefficients*. The idea is to approximate P by an operator P' with constant coefficients in a small open set, and then use results about elliptic operators with constant coefficients proved using the Green's representation. For the approximation of P by P' to be a good approximation, it is necessary that the coefficients of P should not vary too quickly. This can be ensured, for instance, by supposing the coefficients of P to be Hölder continuous with some given bound on their Hölder norm.

As an example, here is a result on Schauder estimates for operators P with Hölder continuous coefficients, part of which will be needed in Chapter 5.

Theorem 1.4.3 *Let B_1, B_2 be the balls of radius 1, 2 in $\mathbb{R}^n$. Let P be a linear elliptic operator of order 2 on functions on B_2, defined by*

$$Pu(x) = a^{ij}(x)\frac{\partial^2 u}{\partial x_i \partial x_j}(x) + b^i(x)\frac{\partial u}{\partial x_i}(x) + c(x)u(x). \tag{1.11}$$

Let $\alpha \in (0,1)$. Suppose the coefficients a^{ij}, b^i and c lie in $C^{0,\alpha}(B_2)$ and there are constants $\lambda, \Lambda > 0$ such that $|a^{ij}(x)\xi_i\xi_j| \geq \lambda |\xi|^2$ for all $x \in B_2$ and $\xi \in \mathbb{R}^n$, and $\|a^{ij}\|_{C^{0,\alpha}} \leq \Lambda$, $\|b^i\|_{C^{0,\alpha}} \leq \Lambda$, and $\|c\|_{C^{0,\alpha}} \leq \Lambda$ on B_2 for all $i,j = 1,\ldots,n$. Then there exist constants C, D depending on n, α, λ and Λ, such that whenever $u \in C^2(B_2)$ and $f \in C^{0,\alpha}(B_2)$ with $Pu = f$, we have $u|_{B_1} \in C^{2,\alpha}(B_1)$ and

$$\|u|_{B_1}\|_{C^{2,\alpha}} \leq C(\|f\|_{C^{0,\alpha}} + \|u\|_{C^0}), \tag{1.12}$$

and whenever $u \in C^2(B_2)$ and f is bounded, then $u|_{B_1} \in C^{1,\alpha}(B_1)$ and

$$\|u|_{B_1}\|_{C^{1,\alpha}} \leq D(\|f\|_{C^0} + \|u\|_{C^0}). \tag{1.13}$$

More generally, let $l \geq 0$ be an integer and $\alpha \in (0,1)$. Suppose the coefficients a^{ij}, b^i and c lie in $C^{l,\alpha}(B_2)$ and there are constants $\lambda, \Lambda > 0$ such that $|a^{ij}(x)\xi_i\xi_j| \geq \lambda|\xi|^2$ for all $x \in B_2$ and $\xi \in \mathbb{R}^n$, and $\|a^{ij}\|_{C^{l,\alpha}} \leq \Lambda$, $\|b^i\|_{C^{l,\alpha}} \leq \Lambda$, and $\|c\|_{C^{l,\alpha}} \leq \Lambda$ on B_2 for all $i,j = 1,\ldots,n$. Then there exists a constant C depending on n, l, α, λ and Λ such that whenever $u \in C^2(B_2)$ and $f \in C^{l,\alpha}(B_2)$ with $Pu = f$, we have $u|_{B_1} \in C^{l+2,\alpha}(B_1)$ and

$$\|u|_{B_1}\|_{C^{l+2,\alpha}} \leq C(\|f\|_{C^{l,\alpha}} + \|u\|_{C^0}). \tag{1.14}$$

Here the estimates (1.12) and (1.14) follow from [83, Th. 6.2, Th. 6.17], and also from [165, Th. 5.6.2]. Estimate (1.13) follows from Morrey [165, Th. 1.5.1]. In fact, Morrey shows that the norm $\|f\|_{L^{n/(1-\alpha)}}$ rather than $\|f\|_{C^0}$ is sufficient in (1.13).

Theorem 1.4.3 specifies exactly what C and D in equations (1.12)–(1.14) depend on, and this is worth looking at. The inequality $|a^{ij}(x)\xi_i\xi_j| \geq \lambda|\xi|^2$ implies that P is elliptic, by definition, so that the constant $\lambda > 0$ represents a sort of lower bound for the ellipticity of P. The constants C and D also depend on Λ, which is a bound for the coefficients of P in $C^{0,\alpha}$ or $C^{k,\alpha}$. Thus, Λ provides a measure of how close P is to being an operator with constant coefficients.

Notice that although u, f exist on B_2, the theorem only gives estimates of u on B_1, and these estimates depend on data on B_2. A result of this sort is called an *interior estimate*, because it estimates u only on the interior of the domain. Here is one reason why we must prove results of this structure. Consider the equation $\Delta u = 0$ on some domain Ω in $\mathbb{R}^n$. The *maximum principle* [83, §3] says that u cannot have a strict maximum at any point in the interior of Ω, roughly because $\Delta u > 0$ at that point. It follows that the maximum of u on Ω must occur at the *boundary* of Ω.

This illustrates the general principle that if P is a linear elliptic operator and $Pu = f$ on Ω, then u is likely to be most badly behaved, and most difficult to bound, near the boundary of Ω. Because of this, it is easier to prove an interior estimate like Theorem 1.4.3, than to estimate u on the whole of its domain.

Now Theorems 1.4.1 and 1.4.2 deal not with subsets of $\mathbb{R}^n$, but with compact manifolds. The final step in the proof of results like these goes as follows. Let M be a compact manifold. Using the compactness of M, we can find a finite set I and sets $\{X_i : i \in I\}$ and $\{Y_i : i \in I\}$, where each X_i, Y_i is an open set in M, the sets X_i form an *open cover* of M, and for each $i \in I$ we have $X_i \subset Y_i$ and the pair (X_i, Y_i) is diffeomorphic to the pair (B_1, B_2), where B_1, B_2 are the balls of radius 1,2 in $\mathbb{R}^n$.

Suppose that we know an interior estimate for linear elliptic equations $Pv = w$ on the balls B_1, B_2 in $\mathbb{R}^n$, analogous to Theorem 1.4.3. Since (X_i, Y_i) is diffeomorphic to (B_1, B_2), we may apply this estimate to (X_i, Y_i), and the result is an estimate of $v|_{X_i}$, depending on norms of $v|_{Y_i}$ and $w|_{Y_i}$. Since the sets X_i form an open cover of M, in this way we estimate v on all of M.

Using this argument, we can use interior estimates for balls in $\mathbb{R}^n$ to prove results for compact manifolds M, that estimate the solution on the whole of M. Therefore, results such as Theorems 1.4.1 and 1.4.2 should be understood as purely *local* results, that do not encode any important global information about M and P.

1.5 Existence of solutions of linear elliptic equations

Now we will use the elliptic regularity results of §1.4 and the Kondrakov Theorem to prove some basic facts about linear elliptic operators. Our first result shows that the kernel of a linear elliptic operator on a compact manifold is very well-behaved.

Theorem 1.5.1 *Let V, W be vector bundles over a compact manifold M, and let P be a smooth linear elliptic operator of order k from V to W. Then P acts by $P : C^\infty(V) \to C^\infty(W)$, $P : C^{k+l,\alpha}(V) \to C^{l,\alpha}(W)$ and $P : L^q_{k+l}(V) \to L^q_l(W)$. The kernel* Ker P

of P is the same for all of these actions, and is a finite-dimensional vector subspace of $C^\infty(V)$.

Proof If $v \in \operatorname{Ker} P$ then $Pv = 0$. Since $0 \in C^\infty(W)$, Theorem 1.4.1 shows that $v \in C^\infty(V)$. Thus $\operatorname{Ker} P$ lies in $C^\infty(V)$, and is therefore the same for all three actions above. Let $\alpha \in (0, 1)$, and define $B = \{v \in \operatorname{Ker} P : \|v\|_{C^{k,\alpha}} \leqslant 1\}$, so that B is the unit ball in $\operatorname{Ker} P$ in the $C^{k,\alpha}$ norm. The Kondrakov Theorem, Theorem 1.2.3, shows that the inclusion $C^{k,\alpha}(V) \hookrightarrow C^k(V)$ is compact. Therefore B lies in a compact subset of $C^k(V)$, and the closure $\overline{B}$ of B in $C^k(V)$ is compact.

But $P : C^k(V) \to C^0(W)$ is continuous, and $P(b) = 0 \in C^0(W)$ if $b \in B$. Thus $P(b') = 0$ if $b' \in \overline{B}$, so $\overline{B} \subset \operatorname{Ker} P$. Since $\operatorname{Ker} P \subset C^{k,\alpha}(V)$, we see that $B = \overline{B}$, and B is a *compact* topological space. Now the only Banach spaces with compact unit balls are finite-dimensional, so $\operatorname{Ker} P$ is finite-dimensional, as we have to prove. □

Now let M be a compact Riemannian manifold and V, W vector bundles over M equipped with metrics in the fibres. Let P be a smooth linear elliptic operator from V to W. Recall from §1.3.5 that $L^2(V)$ has an inner product $\langle \, , \, \rangle_V$. If A is a vector subspace of $L^2(V)$ and $v \in L^2(V)$, we say that $v \perp A$ if $\langle v, a \rangle_V = 0$ for all $a \in A$. Using this notation, we shall prove:

Proposition 1.5.2 *Let V, W be vector bundles over a compact Riemannian manifold M, equipped with metrics in the fibres, and let P be a smooth linear elliptic operator of order k from V to W. Let $l \geqslant 0$ be an integer, and let $\alpha \in (0, 1)$. Then there is a constant $D > 0$ such that if $v \in C^{k+l,\alpha}(V)$ and $v \perp \operatorname{Ker} P$, then $\|v\|_{C^{k+l,\alpha}} \leqslant D\|Pv\|_{C^{l,\alpha}}$.*

Similarly, if $p > 1$ and $l \geqslant 0$ is an integer, there is a constant $D > 0$ such that if $v \in L^p_{k+l}(V)$ and $v \perp \operatorname{Ker} P$, then $\|v\|_{L^p_{k+l}} \leqslant D\|Pv\|_{L^p_l}$.

Proof For simplicity, we will prove only the case $\|v\|_{C^{k,\alpha}} \leqslant D\|Pv\|_{C^{0,\alpha}}$. The proofs in the other cases work in exactly the same way, and are left to the reader. Define a subset S of $C^{k,\alpha}(V)$ by $S = \{v \in C^{k,\alpha}(V) : v \perp \operatorname{Ker} P \text{ and } \|v\|_{C^{k,\alpha}} = 1\}$, and let $\gamma = \inf\{\|Ps\|_{C^{0,\alpha}} : s \in S\}$. Suppose for a contradiction that $\gamma = 0$. Then we can choose a sequence $\{s_j\}_{j=1}^\infty$ in S such that $\|Ps_j\|_{C^{0,\alpha}} \to 0$ as $j \to \infty$. Now S is bounded in $C^{k,\alpha}(V)$ and the inclusion $C^{k,\alpha}(V) \hookrightarrow C^k(V)$ is compact, by the Kondrakov Theorem. Therefore there exists a subsequence $\{s_{i_j}\}_{j=1}^\infty$ that converges in $C^k(V)$ to some $s' \in C^k(V)$.

As $s_{i_j} \to s'$ in C^k we see that $Ps_{i_j} \to Ps'$ in C^0. But $\|Ps_{i_j}\|_{C^{0,\alpha}} \to 0$, and $\|Ps_{i_j}\|_{C^0} \leqslant \|Ps_{i_j}\|_{C^{0,\alpha}}$. Thus $Ps' = 0$ and $s' \in \operatorname{Ker} P$, so that $s' \in C^{k,\alpha}(V)$. Now by Theorem 1.4.1, there is a constant C such that $\|v\|_{C^{k,\alpha}} \leqslant C(\|Pv\|_{C^{0,\alpha}} + \|v\|_{C^0})$ for all $v \in C^{k,\alpha}(V)$. Therefore

$$\|s_{i_j} - s'\|_{C^{k,\alpha}} \leqslant C(\|Ps_{i_j}\|_{C^{0,\alpha}} + \|s_{i_j} - s'\|_{C^0})$$

for each j, since $Ps' = 0$. But $\|Ps_{i_j}\|_{C^{0,\alpha}} \to 0$ as $j \to \infty$, and $\|s_{i_j} - s'\|_{C^0} \to \infty$ as $j \to \infty$ because s_{i_j} converges to s' in C^k and so in C^0. Thus $\|s_{i_j} - s'\|_{C^{k,\alpha}} \to 0$ as $j \to \infty$. But S is closed in $C^{k,\alpha}(V)$, and therefore $s' \in S$.

As $s' \in S$, we have $s' \perp \operatorname{Ker} P$. But also $s' \in \operatorname{Ker} P$, from above. So $s' = 0$. However, $\|s'\|_{C^{k,\alpha}} = 1$ since $s' \in S$, a contradiction. Therefore $\gamma > 0$. Put $D = \gamma^{-1}$. Then for all $s \in S$ we have $\|s\|_{C^{k,\alpha}} = 1 \leqslant D\|Ps\|_{C^{0,\alpha}}$, by definition of γ. But any $v \in C^{k,\alpha}(V)$ with $v \perp \operatorname{Ker} P$ can be written $v = \lambda s$ for some $s \in S$, and so $\|v\|_{C^{k,\alpha}} \leqslant D\|Pv\|_{C^{0,\alpha}}$, as we have to prove. □

From §1.3.5, if V, W are vector bundles, with metrics on the fibres, over a compact Riemannian manifold M, and P is a smooth linear elliptic operator from V to W, then there is a smooth linear elliptic operator P^* from W to V. Our next result is an *existence* result for the equation $Pv = w$, as it gives a simple condition on w, that $w \perp \operatorname{Ker} P^*$, for there to exist a solution v. This is called the *Fredholm alternative*.

Theorem 1.5.3 *Suppose V, W are vector bundles over a compact Riemannian manifold M, equipped with metrics in the fibres, and P is a smooth linear elliptic operator of order k from V to W. Let $l \geqslant 0$ be an integer, let $p > 1$, and let $\alpha \in (0, 1)$. Then the images of the maps*

$$P : C^{k+l,\alpha}(V) \to C^{l,\alpha}(W) \quad \text{and} \quad P : L^p_{k+l}(V) \to L^p_l(W)$$

are closed linear subspaces of $C^{l,\alpha}(W)$ and $L^p_l(W)$ respectively. If $w \in C^{l,\alpha}(W)$ then there exists $v \in C^{k+l,\alpha}(V)$ with $Pv = w$ if and only if $w \perp \operatorname{Ker} P^*$, and if one requires that $v \perp \operatorname{Ker} P$ then v is unique. Similarly, if $w \in L^p_l(W)$ then there exists $v \in L^p_{k+l}(V)$ with $Pv = w$ if and only if $w \perp \operatorname{Ker} P^*$, and if $v \perp \operatorname{Ker} P$ then v is unique.

Proof Let $\{w_j\}_{j=1}^\infty$ be a sequence in $P\big[C^{k+l,\alpha}(V)\big]$ that converges to some w' in $C^{l,\alpha}(W)$. Then for each w_j there exists a unique $v_j \in C^{k+l,\alpha}(V)$ such that $v_j \perp \operatorname{Ker} P$ and $Pv_j = w_j$. Applying Proposition 1.5.2 we see that for all i, j, $\|v_i - v_j\|_{C^{k+l,\alpha}} \leqslant D\|w_i - w_j\|_{C^{l,\alpha}}$, for D some constant. Since $\{w_j\}_{j=1}^\infty$ converges in $C^{l,\alpha}(W)$, $\|w_i - w_j\|_{C^{l,\alpha}} \to 0$ as $i, j \to \infty$, and therefore $\|v_i - v_j\|_{C^{k+l,\alpha}} \to 0$ as $i, j \to \infty$, and $\{v_j\}_{j=1}^\infty$ is a *Cauchy sequence* in $C^{k+l,\alpha}(V)$.

As $C^{k+l,\alpha}(V)$ is a Banach space and therefore complete, the sequence $\{v_j\}_{j=1}^\infty$ converges to some $v' \in C^{k+l,\alpha}(V)$. By continuity, $P(v') = w'$, so that $w' \in P\big[C^{k+l,\alpha}(V)\big]$. Therefore $P\big[C^{k+l,\alpha}(V)\big]$ contains its limit points, and is a *closed* linear subspace of $C^{l,\alpha}(W)$. Similarly, $P\big[L^p_{k+l}(V)\big]$ is closed in $L^p_l(W)$. This proves the first part of the theorem.

By definition of P^*, if $v \in L^2_k(V)$ and $w \in L^2_k(W)$, then $\langle v, P^*w \rangle_V = \langle Pv, w \rangle_W$. It follows that if $w \in C^{l,\alpha}(W)$, then $w \in \operatorname{Ker} P^*$ if and only if $\langle Pv, w \rangle_W = 0$ for all $v \in C^{k+l,\alpha}(V)$. So, $\operatorname{Ker} P^*$ is the orthogonal subspace to $P\big[C^{k+l,\alpha}(V)\big]$. But $P\big[C^{k+l,\alpha}(V)\big]$ is closed. Therefore, if $w \in C^{l,\alpha}(W)$, then $w \perp \operatorname{Ker} P^*$ if and only if $w \in P\big[C^{k+l,\alpha}(V)\big]$, that is, if and only if there exists $v \in C^{k+l,\alpha}(V)$ with $Pv = w$. Clearly, we may add some element of $\operatorname{Ker} P$ to v to make $v \perp \operatorname{Ker} P$, and then v is unique. This proves the second part of the theorem. The last part follows by the same method. □

From elementary linear algebra, if A, B are finite-dimensional inner product spaces and $L : A \to B$ is a linear map, then $\operatorname{Ker} L$ and $\operatorname{Ker} L^*$ are finite-dimensional vector subspaces of A, B. For given $b \in B$, the equation $La = b$ has a solution $a \in A$ if and only if $b \perp \operatorname{Ker} L^*$, and two solutions differ by an element of $\operatorname{Ker} L$. Now by Theorems 1.5.1 and 1.5.3, these properties also hold for linear elliptic operators $P : C^{k+l,\alpha}(V) \to C^{l,\alpha}(W)$ or $P : L^p_{k+l}(V) \to L^p_k(W)$. Thus, linear elliptic operators behave very like linear operators on *finite-dimensional* vector spaces.

This gives us a way of thinking about linear elliptic operators. In the situation of theorem 1.5.3, define $E = \{v \in C^{k+l,\alpha}(V) : v \perp \operatorname{Ker} P\}$ and $F = \{w \in C^{l,\alpha}(W) : w \perp \operatorname{Ker} P^*\}$. Then $C^{k+l,\alpha}(V) = \operatorname{Ker} P \oplus E$ and $C^{l,\alpha}(W) = \operatorname{Ker} P^* \oplus F$, and the theorem implies that $P : E \to F$ is a *linear homeomorphism*, that is, it is both an invertible linear map and an isomorphism of E and F as topological spaces.

Now $\operatorname{Ker} P$ and $\operatorname{Ker} P^*$ are finite-dimensional, and E and F are infinite-dimensional. In some sense, P is close to being an *invertible* map between the infinite-dimensional spaces $C^{k+l,\alpha}(V)$ and $C^{l,\alpha}(W)$, as $P : E \to F$ is invertible, and it is only the finite-dimensional pieces $\operatorname{Ker} P$ and $\operatorname{Ker} P^*$ that cause the problem. Because of this, the existence and uniqueness of solutions of linear elliptic equations can be reduced, more-or-less, to finite-dimensional linear algebra. In contrast, non-elliptic linear differential equations are truly infinite-dimensional problems, and are more difficult to deal with.

Here is another example of the analogy between linear elliptic operators and finite-dimensional linear algebra. If $L : A \to B$ is a linear map of finite-dimensional inner product spaces A, B, then $\dim \operatorname{Ker} L - \dim \operatorname{Ker} L^* = \dim A - \dim B$. Thus the integer $\dim \operatorname{Ker} L - \dim \operatorname{Ker} L^*$ depends only on A and B, and is independent of L. Now let V, W be vector bundles over a compact Riemannian manifold M, with metrics in the fibres, and let P be a smooth linear elliptic operator of order k from V to W. Define the *index of* P, $\operatorname{ind} P$, by

$$\operatorname{ind} P = \dim \operatorname{Ker} P - \dim \operatorname{Ker} P^*.$$

Then $\operatorname{ind} P$ is an integer, and is well-defined. The *Atiyah–Singer Index Theorem* [15] gives a formula for $\operatorname{ind} P$ in terms of topological invariants of the symbol $\sigma(P)$. That is, the index of P is actually a topological invariant. It is unchanged by deformations of P that preserve ellipticity, and in this sense is independent of P.

Finally, here is a version of the results of this section for operators with $C^{l,\alpha}$ coefficients. To prove it, follow the proofs above but apply Theorem 1.4.2 instead of Theorem 1.4.1 wherever it occurs. The reason for requiring $l \geq k$ is in order that P^* should exist.

Theorem 1.5.4 *Let $k > 0$ and $l \geq k$ be integers, and $\alpha \in (0, 1)$. Suppose V, W are vector bundles over a compact Riemannian manifold M, equipped with metrics in the fibres, and P is a linear elliptic operator of order k from V to W with $C^{l,\alpha}$ coefficients. Then P^* is elliptic with $C^{l-k,\alpha}$ coefficients, and $\operatorname{Ker} P$, $\operatorname{Ker} P^*$ are finite-dimensional subspaces of $C^{k+l,\alpha}(V)$ and $C^{l,\alpha}(W)$ respectively. If $w \in C^{l,\alpha}(W)$ then there exists $v \in C^{k+l,\alpha}(V)$ with $Pv = w$ if and only if $w \perp \operatorname{Ker} P^*$, and if one requires that $v \perp \operatorname{Ker} P$ then v is unique.*

2
INTRODUCTION TO CONNECTIONS, CURVATURE AND HOLONOMY GROUPS

In this chapter we will introduce the theory of connections, focussing in particular on two topics, the *curvature* and the *holonomy group* of a connection. Connections can be defined in two different sorts of bundle, that is, *vector bundles* and *principal bundles*. Both definitions will be given in §2.1.

Sections 2.2–2.4 define the holonomy group of a connection in a vector or principal bundle, and explain some of its basic properties, including its relationship with the curvature of the connection. The curvature is a *local* invariant of the connection, since it varies from point to point on the manifold, whereas the holonomy group is a *global* invariant, as it is independent of any base point in the manifold.

Section 2.5 considers connections on the tangent bundle TM of a manifold M, defines the *torsion* of a connection on TM, and discusses the holonomy groups of *torsion-free* connections. Finally, §2.6 defines *G-structures* on a manifold and considers the question of existence and uniqueness of torsion-free connections compatible with a G-structure. For a more detailed introduction to connections and holonomy groups, see Kobayashi and Nomizu [133, Ch. 2, App. 4,5,7].

2.1 Bundles, connections and curvature

We now discuss connections, and their curvature. Connections can be defined in two settings: vector bundles and principal bundles. These two concepts are different, but very closely related. We will define both kinds of connection, and explain the links between them.

2.1.1 Vector bundles and principal bundles

We begin by defining vector bundles and principal bundles.

Definition 2.1.1 Let M be a manifold. A *vector bundle* E over M is a fibre bundle whose fibres are (real or complex) vector spaces. That is, E is a manifold equipped with a smooth projection $\pi : E \to M$. For each $m \in M$ the fibre $E_m = \pi^{-1}(m)$ has the structure of a vector space, and there is an open neighbourhood U_m of m such that $\pi^{-1}(U_m) \cong U_m \times V$, where V is the fibre of E.

Now let M be a manifold, and G a Lie group. A *principal bundle* P over M with fibre G is a manifold P equipped with a smooth projection $\pi : P \to M$, and an action of G on P, which we will write as $p \xmapsto{g} g \cdot p$, for $g \in G$ and $p \in P$. This G-action must be smooth and free, and the projection $\pi : P \to M$ must be a fibration, with fibres the orbits of the G-action, so that for each $m \in M$ the fibre $\pi^{-1}(m)$ is a copy of G.

BUNDLES, CONNECTIONS AND CURVATURE

Vector bundles and principal bundles are basic tools in differential geometry. Many geometric structures can be defined using either vector or principal bundles. Thus vector and principal bundles often provide two different but equivalent approaches to the same problem, and it is useful to understand both.

We shall explain the links between vector and principal bundle methods by showing how to translate from one to the other, and back. First, here is a way to go from vector to principal bundles.

Definition 2.1.2 Let M be a manifold, and $E \to M$ a vector bundle with fibre $\mathbb{R}^k$. Define a manifold F^E by

$$F^E = \{(m, e_1, \ldots, e_k) : m \in M \text{ and } (e_1, \ldots, e_k) \text{ is a basis for } E_m\}.$$

Define $\pi : F^E \to M$ by $\pi : (m, e_1, \ldots, e_k) \mapsto m$. For each $A = (A_{ij})$ in $\text{GL}(k, \mathbb{R})$ and $(m, e_1, \ldots, e_k)$ in F^E, define $A \cdot (m, e_1, \ldots, e_k) = (m, e'_1, \ldots, e'_k)$, where $e'_i = \sum_{j=1}^k A_{ij} e_j$. This gives an action of $\text{GL}(k, \mathbb{R})$ on F^E, which makes F^E into a principal bundle over M, with fibre $\text{GL}(k, \mathbb{R})$. We call F^E the *frame bundle* of E.

One frame bundle is of particular importance. When $E = TM$, the bundle F^{TM} will be written F, and called the *frame bundle of M*.

We can also pass from principal bundles to vector bundles.

Definition 2.1.3 Suppose M is a manifold, and P a principal bundle over M with fibre G, a Lie group. Let ρ be a representation of G on a vector space V. Then G acts on the product space $P \times V$ by the principal bundle action on the first factor, and ρ on the second. Define $\rho(P) = (P \times V)/G$, the quotient of $P \times V$ by this G-action. Now $P/G = M$, so the obvious map from $(P \times V)/G$ to P/G yields a projection from $\rho(P)$ to M. Since G acts freely on P, this projection has fibre V, and thus $\rho(P)$ is a *vector bundle* over M, with fibre V.

These two constructions are inverse, in the sense that if ρ is the canonical representation of $\text{GL}(k, \mathbb{R})$ on $\mathbb{R}^k$ then $E \cong \rho(F^E)$. This gives a 1-1 correspondence between vector bundles over M with fibre $\mathbb{R}^k$, and principal bundles over M with fibre $\text{GL}(k, \mathbb{R})$. But any Lie group G can be the fibre of a principal bundle, and not just $G = \text{GL}(k, \mathbb{R})$, so principal bundles are more general than vector bundles.

Let P be a principal bundle over M with fibre G, let $\mathfrak{g}$ be the Lie algebra of G, and let $\text{ad} : G \to \text{GL}(\mathfrak{g})$ be the adjoint representation of G on $\mathfrak{g}$. Definition 2.1.3 gives a natural vector bundle $\text{ad}(P)$ over M, with fibre $\mathfrak{g}$, called the *adjoint bundle*. This will be important later.

Let ρ be a representation of G on V, and $\pi : P \times V \to \rho(P)$ the natural projection. We may regard $P \times V$ as the trivial vector bundle over P with fibre V. Then if $e \in C^\infty(\rho(P))$ is a smooth section of $\rho(P)$ over M, the pull-back $\pi^*(e)$ is a smooth section of $P \times V$ over P. Moreover, $\pi^*(e)$ is invariant under the action of G on $P \times V$, and this gives a 1-1 correspondence between sections of $\rho(P)$ over M and G-invariant sections of $P \times V$ over P.

2.1.2 Connections on vector bundles

Here is the definition of a connection on a vector bundle.

Definition 2.1.4 Let M be a manifold, and $E \to M$ a vector bundle. A *connection* ∇^E on E is a linear map $\nabla^E : C^\infty(E) \to C^\infty(E \otimes T^*M)$ satisfying the condition

$$\nabla^E(\alpha\, e) = \alpha \nabla^E e + e \otimes d\alpha,$$

whenever $e \in C^\infty(E)$ is a smooth section of E and α is a smooth function on M. If ∇^E is such a connection, $e \in C^\infty(E)$, and $v \in C^\infty(TM)$ is a vector field, then we write $\nabla^E_v e = v \cdot \nabla^E e \in C^\infty(E)$, where '$\cdot$' contracts together the TM and T^*M factors in v and $\nabla^E e$. Then if $v \in C^\infty(TM)$ and $e \in C^\infty(E)$ and α, β are smooth functions on M, we have

$$\nabla^E_{\alpha v}(\beta e) = \alpha\beta \nabla^E_v e + \alpha(v \cdot \beta)e. \tag{2.1}$$

Here $v \cdot \beta$ is the Lie derivative of β by v. It is a smooth function on M, and could also be written $v \cdot d\beta$.

Suppose E is a vector bundle with fibre $\mathbb{R}^k$ over M, and let $e_1, \ldots, e_k$ be smooth sections of E over some open set $U \subset M$, that form a basis of E at each point of U. Then every smooth section of E over U can be written uniquely as $\sum_{i=1}^k \alpha_i e_i$, where $\alpha_1, \ldots, \alpha_k$ are smooth functions on U. Let $f_1, \ldots, f_k$ be any smooth sections of $E \otimes T^*M$ over U, and define

$$\nabla^E \left[\sum_{i=1}^k \alpha_i e_i \right] = \sum_{i=1}^k (\alpha_i f_i + e_i \otimes d\alpha_i) \tag{2.2}$$

for all smooth functions $\alpha_1, \ldots, \alpha_k$ on U. Then ∇^E is a connection on E over U, and moreover, every connection on E over U can be written uniquely in this way.

Next we explain how to define the *curvature* of a connection on a vector bundle. Curvature is a very important topic in geometry, and there are a number of ways to define it. The approach we take uses vector fields, and the Lie bracket of vector fields. Let M be a manifold, and E a vector bundle over M. Write $\text{End}(E) = E \otimes E^*$, where E^* is the dual vector bundle to E. Let ∇^E be a connection on E. Then the curvature $R(\nabla^E)$ of the connection ∇^E is a smooth section of the vector bundle $\text{End}(E) \otimes \Lambda^2 T^*M$, defined as follows.

Proposition 2.1.5 *Let M be a manifold, E a vector bundle over M, and ∇^E a connection on E. Suppose that $v, w \in C^\infty(TM)$ are vector fields and $e \in C^\infty(E)$, and that α, β, γ are smooth functions on M. Then*

$$\nabla^E_{\alpha v} \nabla^E_{\beta w}(\gamma e) - \nabla^E_{\beta w} \nabla^E_{\alpha v}(\gamma e) - \nabla^E_{[\alpha v, \beta w]}(\gamma e) = \\ \alpha\beta\gamma \cdot \left\{ \nabla^E_v \nabla^E_w e - \nabla^E_w \nabla^E_v e - \nabla^E_{[v,w]} e \right\}, \tag{2.3}$$

where $[v, w]$ is the Lie bracket. Thus the expression $\nabla^E_v \nabla^E_w e - \nabla^E_w \nabla^E_v e - \nabla^E_{[v,w]} e$ is pointwise-linear in v, w and e. Also, it is clearly antisymmetric in v and w. Therefore *there exists a unique, smooth section* $R(\nabla^E) \in C^\infty(\text{End}(E) \otimes \Lambda^2 T^*M)$ *called the* curvature *of* ∇^E, *that satisfies the equation*

$$R(\nabla^E) \cdot (e \otimes v \wedge w) = \nabla^E_v \nabla^E_w e - \nabla^E_w \nabla^E_v e - \nabla^E_{[v,w]} e \qquad (2.4)$$

for all $v, w \in C^\infty(TM)$ *and* $e \in C^\infty(E)$.

Proof If $v, w \in C^\infty(TM)$ and α, β are smooth functions on M, then $[\alpha v, \beta w] = \alpha\beta[v, w] + \alpha(v \cdot \beta)w - \beta(w \cdot \alpha)v$. Using this and (2.1) to expand the terms on the left hand side of (2.3), we see that

$$\begin{aligned}
\nabla^E_{\alpha v}\nabla^E_{\beta w}(\gamma e) &= \alpha\beta\gamma\nabla^E_v\nabla^E_w e + \alpha\beta(w\cdot\gamma)\nabla^E_v e + \{\alpha\beta(v\cdot\gamma) + \alpha(v\cdot\beta)\gamma\}\nabla^E_w e \\
&\quad + \{\alpha(v\cdot\beta)(w\cdot\gamma) + \alpha\beta(v\cdot(w\cdot\gamma))\}e, \\
\nabla^E_{\beta w}\nabla^E_{\alpha v}(\gamma e) &= \alpha\beta\gamma\nabla^E_w\nabla^E_v e + \{\alpha\beta(w\cdot\gamma) + (w\cdot\alpha)\beta\gamma\}\nabla^E_v e + \alpha\beta(v\cdot\gamma)\nabla^E_w e \\
&\quad + \{(w\cdot\alpha)\beta(v\cdot\gamma) + \alpha\beta(w\cdot(v\cdot\gamma))\}e, \\
\nabla^E_{[\alpha v,\beta w]}(\gamma e) &= \alpha\beta\gamma\nabla^E_{[v,w]}e - (w\cdot\alpha)\beta\gamma\nabla^E_v e + \alpha(v\cdot\beta)\gamma\nabla^E_w e \\
&\quad + \{\alpha\beta([v,w]\cdot\gamma) + \alpha(v\cdot\beta)(w\cdot\gamma) - (w\cdot\alpha)\beta(v\cdot\gamma)\}e.
\end{aligned}$$

Combining these equations with the identity $v \cdot (w \cdot \gamma) - w \cdot (v \cdot \gamma) = [v, w] \cdot \gamma$, after some cancellation we prove (2.3), and the proposition follows. □

Here is one way to understand the curvature of ∇^E. Let $(x_1, \ldots, x_n)$ be local coordinates on M, and define $v_i = \partial/\partial x_i$ for $i = 1, \ldots, n$. Then v_i is a vector field on M, and $[v_i, v_j] = 0$. Let e be a smooth section of E. Then we may interpret $\nabla^E_{v_i} e$ as a kind of partial derivative $\partial e/\partial x_i$ of e. Using (or abusing) this partial derivative notation, equation (2.4) implies that

$$R(\nabla^E) \cdot (e \otimes v_i \wedge v_j) = \frac{\partial^2 e}{\partial x_i \partial x_j} - \frac{\partial^2 e}{\partial x_j \partial x_i}. \qquad (2.5)$$

Now, partial derivatives of functions commute, so $\partial^2 f/\partial x_i \partial x_j = \partial^2 f/\partial x_j \partial x_i$ if f is a smooth function on M. However, this does not hold for sections of E, as (2.5) shows that *the curvature* $R(\nabla^E)$ *measures how much partial derivatives in E fail to commute*.

2.1.3 *Connections on principal bundles*

Suppose P is a principal bundle over a manifold M, with fibre G and projection $\pi : P \to M$. Let $p \in P$, and set $m = \pi(p)$. Then the derivative of π gives a linear map $d\pi_p : T_pP \to T_mM$. Define a subspace C_p of T_pP by $C_p = \text{Ker}(d\pi_p)$. Then the subspaces C_p form a vector subbundle C of the tangent bundle TP, called the *vertical subbundle*. Note that C_p is $T_p(\pi^{-1}(m))$, the tangent space to the fibre of $\pi : P \to M$ over m. But the fibres of π are the orbits of the free G-action on P. It follows that there is a natural isomorphism $C_p \cong \mathfrak{g}$ between C_p and the Lie algebra $\mathfrak{g}$ of G.

Here is the definition of a connection on P.

Definition 2.1.6 Let M be a manifold, and P a principal bundle over M with fibre G, a Lie group. A *connection* on P is a vector subbundle D of TP called the *horizontal* *subbundle*, that is invariant under the G-action on P, and which satisfies $T_pP = C_p \oplus D_p$ for each $p \in P$. If $\pi(p) = m$, then $\mathrm{d}\pi_p$ maps $T_pP = C_p \oplus D_p$ onto T_mM, and as $C_p = \mathrm{Ker}\, \mathrm{d}\pi_p$, we see that $\mathrm{d}\pi_p$ induces an isomorphism between D_p and T_mM.

Thus the horizontal subbundle D is naturally isomorphic to $\pi^*(TM)$. So if $v \in C^\infty(TM)$ is a vector field on M, there is a unique section $\lambda(v)$ of the bundle $D \subset TP$ over P, such that $\mathrm{d}\pi_p(\lambda(v)|_p) = v|_{\pi(p)}$ for each $p \in P$. We call $\lambda(v)$ the *horizontal lift* of v. It is a vector field on P, and is invariant under the action of G on P.

We now define the *curvature* of a connection on a principal bundle. Let M be a manifold and P a principal bundle over M with fibre G, a Lie group with Lie algebra $\mathfrak{g}$, and let D be a connection on P. If $v, w \in C^\infty(TM)$ and α, β are smooth functions on M, then by a similar argument to the proof of (2.3) in Proposition 2.1.5, we can show that

$$[\lambda(\alpha v), \lambda(\beta w)] - \lambda([\alpha v, \beta w]) = \alpha\beta \cdot \left\{[\lambda(v), \lambda(w)] - \lambda([v, w])\right\},$$

where $[\,,\,]$ is the Lie bracket of vector fields. Thus the expression $[\lambda(v), \lambda(w)] - \lambda([v, w])$ is pointwise-linear and antisymmetric in v, w. Also, as $\mathrm{d}\pi(\lambda(v)) = v$ for all vector fields v on M we see that

$$\mathrm{d}\pi([\lambda(v), \lambda(w)]) = \mathrm{d}\pi(\lambda([v, w])) = [v, w].$$

Therefore, $[\lambda(v), \lambda(w)] - \lambda([v, w])$ lies in the kernel of $\mathrm{d}\pi$, which is the vertical subbundle C of TP. But there is a natural isomorphism $C_p \cong \mathfrak{g}$ for each $p \in P$, and thus we may regard $[\lambda(v), \lambda(w)] - \lambda([v, w])$ as a section of the trivial vector bundle $P \times \mathfrak{g}$ over P.

As $\lambda(v), \lambda(w)$ and $\lambda([v, w])$ are invariant under the action of G on P, this section of $P \times \mathfrak{g}$ is invariant under the natural action of G on $P \times \mathfrak{g}$. But from above there is a 1-1 correspondence between G-invariant sections of $P \times \mathfrak{g}$ over P, and sections of the adjoint bundle $\mathrm{ad}(P)$ over M. We use this to deduce the following result, which defines the *curvature* $R(P, D)$ of a connection D on P.

Proposition 2.1.7 *Let M be a manifold, G a Lie group with Lie algebra $\mathfrak{g}$, P a principal bundle over M with fibre G, and D a connection on P. Then there exists a unique, smooth section $R(P, D)$ of the vector bundle $\mathrm{ad}(P) \otimes \Lambda^2 T^*M$ called the curvature of D, that satisfies*

$$\pi^*\big(R(P, D) \cdot v \wedge w\big) = [\lambda(v), \lambda(w)] - \lambda([v, w]) \qquad (2.6)$$

for all $v, w \in C^\infty(TM)$. Here the left hand side is a $\mathfrak{g}$-valued function on P, the right hand side is a section of the subbundle $C \subset TP$, and the two sides are identified using the natural isomorphism $C_p \cong \mathfrak{g}$ for $p \in P$.

Next we relate connections on vector and principal bundles. Let M, P and G be as above. Let ρ be a representation of G on a vector space V, and define $E \to M$ to be the vector bundle $\rho(P)$ over M. Given a connection D on the principal bundle P, we will explain how to construct a unique connection ∇^E on E. Let $e \in C^\infty(E)$, so that

$\pi^*(e)$ is a section of $P \times V$ over P. Then $\pi^*(e)$ is a function $\pi^*(e) : P \to V$, so its exterior derivative is a linear map $d\pi^*(e)|_p : T_p P \to V$ for each $p \in P$. Thus $d\pi^*(e)$ is a smooth section of the vector bundle $V \otimes T^*P$ over P.

Let D be a connection on P. Then for each $p \in P$ there are isomorphisms

$$T_p P \cong C_p \oplus D_p, \qquad C_p \cong \mathfrak{g} \qquad \text{and} \qquad D_p \cong \pi^*(T_{\pi(p)} M).$$

These give a natural splitting $V \otimes T^*P \cong V \otimes \mathfrak{g}^* \oplus V \otimes \pi^*(T^*M)$. Write $\pi_D(d\pi^*(e))$ for the component of $d\pi^*(e)$ in $C^\infty(V \otimes \pi^*(T^*M))$ in this splitting. Now both $\pi^*(e)$ and the vector bundle splitting are G-invariant, so $\pi_D(d\pi^*(e))$ must be G-invariant. But there is a 1-1 correspondence between G-invariant sections of $V \otimes \pi^*(T^*M)$ over P, and sections of the corresponding vector bundle $E \otimes T^*M$ over M. Therefore $\pi_D(d\pi^*(e))$ is the pull-back of a unique element of $C^\infty(E \otimes T^*M)$. We use this to define ∇^E.

Definition 2.1.8 Suppose M is a manifold, P a principal bundle over M with fibre G, and D a connection on P. Let ρ be a representation of G on a vector space V, and define E to be the vector bundle $\rho(P)$ over M. If $e \in C^\infty(E)$, then $\pi_D(d\pi^*(e))$ is a G-invariant section of $V \otimes \pi^*(T^*M)$ over P. Define $\nabla^E e \in C^\infty(E \otimes T^*M)$ to be the unique section of $E \otimes T^*M$ with pull-back $\pi_D(d\pi^*(e))$ under the natural projection $V \otimes \pi^*(T^*M) \to E$. This defines a connection ∇^E on the vector bundle E over M.

To each connection D on a principal bundle P, we have associated a unique connection ∇^E on the vector bundle $E = \rho(P)$. If $G = \mathrm{GL}(k, \mathbb{R})$ and ρ is the standard representation of G on $\mathbb{R}^k$, so that P is the frame bundle F^E of E, then this gives a 1-1 correspondence between connections on P and E. However, for general G and ρ the map $D \mapsto \nabla^E$ may be neither injective nor surjective.

Our final result, which follows quickly from the definitions, relates the ideas of curvature of connections in vector and principal bundles.

Proposition 2.1.9 *Suppose M is a manifold, G a Lie group with Lie algebra $\mathfrak{g}$, P a principal bundle over M with fibre G, and D a connection on P, with curvature $R(P, D)$. Let ρ be a representation of G on a vector space V, E the vector bundle $\rho(P)$ over M, and ∇^E the connection given in Definition 2.1.8, with curvature $R(\nabla^E)$.*

Now $\mathfrak{g}$ and $\mathrm{End}(V)$ are representations of G, and ρ gives a G-equivariant linear map $d\rho : \mathfrak{g} \to \mathrm{End}(V)$. This induces a map $d\rho : \mathrm{ad}(P) \to \mathrm{End}(E)$ of the vector bundles $\mathrm{ad}(P)$ and $\mathrm{End}(E)$ over M corresponding to $\mathfrak{g}$ and $\mathrm{End}(V)$. Let

$$d\rho \otimes \mathrm{id} : \mathrm{ad}(P) \otimes \Lambda^2 T^*M \to \mathrm{End}(E) \otimes \Lambda^2 T^*M$$

be the product with the identity on $\Lambda^2 T^*M$. Then $(d\rho \otimes \mathrm{id})(R(P, D)) = R(\nabla^E)$.

Thus, the definitions of curvature of connections in vector and principal bundles are essentially equivalent.

2.2 Vector bundles, connections and holonomy groups

We now define the holonomy group of a connection in a vector bundle, and prove some elementary facts about it. Let M be a manifold, $E \to M$ a vector bundle over M, and

∇^E a connection on E. Let $\gamma : [0, 1] \to M$ be a smooth curve in M. Then the pull-back $\gamma^*(E)$ of E to $[0, 1]$ is a vector bundle over $[0, 1]$ with fibre $E_{\gamma(t)}$ over $t \in [0, 1]$, where E_x is the fibre of E over $x \in M$.

Let s be a smooth section of $\gamma^*(E)$ over $[0, 1]$, so that $s(t) \in E_{\gamma(t)}$ for each $t \in [0, 1]$. The connection ∇^E pulls back under γ to give a connection on $\gamma^*(E)$ over $[0, 1]$. We say that s is *parallel* if its derivative under this pulled-back connection is zero, i.e. if $\nabla^E_{\dot\gamma(t)} s(t) = 0$ for all $t \in [0, 1]$, where $\dot\gamma(t)$ is $\frac{d}{dt}\gamma(t)$, regarded as a vector in $T_{\gamma(t)}M$.

Now this is a first-order ordinary differential equation in $s(t)$, and so for each possible inital value $e \in E_{\gamma(0)}$, there exists a unique, smooth solution s with $s(0) = e$. We shall use this to define the idea of *parallel transport* along γ.

Definition 2.2.1 Let M be a manifold, E a vector bundle over M, and ∇^E a connection on E. Suppose $\gamma : [0, 1] \to M$ is smooth, with $\gamma(0) = x$ and $\gamma(1) = y$, where $x, y \in M$. Then for each $e \in E_x$, there exists a unique smooth section s of $\gamma^*(E)$ satisfying $\nabla^E_{\dot\gamma(t)} s(t) = 0$ for $t \in [0, 1]$, with $s(0) = e$. Define $P_\gamma(e) = s(1)$. Then $P_\gamma : E_x \to E_y$ is a well-defined linear map, called the *parallel transport map*. This definition easily generalizes to the case when γ is continuous and piecewise-smooth, by requiring s to be continuous, and differentiable whenever γ is differentiable.

Here are some elementary properties of parallel transport. Let M, E and ∇^E be as above, let $x, y, z \in M$, and let α, β be piecewise-smooth paths in M with $\alpha(0) = x$, $\alpha(1) = y = \beta(0)$, and $\beta(1) = z$. Define paths α^{-1} and $\beta\alpha$ by

$$\alpha^{-1}(t) = \alpha(1-t), \quad \text{and}$$

$$\beta\alpha(t) = \begin{cases} \alpha(2t) & \text{if } 0 \leqslant t \leqslant \tfrac{1}{2}, \\ \beta(2t-1) & \text{if } \tfrac{1}{2} \leqslant t \leqslant 1. \end{cases}$$

Then α^{-1} and $\beta\alpha$ are piecewise-smooth paths in M with $\alpha^{-1}(0) = y$, $\alpha^{-1}(1) = x$, $\beta\alpha(0) = x$ and $\beta\alpha(1) = z$.

Suppose $e_x \in E_x$, and $P_\alpha(e_x) = e_y \in E_y$. Then there is a unique parallel section s of $\alpha^{-1}(E)$ with $s(0) = e_x$ and $s(1) = e_y$. Define $s'(t) = s(1-t)$. Then s' is a parallel section of $(\alpha^{-1})^*(E)$. Since $s'(0) = e_y$ and $s'(1) = e_x$, it follows that $P_{\alpha^{-1}}(e_y) = e_x$. Thus, if $P_\alpha(e_x) = e_y$, then $P_{\alpha^{-1}}(e_y) = e_x$, and so P_α and $P_{\alpha^{-1}}$ are *inverse maps*. In particular, this implies that if γ is any piecewise-smooth path in M, then P_γ is *invertible*. By a similar argument, we can also show that $P_{\beta\alpha} = P_\beta \circ P_\alpha$.

Definition 2.2.2 Let M be a manifold, E a vector bundle over M, and ∇^E a connection on E. Fix a point $x \in M$. We say that γ is a *loop based at* x if $\gamma : [0, 1] \to M$ is a piecewise-smooth path with $\gamma(0) = \gamma(1) = x$. If γ is a loop based at x, then the parallel transport map $P_\gamma : E_x \to E_x$ is an invertible linear map, so that P_γ lies in $\mathrm{GL}(E_x)$, the group of invertible linear transformations of E_x. Define the *holonomy group* $\mathrm{Hol}_x(\nabla^E)$ of ∇^E based at x to be

$$\mathrm{Hol}_x(\nabla^E) = \{ P_\gamma : \gamma \text{ is a loop based at } x \} \subset \mathrm{GL}(E_x). \tag{2.7}$$

If α, β are loops based at x, then α^{-1} and $\beta\alpha$ are too, and from above we have $P_{\alpha^{-1}} = P_\alpha^{-1}$ and $P_{\beta\alpha} = P_\beta \circ P_\alpha$. Thus, if P_α and P_β lie in $\text{Hol}_x(\nabla^E)$, then so do P_α^{-1} and $P_\beta \circ P_\alpha$. This shows that $\text{Hol}_x(\nabla^E)$ is closed under inverses and products in $\text{GL}(E_x)$, and therefore $\text{Hol}_x(\nabla^E)$ is a *subgroup* of $\text{GL}(E_x)$, which justifies calling it a group.

Note that in this book we suppose all manifolds to be *connected*. Suppose $x, y \in M$. Since M is connected, we can find a piecewise-smooth path $\gamma : [0, 1] \to M$ with $\gamma(0) = x$ and $\gamma(1) = y$, so that $P_\gamma : E_x \to E_y$. If α is a loop based at x, then $\gamma\alpha\gamma^{-1}$ is a loop based at y, and $P_{\gamma\alpha\gamma^{-1}} = P_\gamma \circ P_\alpha \circ P_\gamma^{-1}$. Hence, if $P_\alpha \in \text{Hol}_x(\nabla^E)$, then $P_\gamma \circ P_\alpha \circ P_\gamma^{-1} \in \text{Hol}_y(\nabla^E)$. Thus

$$P_\gamma \, \text{Hol}_x(\nabla^E) \, P_\gamma^{-1} = \text{Hol}_y(\nabla^E). \tag{2.8}$$

Now this shows that the holonomy group $\text{Hol}_x(\nabla^E)$ is *independent of the base point* x, in the following sense. Suppose E has fibre $\mathbb{R}^k$, say. Then any identification $E_x \cong \mathbb{R}^k$ induces an isomorphism $\text{GL}(E_x) \cong \text{GL}(k, \mathbb{R})$, and so we may regard $\text{Hol}_x(\nabla^E)$ as a subgroup H of $\text{GL}(k, \mathbb{R})$. If we choose a different identification $E_x \cong \mathbb{R}^k$, we instead get the subgroup aHa^{-1} of $\text{GL}(k, \mathbb{R})$, for some $a \in \text{GL}(k, \mathbb{R})$. Thus, the holonomy group is a subgroup of $\text{GL}(k, \mathbb{R})$, defined up to conjugation. Moreover, (2.8) shows that if $x, y \in M$, then $\text{Hol}_x(\nabla^E)$ and $\text{Hol}_y(\nabla^E)$ yield the same subgroup of $\text{GL}(k, \mathbb{R})$, up to conjugation. This proves:

Proposition 2.2.3 *Let M be a manifold, E a vector bundle over M with fibre $\mathbb{R}^k$, and ∇^E a connection on E. For each $x \in M$, the holonomy group $\text{Hol}_x(\nabla^E)$ may be regarded as a subgroup of $\text{GL}(k, \mathbb{R})$ defined up to conjugation in $\text{GL}(k, \mathbb{R})$, and in this sense it is independent of the base point x.*

Because of this we may omit the subscript x and write the holonomy group of ∇^E as $\text{Hol}(\nabla^E) \subset \text{GL}(k, \mathbb{R})$, implicitly supposing that two subgroups of $\text{GL}(k, \mathbb{R})$ are equivalent if they are conjugate in $\text{GL}(k, \mathbb{R})$. In the same way, if E is a complex vector bundle with fibre $\mathbb{C}^k$, then the holonomy group of ∇^E is a subgroup of $\text{GL}(k, \mathbb{C})$, up to conjugation. The proposition shows that the holonomy group is a *global invariant* of the connection. Next we show that if M is simply-connected, then $\text{Hol}(\nabla^E)$ is a *connected Lie group*.

Proposition 2.2.4 *Let M be a simply-connected manifold, E a vector bundle over M with fibre $\mathbb{R}^k$, and ∇^E a connection on E. Then $\text{Hol}(\nabla^E)$ is a connected Lie subgroup of $\text{GL}(k, \mathbb{R})$.*

Proof Choose a base point $x \in M$, and let γ be a loop in M based at x. Since M is simply-connected, the loop γ can be contracted to the constant loop at x, that is, there exists a family $\{\gamma_s : s \in [0, 1]\}$, where $\gamma_s : [0, 1] \to M$ satisfies $\gamma_s(0) = \gamma_s(1) = x$, $\gamma_0(t) = x$ for $t \in [0, 1]$ and $\gamma_1 = \gamma$, and $\gamma_s(t)$ depends continuously on s and t. In fact, as shown in [133, p. 73–75], one can also suppose that γ_s is piecewise-smooth, and depends on s in a piecewise-smooth way.

Therefore $s \mapsto P_{\gamma_s}$ is a piecewise-smooth map from $[0, 1]$ to $\text{Hol}_x(\nabla^E)$. Since γ_0 is the constant loop at x we see that $P_{\gamma_0} = 1$, and $P_{\gamma_1} = P_\gamma$ as $\gamma_1 = \gamma$. Thus, each P_γ

in $\mathrm{Hol}(\nabla^E)$ can be joined to the identity by a piecewise-smooth path in $\mathrm{Hol}(\nabla^E)$. Now by a theorem of Yamabe [224], every arcwise-connected subgroup of a Lie group is a connected Lie subgroup. Hence, $\mathrm{Hol}(\nabla^E)$ is a connected Lie subgroup of $\mathrm{GL}(k, \mathbb{R})$.

□

When M is not simply-connected, it is convenient to consider the *restricted holonomy group* $\mathrm{Hol}^0(\nabla^E)$, which we now define.

Definition 2.2.5 Let M be a manifold, E a vector bundle over M with fibre $\mathbb{R}^k$, and ∇^E a connection on E. Fix $x \in M$. A loop γ based at x is called *null-homotopic* if it can be deformed to the constant loop at x. Define the *restricted holonomy group* $\mathrm{Hol}_x^0(\nabla^E)$ of ∇^E to be

$$\mathrm{Hol}_x^0(\nabla^E) = \{P_\gamma : \gamma \text{ is a null-homotopic loop based at } x\}. \tag{2.9}$$

Then $\mathrm{Hol}_x^0(\nabla^E)$ is a subgroup of $\mathrm{GL}(E_x)$. As above we may regard $\mathrm{Hol}_x^0(\nabla^E)$ as a subgroup of $\mathrm{GL}(k, \mathbb{R})$ defined up to conjugation, and it is then independent of the base point x, and so is written $\mathrm{Hol}^0(\nabla^E) \subseteq \mathrm{GL}(k, \mathbb{R})$.

Here are some important properties of $\mathrm{Hol}^0(\nabla^E)$.

Proposition 2.2.6 *Let M be a manifold, E a vector bundle over M with fibre $\mathbb{R}^k$, and ∇^E a connection on E. Then $\mathrm{Hol}^0(\nabla^E)$ is a connected Lie subgroup of $\mathrm{GL}(k, \mathbb{R})$. It is the connected component of $\mathrm{Hol}(\nabla^E)$ containing the identity, and is a normal subgroup of $\mathrm{Hol}(\nabla^E)$. There is a natural, surjective group homomorphism $\phi : \pi_1(M) \to \mathrm{Hol}(\nabla^E)/\mathrm{Hol}^0(\nabla^E)$. Thus, if M is simply-connected, then $\mathrm{Hol}(\nabla^E) = \mathrm{Hol}^0(\nabla^E)$.*

Proof The argument used in Proposition 2.2.4 shows that the restricted holonomy group $\mathrm{Hol}^0(\nabla^E)$ is a connected Lie subgroup of $\mathrm{GL}(k, \mathbb{R})$. Fix $x \in M$. If α, β are loops based at x and β is null-homotopic, then $\alpha\beta\alpha^{-1}$ is null-homotopic. Thus, if $P_\alpha \in \mathrm{Hol}_x(\nabla^E)$ and $P_\beta \in \mathrm{Hol}_x^0(\nabla^E)$, then $P_{\alpha\beta\alpha^{-1}} = P_\alpha P_\beta P_\alpha^{-1}$ also lies in $\mathrm{Hol}_x^0(\nabla^E)$, and so $\mathrm{Hol}_x^0(\nabla^E)$ is a normal subgroup of $\mathrm{Hol}_x(\nabla^E)$.

The group homomorphism $\phi : \pi_1(M) \to \mathrm{Hol}_x(\nabla^E)/\mathrm{Hol}_x^0(\nabla^E)$ is given by $\phi([\gamma]) = P_\gamma \cdot \mathrm{Hol}_x^0(\nabla^E)$, where γ is a loop based at x and $[\gamma]$ the corresponding element of $\pi_1(M)$. It is easy to verify that ϕ is a surjective group homomorphism. Since $\pi_1(M)$ is countable, the quotient group $\mathrm{Hol}_x(\nabla^E)/\mathrm{Hol}_x^0(\nabla^E)$ is also countable. Therefore, $\mathrm{Hol}_x^0(\nabla^E)$ is the connected component of $\mathrm{Hol}_x(\nabla^E)$ containing the identity. This completes the proof. □

Now we can define the *Lie algebra* of $\mathrm{Hol}^0(\nabla^E)$.

Definition 2.2.7 Let M be a manifold, E a vector bundle over M with fibre $\mathbb{R}^k$, and ∇^E a connection on E. Then $\mathrm{Hol}^0(\nabla^E)$ is a Lie subgroup of $\mathrm{GL}(k, \mathbb{R})$, defined up to conjugation. Define the *holonomy algebra* $\mathfrak{hol}(\nabla^E)$ to be the Lie algebra of $\mathrm{Hol}^0(\nabla^E)$. It is a Lie subalgebra of $\mathfrak{gl}(k, \mathbb{R})$, defined up to the adjoint action of $\mathrm{GL}(k, \mathbb{R})$. Similarly, $\mathrm{Hol}_x^0(\nabla^E)$ is a Lie subgroup of $\mathrm{GL}(E_x)$ for all $x \in M$. Define $\mathfrak{hol}_x(\nabla^E)$ to be the Lie algebra of $\mathrm{Hol}_x^0(\nabla^E)$. It is a Lie subalgebra of $\mathrm{End}(E_x)$.

Note that because $\mathrm{Hol}^0(\nabla^E)$ is the identity component of $\mathrm{Hol}(\nabla^E)$, the Lie algebras of $\mathrm{Hol}^0(\nabla^E)$ and $\mathrm{Hol}(\nabla^E)$ coincide. Also, although $\mathrm{Hol}^0(\nabla^E)$ is a Lie subgroup of $\mathrm{GL}(k, \mathbb{R})$, it is not necessarily a *closed* subgroup, and so it may not be a submanifold of $\mathrm{GL}(k, \mathbb{R})$ in the strictest sense. (The inclusion of $\mathbb{R}$ in $T^2 = \mathbb{R}^2/\mathbb{Z}^2$ given by $t \mapsto (t + \mathbb{Z}, t\sqrt{2} + \mathbb{Z})$ for $t \in \mathbb{R}$ gives an example of a non-closed Lie subgroup of a Lie group, and this is the sort of behaviour we have in mind.) Even if $\mathrm{Hol}^0(\nabla^E)$ is closed, the full holonomy group $\mathrm{Hol}(\nabla^E)$ may not be closed in $\mathrm{GL}(k, \mathbb{R})$.

The term 'holonomy group' is in some ways misleading, as it suggests that the holonomy group is defined simply as an abstract Lie group. In fact, if ∇^E is a connection on a vector bundle E, then the holonomy group $\mathrm{Hol}(\nabla^E)$ comes equipped with a natural *representation* on the fibre $\mathbb{R}^k$ of E, or equivalently, $\mathrm{Hol}(\nabla^E)$ is embedded as a subgroup of $\mathrm{GL}(k, \mathbb{R})$. Thus, when we describe the holonomy group of a connection, we must specify not only a Lie group, but also a representation of this group. It is important to remember this. We will refer to the representation of $\mathrm{Hol}(\nabla^E)$ on the fibre of E as the *holonomy representation*.

2.3 Holonomy groups and principal bundles

Next we define holonomy groups of connections in principal bundles.

Definition 2.3.1 Let M be a manifold, P a principal bundle over M with fibre G, and D a connection in P. Let $\gamma : [0, 1] \to P$ be a smooth curve in P. Then $\dot\gamma(t) \in T_{\gamma(t)}P$ is tangent to $\gamma([0, 1])$ for each $t \in [0, 1]$. We call γ a *horizontal* curve if its tangent vectors are horizontal, that is, $\dot\gamma(t) \in D_{\gamma(t)}$ for each $t \in [0, 1]$. Similarly, if $\gamma : [0, 1] \to P$ is piecewise-smooth, we say that γ is *horizontal* if $\dot\gamma(t) \in D_{\gamma(t)}$ for t in the open, dense subset of $[0, 1]$ where $\dot\gamma(t)$ is well-defined.

Now, if $\gamma : [0, 1] \to M$ is piecewise-smooth with $\gamma(0) = m$, and $p \in P$ with $\pi(p) = m$, then there exists a unique horizontal, piecewise-smooth map $\gamma' : [0, 1] \to P$ such that $\gamma'(0) = p$ and $\pi \circ \gamma'$ is equal to γ, as maps $[0, 1] \to M$. This follows from existence results for ordinary differential equations, and is analogous to the facts about existence and uniqueness of parallel sections of $\gamma^*(E)$ used in Definition 2.2.1. We call γ' a *horizontal lift* of γ.

Here is the definition of holonomy group.

Definition 2.3.2 Let M be a manifold, P a principal bundle over M with fibre G, and D a connection in P. For $p, q \in P$, write $p \sim q$ if there exists a piecewise-smooth horizontal curve in P joining p to q. Clearly, $\sim$ is an equivalence relation. Fix $p \in P$, and define the *holonomy group* of (P, D) based at p to be $\mathrm{Hol}_p(P, D) = \{g \in G : p \sim g \cdot p\}$. Similarly, define the *restricted holonomy group* $\mathrm{Hol}_p^0(P, D)$ to be the set of $g \in G$ for which there exists a piecewise-smooth, horizontal curve $\gamma : [0, 1] \to P$ such that $\gamma(0) = p$, $\gamma(1) = g \cdot p$, and $\pi \circ \gamma$ is null-homotopic in M.

If $g \in G$ and $p, q \in P$ with $p \sim q$, then there is a horizontal curve γ in P joining p and q. Applying g to γ, we see that $g \cdot \gamma$ is a horizontal curve joining $g \cdot p$ and $g \cdot q$. Therefore if $g \in G$ and $p \sim q$, then $g \cdot p \sim g \cdot q$. If $g \in \mathrm{Hol}_p(P, D)$, then

$p \sim g \cdot p$. Applying g^{-1} gives that $g^{-1} \cdot p \sim g^{-1} \cdot (g \cdot p) = p$. Thus $p \sim g^{-1} \cdot p$ and $g^{-1} \in \text{Hol}_p(P, D)$, so that $\text{Hol}_p(P, D)$ contains inverses of its elements.

Now suppose that $g, h \in \text{Hol}_p(P, D)$. Applying g to $p \sim h \cdot p$ shows that $g \cdot p \sim (gh) \cdot p$. But $p \sim g \cdot p$, so $p \sim (gh) \cdot p$ as $\sim$ is an equivalence relation, and $gh \in \text{Hol}_p(P, D)$. So $\text{Hol}_p(P, D)$ is closed under products, and therefore it is a *subgroup* of G. A similar argument shows that $\text{Hol}^0_p(P, D)$ is a subgroup of G.

Since $\sim$ is an equivalence relation, it is easy to see that if $p, q \in P$ and $p \sim q$, then $\text{Hol}_p(P, D) = \text{Hol}_q(P, D)$. Also, one can show that for all $g \in G$ and $p \in P$, we have $\text{Hol}_{g \cdot p}(P, D) = g\,\text{Hol}_p(P, D) g^{-1}$. Now if $p, q \in P$, then $\pi(p), \pi(q) \in M$. As M is connected, there exists a piecewise-smooth path γ in M with $\gamma(0) = \pi(p)$ and $\gamma(1) = \pi(q)$. There is a unique horizontal lift γ' of γ with $\gamma'(0) = p$ and $\gamma'(1) = q'$, for some $q' \in P$. As $\pi(q') = \pi(q)$, we see that $q' = g \cdot q$ for some $g \in G$, and as γ' is horizontal we have $p \sim q'$. Thus, whenever $p, q \in P$, there exists $g \in G$ with $q \sim g \cdot p$, and so from above

$$\text{Hol}_q(P, D) = \text{Hol}_{g \cdot p}(P, D) = g\,\text{Hol}_p(P, D) g^{-1}. \tag{2.10}$$

This proves the following result, the analogue of Proposition 2.2.3.

Proposition 2.3.3 *Let M be a manifold, P a principal bundle over M with fibre G, and D a connection in P. Then the holonomy group $\text{Hol}_p(P, D)$ depends on the base point $p \in P$ only up to conjugation in G. Thus we may regard the holonomy group as an equivalence class of subgroups of G under conjugation, and it is then independent of p and is written $\text{Hol}(P, D)$. Similarly, we may regard the restricted holonomy group $\text{Hol}^0_p(P, D)$ as an equivalence class of subgroups of G under conjugation, and write it $\text{Hol}^0(P, D)$.*

By following the proofs of Propositions 2.2.4 and 2.2.6, we can show:

Proposition 2.3.4 *Let M be a manifold, P a principal bundle over M with fibre G, and D a connection in P. Then $\text{Hol}^0(P, D)$ is a connected Lie subgroup of G. It is the connected component of $\text{Hol}(P, D)$ containing the identity, and is normal in $\text{Hol}(P, D)$. There is a natural, surjective group homomorphism $\phi : \pi_1(M) \to \text{Hol}(P, D)/\text{Hol}^0(P, D)$. If M is simply-connected, then $\text{Hol}(P, D) = \text{Hol}^0(P, D)$.*

We define the Lie algebra of $\text{Hol}^0(P, D)$.

Definition 2.3.5 Suppose M is a manifold, P a principal bundle over M with fibre G, and D a connection in P. Then $\text{Hol}^0(P, D)$ is a connected Lie subgroup of G, defined up to conjugation. Define the *holonomy algebra* $\mathfrak{hol}(P, D)$ to be the Lie algebra of $\text{Hol}^0(P, D)$. Then $\mathfrak{hol}(P, D)$ is a Lie subalgebra of the Lie algebra $\mathfrak{g}$ of G, and is defined up to the adjoint action of G on $\mathfrak{g}$.

Similarly, $\text{Hol}^0_p(P, D)$ is a Lie subgroup of G for all $p \in P$. Let $\mathfrak{hol}_p(P, D) \subseteq \mathfrak{g}$ be the Lie algebra of $\text{Hol}^0_p(P, D)$. Let $\pi(p) = m \in M$, and define $\mathfrak{hol}_m(P, D) = \pi\big(\mathfrak{hol}_p(P, D)\big)$, where $\pi : P \times \mathfrak{g} \to \text{ad}(P)$ is as in Definition 2.1.3. Then $\mathfrak{hol}_m(P, D)$ is a vector subspace of $\text{ad}(P)_m$. As $\mathfrak{hol}_{g \cdot p}(P, D) = \text{Ad}(g)\big[\mathfrak{hol}_p(P, D)\big]$ for g in G, we see that $\mathfrak{hol}_m(P, D)$ is independent of the choice of $p \in \pi^{-1}(m)$. Thus $\mathfrak{hol}_m(P, D)$ is well-defined.

Now let M, P, G and D be as above, and fix $p \in P$. Write $H = \text{Hol}_p(P, D)$, and suppose H is a closed Lie subgroup of G. Define $Q = \{q \in P : p \sim q\}$. Clearly, Q is preserved by the action of H on P, and thus H acts freely on Q. Also, π restricts to Q giving a projection $\pi : Q \to M$, and it is easy to see that the fibres of $\pi : Q \to M$ are actually the orbits of H. As H is a closed subgroup of G, it is a Lie group, and one can also show that Q is a submanifold of P, and thus a manifold. If H is not closed in G, then Q is not a submanifold of P in the strict sense.

All this shows that Q is a *principal subbundle* of P, with fibre H. A subbundle of this sort is called a *reduction* of P. Let C' be the vertical subbundle of Q. A point q lies in Q if it can be joined to p by a horizontal curve. Therefore, any horizontal curve starting in Q must remain in Q, and so $T_q Q$ must contain all horizontal vectors at q, giving that $D_q \subset T_q Q$. Now $T_q P = C_q \oplus D_q$, and $D_q \subset T_q Q$, and clearly $C'_q = C_q \cap T_q Q$. But these equations imply that $T_q Q = C'_q \oplus D_q$. Therefore, the restriction D' of the distribution D to Q is in fact a *connection* on Q. Thus we have proved:

Theorem 2.3.6. (Reduction theorem) *Let M be a manifold, P a principal bundle over M with fibre G, and D a connection in P. Fix $p \in P$, let $H = \text{Hol}_p(P, D)$, and suppose that H is a closed Lie subgroup of G. Define $Q = \{q \in P : p \sim q\}$. Then Q is a principal subbundle of P with fibre H, and the connection D on P restricts to a connection D' on Q. In other words, P reduces to Q, and the connection D on P reduces to D' on Q.*

The hypothesis that H is closed in G here may be dropped, but then Q may not be closed in P. An example of such a subgroup $H \subset G$ was given in the previous section. Using this theorem, we can interpret holonomy groups in the following way. Suppose P is a principal bundle over M, with fibre G, and a connection D. Then the holonomy group $\text{Hol}(P, D)$ is the smallest subgroup $H \subseteq G$, up to conjugation, for which it is possible to find a reduction Q of P with fibre H, such that the connection D reduces to Q.

Finally, we shall compare the holonomy groups of connections in vector bundles and in principal bundles, using the ideas of §2.1. The relation between the two is given by the following proposition, which is easy to prove.

Proposition 2.3.7 *Let M be a manifold, and P a principal bundle over M with fibre G. Suppose $\rho : G \to GL(V)$ is a representation of G on a vector space V, and set $E = \rho(V)$. Let D be a connection on P, and ∇^E the connection on E given in Definition 2.1.8. Then $\text{Hol}(P, D)$ and $\text{Hol}(\nabla^E)$ are subgroups of G and $GL(V)$ respectively, each defined up to conjugation, and $\rho(\text{Hol}(P, D)) = \text{Hol}(\nabla^E)$.*

Similarly, suppose M is a manifold, E a vector bundle over M with fibre $\mathbb{R}^k$, and F^E the frame bundle of E. Then F^E is a principal bundle with fibre $GL(k, \mathbb{R})$. Let ∇^E be a connection in E, and D^E the corresponding connection in F^E. Then

$\mathrm{Hol}(\nabla^E)$ and $\mathrm{Hol}(F^E, D^E)$ are both subgroups of $GL(k, \mathbb{R})$ defined up to conjugation, and $\mathrm{Hol}(\nabla^E) = \mathrm{Hol}(F^E, D^E)$.

Thus the two definitions of holonomy group are equivalent, or at least very closely related.

2.4 Holonomy groups and curvature

Given a connection on a vector bundle or a principal bundle, there is a fundamental relationship between the holonomy group (or its Lie algebra) and the curvature of the connection. The holonomy algebra both constrains the curvature, and is determined by it. Here are two results showing that the curvature of a connection lies in a vector bundle derived from the holonomy algebra.

Proposition 2.4.1 *Let M be a manifold, E a vector bundle over M, and ∇^E a connection on E. Then for each $m \in M$ the curvature $R(\nabla^E)_m$ of ∇^E at m lies in $\mathfrak{hol}_m(\nabla^E) \otimes \Lambda^2 T_m^* M$, where $\mathfrak{hol}_m(\nabla^E)$ is the vector subspace of $\mathrm{End}(E_m)$ given in Definition 2.2.7.*

Proposition 2.4.2 *Let M be a manifold, P a principal bundle over M with fibre G, and D a connection on P. Then for each $m \in M$ the curvature $R(P, D)_m$ of D at m lies in $\mathfrak{hol}_m(P, D) \otimes \Lambda^2 T_m^* M$, where $\mathfrak{hol}_m(P, D)$ is the vector subspace of $\mathrm{ad}(P)_m$ given in Definition 2.3.5.*

We will only prove the second proposition, as the first follows from it.

Proof of Proposition 2.4.2 It is enough to show that if v, w are vector fields on M then $(R(P, D) \cdot v \wedge w)_m$ lies in $\mathfrak{hol}_m(P, D)$. Choose $p \in M$ with $m = \pi(p)$. Then $(R(P, D) \cdot v \wedge w)_m$ lies in $\mathfrak{hol}_m(P, D)$ if and only if $\pi^*(R(P, D) \cdot v \wedge w)_p$ lies in $\mathfrak{hol}_p(P, D)$.

So by (2.6), we must show that for all $v, w \in C^\infty(TM)$ and $p \in P$, we have

$$[\lambda(v), \lambda(w)]|_p - \lambda([v, w])|_p \in \mathfrak{hol}_p(P, D). \tag{2.11}$$

Here $[\lambda(v), \lambda(w)]|_p - \lambda([v, w])|_p \in C_p$, which is identified with $\mathfrak{g}$, and $\mathfrak{hol}_p(P, D) \subseteq \mathfrak{g}$. Let Q be the subset $\{q \in P : p \sim q\} \subseteq P$ considered in §2.3. Then by Theorem 2.3.6, Q is a principal subbundle of P with fibre $\mathrm{Hol}_p(P, D)$, and the connection D reduces to Q. This means that at $q \in Q$, we have $D_q \subset T_q Q$.

Consider the restriction of $\lambda(v)$ to Q. Since it is horizontal, it lies in D_q and hence in $T_q Q$ at each $q \in Q$. Thus, $\lambda(v)|_Q$ is a vector field on Q. Similarly, $\lambda(w)|_Q$ and $\lambda([v, w])|_Q$ are vector fields on Q, so $[\lambda(v), \lambda(w)]|_Q$ is a vector field on Q. Since $p \in Q$, we see that $[\lambda(v), \lambda(w)]|_p - \lambda([v, w])|_p \in T_p Q$. But we already know this lies in C_p, so it lies in $C_p \cap T_p Q$. However, $C_p \cap T_p Q$ is identified with $\mathfrak{hol}_p(P, D)$ under the isomorphism $C_p \cong \mathfrak{g}$. This verifies equation (2.11), and the proof is complete. □

There is a kind of converse to Propositions 2.4.1 and 2.4.2, known as the *Ambrose–Singer Holonomy Theorem* [8], [133, p. 89]. We state it here in two forms, for connections in vector bundles and principal bundles.

Theorem 2.4.3 (a) *Let M be a manifold, E a vector bundle over M, and ∇^E a connection on E. Fix $x \in M$, so that $\mathfrak{hol}_x(\nabla^E)$ is a Lie subalgebra of $\mathrm{End}(E_x)$. Then $\mathfrak{hol}_x(\nabla^E)$ is the vector subspace of $\mathrm{End}(E_x)$ spanned by all elements of $\mathrm{End}(E_x)$ of the form $P_\gamma^{-1}\big[R(\nabla^E)_y \cdot (v \wedge w)\big]P_\gamma$, where $x \in M$ is a point, $\gamma : [0,1] \to M$ is piecewise smooth with $\gamma(0) = x$ and $\gamma(1) = y$, $P_\gamma : E_x \to E_y$ is the parallel translation map, and $v, w \in T_y M$.*

(b) *Let M be a manifold, P a principal bundle over M with fibre G, and D a connection on P. Fix $p \in P$, and define $Q = \{q \in P : p \sim q\}$, as in §2.3. Then $\mathfrak{hol}_p(P, D)$ is the vector subspace of the Lie algebra $\mathfrak{g}$ of G spanned by the elements of the form $\pi^*(R(P, D) \cdot v \wedge w)_q$ for all $q \in Q$ and $v, w \in C^\infty(TM)$, where π maps $P \times \mathfrak{g}$ to $\mathrm{ad}(P)$.*

This shows that $R(\nabla^E)$ determines $\mathfrak{hol}(\nabla^E)$, and hence $\mathrm{Hol}^0(\nabla^E)$. For instance, if ∇^E is *flat*, so that $R(\nabla^E) = 0$, then $\mathfrak{hol}(\nabla^E) = \{0\}$, and therefore $\mathrm{Hol}^0(\nabla^E) = \{1\}$. The theorem is used by Kobayashi and Nomizu [133, Th. 8.2, p. 90] to prove the next proposition.

Proposition 2.4.4 *Let M be a manifold and P a principal fibre bundle over M with fibre G. If $\dim M \geqslant 2$ and G is connected, then there exists a connection D on P with $\mathrm{Hol}(P, D) = G$.*

As a corollary we have the following result, which can be seen as a sort of converse to Theorem 2.3.6.

Theorem 2.4.5 *Let M be a manifold, and P a principal bundle over M with fibre G. Suppose $\dim M \geqslant 2$. Then for each connected Lie subgroup $H \subset G$, there exists a connection D on P with holonomy group $\mathrm{Hol}(P, D) = H$ if and only if P reduces to a principal bundle Q with fibre H.*

This shows that the question of which groups can appear as the holonomy group of a connection on a general vector or principal bundle is determined entirely by global topological issues: it comes down to asking when the principal bundle admits a reduction to a subgroup, which can be answered using algebraic topology. Therefore, the question of which groups can be the holonomy groups of a connection on a general bundle is not very interesting from the geometrical point of view. To make the question interesting we must impose additional conditions on the connection, as we will see in the next section.

2.5 Connections on the tangent bundle, and torsion

We now consider connections ∇ on the tangent bundle TM of a manifold M. We shall show that ∇ also acts on the tensors on M, and the *constant tensors* on M are determined by the holonomy group $\mathrm{Hol}(\nabla)$. We also define an invariant $T(\nabla)$ called the *torsion*, and discuss the holonomy groups of torsion-free connections.

Suppose M is a manifold of dimension n, and let F be the *frame bundle* of M, as in Definition 2.1.2. Then TM is a vector bundle over M with fibre $\mathbb{R}^n$, and F a principal bundle over M with fibre $\mathrm{GL}(n, \mathbb{R})$. As in §2.1, there is a 1-1 correspondence between connections ∇ on TM, and connections D on F.

Now sections 2.1–2.4 developed the theories of connections in vector bundles and principal bundles in parallel, comparing them, but keeping the two theories distinct. From now on we will not make this distinction. Instead, we will *identify* a connection ∇ on TM with the corresponding connection D on F. We will refer to both as *connections on M*, and we will make use of both vector and principal bundle methods, according to which picture is most helpful.

2.5.1 *Holonomy groups and constant tensors*

Let M be a manifold of dimension n, and ∇ a connection on M. Then ∇ is identified with a connection D on the frame bundle F of M, which is a principal bundle with fibre $\mathrm{GL}(n,\mathbb{R})$. Now if ρ is any representation of $\mathrm{GL}(n,\mathbb{R})$ on a vector space V then Definitions 2.1.3 and 2.1.8 define a vector bundle $\rho(F)$ on M and a connection, ∇^ρ say, on $\rho(F)$.

Let ρ be the usual representation of $\mathrm{GL}(n,\mathbb{R})$ on $V = \mathbb{R}^n$. Then $\rho(F)$ is just TM, and $\nabla^\rho = \nabla$. However, starting from V we can construct many other representations of $\mathrm{GL}(n,\mathbb{R})$ by taking duals, tensor products, exterior products and so on. From each of these representations, we get a vector bundle with a connection. For instance, the representation of $\mathrm{GL}(n,\mathbb{R})$ on V^* gives the cotangent bundle T^*M, and the representation $\bigotimes^k V \otimes \bigotimes^l V^*$ yields the bundle $\bigotimes^k TM \otimes \bigotimes^l T^*M$.

In fact, all of the vector bundles of tensors $\bigotimes^k TM \otimes \bigotimes^l T^*M$, and subbundles of these such as the symmetric tensors $S^k TM$ or the exterior forms $\Lambda^l T^*M$, arise through the construction of Definition 2.1.3. Thus, Definition 2.1.8 yields a connection on each of these bundles. This gives:

Lemma 2.5.1 *Let M be a manifold. Then a connection ∇ on TM induces connections on all the vector bundles of tensors on M, such as $\bigotimes^k TM \otimes \bigotimes^l T^*M$. All of these induced connections on tensors will also be written ∇.*

Let M be a manifold, ∇ a connection on M, and S a tensor on M, so that $S \in C^\infty(\bigotimes^k TM \otimes \bigotimes^l T^*M)$ for some k, l. We say that S is a *constant tensor* if $\nabla S = 0$. Our next result shows that the constant tensors on M are determined entirely by the holonomy group $\mathrm{Hol}(\nabla)$.

Proposition 2.5.2 *Let M be a manifold, and ∇ a connection on TM. Fix $x \in M$, and let $H = \mathrm{Hol}_x(\nabla)$. Then H is a subgroup of $\mathrm{GL}(T_xM)$. Let E be the vector bundle $\bigotimes^k TM \otimes \bigotimes^l T^*M$ over M. Then the connection ∇ on TM induces a connection ∇^E on E, and H has a natural representation on the fibre E_x of E at x.*

Suppose $S \in C^\infty(E)$ is a constant tensor, so that $\nabla^E S = 0$. Then $S|_x$ is fixed by the action of H on E_x. Conversely, if $S_x \in E_x$ is fixed by the action of H, then there exists a unique tensor $S \in C^\infty(E)$ such that $\nabla^E S = 0$ and $S|_x = S_x$.

Proof Let $\rho : H \to \mathrm{GL}(E_x)$ be the natural representation. Then Proposition 2.3.7 shows that $\mathrm{Hol}_x(\nabla^E) = \rho(H)$. Let γ be a loop in M based at x, and $P_\gamma \in \mathrm{GL}(E_x)$ the parallel translation map using ∇^E in E. Then $P_\gamma \in \mathrm{Hol}_x(\nabla^E)$, so $P_\gamma \in \rho(H)$, and $P_\gamma = \rho(h)$ for some $h \in H$. Moreover, for every $h \in H$ we have $P_\gamma = \rho(h)$ for some loop γ in M based at x.

Now $\nabla^E S = 0$, and therefore the pull-back $\gamma^*(S)$ is a *parallel* section of $\gamma^*(E)$ over $[0, 1]$. Therefore $P_\gamma(S|_{\gamma(0)}) = S|_{\gamma(1)}$. But $\gamma(0) = \gamma(1) = x$, so $P_\gamma(S|_x) = S|_x$. Thus $\rho(h)(S|_x) = S|_x$ for all $h \in H$, and $S|_x$ is fixed by the action of H on E_x.

For the second part, suppose $S_x \in E_x$ is fixed by $\rho(H)$. We will define $S \in C^\infty(E)$ with the required properties. Let $y \in M$ be any point. As M is connected, there is a piecewise-smooth path $\alpha : [0, 1] \to M$ with $\alpha(0) = x$ and $\alpha(1) = y$. Let α and β be two such paths, and let $P_\alpha, P_\beta : E_x \to E_y$ be the parallel transport maps, so that $P_{\alpha^{-1}\beta} = P_\alpha^{-1} P_\beta$. But $\alpha^{-1}\beta$ is a loop based at x, and thus $P_{\alpha^{-1}\beta} = P_\alpha^{-1} P_\beta = \rho(h)$ for some $h \in H$.

Now $\rho(h)(S_x) = S_x$ by assumption. Hence $P_\alpha^{-1} P_\beta(S_x) = S_x$, giving $P_\alpha(S_x) = P_\beta(S_x)$. Therefore, if $\alpha, \beta : [0, 1] \to M$ are any two piecewise-smooth paths from x to y then $P_\alpha(S_x) = P_\beta(S_x)$, and the element $P_\alpha(S_x) \in E_y$ depends only on y, and not on α. Define a section S of E by $S|_y = P_\alpha(S_x)$, where α is any piecewise-smooth path from x to y. Then S is well-defined. If γ is any path in M then $\gamma^*(S)$ is parallel, and thus S is differentiable with $\nabla^E S = 0$. Also $S|_x = S_x$ by definition, and clearly $S \in C^\infty(E)$, which finishes the proof. $\square$

In the proposition we wrote ∇^E for the connection on E, in order to distinguish it from the connection ∇ on TM. Usually we will not make this distinction, but will write ∇ for the connections on all the tensor bundles of M. As a corollary we have:

Corollary 2.5.3 *Let M be a manifold and ∇ a connection on TM, and fix $x \in M$. Define $G \subset \mathrm{GL}(T_xM)$ to be the subgroup of $\mathrm{GL}(T_xM)$ that fixes $S|_x$ for all constant tensors S on M. Then $\mathrm{Hol}_x(\nabla)$ is a subgroup of G.*

Now, in nearly all of the geometrical situations that interest us—if, for instance, $\mathrm{Hol}_x(\nabla)$ is compact and connected—we actually have $\mathrm{Hol}_x(\nabla) = G$ in this corollary. This is not true in every case, as for instance G is closed in $\mathrm{GL}(T_xM)$ but $\mathrm{Hol}_x(\nabla)$ is not always closed, but it is a good general rule. The point is that $\mathrm{Hol}(\nabla) = G$ if $\mathrm{Hol}(\nabla)$ can be defined as the subgroup of $\mathrm{GL}(n, \mathbb{R})$ fixing a collection of elements in the finite-dimensional representations of $\mathrm{GL}(n, \mathbb{R})$, and this is true for most of the subgroups of $\mathrm{GL}(n, \mathbb{R})$ of any geometrical interest.

Thus, we have the principle that given a manifold M and a connection ∇ on TM, the holonomy group $\mathrm{Hol}(\nabla)$ determines the constant tensors on M, and the constant tensors on M usually determine the holonomy group $\mathrm{Hol}(\nabla)$. Therefore, studying the holonomy of a connection, and studying its constant tensors, come down to the same thing.

2.5.2 *The torsion of a connection on M*

Let M be a manifold, and ∇ a connection on M. Then the *torsion* $T(\nabla)$ of ∇ is a tensor on M defined in the following proposition. We leave the proof as an easy exercise, as it is similar to that of Proposition 2.1.5.

Proposition 2.5.4 *Let M be a manifold, and ∇ a connection on TM. Suppose $v, w \in C^\infty(TM)$ are vector fields and α, β smooth functions on M. Then*

$$\nabla_{\alpha v}(\beta w) - \nabla_{\beta w}(\alpha v) - [\alpha v, \beta w] = \alpha\beta \cdot \{\nabla_v w - \nabla_w v - [v, w]\},$$

where $[v, w]$ is the Lie bracket. Thus the expression $\nabla_v w - \nabla_w v - [v, w]$ is pointwise-linear in v and w, and it is clearly antisymmetric in v and w. Therefore there exists a unique, smooth section $T(\nabla) \in C^\infty(TM \otimes \Lambda^2 T^*M)$ called the *torsion* of ∇, that satisfies the equation

$$T(\nabla) \cdot (v \wedge w) = \nabla_v w - \nabla_w v - [v, w] \quad \text{for all } v, w \in C^\infty(TM). \tag{2.12}$$

The torsion $T(\nabla)$ of a connection ∇ is a tensor invariant, similar to the curvature $R(\nabla)$. The definition of $T(\nabla)$ uses ∇ once, but that of $R(\nabla)$ uses ∇ twice. In fact, the torsion is a much simpler invariant than the curvature. Note also that we can only define the torsion of a connection on TM, as the definition makes no sense for an arbitrary vector bundle E over M. A connection ∇ on TM with $T(\nabla) = 0$ is called *torsion-free*, or is said to have *zero torsion*. Torsion-free connections are an important class of connections.

Let M be a manifold and ∇ a connection on M. For simplicity we will write T for the torsion $T(\nabla)$ and R for the curvature $R(\nabla)$ of ∇. Then T and R are tensors on M. Using the index notation, we have

$$T = T^a_{bc} \quad \text{with} \quad T^a_{bc} = -T^a_{cb}, \quad \text{and} \quad R = R^a{}_{bcd} \quad \text{with} \quad R^a{}_{bcd} = -R^a{}_{bdc}.$$

For a *torsion-free* connection ∇, the curvature R and its derivative ∇R have certain extra symmetries, known as the *Bianchi identities*.

Proposition 2.5.5 *Let M be a manifold, and ∇ a torsion-free connection on TM. Then the curvature $R^a{}_{bcd}$ of ∇ satisfies the tensor equations*

$$R^a{}_{bcd} + R^a{}_{cdb} + R^a{}_{dbc} = 0, \quad \text{and} \tag{2.13}$$

$$\nabla_e R^a{}_{bcd} + \nabla_c R^a{}_{bde} + \nabla_d R^a{}_{bec} = 0. \tag{2.14}$$

These are known as the first and second Bianchi identities, respectively.

Proof Equation (2.13) is equivalent to the condition that

$$R \cdot (u \otimes v \wedge w) + R \cdot (v \otimes w \wedge u) + R \cdot (w \otimes u \wedge v) = 0 \tag{2.15}$$

for all vector fields $u, v, w \in C^\infty(TM)$. By definition of R we have

$$R \cdot (u \otimes v \wedge w) = \nabla_v \nabla_w u - \nabla_w \nabla_v u - \nabla_{[v,w]} u. \tag{2.16}$$

But ∇ is torsion-free, so $T = 0$, and therefore

$$T \cdot \bigl(u \wedge [v, w]\bigr) = \nabla_u [v, w] - \nabla_{[v,w]} u - [u, [v, w]] = 0.$$

Substituting this into (2.16) gives

$$R \cdot (u \otimes v \wedge w) = \nabla_v \nabla_w u - \nabla_w \nabla_v u - \nabla_u [v, w] + [u, [v, w]].$$

Applying the three cyclic permutations of u, v, w to this equation and adding the results together, we find that the left hand side of (2.15) is equal to

$$\nabla_u\bigl(\nabla_v w - \nabla_w v - [v,w]\bigr) + \nabla_v\bigl(\nabla_w u - \nabla_u w - [w,u]\bigr)$$
$$+\nabla_w\bigl(\nabla_u v - \nabla_v u - [u,v]\bigr) + [u,[v,w]] + [v,[w,u]] + [w,[u,v]].$$

The first term is $\nabla_u\bigl(T\cdot(v\wedge w)\bigr)$, which is zero as $T = 0$, and the two similar terms vanish in the same way. But the remaining three terms sum to zero, by the Jacobi identity for vector fields. Thus (2.15) holds, and this proves (2.13). Equation (2.14) can be proved by similar methods. □

2.5.3 *The holonomy of torsion-free connections*

It was explained in §2.4 that Theorem 2.4.5 describes exactly the possible holonomy groups of a connection on a bundle, and so the problem of which groups can be the holonomy groups of a general connection on a bundle, is not a very fruitful one. However, the problem can be made much more interesting by restricting attention to *torsion-free* connections. We state this as the following question, which is one of the main motivating problems in the field of holonomy groups.

Question 1: *What are the possible holonomy groups* $\mathrm{Hol}(\nabla)$ *of torsion-free connections* ∇ *on a given manifold* M?

In general, the problem of determining which holonomy groups are realized by torsion-free connections on a given manifold M, a compact manifold for instance, is very difficult and depends strongly on the topology of M. So, let us consider instead the corresponding *local* problem, that is:

Question 2: *What are the possible holonomy groups* $\mathrm{Hol}(\nabla)$ *of torsion-free connections* ∇ *on an open ball in* $\mathbb{R}^n$?

This question is still rather difficult, but some powerful algebraic techniques can be applied to the problem, and a fairly complete answer to the question is known.

The classification of Lie groups, and representations of Lie groups, is well understood. Therefore one could, in principle, write down a list of all possible connected Lie subgroups of $\mathrm{GL}(n,\mathbb{R})$ up to conjugation, for each n. This list is of course infinite, and rather complicated. The idea is to test every subgroup on this list, to see whether or not it can be a holonomy group.

Naturally, because of the complexity of the classification of Lie groups and representations, this is a huge task, and so one looks for short cuts. There is an algebraic method that excludes many groups H from being holonomy groups. This is to study the space $\mathfrak{R}^H$ of possible curvature tensors for H. Proposition 2.4.1 and the first Bianchi identity restrict $\mathfrak{R}^H$, making it small. But the Ambrose–Singer Holonomy Theorem shows that $\mathfrak{R}^H$ must be large enough to generate the Lie algebra $\mathfrak{h}$ of H. If these requirements are not consistent, then H cannot be a holonomy group.

The list of groups H that pass this test is shorter and more manageable. Berger [31, Th. 3–Th. 5, p. 318–320] published, without proof, a list of these groups which is substantially complete, but with some omissions. Later, a number of holonomy groups of torsion-free connections that did not appear on Berger's list were discovered by Bryant, Chi, Merkulov and Schwachhöfer, who called them *exotic holonomy groups*. A (hopefully) complete classification of holonomy groups of torsion-free connections, with proof, has been published by Merkulov and Schwachhöfer [161], to which the reader is referred for further details and references.

These algebraic methods eventually yield a list of candidates H for possible holonomy groups, but they do not prove that every such H actually occurs as a holonomy group. There is another approach to the classification problem using the machinery of *Cartan–Kähler theory*, which is a way of describing how many solutions there are to a given partial differential equation. The advantage of this approach is that it systematically determines whether each H occurs as a holonomy group or not. Bryant [45] uses Cartan–Kähler theory to give a unified treatment of the classification of holonomy groups. We shall discuss the problem of classification of *Riemannian holonomy groups* at greater length in §3.4.

2.6 G-structures and intrinsic torsion

We will now discuss G-structures on manifolds, and their torsion. The theory of G-structures gives a different way of looking at connections on M and their holonomy groups, and is a useful framework for studying geometrical structures.

Definition 2.6.1 Let M be a manifold of dimension n, and F the frame bundle of M, as in §2.5. Then F is a principal bundle over M with fibre $GL(n, \mathbb{R})$. Let G be a Lie subgroup of $GL(n, \mathbb{R})$. Then a *G-structure* on M is a principal subbundle P of F, with fibre G.

The G-structures, for the many possible Lie subgroups $G \subseteq GL(n, \mathbb{R})$, provide a large family of interesting geometrical structures on manifolds. Other geometrical objects such as Riemannian metrics and complex structures can also be interpreted as G-structures, as the following example shows.

Example 2.6.2 Let (M, g) be a Riemannian n-manifold, and F the frame bundle of M. Each point of F is $(x, e_1, \ldots, e_n)$, where $x \in M$ and $(e_1, \ldots, e_n)$ is a basis for $T_x M$. Define P to be the subset of F for which $(e_1, \ldots, e_n)$ is *orthonormal* with respect to g. Then P is a principal subbundle of F with fibre $O(n)$, so P is an $O(n)$-*structure* on M. In fact, this gives a 1-1 correspondence between $O(n)$-structures and Riemannian metrics on M.

Now let M be a manifold of dimension n with frame bundle F, let G be a Lie subgroup of $GL(n, \mathbb{R})$, and P a G-structure on M. Suppose D is a connection on P. Then there is a unique connection D' on F that reduces to D on P. Conversely, a connection D' on F reduces to a connection D on P if and only if for each $p \in P$, the subspace $D'|_p$ of $T_p F$ lies in $T_p P$.

As we explained in §2.5, connections D' on the principal bundle F are equivalent to connections ∇ on the vector bundle TM. We call a connection ∇ on TM *compatible with the G-structure* P, if the corresponding connection on F reduces to P. Thus we see that every connection D on P induces a unique connection ∇ on TM, and conversely, a connection ∇ on TM arises from a connection D on P if and only if ∇ is compatible with P.

Our next result shows that if ∇ is a fixed connection on TM, then there is a compatible G-structure P if and only if $\text{Hol}(\nabla) \subseteq G$.

Proposition 2.6.3 *Let M be a connected manifold of dimension n, with frame bundle F, and fix $f \in F$. Let ∇ be a connection on TM. Then for each Lie subgroup $G \subset \mathrm{GL}(n, \mathbb{R})$, there exists a G-structure P on M compatible with ∇ and containing f if and only if $\mathrm{Hol}_f(\nabla) \subseteq G \subseteq \mathrm{GL}(n, \mathbb{R})$. If P exists then it is unique. More generally, there is a 1-1 correspondence between the set of G-structures on M compatible with ∇, but not necessarily containing f, and the homogeneous space $G \backslash \{a \in \mathrm{GL}(n, \mathbb{R}) : a \, \mathrm{Hol}_f(\nabla) a^{-1} \subseteq G\}$.*

Proof The proof is similar to that of Theorem 2.3.6, so we will be brief. If P exists then it contains f and is closed under G, so it contains $g \cdot f$ for each $g \in G$. As P is compatible with ∇, any horizontal curve starting in P remains in P. Thus, if $p \in P$ and $q \in F$ with $p \sim q$ then $q \in P$, where $\sim$ is the equivalence relation defined in §2.3.

Combining these two facts shows that if $p \in F$ and $p \sim g \cdot f$ for any $g \in G$, then $p \in P$. But as M is connected, every $p \in P$ must satisfy $p \sim g \cdot f$ for some $g \in G$. So P must be $\{p \in F : p \sim g \cdot f \text{ for some } g \in G\}$ if it exists. It is easy to show that this set is a principal bundle over M, with fibre the subgroup of $\mathrm{GL}(n, \mathbb{R})$ generated by G and $\mathrm{Hol}_f(\nabla)$. Hence, P exists if and only if $\mathrm{Hol}_f(\nabla) \subseteq G$, and if it exists it is unique.

Now if $a \in \mathrm{GL}(n, \mathbb{R})$, then $\mathrm{Hol}_{a \cdot f}(\nabla) = a \, \mathrm{Hol}_f(\nabla) a^{-1}$. Thus from above, there exists a unique G-structure P containing $a \cdot f$ if and only if $a \, \mathrm{Hol}_f(\nabla) a^{-1} \subseteq G$. But any G-structure containing $a \cdot f$ also contains $(ga) \cdot f$ for all $g \in G$. Therefore, the set of G-structures on M compatible with ∇ is in 1-1 correspondence with the given set. □

This proposition gives a good picture of the set of G-structures compatible with a fixed connection ∇ on TM. So let us turn the problem around, and ask about the set of connections ∇ on TM compatible with a fixed G-structure P on M. We have seen above that these are in 1-1 correspondence with connections on P, of which there are many. So we shall restrict our attention to *torsion-free* connections ∇, and ask the question: given a G-structure P on a manifold M, how many torsion-free connections ∇ are there on TM compatible with P?

If ∇ and ∇' are two connections on P, then the difference $\alpha = \nabla' - \nabla$ is a smooth section of $\mathrm{ad}(P) \otimes T^*M$. But $\mathrm{ad}(P)$ is a vector subbundle of $TM \otimes T^*M$. So α is a tensor, written α^a_{bc} in index notation, and if v, w are vector fields then $(\nabla'_v w - \nabla_v w)^a = \alpha^a_{bc} w^b v^c$. Substituting this into (2.12) we see that

$$T(\nabla')^a_{bc} = T(\nabla)^a_{bc} - \alpha^a_{bc} + \alpha^a_{cb}. \tag{2.17}$$

Let ∇ be an arbitrary, fixed connection on P. Clearly, there exists a *torsion-free* connection ∇' on P if and only if there is an $\alpha \in C^\infty\big(\mathrm{ad}(P) \otimes T^*M\big)$ with $T(\nabla)^a_{bc} = \alpha^a_{bc} - \alpha^a_{cb}$. Moreover, if some such ∇' does exist, then the set of all torsion-free connections ∇' on P is in 1-1 correspondence with the vector space of $\alpha \in C^\infty\big(\mathrm{ad}(P) \otimes T^*M\big)$ for which $\alpha^a_{bc} = \alpha^a_{cb}$.

Here is an alternative way to explain this.

Definition 2.6.4 Let G be a Lie subgroup of $\mathrm{GL}(n, \mathbb{R})$, and let V be $\mathbb{R}^n$. Then G acts faithfully on V, and $\mathfrak{g} \subset V \otimes V^*$. Define $\sigma : \mathfrak{g} \otimes V^* \to V \otimes \Lambda^2 V^*$ by $\sigma(\alpha^a_{bc}) = \alpha^a_{bc} - \alpha^a_{cb}$, in index notation. Define vector spaces $W_1, \ldots, W_4$ by

$W_1 = V \otimes \Lambda^2 V^*$, $W_2 = \operatorname{Im}\sigma$, $W_3 = V \otimes \Lambda^2 V^* / \operatorname{Im}\sigma$ and $W_4 = \operatorname{Ker}\sigma$,

and let $\rho_j : G \to \operatorname{GL}(W_j)$ be the natural representations of G on $W_1, \ldots, W_4$. Now suppose M is a manifold of dimension n, and P a G-structure on M. Then we can associate a vector bundle $\rho(P)$ over M to each representation ρ of G, as in §2.1.1. Thus $\rho_1(P), \ldots, \rho_4(P)$ are vector bundles over M. Clearly, $\rho_2(P)$ is a vector subbundle of $\rho_1(P)$, and the quotient bundle $\rho_1(P)/\rho_2(P)$ is $\rho_3(P)$.

If ∇ is any connection on P, then its torsion $T(\nabla)$ lies in $C^\infty(\rho_1(P))$, and if ∇, ∇' are two connections on P, then $T(\nabla') - T(\nabla)$ lies in the subspace $C^\infty(\rho_2(P))$ of $C^\infty(\rho_1(P))$. Therefore, the projections of $T(\nabla)$ and $T(\nabla')$ to the quotient bundle $\rho_3(P) = \rho_1(P)/\rho_2(P)$ are equal. Define the *intrinsic torsion* $T^i(P)$ *of* P to be the projection to $\rho_3(P)$ of the torsion $T(\nabla)$ of any connection ∇ on P. Then $T^i(P)$ lies in $C^\infty(\rho_3(P))$, and depends only on the G-structure P and not on the choice of connection ∇.

We call the G-structure P *torsion-free* if $T^i(P) = 0$. Clearly, there exists a torsion-free connection ∇ on P if and only if P is torsion-free, and so the intrinsic torsion $T^i(P)$ is the obstruction to finding a torsion-free connection on P. Any two torsion-free connections differ by an element of $C^\infty(\rho_4(P))$. Thus, if $T^i(P) = 0$ then the torsion-free connections ∇ on P are in 1-1 correspondence with $C^\infty(\rho_4(P))$. If $\operatorname{Ker}\sigma = \{0\}$, this set is a single point, so ∇ is unique.

The proof of the next result is similar to that of Proposition 2.6.3.

Proposition 2.6.5 *Let M be a manifold of dimension n, and G a Lie subgroup of $\operatorname{GL}(n, \mathbb{R})$. Then M admits a torsion-free G-structure P if and only if there exists a torsion-free connection ∇ on TM with $\operatorname{Hol}(\nabla) = H$, for some subgroup H of G.*

This shows that torsion-free G-structures on a manifold M are intimately related to torsion-free connections ∇ on TM with $\operatorname{Hol}(\nabla) = G$. However, torsion-free G-structures are simpler, and often easier to work with, than torsion-free connections with prescribed holonomy. This is because a torsion-free G-structure P is defined by a differential equation $T^i(P) = 0$, whereas the condition $\operatorname{Hol}(\nabla) = G$ involves both differentiation and integration, and is rather more complicated. For this reason we will often use the language of torsion-free G-structures in later chapters.

A number of familiar geometric structures are in fact torsion-free G-structures in disguise. Here are some examples. We saw in Example 2.6.2 that a Riemannian metric g is equivalent to an $\operatorname{O}(n)$-structure P. But when $G = \operatorname{O}(n)$ in Definition 2.6.4 it turns out that σ is both injective and surjective. Therefore, every $\operatorname{O}(n)$-structure P is torsion-free, and there is a unique torsion-free connection ∇ on P. This is the *Levi-Civita connection*, and will be discussed at greater length in §3.1.1.

Set $n = 2m$, and let G be the subgroup $\operatorname{GL}(m, \mathbb{C}) \subset \operatorname{GL}(n, \mathbb{R})$. Then it can be shown that an *almost complex structure* J on a manifold M is equivalent to a $\operatorname{GL}(m, \mathbb{C})$-structure on M, and that J is a *complex structure* if and only if this $\operatorname{GL}(m, \mathbb{C})$-structure is torsion-free. (Complex structures and almost complex structures will be defined in §4.1.) Thus, a complex structure is equivalent to a torsion-free $\operatorname{GL}(m, \mathbb{C})$-structure.

Note that in this case, because $\operatorname{Ker} \sigma$ is nonzero, a complex manifold admits infinitely many torsion-free connections preserving the complex structure. In a similar way, a *symplectic structure* on a manifold M of dimension $2m$ is the same thing as a torsion-free $\operatorname{Sp}(m, \mathbb{R})$-structure, where $\operatorname{Sp}(m, \mathbb{R}) \subset \operatorname{GL}(2m, \mathbb{R})$ is the symplectic group, and a *Kähler structure* on M is the same as a torsion-free $\operatorname{U}(m)$-structure.

3

RIEMANNIAN HOLONOMY GROUPS

Let M be a manifold, and g a Riemannian metric on M. Then there is a unique, preferred connection ∇ on TM called the *Levi-Civita connection*, which is torsion-free and satisfies $\nabla g = 0$. The curvature $R(\nabla)$ of the Levi-Civita connection is called the *Riemann curvature*, and its holonomy group $\text{Hol}(\nabla)$ the *Riemannian holonomy group* $\text{Hol}(g)$ of g.

In 1955, Marcel Berger proved that if (M, g) is a Riemannian manifold with M simply-connected and g irreducible and nonsymmetric, then $\text{Hol}(g)$ must be one of $\text{SO}(n)$, $\text{U}(m)$, $\text{SU}(m)$, $\text{Sp}(m)$, $\text{Sp}(m)\,\text{Sp}(1)$, G_2 or $\text{Spin}(7)$. The goal of the first part of this chapter, sections 3.1–3.4, is to explain what this result means and how it is proved. We start with the Levi-Civita connection, Riemann curvature, and Riemannian holonomy groups. After sections on reducible Riemannian manifolds and symmetric spaces we move onto Berger's classification, describing the proof and the groups on Berger's list.

In the second part of the chapter we discuss some general properties of Riemannian manifolds with reduced holonomy, which are common to some or all of the holonomy groups in Berger's classification. Sections 3.5 and 3.6 explore the relationship between the holonomy group $\text{Hol}(g)$ and the topology of the underlying manifold M, in particular when M is compact, and §3.7 discusses *calibrated geometry*, and its connection with holonomy groups.

For more information on the material of the first part of this chapter, see Kobayashi and Nomizu [133, §III, §IV]. The treatments by Besse [33, §10] and Salamon [184, §2, §10] are also helpful, and Bryant [45] approaches the classification of holonomy groups from a different point of view, that of Cartan–Kähler theory.

3.1 Introduction to Riemannian holonomy groups

We define the Levi-Civita connection ∇ and Riemann curvature tensor R of a Riemannian metric g, and prove some symmetries of R and ∇R. Then we discuss the elementary properties of Riemannian holonomy groups and their relation to torsion-free G-structures.

3.1.1 *The Levi-Civita connection*

Let M be a Riemannian manifold, with metric g. Then there is a preferred connection ∇ on TM called the *Levi-Civita connection*, which is torsion-free and satisfies $\nabla g = 0$. This result is called the *Fundamental Theorem of Riemannian Geometry*, and is very important. Here is a proof.

Theorem 3.1.1 *Let M be a manifold and g a Riemannian metric on M. Then there exists a unique, torsion-free connection ∇ on TM with $\nabla g = 0$, called the Levi-Civita connection.*

Proof Suppose first that ∇ is a torsion-free connection on TM with $\nabla g = 0$. Let $u, v, w \in C^\infty(TM)$ be vector fields on M. Then $g(v, w)$ is a smooth function on M, and so u acts on $g(v, w)$ to give another smooth function $u \cdot g(v, w)$ on M. Since $\nabla g = 0$, using the properties of connections we find that

$$u \cdot g(v, w) = g(\nabla_u v, w) + g(v, \nabla_u w).$$

Combining this with similar expressions for $v \cdot g(u, w)$ and $w \cdot g(u, v)$ gives

$$\begin{aligned}
u \cdot g(v, w) &+ v \cdot g(u, w) - w \cdot g(u, v) \\
&= g(\nabla_u v, w) + g(v, \nabla_u w) + g(\nabla_v u, w) + g(u, \nabla_v w) \\
&\quad - g(\nabla_w u, v) - g(u, \nabla_w v) \\
&= g(\nabla_u v + \nabla_v u, w) + g(\nabla_v w - \nabla_w v, u) + g(\nabla_u w - \nabla_w u, v) \\
&= g(2\nabla_u v - [u, v], w) + g([v, w], u) + g([u, w], v).
\end{aligned}$$

Here we have used $\nabla_u v - \nabla_v u = [u, v]$, and two similar equations, which hold because ∇ is torsion-free. Rearranging this equation shows that

$$\begin{aligned}
2g(\nabla_u v, w) = &u \cdot g(v, w) + v \cdot g(u, w) - w \cdot g(u, v) \\
&+ g([u, v], w) - g([v, w], u) - g([u, w], v).
\end{aligned} \quad (3.1)$$

It is easy to show that for fixed u, v, there is a unique vector field $\nabla_u v$ which satisfies (3.1) for all $w \in C^\infty(TM)$. This defines ∇ uniquely, and it turns out that ∇ is indeed a torsion-free connection with $\nabla g = 0$. □

In §2.5 we saw that a connection on the tangent bundle TM of a manifold M induces connections on vector bundles of tensors on M. Thus, the Levi-Civita connection ∇ of a Riemannian metric g on M induces connections on all the tensors on M. These connections will also be written ∇.

3.1.2 *The Riemann curvature*

Suppose M is a Riemannian manifold, with metric g and Levi-Civita connection ∇. Then the curvature $R(\nabla)$ of ∇ is a tensor $R^a{}_{bcd}$ on M. Define $R_{abcd} = g_{ae} R^e{}_{bcd}$. We shall refer to both $R^a{}_{bcd}$ and R_{abcd} as the *Riemann curvature* of g. The following theorem gives a number of symmetries of R_{abcd}. Equations (3.3) and (3.4) are known as the *first* and *second Bianchi identities*, respectively.

Theorem 3.1.2 *Let (M, g) be a Riemannian manifold, ∇ the Levi-Civita connection of g, and R_{abcd} the Riemann curvature of g. Then R_{abcd} and $\nabla_e R_{abcd}$ satisfy the equations*

$$R_{abcd} = -R_{abdc} = -R_{bacd} = R_{cdab}, \quad (3.2)$$

$$R_{abcd} + R_{adbc} + R_{acdb} = 0, \quad (3.3)$$

and $\quad \nabla_e R_{abcd} + \nabla_c R_{abde} + \nabla_d R_{abec} = 0. \quad (3.4)$

Proof Since ∇ is torsion-free, by Proposition 2.5.5 the Bianchi identities (2.13) and (2.14) hold for $R^a{}_{bcd}$. Contracting these with g and substituting $\nabla g = 0$, we get (3.3) and (3.4). Also $R(\nabla) \in C^\infty(\text{End}(TM) \otimes \Lambda^2 T^*M)$, and thus $R^a{}_{bcd} = -R^a{}_{bdc}$, which gives $R_{abcd} = -R_{abdc}$, the first part of (3.2).

Now ∇ also acts on tensors such as g, and by properties of curvature we deduce that

$$\nabla_c \nabla_d g_{ab} - \nabla_d \nabla_c g_{ab} = -R^e{}_{acd} g_{eb} - R^e{}_{bcd} g_{ae}. \tag{3.5}$$

But the left hand side is zero as $\nabla g = 0$, and the right hand side is $-R_{bacd} - R_{abcd}$ by definition. Therefore $R_{abcd} = -R_{bacd}$, the second part of (3.2).

To prove the third part of (3.2), by permuting a, b, c, d in (3.3) we get

$$R_{abcd} + R_{adbc} + R_{acdb} = 0, \qquad R_{dabc} + R_{dcab} + R_{dbca} = 0, \tag{3.6}$$
$$R_{bcda} + R_{bacd} + R_{bdac} = 0, \qquad R_{cdab} + R_{cbda} + R_{cabd} = 0. \tag{3.7}$$

Adding together the equations (3.6), subtracting the equations (3.7), and applying the first two parts of (3.2) we get $2R_{abcd} - 2R_{cdab} = 0$, and thus $R_{abcd} = R_{cdab}$, as we want. □

Next we define two components of the Riemann curvature tensor, the Ricci curvature and the scalar curvature.

Definition 3.1.3 Let (M, g) be a Riemannian manifold, with Riemann curvature $R^a{}_{bcd}$. Then g is called *flat* if $R^a{}_{bcd} = 0$. The *Ricci curvature* of g is $R_{ab} = R^c{}_{acb}$, and the *scalar curvature* of g is $s = g^{ab} R_{ab} = g^{ab} R^c{}_{acb}$. By (3.2), the Ricci curvature satisfies $R_{ab} = R_{ba}$. We say that g is *Einstein* if $R_{ab} = \lambda g_{ab}$ for some constant $\lambda \in \mathbb{R}$, and that g is *Ricci-flat* if $R_{ab} = 0$.

Einstein and Ricci-flat metrics are interesting for a number of reasons. There are of course a huge number of Riemannian metrics on any manifold of dimension at least two. The Einstein and Ricci-flat metrics provide a natural way of picking out a much smaller set of special, 'best' metrics on the manifold. Also, Einstein and Ricci-flat metrics are of great importance to physicists, because in general relativity, empty space is described by a Ricci-flat Lorentzian metric.

3.1.3 Riemannian holonomy groups

Let (M, g) be a Riemannian manifold of dimension n with Levi-Civita connection ∇. Then $\nabla g = 0$, and so g is a *constant tensor* in the sense of §2.5.1. Therefore, by Proposition 2.5.2, if $x \in M$ then the action of $\text{Hol}_x(\nabla)$ on $T_x M$ preserves the metric $g|_x$ on $T_x M$. But the group of transformations of $T_x M$ preserving $g|_x$ is the *orthogonal group* $O(n)$. Therefore, the holonomy group $\text{Hol}(\nabla)$ is a subgroup of $O(n)$.

Here is another way to see this, using the ideas of Theorem 2.3.6. Let F be the frame bundle of M. Then each point of F is a basis $(e_1, \ldots, e_n)$ for one of the tangent spaces $T_x M$ of M. Define P to be the subset of points $(e_1, \ldots, e_n)$ in F such that $e_1, \ldots, e_n$ are orthonormal with respect to the metric g. Then P is a principal subbundle of F with fibre $O(n)$, that is, a reduction of F. Moreover, because the connection ∇ in F satisfies

$\nabla g = 0$, the connection ∇ reduces to P. Again, we see that $\text{Hol}(\nabla)$ is a subgroup of $O(n)$, defined up to conjugation.

Definition 3.1.4 Let (M, g) be a Riemannian manifold with Levi-Civita connection ∇. Define the *holonomy group* $\text{Hol}(g)$ of g to be $\text{Hol}(\nabla)$. Then $\text{Hol}(g)$ is a subgroup of $O(n)$, defined up to conjugation in $O(n)$. We shall refer to the holonomy group of a Riemannian metric as a *Riemannian holonomy group*. Similarly, define the *restricted holonomy group* $\text{Hol}^0(g)$ of g to be $\text{Hol}^0(\nabla)$. Then $\text{Hol}^0(g)$ is a connected Lie subgroup of $SO(n)$ defined up to conjugation in $O(n)$.

Using the results of §2.5 it is easy to prove the following proposition.

Proposition 3.1.5 *Let M be an n-manifold, and ∇ a torsion-free connection on TM. Then ∇ is the Levi-Civita connection of a Riemannian metric g on M if and only if $\text{Hol}(\nabla)$ is conjugate in $GL(n, \mathbb{R})$ to a subgroup of $O(n)$.*

Thus, Riemannian holonomy is really part of the wider subject of holonomy groups of torsion-free connections.

Definition 3.1.6 Let (M, g) be a Riemannian manifold with Levi-Civita connection ∇. Define the *holonomy algebra* $\mathfrak{hol}(g)$ of g to be $\mathfrak{hol}(\nabla)$. Then $\mathfrak{hol}(g)$ is a Lie subalgebra of $\mathfrak{so}(n)$, defined up to the adjoint action of $O(n)$. Let $x \in M$. Then $\mathfrak{hol}_x(\nabla)$ is a vector subspace of $T_x M \otimes T_x^* M$. We may use the metric g to identify $T_x M \otimes T_x^* M$ and $\otimes^2 T_x^* M$, by equating $T^a{}_b$ with $T_{ab} = g_{ac} T^c{}_b$. This identifies $\mathfrak{hol}_x(\nabla)$ with a vector subspace of $\otimes^2 T_x^* M$ that we will write as $\mathfrak{hol}_x(g)$. It is easy to see that $\mathfrak{hol}_x(g)$ actually lies in $\Lambda^2 T_x^* M$.

Now Proposition 2.4.1 shows that $R^a{}_{bcd}$ lies in $\mathfrak{hol}_x(\nabla) \otimes \Lambda^2 T_x^* M$ at x. Lowering the index a to get R_{abcd} as above, we see that R_{abcd} lies in $\mathfrak{hol}_x(g) \otimes \Lambda^2 T_x^* M$ at x. Using this and eqn (3.2), we have:

Theorem 3.1.7 *Let (M, g) be a Riemannian manifold with Riemann curvature R_{abcd}. Then R_{abcd} lies in the vector subspace $S^2 \mathfrak{hol}_x(g)$ in $\Lambda^2 T_x^* M \otimes \Lambda^2 T_x^* M$ at each $x \in M$.*

Combining this theorem with the Bianchi identities, (3.3) and (3.4), gives quite strong restrictions on the curvature tensor R_{abcd} of a Riemannian metric g with a prescribed holonomy group $\text{Hol}(g)$. These restrictions are the basis of the classification of Riemannian holonomy groups, which will be explained in §3.4.

3.1.4 *Riemannian holonomy groups and torsion-free G-structures*

We now apply the ideas of §2.6 to Riemannian holonomy groups. Suppose M is an n-manifold, and g a Riemannian metric on M. Then Example 2.6.2 defines a unique $O(n)$-structure P on M. If G is a Lie subgroup of $O(n)$ and Q a G-structure on M, we say that Q is *compatible with g* if Q is a subbundle of P. Equivalently, Q is compatible with g if each point of Q, which is a basis of some tangent space $T_x M$, is orthonormal with respect to g.

If Q is compatible with g, then $P = O(n) \cdot Q$ as a subset of the frame bundle F of M, and so one can reconstruct P, and hence g, from Q. Thus, if G is a Lie subgroup of O(n) then a G-structure Q on M gives us a Riemannian metric g on M, and some additional geometric data as well. For instance, an SO(n)-structure on M is equivalent to a metric g, together with a choice of orientation on M.

Putting $G = O(n)$ in Definition 2.6.4, the map $\sigma : \mathfrak{o}(n) \otimes (\mathbb{R}^n)^* \to \mathbb{R}^n \otimes \Lambda^2(\mathbb{R}^n)^*$ turns out to be an isomorphism. As σ is surjective, every O(n) structure P on M is torsion-free, and so there exists a torsion-free connection ∇ on TM compatible with P. And since σ is injective, ∇ is unique. Thus, given a Riemannian metric g on M, there is a unique torsion-free connection ∇ on TM compatible with the O(n)-structure P corresponding to g. This is the Levi-Civita connection of g, and we have found an alternative proof of Theorem 3.1.1.

When G is a Lie subgroup of O(n), then the map σ of Definition 2.6.4 is injective, but not in general surjective. Hence, if Q is a G-structure on M, then the condition $T^i(Q) = 0$ for Q to be torsion-free is in general nontrivial. If Q is torsion-free, then there is a unique torsion-free connection ∇ on TM compatible with Q, which is the Levi-Civita connection of the unique metric g compatible with Q.

If Q is a torsion-free G-structure compatible with g, then Hol(g) is a subgroup of G. Because of this, torsion-free G-structures are a useful tool for studying Riemannian holonomy groups. The following result is easily deduced from Proposition 2.6.3, and shows the relationship between torsion-free G-structures and metrics with prescribed holonomy.

Proposition 3.1.8 *Let (M, g) be a connected Riemannian n-manifold. Then* Hol(g) *is a subgroup of* O(n), *defined up to conjugation. Let G be a Lie subgroup of* O(n). *Then M admits a torsion-free G-structure Q compatible with g if and only if* Hol(g) *is conjugate to a subgroup of G. Moreover, there is a 1-1 correspondence between the set of such G-structures Q, and the homogeneous space*

$$G \backslash \{a \in O(n) : a \operatorname{Hol}(g) a^{-1} \subseteq G\}. \tag{3.8}$$

We shall use G-structures later in the book, as a means of constructing Riemannian metrics g with holonomy G. Very briefly, the argument runs as follows. One writes down an explicit G-structure Q on a manifold M with intrinsic torsion $T^i(Q)$ small, in some suitable sense. Then one proves that Q can be deformed to a nearby G-structure $\tilde{Q}$ with $T^i(\tilde{Q}) = 0$. The metric $\tilde{g}$ associated to $\tilde{Q}$ has Hol($\tilde{g}$) $\subseteq G$. Finally, if M satisfies certain topological conditions, it can be shown that Hol($\tilde{g}$) $= G$.

3.2 Reducible Riemannian manifolds

Let M_1, M_2 be manifolds, and $M_1 \times M_2$ the product manifold. Then at each point (p_1, p_2) of $M_1 \times M_2$, we have $T_{(p_1,p_2)}(M_1 \times M_2) \cong T_{p_1} M_1 \oplus T_{p_2} M_2$. Let g_1, g_2 be Riemannian metrics on M_1, M_2. Then $g_1|_{p_1} + g_2|_{p_2}$ is a metric on $T_{p_1} M_1 \oplus T_{p_2} M_2$. Define the *product metric* $g_1 \times g_2$ on $M_1 \times M_2$ by $g_1 \times g_2|_{(p_1,p_2)} = g_1|_{p_1} + g_2|_{p_2}$ for all $p_1 \in M_1$ and $p_2 \in M_2$. Then $g_1 \times g_2$ is a Riemannian metric on $M_1 \times M_2$, and $M_1 \times M_2$ is a Riemannian manifold. We call $(M_1 \times M_2, g_1 \times g_2)$ a *Riemannian product*.

A Riemannian manifold (M, g) is said to be *reducible* if it is isometric to a Riemannian product $(M_1 \times M_2, g_1 \times g_2)$, with dim $M_i > 0$. Also, (M, g) is said to be *locally reducible* if every point has a reducible open neighbourhood. We shall call (M, g) *irreducible* if it is not locally reducible. The following proposition, which is easy to prove, gives the holonomy group of a product metric $g_1 \times g_2$.

Proposition 3.2.1 *Let (M_1, g_1) and (M_2, g_2) be Riemannian manifolds. Then the product metric $g_1 \times g_2$ has holonomy $\mathrm{Hol}(g_1 \times g_2) = \mathrm{Hol}(g_1) \times \mathrm{Hol}(g_2)$.*

In our next three propositions we shall show that if g is a Riemannian metric and the holonomy representation (see the end of §2.2) of g is reducible, then the metric itself is at least locally reducible, and its holonomy group is a product. The first proposition is left as an exercise for the reader.

Proposition 3.2.2 *Let (M, g) be a Riemannian manifold with Levi-Civita connection ∇, and fix $p \in M$, so that $\mathrm{Hol}_p(g)$ acts on T_pM. Suppose that $T_pM = V_p \oplus W_p$, where V_p, W_p are proper vector subspaces of T_pM preserved by $\mathrm{Hol}_p(g)$, and orthogonal with respect to g. Then there are natural vector subbundles V, W of TM with fibres V_p, W_p at p. These subbundles V, W are orthogonal with respect to g and closed under parallel translation, and satisfy $TM = V \oplus W$ and $T^*M = V^* \oplus W^*$.*

Proposition 3.2.3 *In the situation of Proposition 3.2.2, let R_{abcd} be the Riemann curvature of g. Then R_{abcd} is a section of the subbundle $S^2(\Lambda^2 V^*) \oplus S^2(\Lambda^2 W^*)$ of $S^2(\Lambda^2 T^*M)$. Also, the reduced holonomy group $\mathrm{Hol}^0_p(g)$ is a product group $H_V \times H_W$, where H_V is a subgroup of $\mathrm{SO}(V_p)$ and acts trivially on W_p, and H_W is a subgroup of $\mathrm{SO}(W_p)$ and acts trivially on V_p.*

Proof As $T^*M \cong V^* \oplus W^*$, we see that $\Lambda^2 T^*M \cong \Lambda^2 V^* \oplus \Lambda^2 W^* \oplus V^* \otimes W^*$. By Theorem 3.1.7 we know that R_{abcd} lies in $S^2 \mathfrak{hol}_p(g)$ at $p \in M$, where $\mathfrak{hol}_p(g)$ is a vector subspace of $\Lambda^2 T^*M$ identified with the holonomy algebra $\mathfrak{hol}(g)$. But, because the holonomy algebra preserves the splitting $TM = V \oplus W$, we see that $\mathfrak{hol}_p(g)$ lies in the subspace $\Lambda^2 V_p^* \oplus \Lambda^2 W_p^*$ of $\Lambda^2 T_p^*M$. Therefore, R_{abcd} is a section of

$$\Lambda^2 V^* \otimes \Lambda^2 V^* \oplus \Lambda^2 W^* \otimes \Lambda^2 W^* \oplus \Lambda^2 V^* \otimes \Lambda^2 W^* \oplus \Lambda^2 W^* \otimes \Lambda^2 V^*. \tag{3.9}$$

Now, R_{abcd} satisfies the first Bianchi identity (3.3). Using this, it is easy to show that the components of R_{abcd} in the last two components of (3.9) are in fact zero. Therefore, since R_{abcd} is symmetric in the two $\Lambda^2 T^*M$ factors we see that it is a section of $S^2(\Lambda^2 V^*) \oplus S^2(\Lambda^2 W^*)$, as we have to prove. We deduce that $R^a{}_{bcd}$ is a section of the bundle

$$V \otimes V^* \otimes \Lambda^2 V^* \quad \oplus \quad W \otimes W^* \otimes \Lambda^2 W^*. \tag{3.10}$$

Let $q \in M$, let $\gamma : [0, 1] \to M$ be piecewise-smooth with $\gamma(0) = p$ and $\gamma(1) = q$, and let $P_\gamma : T_pM \to T_qM$ be the parallel translation map. Because R lies in the subbundle (3.10), it follows that we can write

$$\langle R_q \cdot (u \wedge v) : u, v \in T_qM \rangle = A_q \oplus B_q, \tag{3.11}$$

where A_q is a subspace of $V_q \otimes V_q^*$ and B_q is a subspace of $W_q \otimes W_q^*$. Now V and W are closed under parallel translation, so P_γ takes V_p to V_q and W_p to W_q, and thus $P_\gamma^{-1} A_q P_\gamma$ lies in $V_p \otimes V_p^*$ and $P_\gamma^{-1} B_q P_\gamma$ lies in $W_p \otimes W_p^*$.

But Theorem 2.4.3 says that $\mathfrak{hol}_p(\nabla)$ is spanned by the elements of $\operatorname{End}(T_pM)$ of the form $P_\gamma^{-1}[R_q \cdot (u \wedge v)]P_\gamma$, for all $q \in M$ and $u, v \in T_qM$, and we have just shown that for each fixed $q \in M$, the subspace generated by these elements splits into a direct sum of a piece in $V_p \otimes V_p^*$, and a piece in $W_p \otimes W_p^*$. Therefore, the span of these elements for all $q \in M$ splits in the same way, so by Theorem 2.4.3 we see that $\mathfrak{hol}_p(\nabla) = \mathfrak{h}_V \oplus \mathfrak{h}_W$, where $\mathfrak{h}_V$ is a subspace of $V_p \otimes V_p^*$ and $\mathfrak{h}_W$ a subspace of $W_p \otimes W_p^*$.

As $\mathfrak{hol}_p(\nabla)$ is the Lie algebra of $\operatorname{Hol}_p^0(g)$, which is a connected Lie group, we see that $\mathfrak{h}_V, \mathfrak{h}_W$ are the Lie algebras of subgroups H_V, H_W of $SO(V_p)$ and $SO(W_p)$ respectively, and $\operatorname{Hol}_p^0(g) = H_V \times H_W$. This completes the proof. □

Proposition 3.2.4 *In the situation of Proposition 3.2.3, there is a connected open neighbourhood N of p in M and a diffeomorphism $N \cong X \times Y$ for manifolds X, Y, such that under the isomorphism $T(X \times Y) \cong TX \oplus TY$, we have $V|_N = TX$ and $W|_N = TY$. There are Riemannian metrics g_X on X and g_Y on Y such that $g|_N$ is isometric to the product metric $g_X \times g_Y$. Thus g is locally reducible, and moreover $\operatorname{Hol}_q^0(g_X) \subseteq H_V$ and $\operatorname{Hol}_r^0(g_Y) \subseteq H_W$, where $p \in N$ is identified with $(q, r) \in X \times Y$.*

Proof Since V is closed under parallel translation, we deduce that if $u \in C^\infty(TM)$ and $v \in C^\infty(V)$, then $\nabla_u v \in C^\infty(V)$. Suppose that $v, v' \in C^\infty(V)$. Then $\nabla_v v'$ and $\nabla_{v'} v \in C^\infty(V)$. But ∇ is torsion-free, so that $[v, v'] = \nabla_v v' - \nabla_{v'} v$. Thus, if $v, v' \in C^\infty(V)$, then $[v, v'] \in C^\infty(V)$. This proves that V is an *integrable distribution*, so that by the Frobenius Theorem [133, p. 10] we see that locally M is fibred by a family of submanifolds of M, with tangent spaces V.

Similarly, we deduce that W is an integrable distribution. But $TM = V \oplus W$, so these two integrable distributions define a *local product structure* on M. This means that we can identify a connected open neighbourhood N of $p \in M$ with a product manifold $X \times Y$, such that the isomorphism $T(X \times Y) \cong TX \oplus TY$ identifies V with TX and W with TY, as we want.

As V and W are orthogonal, we may write $g = g_V + g_W$, where $g_V \in C^\infty(S^2 V^*)$ and $g_W \in C^\infty(S^2 W^*)$. Since ∇ is torsion-free and N is connected, it is not difficult to show that the restriction of g_V to $N = X \times Y$ is independent of the Y directions, and is therefore the pull-back to $X \times Y$ of a metric g_X on X. Similarly, $g_W|_N$ is the pull-back of a metric g_Y on Y. Therefore $g|_N$ is isometric to $g_X \times g_Y$, as we have to prove. The rest of the proposition follows from Proposition 3.2.1 and the definition of local reducibility. □

From Propositions 3.2.2–3.2.4 we immediately deduce:

Corollary 3.2.5 *Let M be an n-manifold, and g an irreducible Riemannian metric on M. Then the representations of $\operatorname{Hol}(g)$ and $\operatorname{Hol}^0(g)$ on $\mathbb{R}^n$ are irreducible.*

More generally, if (M, g) is a Riemannian manifold of dimension n, then $\operatorname{Hol}^0(g)$ is a subgroup of $SO(n)$, and has a natural representation on $\mathbb{R}^n$. By the representation theory of Lie groups, we may decompose $\mathbb{R}^n$ into a finite direct sum of irreducible representations of $\operatorname{Hol}^0(g)$. By applying Propositions 3.2.2–3.2.4 and using induction on k, we easily prove the following theorem.

Theorem 3.2.6 *Let (M, g) be a Riemannian n-manifold, so that $\mathrm{Hol}^0(g)$ is a subgroup of $\mathrm{SO}(n)$ acting on $\mathbb{R}^n$. Then there is a splitting $\mathbb{R}^n = \mathbb{R}^{n_1} \oplus \cdots \oplus \mathbb{R}^{n_k}$, where $n_j > 0$, and a corresponding isomorphism $\mathrm{Hol}^0(g) = H_1 \times \cdots \times H_k$, where H_j is a connected Lie subgroup of $\mathrm{SO}(n_j)$ acting irreducibly on $\mathbb{R}^{n_j}$.*

The theorem shows that if the holonomy representation of $\mathrm{Hol}^0(g)$ is reducible, then $\mathrm{Hol}^0(g)$ is in fact a product group, and the holonomy representation a direct sum of irreducible representations of each factor. Therefore, if G is a connected Lie group and V a representation of G, such that G and V cannot be written $G = G_1 \times \cdots \times G_k$ and $V = V_1 \oplus \cdots \oplus V_k$ where V_j is an irreducible representation of G_j, then G and V cannot be the reduced holonomy group and holonomy representation of any Riemannian metric. This is a strong statement, because there are very many such pairs G, V, and all are excluded as possible holonomy groups.

Notice that Theorem 3.2.6 does *not* claim that the groups H_j are the holonomy groups of metrics on manifolds of dimension n_j. In fact, a careful examination of the proof shows that the group H_j is generated by subgroups which are the holonomy groups of metrics on small open subsets of $\mathbb{R}^{n_j}$, but there are topological difficulties in assembling these patches into a single Riemannian manifold: there is a natural way to do it, but the resulting topological space may not be Hausdorff.

The reason this is important to us, is in the classification of Riemannian holonomy groups. Theorem 3.2.6 comes very close to saying that if the holonomy representation of $\mathrm{Hol}^0(g)$ is reducible, then $\mathrm{Hol}^0(g)$ is a product of holonomy groups of metrics in lower dimensions. If we knew that this was true, then a classification of those groups $\mathrm{Hol}^0(g)$ with irreducible holonomy representations would imply a classification of all groups $\mathrm{Hol}^0(g)$, and thus we could restrict our attention to holonomy groups with irreducible representations.

In fact, this problem turns out not to matter. The classification theory for irreducible holonomy groups that we will summarize in §3.4 also works to classify the groups H_j arising in Theorem 3.2.6, and the classification is the same. This is because the failure of the Hausdorff condition we referred to above does not affect the proof. For further discussion of this point, see Besse [33, §10.42, §10.107].

Next, suppose that the manifold M in Propositions 3.2.2–3.2.4 is *simply-connected*, and the metric g is *complete*. In this case, by a result of de Rham [61], the local product structure constructed in Proposition 3.2.4 is actually a global product structure, so that $M \cong X \times Y$, and g is globally isometric to a Riemannian product metric $g_X \times g_Y$. Therefore, using similar arguments to the previous theorem we may prove the following result, which is a sort of converse to Proposition 3.2.1.

Theorem 3.2.7 *Suppose (M, g) is a complete, simply-connected Riemannian manifold. Then there exist complete, simply-connected Riemannian manifolds $(M_1, g_1), \ldots, (M_k, g_k)$, such that the holonomy representation of $\mathrm{Hol}(g_j)$ is irreducible, (M, g) is isometric to the product $(M_1 \times \cdots \times M_k, g_1 \times \cdots \times g_k)$, and $\mathrm{Hol}(g) = \mathrm{Hol}(g_1) \times \cdots \times \mathrm{Hol}(g_k)$.*

It is shown in [133, App. 5], using the theory of Lie groups, that every connected Lie subgroup of $SO(n)$ that acts irreducibly on $\mathbb{R}^n$ is *closed* in $SO(n)$. From this result and Theorem 3.2.6 we deduce:

Theorem 3.2.8 *Let (M, g) be a Riemannian n-manifold. Then $\mathrm{Hol}^0(g)$ is a closed, connected Lie subgroup of $SO(n)$.*

Since $SO(n)$ is compact, this implies that $\mathrm{Hol}^0(g)$ is also compact. By a study of the fundamental group of a compact, irreducible Riemannian manifold, Cheeger and Gromoll [58, Th. 6] prove the following result.

Theorem 3.2.9 *Let (M, g) be a compact, irreducible Riemannian n-manifold. Then $\mathrm{Hol}(g)$ is a compact Lie subgroup of $O(n)$.*

3.3 Riemannian symmetric spaces

We will now briefly describe the theory of Riemannian symmetric spaces. These were introduced in 1925 by Élie Cartan, who classified them completely, by applying his own classification of irreducible representations of Lie groups. For a very thorough treatment of symmetric spaces, see Helgason's book [100]. Also, the treatments by Kobayashi and Nomizu [134, Chap. XI] and Besse [33, §§7.F, 10.G, 10.K] are helpful. Here is the definition.

Definition 3.3.1 A Riemannian manifold (M, g) is said to be a *Riemannian symmetric space* if for every point $p \in M$ there exists an isometry $s_p : M \to M$ that is an involution (that is, s_p^2 is the identity), such that p is an isolated fixed point of s_p.

Let (M, g) be a *complete* Riemannian manifold, and $p \in M$. Then for each unit vector $u \in T_p M$, there is a unique geodesic $\gamma_u : \mathbb{R} \to M$ parametrized by arc length, such that $\gamma_u(0) = p$ and $\dot{\gamma}_u(0) = u$. Define the *exponential map* $\exp_p : T_p M \to M$ by $\exp_p(tu) = \gamma_u(t)$ for all $t \in \mathbb{R}$ and unit vectors $u \in T_p M$. Then $\exp_p$ induces a diffeomorphism between a neighbourhood of 0 in $T_p M$, and a neighbourhood of p in M.

Identifying $T_p M \cong \mathbb{R}^n$ yields a coordinate system on M near p. These are called *normal coordinates* or *geodesic normal coordinates* at p. If we start with a Riemannian manifold (M, g) that is not complete, then $\exp_p$ can still be defined in a neighbourhood of 0 in $T_p M$. The following lemma shows that the isometries in Definition 3.3.1 assume a particularly simple form in normal coordinates.

Lemma 3.3.2 *Let (M, g) be a complete Riemannian manifold, let $p \in M$, and suppose $s : M \to M$ is an involutive isometry with isolated fixed point p. Then $s\bigl(\exp_p(v)\bigr) = \exp_p(-v)$ for all $v \in T_p M$.*

Proof The derivative $\mathrm{d}s$ of s maps $T_p M$ to itself. Since s^2 is the identity, $(\mathrm{d}s)^2$ is the identity on $T_p M$, and as p is an isolated fixed point, 0 is the sole fixed point of $\mathrm{d}s$ on $T_p M$. Clearly, this implies that $\mathrm{d}s(v) = -v$ for all $v \in T_p M$. Now s preserves g, as it is an isometry, and the map $\exp_p$ depends solely on g. Therefore, $\exp_p$ must commute with s, in the sense that $s\bigl(\exp_p(v)\bigr) = \exp_p\bigl(\mathrm{d}s(v)\bigr)$ for $v \in T_p M$. Substituting $\mathrm{d}s(v) = -v$ gives the result. □

In the next three propositions we explore the geometry of Riemannian symmetric spaces.

Proposition 3.3.3 *Let (M, g) be a connected, simply-connected Riemannian symmetric space. Then g is complete. Let G be the group of isometries of (M, g) generated by elements of the form $s_q \circ s_r$ for $q, r \in M$. Then G is a connected Lie group acting transitively on M. Choose $p \in M$, and let H be the subgroup of G fixing p. Then H is a closed, connected Lie subgroup of G, and M is the homogeneous space G/H.*

Proof Let $\gamma : (-\epsilon, \epsilon) \to M$ be a geodesic segment in M, parametrized by arc length. Then the lemma shows that $s_{\gamma(0)}(\gamma(y)) = \gamma(-y)$ for $y \in (-\epsilon, \epsilon)$. More generally we see that $s_{\gamma(x/2)}(\gamma(y)) = \gamma(x - y)$ whenever $\frac{1}{2}x$, y and $x - y$ lie in $(-\epsilon, \epsilon)$. This gives

$$s_{\gamma(x/2)} \circ s_{\gamma(0)}\bigl(\gamma(y)\bigr) = \gamma(x + y), \tag{3.12}$$

provided $\frac{1}{2}x$, y and $x + y$ are in $(-\epsilon, \epsilon)$. Thus, the map $\alpha_x = s_{\gamma(x/2)} \circ s_{\gamma(0)}$ moves points a distance x along γ. But α_x is defined on the whole of M. By applying α_x and its inverse α_{-x} many times, one can define the geodesic γ not just on $(-\epsilon, \epsilon)$, but on $\mathbb{R}$. Therefore, every geodesic in M can be extended indefinitely, and g is complete.

Now α_x lies in G by definition. Hence, if p, q are two points in M joined by a geodesic segment of length x, there exists an element α_x in G such that $\alpha_x(p) = q$. But M is connected, and so every two points p, q in M can be joined by a finite number of geodesic segments put end to end. Composing the corresponding elements of G, we get an element α of G with $\alpha(p) = q$. Therefore G acts transitively on M.

Let $q, r \in M$. Then, as M is connected, there exists a smooth path $\gamma : [0, 1] \to M$ with $\gamma(0) = q$ and $\gamma(1) = r$. Consider the family of isometries $s_q \circ s_{\gamma(t)}$ for $t \in [0, 1]$. This is a smooth path in G, joining the identity at $t = 0$ with $s_q \circ s_r$ at $t = 1$. Thus, the generating elements $s_q \circ s_r$ of G can be joined to the identity by smooth paths, and so G is arcwise-connected.

By the Myers–Steenrod Theorem [33, p. 39], the isometry group of (M, g) is a Lie group acting smoothly on M. Thus G is an arcwise-connected subgroup of a Lie group, so by the theorem of Yamabe quoted in §2.2, G is a connected Lie group. Also G acts smoothly on M, and so H is a closed Lie subgroup of G. Since G acts transitively on M, we have $M \cong G/H$. Because M is simply-connected, G/H is simply-connected, and this implies that H is connected. □

Note that the group G in the proposition may not be the full isometry group of (M, g), or even the identity component of the isometry group. For example, if M is $\mathbb{R}^n$ with the Euclidean metric, then $G = \mathbb{R}^n$ acting by translations and $H = \{0\}$, but the full isometry group also includes the rotations $O(n)$ acting on $\mathbb{R}^n$.

Proposition 3.3.4 *Let M, g, G, p and H be as above, and let $\mathfrak{g}$ be the Lie algebra of G. Then there is an involutive Lie group isomorphism $\sigma : G \to G$, and a splitting $\mathfrak{g} = \mathfrak{h} \oplus \mathfrak{m}$, where $\mathfrak{h}$ is the Lie algebra of H, and $\mathfrak{h}$ and $\mathfrak{m}$ are the eigenspaces of the involution $d\sigma : \mathfrak{g} \to \mathfrak{g}$ with eigenvalues 1 and -1 respectively. These subspaces satisfy*

$$[\mathfrak{h}, \mathfrak{h}] \subseteq \mathfrak{h}, \qquad [\mathfrak{h}, \mathfrak{m}] \subseteq \mathfrak{m} \qquad \text{and} \qquad [\mathfrak{m}, \mathfrak{m}] \subseteq \mathfrak{h}. \tag{3.13}$$

There is a natural isomorphism $\mathfrak{m} \cong T_pM$. *The adjoint action of H on* $\mathfrak{g}$ *induces a representation of H on* $\mathfrak{m}$, *or equivalently* T_pM, *and this representation is faithful. Also, H is the identity component of the fixed point set of* σ.

Proof Define $\sigma : G \to G$ by $\sigma(\alpha) = s_p \circ \alpha \circ s_p$. Clearly σ does map G to G, and is a group isomorphism, so $d\sigma : \mathfrak{g} \to \mathfrak{g}$ is a Lie algebra isomorphism. Also, σ is an involution, as $s_p^2 = 1$, so $(d\sigma)^2$ is the identity. Therefore $d\sigma$ has eigenvalues ± 1, and $\mathfrak{g}$ is the direct sum of the corresponding eigenspaces.

As M is connected, the isometries s_p in Definition 3.3.1 are unique. Hence, for $p \in M$ and $\alpha \in G$ we have $\alpha \circ s_p \circ \alpha^{-1} = s_{\alpha(p)}$. Therefore $h \circ s_p \circ h^{-1} = s_p$ for $h \in H$, so that $\sigma(h) = s_p \circ h \circ s_p = h$. Thus, H is fixed by σ, and $d\sigma$ is the identity on $\mathfrak{h}$.

The identification of G/H with M identifies p with the coset H and T_pM with $\mathfrak{g}/\mathfrak{h}$. Under this identification, the maps $ds_p : T_pM \to T_pM$ and $d\sigma : \mathfrak{g}/\mathfrak{h} \to \mathfrak{g}/\mathfrak{h}$ coincide. But ds_p multiplies by -1, as in Lemma 3.3.2. Thus, $d\sigma$ is the identity on $\mathfrak{h} \subset \mathfrak{g}$, and acts as -1 on $\mathfrak{g}/\mathfrak{h}$. Therefore there exists a unique splitting $\mathfrak{g} = \mathfrak{h} \oplus \mathfrak{m}$ where $\mathfrak{h}, \mathfrak{m}$ are the 1 and -1 eigenspaces of $d\sigma$, as we want. The relations in (3.13) then follow easily from the fact that $d\sigma$ is a Lie algebra isomorphism.

The splitting $\mathfrak{g} = \mathfrak{h} \oplus \mathfrak{m}$ and the isomorphism $T_pM \cong \mathfrak{g}/\mathfrak{h}$ give an isomorphism $\mathfrak{m} \cong T_pM$, as required. Clearly, $\mathfrak{m}$ is preserved by the adjoint action of H on $\mathfrak{g}$, and this gives a representation of H on $\mathfrak{m}$, and so on T_pM. This action of H on T_pM can also be described as follows: each element $h \in H$ fixes p, and so dh maps T_pM to itself. Now g is complete, so the exponential map $\exp_p : T_pM \to M$ is well-defined, and clearly satisfies $\exp_p(dh(v)) = h \cdot \exp_p(v)$ for all $h \in H$ and $v \in T_pM$.

Thus, $dh : T_pM \to T_pM$ determines the action of h on the subset $\exp_p(T_pM)$ of M, and so on all of M, as M is connected. It follows that if dh is the identity then h is the identity. Therefore, the representation of H on T_pM, or equivalently on $\mathfrak{m}$, is faithful. Finally, we know that H is part of the fixed point set of σ, that H is connected, and that the subspace of $\mathfrak{g}$ fixed by $d\sigma$ is $\mathfrak{h}$. Together these show that H is the identity component of the fixed points of σ. $\square$

Proposition 3.3.5 *In the situation of Proposition 3.3.4, the Riemann curvature* R_p *of* g *at* p *lies in* $T_pM \otimes T_p^*M \otimes \Lambda^2 T_p^*M$. *Identifying* $\mathfrak{m}$ *and* T_pM *in the natural way,* R_p *is given by the equation*

$$R_p \cdot (u \otimes v \wedge w) = [u, [v, w]], \tag{3.14}$$

for all $u, v, w \in \mathfrak{m}$. *Moreover, the holonomy group* $\mathrm{Hol}_p(g)$ *is* H, *with the above representation on* T_pM, *and the Riemann curvature* R *of* g *satisfies* $\nabla R = 0$.

Proof Using the splitting $\mathfrak{g} = \mathfrak{h} \oplus \mathfrak{m}$, one can construct a unique torsion-free, G-invariant connection ∇ on TM. This satisfies $\nabla g = 0$, and so ∇ is the Levi-Civita connection of g. Explicit computation then yields the formula (3.14) for R_p. For more details of this argument, see [134, §X.2, §XI.3]. The formula shows that for $v, w \in \mathfrak{m}$, we have $R_p \cdot (v \wedge w) = -\mathrm{ad}([v, w])$ in $\mathrm{End}(\mathfrak{m})$. But $[\mathfrak{m}, \mathfrak{m}] \subset \mathfrak{h}$ by (3.13), so that $R_p \cdot (v \wedge w) \in \mathrm{ad}(\mathfrak{h})$.

By definition, G is generated as a group by $s_q \circ s_r$ for $q, r \in M$. By fixing $r = p$ and letting q approach p we can prove an infinitesimal version of this statement, which says that $\mathfrak{g}$ is generated by the elements $\frac{d}{dt}(s_{\exp_p(tv)} \circ s_p)|_{t=0}$ of $\mathfrak{g}$, for all $v \in T_pM$. But these are exactly the elements of $\mathfrak{m}$, and thus $\mathfrak{g}$ is generated as a Lie algebra by $\mathfrak{m}$. Therefore (3.13) implies that $[\mathfrak{m}, \mathfrak{m}] = \mathfrak{h}$, and so the equation $R_p \cdot (v \wedge w) = -\mathrm{ad}([v, w])$ gives

$$\langle R_p \cdot (v \wedge w) : v, w \in \mathfrak{m} \rangle = \mathrm{ad}(\mathfrak{h}). \tag{3.15}$$

Proposition 2.4.1 then shows that $\mathrm{ad}(\mathfrak{h})$ is a subset of $\mathfrak{hol}_p(g)$, the Lie algebra of $\mathrm{Hol}_p(g)$.

Let $q \in M$. Then $s_p : M \to M$ is an isometry, and so preserves R and ∇R. But ds_q acts as -1 on $T_q M$, and thus $ds_q(\nabla R|_q) = -\nabla R|_q$. Therefore $\nabla R|_q = 0$ for all $q \in M$, giving and $\nabla R = 0$ as we want. Let $\gamma : [0, 1] \to M$ be piecewise-smooth with $\gamma(0) = p$ and $\gamma(1) = q$, and let $P_\gamma : T_pM \to T_qM$ be the parallel translation map. Since $\nabla R = 0$, it follows that

$$P_\gamma^{-1}[R_q \cdot (v \wedge w)]P_\gamma = R_p \cdot (P_\gamma^{-1}v \wedge P_\gamma^{-1}w), \tag{3.16}$$

for all $v, w \in T_qM$.

Now Theorem 2.4.3 shows that $\mathfrak{hol}_p(\nabla)$ is spanned by elements of $\mathrm{End}(T_pM)$ of the form $P_\gamma^{-1}[R_q \cdot (v \wedge w)]P_\gamma$. Therefore (3.15) implies that $\mathfrak{hol}_p(\nabla) = \mathrm{ad}(\mathfrak{h})$. But H is connected by Proposition 3.3.3, and $\mathrm{Hol}_p(g)$ is connected as M is simply-connected, and so $\mathrm{Hol}_p(g) = \mathrm{Ad}(H)$, where Ad is the adjoint representation of H on $\mathfrak{m}$, and we identify $\mathfrak{m}$ and T_pM. Since the representation is faithful, we have $\mathrm{Hol}_p(g) \cong H$, and the proof is complete. $\square$

Propositions 3.3.3–3.3.5 reduce the problem of understanding and classifying simply-connected Riemannian symmetric spaces to a problem in the theory of Lie groups. This rather difficult problem was solved completely by E. Cartan in 1926–7, who was able to write down a complete list of all simply-connected Riemannian symmetric spaces. Helgason [100, Chap. IX] discusses Cartan's proof, and Besse [33, §7.H, §10.K] gives tables of all the possibilities.

Using the results above, the holonomy group of a Riemannian symmetric space is easily found. Therefore, Cartan's classification implies the classification of the holonomy groups of Riemannian symmetric spaces. A considerable number of Riemannian holonomy groups arise in this way: for example, every connected, compact, simple Lie group is (up to a finite cover) the holonomy group of an irreducible Riemannian symmetric space, with holonomy representation the adjoint representation.

Some well-known examples of Riemannian symmetric spaces are $\mathbb{R}^n$ with the Euclidean metric, $\mathcal{S}^n$ with its round metric, $\mathcal{H}^n$ with the hyperbolic metric, and $\mathbb{CP}^n$ with the Fubini–Study metric. The corresponding groups are $G = \mathrm{SO}(n+1)$ and $H = \mathrm{SO}(n)$ for $\mathcal{S}^n$, and $G = \mathrm{SO}(n, 1)$ and $H = \mathrm{SO}(n)$ for $\mathcal{H}^n$, so that $\mathcal{S}^n$ and $\mathcal{H}^n$ both have holonomy group $\mathrm{SO}(n)$, and $G = \mathrm{U}(n+1)/\mathrm{U}(1)$ and $H = \mathrm{U}(n)$ for $\mathbb{CP}^n$, so that $\mathbb{CP}^n$ has holonomy $\mathrm{U}(n)$.

Next we shall discuss *locally symmetric* Riemannian manifolds, which satisfy a local version of the symmetric space condition.

Definition 3.3.6 We say that a Riemannian manifold (M, g) is *locally symmetric* if every point $p \in M$ admits an open neighbourhood U_p in M, and an involutive isometry $s_p : U_p \to U_p$, with unique fixed point p. We say that (M, g) is *nonsymmetric* if it is not locally symmetric.

Clearly, every Riemannian symmetric space is locally symmetric. Conversely, we can show that every locally symmetric Riemannian manifold is locally isometric to a Riemannian symmetric space.

Theorem 3.3.7 *Suppose that (M, g) is a locally symmetric Riemannian manifold. Then there is a unique simply-connected Riemannian symmetric space (N, h), such that (M, g) is locally isometric to (N, h). In other words, given any points $p \in M$ and $q \in N$, there exist isometric open neighbourhoods U of p in M and V of q in N.*

A proof of this theorem can be found in [100, p. 183], but we will not give it. Here is one way the result can be proved. The problem is that because the isometries s_p are defined only locally, they cannot be put together to form Lie groups of isometries G, H, as in Proposition 3.3.3. However, the Lie algebras $\mathfrak{g},\mathfrak{h}$ of G, H can be defined in the locally symmetric case, as Lie algebras of Killing vector fields, defined locally. Let G be the unique connected, simply-connected Lie group with Lie algebra $\mathfrak{g}$, and let H be the unique connected Lie subgroup of G with Lie algebra $\mathfrak{h}$. Then $N = G/H$ is a Riemannian symmetric space, with the desired properties.

Let (M, g) be a Riemannian manifold. If (M, g) is locally symmetric, then by Theorem 3.3.7 it is locally isometric to a Riemannian symmetric space, and so Proposition 3.3.5 shows that $\nabla R = 0$, where ∇ is the Levi-Civita connection, and R the Riemann curvature. Surprisingly, the converse is also true: if $\nabla R = 0$, then (M, g) is locally symmetric.

Theorem 3.3.8 *Let (M, g) be a Riemannian manifold, with Levi-Civita connection ∇ and Riemann curvature R. Then (M, g) is locally symmetric if and only if $\nabla R = 0$.*

See [134, p. 244] for a proof of this. Here is how the proof works. Let (M, g) be a Riemannian manifold, and let $p \in M$. Then $\exp_p : T_pM \to M$ is well-defined and injective in a small ball $B_\epsilon(0)$ about the origin in T_pM. Define $U_p = \exp_p\bigl(B_\epsilon(0)\bigr)$, and define $s_p : U_p \to U_p$ by $s_p\bigl(\exp_p(v)\bigr) = \exp_p(-v)$. This map s_p is called the *geodesic symmetry* about p, and is clearly an involution.

Now, if $\nabla R = 0$ it can be shown that s_p is an *isometry* on U_p. This is because the Jacobi fields along a geodesic are the solutions of a differential equation with constant coefficients, and therefore the metric g on U_p assumes a simple form in normal coordinates at p, determined entirely by the metric and curvature at p. A similar argument is given in detail in [133, §VI, Th. 7.2, Th. 7.4].

Theorem 3.3.8 is important in the classification of Riemannian holonomy groups. For suppose (M, g) is a Riemannian manifold with $\nabla R = 0$. Then (M, g) is locally isometric to a simply-connected Riemannian symmetric space (N, h) by Theorems 3.3.7 and 3.3.8, and therefore $\mathrm{Hol}^0(g) = \mathrm{Hol}(h)$. But, as we have seen above, the classification of holonomy groups of Riemannian symmetric spaces comes out of Cartan's classification of Riemannian symmetric spaces, and is already well understood.

Therefore, we may restrict our attention to holonomy groups of Riemannian metrics that are nonsymmetric, for which $\nabla R \neq 0$. Now, this condition $\nabla R \neq 0$ can be used to exclude many candidate holonomy groups, in the following way. Theorems 3.1.2 and 3.1.7 show that if a metric g has a prescribed holonomy group $H \subset O(n)$, then the Riemann curvature R and its derivative ∇R have certain symmetries, and also lie in vector subspaces determined by the Lie algebra $\mathfrak{h}$ of H. For some groups H, these conditions force $\nabla R = 0$, so that H cannot be a nonsymmetric holonomy group.

3.4 The classification of Riemannian holonomy groups

In this section we shall discuss the question: which subgroups of $O(n)$ can be the holonomy group of a Riemannian n-manifold (M, g)? To simplify the answer, it is convenient to restrict the question in three ways. Firstly, we suppose that M is simply-connected, or equivalently, we study the restricted holonomy group $\mathrm{Hol}^0(g)$ instead of the holonomy group $\mathrm{Hol}(g)$. This eliminates issues to do with the fundamental group and global topology of M.

Secondly, we know from §3.2 that if g is locally reducible then $\mathrm{Hol}^0(g)$ is a product of holonomy groups in lower dimensions. Therefore, we suppose that g is irreducible. And thirdly, from §3.3, if g is locally symmetric then $\mathrm{Hol}^0(g)$ lies on the list of holonomy groups of Riemannian symmetric spaces, which are already known. So we suppose that g is not locally symmetric, and ask another question: which subgroups of $SO(n)$ can be the holonomy group of an irreducible, nonsymmetric Riemannian metric g on a simply-connected n-manifold M? In 1955, Berger [31, Th. 3, p. 318] proved the following result, which is the first part of the answer to this question.

Theorem 3.4.1. (Berger) *Suppose M is a simply-connected manifold of dimension n, and that g is a Riemannian metric on M, that is irreducible and nonsymmetric. Then exactly one of the following seven cases holds.*

(i) $\mathrm{Hol}(g) = SO(n)$,
(ii) $n = 2m$ with $m \geqslant 2$, and $\mathrm{Hol}(g) = U(m)$ in $SO(2m)$,
(iii) $n = 2m$ with $m \geqslant 2$, and $\mathrm{Hol}(g) = SU(m)$ in $SO(2m)$,
(iv) $n = 4m$ with $m \geqslant 2$, and $\mathrm{Hol}(g) = Sp(m)$ in $SO(4m)$,
(v) $n = 4m$ with $m \geqslant 2$, and $\mathrm{Hol}(g) = Sp(m)\,Sp(1)$ in $SO(4m)$,
(vi) $n = 7$ and $\mathrm{Hol}(g) = G_2$ in $SO(7)$, or
(vii) $n = 8$ and $\mathrm{Hol}(g) = \mathrm{Spin}(7)$ in $SO(8)$.

In fact Berger also included the eighth case $n = 16$ and $\mathrm{Hol}(g) = \mathrm{Spin}(9)$ in $SO(16)$, but it was shown by Alekseevskii [6] and also by Brown and Gray [41] that any Riemannian metric with holonomy group $\mathrm{Spin}(9)$ is symmetric. We shall refer to Theorem 3.4.1 as *Berger's Theorem*, and to the groups in parts (i)–(vii) as *Berger's list*. Berger's proof will be discussed in §3.4.3. It is rather algebraic, and uses the classification of Lie groups and their representations, and the symmetry properties of the curvature tensor.

Berger proved that the groups on his list were the only possibilities, but he did not show whether the groups actually do occur as holonomy groups. It is now known (but this took another thirty years to find out) that all of the groups on Berger's list do occur as the holonomy groups of irreducible, nonsymmetric metrics.

Here are a few remarks about Berger's Theorem.

- Theorem 3.4.1 gives the holonomy group Hol(g) not just as an abstract group, but as a particular subgroup of SO(n). In other words, the holonomy representation of Hol(g) on $\mathbb{R}^n$ is completely specified.
- Combining Theorem 3.4.1 with the results of §3.2 and §3.3, we see that the restricted holonomy group Hol^0(g) of any Riemannian manifold (M, g) is a product of groups from Berger's list and the holonomy groups of Riemannian symmetric spaces, which are known from Cartan's classification.
- In cases (ii)–(v) of Theorem 3.4.1, we require that $m \geqslant 2$. The reason for this is to avoid repeating holonomy groups. In case (ii), when $m = 1$ we have U(1) = SO(2), coinciding with $n = 2$ in case (i). In case (iii) we have that SU(1) = {1}, which acts reducibly, and can be regarded as SO(1) × SO(1) acting on $\mathbb{R}^2 = \mathbb{R} \oplus \mathbb{R}$. In case (iv) we have Sp(1) = SU(2) in SO(4), and in case (v) we have Sp(1) Sp(1) = SO(4).

In §3.4.1, we will say a little bit about each of the groups on Berger's list. The rest of the book will discuss cases (ii)–(vii) in much more detail. Then §3.4.2 will discuss Berger's list as a whole, bringing out various common features and themes. Finally, §3.4.3 explains the principles behind the proofs by Berger and Simons of Theorem 3.4.1.

3.4.1 The groups on Berger's list

We make some brief remarks, with references, about each group on Berger's list.

(i) SO(n) is the holonomy group of generic Riemannian metrics.

(ii) Riemannian metrics g with Hol(g) ⊆ U(m) are called *Kähler metrics*. Kähler metrics are a natural class of metrics on complex manifolds, and generic Kähler metrics on a given complex manifold have holonomy U(m). Kähler geometry is covered by Griffiths and Harris [93], and Kobayashi and Nomizu [134, §IX].

(iii) Metrics g with Hol(g) ⊆ SU(m) are called *Calabi–Yau metrics*. Since SU(m) is a subgroup of U(m), all Calabi–Yau metrics are Kähler. If g is Kähler, then Hol^0(g) ⊆ SU(m) if and only if g is Ricci-flat. Thus Calabi–Yau metrics are locally the same as Ricci-flat Kähler metrics.

Explicit examples of *complete* metrics with holonomy SU(m) were given by Calabi [51]. The existence of metrics with holonomy SU(m) on *compact* manifolds follows from Yau's solution of the Calabi conjecture, [226]. The most well-known example is the $K3$ surface, which admits a family of metrics with holonomy SU(2).

(iv) Metrics g with Hol(g) ⊆ Sp(m) are called *hyperkähler metrics*. As Sp(m) ⊆ SU(2m) ⊂ U(2m), hyperkähler metrics are Ricci-flat and Kähler. Explicit examples of *complete* metrics with holonomy Sp(m) were found by Calabi [51]. Yau's solution of the Calabi conjecture can be used to construct metrics with holonomy

Sp(m) on *compact* manifolds; examples were given by Fujiki [77] in the case Sp(2), and Beauville [28] in the case Sp(m).

(v) Metrics g with holonomy group Sp(m) Sp(1) for $m \geqslant 2$ are called *quaternionic Kähler metrics*. (Note that quaternionic Kähler metrics are not in fact Kähler.) They are Einstein, but not Ricci-flat. For the theory of quaternionic Kähler manifolds, see Salamon [184], and for explicit examples, see Galicki and Lawson [79, 80]. It is (the author believes) still an important open question whether there exist quaternionic Kähler metrics with positive scalar curvature on compact manifolds, that are not locally symmetric. For some progress on this problem, see LeBrun and Salamon [149].

(vi) and (vii) The holonomy groups G_2 and Spin(7) are called the *exceptional holonomy groups*. The existence of metrics with holonomy G_2 and Spin(7) was first established in 1985 by Bryant [43], using the theory of exterior differential systems. Explicit examples of complete metrics with holonomy G_2 and Spin(7) were found by Bryant and Salamon [47]. Metrics with holonomy G_2 and Spin(7) on compact manifolds were constructed by the author in [121, 122] for the case of G_2, and [120] for the case of Spin(7).

3.4.2 *A discussion of Berger's list*

Attempts to generalize the concept of number from real numbers to complex numbers and beyond led to the discovery of the four *division algebras*: the *real numbers* $\mathbb{R}$, the *complex numbers* $\mathbb{C}$, the *quaternions* $\mathbb{H}$, and the *octonions* or *Cayley numbers* $\mathbb{O}$. At each step in this sequence, the dimension doubles, and one algebraic property is lost. So, the complex numbers have dimension 2 (over $\mathbb{R}$) and are not ordered; the quaternions have dimension 4 and are not commutative; and the octonions have dimension 8 and are not associative. The sequence stops here, possibly because there are no algebraic properties left to lose.

The groups on Berger's list correspond to the division algebras. First consider cases (i)–(v). The group SO(n) is a group of automorphisms of $\mathbb{R}^n$. Both U(m) and SU(m) are groups of automorphisms of $\mathbb{C}^m$, and Sp(m) and Sp(m) Sp(1) are automorphism groups of $\mathbb{H}^m$. To make the analogy between $\mathbb{R}$, $\mathbb{C}$ and $\mathbb{H}$ more complete, we add the holonomy group O(n). Then O(n), U(m) and Sp(m) Sp(1) are automorphism groups of $\mathbb{R}^n$, $\mathbb{C}^m$ and $\mathbb{H}^m$ respectively, preserving a metric. And SO(n), SU(m) and Sp(m) are the subgroups of O(n), U(m) and Sp(m) Sp(1) with 'determinant 1' in an appropriate sense.

The exceptional cases (vi) and (vii) also fit into this pattern, although not as neatly. One can regard G_2 and Spin(7) as automorphism groups of $\mathbb{O}$. The octonions split as $\mathbb{O} \cong \mathbb{R} \oplus \operatorname{Im} \mathbb{O}$, where $\operatorname{Im} \mathbb{O} \cong \mathbb{R}^7$ is the *imaginary octonions*. The automorphism group of $\operatorname{Im} \mathbb{O}$ is G_2.

In some sense, G_2 is the group of 'determinant 1' automorphisms of $\mathbb{O}$, so it fits into the sequence SO(n), SU(m), Sp(m), G_2. Similarly, Spin(7) is the group of automorphisms of $\mathbb{O} \cong \mathbb{R}^8$ which preserve a certain part of the multiplicative structure of $\mathbb{O}$, and it fits into the sequence O(n), U(m), Sp(m) Sp(1), Spin(7).

Here are three ways in which we can gather together the holonomy groups on Berger's list into subsets with common features.

- The *Kähler holonomy groups* are U(m), SU(m) and Sp(m). Any Riemannian manifold with one of these holonomy groups is a Kähler manifold, and thus a complex manifold. Therefore one can use complex geometry to study the Kähler holonomy groups, and this is a tremendous advantage.

  As complex manifolds are locally trivial, complex geometry has a very different character to Riemannian geometry, and a great deal is known about the *global* geometry of complex manifolds, particularly through complex algebraic geometry, which has no real parallel in Riemannian geometry.

  Although metrics with holonomy Sp(m) Sp(1) for $m > 1$ are not Kähler, they should be considered along with the Kähler holonomy groups. A Riemannian manifold M with holonomy Sp(m) Sp(1) has a *twistor space* [184], a complex manifold Z of real dimension $4m + 2$, which fibres over M with fibre $\mathbb{CP}^1$. If M has positive scalar curvature, then Z is Kähler. Thus, metrics with this holonomy group can also be studied using complex and Kähler geometry.

- The *Ricci-flat holonomy groups* are SU(m), Sp(m), G_2 and Spin(7). Any metric with one of these holonomy groups is Ricci-flat. They are the main focus of this book. Because irreducible symmetric spaces (other than $\mathbb{R}$) are Einstein with nonzero scalar curvature, none of the Ricci-flat holonomy groups can be the holonomy group of a symmetric space, or more generally of a homogeneous space. Because of this, simple examples of metrics with the Ricci-flat holonomy groups are difficult to find, and one has to work harder to get a feel for what the geometry of these metrics is like.

- The *exceptional holonomy groups* are G_2 and Spin(7). They are the exceptional cases in Berger's classification, and they are rather different from the other holonomy groups. The holonomy groups U(m), SU(m), Sp(m) and Sp(m) Sp(1) can all be approached through complex geometry, and SO(n) is uninteresting for obvious reasons.

  This leaves G_2 and Spin(7), which are similar to one another but stand out from the rest. Since we cannot use complex manifold theory to tell us about the *global* geometry of manifolds with holonomy G_2 and Spin(7), at present our understanding of them is essentially *local* in nature.

3.4.3 *A sketch of the proof of Berger's Theorem*

Here is a sketch of Berger's proof of Theorem 3.4.1. As M is simply-connected, Theorem 3.2.8 shows that Hol(g) is a closed, connected Lie subgroup of SO(n), and since g is irreducible, Corollary 3.2.5 shows that the representation of Hol(g) on $\mathbb{R}^n$ is irreducible. So, suppose that H is a closed, connected subgroup of SO(n) with irreducible representation on $\mathbb{R}^n$, and Lie algebra $\mathfrak{h}$. The classification of all such groups H follows from the classification of Lie groups (and is of considerable complexity).

Berger's method was to take the list of all such groups H, and to apply two tests to each possibility to find out if it could be a holonomy group. The only groups H

which passed both tests are those in the theorem. Berger's tests are algebraic and involve the curvature tensor. Suppose that R_{abcd} is the Riemann curvature of a metric g with $\text{Hol}(g) = H$. Then Theorem 3.1.2 shows that $R_{abcd} \in S^2\mathfrak{h}$, and the first Bianchi identity (3.3) applies.

If $\mathfrak{h}$ has large codimension in $\mathfrak{so}(n)$, then the vector space $\mathfrak{R}^H$ of elements of $S^2\mathfrak{h}$ satisfying (3.3) will be small, or even zero. But Theorem 2.4.3 shows that $\mathfrak{R}^H$ must be big enough to generate $\mathfrak{h}$. For many of the candidate groups H this does not hold, and so H cannot be a holonomy group. This is the first test. Now $\nabla_e R_{abcd}$ lies in $(\mathbb{R}^n)^* \otimes \mathfrak{R}^H$, and also satisfies the second Bianchi identity (3.4). Frequently these requirements imply that $\nabla R = 0$, so that g is locally symmetric. Therefore we may exclude such H, and this is Berger's second test.

Later, with the benefit of hindsight, Simons [194] found a shorter (but still difficult) proof of Theorem 3.4.1 based on showing that if g is irreducible and nonsymmetric, then $\text{Hol}(g)$ must act transitively on the unit sphere in $\mathbb{R}^n$. But the list of compact, connected Lie groups acting transitively and effectively on spheres had already been found by Montgomery and Samelson [163], [37].

They turn out to be the groups on Berger's list, plus two others, the group $\text{Sp}(m)\text{U}(1)$ acting on S^{4m-1} for $m > 1$, and the group $\text{Spin}(9)$ acting on S^{15}. Thus, to complete the second proof of Berger's Theorem one must show that these two cannot occur as holonomy groups. Short accounts of Simons' proof are given by Besse [33, p. 303–305] and Salamon [184, p. 149–151].

3.5 Holonomy groups, exterior forms and de Rham cohomology

Let (M, g) be a compact Riemannian manifold. In this section we explore the links between $\text{Hol}(g)$ and the de Rham cohomology $H^*(M, \mathbb{R})$. We explain how a G-structure on M divides the bundle of k-forms on M into a sum of vector subbundles, corresponding to irreducible representations of G. If the G-structure is torsion-free, then $H^k(M, \mathbb{R})$ has an analogous decomposition into vector subspaces by Hodge theory. Finally we show that if $\text{Hol}(g)$ is one of the Ricci-flat holonomy groups then $H^1(M, \mathbb{R}) = \{0\}$, and $\pi_1(M)$ is finite.

3.5.1 *Decomposition of exterior forms*

Let M be an n-manifold, G a Lie subgroup of $\text{GL}(n, \mathbb{R})$, and Q a G-structure on M. Then from Definition 2.1.3, to each representation ρ of G on a vector space V we can associate a vector bundle $\rho(Q)$ over M, with fibre V. In particular, $\text{GL}(n, \mathbb{R})$ has a natural representation on $V = \mathbb{R}^n$, and if ρ is the restriction of this to G, then $\rho(Q) = TM$.

The representation of G on V induces representations of G on the dual vector space V^* and its exterior powers $\Lambda^k V^*$. Write ρ^k for the representation of G on $\Lambda^k V^*$. Then $\rho^k(Q) = \Lambda^k T^*M$, the bundle of k-forms on M. Now $\Lambda^k V^*$ is an *irreducible* representation of $\text{GL}(n, \mathbb{R})$ for every k. However, if G is a proper subgroup of $\text{GL}(n, \mathbb{R})$, it can happen that the representation ρ^k of G on $\Lambda^k V^*$ is *reducible*. Then we can write $\Lambda^k V^* = \bigoplus_{i \in I^k} W_i^k$ and $\rho^k = \bigoplus_{i \in I^k} \rho_i^k$, where each (ρ_i^k, W_i^k) is an irreducible representation of G, and I^k is a finite indexing set.

But then $\Lambda^k T^*M = \rho^k(Q) = \bigoplus_{i \in I^k} \rho_i^k(Q)$. This means that a G-structure on M induces a splitting of the vector bundle $\Lambda^k T^*M$ of k-forms on M, into a direct sum of vector subbundles $\rho_i^k(Q)$ corresponding to irreducible representations of G. We shall use the notation Λ_i^k for $\rho_i^k(Q)$, and $\pi_i : \Lambda^k T^*M \to \Lambda_i^k$ for the projection to Λ_i^k in the decomposition $\Lambda^k T^*M = \bigoplus_{i \in I^k} \Lambda_i^k$.

Note that analogous decompositions also hold for tensor bundles $\bigotimes^k TM \otimes \bigotimes^l T^*M$ on a manifold with a G-structure Q, and also, if G is a subgroup of $O(n)$ and M is spin, for the spin bundles of M with respect to the Riemannian metric induced by Q. We will not make much use of these, though. The following proposition, which is trivial to prove, summarizes the material above.

Proposition 3.5.1 *Let G be a Lie subgroup of* $\mathrm{GL}(n, \mathbb{R})$. *Write (ρ, V) for the natural representation of G on $\mathbb{R}^n$, and let ρ^k be the induced representation of G on $\Lambda^k V^*$. Then $(\rho^k, \Lambda^k V^*)$ is a direct sum of irreducible representations (ρ_i^k, W_i^k) of G, for $i \in I^k$, an indexing set.*

Suppose M is an n-manifold, and Q a G-structure on M. Then there is a natural decomposition

$$\Lambda^k T^*M = \bigoplus_{i \in I^k} \Lambda_i^k, \tag{3.17}$$

where Λ_i^k is a vector subbundle of $\Lambda^k T^*M$ with fibre W_i^k. If two representations (ρ_i^k, W_i^k) and (ρ_j^l, W_j^l) are isomorphic, then Λ_i^k and Λ_j^l are isomorphic. If $\phi : \Lambda^k V^* \to \Lambda^l V^*$ is a G-equivariant linear map, there is a corresponding map $\Phi : \Lambda^k T^*M \to \Lambda^l T^*M$ of vector bundles.

As an example of these ideas, we explain the Hodge star of §1.1.2.

Example 3.5.2 Let G be the subgroup $\mathrm{SO}(n)$ of $\mathrm{GL}(n, \mathbb{R})$. Then a G-structure on an n-manifold M is equivalent to a Riemannian metric g and an orientation on M. There is an isomorphism $* : \Lambda^k V^* \to \Lambda^{n-k} V^*$ between the representations of $\mathrm{SO}(n)$. By Proposition 3.5.1, this induces an isomorphism $* : \Lambda^k T^*M \to \Lambda^{n-k} T^*M$ called the *Hodge star*.

In the case $n = 4m$, the map $* : \Lambda^{2m} V^* \to \Lambda^{2m} V^*$ satisfies $*^2 = 1$, and so $\Lambda^{2m} V^*$ splits as $\Lambda^{2m} V^* = W_+^{2m} \oplus W_-^{2m}$, where $W_\pm^{2m}$ are the eigenspaces of $*$ with eigenvalues ± 1. Here W_+^{2m} and W_-^{2m} are in fact irreducible representations of $\mathrm{SO}(4m)$ of equal dimension, and we choose the indexing set $I^{2m} = \{+, -\}$. Thus, there is a corresponding splitting $\Lambda^{2m} T^*M \cong \Lambda_+^{2m} \oplus \Lambda_-^{2m}$.

We are principally interested in the case in which G is a Lie subgroup of $O(n)$ in $\mathrm{GL}(n, \mathbb{R})$, and Q a torsion-free G-structure. The condition that Q be torsion-free implies that the exterior derivative d on k-forms, and its formal adjoint d*, behave in a special way with regard to the splitting $\Lambda^k T^*M = \bigoplus_{i \in I^k} \Lambda_i^k$, and this has important consequences for the de Rham cohomology of M.

3.5.2 *Hodge theory and the splitting of the de Rham cohomology*

Suppose (M, g) is a compact Riemannian n-manifold. Let G be a Lie subgroup of $O(n)$, and suppose that $\mathrm{Hol}(g) \subseteq G$. Then from §3.1.4, there is a unique, torsion-free $O(n)$-structure P on M induced by g, and a torsion-free G-structure Q on M contained in P. Moreover, the Levi-Civita connection ∇ of g reduces to Q.

Our goal is to study the *Hodge theory* of M. Hodge theory is described in §1.1.3, and concerns the *Laplacian* $\Delta = dd^* + d^*d$ acting on k-forms on M, and its kernel $\mathcal{H}^k$, which is a finite-dimensional vector space of k-forms. The *Weitzenbock formula* for k-forms ([33, §1.I], [186, Prop. 4.10]) is

$$(dd^* + d^*d)\xi = \nabla^*\nabla\xi - 2\tilde{R}(\xi), \tag{3.18}$$

where, using the index notation and writing R_{ab} for the Ricci curvature and $R^a{}_{bcd}$ for the Riemann curvature of g, we have

$$\begin{aligned}\tilde{R}(\xi)_{r_1\ldots r_k} &= \sum_{1 \leqslant i < j \leqslant k} g^{bc} R^a{}_{r_i c r_j} \xi_{r_1 \ldots r_{i-1} a r_{i+1} \ldots r_{j-1} b r_{j+1} \ldots r_k} \\ &\quad - \frac{1}{2} \sum_{j=1}^k g^{ab} R_{r_j b} \xi_{r_1 \ldots r_{j-1} a r_{j+1} \ldots r_k}.\end{aligned} \tag{3.19}$$

By Proposition 3.5.1 there is a splitting (3.17) induced by Q. This gives an isomorphism

$$C^\infty(\Lambda^k T^*M) = \bigoplus_{i \in I^k} C^\infty(\Lambda_i^k). \tag{3.20}$$

Suppose that ξ lies in $C^\infty(\Lambda_i^k)$ for some $i \in I^k$. Because ∇ preserves the G-structure Q, it preserves the decomposition (3.17). Thus $\nabla\xi \in C^\infty(T^*M \otimes \Lambda_i^k)$, and so $\nabla^*\nabla\xi \in C^\infty(\Lambda_i^k)$.

Now Theorem 3.1.7 shows that the Riemann curvature R_{abcd} lies in $S^2\mathfrak{hol}_x(g)$ at each $x \in M$, where $\mathfrak{hol}_x(g)$ is a vector subspace of $\Lambda^2 T_x^*M$ isomorphic to the holonomy algebra $\mathfrak{hol}(g)$. As $\mathrm{Hol}(g) \subseteq G$, we have $\mathfrak{hol}(g) \subset \mathfrak{g}$, where $\mathfrak{g}$ is the Lie algebra of G. Using this, one can show that the linear map $\tilde{R} : \Lambda^k T^*M \to \Lambda^k T^*M$ defined by (3.19) preserves the splitting (3.17), and thus $\tilde{R}(\xi) \in C^\infty(\Lambda_i^k)$. Therefore we see that if $\xi \in C^\infty(\Lambda_i^k)$, then both $\nabla^*\nabla\xi$ and $\tilde{R}(\xi)$ lie in $C^\infty(\Lambda_i^k)$, and so $(dd^* + d^*d)\xi$ lies in $C^\infty(\Lambda_i^k)$ by (3.18). This proves that the Laplacian $\Delta = dd^* + d^*d$ maps $C^\infty(\Lambda_i^k)$ into itself for each $i \in I^k$.

We have shown that in the splitting (3.20), the Laplacian Δ takes each factor $C^\infty(\Lambda_i^k)$ to itself. Therefore we have

$$\mathcal{H}^k = \bigoplus_{i \in I^k} \mathcal{H}_i^k, \text{ where } \mathcal{H}^k = \mathrm{Ker}\,\Delta \text{ and } \mathcal{H}_i^k = \mathrm{Ker}(\Delta|_{\Lambda_i^k}). \tag{3.21}$$

In Proposition 3.5.1 we saw that if W_i^k and W_j^l are isomorphic representations of G, then the vector bundles Λ_i^k and Λ_j^l are isomorphic. Now it turns out that $\nabla^*\nabla$ and $\tilde{R}$ depend only on the representation of G, and not on its particular embedding in $\Lambda^k V^*$. By (3.18) this is true also of $dd^* + d^*d$, and so if W_i^k and W_j^l are isomorphic representations of G, it follows that $\mathcal{H}_i^k$ and $\mathcal{H}_j^l$ are isomorphic.

But $H^k(M, \mathbb{R})$ is isomorphic to $\mathcal{H}^k$ by Theorem 1.1.3. Thus, (3.21) gives a decomposition of $H^k(M, \mathbb{R})$ corresponding to the splitting (3.17). This gives the following theorem.

Theorem 3.5.3 *Suppose M is a compact n-manifold and Q a torsion-free G-structure on M, where G is a Lie subgroup of $O(n)$, and let g be the metric associated to Q. Then Proposition 3.5.1 gives a splitting $\Lambda^k T^* M = \bigoplus_{i \in I^k} \Lambda_i^k$, corresponding to the decomposition of $\Lambda^k (\mathbb{R}^n)^*$ into irreducible representations of G.*

The Laplacian $\Delta = dd^* + d^*d$ associated to g maps $C^\infty(\Lambda_i^k)$ to itself. Define $\mathcal{H}_i^k = \text{Ker}\, \Delta|_{\Lambda_i^k}$, and let $H_i^k(M, \mathbb{R})$ be the vector subspace of the de Rham cohomology group $H^k(M, \mathbb{R})$ with representatives in $\mathcal{H}_i^k$. Then $H_i^k(M, \mathbb{R}) \cong \mathcal{H}_i^k$, and we have the direct sum

$$H^k(M, \mathbb{R}) = \bigoplus_{i \in I^k} H_i^k(M, \mathbb{R}). \tag{3.22}$$

If W_i^k and W_j^l are isomorphic as representations of G, then $H_i^k(M, \mathbb{R}) \cong H_j^l(M, \mathbb{R})$.

The theorem shows that if a Riemannian metric g on a compact manifold M has $\text{Hol}(g) = G$, then the de Rham cohomology $H^*(M, \mathbb{R})$ has a natural decomposition into smaller pieces, which depend on G and its representations. The Betti numbers of M are $b^k(M) = \dim H^k(M, \mathbb{R})$. Define the *refined Betti numbers* $b_i^k(M)$ to be $b_i^k(M) = \dim H_i^k(M, \mathbb{R})$, for $i \in I^k$. Then (3.22) shows that $b^k(M) = \sum_{i \in I^k} b_i^k(M)$. The refined Betti numbers carry both *topological* information about M, and *geometrical* information about the G-structure Q.

If a compact manifold M admits a metric g with $\text{Hol}(g) = G$, then the theorem forces its cohomology $H^*(M, \mathbb{R})$ to assume a certain form. Conversely, if one can show that $H^*(M, \mathbb{R})$ cannot be written in this way, then M cannot admit any metric g with $\text{Hol}(g) = G$. Thus, one can prove that some compact manifolds do not admit metrics with a given holonomy group, for purely topological reasons. For example, when $G = U(m)$ one finds that $b^k(M)$ must be even when k is odd; roughly speaking, this is because each irreducible representation occurs twice.

Suppose that for some k and $i \in I^k$, we have $W_i^k \cong \mathbb{R}$, the trivial representation of G. Now $\Lambda^0(\mathbb{R}^n)^* = W_1^0$ is also a copy of $\mathbb{R}$, the trivial representation. So W_i^k and W_1^0 are isomorphic as representations of G, and $H_i^k(M, \mathbb{R}) \cong H_1^0(M, \mathbb{R}) = H^0(M, \mathbb{R})$ by Theorem 3.5.3. But $H^0(M, \mathbb{R}) = \mathbb{R}$, as M is connected. Thus, if W_i^k is the trivial representation $\mathbb{R}$, then there is a natural isomorphism $H_i^k(M, \mathbb{R}) \cong \mathbb{R}$.

The explanation for this is simple. Proposition 2.5.2 shows that there is a 1-1 correspondence between *constant tensors* on M, and invariant elements of the corresponding representation of $\text{Hol}(g)$. Since $\text{Hol}(g) \subseteq G$, if W_i^k is a trivial representation its elements are invariant under $\text{Hol}(g)$, and so correspond to *constant k-forms*. Now if ξ

is a constant k-form, then $\nabla \xi = 0$. But $d\xi$ and $d^*\xi$ are components of $\nabla \xi$, and so $d\xi = d^*\xi = 0$. Hence $(dd^* + d^*d)\xi = 0$, and ξ lies in $\mathcal{H}^k$. Thus, if W_i^k is a trivial representation, then $\mathcal{H}_i^k$ is a vector space of constant k-forms isomorphic to W_i^k.

We have shown that if Q is a torsion-free G-structure on a compact manifold M, then to each G-invariant element of $\Lambda^k V^*$ there corresponds a constant k-form ξ, and this defines a cohomology class $[\xi] \in H^k(M, \mathbb{R})$. So, to each torsion-free G-structure Q we associate a collection of cohomology classes in $H^k(M, \mathbb{R})$, corresponding to the G-invariant part of $\Lambda^k V^*$. Now $H^k(M, \mathbb{R})$ is a topological invariant of M, and does not depend on Q. Thus we can compare the cohomology classes associated to *different* torsion-free G-structures Q, Q' on M.

Often, the family of torsion-free G-structures on M, when divided by the group of diffeomorphisms of M isotopic to the identity, forms a finite-dimensional manifold $\mathcal{M}_G$ called a *moduli space*. The cohomology classes associated to a torsion-free G-structure give maps $\mathcal{M}_G \to H^k(M, \mathbb{R})$, and these provide a natural coordinate system on $\mathcal{M}_G$. More generally, the splitting (3.22) may also be regarded in this light. This topological information can be exploited to give a local (and, in good cases, global) description of the moduli spaces $\mathcal{M}_G$, when G is one of the Ricci-flat holonomy groups.

3.5.3 *One-forms and the Ricci-flat holonomy groups*

Let (M, g) be a compact Riemannian n-manifold. From eqns (3.18) and (3.19) we see that if ξ is a 1-form, then

$$(dd^* + d^*d)\xi_a = \nabla^*\nabla \xi_a + R_{ab}g^{bc}\xi_c, \quad (3.23)$$

where R_{ab} is the Ricci curvature of g. Suppose now that $\xi \in \mathcal{H}^1$, the kernel of $dd^* + d^*d$ on 1-forms. Then we have $\nabla^*\nabla \xi_a + R_{ab}g^{bc}\xi_c = 0$. Taking the inner product of this equation with ξ and integrating by parts yields

$$\|\nabla \xi\|_{L^2}^2 + \int_M R_{ab}g^{bc}g^{ad}\xi_c\xi_d dV = 0. \quad (3.24)$$

If the Ricci curvature R_{ab} is zero, this shows that $\|\nabla \xi\|_{L^2} = 0$, so that $\nabla \xi = 0$.

More generally, if R_{ab} is *nonnegative*, then the second term in (3.24) is nonnegative. But the first term is also nonnegative, so both must be zero. Thus, if R_{ab} is nonnegative, we have $\nabla \xi = 0$. If R_{ab} is *positive definite*, then either $\xi \equiv 0$, or the second term in (3.24) is positive, which is a contradiction. This shows that if R_{ab} is zero or nonnegative, then all 1-forms ξ in $\mathcal{H}^1$ are constant, and if R_{ab} is positive definite, then all 1-forms ξ in $\mathcal{H}^1$ are zero.

Suppose that R_{ab} is nonnegative, and that $\dim \mathcal{H}^1 = k > 0$. Choose a basis $\xi_1, \ldots, \xi_k$ for $\mathcal{H}^1$. Then $\xi_1, \ldots, \xi_k$ are constant 1-forms, and so by Proposition 2.5.2 they correspond to elements of $(\mathbb{R}^n)^*$ fixed by $\mathrm{Hol}(g)$. Clearly, this implies that $k \leqslant n$ and that $\mathbb{R}^n$ splits orthogonally as $\mathbb{R}^n = \mathbb{R}^k \oplus \mathbb{R}^{n-k}$, where $\mathrm{Hol}(g)$ preserves the splitting and acts trivially on $\mathbb{R}^k$.

Thus, the action of $\mathrm{Hol}(g)$ on $\mathbb{R}^n$ is *reducible*, and so by Corollary 3.2.5 the metric g is locally reducible. Moreover, let $(\tilde{M}, \tilde{g})$ be the universal cover of (M, g). Then

Hol($\tilde{g}$) = Hol^0(g) $\subseteq$ Hol(g), and $\tilde{g}$ is *complete* as M is compact. So Theorem 3.2.7 applies to show that ($\tilde{M}, \tilde{g}$) is globally reducible. In fact ($\tilde{M}, \tilde{g}$) is isometric to a product $\mathbb{R}^k \times N$, where $\mathbb{R}^k$ carries the Euclidean metric and N is a Riemannian ($n-k$)-manifold.

But $\mathcal{H}^1 \cong H^1(M, \mathbb{R})$ by Theorem 1.1.3. Thus we have proved the following result, known as the *Bochner Theorem*.

Theorem 3.5.4. (Bochner [35]) *Let (M, g) be a compact Riemannian manifold. If the Ricci curvature of g is nonnegative, then* dim $H^1(M, \mathbb{R}) = k \leqslant$ dim M, *and the universal cover $(\tilde{M}, \tilde{g})$ of (M, g) is isometric to a product $\mathbb{R}^k \times N$, where $\mathbb{R}^k$ has the Euclidean metric. If the Ricci curvature of g is positive definite, then $H^1(M, \mathbb{R}) = \{0\}$.*

One can also prove strong results on the fundamental group of a compact manifold with nonnegative Ricci curvature. As a consequence of the *Cheeger–Gromoll splitting Theorem* [58], [33, §6.G] we get the following result, taken from Besse [33, Cor. 6.67].

Theorem 3.5.5 *Suppose (M, g) is a compact Riemannian manifold. If g is Ricci-flat then M admits a finite cover isometric to $T^k \times N$, where T^k carries a flat metric and N is a compact, simply-connected Riemannian manifold. If the Ricci curvature of M is positive definite then $\pi_1(M)$ is finite.*

Now suppose that Hol(g) is one of the Ricci-flat holonomy groups SU(m), Sp(m), G_2 and Spin(7). Then $R_{ab} = 0$, so that $\mathcal{H}^1$ consists of constant 1-forms, from above. But by Proposition 2.5.2, the constant tensors on M are entirely determined by Hol(g). Since SU(m), Sp(m), G_2 and Spin(7) all fix no nonzero elements in $(\mathbb{R}^n)^*$, there are no nonzero constant 1-forms. Thus $\mathcal{H}^1 = 0$, and so $H^1(M, \mathbb{R}) = \{0\}$. Moreover, it follows from Theorem 3.5.5 that $\pi_1(M)$ is finite. Therefore we have:

Corollary 3.5.6 *Let (M, g) be a compact Riemannian manifold, and suppose that* Hol(g) *is one of the Ricci-flat holonomy groups* SU(m), Sp(m), G_2 *and* Spin(7). *Then $H^1(M, \mathbb{R}) = \{0\}$, so that $b^1(M) = 0$, and the fundamental group $\pi_1(M)$ is finite.*

This is an example of how the topology of a compact manifold M can impose constraints on the possible holonomy groups Hol(g) of Riemannian metrics g on M.

3.6 Spinors and holonomy groups

If (M, g) is a Riemannian spin manifold, then there is a natural vector bundle S over M called the *spin bundle*, and sections of S are called *spinors*. Spinors are closely related to tensors, and have similar properties. In particular, just as the constant tensors on M are determined by Hol(g), so the constant spinors on M are also determined by Hol(g), with the right choice of spin structure. In this section we explore the link between spin geometry and holonomy groups, and deduce some topological information about compact $4m$-manifolds M with the Ricci-flat holonomy groups.

3.6.1 *Introduction to spin geometry*

Here is a brief explanation of some ideas from spin geometry that we will need later. Some general references on this material are Lawson and Michelson [145], and Harvey [97].

For each $n \geq 3$, the Lie group $SO(n)$ is connected and has fundamental group $\pi_1(SO(n)) = \mathbb{Z}_2$. Therefore it has a double cover, the Lie group $Spin(n)$, which is a compact, connected, simply-connected Lie group. The covering map $\pi : Spin(n) \to SO(n)$ is a Lie group homomorphism. There is a natural representation Δ^n of $Spin(n)$, called the *spin representation*. It has the following properties:

- Δ^{2m} is a complex representation of $Spin(2m)$, with complex dimension 2^m. It splits into a direct sum $\Delta^{2m} = \Delta_+^{2m} \oplus \Delta_-^{2m}$, where $\Delta_\pm^{2m}$ are irreducible representations of $Spin(2m)$ with complex dimension 2^{m-1}.
- Δ^{2m+1} is a complex representation of $Spin(2m+1)$, with complex dimension 2^m. It is irreducible, and does not split into positive and negative parts.
- When $n = 8k-1$, $8k$ or $8k+1$, $\Delta^n = \Delta_\mathbb{R}^n \otimes_\mathbb{R} \mathbb{C}$, where $\Delta_\mathbb{R}^n$ is a *real* representation of $Spin(n)$.

Now let (M, g) be an oriented Riemannian n-manifold. The metric and orientation on M induce a unique $SO(n)$-structure P on M. A *spin structure* $(\tilde{P}, \pi)$ on M is a principal bundle $\tilde{P}$ over M with fibre $Spin(n)$, together with a map of bundles $\pi : \tilde{P} \to P$, that is locally modelled on the projection $\pi : Spin(n) \to SO(n)$. We may regard $\tilde{P}$ as a double cover of P, and π as the covering map.

Spin structures do not exist on every manifold. In fact, an oriented Riemannian manifold M admits a spin structure if and only if $w_2(M) = 0$, where $w_2(M) \in H^2(M, \mathbb{Z}_2)$ is the second Stiefel–Whitney class of M. Also, if a spin structure exists it may not be unique: when $w_2(M) = 0$, the family of spin structures on M is parametrized by $H^1(M, \mathbb{Z}_2)$. This is finite if M is compact, and zero if M is simply-connected. We call M a *spin manifold* if $w_2(M) = 0$, that is, if M admits a spin structure.

Let (M, g) be an oriented, spin Riemannian n-manifold, and choose a spin structure $(\tilde{P}, \pi)$ on M. Define the *(complex) spin bundle* $S \to M$ to be $S = \tilde{P} \times_{Spin(n)} \Delta^n$. Then S is a complex vector bundle over M, with fibre Δ^n. Sections of S are called *spinors*. If $n = 2m$, then Δ^n splits as $\Delta^n = \Delta_+^n \oplus \Delta_-^n$, and so S also splits as $S = S_+ \oplus S_-$, where $S_\pm$ are vector subbundles of S with fibre $\Delta_\pm^n$. Sections of S_+, S_- are called *positive* and *negative spinors* respectively. In dimensions $8k-1$, $8k$ and $8k+1$ there is a real spin representation $\Delta_\mathbb{R}^n$ as well as a complex one Δ^n. In this case one defines the *real spin bundle* $S_\mathbb{R} = \tilde{P} \times_{Spin(n)} \Delta_\mathbb{R}^n$. We shall always work with complex spinors, unless we explicitly say otherwise.

The $SO(n)$-bundle P over M has a natural connection, the Levi-Civita connection ∇ of g. Because $\pi : \tilde{P} \to P$ is locally an isomorphism, we may lift ∇ to $\tilde{P}$. Thus, $\tilde{P}$ also carries a natural connection, and as in §2.1, this induces a connection $\nabla^S : C^\infty(S) \to C^\infty(T^*M \otimes S)$ on S, called the *spin connection*. Now, there is a natural linear map from $T^*M \otimes S$ to S, defined by *Clifford multiplication*. Composing this map with ∇^S gives a first-order, linear partial differential operator $D : C^\infty(S) \to C^\infty(S)$ called the *Dirac operator*.

The Dirac operator is self-adjoint and elliptic. In even dimensions, it splits as a sum $D = D_+ \oplus D_-$, where D_+ maps $C^\infty(S_+) \to C^\infty(S_-)$ and D_- maps $C^\infty(S_-) \to C^\infty(S_+)$. Here $D_\pm$ are both first-order linear elliptic operators, and D_- is the formal adjoint of D_+, and vice versa. The result of changing the orientation of M is, in even dimensions, to exchange S_+ and S_-, and D_+ and D_-.

3.6.2 Parallel spinors and holonomy groups

Let (M, g) be an oriented Riemannian n-manifold with a spin structure. Then the holonomy group $\text{Hol}(\nabla^S)$ is a subgroup of $\text{Spin}(n)$. Moreover, under the projection $\pi : \text{Spin}(n) \to \text{SO}(n)$, the image of $\text{Hol}(\nabla^S)$ is exactly $\text{Hol}(g)$. The projection $\pi : \text{Hol}(\nabla^S) \to \text{Hol}(g)$ may be an isomorphism, or it may be a double cover; in general, this depends on the choice of spin structure. However, if M is simply-connected, then both $\text{Hol}(g)$ and $\text{Hol}(\nabla^S)$ are connected, which forces $\text{Hol}(\nabla^S)$ to be the identity component of $\pi^{-1}(\text{Hol}(g))$ in $\text{Spin}(n)$. Thus, for simply-connected spin manifolds, the classification of holonomy groups of spin connections ∇^S follows from that of Riemannian holonomy groups.

Suppose that $\sigma \in C^\infty(S)$ satisfies $\nabla^S \sigma = 0$, so that σ is a *parallel spinor*, or *constant spinor*. Just as in §2.5.1 we found a 1-1 correspondence between constant tensors and elements of the appropriate representation invariant under $\text{Hol}(\nabla)$, so there is a 1-1 correspondence between parallel spinors and elements of Δ^n invariant under $\text{Hol}(\nabla^S)$. Therefore, one can apply Berger's classification of Riemannian holonomy groups to classify the holonomy groups of metrics with parallel spinors. This has been done by Wang [218, p. 59], in the following result.

Theorem 3.6.1 *Let M be an orientable, connected, simply-connected spin n-manifold for $n \geq 3$, and g an irreducible Riemannian metric on M. Define N to be the dimension of the space of parallel spinors on M. If n is even, define $N_\pm$ to be the dimensions of the spaces of parallel spinors in $C^\infty(S_\pm)$, so that $N = N_+ + N_-$.*

Suppose $N \geq 1$. Then, after making an appropriate choice of orientation for M, exactly one of the following holds:

  (i) *$n = 4m$ for $m \geq 1$ and $\text{Hol}(g) = \text{SU}(2m)$, with $N_+ = 2$ and $N_- = 0$,*
  (ii) *$n = 4m$ for $m \geq 2$ and $\text{Hol}(g) = \text{Sp}(m)$, with $N_+ = m + 1$ and $N_- = 0$,*
  (iii) *$n = 4m + 2$ for $m \geq 1$ and $\text{Hol}(g) = \text{SU}(2m + 1)$, with $N_+ = 1$ and $N_- = 1$,*
  (iv) *$n = 7$ and $\text{Hol}(g) = G_2$, with $N = 1$, and*
  (v) *$n = 8$ and $\text{Hol}(g) = \text{Spin}(7)$, with $N_+ = 1$ and $N_- = 0$.*

With the opposite orientation, the values of $N_\pm$ are exchanged.

Notice that the holonomy groups appearing here are exactly the Ricci-flat holonomy groups. Hence, every Riemannian spin manifold that admits a nonzero parallel spinor is Ricci-flat. (In fact, this can be proved directly.) Conversely, it is natural to ask whether every Riemannian manifold with one of the Ricci-flat holonomy groups is in fact a spin manifold, and possesses constant spinors. The answer to this is yes, and it follows from the next proposition.

Proposition 3.6.2 *Suppose M is an n-manifold that admits a G-structure Q, where $n \geq 3$ and G is a connected, simply-connected subgroup of $\text{SO}(n)$. Then M is spin, and has a natural spin structure $\tilde{P}$ induced by Q.*

Proof Since G and $SO(n)$ are connected, the embedding $\iota : G \hookrightarrow SO(n)$ lifts to a homomorphism $\tilde{\iota}$ between the universal covers of G and $SO(n)$. As G is simply-connected and the universal cover of $SO(n)$ is $\text{Spin}(n)$, as $n \geqslant 3$, this gives an injective Lie group homomorphism $\tilde{\iota} : G \hookrightarrow \text{Spin}(n)$ such that $\pi \circ \tilde{\iota} = \iota$, where $\pi : \text{Spin}(n) \to SO(n)$ is the covering map.

Now the G-structure Q on M induces an $SO(n)$-structure $P = SO(n) \cdot Q$ on M. By definition, M is spin if and only if P admits a double cover $\tilde{P}$, which fibres over M with fibre $\text{Spin}(n)$. But using the embedding $\tilde{\iota} : G \hookrightarrow \text{Spin}(n)$ we may define $\tilde{P} = Q \times_G \text{Spin}(n)$, and this is indeed a double cover of P that fibres over M with fibre $\text{Spin}(n)$. Thus M is spin. □

Since all of the Ricci-flat holonomy groups $SU(m)$, $Sp(m)$, G_2 and $\text{Spin}(7)$ are connected and simply-connected, the next corollary quickly follows.

Corollary 3.6.3 *Let (M, g) be a Riemannian manifold, and suppose that $\text{Hol}(g)$ is one of the Ricci-flat holonomy groups $SU(m)$, $Sp(m)$, G_2 and $\text{Spin}(7)$. Then M is spin, with a preferred spin structure. With this spin structure, the spaces of parallel spinors on M are nonzero and have the dimensions prescribed by Theorem 3.6.1.*

Thus, an irreducible metric has one of the Ricci-flat holonomy groups if and only if it admits a nonzero constant spinor.

3.6.3 *Harmonic spinors and the $\hat{A}$-genus*

Let (M, g) be a compact Riemannian spin manifold, with spin bundle S and Dirac operator D. The Weitzenbock formula of Lichnerowicz [153, p. 8, eqn (7)] states that if $\sigma \in C^\infty(S)$ then

$$D^2 \sigma = (\nabla^S)^* \nabla^S \sigma + \tfrac{1}{4} s \, \sigma, \tag{3.25}$$

where s is the *scalar curvature* of g. We call σ a *harmonic spinor* if $D\sigma = 0$, that is, if $\sigma \in \text{Ker } D$. Using the 'Bochner argument' used to prove Theorem 3.5.4, one can easily show:

Proposition 3.6.4 *Let (M, g) be a compact Riemannian spin manifold. If the scalar curvature s of g is zero, then every harmonic spinor on M is parallel. If the scalar curvature s of g is positive, then there are no nonzero harmonic spinors.*

Suppose (M, g) is a compact Riemannian spin manifold of dimension $4m$, for $m \geqslant 1$. Then D_+ is a linear elliptic operator, with adjoint D_-, which has *index* $\text{ind } D_+ = \dim \text{Ker } D_+ - \dim \text{Ker } D_-$. Now the *Atiyah–Singer Index Theorem* [15] gives a topological formula for $\text{ind } D_+$, and by [15, Th. 5.3], $\text{ind } D_+$ is equal to $\hat{A}(M)$, a characteristic class of M called the $\hat{A}$-*genus*. From Proposition 3.6.4, if M is a compact Riemannian spin manifold of dimension $4m$ with positive scalar curvature, then $\text{Ker } D_\pm = \{0\}$, and so $\hat{A}(M) = 0$. Conversely, if M is a compact spin manifold of dimension $4m$ and $\hat{A}(M) \neq 0$, then there are no Riemannian metrics with positive scalar curvature on M.

If M is a compact Riemannian spin manifold with zero scalar curvature, Proposition 3.6.4 shows that $\text{Ker } D_\pm$ are the spaces of parallel positive and negative spinors, which are determined by $\mathrm{Hol}(\nabla^s)$. Thus, ind D_+ is determined by $\mathrm{Hol}(\nabla^s)$, and in fact by $\mathrm{Hol}(g)$. When $\mathrm{Hol}(g)$ is $\mathrm{SU}(2m)$, $\mathrm{Sp}(m)$ or $\mathrm{Spin}(7)$, by Corollary 3.6.3 the dimensions of $\mathrm{Ker}\, D_\pm$ are those given in Theorem 3.6.1, and this gives $\hat{A}(M)$ explicitly. Thus we have proved the following result.

Theorem 3.6.5 *Let (M, g) be a compact Riemannian spin manifold of dimension $4m$, for $m \geqslant 1$. If the scalar curvature of g is positive, then $\hat{A}(M) = 0$. If the scalar curvature of g is zero, then $\hat{A}(M)$ is an integer determined by the holonomy group $\mathrm{Hol}(g)$ of g. In particular, if $\mathrm{Hol}(g) = \mathrm{SU}(2m)$ then $\hat{A}(M) = 2$. If $\mathrm{Hol}(g) = \mathrm{SU}(2k) \times \mathrm{SU}(2m-2k)$ for $0 < k < m$ then $\hat{A}(M) = 4$. If $\mathrm{Hol}(g) = \mathrm{Sp}(m)$ then $\hat{A}(M) = m+1$, and if $m = 2$ and $\mathrm{Hol}(g) = \mathrm{Spin}(7)$ then $\hat{A}(M) = 1$.*

Here is how we will apply this theorem later in the book. Let G be one of $\mathrm{SU}(2m)$, $\mathrm{Sp}(m)$ or $\mathrm{Spin}(7)$. Then one can use analytic methods to construct a torsion-free G-structure Q on a compact manifold M. The Riemannian metric g associated to Q must then have $\mathrm{Hol}(g) \subseteq G$, but it does not immediately follow that $\mathrm{Hol}(g) = G$. However, by studying the topology of M we may work out $\hat{A}(M)$, and use this to distinguish between the different possibilities for $\mathrm{Hol}(g)$.

3.7 Calibrated geometry and holonomy groups

The theory of *calibrated geometry* was invented by Harvey and Lawson [99]. It concerns calibrated submanifolds, a special kind of minimal submanifold of a Riemannian manifold M, which are defined using a closed form on M called a calibration. It is closely connected with the theory of Riemannian holonomy groups because Riemannian manifolds with reduced holonomy usually come equipped with one or more natural calibrations. Two good references on calibrated geometry are Harvey and Lawson [99] and Harvey [97].

Definition 3.7.1 Let (M, g) be a Riemannian manifold. An *oriented tangent k-plane* V on M is a vector subspace V of some tangent space $T_x M$ to M with $\dim V = k$, equipped with an orientation. If V is an oriented tangent k-plane on M then $g|_V$ is a Euclidean metric on V, so combining $g|_V$ with the orientation on V gives a natural *volume form* vol_V on V, which is a k-form on V.

Now let φ be a closed k-form on M. We say that φ is a *calibration* on M if for every oriented k-plane V on M we have $\varphi|_V \leqslant \mathrm{vol}_V$. Here $\varphi|_V = \alpha \cdot \mathrm{vol}_V$ for some $\alpha \in \mathbb{R}$, and $\varphi|_V \leqslant \mathrm{vol}_V$ if $\alpha \leqslant 1$. Let N be an oriented submanifold of M with dimension k. Then each tangent space $T_x N$ for $x \in N$ is an oriented tangent k-plane. We say that N is a *calibrated submanifold* or *φ-submanifold* if $\varphi|_{T_x N} = \mathrm{vol}_{T_x N}$ for all $x \in N$.

All calibrated submanifolds are *minimal submanifolds*. We prove this for compact calibrated submanifolds, but noncompact calibrated submanifolds are also locally volume-minimizing.

Proposition 3.7.2 *Let (M, g) be a Riemannian manifold, φ a calibration on M, and N a compact φ-submanifold in M. Then N is volume-minimizing in its homology class.*

Proof Let dim $N = k$, and let $[N] \in H_k(M, \mathbb{R})$ and $[\varphi] \in H^k(M, \mathbb{R})$ be the homology and cohomology classes of N and φ. Then

$$[\varphi] \cdot [N] = \int_{x \in N} \varphi|_{T_x N} = \int_{x \in N} \text{vol}_{T_x N} = \text{vol}(N),$$

since $\varphi|_{T_x N} = \text{vol}_{T_x N}$ for each $x \in N$, as N is a calibrated submanifold. If N' is any other compact k-submanifold of M with $[N'] = [N]$ in $H_k(M, \mathbb{R})$, then

$$[\varphi] \cdot [N] = [\varphi] \cdot [N'] = \int_{x \in N'} \varphi|_{T_x N'} \leqslant \int_{x \in N'} \text{vol}_{T_x N'} = \text{vol}(N'),$$

since $\varphi|_{T_x N'} \leqslant \text{vol}_{T_x N'}$ because φ is a calibration. The last two equations give $\text{vol}(N) \leqslant \text{vol}(N')$. Thus N is volume-minimizing in its homology class. □

Note that a calibration φ on (M, g) can only have nontrivial calibrated submanifolds if there exist oriented tangent k-planes V on M with $\varphi|_V = \text{vol}_V$. For instance, $\varphi = 0$ is a calibration on M, but has no calibrated submanifolds. This means that a calibration φ is only interesting if the set of oriented tangent k-planes V on M with $\varphi|_V = \text{vol}_V$ has reasonably large dimension.

Next we explain the connection with Riemannian holonomy. Let $G \subset O(n)$ be a possible holonomy group of a Riemannian metric. In particular, we can take G to be one of the holonomy groups $U(m)$, $SU(m)$, $Sp(m)$, G_2 or $Spin(7)$ from Berger's list. Then G acts on the k-forms $\Lambda^k (\mathbb{R}^n)^*$ on $\mathbb{R}^n$, so we can look for G-invariant k-forms on $\mathbb{R}^n$.

Suppose φ_0 is a nonzero, G-invariant k-form on $\mathbb{R}^n$. By rescaling φ_0 we can arrange that for each oriented k-plane $U \subset \mathbb{R}^n$ we have $\varphi_0|_U \leqslant \text{vol}_U$, and that $\varphi_0|_U = \text{vol}_U$ for at least one such U. Let $\mathcal{F}$ be the family of oriented k-planes U in $\mathbb{R}^n$ with $\varphi_0|_U = \text{vol}_U$, and let $l = \dim \mathcal{F}$. Then $\mathcal{F}$ is nonempty. Since φ_0 is G-invariant, if $U \in \mathcal{F}$ then $\gamma \cdot U \in \mathcal{F}$ for all $\gamma \in G$. This usually means that $l > 0$.

Let M be a manifold of dimension n, and g a metric on M with Levi-Civita connection ∇ and holonomy group G. Then by Proposition 2.5.2 there is a k-form φ on M with $\nabla \varphi = 0$, corresponding to φ_0. Hence $d\varphi = 0$, and φ is closed. Also, the condition $\varphi_0|_U \leqslant \text{vol}_U$ for all oriented k-planes U in $\mathbb{R}^n$ implies that $\varphi|_V \leqslant \text{vol}_V$ for all oriented tangent k-planes in M. Thus φ is a *calibration* on M.

At each point $x \in M$ there is an l-dimensional family $\mathcal{F}_x$ of oriented tangent k-planes V with $\varphi|_V = \text{vol}_V$, isomorphic to $\mathcal{F}$. Hence, the set of oriented tangent k-planes V in M with $\varphi|_V = \text{vol}_V$ has dimension $l + n$, which is reasonably large. This suggests that locally there should exist many φ-submanifolds N in M, so the calibrated geometry of φ on (M, g) is nontrivial.

This gives us a general method for finding interesting calibrations on manifolds with reduced holonomy. Here are some examples of this, taken from [99].

- Let $G = U(m) \subset O(2m)$. Then G preserves a 2-form ω_0 on $\mathbb{R}^{2m}$. If g is a metric on M with holonomy $U(m)$ then g is *Kähler* with complex structure J, and the 2-form ω on M associated to ω_0 is the *Kähler form* of g. These terms are explained in Chapter 4.

One can show that ω is a calibration on (M, g), and the calibrated submanifolds are exactly the *holomorphic curves* in (M, J). More generally $\omega^k/k!$ is a calibration on M for $1 \leqslant k \leqslant m$, and the corresponding calibrated submanifolds are the complex k-dimensional submanifolds of (M, J).

- Let $G = \mathrm{SU}(m) \subset \mathrm{O}(2m)$. Riemannian manifolds (M, g) with holonomy $\mathrm{SU}(m)$ are called *Calabi–Yau manifolds*, and are the subject of Chapter 6. A Calabi–Yau manifold comes equipped with a complex m-form θ called a *holomorphic volume form*. The real part $\mathrm{Re}\,\theta$ is a calibration on M, and the corresponding calibrated submanifolds are called *special Lagrangian submanifolds*.

- The group $G_2 \subset \mathrm{O}(7)$ preserves a 3-form φ_0 and a 4-form $*\varphi_0$ on $\mathbb{R}^7$, which will be given explicitly in §10.1. Thus a Riemannian 7-manifold (M, g) with holonomy G_2 comes with a 3-form φ and 4-form $*\varphi$, which are both calibrations. We call φ-submanifolds *associative 3-folds*, and $*\varphi$-submanifolds *coassociative 4-folds*.

- The group $\mathrm{Spin}(7) \subset \mathrm{O}(8)$ preserves a 4-form Ω_0 on $\mathbb{R}^8$, which will be given explicitly in §10.5. Thus a Riemannian 8-manifold (M, g) with holonomy $\mathrm{Spin}(7)$ has a 4-form Ω, which is a calibration. We call Ω-submanifolds *Cayley 4-folds*.

4

KÄHLER MANIFOLDS

In this chapter we shall introduce the very rich geometry of *complex* and *Kähler manifolds*. Complex manifolds are manifolds with a geometric structure called a *complex structure*, which gives every tangent space the structure of a complex vector space. They are defined in §4.1, together with complex submanifolds and holomorphic maps. Section 4.2 discusses tensors on complex manifolds, and their decomposition into components using the complex structure, and §4.3 defines holomorphic vector bundles over a complex manifold.

Sections 4.4–4.7 deal with *Kähler metrics* on complex manifolds. A Kähler metric is a Riemannian metric on a complex manifold, that is compatible with the complex structure in a natural way. Also, Kähler metrics have special holonomy groups: if g is a Kähler metric on a complex manifold of dimension m, then the holonomy group $\text{Hol}(g)$ is a subgroup of $\text{U}(m)$. Quite a lot of the geometry of Kähler metrics that we will describe has parallels in the geometry of other holonomy groups. We discuss Kähler potentials, the curvature of Kähler metrics, and exterior forms on Kähler manifolds.

In §4.1–§4.7 we treat complex and Kähler manifolds using differential and Riemannian geometry. But one can also study complex and Kähler manifolds using *complex algebraic geometry*, and we give an introduction to this in §4.8–§4.10. Algebraic geometry is a very large subject and we cannot do it justice in a few pages, so we aim only to provide enough background for the reader to understand the algebraic parts of the rest of the book, which occur mostly in Chapters 6 and 7.

Section 4.8 introduces *complex algebraic varieties*, the objects studied in complex algebraic geometry, and briefly describes some of the fundamental ideas—morphisms, rational maps, sheaves and so on. We then cover two areas in more detail: singularities, resolutions and deformations in §4.9, and holomorphic line bundles and divisors in §4.10.

And now a word about notation. Unfortunately, the literature on Kähler geometry is rather inconsistent about the notation it uses. For example, while writing this chapter I found four different definitions of the Kähler form ω of §4.4 in various books, that differ from the definition we shall give by constant factors. In the same way, the operator d^c of §4.2, the Ricci form ρ of §4.6 and the Laplacian Δ on a Kähler manifold all have several definitions, differing by constant factors. I have done my best to make the formulae in this book consistent with each other, but readers are warned that other books and papers have other conventions.

4.1 Introduction to complex manifolds

A complex manifold is a real, even-dimensional manifold equipped with a geometric structure called a *complex structure*. Here are three ways to define the idea of a complex structure.

First Definition Let M be a real manifold of dimension $2m$. A *complex chart* on M is a pair (U, ψ), where U is open in M and $\psi : U \to \mathbb{C}^m$ is a diffeomorphism between U and some open set in $\mathbb{C}^m$. Equivalently, ψ gives a set of complex coordinates $z_1, \ldots, z_m$ on U. If (U_1, ψ_1) and (U_2, ψ_2) are two complex charts, then the *transition function* is $\psi_{12} : \psi_1(U_1 \cap U_2) \to \psi_2(U_1 \cap U_2)$, given by $\psi_{12} = \psi_2 \circ \psi_1^{-1}$. We say M is a *complex manifold* if it has an atlas of complex charts (U, ψ), such that all the transition functions are *holomorphic*, as maps from $\mathbb{C}^m$ to itself.

This is the traditional definition of complex manifold, using holomorphic coordinates. However, in this book we prefer to take a more differential geometric point of view, and to define geometric structures using tensors. So, here are the preliminaries to our second definition of complex manifold.

Let M be a real manifold of dimension $2m$. We define an *almost complex structure* J on M to be a smooth tensor J_a^b on M satisfying $J_a^b J_b^c = -\delta_a^c$. Let v be a smooth vector field on M, written v^a in index notation, and define a new vector field Jv by $(Jv)^b = J_a^b v^a$. Thus J acts linearly on vector fields. The equation $J_a^b J_b^c = -\delta_a^c$ implies that $J(Jv) = -v$, so that $J^2 = -1$. Observe that J gives each tangent space T_pM the structure of a *complex vector space*.

For all smooth vector fields v, w on M, define a vector field $N_J(v, w)$ by

$$N_J(v, w) = [v, w] + J\bigl([Jv, w] + [v, Jw]\bigr) - [Jv, Jw],$$

where $[,]$ is the Lie bracket of vector fields. It turns out that N_J is a *tensor*, meaning that $N_J(v, w)$ is pointwise bilinear in v and w. We call N_J the *Nijenhuis tensor* of J.

Second Definition Let M be a real manifold of dimension $2m$, and J an almost complex structure on M. We call J a *complex structure* if $N_J \equiv 0$ on M. A *complex manifold* is a manifold M equipped with a complex structure J. We shall often use the notation (M, J) to refer to a manifold and its complex structure.

Here is why the first two definitions are equivalent. Let $f : M \to \mathbb{C}$ be a smooth complex function on M. We say that f is *holomorphic* if $J_a^b(\mathrm{d}f)_b \equiv i(\mathrm{d}f)_a$ on M. These are called the Cauchy–Riemann equations. It turns out that if $m > 1$, the equations are overdetermined, and the Nijenhuis tensor N_J is an obstruction to the existence of holomorphic functions. Simply put, if $N_J \equiv 0$ there are many holomorphic functions locally, but if $N_J \not\equiv 0$ there are few holomorphic functions.

Let (U, ψ) be a complex chart on M. Then ψ is a set of complex coordinates $(z_1, \ldots, z_m)$ on U, where $z_j : U \to \mathbb{C}$ is a smooth complex function. We call (U, ψ) a *holomorphic chart* if each of the functions $z_1, \ldots, z_m$ is holomorphic in the above sense.

The *Newlander–Nirenberg Theorem* shows that a necessary and sufficient condition for there to exist a holomorphic chart around each point of M, is the vanishing of the

Nijenhuis tensor N_J of J. Therefore, if (M, J) is a complex manifold in the sense of the second definition, then M has an atlas of holomorphic charts. This atlas makes M into a complex manifold in the sense of the first definition.

Our third way to define the idea of complex structure uses the language of G-structures and intrinsic torsion, that was defined in §2.6. Let M be a real $2m$-manifold with frame bundle F, and let J be an almost complex structure on M. Define P to be the subbundle of frames in F in which the components of J assume the standard form

$$J_b^a = \begin{cases} 1 & \text{if } a = b + m, \\ -1 & \text{if } a = b - m, \\ 0 & \text{otherwise,} \end{cases}$$

for $a, b = 1, \ldots, 2m$. Then P is a principal subbundle of F with fibre $\mathrm{GL}(m, \mathbb{C})$, which is a Lie subgroup of $\mathrm{GL}(2m, \mathbb{R})$ in the obvious way, and so P is a $\mathrm{GL}(m, \mathbb{C})$-structure on M. Clearly, this defines a 1-1 correspondence between almost complex structures J and $\mathrm{GL}(m, \mathbb{C})$-structures P on M.

It turns out that the Nijenhuis tensor N_J of J is equivalent to the *intrinsic torsion* $T^i(P)$ of P, as defined in Definition 2.6.4. Thus, $N_J \equiv 0$ if and only if P is *torsion-free*. So, the second definition of complex structure is equivalent to the following:

Third Definition Let M be a real manifold of dimension $2m$. Then a *complex structure* on M is a torsion-free $\mathrm{GL}(m, \mathbb{C})$-structure on M.

All three definitions are useful for different purposes. The second definition is convenient for differential geometric calculations, and we will use it most often. But the first definition is best for defining complex manifolds explicitly. Here is a simple example.

Example 4.1.1 A very important family of complex manifolds are the *complex projective spaces* $\mathbb{CP}^m$. Define $\mathbb{CP}^m$ to be the set of one-dimensional vector subspaces of $\mathbb{C}^{m+1}$. Let $(z_0, \ldots, z_m)$ be a point in $\mathbb{C}^{m+1} \setminus \{0\}$. Then we write

$$[z_0, \ldots, z_m] = \{(\alpha z_0, \ldots, \alpha z_m) : \alpha \in \mathbb{C}\} \in \mathbb{CP}^m.$$

Every point in $\mathbb{CP}^m$ is of the form $[z_0, \ldots, z_m]$ for some $(z_0, \ldots, z_m) \in \mathbb{C}^{m+1} \setminus \{0\}$. This notation is called *homogeneous coordinates* for $\mathbb{CP}^m$. The homogeneous coordinates of a point are not unique, since if $\lambda \in \mathbb{C}$ is nonzero then $[z_0, \ldots, z_m]$ and $[\lambda z_0, \ldots, \lambda z_m]$ represent the same point in $\mathbb{CP}^m$.

We will define a set of complex charts on $\mathbb{CP}^m$, that make it into a compact complex manifold of dimension m, in the sense of the first definition above. For each $j = 0, 1, \ldots, m$, define $U_j = \{[z_0, \ldots, z_m] \in \mathbb{CP}^m : z_j \neq 0\}$. Then U_j is an open set in $\mathbb{CP}^m$, and every point in U_j can be written uniquely as $[z_0, \ldots, z_{j-1}, 1, z_{j+1}, \ldots, z_m]$, that is, with $z_j = 1$. Define a map $\psi_j : U_j \to \mathbb{C}^m$ by

$$\psi_j([z_0, \ldots, z_{j-1}, 1, z_{j+1}, \ldots, z_m]) = (z_0, \ldots, z_{j-1}, z_{j+1}, \ldots, z_m).$$

Then ψ_j is a diffeomorphism, and (U_j, ψ_j) is a *complex chart* on $\mathbb{CP}^m$. It is easy to show that the charts (U_j, ψ_j) for $j = 0, 1, \ldots, m$ form a *holomorphic atlas* for $\mathbb{CP}^m$, and thus $\mathbb{CP}^m$ is a *complex manifold*.

4.1.1 Holomorphic maps and complex submanifolds

Many concepts in real differential geometry have natural analogues in the world of complex manifolds. We will now examine the ideas of holomorphic maps and complex submanifolds, which are the complex analogues of smooth maps and submanifolds respectively.

Let M, N be complex manifolds with complex structures J_M, J_N, and let $f : M \to N$ be a smooth map. At each point $p \in M$, the derivative of f is a linear map $d_p f : T_p M \to T_{f(p)} N$. We say that f is a *holomorphic map* if, for each point $p \in M$ and each $v \in T_p M$, the equation $J_N(d_p f(v)) = d_p f(J_M(v))$ holds. In other words, we must have $J_N \circ df = df \circ J_M$ as maps between the vector bundles TM and $f^*(TN)$ over M.

A map $f : M \to N$ between complex manifolds is called *biholomorphic*, or a *biholomorphism*, if an inverse map $f^{-1} : N \to M$ exists, and both f and f^{-1} are holomorphic maps. Biholomorphic maps are the natural notion of isomorphism of complex manifolds, just as diffeomorphisms are the natural notion of isomorphism of smooth manifolds.

Now let M be a complex manifold with complex structure J, and let N be a submanifold of M. Then for each $p \in N$, the tangent space $T_p N$ is a vector subspace of the tangent space $T_p M$. We say that N is a *complex submanifold* of M if $J(T_p N) = T_p N$ for each $p \in N$, that is, if the tangent spaces of N are closed under J.

If N is a complex submanifold, then the restriction of J to TN is a complex structure on N, so that N is a complex manifold, and the inclusion map $i : N \to M$ is holomorphic. It can be shown that a submanifold $N \subset M$ is a complex submanifold if and only if it can locally be written as the zeros of a finite number of holomorphic functions.

The complex projective spaces $\mathbb{CP}^m$ have many complex submanifolds, which are defined as the set of zeros of a collection of polynomials. Such submanifolds are called *complex algebraic varieties*, and will be the subject of §4.8. Here are two examples, to illustrate the ideas of holomorphic map and complex submanifold.

Example 4.1.2 Define a map $f : \mathbb{CP}^1 \to \mathbb{CP}^2$ by

$$f([x, y]) = [x^2, xy, y^2].$$

As above, if $\lambda \in \mathbb{C}$ is nonzero then the points $[\lambda x, \lambda y]$ and $[x, y]$ are the same. To see that f is well-defined, we must show that the definition is independent of the choice of homogeneous coordinates. But this is obvious because

$$[(\lambda x)^2, (\lambda x)(\lambda y), (\lambda y)^2] = [\lambda^2 x^2, \lambda^2 xy, \lambda^2 y^2] = [x^2, xy, y^2],$$

and so f is well-defined. The reason that this works is that the polynomials x^2, xy and y^2 are all *homogeneous of the same degree*.

This map f is a *holomorphic map* between the complex manifolds $\mathbb{CP}^1$ and $\mathbb{CP}^2$. Its image $N = \text{Im } f$ is a *complex submanifold* of $\mathbb{CP}^2$, which is isomorphic as a complex manifold to $\mathbb{CP}^1$. There is another way to define N as a subset of $\mathbb{CP}^2$. It can be shown that

$$N = \{[z_0, z_1, z_2] \in \mathbb{CP}^2 : z_0 z_2 - z_1^2 = 0\}.$$

Here we define N as the zeros of the homogeneous quadratic polynomial $z_0 z_2 - z_1^2$. It is an example of a *conic* in $\mathbb{CP}^2$.

Example 4.1.3 Let C be the *cubic* in $\mathbb{CP}^2$ given by

$$C = \{[z_0, z_1, z_2] \in \mathbb{CP}^2 : z_0^3 + z_1^3 + z_2^3 = 0\}.$$

This is a compact, complex submanifold of $\mathbb{CP}^2$. As a real manifold, C is diffeomorphic to the torus T^2. More generally, let $P(z_0, z_1, z_2)$ be a homogeneous complex polynomial of degree $d \geqslant 1$, and define a set $C_P \subset \mathbb{CP}^2$ by

$$C_P = \{[z_0, z_1, z_2] \in \mathbb{CP}^2 : P(z_0, z_1, z_2) = 0\}.$$

For *generic* polynomials P the curve C_P is a compact complex submanifold of $\mathbb{CP}^2$, which is diffeomorphic as a real manifold to a surface of genus $g = \frac{1}{2}(d-1)(d-2)$. For more details, see [93, §2.1].

4.2 Tensors on complex manifolds

We showed in §3.5.1 that if G is a Lie subgroup of $\mathrm{GL}(n, \mathbb{R})$ and Q is a G-structure on an n-manifold M, then the bundles of tensors and exterior forms on M decompose into a direct sum of subbundles corresponding to irreducible representations of G. Now §4.1 defined a complex structure to be a torsion-free $\mathrm{GL}(m, \mathbb{C})$-structure on a $2m$-manifold. Thus, the bundles of tensors and exterior forms on a complex manifold split into subbundles corresponding to irreducible representations of $\mathrm{GL}(m, \mathbb{C})$. In this section we will explain these splittings, and some of their consequences.

Let M be a manifold of dimension $2m$, and J a complex structure on M. Then J acts linearly on vector fields v by $v \mapsto Jv$, such that $J(Jv) = -v$. As vector fields are sections of the tangent bundle TM of M, we may regard J as a bundle-linear map $J : TM \to TM$. At a point p in M, this gives a linear map $J_p : T_p M \to T_p M$. Now $T_p M$ is a real vector space isomorphic to $\mathbb{R}^{2m}$. It is convenient to *complexify* $T_p M$ to get $T_p M \otimes_\mathbb{R} \mathbb{C}$, which is a complex vector space isomorphic to $\mathbb{C}^{2m}$. (Note that this operation of complexification is independent of J.) The map J_p extends naturally to a map $J_p : T_p M \otimes_\mathbb{R} \mathbb{C} \to T_p M \otimes_\mathbb{R} \mathbb{C}$, which is linear over $\mathbb{C}$.

Consider the eigenvalues and eigenvectors of J_p in $T_p M \otimes_\mathbb{R} \mathbb{C}$. Since $J_p^2 = -\mathrm{id}$, where id is the identity, any eigenvalue λ of J_p must satisfy $\lambda^2 = -1$. Hence $\lambda = \pm i$. Define $T_p^{(1,0)} M$ to be the eigenspace of J_p in $T_p M \otimes_\mathbb{R} \mathbb{C}$ with eigenvalue i, and $T_p^{(0,1)} M$ to be the eigenspace with eigenvalue $-i$. It is easy to show that $T_p^{(1,0)} M \cong \mathbb{C}^m \cong T_p^{(0,1)} M$, that $T_p M \otimes_\mathbb{R} \mathbb{C} = T_p^{(1,0)} M \oplus T_p^{(0,1)} M$, and that $T_p^{(1,0)} M$ and $T_p^{(0,1)} M$ are complex conjugate subspaces under the natural complex conjugation on $T_p M \otimes_\mathbb{R} \mathbb{C}$.

As this works at every point p in M, we have defined two subbundles $T^{(1,0)}M$ and $T^{(0,1)}M$ of $TM \otimes_{\mathbb{R}} \mathbb{C}$, with $TM \otimes_{\mathbb{R}} \mathbb{C} = T^{(1,0)}M \oplus T^{(0,1)}M$. What we have shown is that a complex structure on M splits the complexified tangent bundle into two subbundles. In a similar way, the complexified cotangent bundle, and in fact complexified tensors of all kinds on M, are split into subbundles by the complex structure.

This is an important idea in complex geometry, and to make use of it we will usually work with *complex-valued tensors* on complex manifolds, that is, all vector bundles will be complexified, as above with the tangent bundle. We will now develop the idea in two different ways. Firstly, a notation for tensors on complex manifolds will be defined, and secondly, the decomposition of exterior forms by the complex structure is explained, and the ∂ and $\bar{\partial}$ operators are defined.

4.2.1 *The decomposition of complex tensors*

Let M be a complex manifold with complex structure J, which will be written with indices as J^k_j. Let $S = S^{a\cdots}_{\cdots}$ be a tensor on M, taking values in $\mathbb{C}$. Here a is a contravariant index of S, and any other indices of S are represented by dots. The Greek characters $\alpha, \beta, \gamma, \delta, \epsilon$ and their conjugates $\bar{\alpha}, \bar{\beta}, \bar{\gamma}, \bar{\delta}, \bar{\epsilon}$, will be used in place of the Roman indices a, b, c, d, e respectively. They are tensor indices in the normal sense, and their use is actually a shorthand indicating a modification to the tensor itself.

Define $S^{\alpha\cdots}_{\cdots} = \frac{1}{2}(S^{a\cdots}_{\cdots} - iJ^a_j S^{j\cdots}_{\cdots})$ and $S^{\bar{\alpha}\cdots}_{\cdots} = \frac{1}{2}(S^{a\cdots}_{\cdots} + iJ^a_j S^{j\cdots}_{\cdots})$. In the same way, if b is a covariant index on a complex-valued tensor $T^{\cdots}_{b\cdots}$, define $T^{\cdots}_{\beta\cdots} = \frac{1}{2}(T^{\cdots}_{b\cdots} - iJ^j_b T^{\cdots}_{j\cdots})$ and $T^{\cdots}_{\bar{\beta}\cdots} = \frac{1}{2}(T^{\cdots}_{b\cdots} + iJ^j_b T^{\cdots}_{j\cdots})$. These operations on tensors are projections, and satisfy $S^{a\cdots}_{\cdots} = S^{\alpha\cdots}_{\cdots} + S^{\bar{\alpha}\cdots}_{\cdots}$ and $T^{\cdots}_{b\cdots} = T^{\cdots}_{\beta\cdots} + T^{\cdots}_{\bar{\beta}\cdots}$.

Let δ^b_a be the Kronecker delta, regarded as a tensor on M. Then $\delta^b_a = \delta^\beta_\alpha + \delta^{\bar{\beta}}_{\bar{\alpha}}$ in this notation. It is also easy to show that $J^b_a = i\delta^\beta_\alpha - i\delta^{\bar{\beta}}_{\bar{\alpha}}$. Thus J acts on tensor indices of the form $\alpha, \beta, \ldots$ by multiplication by i, so we may think of these as *complex linear* components with respect to J. Similarly, J acts on indices of the form $\bar{\alpha}, \bar{\beta}, \ldots$ by multiplication by $-i$, and we may think of these as *complex antilinear* components.

4.2.2 *Exterior forms on complex manifolds*

By the argument used above, the complexified cotangent bundle $T^*M \otimes_{\mathbb{R}} \mathbb{C}$ splits into pieces: $T^*M \otimes_{\mathbb{R}} \mathbb{C} = T^{*(1,0)}M \oplus T^{*(0,1)}M$. Now if U, V, W are vector spaces with $U = V \oplus W$, then the exterior powers of U, V and W are related by $\Lambda^k U = \bigoplus_{j=0}^k \Lambda^j V \otimes \Lambda^{k-j} W$. Using the splitting of $T^*M \otimes_{\mathbb{R}} C$, it follows that

$$\Lambda^k T^*M \otimes_{\mathbb{R}} \mathbb{C} = \bigoplus_{j=0}^k \Lambda^j T^{*(1,0)}M \otimes \Lambda^{k-j}T^{*(0,1)}M. \tag{4.1}$$

Define $\Lambda^{p,q}M$ to be the bundle $\Lambda^p T^{*(1,0)}M \otimes \Lambda^q T^{*(0,1)}M$. Then (4.1) gives

$$\Lambda^k T^*M \otimes_{\mathbb{R}} \mathbb{C} = \bigoplus_{j=0}^k \Lambda^{j,k-j}M. \tag{4.2}$$

This is the decomposition of the exterior k-forms on M induced by the complex structure J. A section of $\Lambda^{p,q} M$ is called a *(p, q)-form*.

We may use the splittings of $\Lambda^k T^* M \otimes_{\mathbb{R}} \mathbb{C}$ and $\Lambda^{k+1} T^* M \otimes_{\mathbb{R}} \mathbb{C}$ to divide the exterior derivative d on complex k-forms into components, each component mapping sections of $\Lambda^{p,q} M$ to sections of $\Lambda^{r,s} M$, where $p + q = k$ and $r + s = k + 1$. Provided J is a complex structure (not just an almost complex structure), the only nonzero components are those that map $\Lambda^{p,q} M$ to $\Lambda^{p+1,q} M$ and to $\Lambda^{p,q+1} M$.

Define ∂ to be the component of d mapping $C^\infty(\Lambda^{p,q} M)$ to $C^\infty(\Lambda^{p+1,q} M)$, and $\bar\partial$ to be the component of d mapping $C^\infty(\Lambda^{p,q} M)$ to $C^\infty(\Lambda^{p,q+1} M)$. Then $\partial, \bar\partial$ are first-order partial differential operators on complex k-forms which satisfy $d = \partial + \bar\partial$. The identity $d^2 = 0$ implies that $\partial^2 = \bar\partial^2 = 0$ and $\partial\bar\partial + \bar\partial\partial = 0$. As $\bar\partial^2 = 0$, we may define the *Dolbeault cohomology groups* $H^{p,q}_{\bar\partial}(M)$ of a complex manifold, by

$$H^{p,q}_{\bar\partial}(M) = \frac{\mathrm{Ker}\left(\bar\partial : C^\infty(\Lambda^{p,q} M) \to C^\infty(\Lambda^{p,q+1} M)\right)}{\mathrm{Im}\left(\bar\partial : C^\infty(\Lambda^{p-1,q} M) \to C^\infty(\Lambda^{p,q} M)\right)}. \tag{4.3}$$

The Dolbeault cohomology groups depend on the complex structure of M.

Now define an operator $d^c : C^\infty(\Lambda^k T^* M \otimes_{\mathbb{R}} \mathbb{C}) \to C^\infty(\Lambda^{k+1} T^* M \otimes_{\mathbb{R}} \mathbb{C})$ by $d^c = i(\bar\partial - \partial)$. It is easy to show that

$$dd^c + d^c d = 0, \quad (d^c)^2 = 0, \quad \partial = \tfrac{1}{2}(d + id^c), \quad \bar\partial = \tfrac{1}{2}(d - id^c) \text{ and } dd^c = 2i\partial\bar\partial.$$

Also d^c is a *real* operator, that is, if α is a real k-form then $d^c \alpha$ is a real $(k+1)$-form.

4.3 Holomorphic vector bundles

Next we define holomorphic vector bundles over a complex manifold, which are the analogues in complex geometry of smooth vector bundles over real manifolds. A good reference for the material in this section is [93, §0.5 & §1.1].

Definition 4.3.1 Let M be a complex manifold. Let $\{E_p : p \in M\}$ be a family of complex vector spaces of dimension k, parametrized by M. Let E be the total space of this family, and $\pi : E \to M$ be the natural projection. Suppose also that E has the structure of a complex manifold. This collection of data (the family of complex vector spaces, with a complex structure on its total space) is called a *holomorphic vector bundle* with fibre $\mathbb{C}^k$, if the following conditions hold.

  (i) The map $\pi : E \to M$ is a holomorphic map of complex manifolds.
  (ii) For each $p \in M$ there exists an open neighbourhood $U \subset M$, and a biholomorphic map $\varphi_U : \pi^{-1}(U) \to U \times \mathbb{C}^k$.
  (iii) In part (ii), for each $u \in U$ the map φ_U takes E_u to $\{u\} \times \mathbb{C}^k$, and this is an isomorphism between E_u and $\mathbb{C}^k$ as complex vector spaces.

The vector space E_p is called the *fibre* of E over p. Usually we will refer to E as the holomorphic vector bundle, implicitly assuming that the rest of the structure is given.

Let E and F be holomorphic vector bundles over M. Then E^* and $E \otimes F$ are also holomorphic bundles in a natural way, where E^* is the *dual vector bundle* to E with fibre E_p^* at $p \in M$, and $E \otimes F$ is the *tensor product bundle*, with fibre $E_p \otimes F_p$ at $p \in M$.

Suppose that E is a holomorphic vector bundle over M, with projection $\pi : E \to M$. A *holomorphic section s* of E is a holomorphic map $s : M \to E$, such that $\pi \circ s$ is the identity map on M. Because the fibres of E are complex vector spaces, holomorphic sections of E can be added together and multiplied by complex constants. Thus the holomorphic sections of E form a complex vector space, which is finite-dimensional if M is compact.

Now, every complex manifold M comes equipped with a number of natural holomorphic vector bundles. For example, the product $M \times \mathbb{C}^k$ is a holomorphic vector bundle over M, called the *trivial* vector bundle with fibre $\mathbb{C}^k$. Also, the tangent bundle TM and the cotangent bundle T^*M are both real vector bundles over M, but we may make them into complex vector bundles by identifying J with multiplication by $i \in \mathbb{C}$. The total spaces of TM and T^*M both have natural complex structures, which make them into *holomorphic vector bundles*.

From TM and T^*M we can make other holomorphic vector bundles, using the tensor product. We will consider the bundles of exterior forms. The vector bundles $\Lambda^{p,q}M$ defined in §4.2.2 are *complex vector bundles*, smooth vector bundles with complex vector spaces as fibres. But which of these are holomorphic vector bundles? It turns out that $\Lambda^{p,q}M$ is a holomorphic vector bundle in a natural way if and only if $q = 0$, so that $\Lambda^{p,0}M$ is a holomorphic vector bundle for $p = 0, 1, \ldots, m$. There are natural isomorphisms

$$\Lambda^{0,0}M \cong M \times \mathbb{C}, \quad \Lambda^{1,0}M \cong T^*M, \quad \text{and} \quad \Lambda^{p,0}M \cong \Lambda_{\mathbb{C}}^p T^*M,$$

as holomorphic vector bundles.

Now let $s \in C^\infty(\Lambda^{p,0}M)$, so that s is a smooth section of $\Lambda^{p,0}M$. It can be shown that s is a *holomorphic section* of $\Lambda^{p,0}M$ if and only if $\bar{\partial}s = 0$ in $C^\infty(\Lambda^{p,1}M)$. A holomorphic section of $\Lambda^{p,0}M$ is called a *holomorphic p-form*. From eqn (4.3) we see that the Dolbeault group $H_{\bar{\partial}}^{p,0}(M)$ is actually the vector space of holomorphic p-forms on M.

4.4 Introduction to Kähler manifolds

Let (M, J) be a complex manifold, and let g be a Riemannian metric on M. We call g a *Hermitian metric* if three equivalent conditions hold:

(i) $g(v, w) = g(Jv, Jw)$ for all vector fields v, w on M,
(ii) In index notation, $g_{ab} \equiv J_a^c J_b^d g_{cd}$,
(iii) In the notation of §4.2, $g_{ab} \equiv g_{\alpha\bar{\beta}} + g_{\bar{\alpha}\beta}$. That is, $g_{\alpha\beta} \equiv g_{\bar{\alpha}\bar{\beta}} \equiv 0$.

This is a natural compatibility condition between a complex structure and a Riemannian metric. If g is a Hermitian metric, we define a 2-form ω on M called the *Hermitian form of g* in three equivalent ways:

(i) $\omega(v, w) = g(Jv, w)$ for all vector fields v, w on M,
(ii) In index notation, $\omega_{ac} = J_a^b g_{bc}$,
(iii) In the notation of §4.2, $\omega_{ab} = ig_{\alpha\bar{\beta}} - ig_{\bar{\alpha}\beta}$.

Then ω is a (1,1)-form, and we may reconstruct g from ω using the equation $g(v, w) = \omega(v, Jw)$. Let us define a (1,1)-form ω on a complex manifold to be *positive* if $\omega(v, Jv) > 0$ for all nonzero vectors v. It is easy to see that if ω is a (1,1)-form on a complex manifold, then ω is the Hermitian form of a Hermitian metric if and only if ω is positive. The idea of Hermitian metric also makes sense for J an almost complex structure.

Definition 4.4.1 Let M be a complex manifold, and g a Hermitian metric on M, with Hermitian form ω. We say g is a *Kähler metric* if $d\omega = 0$. In this case we call ω the *Kähler form*.

Here are some important facts about Kähler metrics.

Proposition 4.4.2 *Let M be a manifold of dimension $2m$, J an almost complex structure on M, and g a Hermitian metric, with Hermitian form ω. Let ∇ be the Levi-Civita connection of g. Then the following conditions are equivalent:*

(i) *J is a complex structure and g is Kähler,*
(ii) $\nabla J = 0$,
(iii) $\nabla \omega = 0$,
(iv) *The holonomy group of g is contained in $U(m)$, and J is associated to the corresponding $U(m)$-structure.*

In §4.1 we saw that one way to define a complex structure is as a torsion-free $GL(m, \mathbb{C})$-structure on a $2m$-manifold M. In the same way, a Kähler structure can be defined to be a *torsion-free $U(m)$-structure* on M, as in §2.6.

Notice that if g is a Hermitian metric on a complex manifold, then the rather weak condition $d\omega = 0$ implies the much stronger conditions that $\nabla \omega = \nabla J = 0$. One moral is that Kähler metrics are easy to construct, as closed 2-forms are easy to find, but they have many interesting properties following from $\nabla \omega = \nabla J = 0$. Also, note that a Kähler metric is essentially the same thing as a Riemannian metric with holonomy $U(m)$, or some subgroup of $U(m)$. The highest exterior power ω^m of ω is proportional to the volume form dV_g of g, and with the conventions used in this book, the relationship is

$$\omega^m = m! \cdot dV_g. \tag{4.4}$$

Let M be a Kähler manifold with Kähler metric g and Kähler form ω. If N is a *complex submanifold* of M in the sense of §4.1.1, then the restriction of g to N is also Kähler. (One way to see this is that the restriction of ω to N is clearly a closed, positive (1,1)-form.) Thus, any complex submanifold of a Kähler manifold is a Kähler manifold in its own right.

Example 4.4.3 The complex manifold $\mathbb{CP}^m$, described in Example 4.1.1, carries a natural Kähler metric. Here is one way to define it. There is a natural projection

$$\pi : \mathbb{C}^{m+1} \setminus \{0\} \to \mathbb{CP}^m, \quad \text{defined by} \quad \pi : (z_0, \ldots, z_m) \mapsto [z_0, \ldots, z_m].$$

Define a real function $u : \mathbb{C}^{m+1} \setminus \{0\} \to (0, \infty)$ by $u(z_0, \ldots, z_m) = |z_0|^2 + \cdots + |z_m|^2$. Define a closed (1,1)-form α on $\mathbb{C}^{m+1} \setminus \{0\}$ by $\alpha = \mathrm{dd}^c(\log u)$. Now α is not the Kähler form of any Kähler metric on $\mathbb{C}^{m+1} \setminus \{0\}$, because it is not positive. However, there does exist a unique positive (1,1)-form ω on $\mathbb{CP}^m$, such that $\alpha = \pi^*(\omega)$. The Kähler metric g on $\mathbb{CP}^m$ with Kähler form ω is the *Fubini–Study metric*.

The idea used here, of using dd^c to make Kähler forms, will be explored in §4.5. Since $\mathbb{CP}^m$ is Kähler, it follows that any complex submanifold of $\mathbb{CP}^m$ is also a Kähler manifold. Now there are lots of complex submanifolds in the complex projective spaces $\mathbb{CP}^m$. They are studied in the subject of *complex algebraic geometry*, which will be introduced in §4.8. This gives a great number of examples of Kähler manifolds.

4.5 Kähler potentials

Let (M, J) be a complex manifold. We have seen that to each Kähler metric g on M there is associated a closed real (1,1)-form ω, called the Kähler form. Conversely, if ω is a closed real (1,1)-form on M, then ω is the Kähler form of a Kähler metric if and only if ω is *positive* (that is, $\omega(v, Jv) > 0$ for all nonzero vectors v). Positivity is an *open condition* on closed real (1,1)-forms, meaning that it holds on an open set in the space of closed real (1,1)-forms.

Let ϕ be a smooth real function on M. Then $\mathrm{dd}^c\phi$ is clearly a closed (and in fact exact) real 2-form, as both d and d^c are real operators. But since $\mathrm{dd}^c = 2i\partial\bar{\partial}$, it follows that $\mathrm{dd}^c\phi$ is also a (1,1)-form. Thus if ϕ is a real function then $\mathrm{dd}^c\phi$ is a closed real (1,1)-form. The following results are converses to this.

The Local dd^c-Lemma *Let η be a smooth, closed, real (1, 1)-form on the unit disc in $\mathbb{C}^m$. Then there exists a smooth real function ϕ on the unit disc such that $\eta = \mathrm{dd}^c\phi$.*

The Global dd^c-Lemma *Let M be a compact Kähler manifold, and η a smooth, exact, real (1, 1)-form on M. Then there exists a smooth real function ϕ on M such that $\eta = \mathrm{dd}^c\phi$.*

Note that in the second result it is necessary that M should be Kähler, rather than just complex, for the result to hold. From the local dd^c-Lemma it follows that if g is a Kähler metric on M with Kähler form ω, then locally in M we may write $\omega = \mathrm{dd}^c\phi$ for some real function ϕ. Such a function ϕ is called a *Kähler potential* for the metric g. However, in general we cannot find a global Kähler potential for g, for the following reason.

Suppose M is a compact Kähler manifold of dimension $2m$, with Kähler form ω. As ω is closed, it defines a de Rham cohomology class $[\omega] \in H^2(M, \mathbb{R})$, called the *Kähler class*. Now ω^m, the wedge product of m copies of ω, is equal to $m!$ times the volume form of g. Hence $\int_M \omega^m = m!\,\mathrm{vol}(M)$. But $\mathrm{vol}(M) > 0$, and $\int_M \omega^m$ depends only on the cohomology class $[\omega]$. It follows that $[\omega]$ must be nonzero. However, $dd^c\phi$ is exact, so that $[dd^c\phi]$ is zero in $H^2(M,\mathbb{R})$. Therefore, on a compact Kähler manifold it is impossible to find a global Kähler potential; but we do have the following useful result.

Lemma 4.5.1 *Let M be a compact, complex manifold, and let g, g' be Kähler metrics on M with Kähler forms ω, ω'. Suppose that $[\omega] = [\omega'] \in H^2(M,\mathbb{R})$. Then there exists a smooth, real function ϕ on M such that $\omega' = \omega + dd^c\phi$. This function ϕ is unique up to the addition of a constant.*

Proof Since $[\omega] = [\omega']$, $\omega' - \omega$ is an exact, real (1,1)-form. So, by the global dd^c-Lemma, a function ϕ exists with $\omega' - \omega = dd^c\phi$, and $\omega' = \omega + dd^c\phi$ as we want. If ϕ_1 and ϕ_2 are both solutions, then by subtraction $dd^c(\phi_1 - \phi_2) = 0$ on M, which implies that $\phi_1 - \phi_2$ is constant, as M is compact. Therefore ϕ is unique up to a constant. □

The lemma gives a parametrization of the Kähler metrics with a fixed Kähler class, by smooth functions on the manifold. We may also express the metric g' in terms of g and ϕ. As $\omega' = \omega + dd^c\phi = \omega + 2i\partial\bar{\partial}\phi$, we have

$$\omega'_{\alpha\bar{\beta}} = \omega_{\alpha\bar{\beta}} + i\partial_\alpha\bar{\partial}_{\bar{\beta}}\phi \quad \text{and} \quad \omega'_{\bar{\alpha}\beta} = \omega_{\bar{\alpha}\beta} - i\bar{\partial}_{\bar{\alpha}}\partial_\beta\phi.$$

But $g_{\alpha\bar{\beta}} = -i\omega_{\alpha\bar{\beta}}$, $g_{\bar{\alpha}\beta} = i\omega_{\bar{\alpha}\beta}$, $g'_{\alpha\bar{\beta}} = -i\omega'_{\alpha\bar{\beta}}$ and $g'_{\bar{\alpha}\beta} = i\omega'_{\bar{\alpha}\beta}$, and therefore

$$g'_{\alpha\bar{\beta}} = g_{\alpha\bar{\beta}} + \partial_\alpha\bar{\partial}_{\bar{\beta}}\phi \quad \text{and} \quad g'_{\bar{\alpha}\beta} = g_{\bar{\alpha}\beta} + \bar{\partial}_{\bar{\alpha}}\partial_\beta\phi.$$

4.6 Curvature of Kähler manifolds

Let M be a $2m$-manifold, and g a Kähler metric on M. Then $\mathrm{Hol}(g) \subseteq \mathrm{U}(m)$, by Proposition 4.4.2. Applying Theorem 3.1.7, one can show that in the notation of §4.2, the Riemann curvature tensor of g satisfies

$$R^a{}_{bcd} = R^\alpha{}_{\beta\gamma\bar{\delta}} + R^\alpha{}_{\beta\bar{\gamma}\delta} + R^{\bar{\alpha}}{}_{\bar{\beta}\gamma\bar{\delta}} + R^{\bar{\alpha}}{}_{\bar{\beta}\bar{\gamma}\delta}. \tag{4.5}$$

Now a general tensor $T^a{}_{bcd}$ has 16 components in its complex decomposition. Equation (4.5) says that 12 of these components vanish for the curvature tensor of a Kähler manifold, leaving only 4 components. However, using symmetries of Riemann curvature, and complex conjugation, we may identify $R^\alpha{}_{\beta\bar{\gamma}\delta}$ with $R^\alpha{}_{\beta\gamma\bar{\delta}}$, and identify both $R^{\bar{\alpha}}{}_{\bar{\beta}\gamma\bar{\delta}}$ and $R^{\bar{\alpha}}{}_{\bar{\beta}\bar{\gamma}\delta}$ with the complex conjugate of $R^\alpha{}_{\beta\gamma\bar{\delta}}$. Thus the Kähler curvature is determined by the single component $R^\alpha{}_{\beta\gamma\bar{\delta}}$.

The *Ricci curvature* is $R_{bd} = R^a{}_{bad}$. From (4.5) we see that

$$R_{bd} = R^\alpha{}_{\beta\alpha\bar{\delta}} + R^{\bar{\alpha}}{}_{\bar{\beta}\bar{\alpha}\delta}. \tag{4.6}$$

Hence $R_{ab} = R_{\alpha\bar{\beta}} + R_{\bar{\alpha}\beta}$, and $R_{\alpha\beta} = R_{\bar{\alpha}\bar{\beta}} = 0$. Also, $R_{ab} = R_{ba}$ by symmetries of curvature. Therefore, the Ricci curvature satisfies the same conditions as a Hermitian metric. From a Hermitian metric we can make a Hermitian form, so we will try the same trick with the Ricci curvature. Define the *Ricci form* ρ by

$$\rho_{ab} = iR_{\alpha\bar{\beta}} - iR_{\bar{\alpha}\beta}, \quad \text{or equivalently} \quad \rho_{ac} = J_a^b R_{bc}.$$

Then ρ is a real (1,1)-form, and we may recover the Ricci curvature from ρ using the equation $R_{ab} = \rho_{ac} J_b^c$. It is a remarkable fact that ρ is a *closed* 2-form. The cohomology class $[\rho] \in H^2(M, \mathbb{R})$ depends only on the complex structure of M, and is equal to $2\pi\, c_1(M)$, where $c_1(M)$ is the first Chern class of M.

To see why this is so, we will give an explicit expression for the Ricci curvature in coordinates. Let $(z_1, \ldots, z_m)$ be holomorphic coordinates on an open set in M. Let $g_{ab} = g_{\alpha\bar{\beta}} + g_{\bar{\alpha}\beta}$ be the Kähler metric. We may regard α as an index for $dz_1, \ldots, dz_m$, and $\bar{\beta}$ as an index for $d\bar{z}_1, \ldots, d\bar{z}_m$. Hence, α and $\bar{\beta}$ are both indices running from 1 to m, and $g_{\alpha\bar{\beta}}$ is an $m \times m$ complex matrix.

It is easy to see that $g_{\alpha\bar{\beta}}$ is a *Hermitian* matrix (that is, $g_{\alpha\bar{\beta}} = \overline{g_{\beta\bar{\alpha}}}$), and so it has real eigenvalues, and $\det(g_{\alpha\bar{\beta}})$ is a *real* function. This determinant is also given by the equation

$$\omega^m = i^m m!\, \det(g_{\alpha\bar{\beta}})\, dz_1 \wedge d\bar{z}_1 \wedge dz_2 \wedge d\bar{z}_2 \wedge \cdots \wedge dz_m \wedge d\bar{z}_m. \tag{4.7}$$

Here ω^m is the m-fold wedge product of ω. We see from (4.7) that $\det(g_{\alpha\bar{\beta}})$ is *positive*, as ω^m is $m!$ times the volume form of g and so is a positive $2m$-form.

It can be shown that the Ricci curvature is given by

$$R_{\alpha\bar{\beta}} = -\partial_\alpha \bar{\partial}_{\bar{\beta}}\left[\log \det(g_{\gamma\bar{\delta}})\right],$$

and therefore the Ricci form is

$$\rho = -i\partial\bar{\partial}\left[\log \det(g_{\gamma\bar{\delta}})\right] = -\tfrac{1}{2}\,dd^c\left[\log \det(g_{\gamma\bar{\delta}})\right]. \tag{4.8}$$

Thus locally we may write $\rho = -\tfrac{1}{2} dd^c f$ for a smooth real function f, so ρ is closed. As the determinant only makes sense in a holomorphic coordinate system, and we cannot find holomorphic coordinates on the whole of M, this is only a local expression for ρ.

As we remarked in §3.4.1, a Kähler metric g on M has $\mathrm{Hol}^0(g) \subseteq \mathrm{SU}(m)$ if and only if it is *Ricci-flat*, and thus if and only if it has Ricci form $\rho = 0$. Such metrics are called *Calabi–Yau metrics*, because they can be constructed using Yau's solution of the Calabi conjecture, as in Chapter 5.

4.7 Exterior forms on Kähler manifolds

Section 1.1.2 defined the Hodge star $*$ and the operators d^* and Δ_d on an oriented Riemannian manifold. We begin by defining analogues of these on a Kähler manifold. Let M be a Kähler manifold of real dimension $2m$, with Kähler metric g. The complex structure induces a natural orientation on M, and the metric and the orientation combine to give a volume form dV_g on M, which is a real $2m$-form.

Let α, β be *complex k-forms* on M. Define a pointwise inner product (α, β) on M by

$$(\alpha, \beta) = \alpha_{a_1...a_k} \overline{\beta_{b_1...b_k}} g^{a_1 b_1} \ldots g^{a_k b_k} \tag{4.9}$$

in index notation. Here (α, β) is a *complex* function on M, which is linear in α and antilinear in β, that is, linear in the complex conjugate $\overline{\beta}$ of β. Note that $(\beta, \alpha) = \overline{(\alpha, \beta)}$, and that (α, α) is a nonnegative real function on M.

When M is compact, define the L^2 inner product of complex k-forms α, β by $\langle \alpha, \beta \rangle = \int_M (\alpha, \beta) dV_g$. Then $\langle\,,\,\rangle$ is a *Hermitian inner product* on the space of complex k-forms. That is, $\langle \alpha, \beta \rangle$ is a complex number, bilinear in α and $\overline{\beta}$, such that $\langle \beta, \alpha \rangle = \overline{\langle \alpha, \beta \rangle}$, and $\langle \alpha, \alpha \rangle = \|\alpha\|_{L^2}^2$ is real and nonnegative.

Now let the *Hodge star on Kähler manifolds* be the unique map $* : \Lambda^k T^* M \otimes_{\mathbb{R}} \mathbb{C} \to \Lambda^{2m-k} T^* M \otimes_{\mathbb{R}} \mathbb{C}$, satisfying the equation $\alpha \wedge (*\beta) = (\alpha, \beta) dV_g$ for all complex k-forms α, β. Then $*\beta$ is antilinear in β. The relation to the Hodge star on real forms defined in §1.1.2, is that if $\beta = \beta_1 + i\beta_2$ for β_1, β_2 real k-forms, then $*\beta = *\beta_1 - i*\beta_2$. The Hodge star on Kähler manifolds satisfies $*1 = dV_g$ and $*(*\beta) = (-1)^k \beta$, for β a complex k-form, so that $*^{-1} = (-1)^k *$.

Since M is complex, we have operators d, ∂ and $\bar{\partial}$ taking complex k-forms to complex $(k+1)$-forms. Define operators d^*, ∂^* and $\bar{\partial}^*$ by

$$d^*\alpha = -*d(*\alpha), \quad \partial^*\alpha = -*\partial(*\alpha) \text{ and } \bar{\partial}^*\alpha = -*\bar{\partial}(*\alpha). \tag{4.10}$$

Then d^*, ∂^* and $\bar{\partial}^*$ all take complex k-forms to complex $(k-1)$-forms. Moreover, the argument used in §1.1.2 to show that $\langle \alpha, d^*\beta \rangle = \langle d\alpha, \beta \rangle$ for α a $(k-1)$-form and β a k-form on a compact oriented Riemannian manifold, also shows that

$$\langle \alpha, d^*\beta \rangle = \langle d\alpha, \beta \rangle, \quad \langle \alpha, \partial^*\beta \rangle = \langle \partial\alpha, \beta \rangle \text{ and } \langle \alpha, \bar{\partial}^*\beta \rangle = \langle \bar{\partial}\alpha, \beta \rangle,$$

where α is a complex $(k-1)$-form and β a complex k-form on a compact Kähler manifold.

In §1.1 we defined the Laplacian $\Delta = dd^* + d^*d$ on Riemannian manifolds. By analogy, from the six operators d, ∂, $\bar{\partial}$, d^*, ∂^* and $\bar{\partial}^*$ we can make three different Laplacians on complex k-forms:

$$\Delta_d = dd^* + d^*d, \quad \Delta_\partial = \partial\partial^* + \partial^*\partial \text{ and } \Delta_{\bar{\partial}} = \bar{\partial}\bar{\partial}^* + \bar{\partial}^*\bar{\partial}.$$

We call Δ_d the d-*Laplacian*, Δ_∂ the ∂-*Laplacian* and $\Delta_{\bar{\partial}}$ the $\bar{\partial}$-*Laplacian*. It can be shown (see [93, p. 115]) that these satisfy

$$\Delta_\partial = \Delta_{\bar{\partial}} = \tfrac{1}{2}\Delta_d. \tag{4.11}$$

Now it is easy to show that $*$ takes $\Lambda^{p,q}M$ to $\Lambda^{m-p,m-q}M$. Since ∂ maps $C^\infty(\Lambda^{p,q}M)$ to $C^\infty(\Lambda^{p+1,q}M)$ and $\bar{\partial}$ maps $C^\infty(\Lambda^{p,q}M)$ to $C^\infty(\Lambda^{p,q+1}M)$, we deduce from (4.10) that ∂^* maps $C^\infty(\Lambda^{p,q}M)$ to $C^\infty(\Lambda^{p-1,q}M)$ and $\bar{\partial}^*$ maps

$C^\infty(\Lambda^{p,q}M)$ to $C^\infty(\Lambda^{p,q-1}M)$. It follows easily that Δ_∂ (and hence $\Delta_{\bar\partial}$ and Δ_d) maps $C^\infty(\Lambda^{p,q}M)$ to $C^\infty(\Lambda^{p,q}M)$.

It is conventional to call the $\bar\partial$-Laplacian on a Kähler manifold the Laplacian, and to write it Δ rather than $\Delta_{\bar\partial}$. This can lead to confusion, because on a Riemannian manifold we call the d-Laplacian Δ_d the Laplacian, so that the Laplacian on a Kähler manifold is half the Laplacian on a Riemannian manifold.

4.7.1 Hodge theory on Kähler manifolds

In §1.1.3 we summarized the ideas of Hodge theory for a compact Riemannian manifold (M, g). Then in §3.5.2 we showed that the space $\mathcal{H}^k$ of Hodge k-forms is a direct sum of subspaces $\mathcal{H}^k_i$ corresponding to irreducible representations of the holonomy group $\mathrm{Hol}(g)$ of g, and deduced that the de Rham cohomology group $H^k(M, \mathbb{R})$ decomposes in the same way.

Since a Kähler metric g has $\mathrm{Hol}(g) \subseteq \mathrm{U}(m)$, these ideas apply to compact Kähler manifolds, and we will now work them out in more detail. Our notation differs slightly from §1.1.3 and §3.5.2, in that we deal with complex rather than real k-forms, and write the summands $\mathcal{H}^{p,q}$ rather than $\mathcal{H}^k_i$. Let M be a compact Kähler manifold, and define

$$\mathcal{H}^{p,q} = \mathrm{Ker}\Big(\Delta : C^\infty(\Lambda^{p,q}M) \to C^\infty(\Lambda^{p,q}M)\Big)$$

so that $\mathcal{H}^{p,q}$ is the vector space of harmonic (p, q)-forms on M. It is easy to show by (4.11) and integration by parts that $\alpha \in \mathcal{H}^{p,q}$ if and only if $\partial\alpha = \bar\partial\alpha = \partial^*\alpha = \bar\partial^*\alpha = 0$. Here is a version of the Hodge decomposition theorem for the $\bar\partial$ operator, proved in [93, p. 84].

Theorem 4.7.1 *Let M be a compact Kähler manifold. Then*

$$C^\infty(\Lambda^{p,q}M) = \mathcal{H}^{p,q} \oplus \bar\partial\big[C^\infty(\Lambda^{p-1,q}M)\big] \oplus \bar\partial^*\big[C^\infty(\Lambda^{p+1,q}M)\big], \text{ where}$$
$$\mathrm{Ker}\,\bar\partial = \mathcal{H}^{p,q} \oplus \bar\partial\big[C^\infty(\Lambda^{p-1,q}M)\big] \text{ and } \mathrm{Ker}\,\bar\partial^* = \mathcal{H}^{p,q} \oplus \bar\partial^*\big[C^\infty(\Lambda^{p+1,q}M)\big].$$

Comparing Theorem 4.7.1 with the definition (4.3) of the Dolbeault groups $H^{p,q}_{\bar\partial}(M)$ of M we see that $H^{p,q}_{\bar\partial}(M) \cong \mathcal{H}^{p,q}$. Now define

$$\mathcal{H}^k = \mathrm{Ker}\Big(\Delta : C^\infty(\Lambda^k T^*M \otimes_\mathbb{R} \mathbb{C}) \to C^\infty(\Lambda^k T^*M \otimes_\mathbb{R} \mathbb{C})\Big).$$

As $\Delta = \frac{1}{2}\Delta_d$ by (4.11), Theorem 1.1.4 implies that there is a natural isomorphism between $\mathcal{H}^k$ and the complex cohomology $H^k(M, \mathbb{C})$ of M. But

$$\mathcal{H}^k = \bigoplus_{j=0}^{k} \mathcal{H}^{j,k-j}.$$

Define $H^{p,q}(M)$ to be the vector subspace of $H^{p+q}(M, \mathbb{C})$ with representatives in $\mathcal{H}^{p,q}$. Then we have:

Theorem 4.7.2 *Let M be a compact Kähler manifold of real dimension $2m$. Then $H^k(M, \mathbb{C})$ decomposes as $H^k(M, \mathbb{C}) = \bigoplus_{j=0}^{k} H^{j,k-j}(M)$. Every element of $H^{p,q}(M)$ is represented by a unique harmonic (p, q)-form. Moreover for all p, q we have $H^{p,q}(M) \cong H_{\bar{\partial}}^{p,q}(M)$,*

$$H^{p,q}(M) \cong \overline{H^{q,p}(M)} \quad \text{and} \quad H^{p,q}(M) \cong \left(H^{m-p,m-q}(M)\right)^*.$$

Note that if M is a compact complex manifold admitting Kähler metrics, then the decomposition $H^k(M, \mathbb{C}) = \bigoplus_{j=0}^{k} H^{j,k-j}(M)$ given in the theorem depends only on the complex structure of M, and not on the choice of a particular Kähler structure.

Define the *Hodge numbers* $h^{p,q}$ by $h^{p,q} = \dim H^{p,q}(M)$. From Theorem 4.7.2 we see that

$$b^k = \sum_{j=0}^{k} h^{j,k-j} \quad \text{and} \quad h^{p,q} = h^{q,p} = h^{m-p,m-q} = h^{m-q,m-p}. \tag{4.12}$$

From these equations one can deduce that some compact manifolds, even complex manifolds, cannot admit a Kähler metric for topological reasons. Here are two ways in which this can happen. First, if $k = 2l + 1$ then $b^k = 2\sum_{j=0}^{l} h^{j,k-j}$, so that if M is a compact Kähler manifold, then b^k is even when k is odd. Thus, any compact manifold that admits a Kähler metric must satisfy this topological condition. For instance, the 4-manifold $\mathcal{S}^3 \times \mathcal{S}^1$ has a complex structure, but as $b^1 = 1$ it can have no Kähler metric.

Secondly, if M is a compact complex manifold, then it has Dolbeault groups $H_{\bar{\partial}}^{p,q}(M)$ depending on its complex structure. Theorem 4.7.2 shows that a necessary condition for M to admit a Kähler metric with this complex structure is that $\dim H_{\bar{\partial}}^{p,q}(M) = \dim H_{\bar{\partial}}^{q,p}(M)$ for all p, q, which is not always the case.

Now in the splitting (4.2) of complex k-forms on a complex manifold into (p, q)-forms, the summands $\Lambda^{p,q} M$ correspond to irreducible representations of $\text{GL}(m, \mathbb{C})$. However, when $1 \leqslant p, q \leqslant m-1$, the corresponding representation of $\text{U}(m)$ is *not* irreducible, but is the sum of several irreducible subrepresentations. This means that $\mathcal{H}^{p,q}$ and $H^{p,q}(M)$ can be split into smaller pieces, using the $\text{U}(m)$-structure. The simplest example of this is that $\Lambda^{1,1} M = \langle \omega \rangle \oplus \Lambda_0^{1,1} M$, where ω is the Kähler form and $\Lambda_0^{1,1} M$ is the bundle of $(1,1)$-forms orthogonal to ω. However, we prefer to work with the splitting into (p, q)-forms, as it is simpler and loses little information.

Finally we consider *Kähler classes* and the *Kähler cone*.

Definition 4.7.3 Let (M, J) be a compact complex manifold admitting Kähler metrics. If g is a Kähler metric on M, then the Kähler form ω of g is a closed real 2-form, and so defines a de Rham cohomology class $[\omega] \in H^2(M, \mathbb{R})$, called the *Kähler class* of g. It is a *topological invariant* of g. Since ω is also a $(1,1)$-form, $[\omega]$ lies in the intersection $H^{1,1}(M) \cap H^2(M, \mathbb{R})$, regarding $H^{1,1}(M)$ and $H^2(M, \mathbb{R})$ as vector subspaces of $H^2(M, \mathbb{C})$. Define the *Kähler cone* $\mathcal{K}$ of M to be the set of Kähler classes $[\omega] \in H^{1,1}(M) \cap H^2(M, \mathbb{R})$ of Kähler metrics on M.

If g_1, g_2 are Kähler metrics on M and $t_1, t_2 > 0$, then $t_1 g_1 + t_2 g_2$ is also Kähler. Thus, if $\alpha_1, \alpha_2 \in \mathcal{K}$ and $t_1, t_2 > 0$ then $t_1 \alpha_1 + t_2 \alpha_2 \in \mathcal{K}$, so that $\mathcal{K}$ is a *convex cone*. Furthermore, if ω is the Kähler form of a Kähler metric and η is a smooth, closed real (1,1)-form on M with $|\eta| < 1$ on M, then $\omega + \eta$ is also the Kähler form of a Kähler metric on M. As M is compact, this implies that $\mathcal{K}$ is *open* in $H^{1,1}(M) \cap H^2(M, \mathbb{R})$.

Suppose that Σ is a compact complex curve in a complex manifold (M, J), and that ω is the Kähler form of a Kähler metric g on M. Then Σ defines a homology class $[\Sigma] \in H_2(M, \mathbb{R})$, and the area of Σ with respect to g is $[\omega] \cdot [\Sigma] \in \mathbb{R}$. But this area must be positive. Therefore, each α in the Kähler cone $\mathcal{K}$ of M must satisfy $\alpha \cdot [\Sigma] > 0$, for each compact complex curve $\Sigma \subset M$. In simple cases $\mathcal{K}$ is exactly the subset of $H^{1,1}(M) \cap H^2(M, \mathbb{R})$ satisfying these inequalities, and is a polyhedral cone bounded by a finite number of hyperplanes, but this is not always true.

4.8 Complex algebraic varieties

This section is designed as a rather brief introduction to complex algebraic geometry. We shall define complex algebraic varieties, and discuss related ideas such as the Zariski topology, sheaves, and schemes. Before beginning, here are some introductory books on algebraic geometry. Griffiths and Harris [93] cover complex algebraic geometry, taking quite a differential geometric point of view, and discussing manifolds, Kähler metrics, Hodge theory, and so on. Hartshorne [96] has a more algebraic approach, and this section is largely based on [96, §1 & §2]. Two other books, both rather algebraic, are Harris [95] and Iitaka [111]. Harris' book is more elementary and contains lots of examples.

Let $\mathbb{C}^m$ have complex coordinates $(z_1, \ldots, z_m)$, and let $\mathbb{C}[z_1, \ldots, z_m]$ denote the ring of polynomials in the variables $z_1, \ldots, z_m$, with complex coefficients. Then $\mathbb{C}[z_1, \ldots, z_m]$ is a *ring of complex functions* on $\mathbb{C}^m$. Now $\mathbb{C}^m$ is a topological space, with the usual manifold topology. However, there is another natural topology on $\mathbb{C}^m$ called the *Zariski topology*, which is more useful for the purposes of algebraic geometry. Here is its definition.

Definition 4.8.1 An *algebraic set* in $\mathbb{C}^m$ is the set of common zeros of a finite number of polynomials in $\mathbb{C}[z_1, \ldots, z_m]$. It is easy to show that if X, Y are algebraic sets in $\mathbb{C}^m$, then $X \cap Y$ and $X \cup Y$ are algebraic sets. Also, $\emptyset$ and $\mathbb{C}^m$ are algebraic sets.

Define the *Zariski topology* on $\mathbb{C}^m$ by taking the open subsets to be $\mathbb{C}^m \setminus X$, for all algebraic sets X. This gives a topology on $\mathbb{C}^m$, in which a subset $X \subset \mathbb{C}^m$ is closed if and only if it is algebraic.

In this section, we shall regard $\mathbb{C}^m$ as a topological space with the Zariski topology, rather than the usual topology. The simplest sort of complex algebraic varieties are affine varieties, which we define now.

Definition 4.8.2 An algebraic set X in $\mathbb{C}^m$ is said to be *irreducible* if it is not the union $X_1 \cup X_2$ of two proper subsets, which are also algebraic sets in $\mathbb{C}^m$.

An *affine algebraic variety*, or simply *affine variety*, is an irreducible algebraic set in $\mathbb{C}^m$. It is considered to be a topological space, with the induced (Zariski) topology. A *quasi-affine variety* is an open set in an affine variety.

Let X be an affine variety in $\mathbb{C}^m$. Let $I(X)$ be the set of polynomials in $\mathbb{C}[z_1, \ldots, z_m]$ that vanish on X, and $A(X)$ the set $\{f|_X : f \in \mathbb{C}[z_1, \ldots, z_m]\}$ of functions on X. It is easy to see that $A(X)$ is a ring of functions on X, that $I(X)$ is an ideal in the ring $\mathbb{C}[z_1, \ldots, z_m]$, and that $A(X) \cong \mathbb{C}[z_1, \ldots, z_m]/I(X)$. We call $A(X)$ the *affine coordinate ring* of X.

Choose $x \in X$, and define I_x to be the set of functions in $A(X)$ that are zero at x. Clearly, I_x is an ideal in $A(X)$. In fact, I_x is a *maximal ideal* in $A(X)$. Moreover, every maximal ideal in $A(X)$ is of the form I_x for some $x \in X$, and if $x_1, x_2 \in X$, then $I_{x_1} = I_{x_2}$ if and only if $x_1 = x_2$. It follows that there is a 1-1 correspondence between the points of X, and the set of maximal ideals in $A(X)$. This is the beginning of the subject of *affine algebraic geometry*. The idea is that the ring $A(X)$ is regarded as the primary object, and then X is derived from $A(X)$. The philosophy is to investigate affine varieties by using the algebraic properties of rings of functions on them.

Next, we will discuss *projective varieties*. Projective varieties are subsets of $\mathbb{CP}^m$, just as affine varieties are subsets of $\mathbb{C}^m$. Since we cannot define polynomials on $\mathbb{CP}^m$, instead we use *homogeneous polynomials* on $\mathbb{C}^{m+1}$.

Definition 4.8.3 Let $\mathbb{CP}^m$ be the complex projective space, and let $[z_0, \ldots, z_m]$ be homogeneous coordinates on $\mathbb{CP}^m$, as described in Example 4.1.1. Let d be a nonnegative integer. A polynomial $f \in \mathbb{C}[z_0, \ldots, z_m]$ is called *homogeneous of degree* d if $f(\lambda z_0, \ldots, \lambda z_m) = \lambda^d f(z_0, \ldots, z_m)$ for all λ and $z_0, \ldots, z_m \in \mathbb{C}$. Let f be a homogeneous polynomial in $\mathbb{C}[z_0, \ldots, z_m]$, and let $[z_0, \ldots, z_m] \in \mathbb{CP}^m$. We say that $[z_0, \ldots, z_m]$ is a *zero* of f if $f(z_0, \ldots, z_m) = 0$. As f is homogeneous, this definition does not depend on the choice of homogeneous coordinates for $[z_0, \ldots, z_m]$.

Here are the analogous definitions of the Zariski topology on $\mathbb{CP}^m$, and projective and quasi-projective varieties.

Definition 4.8.4 Define an *algebraic set* in $\mathbb{CP}^m$ to be the set of common zeros of a finite number of homogeneous polynomials in $\mathbb{C}[z_0, \ldots, z_m]$. Define the *Zariski topology* on $\mathbb{CP}^m$ by taking the open subsets to be $\mathbb{CP}^m \setminus X$, for all algebraic sets $X \subset \mathbb{CP}^m$. An algebraic set X in $\mathbb{CP}^m$ is said to be *irreducible* if it is not the union of two proper algebraic subsets.

A *projective algebraic variety*, or simply *projective variety*, is defined to be an irreducible algebraic set in $\mathbb{CP}^m$. A *quasi-projective variety* is defined to be an open subset of a projective variety, in the Zariski topology.

Now $\mathbb{C}^m$ can be identified with the Zariski open set $\{[z_0, \ldots, z_m] \in \mathbb{CP}^m : z_0 \neq 0\}$ in $\mathbb{CP}^m$. Making this identification, we see that affine, quasi-affine and projective varieties are all examples of quasi-projective varieties. Because of this, quasi-projective varieties are often referred to as *algebraic varieties*, or simply *varieties*. In this section we consider varieties to be topological spaces with the Zariski topology, unless we specify otherwise.

An affine variety is studied using the ring of polynomials on it. On projective varieties we cannot consider polynomials, so we consider two other classes of functions, the *rational functions* and *regular functions*.

Definition 4.8.5 Let X be a quasi-projective variety in $\mathbb{CP}^m$. Let $g, h \in \mathbb{C}[z_0, \ldots, z_m]$ be homogeneous polynomials of the same degree d. Define a subset U_h in X by $U_h = \{[z_0, \ldots, z_m] \in X : h(z_0, \ldots, z_m) \neq 0\}$. Then U_h is a Zariski open set. Define the *rational function* $f : U_h \to \mathbb{C}$ by

$$f([z_0, \ldots, z_m]) = \frac{g(z_0, \ldots, z_m)}{h(z_0, \ldots, z_m)}.$$

As g and h are both homogeneous of the same degree, f is independent of the choice of homogeneous coordinates for each point, and so is well-defined.

Let U be open in X and let $f : U \to \mathbb{C}$ be a function. If $p \in U$, we say that f is *regular* at p if there is an open set $U' \subset U$ containing p, and f is equal to a rational function on U'. If f is regular at every point $p \in U$, we say f is *regular*. A regular function is one that is *locally* equal to a rational function.

Next we define two natural notions of map between varieties, *morphisms* and *rational maps*.

Definition 4.8.6 Let X and Y be varieties. A *morphism* $\phi : X \to Y$ is a continuous map (with the Zariski topologies) such that whenever V is open in Y and $f : V \to \mathbb{C}$ is regular, then $f \circ \phi : \phi^{-1}(V) \to \mathbb{C}$ is also regular. Clearly, if $\phi : X \to Y$ and $\psi : Y \to Z$ are morphisms of varieties, then $\psi \circ \phi : X \to Z$ is also a morphism. A map $\phi : X \to Y$ is called an *isomorphism* if ϕ is bijective, so that it has an inverse $\phi^{-1} : Y \to X$, and both ϕ and ϕ^{-1} are morphisms.

Definition 4.8.7 Let X and Y be varieties. A *rational map* $\phi : X \dashrightarrow Y$ is an equivalence class of morphisms $\phi_U : U \to Y$, where U is a dense open set in X, and morphisms $\phi_U : U \to Y$ and $\phi_V : V \to Y$ are equivalent if $\phi_U|_{U \cap V} = \phi_V|_{U \cap V}$. Note that a rational map is *not* in general a map of the set X to the set Y.

A *birational map* $\phi : X \dashrightarrow Y$ is a rational map which admits a rational inverse. That is, ϕ is an equivalence class of maps $\phi_U : U \to V$, where U, V are dense open sets in X, Y respectively, and ϕ_U is an isomorphism of varieties. If there is a birational map between X and Y, we say X and Y are *birationally equivalent*, or simply *birational*. This is an equivalence relation.

A *birational morphism* $\phi : X \to Y$ is a morphism of varieties which is also a birational map. That is, there should exist dense open subsets $U \subset X$ and $V \subset Y$ such that $\phi(U) = V$ and $\phi|_U : U \to V$ is an isomorphism. A birational morphism $\phi : X \to Y$ is a genuine map from the set X to Y.

If X and Y are isomorphic varieties, then they are birational. However, if X and Y are birational, they need not be isomorphic. Thus, birationality is a cruder equivalence relation on varieties than isomorphism.

Let X be a variety, and U be open in X. Define A_U to be the set of regular functions on U. Then A_U is a *ring of functions on U*. If U, V are open sets with $U \subseteq V$, then restriction from V to U gives a natural map $r_{V,U} : A_V \to A_U$, which is a ring homomorphism. All this information is packaged together in a composite mathematical object called a *sheaf of rings* on X (see [93, p. 35] or [111, p. 27]), which we now define.

Definition 4.8.8 Let X be a topological space with topology $\mathcal{T}$. A *sheaf of rings* $\mathcal{F}$ on X associates to each open set $U \in \mathcal{T}$ a ring $\mathcal{F}(U)$, called the *sections* of $\mathcal{F}$ over U, and to each pair $U \subset V$ in $\mathcal{T}$ a ring homomorphism $r_{V,U} : \mathcal{F}(V) \to \mathcal{F}(U)$, such that conditions (i)–(v) below are satisfied. The map $r_{V,U}$ is called the *restriction map*, and for $\sigma \in \mathcal{F}(V)$ we write $r_{V,U}(\sigma) = \sigma|_U$. Here are the necessary axioms.

(i) $\mathcal{F}(\emptyset) = \{0\}$.
(ii) $r_{U,U} : \mathcal{F}(U) \to \mathcal{F}(U)$ is the identity for all $U \in \mathcal{T}$.
(iii) If $U, V, W \in \mathcal{T}$ with $U \subset V \subset W$, then $r_{W,U} = r_{W,V} \circ r_{V,U}$.
(iv) If $U, V \in \mathcal{T}$ and $\sigma \in \mathcal{F}(U)$, $\tau \in \mathcal{F}(V)$ satisfy $\sigma|_{U \cap V} = \tau|_{U \cap V}$, then there exists $\rho \in \mathcal{F}(U \cup V)$ such that $\rho|_U = \sigma$ and $\rho|_V = \tau$.
(v) If $U, V \in \mathcal{T}$ and $\sigma \in \mathcal{F}(U \cup V)$ satisfies $\sigma|_U = 0$ and $\sigma|_V = 0$, then $\sigma = 0$.

A *sheaf of groups* on X is defined in exactly the same way, except that $\mathcal{F}(U)$ should be a group rather than a ring, and the restrictions $r_{V,U}$ should be group homomorphisms. A *ringed space* is defined to be a pair $(X, \mathcal{O})$, where X is a topological space and $\mathcal{O}$ a sheaf of rings on X. We call X the *base space*, and $\mathcal{O}$ the *structure sheaf*.

If $\mathcal{F}$ is a sheaf of groups or rings over a topological space X, then one can define the *sheaf cohomology groups* $H^j(X, \mathcal{F})$ for $j = 0, 1, 2, \ldots$. They are an important tool in algebraic geometry. The group $H^0(X, \mathcal{F})$ is $\mathcal{F}(X)$, the group of global sections of $\mathcal{F}$ over X, but the groups $H^k(X, \mathcal{F})$ for $k \geqslant 1$ are more difficult to interpret. For more details, see [93, §0.3], [96, §3] or [111, §4].

Let X be a projective variety. Then X is a topological space, with the Zariski topology, and the regular functions on X form a sheaf $\mathcal{O}$ of rings on X, called the *sheaf of regular functions on X*. Thus $(X, \mathcal{O})$ is a *ringed space*. Moreover, for each open set U, the ring $\mathcal{O}(U)$ is actually a ring of complex functions on U.

Complex algebraic geometry can be described as *the study of complex algebraic varieties up to isomorphisms*. Let X and Y be varieties, and $\phi : X \to Y$ an isomorphism of varieties. Consider the question: if X and Y are isomorphic, what features of X and Y have to be 'the same'? Well, from the definition we see that X and Y have to be isomorphic as topological spaces, with the Zariski topologies, and the sheaves of regular functions must also agree. However, there is no need for the embeddings $X \hookrightarrow \mathbb{CP}^m$ and $Y \hookrightarrow \mathbb{CP}^n$ to be related at all.

Because of this, it is useful to think of a variety not as a particular subset of $\mathbb{CP}^m$, but as a topological space X equipped with a sheaf of rings $\mathcal{O}$. Following this idea, one can define the concept of an *abstract variety*, which is a variety without a given embedding in $\mathbb{CP}^m$. In affine algebraic geometry, the primary object is a ring A. The topological space X is derived from A as the set of maximal ideals, and is studied using algebraic tools and a lot of ring theory.

In more general algebraic geometry, the primary object is often an abstract variety, regarded as a topological space X equipped with a sheaf of rings $\mathcal{O}$. It is also studied from a very algebraic point of view. In fact, a lot of algebraic geometry is written in terms of *schemes* [96, §2], which are closely related to varieties. Recall that in an affine variety, the points of the topological space are the maximal ideals of a ring. In an *affine scheme* the points of the topological space are instead the *prime ideals* of a ring. A *scheme* is a ringed space $(X, \mathcal{O})$ that is locally isomorphic to an affine scheme.

4.9 Singular varieties, resolutions, and deformations

Let X be a variety in $\mathbb{CP}^m$, and let $x \in X$. We say that x is a *nonsingular point* if X is a complex submanifold of $\mathbb{CP}^m$ in a neighbourhood of x. We say that x is *singular* if it is not nonsingular. The variety X is called *singular* if it has singular points, and *nonsingular* otherwise. In general, the nonsingular points form a dense open subset of X, and the singular points are a finite union of subvarieties of X. There is also an equivalent, algebraic way to define the idea of singular point, using the idea of *local ring*.

For example, let $p(z_1, \ldots, z_m)$ be a complex polynomial that is not constant and has no repeated factors, and let X be the hypersurface $\{(z_1, \ldots, z_m) \in \mathbb{C}^m : p(z_1, \ldots, z_m) = 0\}$ in $\mathbb{C}^m$. Then a point $x \in X$ is singular if and only if $\partial p / \partial z_j = 0$ at x for $j = 1, \ldots, m$. So, for instance, the quadric $z_1^2 + z_2^2 + z_3^2 = 0$ in $\mathbb{C}^3$ has just one singular point at $(0, 0, 0)$.

Clearly, a complex algebraic variety X is a *complex manifold* if and only if it is nonsingular. The converse, however, is not true: not every complex manifold is an algebraic variety. Let X be a compact complex manifold. A *meromorphic function* f on X is a singular holomorphic function, that can be written locally as the quotient of two holomorphic functions. On an algebraic variety, all the regular functions are meromorphic. Therefore a variety must have a lot of meromorphic functions—enough to form a holomorphic coordinate system near each point, for instance.

So, if a compact complex manifold has only a few meromorphic functions, then it cannot be an algebraic variety. There are many compact complex manifolds that admit no nonconstant meromorphic functions at all, and these are not algebraic varieties. However, *Chow's Theorem* [93, p. 167] states that any compact complex submanifold of $\mathbb{CP}^m$ is algebraic. The study of nonalgebraic complex manifolds is sometimes called *transcendental complex geometry*.

There is a natural generalization of the idea of complex manifold to include singularities, called a *complex analytic variety*.

Definition 4.9.1 Let U be an open set in $\mathbb{C}^m$, in the usual topology, rather than the Zariski topology. An *analytic subset of* U is a subset $S \subseteq U$ defined by the vanishing of a finite number of holomorphic functions on U. The restriction to S of the sheaf $\mathcal{O}_U$ of holomorphic functions on U is a sheaf of rings $\mathcal{O}_S$ on S. We define a *(complex) analytic variety* to be a ringed space $(X, \mathcal{O}_X)$ such that X is Hausdorff, and $(X, \mathcal{O}_X)$ is locally isomorphic to $(S, \mathcal{O}_S)$ for analytic subsets $S \subseteq U \subseteq \mathbb{C}^m$. Here sheaves and ringed spaces are defined in Definition 4.8.8.

We call a point $x \in X$ *nonsingular* if $(X, \mathcal{O}_X)$ is locally isomorphic to $(\mathbb{C}^k, \mathcal{O}_{\mathbb{C}^k})$ near x, where k is the dimension of X near x. We call x *singular* if it is not nonsingular. If X contains no singular points then it is a *complex manifold*. Otherwise we call X a *singular complex manifold*. For more information about analytic varieties, see [93, p. 12–14].

Complex algebraic varieties are examples of complex analytic varieties. Conversely, complex analytic varieties are locally isomorphic to complex algebraic varieties, but not necessarily globally isomorphic. All the ideas in the rest of this section work equally well in the setting of algebraic varieties and in the setting of analytic varieties, but we will only give definitions for one of the two.

Now a singular point in a variety X is a point where the manifold structure of X breaks down in some way. Given a singular variety X, it is an important problem in algebraic geometry to understand how to repair the singularities of X, and make a new, nonsingular variety $\tilde{X}$ closely related to X. There are two main strategies used to do this, called *resolution* and *deformation*.

4.9.1 *Resolutions of singular varieties*

Definition 4.9.2 Let X be a singular variety. A *resolution* $(\tilde{X}, \pi)$ of X is a normal, nonsingular variety $\tilde{X}$ with a proper birational morphism $\pi : \tilde{X} \to X$. Here *normal varieties* are defined by Griffiths and Harris [93, p. 177] and Iitaka [111, §2], and *proper morphisms* by Hartshorne [96, p. 95–105]. From [96, p. 95], a morphism $f : X \to Y$ of complex algebraic varieties is proper if the preimage of any compact set in Y is compact in X, using the manifold topologies on X and Y, and not the Zariski topologies.

This means that $\tilde{X}$ is a complex manifold, and the map $\pi : \tilde{X} \to X$ is surjective. There are dense open sets of X and $\tilde{X}$ on which π is also injective, and in fact biholomorphic. But if x is a singular point of X, then $\pi^{-1}(x)$ is in general a compact subvariety of $\tilde{X}$, rather than a single point. Often $\pi^{-1}(x)$ is a submanifold of $\tilde{X}$, or a finite union of submanifolds. Thus, in a resolution we repair the singularities by replacing each singular point by a submanifold, or more general subvariety.

One way to construct resolutions is to use a technique called *blowing up*, which we define first for affine varieties.

Definition 4.9.3 Let $X \subset \mathbb{C}^m$ be an affine variety with affine coordinate ring $A(X)$, let Y be a closed subvariety of X, and let $I_Y \subset A(X)$ be the ideal of functions in $A(X)$ that are zero on Y. Then I_Y is finitely generated, and we can choose a set of generators $f_0, \ldots, f_n \in I_Y$ for I_Y. Define a map $\phi : X \setminus Y \to \mathbb{CP}^n$ by $\phi(x) = [f_0(x), \ldots, f_n(x)]$.

The *(algebraic) blow-up* $\tilde{X}$ of X along Y is the closure in $X \times \mathbb{CP}^n$ of the graph of ϕ, that is,

$$\tilde{X} = \overline{\left\{(x, \phi(x)) : x \in X \setminus Y\right\}} \subset X \times \mathbb{CP}^n.$$

The *projection* $\pi : \tilde{X} \to X$ is the map $\pi : (x, z) \mapsto x$. Then $\tilde{X}$ is a variety, and $\pi : \tilde{X} \to X$ is a birational morphism. In a similar way, one can define the blow-up $(\tilde{X}, \pi)$ of a general algebraic or analytic variety X along a closed subvariety Y. For more information, see [93, p. 182, p. 602] and [95, p. 82].

Here $\pi : \tilde{X} \to X$ is surjective, and $\pi : \tilde{X} \setminus \pi^{-1}(Y) \to X \setminus Y$ is an isomorphism. The pull-back $\pi^{-1}(Y)$ is a finite union of closed subvarieties of $\tilde{X}$ of codimension one, called the *exceptional divisor*. If X is nonsingular and Y is a submanifold of X, then

$\pi^{-1}(Y)$ is the projectivized normal bundle of Y in X. In particular, if Y is the single point y, then $\pi^{-1}(y)$ is the complex projective space $P(T_y X)$.

Now suppose X is a singular variety, and let $Y \subset X$ be the set of singular points in X. Then Y is a finite union of subvarieties of X, and we can consider the blow-up $\tilde{X}$ of X along Y. Although $\tilde{X}$ may not be nonsingular, it is a general principle that the singularities of $\tilde{X}$ are usually of a less severe kind, and easier to resolve, than those of X. Our next result, sometimes called the *Resolution of Singularities Theorem*, shows that the singularities of any variety can be resolved by a finite number of blow-ups.

Theorem 4.9.4. (Hironaka [101]) *Let X be a complex algebraic variety. Then there exists a resolution $\pi : \tilde{X} \to X$, which is the result of a finite sequence of blow-ups of X. That is, there are varieties $X = X_0, X_1, \ldots, X_n = \tilde{X}$, such that X_j is a blow-up of X_{j-1} along some subvariety, with projection $\pi_j : X_j \to X_{j-1}$, and the map $\pi : \tilde{X} \to X$ is $\pi = \pi_1 \circ \cdots \circ \pi_n$.*

4.9.2 Deformations of singular and nonsingular varieties

Definition 4.9.5 Let X be a complex analytic variety of dimension m. A 1-*parameter family of deformations of X* is a complex analytic variety $\mathcal{X}$ of dimension $m+1$, together with a proper holomorphic map $f : \mathcal{X} \to \Delta$ where Δ is the unit disc in $\mathbb{C}$, such that $X_0 = f^{-1}(0)$ is isomorphic to X. The other fibres $X_t = f^{-1}(t)$ for $t \neq 0$ are called *deformations* of X.

If the deformations X_t are nonsingular for $t \neq 0$, they are called *smoothings* of X. By a *small deformation* of X we mean a deformation X_t where t is small. That is, when we say something is true for all small deformations of X, we mean that in any 1-parameter family of deformations $\{X_t : t \in \Delta\}$ of X, the statement holds for all sufficiently small t. We say that X is *rigid* if all small deformations X_t of X are biholomorphic to X.

We shall be interested in deformations of complex analytic varieties for two reasons. Firstly, a singular variety X may admit a family of nonsingular deformations X_t. Thus, as with resolutions, deformation gives a way of repairing the singularities of X to get a nonsingular variety.

From the point of view of algebraic geometry, there is a big difference between resolution and deformation. If X is a singular variety and $\tilde{X}$ a resolution of X, then X and $\tilde{X}$ are birationally equivalent, and share the same field of meromorphic functions. So to algebraic geometers, who often try to classify varieties up to birational equivalence, X and $\tilde{X}$ are nearly the same thing. But a variety X and its deformations X_t can be algebraically very different.

The second reason we will be interested in deformations is when we wish to describe the family of all integrable complex structures upon a particular compact manifold, up to isomorphism. Suppose that (X, J) is a compact complex manifold. Then all small deformations X_t of X are nonsingular, and are diffeomorphic to X as real manifolds. Thus small deformations of X are equivalent to complex structures J_t on X that are close to the complex structure J, in a suitable sense.

So, to understand the local geometry of the moduli space of complex structures on X, we need a way to study the collection of *all* small deformations of a complex analytic variety.

Definition 4.9.6 Let X be a complex analytic variety. A *family of deformations of* X consists of a (possibly singular) complex analytic variety T called the *base space* containing a *base point* t_0, and a complex analytic variety $\mathcal{X}$ with a flat holomorphic map $f : \mathcal{X} \to T$, such that $X_{t_0} = f^{-1}(t_0)$ is isomorphic to X. The other fibres $X_t = f^{-1}(t_0)$ are then deformations of X. Here *flatness* is a technical condition upon morphisms of algebraic or analytic varieties defined in [96, §III.9]. It implies, in particular, that $\dim X_t = \dim X$ for all $t \in T$.

If (S, s_0) is another complex analytic variety with base point and $F : S \to T$ is a holomorphic map with $F(s_0) = t_0$, then we get an induced family of deformations $F^*(\mathcal{X})$ of X over S. A family of deformations of X is called *versal* or *semi-universal* if any other family of small deformations of X can be induced from it by a suitable map F. It is called *universal* if this map F is unique.

Note that some authors (e.g. Slodowy [196, p. 7]) define semi-universality differently, and distinguish between versal and semi-universal deformations. If $\{X_t : t \in T\}$ is a *universal* family of deformations of X, then every small deformation of X appears exactly once in the family. Thus the collection of all deformations of X is locally isomorphic to the base space T of the universal family, and has the structure of a complex analytic variety. In particular, if (X, J) is a compact complex manifold and $\{X_t : t \in T\}$ is a universal family of deformations of (X, J), then the moduli space of all complex structures J_t on X is locally isomorphic to T.

However, there exist compact complex manifolds (X, J) which have no universal family of deformations. The moduli space of complex structures on X up to isomorphism has a natural topology. If this topology is not Hausdorff near J then (X, J) cannot have a universal family of deformations, because the base space T would be non-Hausdorff, contradicting its definition as a complex analytic variety.

Rather than working with moduli spaces of pathological topology, it is helpful in this case to consider a *versal* family of deformations of X. In a versal family every small deformation of X is represented at least once, but some may appear many times. Now the theory of deformations of compact complex manifolds was developed by Kodaira, Spencer and Kuranishi, and is described in Kodaira [136]. The main result in this theory is that a versal family of deformations exists for any compact complex manifold, and can be constructed using sheaf cohomology.

Let X be a compact complex manifold, and let Θ_X be the sheaf of holomorphic vector fields of X. Then the *sheaf cohomology groups* $H^*(X, \Theta_X)$ are the cohomology of the complex

$$0 \to C^\infty(T^{1,0}X) \xrightarrow{\bar{\partial}} C^\infty(T^{1,0}X \otimes \Lambda^{0,1}X) \xrightarrow{\bar{\partial}} C^\infty(T^{1,0}X \otimes \Lambda^{0,2}X) \xrightarrow{\bar{\partial}} \cdots .$$

Here we interpret $H^1(X, \Theta_X)$ as the space of infinitesimal deformations of the complex structure of X, and $H^2(X, \Theta_X)$ as the space of obstructions to lifting an infinitesimal deformation to an actual deformation of the complex structure of X. Kodaira,

Spencer and Kuranishi prove that there is an open neighbourhood U of 0 in $H^1(X, \Theta_X)$ and a holomorphic map $\Phi : U \to H^2(X, \Theta_X)$ with $\Phi(0) = d\Phi(0) = 0$, such that $T = \Phi^{-1}(0)$ is the base of a versal family of deformations of X, with base point 0, called the *Kuranishi family* of X. The group $H^0(X, \Theta_X)$ also has an interpretation as the Lie algebra of the group of holomorphic automorphisms of X, and if $H^0(X, \Theta_X) = 0$ then the Kuranishi family is universal.

4.10 Line bundles and divisors

Let M be a complex manifold. A *holomorphic line bundle* over M is a holomorphic vector bundle with fibre $\mathbb{C}$, the complex line. Holomorphic lines bundles are important in algebraic geometry. If L, L' and L'' are holomorphic line bundles over M, then the dual bundle L^* and the tensor product $L \otimes L'$ are also holomorphic line bundles. These operations satisfy the equations

$$L \otimes L' \cong L' \otimes L, \quad (L \otimes L') \otimes L'' \cong L \otimes (L' \otimes L''), \quad \text{and} \quad L \otimes L^* \cong \tilde{\mathbb{C}},$$

where $\tilde{\mathbb{C}}$ is the trivial line bundle $M \times \mathbb{C}$.

Define $\mathcal{P}_M$ to be the set of isomorphism classes of holomorphic line bundles over M. From the equation above we see that $\mathcal{P}_M$ is an *abelian group*, where multiplication is given by the tensor product, inverses are dual bundles, and the identity is the trivial bundle $\tilde{\mathbb{C}}$. This group is called the *Picard group* of M. It can be identified with the *sheaf cohomology group* $H^1(M, \mathcal{O}^*)$ [93, p. 133], but we will not explain this.

Because of the group structure on $\mathcal{P}_M$, it is convenient to use a multiplicative notation for line bundles. Let M be a complex manifold, L a holomorphic line bundle over M, and $k \in \mathbb{Z}$. Then we write

$$L^k = \bigotimes^k L \quad \text{if } k > 0, \quad L^k = \bigotimes^{-k} L^* \quad \text{if } k < 0, \text{ and} \quad L^0 = \tilde{\mathbb{C}}.$$

In particular, the dual L^* is written L^{-1}.

If M is a complex manifold of dimension m, then $\Lambda^{p,0}M$ is a holomorphic vector bundle with fibre dimension $\binom{m}{p}$. Thus, when $p = m$, the fibre of $\Lambda^{m,0}M$ is $\mathbb{C}$, and $\Lambda^{m,0}M$ is a *holomorphic line bundle*. This is called the *canonical bundle* of M, and is written K_M. It is the bundle of *complex volume forms* on M, and is an important tool in algebraic geometry.

Let L be a holomorphic line bundle over a complex manifold M. The *first Chern class* $c_1(L)$ of L is a topological invariant of L called a *characteristic class*, which lies in the cohomology group $H^2(M, \mathbb{Z})$. Characteristic classes are described in [162]. The first Chern class classifies line bundles as smooth vector bundles. It satisfies

$$c_1(L^*) = -c_1(L) \quad \text{and} \quad c_1(L \otimes L') = c_1(L) + c_1(L').$$

Thus $c_1 : \mathcal{P}_M \to H^2(M, \mathbb{Z})$ is a homomorphism of abelian groups.

Let M be a complex manifold, and L a holomorphic line bundle over M. For each open set $U \subset M$, define $\mathcal{O}_L(U)$ to be the vector space of holomorphic sections of L over U, and if U, V are open in M with $U \subset V$, let $r_{V,U} : \mathcal{O}_L(V) \to \mathcal{O}_L(U)$ be the restriction map. Then $\mathcal{O}_L$ is a sheaf of groups over M, the *sheaf of holomorphic sections of L*. Now we describe the line bundles over the projective space $\mathbb{CP}^m$.

Example 4.10.1 Recall that in Example 4.1.1, $\mathbb{CP}^m$ was defined to be the set of one-dimensional vector spaces of $\mathbb{C}^{m+1}$. Define the *tautological line bundle* L^{-1} over $\mathbb{CP}^m$ to be the subbundle of the trivial bundle $\mathbb{CP}^m \times \mathbb{C}^{m+1}$, whose fibre at $x \in \mathbb{CP}^m$ is the line in $\mathbb{C}^{m+1}$ represented by x. Then L^{-1} is a vector bundle over $\mathbb{CP}^m$ with fibre $\mathbb{C}$. The total space of L^{-1} is a complex submanifold of $\mathbb{CP}^m \times \mathbb{C}^{m+1}$, and has the structure of a complex manifold. Thus, L^{-1} is a holomorphic line bundle over $\mathbb{CP}^m$.

Define L to be the dual of L^{-1}. Then L is a holomorphic line bundle over $\mathbb{CP}^m$, called the *hyperplane bundle*. So, L^k is a holomorphic line bundle over $\mathbb{CP}^m$ for each $k \in \mathbb{Z}$. It can be shown [93, p. 145] that every holomorphic line bundle over $\mathbb{CP}^m$ is isomorphic to L^k for some $k \in \mathbb{Z}$. There is an isomorphism $H^2(\mathbb{CP}^m, \mathbb{Z}) \cong \mathbb{Z}$, and making this identification we find that $c_1(L^k) = k \in \mathbb{Z}$. Thus $c_1 : \mathcal{P}_{\mathbb{CP}^m} \to H^2(\mathbb{CP}^m, \mathbb{Z}) \cong \mathbb{Z}$ is a group isomorphism. The canonical bundle $K_{\mathbb{CP}^m}$ of $\mathbb{CP}^m$ is isomorphic to L^{-m-1}.

The sheaf of holomorphic sections $\mathcal{O}_{L^k}$ of L^k over $\mathbb{CP}^m$ is written $\mathcal{O}(k)$. (Also, by an abuse of notation, $\mathcal{O}(k)$ often denotes the line bundle L^k.) The vector space of holomorphic sections of L^k is $H^0(\mathbb{CP}^m, \mathcal{O}(k))$, in the notation of sheaf cohomology. If $k < 0$ then $H^0(\mathbb{CP}^m, \mathcal{O}(k))$ is zero, and if $k \geq 0$ it is canonically identified with the set of homogeneous polynomials of degree k on $\mathbb{C}^{m+1}$, which is a vector space of dimension $\binom{m+k}{m}$.

Let M be a compact complex manifold, and L a holomorphic line bundle over M. Let V be the vector space $H^0(M, \mathcal{O}_L)$ of holomorphic sections of L over M. Then V is a finite-dimensional vector space over $\mathbb{C}$, of dimension $m+1$, say. For each point $p \in M$, define a map $\phi_p : V \to L_p$ by $\phi_p(s) = s(p)$. Then ϕ_p is linear, so that $\phi_p \in V^* \otimes L_p$. Define p to be a *base point* of L if $\phi_p = 0$, and let $B \subset M$ be the set of base points of L. Then if $p \in M \setminus B$, then ϕ_p is nonzero in $V^* \otimes L_p$, and thus $[\phi_p] \in P(V^* \otimes L_p)$.

But L is a line bundle, so $L_p \cong \mathbb{C}$ as complex vector spaces. Therefore the projective spaces $P(V^*)$ and $P(V^* \otimes L_p)$ are naturally isomorphic, and we can regard $[\phi_p]$ as a point in $P(V^*)$. So, define a map $\iota_L : M \setminus B \to P(V^*)$ by $\iota_L(p) = [\phi_p]$. Now $P(V^*)$ is a complex projective space $\mathbb{CP}^m$, and thus a complex manifold, and B is closed in M, so that $M \setminus B$ is also a complex manifold. It is easy to show that $\iota_L : M \setminus B \to \mathbb{CP}^m$ is a *holomorphic map* of complex manifolds. When $L = K_M^r$, a power of the canonical bundle, the maps $\iota_{K_M^r}$ are called the *pluricanonical maps*, and are important in algebraic geometry.

A line bundle L over a compact complex manifold M is called *very ample* if L has no base points in M, and the map $\iota_L : M \to \mathbb{CP}^m$ is an *embedding* of M in $\mathbb{CP}^m$. Also, L is called *ample* if L^k is very ample for some $k > 0$. Thus, if L is very ample, then ι_L identifies M with a complex submanifold of $\mathbb{CP}^m$, its image $\iota_L(M)$. Now by Chow's Theorem [93, p. 167], every compact complex submanifold of $\mathbb{CP}^m$ is a nonsingular projective variety. Thus, if M is a compact complex manifold with an ample line bundle, then M is a projective variety.

The remarkable Kodaira Embedding Theorem [93, p. 181] gives a simple criterion for a holomorphic line bundle L over a compact complex manifold M to be ample. A line bundle L is called *positive* if its first Chern class $c_1(L)$ can be represented, as a de Rham cohomology class, by a closed (1, 1)-form α which is positive in the sense of §4.4. The Kodaira Embedding Theorem says that L is ample if and only if it is positive. Therefore, a compact complex manifold with a positive line bundle is a projective variety. Because of this, many problems on compact complex and Kähler manifolds become problems about projective varieties, and can be attacked algebraically.

4.10.1 Divisors

Now we shall explore the connections between line bundles and divisors, which are formal sums of hypersurfaces in complex manifolds.

Definition 4.10.2 Let M be a complex manifold. A closed subset $N \subset M$ is said to be a *hypersurface* in M if for each $p \in N$ there is an open neighbourhood U of p in M and a nonzero holomorphic function $f : U \to \mathbb{C}$, such that $N \cap U = \{u \in U : f(u) = 0\}$. A hypersurface $N \subset M$ is called *irreducible* if it is not the union of two hypersurfaces N_1, N_2 with $N_1, N_2 \neq N$.

In general, a hypersurface $N \subset M$ is a singular submanifold of M, of codimension one. Every hypersurface in M can be written uniquely as a union of irreducible hypersurfaces, and if M is compact then this union is finite. Suppose now that M is a complex manifold, L a holomorphic line bundle over M, and s a nonzero holomorphic section of L. Define $N \subset M$ to be the set $\{m \in M : s(m) = 0\}$. Then N is a hypersurface in M. Thus, there is a link between holomorphic line bundles, holomorphic sections, and hypersurfaces. To render this link more explicit, we make another definition.

Definition 4.10.3 Let M be a complex manifold. An irreducible hypersurface $N \subset M$ is called a *prime divisor* on M. A *divisor* D on M is a locally finite formal linear combination

$$D = \sum_i a_i N_i, \tag{4.13}$$

where $a_i \in \mathbb{Z}$, and each N_i is a prime divisor. Here 'locally finite' means that each compact subset of M meets only a finite number of the hypersurfaces N_i. The divisor D is called *effective* if $a_i \geq 0$ for all i.

Suppose as before that M is a complex manifold, L a holomorphic line bundle over M, and s a nonzero holomorphic section of L. Let N be the hypersurface $N = \{m \in M : s(m) = 0\}$. Then N may be written in a unique way as a locally finite union $N = \bigcup_i N_i$, where the N_i are prime divisors. For each i, there is a unique positive integer a_i, such that s *vanishes to order* a_i *along* N_i. Define $D = \sum_i a_i N_i$. Then D is an effective divisor. In this way, whenever we have a nonzero holomorphic section of a holomorphic line bundle over M, we construct an effective divisor on M.

This construction is reversible, in the following sense. Suppose that L_1, L_2 are holomorphic line bundles over M, and s_1, s_2 are nonzero holomorphic sections of L_1, L_2 respectively. Let D_1, D_2 be the effective divisors constructed from s_1, s_2. It can be shown that $D_1 = D_2$ if and only if there exists an isomorphism $\phi : L_1 \to L_2$ of holomorphic line bundles, such that $\phi(s_1) = s_2$, and the isomorphism ϕ is then unique.

Moreover, if D is an effective divisor on M, then there exists a holomorphic line bundle L over M, and a nonzero holomorphic section s of L, that yields the divisor D.

Thus there is a 1-1 correspondence between effective divisors on M, and isomorphism classes of holomorphic line bundles equipped with nonzero holomorphic sections. In the same way, there is a 1-1 correspondence between divisors on M, and isomorphism classes of line bundles equipped with nonzero *meromorphic* sections. In this case, the divisor $\sum_i a_i N_i$ corresponds to a section s with a *zero* of order a_i along N_i if $a_i > 0$, and a *pole* of order $-a_i$ along N_i if $a_i < 0$.

Let M be a compact complex manifold of complex dimension m, and L a holomorphic line bundle over M, with a nonzero meromorphic section associated to a divisor D. Then D defines a homology class $[D] \in H_{2m-2}(M, \mathbb{Z})$. Under the natural isomorphism $H^2(M, \mathbb{Z}) \cong H_{2m-2}(M, \mathbb{Z})$, this homology class is identified with $c_1(L)$, the first Chern class of L. This gives one way to understand $c_1(L)$.

Here is a result on the topology of a hypersurface N in M associated to a *positive* line bundle L.

Theorem 4.10.4. (Lefschetz Hyperplane Theorem) *Let M be a compact, m-dimensional complex manifold, N a nonsingular hypersurface in M, and L the holomorphic line bundle over M associated to the divisor N. Suppose L is positive. Then*

(a) *the map $H^k(M, \mathbb{C}) \to H^k(N, \mathbb{C})$ induced by the inclusion $N \hookrightarrow M$ is an isomorphism for $0 \leqslant k \leqslant m - 2$ and injective for $k = m - 1$, and*

(b) *the map of homotopy groups $\pi_k(N) \to \pi_k(M)$ induced by the inclusion $N \hookrightarrow M$ is an isomorphism for $0 \leqslant k \leqslant m - 2$ and surjective for $k = m - 1$.*

The result also holds if M and N are orbifolds instead of manifolds, and N is a nonsingular hypersurface in the orbifold sense.

This is known as the *Lefschetz Hyperplane Theorem*, as we can take M to be a submanifold of $\mathbb{CP}^n$ and L the restriction to M of the line bundle $\mathcal{O}(1)$ over $\mathbb{CP}^n$, and then N is the intersection of M with a hyperplane H in $\mathbb{CP}^n$. Part (a) is proved in Griffiths and Harris [93, p. 156] for complex manifolds, and rather more general and more complicated versions of (b) are proved by Goresky and MacPherson [84, p. 153] and Hamm [94], in which M and N can be singular complex varieties, and not just orbifolds.

5

THE CALABI CONJECTURE

Let (M, J) be a compact, complex manifold, and g a Kähler metric on M, with Ricci form ρ. From §4.6 we know that ρ is a closed (1,1)-form and $[\rho] = 2\pi c_1(M) \in H^2(M, \mathbb{R})$. It is natural to ask which closed (1,1)-forms can be the Ricci forms of a Kähler metric on M. The Calabi conjecture [49, 50] answers this question.

The Calabi conjecture *Let (M, J) be a compact, complex manifold, and g a Kähler metric on M, with Kähler form ω. Suppose that ρ' is a real, closed $(1, 1)$-form on M with $[\rho'] = 2\pi c_1(M)$. Then there exists a unique Kähler metric g' on M with Kähler form ω', such that $[\omega'] = [\omega] \in H^2(M, \mathbb{R})$, and the Ricci form of g' is ρ'.*

The conjecture was posed by Calabi in 1954, who also showed that if g' exists it must be unique. It was eventually proved by Yau in 1976, [225, 226]. Before this, Aubin [16] had made significant progress towards a proof. In this chapter we will give a proof of the Calabi conjecture that broadly follows Yau's own proof, with some differences. My main references for this chapter are Yau's paper [226], and the treatment given in Aubin's book [17, §7]. The proof is also explained, in French, by Bourguignon et al. [39].

In §5.1 the Calabi conjecture is reformulated as a nonlinear, elliptic partial differential equation in a real function ϕ. Section 5.2 states four results, Theorems C1–C4, and then proves the Calabi conjecture assuming these theorems. After some preparatory work in §5.3, Theorems C1–C4 are proved in §5.4–§5.7 respectively, and the proof of the Calabi conjecture is complete. Finally, section 5.8 discusses some analytic issues from the proof.

The proof of the Calabi conjecture is very important in the subject of Riemannian holonomy groups, for the following reason. Suppose M is a compact Kähler manifold with $c_1(M) = 0$. Then we may choose the 2-form ρ' in the Calabi conjecture to be zero, and so the proof of the conjecture guarantees the existence of a Kähler metric g' on M with zero Ricci form. Thus, we construct families of *Ricci-flat Kähler metrics* on compact complex manifolds.

Now from §3.4.1, a generic Kähler metric g has $\text{Hol}^0(g) = \text{U}(m)$, but if g is Ricci-flat then $\text{Hol}^0(g) \subseteq \text{SU}(m)$. If g is irreducible, Berger's Theorem implies that either $\text{Hol}^0(g) = \text{SU}(m)$, or $m = 2k$ and $\text{Hol}^0(g) = \text{Sp}(k)$. Therefore, the Calabi conjecture proof yields examples of compact Riemannian manifolds with holonomy $\text{SU}(m)$ and $\text{Sp}(k)$. These manifolds, called *Calabi–Yau manifolds* and *hyperkähler manifolds* respectively, will be the subject of Chapters 6 and 7.

There are other applications of the proof of the Calabi conjecture which we shall not discuss; for instance, it can be used to find Kähler metrics with positive or negative definite Ricci curvature on some compact complex manifolds M, and this has consequences for the fundamental group $\pi_1(M)$, and the group of biholomorphisms of M. There are also results on the existence of Kähler–Einstein metrics on complex manifolds that are closely related to the Calabi conjecture proof. For more information on these topics, see Besse [33, §11].

5.1 Reformulating the Calabi conjecture

We shall rewrite the Calabi conjecture in terms of a partial differential equation. Let (M, J) be a compact, complex manifold, g a Kähler metric on M with Kähler form ω, and ρ the Ricci form of g. Let ρ' be a real, closed (1,1)-form on M with $[\rho'] = 2\pi\, c_1(M)$. To solve the Calabi conjecture we must find a Kähler metric g', with Kähler form ω', such that $[\omega] = [\omega']$ and g' has Ricci form ρ'.

As $[\rho'] = 2\pi\, c_1(M) = [\rho]$ we have $[\rho' - \rho] = 0$ in $H^2(M, \mathbb{R})$, so by the proof of Lemma 4.5.1 there exists a smooth real function f on M, unique up to addition of a constant, such that

$$\rho' = \rho - \tfrac{1}{2}\mathrm{d}\mathrm{d}^c f. \tag{5.1}$$

Define a smooth, positive function F on M by $(\omega')^m = F \cdot \omega^m$. Using eqns (4.7) and (4.8) of §4.6 we deduce that $\tfrac{1}{2}\mathrm{d}\mathrm{d}^c(\log F) = \rho - \rho' = \tfrac{1}{2}\mathrm{d}\mathrm{d}^c f$. Thus $\mathrm{d}\mathrm{d}^c(f - \log F) = 0$, so that $f - \log F$ is constant on M.

Define $A > 0$ by $f - \log F = -\log A$. Then $F = A \mathrm{e}^f$, and g' must satisfy

$$(\omega')^m = A \mathrm{e}^f \omega^m. \tag{5.2}$$

As $[\omega'] = [\omega] \in H^2(M, \mathbb{R})$, and M is compact, we see that $\int_M (\omega')^m = \int_M \omega^m$. Substituting (5.2) in and applying (4.4), we deduce that

$$A \int_M \mathrm{e}^f\, \mathrm{d}V_g = \int_M \mathrm{d}V_g = \mathrm{vol}_g(M), \tag{5.3}$$

where $\mathrm{d}V_g$ is the volume form on M induced by g, and $\mathrm{vol}_g(M)$ the volume of M with this volume form. This determines the constant A.

Note that in this book, all manifolds are by definition assumed to be *connected*. If M were not connected then we would have to choose a different constant A for each connected component of M. We have shown that the Calabi conjecture is equivalent to the following:

The Calabi conjecture (second version) *Let (M, J) be a compact, complex manifold, and g a Kähler metric on M, with Kähler form ω. Let f be a smooth real function on M, and define $A > 0$ by $A \int_M \mathrm{e}^f\, \mathrm{d}V_g = \mathrm{vol}_g(M)$. Then there exists a unique Kähler metric g' on M with Kähler form ω', such that $[\omega'] = [\omega] \in H^2(M, \mathbb{R})$, and $(\omega')^m = A\mathrm{e}^f \omega^m$.*

Here is a way to understand this. The conjecture is about the existence of metrics with *prescribed volume forms*. Every volume form on M may be written as $F \mathrm{d}V_g$, for F a smooth real function. We impose two conditions on this volume form: firstly that it should be positive, so that $F > 0$, and secondly that it should have the same total volume as dV_g, so that $\int_M F\,dV_g = \int_M dV_g$. Then the Calabi conjecture says that there is a unique Kähler metric g' with the same Kähler class, such that $dV_{g'} = F\,dV_g$, that is, with the chosen volume form.

This is in fact a considerable simplification. The first statement of the conjecture prescribed the Ricci curvature of g', which depends on g' and its second derivatives, and was in effect m^2 real equations on g'. But this second statement depends only on g', not on its derivatives, and imposes only one real equation on g'.

Next, observe that as $[\omega'] = [\omega]$, by Lemma 4.5.1 there exists a smooth real function ϕ on M, unique up to addition of a constant, such that

$$\omega' = \omega + dd^c\phi. \tag{5.4}$$

Suppose also that ϕ satisfies the equation $\int_M \phi\,dV_g = 0$. This then specifies ϕ uniquely. So, we deduce that the Calabi conjecture is equivalent to the following:

The Calabi conjecture (third version) *Let (M, J) be a compact, complex manifold, and g a Kähler metric on M, with Kähler form ω. Let f be a smooth real function on M, and define $A > 0$ by $A\int_M e^f\,dV_g = \mathrm{vol}_g(M)$. Then there exists a unique smooth real function ϕ such that*

(i) *$\omega + dd^c\phi$ is a positive $(1, 1)$-form,*
(ii) *$\int_M \phi\,dV_g = 0$, and*
(iii) *$(\omega + dd^c\phi)^m = Ae^f\omega^m$ on M.*

Moreover, part (iii) is equivalent to the following:

(iii)' *Choose holomorphic coordinates $z_1, \ldots, z_m$ on an open set U in M. Then $g_{\alpha\bar{\beta}}$ may be interpreted as an $m \times m$ Hermitian matrix indexed by $\alpha, \bar{\beta} = 1, 2, \ldots, m$ in U. The condition on ϕ is*

$$\det\left(g_{\alpha\bar{\beta}} + \frac{\partial^2\phi}{\partial z_\alpha \partial \bar{z}_{\bar{\beta}}}\right) = Ae^f \det(g_{\alpha\bar{\beta}}). \tag{5.5}$$

For part (iii)', eqn (5.4) gives $g'_{\alpha\bar{\beta}} = g_{\alpha\bar{\beta}} + \partial_\alpha\bar{\partial}_{\bar{\beta}}\phi$, and the result follows from eqn (4.7) of §4.6. Equation (5.5) is a nonlinear, elliptic, second-order partial differential equation in ϕ, of a kind known as a *Monge–Ampère equation*. We have reduced the Calabi conjecture to a problem in analysis, that of showing that a particular p.d.e. has a unique, smooth solution.

The difficulty of the Calabi conjecture, and the reason it took twenty years to complete, is that nonlinear equations in general are difficult to solve, and the nonlinearities of (5.5) are of a particularly severe kind, as they are nonlinear in the derivatives of highest order.

In fact part (i) follows from part (iii), as the following lemma shows.

Lemma 5.1.1 *Let (M, J) be a compact, complex manifold, and g a Kähler metric on M, with Kähler form ω. Let $f \in C^0(M)$, and define A by $A\int_M e^f\,dV_g = \mathrm{vol}_g(M)$.*

Suppose that $\phi \in C^2(M)$ satisfies the equation $(\omega + dd^c\phi)^m = Ae^f\omega^m$ on M. Then $\omega + dd^c\phi$ is a positive $(1, 1)$-form.

Proof Choose holomorphic coordinates $z_1, \ldots, z_m$ on a connected open set U in M. Then in U, the new metric g' is

$$g'_{\alpha\bar{\beta}} = g_{\alpha\bar{\beta}} + \frac{\partial^2\phi}{\partial z_\alpha \partial \bar{z}_{\bar{\beta}}}.$$

As usual, we may interpret $g'_{\alpha\bar{\beta}}$ as an $m \times m$ Hermitian matrix indexed by $\alpha, \bar{\beta} = 1, 2, \ldots, m$ in U. A Hermitian matrix has real eigenvalues. From §4.4, $\omega + dd^c\phi$ is a positive (1,1)-form if and only if g' is a Hermitian metric, that is, if and only if the eigenvalues of the matrix $g'_{\alpha\bar{\beta}}$ are all *positive*.

But from (5.5), $\det(g'_{\alpha\bar{\beta}}) > 0$ on U, so $g'_{\alpha\bar{\beta}}$ has no zero eigenvalues. Therefore by continuity of $g'_{\alpha\bar{\beta}}$, if the eigenvalues of $g'_{\alpha\bar{\beta}}$ are positive at some point $p \in U$ then they are positive everywhere in U. So by covering M with such open sets U and using the connectedness of M, we can show that if $\omega + dd^c\phi$ is positive at some point $p \in M$, then it is positive on all of M.

Since M is compact and ϕ is continuous, ϕ has a minimum on M. Let $p \in M$ be a minimum point of ϕ, and U a coordinate patch containing p. It is easy to show that at p the matrix $\partial^2\phi/\partial z_\alpha \partial \bar{z}_{\bar{\beta}}$ has nonnegative eigenvalues, and so $g'_{\alpha\bar{\beta}}$ has positive eigenvalues at p. Thus $\omega + dd^c\phi$ is positive at p, and everywhere on M. $\square$

5.2 Overview of the proof of the Calabi conjecture

We begin by stating four results, Theorems C1–C4, which will be proved later in the chapter. These are the four main theorems which make up our proof of the Calabi conjecture. After this, we will prove the Calabi conjecture assuming Theorems C1–C4, and make some comments on the proof.

Theorem C1 *Let (M, J) be a compact, complex manifold, and g a Kähler metric on M, with Kähler form ω. Let $Q_1 \geq 0$. Then there exist $Q_2, Q_3, Q_4 \geq 0$ depending only on M, J, g and Q_1, such that the following holds.*

Suppose $f \in C^3(M)$, $\phi \in C^5(M)$ and $A > 0$ satisfy the equations

$$\|f\|_{C^3} \leq Q_1, \quad \int_M \phi\, dV_g = 0, \quad \text{and} \quad (\omega + dd^c\phi)^m = Ae^f\omega^m.$$

Then $\|\phi\|_{C^0} \leq Q_2$, $\|dd^c\phi\|_{C^0} \leq Q_3$ and $\|\nabla dd^c\phi\|_{C^0} \leq Q_4$.

Theorem C2 *Let (M, J) be a compact, complex manifold, and g a Kähler metric on M, with Kähler form ω. Let $Q_1, \ldots, Q_4 \geq 0$ and $\alpha \in (0, 1)$. Then there exists $Q_5 \geq 0$ depending only on $M, J, g, Q_1, \ldots, Q_4$ and α, such that the following holds.*

Suppose $f \in C^{3,\alpha}(M)$, $\phi \in C^5(M)$ and $A > 0$ satisfy $(\omega + dd^c\phi)^m = Ae^f\omega^m$ and the inequalities

$$\|f\|_{C^{3,\alpha}} \leq Q_1, \quad \|\phi\|_{C^0} \leq Q_2, \quad \|dd^c\phi\|_{C^0} \leq Q_3 \quad \text{and} \quad \|\nabla dd^c\phi\|_{C^0} \leq Q_4.$$

Then $\phi \in C^{5,\alpha}(M)$ and $\|\phi\|_{C^{5,\alpha}} \leq Q_5$. Also, if $f \in C^{k,\alpha}(M)$ for $k \geq 3$ then $\phi \in C^{k+2,\alpha}(M)$, and if $f \in C^\infty(M)$ then $\phi \in C^\infty(M)$.

Theorem C3 *Let (M, J) be a compact complex manifold, and g a Kähler metric on M, with Kähler form ω. Fix $\alpha \in (0, 1)$, and suppose that $f' \in C^{3,\alpha}(M)$, $\phi' \in C^{5,\alpha}(M)$ and $A' > 0$ satisfy the equations*

$$\int_M \phi' \, dV_g = 0 \quad \text{and} \quad (\omega + dd^c \phi')^m = A' e^{f'} \omega^m.$$

Then whenever $f \in C^{3,\alpha}(M)$ and $\|f - f'\|_{C^{3,\alpha}}$ is sufficiently small, there exist $\phi \in C^{5,\alpha}(M)$ and $A > 0$ such that

$$\int_M \phi \, dV_g = 0 \quad \text{and} \quad (\omega + dd^c \phi)^m = A e^{f} \omega^m.$$

Theorem C4 *Let (M, J) be a compact complex manifold, and g a Kähler metric on M, with Kähler form ω. Let $f \in C^1(M)$. Then there is at most one function $\phi \in C^3(M)$ such that $\int_M \phi \, dV_g = 0$, and $(\omega + dd^c \phi)^m = A e^{f} \omega^m$ on M.*

Here are some remarks on these results. The positive constants A, A' above are determined entirely by f, f' using (5.3). Theorem C1 is due to Yau [226, §2]. Results of this type are called *a priori estimates*, because it tells us in advance (a priori) that any solution to a given equation must satisfy a certain bound. Finding such a priori estimates was the most difficult part of the Calabi conjecture, and was Yau's biggest contribution to the proof.

As eqn (5.5) is a nonlinear, elliptic p.d.e., we can draw on the fruit of decades of hard work on the properties of solutions of elliptic equations. Theorem C2 uses results about the differentiability of solutions of elliptic equations, and Theorem C3 uses results on the existence of solutions of elliptic equations. Theorem C4 shows that if ϕ exists, then it is unique. It has an elementary proof found by Calabi [50, p. 86]. Broadly speaking, Theorems C1–C3 concern the *existence* of the function ϕ, Theorem C2 is about the *smoothness* of ϕ, and Theorem C4 is about the *uniqueness* of ϕ. Using Theorems C1–C4 we will now prove the Calabi conjecture.

5.2.1 The proof of the Calabi conjecture

We start with a definition, the purpose of which will become clear soon.

Definition 5.2.1 Let (M, J) be a compact complex manifold, and g a Kähler metric on M, with Kähler form ω. Fix $\alpha \in (0, 1)$ and $f \in C^{3,\alpha}(M)$. Define S to be the set of all $t \in [0, 1]$ for which there exists $\phi \in C^{5,\alpha}(M)$ with $\int_M \phi \, dV_g = 0$ and $A > 0$, such that $(\omega + dd^c \phi)^m = A e^{tf} \omega^m$ on M.

Now, using Theorems C1 and C2 we will show that this set S is *closed*, and using Theorem C3 we will show that S is *open*.

Theorem 5.2.2 *In Definition 5.2.1, the set S is a closed subset of $[0, 1]$.*

Proof It must be shown that S contains its limit points, and therefore is closed. Let $\{t_j\}_{j=0}^{\infty}$ be a sequence in S, which converges to some $t' \in [0, 1]$. We will prove that $t' \in S$. Since $t_j \in S$, by definition there exists $\phi_j \in C^{5,\alpha}(M)$ and $A_j > 0$ such that

$$\int_M \phi_j \, dV_g = 0 \quad \text{and} \quad (\omega + dd^c \phi_j)^m = A_j e^{t_j f} \omega^m. \tag{5.6}$$

Define Q_1 by $Q_1 = \|f\|_{C^{3,\alpha}}$. Let Q_2, Q_3, Q_4 be the constants given by Theorem C1, which depend on Q_1, and Q_5 the constant given by Theorem C2, which depends on $Q_1, \ldots, Q_4$.

As $t_j \in [0, 1]$, $\|t_j f\|_{C^3} \leq Q_1$. So, applying Theorem C1 with ϕ_j in place of ϕ and $t_j f$ in place of f, we see that $\|\phi_j\|_{C^0} \leq Q_2$, $\|dd^c \phi_j\|_{C^0} \leq Q_3$ and $\|\nabla dd^c \phi_j\|_{C^0} \leq Q_4$ for all j. Thus, by Theorem C2, $\phi_j \in C^{5,\alpha}(M)$ and $\|\phi_j\|_{C^{5,\alpha}} \leq Q_5$ for all j. Now the Kondrakov Theorem, Theorem 1.2.3, says that the inclusion $C^{5,\alpha}(M) \to C^5(M)$ is *compact*, in the sense of Definition 1.2.2. It follows that as the sequence $\{\phi_j\}_{j=0}^\infty$ is bounded in $C^{5,\alpha}(M)$, it lies in a compact subset of $C^5(M)$. Therefore there exists a subsequence $\{\phi_{i_j}\}_{j=0}^\infty$ which converges in $C^5(M)$. Let $\phi' \in C^5(M)$ be the limit of this subsequence.

Define $A' > 0$ by $A' \int_M e^{t'f} dV_g = \text{vol}_g(M)$. Then $A_{i_j} \to A'$ as $j \to \infty$, because $t_{i_j} \to t'$ as $j \to \infty$. Since $\{\phi_{i_j}\}_{j=0}^\infty$ converges in C^2 we may take the limit in (5.6), giving

$$\int_M \phi' \, dV_g = 0 \quad \text{and} \quad (\omega + dd^c \phi')^m = A' e^{t'f} \omega^m. \tag{5.7}$$

Theorems C1 and C2 then show that $\phi' \in C^{5,\alpha}(M)$. Therefore $t' \in S$. So S contains its limit points, and is closed. □

Theorem 5.2.3 *In Definition 5.2.1, the set S is an open subset of $[0, 1]$.*

Proof Suppose $t' \in S$. Then by definition there exist $\phi' \in C^{5,\alpha}(M)$ with $\int_M \phi' \, dV_g = 0$ and $A' > 0$, such that $(\omega + dd^c \phi')^m = A' e^{t'f} \omega^m$ on M. Apply Theorem C3, with $t'f$ in place of f', and tf in place of f, for $t \in [0, 1]$. The theorem shows that whenever $|t - t'| \cdot \|f\|_{C^{3,\alpha}}$ is sufficiently small, there exist $\phi \in C^{5,\alpha}(M)$ and $A > 0$ such that

$$\int_M \phi \, dV_g = 0 \quad \text{and} \quad (\omega + dd^c \phi)^m = A e^{tf} \omega^m.$$

But then $t \in S$. Thus, if $t \in [0, 1]$ is sufficiently close to t' then $t \in S$, and S contains an open neighbourhood in $[0, 1]$ of each t' in S. So S is open. □

Using Lemma 5.1.1 and Theorems 5.2.2 and 5.2.3 we shall prove an existence result for the function ϕ. Notice that parts (i)–(iii) come from the third version of the Calabi conjecture.

Theorem 5.2.4 *Let (M, J) be a compact complex manifold, and g a Kähler metric on M with Kähler form ω. Choose $\alpha \in (0, 1)$, and let $f \in C^{3,\alpha}(M)$. Then there exist $\phi \in C^{5,\alpha}(M)$ and $A > 0$ such that*

(i) $\omega + dd^c \phi$ *is a positive $(1, 1)$-form,*
(ii) $\int_M \phi \, dV_g = 0$, *and*
(iii) $(\omega + dd^c \phi)^m = A e^f \omega^m$ *on M.*

Proof Theorems 5.2.2 and 5.2.3 imply that S is an open and closed subset of $[0, 1]$. Since $[0, 1]$ is connected, either $S = \emptyset$ or $S = [0, 1]$. But when $t = 0$, the function $\phi \equiv 0$ satisfies the conditions in Definition 5.2.1, so that $0 \in S$. Thus S cannot be empty, and $S = [0, 1]$. It follows that $1 \in S$. So, setting $t = 1$, there exists $\phi \in C^{5,\alpha}(M)$ with $\int_M \phi \, dV_g = 0$, such that $(\omega + dd^c\phi)^m = Ae^f\omega^m$ on M, as $A = A_1$. Therefore parts (ii) and (iii) of the theorem hold for ϕ. By Lemma 5.1.1, part (i) holds as well. This completes the proof. □

Finally, using Theorem 5.2.4 and Theorems C2 and C4, we show:

Theorem 5.2.5 *The Calabi conjecture is true.*

Proof Suppose (M, J) is a compact, complex manifold, and g a Kähler metric on M, with Kähler form ω. Let $f \in C^\infty(M)$. Then Theorem 5.2.4 constructs $\phi \in C^{5,\alpha}(M)$ and $A > 0$ for which conditions (i)–(iii) of the third version of the Calabi conjecture hold. Theorems C2 and C4 show that ϕ is smooth and unique. This proves the third version of the Calabi conjecture. □

5.2.2 The continuity method

The idea used in the proofs above is known as the *continuity method*, and it works like this. The goal is to prove that a particular nonlinear equation, in our case the equation

$$(\omega + dd^c\phi)^m = Ae^f\omega^m,$$

has a solution ϕ. The first step is to think of a similar equation which we already know has a solution. In this case we choose the equation

$$(\omega + dd^c\phi)^m = \omega^m,$$

which has the obvious solution $\phi = 0$.

The second step is to write down a 1-parameter family of equations depending continuously on $t \in [0, 1]$, such that when $t = 0$ the equation is the one we know has a solution, and when $t = 1$ the equation is the one which must be solved. In our case this family of equations is

$$(\omega + dd^c\phi_t)^m = A_t e^{tf}\omega^m.$$

To complete the proof one must show that the set S of $t \in [0, 1]$ for which the corresponding equation has a solution ϕ_t, is both open and closed in $[0, 1]$. For then, as the equation is soluble when $t = 0$, it is also soluble when $t = 1$ by the argument in Theorem 5.2.4, which is what we want.

Here are two standard arguments that are used to show S is open and closed. To show S is open, suppose that $t' \in S$, so a solution $\phi_{t'}$ exists. Then, one tries to show that when t is close to t' in $[0, 1]$, there is a solution ϕ_t that is close to $\phi_{t'}$ (in some Banach space). To do this it is usually enough to consider the *linearization* of the equation about $\phi_{t'}$, which simplifies the problem.

To show S is closed, one shows that S contains its limit points. Suppose $\{t_j\}_{j=0}^\infty$ is a sequence in S that converges to t'. Then there is a corresponding sequence of solutions

$\{\phi_{t_j}\}_{j=0}^{\infty}$. Now by establishing *a priori bounds* on all solutions ϕ_t in some Banach norm, it may be possible to show that they lie in some *compact* subset in a Banach space. If this is so, the sequence $\{\phi_{t_j}\}_{j=0}^{\infty}$ contains a convergent subsequence. One then shows that the limit of this subsequence is a solution $\phi_{t'}$ for $t = t'$, so $t' \in S$, and S is closed. This is the continuity method.

5.3 Calculations at a point

Let (M, J) be a compact, complex manifold of dimension m, and g a Kähler metric on M with Kähler form ω. Let $f \in C^0(M)$, $\phi \in C^2(M)$ and $A > 0$. Set $\omega' = \omega + dd^c\phi$, and suppose $(\omega')^m = Ae^f \omega^m$ on M. Lemma 5.1.1 then shows that ω' is a real, positive (1,1)-form, which therefore determines a Kähler metric g'. Let p be a point in M. In this section we shall find expressions for ω and ω' at p, and derive several inequalities that will be useful later.

Lemma 5.3.1 *In the situation above, we may choose holomorphic coordinates $z_1, \ldots, z_m$ on M near p, such that g, g', ω and ω' are given at p by*

$$g_p = 2|dz_1|^2 + \cdots + 2|dz_m|^2, \quad g'_p = 2a_1|dz_1|^2 + \cdots + 2a_m|dz_m|^2, \tag{5.8}$$

$$\omega_p = i(dz_1 \wedge d\bar{z}_1 + \cdots + dz_m \wedge d\bar{z}_m),$$
$$\text{and} \quad \omega'_p = i(a_1 dz_1 \wedge d\bar{z}_1 + \cdots + a_m dz_m \wedge d\bar{z}_m), \tag{5.9}$$

where $a_1, \ldots, a_m$ are positive real numbers.

Proof As $T_p^{(1,0)} M$ is isomorphic to $\mathbb{C}^m$ as a complex vector space, if we fix a basis for $T_p^{(1,0)} M$ over $\mathbb{C}$, then we may regard $(g_p)_{\alpha\bar{\beta}}$ and $(g'_p)_{\alpha\bar{\beta}}$ as invertible, Hermitian $m \times m$ complex matrices. By elementary linear algebra, using a process called simultaneous diagonalization, one can choose a basis $(v_1, \ldots, v_m)$ for $T_p M$ over $\mathbb{C}$, with respect to which

$$(g_p)_{\alpha\bar{\beta}} = \begin{cases} 1 & \text{if } \alpha = \bar{\beta} \\ 0 & \text{if } \alpha \neq \bar{\beta}, \end{cases} \quad \text{and} \quad (g'_p)_{\alpha\bar{\beta}} = \begin{cases} a_j & \text{if } \alpha = \bar{\beta} = j \\ 0 & \text{if } \alpha \neq \bar{\beta}, \end{cases} \tag{5.10}$$

where $a_1, \ldots, a_m$ are real numbers, and $a_j > 0$ for $j = 1, \ldots, m$ as g' is a metric.

Clearly, it is possible to find holomorphic coordinates $z_1, \ldots, z_m$ on M near p, such that $v_j = \partial/\partial z_j$ at p. Equations (5.8) and (5.9) then follow immediately from (5.10) and the equation $\omega_{ac} = J_a^b g_{bc}$. □

Next, we relate $a_1, \ldots, a_m$ to Ae^f and $\Delta\phi$. In order to be consistent with [226] and [17], we define the Laplacian Δ on a Kähler manifold by $\Delta\phi = -g^{\alpha\bar{\beta}} \partial_\alpha \bar{\partial}_{\bar{\beta}} \phi$. Be warned that this is equal to half of the usual d-Laplacian on a Riemannian manifold.

Lemma 5.3.2 *In the situation of the previous lemma, we have*

$$\prod_{j=1}^m a_j = Ae^{f(p)}, \quad \frac{\partial^2 \phi}{\partial z_j \partial \bar{z}_j}(p) = a_j - 1 \quad \text{and} \quad (\Delta\phi)(p) = m - \sum_{j=1}^m a_j. \tag{5.11}$$

Proof From (5.9) we see that $\omega_p^m = i^m m! \, dz_1 \wedge d\bar{z}_1 \wedge \cdots \wedge dz_m \wedge d\bar{z}_m$ and $(\omega_p')^m = \prod_{j=1}^m a_j \cdot i^m m! \, dz_1 \wedge d\bar{z}_1 \wedge \cdots \wedge dz_m \wedge d\bar{z}_m$. But $(\omega')^m = Ae^f \omega^m$, and therefore $\prod_{j=1}^m a_j = Ae^{f(p)}$, the first equation of (5.11). As $\omega' = \omega + dd^c\phi$, we have

$$(g_p')_{\alpha\bar{\beta}} = (g_p)_{\alpha\bar{\beta}} + \frac{\partial^2 \phi}{\partial z_\alpha \partial \bar{z}_{\bar{\beta}}}.$$

Putting $\alpha = \bar{\beta} = j$ and substituting (5.10) in gives the second equation of (5.11). Now $g^{\alpha\bar{\beta}}$ is the inverse of $g_{\alpha\bar{\beta}}$ as an $m \times m$ matrix. Hence by (5.10), $g_p^{\alpha\bar{\beta}}$ is 1 if $\alpha = \bar{\beta}$, and 0 otherwise. So, as $\Delta\phi = -g^{\alpha\bar{\beta}}\partial_\alpha\partial_{\bar{\beta}}\phi$, the third equation of (5.11) follows from the second. □

Let $T = T_{c_1 \cdots c_l}^{a_1 \cdots a_k}$ be a tensor on M, and g a Riemannian metric on M. We define $|T|_g$ by

$$|T|_g^2 = T_{c_1 \cdots c_l}^{a_1 \cdots a_k} T_{d_1 \cdots d_l}^{b_1 \cdots b_k} g_{a_1 b_1} \cdots g_{a_k b_k} g^{c_1 d_1} \cdots g^{c_l d_l}. \tag{5.12}$$

Using this notation, and the material of the previous two lemmas, the following result is very easy, so we omit the proof.

Lemma 5.3.3 *In the situation of Lemma 5.3.1, at p we have*

$$\left|dd^c\phi\right|_g^2 = 2\sum_{j=1}^m (a_j - 1)^2, \quad \left|g_{ab}'\right|_g^2 = 2\sum_{j=1}^m a_j^2 \quad \text{and} \quad \left|g'^{ab}\right|_g^2 = 2\sum_{j=1}^m a_j^{-2}.$$

Here g'^{ab} is the matrix inverse of g'_{ab} in coordinates. Now we shall prove some inequalities that will be useful in the next few sections.

Proposition 5.3.4 *Let (M, J) be a compact, complex manifold and g a Kähler metric on M, with Kähler form ω. Let $f \in C^0(M)$, $\phi \in C^2(M)$ and $A > 0$. Set $\omega' = \omega + dd^c\phi$, suppose that $(\omega')^m = Ae^f \omega^m$, and let g' be the metric with Kähler form ω'. Then*

$$\Delta\phi \leqslant m - mA^{1/m}e^{f/m} < m, \tag{5.13}$$

and there are constants c_1, c_2 and c_3 depending only on m and upper bounds for $\|f\|_{C^0}$ and $\|\Delta\phi\|_{C^0}$, such that

$$\left\|g_{ab}'\right\|_{C^0} \leqslant c_1, \quad \left\|g'^{ab}\right\|_{C^0} \leqslant c_2 \quad \text{and} \quad \left\|dd^c\phi\right\|_{C^0} \leqslant c_3. \tag{5.14}$$

Here all norms are with respect to the metric g.

Proof Inequality (5.13) follows immediately from the first and last equations of (5.11), and the fact that the geometric mean $(a_1 \cdots a_m)^{1/m}$ is less than or equal to the arithmetic mean $\frac{1}{m}(a_1 + \cdots + a_m)$. As $\Delta\phi(p) = m - \sum_{j=1}^m a_j$ by (5.11), and the a_j are positive, from Lemma 5.3.3 one can show that at p, $|dd^c\phi|_g^2 \leqslant 2m + 2(m - \Delta\phi)^2$, and $|g_{ab}'|_g^2 \leqslant$

$2(m - \Delta\phi)^2$. As these hold for all $p \in M$, the first and last inequalities of (5.14) hold with constants c_1, c_3 depending only on m and $\|\Delta\phi\|_{C^0}$.

Now if $\log A > -\inf_M f$ then $Ae^f > 1$ on M, which contradicts the equation $\int_M Ae^f dV_g = \int_M dV_g$. Similarly $\log A < -\sup_M f$ leads to a contradiction. Hence, $-\sup_M f \leq \log A \leq -\inf_M f$, and $|\log A| \leq \|f\|_{C^0}$. It follows that

$$e^{-2\|f\|_{C^0}} \leq Ae^f \leq e^{2\|f\|_{C^0}} \tag{5.15}$$

on M. Since $\prod_{j=1}^m a_j = Ae^{f(p)}$ by (5.11), we see that

$$a_j^{-1} = A^{-1}e^{-f(p)} \prod_{\substack{1 \leq k \leq m, \\ j \neq k}} a_k.$$

Using (5.15) to estimate $A^{-1}e^{-f(p)}$ and the inequality $a_k \leq m - \Delta\phi(p)$ derived from the third equation of (5.11), we see that

$$a_j^{-2} \leq e^{4\|f\|_{C^0}}(m - \Delta\phi(p))^{2m-2}.$$

From this and Lemma 5.3.3 the second inequality of (5.14) follows for some c_2 depending on m, $\|f\|_{C^0}$ and $\|\Delta\phi\|_{C^0}$, and the proof is complete. □

The proposition shows that an a priori bound for $\Delta\phi$ yields a priori bounds for g' and $dd^c\phi$.

Lemma 5.3.5 *In the situation of Proposition 5.3.4, we have*

$$d\phi \wedge d^c\phi \wedge \omega^{m-1} = \frac{1}{m}|\nabla\phi|_g^2 \omega^m \quad \text{and} \quad d\phi \wedge d^c\phi \wedge \omega^{m-j-1} \wedge (\omega')^j = F_j \omega^m$$

for $j = 1, 2, \ldots, m-1$, where F_j is a nonnegative real function on M.

Proof We have $d\phi \wedge d^c\phi \wedge \omega^{m-1} = (d\phi \wedge d^c\phi, *(\omega^{m-1}))dV_g$ by properties of the Hodge star. But $*(\omega^{m-1}) = (m-1)!\,\omega$, and $\omega^m = m!\,dV_g$ by (4.4). Therefore

$$d\phi \wedge d^c\phi \wedge \omega^{m-1} = \frac{(m-1)!}{m!}(d\phi \wedge d^c\phi, \omega) \cdot \omega^m. \tag{5.16}$$

Now $(d^c\phi)_a = -J_a^b(d\phi)_b$, so $(d\phi \wedge d^c\phi, \omega) = -(d\phi)_a J_b^e(d\phi)_e \omega_{cd} g^{ac} g^{bd} = |\nabla\phi|_g^2$, since $-J_b^e \omega_{cd} g^{ac} g^{bd} = g^{ae}$. Substituting this into (5.16) gives the first equation of the lemma.

Looking carefully at the proof, one can see that the reason $d\phi \wedge d^c\phi \wedge \omega^{m-1}$ turns out to be a nonnegative multiple of ω^m is that $*(\omega^{m-1})$ is a *positive* (1,1)-form, in the sense of §4.4. In the same way, if we can show that $*(\omega^{m-j-1} \wedge (\omega')^j)$ is a positive (1,1)-form, it will follow that $d\phi \wedge d^c\phi \wedge \omega^{m-j-1} \wedge (\omega')^j$ is a nonnegative multiple of ω^m, which is what we have to prove. But using the explicit expressions (5.9) for ω and ω' at a point, one can readily show that $*(\omega^{m-j-1} \wedge (\omega')^j)$ is positive, and the proof is finished. □

5.4 The proof of Theorem C1

We will now prove Theorem C1 of §5.2.

Theorem C1 *Let (M, J) be a compact, complex manifold, and g a Kähler metric on M, with Kähler form ω. Let $Q_1 \geq 0$. Then there exist $Q_2, Q_3, Q_4 \geq 0$ depending only on M, J, g and Q_1, such that the following holds.*

Suppose $f \in C^3(M)$, $\phi \in C^5(M)$ and $A > 0$ satisfy the equations

$$\|f\|_{C^3} \leq Q_1, \quad \int_M \phi\, dV_g = 0, \quad \text{and} \quad (\omega + dd^c\phi)^m = A e^f \omega^m.$$

Then $\|\phi\|_{C^0} \leq Q_2$, $\|dd^c\phi\|_{C^0} \leq Q_3$ and $\|\nabla dd^c\phi\|_{C^0} \leq Q_4$.

The result was first proved by Yau [226, Prop. 2.1]. Parts of the proof (in particular, the a priori estimate of $\|\phi\|_{C^0}$ in §5.4.1) were simplified by Kazdan, Bourguignon and Aubin, and the treatment we will give is based on that in Aubin's book [17, p. 148–153]. The way the proof works is to make a sequence of a priori estimates of ϕ in different Banach norms. Each estimate depends on the previous estimate, and each estimate is slightly stronger than the last.

First an estimate of $\|\phi\|_{L^2}$ is found. Then we show that an estimate of $\|\phi\|_{L^p}$ leads to an estimate of $\|\phi\|_{L^{p\epsilon}}$, where $\epsilon = \frac{m}{m-1}$, and so by an induction argument we estimate $\|\phi\|_{L^p}$ for all p. Taking the limit as $p \to \infty$ yields an estimate for $\|\phi\|_{C^0}$, proving the first part of the theorem. Using this estimate we are then able to bound $\|\Delta\phi\|_{C^0}$, and from this we bound $\|dd^c\phi\|_{C^0}$. Finally we bound $\|\nabla dd^c\phi\|_{C^0}$ and the theorem follows.

For the rest of this section, we study the following situation. Let M be a compact Kähler manifold, with complex structure J, Kähler metric g and Kähler form ω. Let $f \in C^3(M)$ be a fixed function satisfying $\|f\|_{C^3} \leq Q_1$, where Q_1 is a given constant. For simplicity we shall assume, by adding a constant to f if necessary, that $\int_M e^f dV_g = \text{vol}_g(M)$, so that the constant A is 1. (As A can be estimated in terms of $\|f\|_{C^0}$, this assumption does not matter.) Let $\phi \in C^5(M)$ satisfy the equations

$$\int_M \phi\, dV_g = 0 \quad \text{and} \quad (\omega + dd^c\phi)^m = e^f \omega^m,$$

define $\omega' = \omega + dd^c\phi$, and let g' be the associated Kähler metric.

5.4.1 Estimates of order zero

We begin with the following proposition.

Proposition 5.4.1 *Let $p > 1$ be a real number. Then in the situation above,*

$$\int_M \left|\nabla |\phi|^{p/2}\right|_g^2 dV_g \leq \frac{mp^2}{4(p-1)} \int_M (1 - e^f)\phi|\phi|^{p-2} dV_g. \tag{5.17}$$

Proof As $(\omega')^m = e^f \omega^m$ and $\omega - \omega' = -dd^c\phi$, we see that

$$(1 - e^f)\omega^m = \omega^m - (\omega')^m = -dd^c\phi \wedge \left(\omega^{m-1} + \cdots + (\omega')^{m-1}\right). \tag{5.18}$$

Now M is compact, and so Stokes' Theorem shows that

$$\int_M d\left[\phi|\phi|^{p-2} d^c\phi \wedge \left(\omega^{m-1} + \omega^{m-2} \wedge \omega' + \cdots + (\omega')^{m-1}\right)\right] = 0. \tag{5.19}$$

Expanding this equation, remembering that $d\omega = d\omega' = 0$, and substituting in (5.18) and the equation $d(\phi|\phi|^{p-2}) = (p-1)|\phi|^{p-2}d\phi$, we find

$$\int_M \phi|\phi|^{p-2}(1 - e^f)\omega^m = (p-1)\int_M |\phi|^{p-2}d\phi \wedge d^c\phi \wedge$$
$$\left(\omega^{m-1} + \omega^{m-2} \wedge \omega' + \cdots + (\omega')^{m-1}\right).$$

Now Lemma 5.3.5 gives expressions for $d\phi \wedge d^c\phi \wedge \omega^{m-1}$ and $d\phi \wedge d^c\phi \wedge \omega^{m-j-1} \wedge (\omega')^j$, and substituting these in shows that

$$\int_M (1 - e^f)\phi|\phi|^{p-2}\omega^m = \frac{p-1}{m}\int_M |\phi|^{p-2}\left(|\nabla\phi|_g^2 + F_1 + \cdots + F_{m-1}\right)\omega^m,$$

where $F_1, \ldots, F_{m-1}$ are nonnegative functions on M. However, $\omega^m = m! \, dV_g$ by (4.4), so

$$\int_M |\phi|^{p-2}\left(|\nabla\phi|_g^2 + F_1 + \cdots + F_{m-1}\right)dV_g = \frac{m}{p-1}\int_M (1 - e^f)\phi|\phi|^{p-2}dV_g.$$

Combining this with the equation $\frac{1}{4}p^2|\phi|^{p-2}|\nabla\phi|_g^2 = \left|\nabla|\phi|^{p/2}\right|_g^2$ and the inequality $0 \leq |\phi|^{p-2}F_j$ gives (5.17), as we have to show. □

For the rest of this section, let $\epsilon = \frac{m}{m-1}$. Next we prove

Lemma 5.4.2 *There are constants C_1, C_2 depending on M and g, such that if $\psi \in L_1^2(M)$ then $\|\psi\|_{L^{2\epsilon}}^2 \leq C_1(\|\nabla\psi\|_{L^2}^2 + \|\psi\|_{L^2}^2)$, and if also $\int_M \psi \, dV_g = 0$ then $\|\psi\|_{L^2} \leq C_2\|\nabla\psi\|_{L^2}$.*

Proof By the Sobolev Embedding Theorem, Theorem 1.2.1, $L_1^2(M)$ is continuously embedded in $L^{2\epsilon}(M)$, and the first inequality of the lemma easily follows. Now consider the operator $d^*d : C^\infty(M) \to C^\infty(M)$. It is well-known that $\text{Ker}(d^*d)$ is the constant functions, and the nonzero eigenvalues of d^*d are positive and form a discrete spectrum. Let $\lambda_1 > 0$ be the smallest positive eigenvalue of d^*d. If $\psi \in C^\infty(M)$ satisfies $\int_M \psi \, dV_g = 0$, then ψ is L^2-orthogonal to $\text{Ker}(d^*d)$, so ψ is a sum of eigenvectors of d^*d with eigenvalues not less than λ_1.

It follows that $\langle\psi, d^*d\psi\rangle \geq \lambda_1\langle\psi, \psi\rangle$, so $\|d\psi\|_{L^2}^2 \geq \lambda_1\|\psi\|_{L^2}^2$ by integration by parts. As $C^\infty(M)$ is dense in $L_1^2(M)$, this inequality extends to $\psi \in L_1^2(M)$, by continuity. Thus if $\psi \in L_1^2(M)$ and $\int_M \psi \, dV_g = 0$ then $\|\psi\|_{L^2} \leq \lambda_1^{-1/2}\|\nabla\psi\|_{L^2}$, giving the second inequality of the lemma with $C_2 = \lambda_1^{-1/2}$. □

In the next two results we find a priori estimates of $\|\phi\|_{L^p}$ for $p \in [2, 2\epsilon]$ and $p \in [2, \infty)$ respectively.

Lemma 5.4.3 *There is a constant C_3 depending on M, g and Q_1 such that if $p \in [2, 2\epsilon]$ then $\|\phi\|_{L^p} \leq C_3$.*

Proof Putting $p = 2$ in Proposition 5.4.1 and using the inequality $|1 - e^f| \leq e^{Q_1}$, we see that $\|\nabla \phi\|_{L^2}^2 \leq m\, e^{Q_1} \|\phi\|_{L^1}$. As $\int_M \phi\, dV_g = 0$, Lemma 5.4.2 shows that $\|\phi\|_{L^2} \leq C_2 \|\nabla \phi\|_{L^2}$, and $\|\phi\|_{L^1} \leq \mathrm{vol}_g(M)^{1/2} \|\phi\|_{L^2}$ by Hölder's inequality. Combining these three shows that

$$\|\nabla \phi\|_{L^2}^2 \leq m\, C_2 e^{Q_1} \mathrm{vol}_g(M)^{1/2} \cdot \|\nabla \phi\|_{L^2},$$

and so $\|\nabla \phi\|_{L^2} \leq c$, where $c = m\, C_2 e^{Q_1} \mathrm{vol}_g(M)^{1/2}$. Therefore

$$\|\phi\|_{L^2} \leq c\, C_2 \quad \text{and so} \quad \|\phi\|_{L^{2\epsilon}}^2 \leq C_1(c^2 + c^2 C_2^2),$$

using Lemma 5.4.2. Define C_3 by $C_3 = \max\bigl(c\, C_2, c\, C_1^{1/2}(1 + C_2^2)^{1/2}\bigr)$. Then $\|\phi\|_{L^2} \leq C_3$ and $\|\phi\|_{L^{2\epsilon}} \leq C_3$, and it follows by Hölder's inequality that if $p \in [2, 2\epsilon]$ then $\|\phi\|_{L^p} \leq C_3$. □

Proposition 5.4.4 *There are constants Q_2, C_4 depending on M, g and Q_1 such that for each $p \geq 2$, we have $\|\phi\|_{L^p} \leq Q_2(C_4 p)^{-m/p}$.*

Proof Define a constant $C_4 > 0$ by $C_4 = C_1 \epsilon^{m-1}\bigl(m\, e^{Q_1} + \tfrac{1}{2}\bigr)$, and choose a constant $Q_2 > 0$ such that

$$\begin{aligned} Q_2 &\geq C_3(C_4 p)^{m/p} \quad \text{for } 2 \leq p \leq 2\epsilon, \text{ and} \\ Q_2 &\geq (C_4 p)^{m/p} \quad \text{for } 2 \leq p < \infty. \end{aligned} \quad (5.20)$$

Such a constant must exist, as $\lim_{p \to \infty}(C_4 p)^{m/p} = 1$.

We will prove the theorem by a form of induction on p. The first step is that if $p \in [2, 2\epsilon]$, then $\|\phi\|_{L^p} \leq C_3$ by Lemma 5.4.3, and $C_3 \leq Q_2(C_4 p)^{-m/p}$ by the first inequality of (5.20), and so $\|\phi\|_{L^p} \leq Q_2(C_4 p)^{-m/p}$ as we want. For the inductive step, suppose that $\|\phi\|_{L^p} \leq Q_2(C_4 p)^{-m/p}$ holds for all p with $2 \leq p \leq k$, where $k \geq 2\epsilon$. We shall show that $\|\phi\|_{L^q} \leq Q_2(C_4 q)^{-m/q}$ holds for all q with $2 \leq q \leq \epsilon k$, and therefore by induction, the inequality holds for all $p \in [1, \infty)$.

Let $p \in [2, k]$. Then $p^2/4(p-1) \leq p$ as $p \geq 2$, and since $|1 - e^f| \leq e^{Q_1}$ we see from Proposition 5.4.1 that $\|\nabla |\phi|^{p/2}\|_{L^2}^2 \leq mp\, e^{Q_1} \|\phi\|_{L^{p-1}}^{p-1}$. Applying Lemma 5.4.2 with $\psi = |\phi|^{p/2}$ gives $\|\phi\|_{L^{\epsilon p}}^p \leq C_1\bigl(\|\nabla |\phi|^{p/2}\|_{L^2}^2 + \|\phi\|_{L^p}^p\bigr)$. Combining these two equations shows that

$$\|\phi\|_{L^{\epsilon p}}^p \leq mp\, C_1 e^{Q_1} \|\phi\|_{L^{p-1}}^{p-1} + C_1 \|\phi\|_{L^p}^p.$$

Let $q = \epsilon p$. As $p \in [2, k]$ we have $\|\phi\|_{L^p} \leq Q_2(C_4 p)^{-m/p}$, and by the second part of (5.20) we see that $Q_2(C_4 p)^{-m/p} \geq 1$, and thus

$$\|\phi\|_{L^q}^p \leq Q_2^p (C_4 p)^{-m} (mp\, C_1 e^{Q_1} + C_1),$$
and also $\quad (Q_2(C_4 q)^{-m/q})^p = Q_2^p (C_4 p\epsilon)^{1-m}.$

However, as $p \geq 2$ the inequality $mp\, C_1 e^{Q_1} + C_1 \leq C_4 p \epsilon^{1-m}$ follows from the definition of C_4, so comparing the right hand sides of the above equations we see that $\|\phi\|_{L^q}^p \leq (Q_2(C_4 q)^{-m/q})^p$, and therefore $\|\phi\|_{L^q} \leq Q_2(C_4 q)^{-m/q}$. This holds for all $q \in [2\epsilon, \epsilon k]$, and the inductive step is complete. □

Now we can prove the first part of Theorem C1.

Corollary 5.4.5 *The function ϕ satisfies $\|\phi\|_{C^0} \leq Q_2$.*

Proof As ϕ is continuous on a compact manifold, $\|\phi\|_{C^0} = \lim_{p \to \infty} \|\phi\|_{L^p}$. But $\|\phi\|_{L^p} \leq Q_2(C_4 p)^{-m/p}$ by Proposition 5.4.4, and $\lim_{p \to \infty} (C_4 p)^{-m/p} = 1$, so the result follows. □

5.4.2 Second-order estimates

Here is some notation that will be used for the next calculations. We have two Kähler metrics g and g' on M. Let ∇ be the Levi-Civita connection of g. If T is a tensor on M, we will write $\nabla_{a_1 \cdots a_k} T$ in place of $\nabla_{a_1} \nabla_{a_2} \cdots \nabla_{a_k} T$, the kth derivative of T using ∇. This notation will be used together with the notation for complex tensors introduced in §4.2. Let $R^a{}_{bcd}$ be the Riemann curvature of g. Also, for $\psi \in C^2(M)$, let $\Delta \psi$ be the Laplacian of ψ with respect to g, and let $\Delta' \psi$ be the Laplacian of ψ with respect to g'. Then in this notation,

$$\Delta \psi = -g^{\alpha \bar{\beta}} \nabla_{\alpha \bar{\beta}} \psi \quad \text{and} \quad \Delta' \psi = -g'^{\alpha \bar{\beta}} \nabla_{\alpha \bar{\beta}} \psi.$$

The second formula defines the Laplacian with respect to g' using the Levi-Civita connection of g. It is valid because this component of $\nabla^2 \psi$ is independent of the choice of Kähler metric.

Using this notation we shall prove:

Proposition 5.4.6 *In the situation above, we have*

$$\Delta'(\Delta \phi) = -\Delta f + g^{\alpha \bar{\lambda}} g'^{\mu \bar{\beta}} g'^{\gamma \bar{\nu}} \nabla_{\alpha \bar{\beta} \gamma} \phi \, \nabla_{\bar{\lambda} \mu \bar{\nu}} \phi \qquad (5.21)$$
$$+ g'^{\alpha \bar{\beta}} g^{\gamma \bar{\delta}} \big(R^{\bar{\epsilon}}{}_{\bar{\delta} \gamma \bar{\beta}} \nabla_{\alpha \bar{\epsilon}} \phi - R^{\bar{\epsilon}}{}_{\bar{\beta} \alpha \bar{\delta}} \nabla_{\gamma \bar{\epsilon}} \phi \big).$$

Proof Taking the log of eqn (5.5) and applying ∇ gives $\nabla_{\bar{\lambda}} f = g'^{\mu \bar{\nu}} \nabla_{\bar{\lambda} \mu \bar{\nu}} \phi$. Therefore, as $\Delta f = -g^{\alpha \bar{\lambda}} \nabla_{\alpha \bar{\lambda}} f$, we have

$$\Delta f = -g^{\alpha \bar{\lambda}} (\nabla_\alpha g'^{\mu \bar{\nu}}) \nabla_{\bar{\lambda} \mu \bar{\nu}} \phi - g^{\alpha \bar{\lambda}} g'^{\mu \bar{\nu}} \nabla_{\alpha \bar{\lambda} \mu \bar{\nu}} \phi. \qquad (5.22)$$

But $g'^{\mu \bar{\beta}} g'_{\bar{\beta} \gamma} = \delta^\mu_\gamma$, and $\nabla_\alpha g'_{\bar{\beta} \gamma} = \nabla_{\alpha \bar{\beta} \gamma} \phi$ as $g'_{\bar{\beta} \gamma} = g_{\bar{\beta} \gamma} + \nabla_{\bar{\beta} \gamma} \phi$, so that

$$0 = \nabla_\alpha \delta^\mu_\gamma = g'_{\bar{\beta} \gamma} \nabla_\alpha g'^{\mu \bar{\beta}} + g'^{\mu \bar{\beta}} \nabla_{\alpha \bar{\beta} \gamma} \phi.$$

Contracting with $g'^{\gamma\bar{\nu}}$ shows that $\nabla_\alpha g'^{\mu\bar{\nu}} = -g'^{\mu\bar{\beta}} g'^{\gamma\bar{\nu}} \nabla_{\alpha\bar{\beta}\gamma}\phi$. Substituting this into (5.22) yields

$$\Delta f = g^{\alpha\bar{\lambda}} g'^{\mu\bar{\beta}} g'^{\gamma\bar{\nu}} \nabla_{\alpha\bar{\beta}\gamma}\phi \nabla_{\bar{\lambda}\mu\bar{\nu}}\phi - g^{\alpha\bar{\lambda}} g'^{\mu\bar{\nu}} \nabla_{\alpha\bar{\lambda}\mu\bar{\nu}}\phi, \tag{5.23}$$

and rearranging and changing some indices gives

$$g'^{\alpha\bar{\beta}} g^{\gamma\bar{\delta}} \nabla_{\gamma\bar{\delta}\alpha\bar{\beta}}\phi = -\Delta f + g^{\alpha\bar{\lambda}} g'^{\mu\bar{\beta}} g'^{\gamma\bar{\nu}} \nabla_{\alpha\bar{\beta}\gamma}\phi \nabla_{\bar{\lambda}\mu\bar{\nu}}\phi. \tag{5.24}$$

However, it can be shown that

$$g'^{\alpha\bar{\beta}} g^{\gamma\bar{\delta}} \nabla_{\alpha\bar{\beta}\gamma\bar{\delta}}\phi - g'^{\alpha\bar{\beta}} g^{\gamma\bar{\delta}} \nabla_{\gamma\bar{\delta}\alpha\bar{\beta}}\phi = g'^{\alpha\bar{\beta}} g^{\gamma\bar{\delta}} \left(R^{\bar{\epsilon}}_{\bar{\delta}\gamma\bar{\beta}} \nabla_{\alpha\bar{\epsilon}}\phi - R^{\bar{\epsilon}}_{\bar{\beta}\alpha\bar{\delta}} \nabla_{\gamma\bar{\epsilon}}\phi \right),$$

and this combines with (5.24) and $\Delta'\Delta\phi = g'^{\alpha\bar{\beta}} g^{\gamma\bar{\delta}} \nabla_{\alpha\bar{\beta}\gamma\bar{\delta}}\phi$ to give (5.21), as we want. $\square$

The next result gives the second part of Theorem C1.

Proposition 5.4.7 *There are constants c_1, c_2 and Q_3 depending on M, J, g and Q_1, such that*

$$\|g'_{ab}\|_{C^0} \leqslant c_1, \quad \|g'^{ab}\|_{C^0} \leqslant c_2 \quad \text{and} \quad \|dd^c\phi\|_{C^0} \leqslant Q_3. \tag{5.25}$$

Proof Define a function F on M by $F = \log(m - \Delta\phi) - \kappa\phi$, where κ is a constant to be chosen later. Note that $m - \Delta\phi > 0$ by Proposition 5.3.4, so F is well-defined. We shall find an expression for $\Delta'F$. It is easy to show that

$$\Delta'F = -(m - \Delta\phi)^{-1} \Delta'\Delta\phi + (m - \Delta\phi)^{-2} g'^{\alpha\bar{\lambda}} g^{\mu\bar{\beta}} g'^{\gamma\bar{\nu}} \nabla_{\alpha\bar{\beta}\gamma}\phi \nabla_{\bar{\lambda}\mu\bar{\nu}}\phi - \kappa\Delta'\phi.$$

Now $\Delta'\phi = -g'^{\alpha\bar{\beta}} \nabla_{\alpha\bar{\beta}}\phi = g'^{\alpha\bar{\beta}}(g_{\alpha\bar{\beta}} - g'_{\alpha\bar{\beta}}) = g'^{\alpha\bar{\beta}} g_{\alpha\bar{\beta}} - m$, so (5.21) and the equation above give

$$\Delta'F = (m - \Delta\phi)^{-1} \Delta f + \kappa\left(m - g'^{\alpha\bar{\beta}} g_{\alpha\bar{\beta}}\right) - (m - \Delta\phi)^{-1}(G + H), \tag{5.26}$$

where G and H are defined by

$$G = g^{\alpha\bar{\lambda}} g'^{\mu\bar{\beta}} g'^{\gamma\bar{\nu}} \nabla_{\alpha\bar{\beta}\gamma}\phi \nabla_{\bar{\lambda}\mu\bar{\nu}}\phi - (m - \Delta\phi)^{-1} g'^{\alpha\bar{\lambda}} g^{\mu\bar{\beta}} g'^{\gamma\bar{\nu}} \nabla_{\alpha\bar{\beta}\gamma}\phi \nabla_{\bar{\lambda}\mu\bar{\nu}}\phi,$$
$$H = g'^{\alpha\bar{\beta}} g^{\gamma\bar{\delta}} \left(R^{\bar{\epsilon}}_{\bar{\delta}\gamma\bar{\beta}} \nabla_{\alpha\bar{\epsilon}}\phi - R^{\bar{\epsilon}}_{\bar{\beta}\alpha\bar{\delta}} \nabla_{\gamma\bar{\epsilon}}\phi\right).$$

Now expanding the inequality

$$g^{\alpha\bar{\lambda}} g'^{\mu\bar{\beta}} g'^{\gamma\bar{\nu}} \left[(m - \Delta\phi) \nabla_{\alpha\bar{\beta}\gamma}\phi - g'_{\alpha\bar{\beta}} \nabla_\gamma\Delta\phi\right] \cdot \left[(m - \Delta\phi) \nabla_{\bar{\lambda}\mu\bar{\nu}}\phi - g'_{\bar{\lambda}\mu} \nabla_{\bar{\nu}}\Delta\phi\right] \geqslant 0$$

and dividing by $(m - \Delta\phi)^2$ shows that $G \geqslant 0$. Also, using the inequalities $|\nabla_{\alpha\bar{\beta}}\phi|_g \leqslant (m - \Delta\phi)$ and $|g'^{\alpha\bar{\beta}}|_g \leqslant g'^{\alpha\bar{\beta}} g_{\alpha\bar{\beta}}$ one can see that there is a constant $C_5 \geqslant 0$ depending on m and $\|R\|_{C^0}$ such that $|H| \leq C_5(m - \Delta\phi)g'^{\alpha\bar{\beta}}g_{\alpha\bar{\beta}}$. Substituting these inequalities and $\Delta f \leq Q_1$ into (5.26) gives

$$\Delta' F \leq (m - \Delta\phi)^{-1}Q_1 + \kappa\left(m - g'^{\alpha\bar{\beta}}g_{\alpha\bar{\beta}}\right) + C_5 g'^{\alpha\bar{\beta}}g_{\alpha\bar{\beta}}. \tag{5.27}$$

At a point p where F is maximum, $\Delta' F \geq 0$, and so by (5.27) at p we get

$$(\kappa - C_5)g'^{\alpha\bar{\beta}}g_{\alpha\bar{\beta}} \leq m\kappa + Q_1(m - \Delta\phi)^{-1}. \tag{5.28}$$

Rearranging equation (5.13) of Proposition 5.3.4 gives $m - \Delta\phi \geq m e^{f/m}$, and as $\|f\|_{C^0} \leq Q_1$ this shows that $(m - \Delta\phi)^{-1} \leq \frac{1}{m}e^{Q_1/m}$. Now choose $\kappa = C_5 + 1$. Then (5.28) implies that $g'^{\alpha\bar{\beta}}g_{\alpha\bar{\beta}} \leq C_6$ at p, where $C_6 = m\kappa + \frac{1}{m}Q_1 e^{Q_1/m}$.

Let us apply Lemma 5.3.1 to find expressions for g and g' at p in terms of positive constants $a_1, \ldots, a_m$. In this notation, using Lemma 5.3.2 we have

$$m - \Delta\phi = \sum_{j=1}^{m} a_j, \quad g'^{\alpha\bar{\beta}}g_{\alpha\bar{\beta}} = \sum_{j=1}^{m} a_j^{-1}, \quad \text{and} \quad \prod_{j=1}^{m} a_j = e^{f(p)}.$$

It easily follows that at p we have

$$m - \Delta\phi \leq e^{f(p)}\left(g'^{\alpha\bar{\beta}}g_{\alpha\bar{\beta}}\right)^{m-1} \leq e^{Q_1}C_6^{m-1}.$$

Therefore at a maximum of F, we see that $F \leq Q_1 + (m - 1)\log C_6 - \kappa \inf \phi$, and as $\|\phi\|_{C^0} \leq Q_2$ by Corollary 5.4.5 this shows that $F \leq Q_1 + (m - 1)\log C_6 + \kappa Q_2$ everywhere on M. Substituting for F and exponentiating gives

$$0 < m - \Delta\phi \leq C_6^{m-1}\exp(Q_1 + \kappa Q_2 + \kappa\phi) \leq C_6^{m-1}\exp(Q_1 + 2\kappa Q_2)$$

on M. This gives an a priori estimate for $\|\Delta\phi\|_{C^0}$. But Proposition 5.3.4 gives estimates for $\|g'_{ab}\|_{C^0}$, $\|g'^{ab}\|_{C^0}$ and $\|dd^c\phi\|_{C^0}$ in terms of upper bounds for $\|f\|_{C^0}$ and $\|\Delta\phi\|_{C^0}$, so (5.25) holds for some constants c_1, c_2 and Q_3. All the constants in this proof, including c_1, c_2 and Q_3, depend only on M, J, g and Q_1. $\square$

5.4.3 Third-order estimates

Define a function $S \geq 0$ on M by $4S^2 = |\nabla dd^c\phi|^2_{g'}$, so that in index notation

$$S^2 = g'^{\alpha\bar{\lambda}}g'^{\mu\bar{\beta}}g'^{\gamma\bar{\nu}}\nabla_{\alpha\bar{\beta}\gamma}\phi\,\nabla_{\bar{\lambda}\mu\bar{\nu}}\phi.$$

Our goal is to find an a priori upper bound for S, and this will be done by finding a formula for $\Delta'(S^2)$ and then using an argument similar to that used in Proposition 5.4.7. First, here is some notation that will be used for the next result. Suppose A, B, C are tensors on M. Let us write $P^{a,b,c}(A, B, C)$ for any polynomial in the tensors A, B, C alone, that is homogeneous of degree a in A, degree b in B and degree c in C.

Using this notation, we shall give in the next proposition an expression for $\Delta'(S^2)$. The proof of the proposition is a straightforward but rather long and tedious calculation, and we will omit it. Readers can find the calculation in [226, App. A] and [16, p. 410–411].

Proposition 5.4.8 *In the notation above, we have:*

$$-\Delta'(S^2) = \left|\nabla_{\bar{\alpha}\beta\bar{\gamma}\delta}\phi - g'^{\lambda\bar{\mu}}\nabla_{\bar{\alpha}\lambda\bar{\gamma}}\phi\nabla_{\beta\bar{\mu}\delta}\phi\right|^2_{g'}$$
$$+ \left|\nabla_{\alpha\beta\bar{\gamma}\delta}\phi - g'^{\lambda\bar{\mu}}\nabla_{\alpha\bar{\gamma}\lambda}\phi\nabla_{\beta\bar{\mu}\delta}\phi - g'^{\lambda\bar{\mu}}\nabla_{\alpha\bar{\mu}\delta}\phi\nabla_{\lambda\bar{\gamma}\beta}\phi\right|^2_{g'}$$
$$+ P^{4,2,1}\!\left(g'^{\alpha\bar{\beta}}, \nabla_{\alpha\bar{\beta}\gamma}\phi, \nabla_{\alpha\bar{\beta}}f\right) + P^{4,2,1}\!\left(g'^{\alpha\bar{\beta}}, \nabla_{\alpha\bar{\beta}\gamma}\phi, R^a{}_{bcd}\right)$$
$$+ P^{3,1,1}\!\left(g'^{\alpha\bar{\beta}}, \nabla_{\alpha\bar{\beta}\gamma}\phi, \nabla_{\bar{\alpha}\beta\bar{\gamma}}f\right) + P^{3,1,1}\!\left(g'^{\alpha\bar{\beta}}, \nabla_{\alpha\bar{\beta}\gamma}\phi, \nabla_e R^a{}_{bcd}\right). \tag{5.29}$$

In (5.29), we use the notation defined in (5.12) that for T a tensor on M, $|T|_{g'}$ is the modulus of T calculated with respect to the metric g'. Also, the four polynomials $P^{a,b,c}$ in (5.29) are not the same but different polynomials. Now there is a certain similarity between Proposition 5.4.8 and Proposition 5.4.6. Here is one way to see it.

In (5.21), the left hand side $\Delta'(\Delta\phi)$ involves $\nabla^4\phi$. However, the right hand side involves only $\nabla^2\phi$ and $\nabla^3\phi$, together with g', Δf and R. Moreover, the terms on the right hand side involving $\nabla^3\phi$ are nonnegative. In the same way, the left hand side of (5.29) involves $\nabla^5\phi$, but the right hand side involves only $\nabla^3\phi$ and $\nabla^4\phi$, and the terms on the right hand side involving $\nabla^4\phi$ are nonnegative.

So, both (5.21) and (5.29) express a derivative of ϕ in terms of lower derivatives of ϕ, and g', f and R, and the highest derivative of ϕ on the right hand side contributes only nonnegative terms. Now Proposition 5.4.7 used (5.21) to find an a priori bound for $\|\Delta\phi\|_{C^0}$. Because of the similarities between (5.21) and (5.29), we will be able to use the same method to find an a priori bound for S, and hence for $\|\nabla dd^c\phi\|_{C^0}$.

It follows quickly from (5.29) that

Corollary 5.4.9 *There is a constant C_7 depending on Q_1, c_1, c_2 and $\|R^a{}_{bcd}\|_{C^1}$, such that*

$$\Delta'(S^2) \leqslant C_7(S^2 + S).$$

Here c_1, c_2 are the constants from Proposition 5.4.7. To prove the corollary, observe that the first two terms on the right hand side of (5.29) are nonnegative, and can be neglected. Of the four terms of type $P^{a,b,c}$, the first two are quadratic in $\nabla_{\alpha\bar{\beta}\gamma}\phi$ and must be estimated by a multiple of S^2, and the second two are linear in $\nabla_{\alpha\bar{\beta}\gamma}\phi$ and must be estimated by a multiple of S. As $\|f\|_{C^3} \leqslant Q_1$, the factors of $\nabla_{\alpha\bar{\beta}}f$ and $\nabla_{\bar{\alpha}\beta\bar{\gamma}}f$ are estimated using Q_1, and by using the estimates (5.25) of Proposition 5.4.7 for g'_{ab} and g'^{ab}, the corollary quickly follows.

The next result, together with Corollary 5.4.5 and Proposition 5.4.7, completes the proof of Theorem C1.

Proposition 5.4.10 *There exists Q_4 depending only on M, J, g and Q_1, with $\|\nabla dd^c\phi\|_{C^0} \leqslant Q_4$.*

Proof Using the formulae for g' in §5.3, it is easy to show that

$$g^{\alpha\bar{\lambda}}g'^{\mu\bar{\beta}}g'^{\gamma\bar{\nu}}\nabla_{\alpha\bar{\beta}\gamma}\phi\nabla_{\bar{\lambda}\mu\bar{\nu}}\phi \geqslant c_2^{-1} S^2,$$

where $c_2 > 0$ comes from Proposition 5.4.7. Therefore, from Proposition 5.4.6 it follows that

$$\Delta'(\Delta\phi) \geq c_2^{-1} S^2 - C_8, \tag{5.30}$$

where C_8 depends on M, J, g and Q_1. Consider the function $S^2 - 2c_2 C_7 \Delta\phi$ on M. From Corollary 5.4.9 and (5.30) it follows that

$$\Delta'(S^2 - 2c_2 C_7 \Delta\phi) \leq C_7(S^2 + S) - 2c_2 C_7 (c_2^{-1} S^2 - C_8)$$
$$= -C_7(S - \tfrac{1}{2})^2 + 2c_2 C_7 C_8 + \tfrac{1}{4} C_7.$$

At a maximum point p of $S^2 - 2c_2 C_7 \Delta\phi$, we have $\Delta'(S^2 - 2c_2 C_7 \Delta\phi) \geq 0$, and therefore

$$(S - \tfrac{1}{2})^2 \leq 2c_2 C_8 + \tfrac{1}{4}$$

at p. So, there is a constant $C_9 > 0$ depending on c_2, C_8 and the a priori estimate for $\|\Delta\phi\|_{C^0}$ found in Proposition 5.4.7, such that $S^2 - 2c_2 C_7 \Delta\phi \leq C_9$ at p. As p is a maximum, $S^2 - 2c_2 C_7 \Delta\phi \leq C_9$ holds on M. Using the estimate on $\|\Delta\phi\|_{C^0}$ again, we find an a priori estimate for $\|S\|_{C^0}$.

Now $2S = |\nabla dd^c\phi|_{g'}$. It can be seen that $|\nabla dd^c\phi|_g \leq c_1^{3/2} |\nabla dd^c\phi|_{g'}$, where c_1 comes from Proposition 5.4.7. Thus, an a priori bound for $\|S\|_{C^0}$ gives one for $\|\nabla dd^c\phi\|_{C^0}$, and there is a constant Q_4 depending on M, J, g and Q_1 such that $\|\nabla dd^c\phi\|_{C^0} \leq Q_4$, which completes the proof. □

5.5 The proof of Theorem C2

Here is the proof of Theorem C2 of §5.2.

Theorem C2 *Let (M, J) be a compact, complex manifold, and g a Kähler metric on M, with Kähler form ω. Let $Q_1, \ldots, Q_4 \geq 0$ and $\alpha \in (0, 1)$. Then there exists $Q_5 \geq 0$ depending only on $M, J, g, Q_1, \ldots, Q_4$ and α, such that the following holds.*

Suppose $f \in C^{3,\alpha}(M)$, $\phi \in C^5(M)$ and $A > 0$ satisfy $(\omega + dd^c\phi)^m = Ae^f \omega^m$ and the inequalities

$$\|f\|_{C^{3,\alpha}} \leq Q_1, \quad \|\phi\|_{C^0} \leq Q_2, \quad \|dd^c\phi\|_{C^0} \leq Q_3 \quad \text{and} \quad \|\nabla dd^c\phi\|_{C^0} \leq Q_4.$$

Then $\phi \in C^{5,\alpha}(M)$ and $\|\phi\|_{C^{5,\alpha}} \leq Q_5$. Also, if $f \in C^{k,\alpha}(M)$ for $k \geq 3$ then $\phi \in C^{k+2,\alpha}(M)$, and if $f \in C^\infty(M)$ then $\phi \in C^\infty(M)$.

We shall continue to use most of the notation of §5.4. In particular, the metrics g and g' and the Levi-Civita connection ∇ of g will be the same, repeated derivatives using ∇ will be written $\nabla_{a_1 \ldots a_k}$, and the operators $\Delta = g^{\alpha\bar\beta}\nabla_{\alpha\bar\beta}$ and $\Delta' = g'^{\alpha\bar\beta}\nabla_{\alpha\bar\beta}$ will be used. All norms in this section will be with respect to g.

Let us begin by stating three elliptic regularity results for Δ and Δ', which come from §1.4. Lemma 5.5.1 follows from (1.8) of Theorem 1.4.1. Lemmas 5.5.2 and 5.5.3 follow from estimates (1.13) and (1.14) of Theorem 1.4.3 respectively. To prove Lemmas 5.5.2 and 5.5.3, we cover M by a finite number of overlapping coordinate patches identified with the unit ball in $\mathbb{R}^{2m}$, using the argument described in §1.4.1.

Lemma 5.5.1 *Let $k \geqslant 0$ and $\alpha \in (0, 1)$. Then there exists a constant $E_{k,\alpha} > 0$ depending on k, α, M and g, such that if $\psi \in C^2(M)$ and $\xi \in C^{k,\alpha}(M)$ and $\Delta \psi = \xi$, then $\psi \in C^{k+2,\alpha}(M)$ and $\|\psi\|_{C^{k+2,\alpha}} \leqslant E_{k,\alpha}(\|\xi\|_{C^{k,\alpha}} + \|\psi\|_{C^0})$.*

Lemma 5.5.2 *Let $\alpha \in (0, 1)$. Then there exists a constant $E'_\alpha > 0$ depending on α, M, g and the norms $\|g'_{ab}\|_{C^0}$ and $\|g'^{ab}\|_{C^{0,\alpha}}$, such that if $\psi \in C^2(M)$ and $\xi \in C^0(M)$ and $\Delta'\psi = \xi$, then $\psi \in C^{1,\alpha}(M)$ and $\|\psi\|_{C^{1,\alpha}} \leqslant E'_\alpha(\|\xi\|_{C^0} + \|\psi\|_{C^0})$.*

Lemma 5.5.3 *Let $k \geqslant 0$ be an integer, and $\alpha \in (0, 1)$. Then there exists a constant $E'_{k,\alpha} > 0$ depending on k, α, M, g and the norms $\|g'_{ab}\|_{C^0}$ and $\|g'^{ab}\|_{C^{k,\alpha}}$, such that if $\psi \in C^2(M)$ and $\xi \in C^{k,\alpha}(M)$ and $\Delta'\psi = \xi$, then $\psi \in C^{k+2,\alpha}(M)$ and $\|\psi\|_{C^{k+2,\alpha}} \leqslant E'_{k,\alpha}(\|\xi\|_{C^{k,\alpha}} + \|\psi\|_{C^0})$.*

The proof of the theorem is based on eqn (5.21) of Proposition 5.4.6, which we reproduce here.

$$\Delta'(\Delta\phi) = -\Delta f + g^{\alpha\bar{\lambda}} g'^{\mu\bar{\beta}} g'^{\gamma\bar{\nu}} \nabla_{\alpha\bar{\beta}\gamma}\phi \nabla_{\bar{\lambda}\mu\bar{\nu}}\phi \\ + g'^{\alpha\bar{\beta}} g^{\gamma\bar{\delta}} \big(R^{\bar{\epsilon}}_{\bar{\delta}\gamma\bar{\beta}} \nabla_{\alpha\bar{\epsilon}}\phi - R^{\bar{\epsilon}}_{\bar{\beta}\alpha\bar{\delta}} \nabla_{\gamma\bar{\epsilon}}\phi \big). \quad (5.31)$$

Applying the three lemmas above to this equation, we shall prove the theorem by an inductive process known as 'bootstrapping'. First we will find an a priori estimate for $\|\phi\|_{C^{3,\alpha}}$.

Proposition 5.5.4 *There is a constant D_1 depending on $M, g, J, Q_1, \ldots, Q_4$ and α such that $\|\phi\|_{C^{3,\alpha}} \leqslant D_1$.*

Proof In this proof, all estimates and all constants will depend only on $M, g, J, Q_1, \ldots, Q_4$ and α. By Proposition 5.4.7, the estimate $\|dd^c\phi\|_{C^0} \leqslant Q_3$ implies the estimates $\|g'_{ab}\|_{C^0} \leqslant c_1$ and $\|g'^{ab}\|_{C^0} \leqslant c_2$. Also, $\nabla g'^{ab}$ can be expressed in terms of $\nabla dd^c\phi$, g'^{ab} and J, so that the estimates for $\|g'^{ab}\|_{C^0}$ and $\|\nabla dd^c\phi\|_{C^0}$ yield an estimate for $\|\nabla g'^{ab}\|_{C^0}$. Combining the estimates for $\|g'^{ab}\|_{C^0}$ and $\|\nabla g'^{ab}\|_{C^0}$, we can find an estimate for $\|g'^{ab}\|_{C^{0,\alpha}}$.

Thus there are a priori estimates for $\|g'_{ab}\|_{C^0}$ and $\|g'^{ab}\|_{C^{0,\alpha}}$. Therefore, Lemma 5.5.2 holds with a constant E'_α depending on $M, g, J, Q_1, \ldots, Q_4$ and α. Put $\psi = \Delta\phi$, and let ξ be the right hand side of (5.31), so that $\Delta\psi = \xi$. Now, combining a priori estimates of the C^0 norms of g'^{ab}, $\nabla_{\alpha\bar{\beta}}\phi$ and $\nabla_{\alpha\bar{\beta}\gamma}\phi$, the inequality $\|f\|_{C^2} \leqslant Q_1$ and a bound for R^a_{bcd}, we can find a constant D_2 such that $\|\xi\|_{C^0} \leqslant D_2$.

So, by Lemma 5.5.2, $\Delta\phi = \psi$ lies in $C^{1,\alpha}(M)$ and $\|\Delta\phi\|_{C^{1,\alpha}} \leqslant E'_\alpha(D_2 + Q_3)$, as $\|\Delta\phi\|_{C^0} \leqslant \|dd^c\phi\|_{C^0} \leqslant Q_3$. Therefore by Lemma 5.5.1, $\phi \in C^{3,\alpha}(M)$ and

$$\|\phi\|_{C^{3,\alpha}} \leqslant E_{1,\alpha}(\|\Delta\phi\|_{C^{1,\alpha}} + \|\phi\|_{C^0}) \leqslant E_{1,\alpha}(E'_\alpha(D_2 + Q_3) + Q_2).$$

Thus, putting $D_1 = E_{1,\alpha}(E'_\alpha(D_2 + Q_3) + Q_2)$, the proposition is complete. □

Theorem C2 will follow from the next proposition.

Proposition 5.5.5 *For each $k \geqslant 3$, if $f \in C^{k,\alpha}(M)$ then $\phi \in C^{k+2,\alpha}(M)$, and there exists an a priori bound for $\|\phi\|_{C^{k+2,\alpha}}(M)$ depending on $M, g, J, Q_1, \ldots, Q_4, k, \alpha$, and a bound for $\|f\|_{C^{k,\alpha}}$.*

Proof The proof is by induction on k. The result is stated for $k \geq 3$ only, because Theorem C1 uses $\|f\|_{C^3}$ to bound $\|\nabla \mathrm{dd}^c\phi\|_{C^0}$. However, it is convenient to start the induction at $k = 2$. In this proof, we say a constant 'depends on the k-data' if it depends only on $M, g, J, Q_1, \ldots, Q_4, k, \alpha$ and bounds for $\|f\|_{C^{k,\alpha}}$ and $\|f\|_{C^3}$. Our inductive hypothesis is that $f \in C^{k,\alpha}(M)$ and $\phi \in C^{k+1,\alpha}(M)$, and that there is an a priori bound for $\|\phi\|_{C^{k+1,\alpha}}$ depending on the $(k-1)$-data. By Proposition 5.5.4, this holds for $k = 2$, and this is the first step in the induction.

Write $\psi = \Delta\phi$, and let ξ be the right hand side of (5.31), so that $\Delta'\psi = \xi$. Now the term $-\Delta f$ on the right hand side of (5.31) is bounded in $C^{k-2,\alpha}(M)$ by $\|f\|_{C^{k,\alpha}}$, and hence in terms of the k-data. It is easy to see that every other term on the right hand side of (5.31) can be bounded in $C^{k-2,\alpha}(M)$ in terms of M, g, J and bounds for $\|g'^{ab}\|_{C^0}$ and $\|\phi\|_{C^{k+1,\alpha}}$. By the inductive hypothesis, these are all bounded in terms of the $(k-1)$-data. Therefore, we can find a bound $F_{k,\alpha}$ depending on the k-data, such that $\|\xi\|_{C^{k-2,\alpha}} \leq F_{k,\alpha}$.

Using the inductive hypothesis again, we may bound $\|g'_{ab}\|_{C^0}$ and $\|g'^{ab}\|_{C^{k,\alpha}}$ in terms of the $(k-1)$-data. Therefore we may apply Lemma 5.5.3 to $\psi = \Delta\phi$, which shows that $\Delta\phi \in C^{k,\alpha}(M)$ and

$$\|\Delta\phi\|_{C^{k,\alpha}} \leq E'_{k-2,\alpha}\big(\|\xi\|_{C^{k-2,\alpha}} + \|\Delta\phi\|_{C^0}\big) \leq E'_{k-2,\alpha}\big(F_{k,\alpha} + Q_3\big),$$

since $\|\Delta\phi\|_{C^0} \leq Q_3$, where $E'_{k-2,\alpha}$ depends on the $(k-1)$-data. Thus by Lemma 5.5.1, $\phi \in C^{k+2,\alpha}(M)$ as we have to prove, and

$$\|\phi\|_{C^{k+2,\alpha}} \leq E_{k,\alpha}\big(\|\Delta\phi\|_{C^{k,\alpha}} + \|\phi\|_{C^0}\big) \leq E_{k,\alpha}\big(E'_{k-2,\alpha}(F_{k,\alpha} + Q_3) + Q_2\big),$$

since $\|\phi\|_{C^0} \leq Q_2$. This is an a priori bound for $\|\phi\|_{C^{k+2,\alpha}}$ depending only on the k-data. Therefore by induction, the result holds for all k. $\square$

Now, to prove Theorem C2, first put $k = 3$ in Proposition 5.5.5. This shows that $\phi \in C^{5,\alpha}(M)$ and gives an a priori bound for $\|\phi\|_{C^{5,\alpha}}$. Let Q_5 be this bound. Then $\|\phi\|_{C^{5,\alpha}} \leq Q_5$, and Q_5 depends only on $M, J, g, Q_1, \ldots, Q_4$ and α, since Q_1 is a bound for $\|f\|_{C^{3,\alpha}}$. Also, if $k \geq 3$ and $f \in C^{k,\alpha}(M)$, Proposition 5.5.5 shows that $\phi \in C^{k+2,\alpha}(M)$. Since this holds for all k, if $f \in C^{\infty}(M)$ then $\phi \in C^{\infty}(M)$, and the theorem is proved.

5.6 The proof of Theorem C3

Now we prove Theorem C3 of §5.2.

Theorem C3 *Let (M, J) be a compact complex manifold, and g a Kähler metric on M, with Kähler form ω. Fix $\alpha \in (0, 1)$, and suppose that $f' \in C^{3,\alpha}(M)$, $\phi' \in C^{5,\alpha}(M)$ and $A' > 0$ satisfy the equations*

$$\int_M \phi' \, \mathrm{d}V_g = 0 \quad \text{and} \quad (\omega + \mathrm{dd}^c\phi')^m = A'e^{f'}\omega^m. \tag{5.32}$$

Then whenever $f \in C^{3,\alpha}(M)$ and $\|f - f'\|_{C^{3,\alpha}}$ is sufficiently small, there exist $\phi \in C^{5,\alpha}(M)$ and $A > 0$ such that

$$\int_M \phi \, dV_g = 0 \quad \text{and} \quad (\omega + dd^c \phi)^m = A e^f \omega^m. \tag{5.33}$$

Proof Define X to be the vector subspace of $\phi \in C^{5,\alpha}(M)$ for which $\int_M \phi \, dV_g = 0$. Let U be the subset of $\phi \in X$ such that $\omega + dd^c \phi$ is a *positive* (1,1)-form on M. Then U is open in X. Let Y be the Banach space $C^{3,\alpha}(M)$. Suppose that $\phi \in U$ and $a \in \mathbb{R}$. Then $\omega + dd^c \phi$ is a positive (1,1)-form, so that $(\omega + dd^c \phi)^m$ is a positive multiple of ω^m at each point. Therefore there exists a unique real function f on M such that $(\omega + dd^c \phi)^m = e^{a+f} \omega^m$ on M, and as $\phi \in C^{5,\alpha}(M)$, it follows that $f \in C^{3,\alpha}(M)$.

Define a function $F : U \times \mathbb{R} \to Y$ by $F(\phi, a) = f$, where $(\omega + dd^c \phi)^m = e^{a+f} \omega^m$ on M. We have just shown that F is well-defined, and it is easy to see that F is a smooth map of Banach spaces. Now let f', ϕ' and A' be as in the theorem, and let $a' = \log A'$. By (5.32), ϕ' lies in U and $F(\phi', a') = f'$. We shall evaluate the first derivative $dF_{(\phi', a')}$ of F at (ϕ', a'). Calculation shows that

$$\bigl(\omega + dd^c(\phi' + \epsilon \psi)\bigr)^m = \exp\bigl(a' + \epsilon b + f_\epsilon\bigr) \omega^m,$$

where $f_\epsilon = f' - \epsilon b - \epsilon \Delta' \psi + O(\epsilon^2)$. Here g' is the $C^{3,\alpha}$ Kähler metric with Kähler form $\omega + dd^c \phi'$, and $\Delta' = -g'^{\alpha \bar\beta} \nabla_{\alpha \bar\beta}$ is the Laplacian with respect to g'. Therefore $F(\phi' + \epsilon \psi, a' + \epsilon b) = f_\epsilon$. From the first order term in ϵ in the expression for f_ϵ, we see that the first derivative of F is

$$dF_{(\phi', a')} : X \times \mathbb{R} \to Y, \quad \text{given by} \quad dF_{(\phi', a')}(\psi, b) = -b - \Delta' \psi. \tag{5.34}$$

Now the operator $\Delta' : C^{5,\alpha}(M) \to C^{3,\alpha}(M)$ is a linear elliptic operator of order 2 with $C^{3,\alpha}$ coefficients. As M is connected, the kernel $\operatorname{Ker} \Delta'$ is the set of constant functions on M. Calculated with respect to the metric g, the dual $(\Delta')^*$ of Δ' is given by $(\Delta')^* \psi = e^{f'} \Delta'(e^{-f'} \psi)$. Therefore $\operatorname{Ker}(\Delta')^*$ is the set of constant multiples of $e^{-f'}$. It follows that $\psi \perp \operatorname{Ker} \Delta'$ if and only if $\int_M \psi \, dV_g = 0$, and $\chi \perp \operatorname{Ker}(\Delta')^*$ if and only if $\langle \chi, e^{-f'} \rangle = 0$.

Let us apply Theorem 1.5.4 of §1.5 to Δ'. The theorem shows that if $\chi \in C^{3,\alpha}(M)$, then there exists $\psi \in C^{5,\alpha}(M)$ with $\Delta' \psi = f$ if and only if $\langle \chi, e^{-f'} \rangle = 0$. Also, if in addition we require that $\int_M \psi \, dV_g = 0$, then ψ is unique. Let $\chi \in Y = C^{3,\alpha}(M)$. Then there is a unique $b \in \mathbb{R}$ such that $\langle \chi + b, e^{-f'} \rangle = 0$. Hence, there exists a unique $\psi \in C^{5,\alpha}(M)$ with $\int_M \psi \, dV_g = 0$, such that $\Delta' \psi = -b - \chi$. But then $\psi \in X$, and (5.34) shows that $dF_{(\phi', a')}(\psi, b) = \chi$.

Thus, if $\chi \in Y$, there exist unique $\psi \in X$ and $b \in \mathbb{R}$, such that $dF_{(\phi', a')}(\psi, b) = \chi$. So $dF_{(\phi', a')} : X \times \mathbb{R} \to Y$ is an *invertible* linear map. It is continuous and has continous inverse, and thus is an isomorphism of $X \times \mathbb{R}$ and Y as both vector spaces and topological spaces. Therefore, applying the Inverse Mapping Theorem for Banach spaces, Theorem 1.2.4, there is an open neighbourhood $U' \subset U \times \mathbb{R}$ of (ϕ', a') in $X \times \mathbb{R}$ and an open neighbourhood $V' \subset Y$ of f' in Y, such that $F : U' \to V'$ is a homeomorphism.

So, whenever $f \in C^{3,\alpha}(M) = Y$ and $\|f - f'\|_{C^{3,\alpha}}$ is sufficiently small, we have $f \in V'$, and there is a unique $(\phi, a) \in U'$ with $F(\phi, a) = f$. Since $\phi \in X$, the first equation of (5.33) holds, and $\phi \in C^{5,\alpha}(M)$. Putting $A = e^a$ gives $A > 0$, and as

$F(\phi, a) = f$, the second equation of (5.33) holds by definition of F. This completes the proof of Theorem C3. □

5.7 The proof of Theorem C4

We now prove Theorem C4 of §5.2, which completes our proof of the Calabi conjecture. Note that the proof is very similar to the proof of Proposition 5.4.1.

Theorem C4 *Let (M, J) be a compact complex manifold, and g a Kähler metric on M, with Kähler form ω. Let $f \in C^1(M)$. Then there is at most one function $\phi \in C^3(M)$ such that $\int_M \phi \, dV_g = 0$, and $(\omega + dd^c \phi)^m = Ae^f \omega^m$ on M.*

Proof Suppose $\phi_1, \phi_2 \in C^3(M)$ satisfy $\int_M \phi_j \, dV_g = 0$ and $(\omega + dd^c \phi_j)^m = Ae^f \omega^m$ for $j = 1, 2$. We shall show that $\phi_1 = \phi_2$, so that any solution ϕ of these two equations is unique in $C^3(M)$. Write $\omega_1 = \omega + dd^c \phi_1$ and $\omega_2 = \omega + dd^c \phi_2$. By Lemma 5.1.1, both ω_1 and ω_2 are positive (1,1)-forms. Let g_1 and g_2 be the C^1 Kähler metrics associated to ω_1 and ω_2.

As $\omega_1^m = Ae^f \omega^m = \omega_2^m$ and $\omega_1 - \omega_2 = dd^c(\phi_1 - \phi_2)$, we see that

$$0 = \omega_1^m - \omega_2^m = dd^c(\phi_1 - \phi_2) \wedge \left(\omega_1^{m-1} + \omega_1^{m-2} \wedge \omega_2 + \cdots + \omega_2^{m-1} \right).$$

Now M is compact, and so Stokes' Theorem gives

$$\int_M d\left[(\phi_1 - \phi_2) d^c(\phi_1 - \phi_2) \wedge \left(\omega_1^{m-1} + \cdots + \omega_2^{m-1} \right) \right] = 0. \tag{5.35}$$

Combining the previous two equations and using $d\omega_1 = d\omega_2 = 0$, we find

$$\int_M d(\phi_1 - \phi_2) \wedge d^c(\phi_1 - \phi_2) \wedge \left(\omega_1^{m-1} + \cdots + \omega_2^{m-1} \right) = 0. \tag{5.36}$$

Following the proof of Lemma 5.3.5, we can prove that

$$d(\phi_1 - \phi_2) \wedge d^c(\phi_1 - \phi_2) \wedge \omega_1^{m-1} = \frac{1}{m} \left| d(\phi_1 - \phi_2) \right|_{g_1}^2 \omega_1^m$$

and $\quad d(\phi_1 - \phi_2) \wedge d^c(\phi_1 - \phi_2) \wedge \omega_1^{m-1-j} \wedge \omega_2^j = F_j \omega_1^m$,

where F_j is a nonnegative function on M. Substituting these into (5.36), and using the fact that $d(\phi_1 - \phi_2)$ is continuous, we deduce that $d(\phi_1 - \phi_2) = 0$. Since M is connected, $\phi_1 - \phi_2$ is constant. But $\int_M \phi_1 \, dV_g = 0 = \int_M \phi_2 \, dV_g$, so $\phi_1 - \phi_2 = 0$, and $\phi_1 = \phi_2$. This completes the proof. □

5.8 A discussion of the proof

One question that preoccupied the author during the writing of this chapter is: can the proof be simplified? In particular, is it really necessary to find a priori estimates for the first three derivatives of ϕ, and must one assume that ϕ is five times differentiable, or can one work with fewer derivatives? Here are the reasons for using this many derivatives of ϕ and f, and for choosing the Banach spaces such as $C^{5,\alpha}(M)$ that appear in the proof.

Clearly ϕ must be at least twice differentiable, or else the equation (5.5) that ϕ must satisfy makes no sense. So one must find a priori estimates for at least the first two derivatives of ϕ. However, it is less obvious that one really needs the third-order estimates of §5.4.3. There are results that show that if ϕ is a C^2 solution of a smooth, nonlinear, elliptic, second-order differential equation, then ϕ is C^∞. For example, this follows from Morrey [165, Th. 6.8.1]. These suggest that one could make do with only the second-order estimates in §5.4.2.

However, a careful examination shows that Morrey's results, and others of the same kind, tend to rely not only on bounds on $\nabla^2\phi$, but also on a *modulus of continuity* for $\nabla^2\phi$. Every continuous function on a compact manifold has a modulus of continuity, which is a real function that bounds how quickly the function varies from point to point. To apply these smoothness results in our situation would require an a priori bound for the modulus of continuity of $dd^c\phi$. Finding one is probably no easier than making the third-order estimates of §5.4.3, and having made them we are then able to complete the proof using more elementary results on linear elliptic equations.

Now, about the spaces used in the proofs. In Theorem C1, we need $\phi \in C^5(M)$ so that $\Delta'(S^2)$ should exist in Proposition 5.4.8 of §5.4.3. Theorems C2 and C3 require $\phi \in C^{5,\alpha}(M)$ for several reasons: ϕ must be in $C^5(M)$ for Theorem C1, Hölder spaces have good elliptic regularity properties, and the embedding $C^{5,\alpha}(M) \hookrightarrow C^5(M)$ is compact, which is used in Theorem 5.2.2. In Theorem C4, we take $\phi \in C^3(M)$ because in applying Stokes' Theorem in §5.7 we suppose that $d\omega_1 = d\omega_2 = 0$, which involves the third derivatives of ϕ.

Here is another issue. In §5.4.2, the bound computed for $\|dd^c\phi\|_{C^0}$ depends on $\|f\|_{C^2}$, that is, on C^0 bounds for f and Δf. However, if $\phi \in C^2(M)$, it is only necessary that $f \in C^0(M)$, and experience with linear elliptic equations suggests that if f is bounded in $C^{0,\alpha}(M)$, then ϕ should be bounded in $C^{2,\alpha}(M)$. So, the proof seems to use an unnecessarily strong bound on f. Similarly, the third-order estimates for ϕ in §5.4.3 involve $\|f\|_{C^3}$, when one would expect to use only $\|f\|_{C^{1,\alpha}}$. I do not know whether it is possible to make the a priori estimates of ϕ without using these strong bounds on f, but I suspect the answer may be no.

6
CALABI–YAU MANIFOLDS

In this chapter we study compact Riemannian manifolds with holonomy $SU(m)$. A *Calabi–Yau manifold* is a compact Kähler manifold (M, J, g) of dimension $m \geqslant 2$, with $\text{Hol}(g) = SU(m)$. Calabi–Yau manifolds are Ricci-flat. They are called Calabi–Yau manifolds because one can use Yau's solution of the Calabi conjecture to prove the existence of Kähler metrics g with holonomy $SU(m)$ on suitable compact complex manifolds (M, J).

The holonomy groups $Sp(m)$ for $m \geqslant 1$ are very closely related to the holonomy groups $SU(m)$, since $Sp(m)$ is a subgroup of $SU(2m)$. Thus any metric g with $\text{Hol}(g) = Sp(m)$ is also Kähler and Ricci-flat. The inclusion $Sp(m) \subseteq SU(2m)$ is proper for $m > 1$, but when $m = 1$ we find that $SU(2) = Sp(1)$, so the holonomy groups $SU(2)$ and $Sp(1)$ are the same.

The holonomy groups $Sp(m)$ will be the subject of Chapter 7, so we will not say much about them in this chapter. Also, we have chosen to leave nearly all our discussion of the holonomy group $SU(2) = Sp(1)$ until Chapter 7, because it fits in better with the material there. In particular, we postpone the treatment of $K3$ surfaces (compact manifolds with holonomy $SU(2)$) until §7.3. This means that much of the focus of this chapter is on the holonomy group $SU(3)$.

If (M, J, g) is a Calabi–Yau manifold of dimension $m \geqslant 3$, then the complex manifold (M, J) is a projective manifold with $c_1(M) = 0$. Thus the complex manifolds underlying Calabi–Yau manifolds are algebraic objects, and it is natural to study them using *complex algebraic geometry*. Projective manifolds exist in huge numbers, and the condition $c_1(M) = 0$ is comparatively easy to test for, especially for complete intersections. Many examples of Calabi–Yau manifolds can be constructed simply by considering some class of projective manifolds, and calculating which ones satisfy $c_1(M) = 0$.

Although much of the chapter is algebraic geometry, a rather technical subject, we will try to present it in an elementary way and keep the technicalities to a minimum. One thing we will not say much about is what the metrics on the Calabi–Yau manifolds are like: here we are mostly interested in the underlying complex manifolds. In fact little is known about general Calabi–Yau metrics.

The subject of Calabi–Yau manifolds is too big to be properly described in one chapter of a book. We aim to cover three main areas in detail: firstly, the basic differential geometry of Calabi–Yau manifolds; secondly, orbifolds and crepant resolutions; and thirdly, ways of constructing Calabi–Yau manifolds. This choice of topics is dictated in part by the needs of the rest of the book, because the material on orbifolds and crepant resolutions will form a central part of later chapters on the holonomy groups G_2 and $Spin(7)$. There are other important areas, such as rational curves on Calabi–Yau

3-folds, that we shall not have space to discuss. Some suggestions for further reading will be given at the end of the chapter.

Sections 6.1 and 6.2 introduce the holonomy groups $SU(m)$, and the differential geometry of compact Ricci-flat Kähler manifolds. Then §6.3–§6.6 discuss a special class of resolutions of singular complex manifolds, called *crepant resolutions*. Their importance to Calabi–Yau geometry is that if X is a singular Calabi–Yau manifold and $(\tilde{X}, \pi)$ a Kähler crepant resolution of X, then $\tilde{X}$ is a nonsingular Calabi–Yau manifold. This plays a part in a number of constructions of Calabi–Yau manifolds. We focus in particular on crepant resolutions of *complex orbifolds*.

Section 6.7 describes constructions of Calabi–Yau manifolds, and Calabi–Yau 3-folds in particular, using algebraic geometry. For example, a nonsingular hypersurface of degree $m+1$ in $\mathbb{CP}^m$ is a Calabi–Yau manifold. By replacing $\mathbb{CP}^m$ by a weighted projective space, or more generally by a compact toric variety, one can construct many examples of Calabi–Yau manifolds.

Next, §6.8 explains the deformation theory of the complex manifolds underlying Calabi–Yau manifolds. Section 6.9 gives a very quick introduction to *String Theory*, a branch of theoretical physics that involves Calabi–Yau manifolds, and *Mirror Symmetry*, an exciting set of ideas and conjectures about Calabi–Yau manifolds that has emerged from String Theory. Finally, §6.10 gives a reading list on various topics to do with Calabi–Yau manifolds.

6.1 The holonomy groups $SU(m)$

Identify $\mathbb{R}^{2m}$ with $\mathbb{C}^m$ with complex coordinates $(z_1, \ldots, z_m)$, and define a metric g, a real 2-form ω and a complex m-form θ on $\mathbb{C}^m$ by

$$g = |dz_1|^2 + \cdots + |dz_m|^2, \quad \omega = \frac{i}{2}(dz_1 \wedge d\bar{z}_1 + \cdots + dz_m \wedge d\bar{z}_m),$$
$$\text{and} \quad \theta = dz_1 \wedge \cdots \wedge dz_m. \tag{6.1}$$

The subgroup of $GL(2m, \mathbb{R})$ preserving g, ω and θ is $SU(m)$. Therefore, by the results of §2.5.1, every Riemannian manifold (M, g) with holonomy $SU(m)$ admits natural forms ω and θ, constant under the Levi-Civita connection, such that g, ω and θ can be written in the form (6.1) at each point of M.

There is a unique complex structure J on M such that $\omega_{ac} = J_a^b g_{bc}$. The triple (g, J, ω) is a *Kähler structure* on M, so we call ω the *Kähler form*. Also, θ is a holomorphic $(m, 0)$-form with respect to J, so we call θ the *holomorphic volume form* on M. Thus, every Riemannian manifold with holonomy $SU(m)$ is a Kähler manifold and is equipped with a constant, holomorphic volume form. Conversely, if (M, J, g) is Kähler and θ is a holomorphic volume form on M with $\nabla \theta = 0$, then $\text{Hol}(g) \subseteq SU(m)$.

Now $\Lambda^{m,0}$ is the *canonical bundle* K_M of M, a holomorphic line bundle, and a holomorphic volume form θ is a nonvanishing holomorphic section of K_M. Such a θ exists if and only if K_M is *trivial*, that is, if it is isomorphic to the trivial holomorphic line bundle $M \times \mathbb{C}$ over M. Thus, if (M, J, g) is Kähler and $\text{Hol}(g) \subseteq SU(m)$, then K_M is trivial.

Also, since the *first Chern class* $c_1(M)$ of M is a characteristic class of K_M, we see that $c_1(M) = 0$ in $H^2(M, \mathbb{Z})$. Hence, any complex manifold M admitting Kähler metrics with holonomy in $\mathrm{SU}(m)$ must have $c_1(M) = 0$. Another important property of metrics with holonomy $\mathrm{SU}(m)$ is that they are *Ricci-flat*.

Proposition 6.1.1 *Let (M, J, g) be a Kähler manifold. Then $\mathrm{Hol}^0(g) \subseteq \mathrm{SU}(m)$ if and only if g is Ricci-flat.*

Proof As g is Kähler, the Levi-Civita connection ∇ of g induces a connection ∇^K on $K_M = \Lambda^{m,0} M$, which has holonomy group $\mathrm{Hol}(\nabla^K) \subseteq \mathrm{U}(1)$. Now if $A \in \mathrm{U}(m)$ acts on $\mathbb{C}^m$, the induced map on $\Lambda^{m,0}\mathbb{C}^m$ is multiplication by $\det A$. Therefore, the relationship between $\mathrm{Hol}^0(\nabla) = \mathrm{Hol}^0(g)$ and $\mathrm{Hol}^0(\nabla^K)$ is given by $\mathrm{Hol}^0(\nabla^K) = \det(\mathrm{Hol}^0(\nabla))$, where $\det : \mathrm{U}(m) \to \mathrm{U}(1)$ is the determinant map. It follows that $\mathrm{Hol}^0(\nabla^K) = \{1\}$ if and only if $\mathrm{Hol}^0(g) \subseteq \mathrm{SU}(m)$.

But by the Frobenius Theorem, $\mathrm{Hol}^0(\nabla^K) = \{1\}$ if and only if ∇^K is *flat*, that is, if the curvature of ∇^K is identically zero. Since the gauge group of ∇^K is $\mathrm{U}(1)$, the curvature of ∇^K is just a closed 2-form. It can be shown using (4.8) that this 2-form is exactly the *Ricci form* ρ of g. Thus, we have shown that $\mathrm{Hol}^0(g) \subseteq \mathrm{SU}(m)$ if and only if $\rho \equiv 0$, that is, if and only if g is Ricci-flat. $\square$

Our main interest in this chapter is in *Calabi–Yau manifolds*. The term Calabi–Yau manifold is used to mean several slightly different things. Here are five definitions the author found in the literature, which are all inequivalent.

- a compact complex manifold (M, J) with $c_1(M) = 0$ admitting Kähler metrics,
- a projective manifold M with $c_1(M) = 0$,
- a compact Ricci-flat Kähler manifold,
- a compact Kähler manifold (M, J, g) with $\mathrm{Hol}(g) \subseteq \mathrm{SU}(m)$, and
- a compact Kähler manifold (M, J, g) with $\mathrm{Hol}(g) = \mathrm{SU}(m)$.

Manifolds such as Enriques surfaces or complex tori satisfy some of these definitions, but not others. As we are principally concerned with holonomy groups we adopt the following definition, but readers are warned that conventions vary.

Definition 6.1.2 A *Calabi–Yau manifold* is a compact Kähler manifold (M, J, g) of dimension $m \geqslant 2$, with $\mathrm{Hol}(g) = \mathrm{SU}(m)$.

6.2 Compact Ricci-flat Kähler manifolds

One important consequence of the proof of the Calabi conjecture in Chapter 5, is the existence of families of Ricci-flat Kähler metrics on suitable compact complex manifolds. As we explained in the introduction to Chapter 5, if M is Kähler and $c_1(M) = 0$ in $H^2(M, \mathbb{R})$ then the Calabi conjecture yields a Kähler metric g' on M with Ricci form $\rho' = 0$, so that g' is Ricci-flat. We state this in the following theorem.

Theorem 6.2.1 *Let (M, J) be a compact complex manifold admitting Kähler metrics, with $c_1(M) = 0$. Then there is a unique Ricci-flat Kähler metric in each Kähler class on M. The Ricci-flat Kähler metrics on M form a smooth family of dimension $h^{1,1}(M)$, isomorphic to the Kähler cone $\mathcal{K}$ of M.*

Now, a generic Kähler metric g on a complex m-manifold M has holonomy group $\text{Hol}(g) = \text{U}(m)$. But by Proposition 6.1.1 we know that $\text{Hol}^0(g) \subseteq \text{SU}(m)$ if and only if g is Ricci-flat. Thus Theorem 6.2.1 constructs metrics with special holonomy on compact manifolds. In our next three results we pin down their holonomy groups more precisely.

By Theorem 3.5.5 any compact Ricci-flat Riemannian manifold (M, g) admits a finite cover isometric to $(T^n \times N, g' \times g'')$, where (T^n, g') is a flat Riemannian torus, and (N, g'') is a compact, simply-connected Ricci-flat Riemannian manifold. If g is also Kähler, the factors (T^n, g') and (N, g'') are Kähler, so that $n = 2l$. But N is simply-connected, and g'' is complete as N is compact. Therefore Theorem 3.2.7 applies, and we may write (N, g'') as a Riemannian product $(M_1 \times \cdots \times M_k, g_1 \times \cdots \times g_k)$, where each (M_j, g_j) is irreducible. If (N, g'') is Kähler then the factors (M_j, g_j) are Kähler. This gives

Proposition 6.2.2 *Let (M, J, g) be a compact Ricci-flat Kähler manifold. Then (M, J, g) admits a finite cover isomorphic to the product Kähler manifold*

$$(T^{2l} \times M_1 \times \cdots \times M_k, J_0 \times \cdots \times J_k, g_0 \times \cdots \times g_k), \tag{6.2}$$

where (T^{2l}, J_0, g_0) is a flat Kähler torus, and (M_j, J_j, g_j) is a compact, simply-connected, irreducible, Ricci-flat Kähler manifold for $j = 1, \ldots, k$.

Now suppose (M, J, g) is a compact, simply-connected, irreducible, Ricci-flat Kähler manifold. As g is Ricci-flat, but not flat, it must be nonsymmetric. So M is simply-connected and g is nonsymmetric and irreducible, and Theorem 3.4.1 shows that $\text{Hol}(g)$ lies on Berger's list. Because g is also Kähler, we see that $\text{Hol}(g)$ must be one of $\text{U}(m)$, $\text{SU}(m)$ or $\text{Sp}(m/2)$. But $\text{Hol}^0(g) \subseteq \text{SU}(m)$ as g is Ricci-flat, which excludes $\text{U}(m)$, so $\text{Hol}(g) = \text{SU}(m)$ or $\text{Sp}(m/2)$.

On the other hand, if (M, J, g) is compact and Kähler and $\text{Hol}(g)$ is $\text{SU}(m)$ or $\text{Sp}(m/2)$ then g is Ricci-flat as $\text{Hol}^0(g) \subseteq \text{SU}(m)$, and irreducible as $\text{Hol}^0(g)$ does not split. By the previous proposition M has a simply-connected finite cover, and so $\pi_1(M)$ is finite. Thus we have proved:

Proposition 6.2.3 *Let (M, J, g) be a compact, simply-connected, irreducible, Ricci-flat Kähler manifold, of dimension m. Then either $m \geqslant 2$ and $\text{Hol}(g) = \text{SU}(m)$, or $m \geqslant 4$ is even and $\text{Hol}(g) = \text{Sp}(m/2)$. Conversely, if (M, J, g) is a compact Kähler manifold and $\text{Hol}(g)$ is $\text{SU}(m)$ or $\text{Sp}(m/2)$, then g is Ricci-flat and irreducible and M has finite fundamental group.*

Next we show that any closed $(p, 0)$-form on a compact Ricci-flat manifold is constant. The proof uses the 'Bochner argument' used to prove Theorem 3.5.4.

Proposition 6.2.4 *Let (M, J, g) be a compact Ricci-flat Kähler manifold, and let ξ be a smooth $(p, 0)$-form on M. Then $d\xi = 0$ if and only if $\nabla \xi = 0$, where ∇ is the Levi-Civita connection of g. Hence $H^{p,0}(M)$ is isomorphic to the vector space of constant $(p, 0)$-forms ξ on M.*

Proof Suppose $d\xi = 0$. Then $\partial \xi = \bar{\partial}\xi = 0$. But for $(p,0)$-forms the $\bar{\partial}$-Laplacian $\Delta_{\bar{\partial}} = \bar{\partial}^*\bar{\partial}$, and so $\Delta_{\bar{\partial}}\xi = 0$. Hence by (4.11) we see that $(dd^* + d^*d)\xi = 0$. Taking the inner product with ξ and integrating by parts then shows that $d^*\xi = 0$, as M is compact, and thus ξ lies in the space $\mathcal{H}^{p,0}$ of harmonic $(p,0)$-forms defined in §4.7.1, which is isomorphic to $H^{p,0}(M)$.

Now the Weitzenbock formula for p-forms (3.18) shows that $(dd^* + d^*d)\xi = \nabla^*\nabla\xi - 2\tilde{R}(\xi)$, where $\tilde{R}$ is defined in (3.19) and depends on the Riemann curvature $R^a{}_{bcd}$ of g. Combining the symmetries (3.2) and (3.3) of $R^a{}_{bcd}$ with the decomposition (4.5) of the curvature tensor of a Kähler metric, we may show that $R^\alpha{}_{\beta\bar{\gamma}\delta} = R^\alpha{}_{\delta\bar{\gamma}\beta}$, in the notation of §4.2.1.

Since ξ is a $(p,0)$-form, this implies that the first term on the right hand side of (3.19) is zero. However, the second term on the right hand side of (3.19) depends on the Ricci curvature, and g is Ricci-flat. Thus $\tilde{R}(\xi) = 0$, so that $(dd^* + d^*d)\xi = \nabla^*\nabla\xi$ by the Weitzenbock formula. So $\nabla^*\nabla\xi = 0$, and integrating by parts shows that $\nabla\xi = 0$, as M is compact. This proves that if $d\xi = 0$ then $\nabla\xi = 0$. But clearly $\nabla\xi = 0$ implies that $d\xi = 0$, and the proof is complete. □

Applying the proposition to $(m,0)$-forms when $\dim M = m$, we get:

Corollary 6.2.5 *Let (M, J, g) be a compact Ricci-flat Kähler m-manifold. Then $\mathrm{Hol}(g) \subseteq \mathrm{SU}(m)$ if and only if the canonical bundle K_M of M is trivial.*

Now as we explained in §2.5.1, the tensors constant under a connection ∇ are entirely determined by the holonomy group $\mathrm{Hol}(\nabla)$. Combining this with Proposition 6.2.4, we see that if (M, J, g) is a compact Ricci-flat Kähler manifold, then $H^{p,0}(M)$ is determined by $\mathrm{Hol}(g)$. That is, $\mathrm{Hol}(g)$ is (up to conjugation) a subgroup of $\mathrm{U}(m)$, which acts naturally on $\Lambda^{p,0}\mathbb{C}^m$, and $H^{p,0}(M)$ is isomorphic to the subspace of $\Lambda^{p,0}\mathbb{C}^m$ invariant under $\mathrm{Hol}(g)$.

In particular, the group $\mathrm{SU}(m)$ fixes $\Lambda^{0,0}\mathbb{C}^m$ and $\Lambda^{m,0}\mathbb{C}^m$, but fixes no nonzero elements of $\Lambda^{p,0}\mathbb{C}^m$ for $0 < p < m$. Hence if $\mathrm{Hol}(g) = \mathrm{SU}(m)$ then $H^{p,0}(M)$ is $\mathbb{C}$ if $p = 0, m$ and zero otherwise. We have proved:

Proposition 6.2.6 *Let (M, J, g) be a Calabi–Yau m-fold with Hodge numbers $h^{p,q}$. Then $h^{0,0} = h^{m,0} = 1$ and $h^{p,0} = 0$ for $p \neq 0, m$.*

Next we show that Calabi–Yau manifolds of dimension $m \geqslant 3$ are *algebraic*.

Theorem 6.2.7 *Let (M, J, g) be a Calabi–Yau manifold of dimension $m \geqslant 3$. Then M is projective. That is, (M, J) is isomorphic as a complex manifold to a complex submanifold of $\mathbb{CP}^N$, and is an algebraic variety.*

Proof From Proposition 6.2.6 we see that $h^{2,0}(M) = h^{0,2}(M) = 0$. Thus $H^{1,1}(M) = H^2(M, \mathbb{C})$, and $H^{1,1}(M) \cap H^2(M, \mathbb{R}) = H^2(M, \mathbb{R})$. Now the Kähler cone $\mathcal{K}$ of M is a nonempty open subset of $H^{1,1}(M) \cap H^2(M, \mathbb{R})$, so that $\mathcal{K}$ is open in $H^2(M, \mathbb{R})$. But $H^2(M, \mathbb{Q})$ is dense in $H^2(M, \mathbb{R})$, so $\mathcal{K} \cap H^2(M, \mathbb{Q})$ is nonempty. Choose α in $\mathcal{K} \cap H^2(M, \mathbb{Q})$. Let k be a positive integer such that $k\alpha \in H^2(M, \mathbb{Z})$. Then $k\alpha \in \mathcal{K}$, as $\mathcal{K}$ is a cone, so there exists a closed positive $(1,1)$-form β on M such that $[\beta] = k\alpha$.

Using the theory of holomorphic line bundles explained in §4.10, one can show that there exists a holomorphic line bundle L over M with $c_1(L) = k\alpha$. Since $c_1(L)$ is represented by a positive (1,1)-form β, the Kodaira Embedding Theorem [93, p. 181] shows that L is *ample*. That is, for large n the holomorphic sections of L^n embed M as a complex submanifold of $\mathbb{CP}^N$. But by Chow's Theorem [93, p. 167], any complex submanifold of $\mathbb{CP}^N$ is an algebraic variety. □

This theorem shows that if (M, J, g) is a Calabi–Yau manifold of dimension $m \geqslant 3$, then the underlying complex manifold (M, J) is a projective manifold with $c_1(M) = 0$. Conversely, every projective manifold with $c_1(M) = 0$ admits a family of Ricci-flat Kähler metrics g. Under appropriate conditions on $\pi_1(M)$ and $h^{p,0}(M)$, it follows that (M, J, g) is a Calabi–Yau manifold. Therefore, to answer many questions about Calabi–Yau manifolds in dimension $m \geqslant 3$, it is enough to know about the underlying complex manifolds, which are *algebraic* objects, and it is natural to study them using *complex algebraic geometry*.

6.3 Crepant resolutions, small resolutions, and flops

Let X be a singular complex algebraic variety, and $(\tilde{X}, \pi)$ a resolution of X, as in §4.9.1. We say that $\tilde{X}$ is a *crepant resolution* of X if $\pi^*(K_X) \cong K_{\tilde{X}}$, where K_X and $K_{\tilde{X}}$ are the *canonical bundles* of X and $\tilde{X}$. However, this statement is not as simple as it seems. What do we mean by a holomorphic line bundle such as K_X over a singular variety? How do we define the pull-back $\pi^*(K_X)$, and is it really a holomorphic line bundle on $\tilde{X}$, or some more singular object?

There is an extensive and complicated theory of line bundles over singular varieties, involving Weil divisors, Cartier divisors, and so on. But as this is not an algebraic geometry text we will bypass most of these ideas to simplify things. We shall interpret holomorphic line bundles on singular varieties in terms of *invertible sheaves*.

Definition 6.3.1 Let X be a complex algebraic variety, and let $\mathcal{O}$ be the sheaf of regular functions on X. An *invertible sheaf* $\mathcal{F}$ on X is a sheaf of modules over $\mathcal{O}$ that is locally isomorphic to $\mathcal{O}$.

If X is nonsingular, L is a holomorphic line bundle over X, and $\mathcal{O}(L)$ is the sheaf of holomorphic sections of L, then $\mathcal{O}(L)$ is an invertible sheaf. Conversely, each invertible sheaf over a nonsingular variety comes from a line bundle. Thus, invertible sheaves are a natural generalization to singular varieties of the idea of holomorphic line bundle.

A useful property of invertible sheaves is that if $\mathcal{F}$ is an invertible sheaf over Y and $\pi : X \to Y$ is a morphism, then by [96, §II.5] one can define the *inverse image sheaf* $\pi^*(\mathcal{F})$, which is an invertible sheaf on X. Hence, if X is a singular variety with $\mathcal{O}(K_X)$ invertible, and $(\tilde{X}, \pi)$ is a resolution of X, then $\pi^*(K_X)$ is a well-defined holomorphic line bundle on $\tilde{X}$.

Definition 6.3.2 Let X be a complex algebraic variety, and suppose that the sheaf of regular sections $\mathcal{O}(K_X)$ of K_X is an invertible sheaf. Let $(\tilde{X}, \pi)$ be a *resolution* of X. Then the inverse image sheaf $\pi^*(\mathcal{O}(K_X))$ is an invertible sheaf on $\tilde{X}$, and so represents a holomorphic line bundle on $\tilde{X}$, written $\pi^*(K_X)$.

We say that a prime divisor E in $\tilde{X}$ is *exceptional* if $\pi(E)$ has codimension at least 2 in X. Let the exceptional divisors be $E_1, \ldots, E_n$. Then there exist unique integers $a_1, \ldots, a_n$ such that in divisors on $\tilde{X}$, we have

$$[K_{\tilde{X}}] = [\pi^*(K_X)] + \sum_{i=1}^{n} a_i E_i. \tag{6.3}$$

We call the divisor $\sum_{i=1}^{n} a_i E_i$ on $\tilde{X}$ the *discrepancy* of $(\tilde{X}, \pi)$.
  (i) If $a_i \geqslant 0$ for all i, we say X has *canonical singularities*.
  (ii) if $a_i > 0$ for all i, we say X has *terminal singularities*.
  (iii) If $a_i = 0$ for all i, we say $\pi : \tilde{X} \to X$ is a *minimal resolution* or *crepant resolution*, because it has zero discrepancy. Then $K_{\tilde{X}} \cong \pi^*(K_X)$.

Canonical and terminal singularities were defined by Reid [176, 178]. It can be shown that their definition is independent of the choice of resolution $\pi : \tilde{X} \to X$. They are of interest because they are the singularities which occur in the pluricanonical models of varieties of general type.

Next we consider another special class of resolutions, called *small resolutions*.

Definition 6.3.3 Let X be a singular algebraic variety, and let $(\tilde{X}, \pi)$ be a resolution of X. We say that $\tilde{X}$ is a *small resolution* if it has no exceptional divisors. Here is another way to say this. Define the *exceptional set* E of the resolution to be the set of points $e \in \tilde{X}$ such that $\dim[\pi^{-1}(\pi(e))] \geqslant 1$. Usually $E = \pi^{-1}(S)$, where S is the singular set of X. Then $\tilde{X}$ is a small resolution if E is of codimension at least 2 in $\tilde{X}$.

Clearly, any small resolution is a *crepant resolution*, as the condition in Definition 6.3.2 that $a_i = 0$ for all exceptional prime divisors E_i holds trivially because there are no E_i. For the same reason, any singular manifold X which admits a small resolution has *terminal singularities*. Conversely, if X has terminal singularities then any crepant resolution of X must be a small resolution.

Because blowing up always introduces new exceptional divisors, small resolutions cannot be constructed by the usual strategy of blowing up singular points. This makes small resolutions difficult to find, for general singularities. One common situation in which small resolutions arise is if X is a complex 3-fold with isolated singularities, and $\tilde{X}$ is a resolution of X in which each singular point is replaced by a rational curve $\mathbb{CP}^1$, or a finite union of rational curves. Such singularities are called *double points*, and are studied in [74]. Here is an example.

Example 6.3.4 Define a hypersurface X in $\mathbb{C}^4$ by

$$X = \{(z_1, \ldots, z_4) \in \mathbb{C}^4 : z_1 z_2 = z_3 z_4\}.$$

Then X has a single, isolated singular point at 0, called an *ordinary double point*, or *node*. Define $\tilde{X}_1 \subset \mathbb{C}^4 \times \mathbb{CP}^1$ by

$$\tilde{X}_1 = \{((z_1, \ldots, z_4), [x_1, x_2]) \in \mathbb{C}^4 \times \mathbb{CP}^1 : z_1 x_2 = z_4 x_1, \quad z_3 x_2 = z_2 x_1\},$$

and define $\pi_1 : \tilde{X}_1 \to \mathbb{C}^4$ by $\pi_1 : \big((z_1, \ldots, z_4), [x_1, x_2]\big) \mapsto (z_1, \ldots, z_4)$. Now since x_1, x_2 are not both zero, the equations $z_1 x_2 = z_4 x_1$, $z_3 x_2 = z_2 x_1$ together imply that $z_1 z_2 = z_3 z_4$. Therefore π_1 maps $\tilde{X}_1$ to X. It is easy to see that π_1 is surjective, that $\pi_1^{-1}(0) = \mathbb{CP}^1$, and that π_1 is injective except at $0 \in X$. So $\pi_1 : \tilde{X}_1 \to X$ is a *small resolution* of X.

Similarly, define $\tilde{X}_2 \subset \mathbb{C}^4 \times \mathbb{CP}^1$ by

$$\tilde{X}_2 = \Big\{ \big((z_1, \ldots, z_4), [y_1, y_2]\big) \in \mathbb{C}^4 \times \mathbb{CP}^1 : z_1 y_2 = z_3 y_1, \quad z_4 y_2 = z_2 y_1 \Big\},$$

and define $\pi_2 : \tilde{X}_2 \to X$ as above. Then $\pi_2 : \tilde{X}_2 \to X$ is also a small resolution of X. It can be shown (using toric geometry) that $\tilde{X}_1$ and $\tilde{X}_2$ are the only two crepant resolutions of X, and that they are topologically distinct. Thus, a singularity can admit more than one crepant resolution.

This example is the basis of an important construction in the algebraic geometry of 3-folds, called a *flop*. Suppose that Y is a 3-fold with a single node. Then we can resolve the singularity with a small resolution in two different ways as above, to get two different nonsingular 3-folds Y_1, Y_2, each containing a rational curve $\mathbb{CP}^1$. Conversely, if we have a nonsingular 3-fold Y_1 containing a suitable $\mathbb{CP}^1$, with normal bundle $\mathcal{O}(-1) \oplus \mathcal{O}(-1)$, we can contract the $\mathbb{CP}^1$ to a point to get a 3-fold Y with a single node, and then resolve it in the other way to get a nonsingular 3-fold Y_2 that is different from Y_1.

This process of passing from Y_1 to Y_2 is called a *flop*. Now $c_1(Y_1) = 0$ if and only if $c_1(Y_2) = 0$. Thus, if Y_1 is a Calabi–Yau 3-fold and Y_2 is a flop of Y_1, then Y_2 will also be a Calabi–Yau 3-fold, provided that it is Kähler (which is not always the case). So given one Calabi–Yau 3-fold, one can often construct many more by flopping.

6.4 Crepant resolutions of quotient singularities

Let G be a nontrivial finite subgroup of $\mathrm{GL}(m, \mathbb{C})$. Then G acts on $\mathbb{C}^m$, and the quotient $\mathbb{C}^m/G$ is a singular complex manifold called a *quotient singularity*. It can be made into an algebraic variety, using the algebra of G-invariant polynomials on $\mathbb{C}^m$. If $x \in \mathbb{C}^m$, then xG is a singular point of $\mathbb{C}^m/G$ if and only if the stabilizer subgroup $\mathrm{Stab}(x) = \{\gamma \in G : \gamma \cdot x = x\}$ of x is nontrivial. Thus 0 is always a singular point of $\mathbb{C}^m/G$. If G acts freely on $\mathbb{C}^m \setminus \{0\}$, then 0 is the unique singular point of $\mathbb{C}^m/G$.

For reasons to be explained in §6.6, we are interested in crepant resolutions of quotient singularities $\mathbb{C}^m/G$. Now $\gamma \in G$ acts on $\Lambda^{m,0}\mathbb{C}^m$ by multiplication by $\det \gamma$. Thus the canonical bundle of $\mathbb{C}^m/G$ is only well-defined at 0 if $\det \gamma = 1$ for all $\gamma \in G$, that is, if $G \subset \mathrm{SL}(m, \mathbb{C})$. It follows that $\mathbb{C}^m/G$ can have a crepant resolution only if $G \subset \mathrm{SL}(m, \mathbb{C})$.

Therefore we will restrict our attention to finite subgroups $G \subset \mathrm{SL}(m, \mathbb{C})$. Since any such G is conjugate to a subgroup of $\mathrm{SU}(m)$, we may take $G \subset \mathrm{SU}(m)$ if we wish. In §6.4.1 we discuss crepant resolutions of $\mathbb{C}^2/G$. Then §6.4.2 summarizes what is known about the existence of crepant resolutions of $\mathbb{C}^m/G$ for $m \geq 3$ and $G \subset \mathrm{SL}(m, \mathbb{C})$ a finite subgroup. Finally §6.4.3 explains the McKay correspondence, which relates the topology of a crepant resolution of $\mathbb{C}^m/G$ to the group theory of G.

6.4.1 The Kleinian singularities $\mathbb{C}^2/G$ and their resolutions

The quotient singularities $\mathbb{C}^2/G$, for G a finite subgroup of SU(2), were first classified by Klein in 1884 and are called *Kleinian singularities;* they are also called *Du Val surface singularities*, or *rational double points*. The theory of these singularities and their resolutions (see for instance Slodowy [196]) is very rich, and has many connections to other areas of mathematics.

There is a 1-1 correspondence between nontrivial finite subgroups $G \subset$ SU(2) and the *Dynkin diagrams* of type A_r $(r \geqslant 1)$, D_r $(r \geqslant 4)$, E_6, E_7 and E_8. Let Γ be the Dynkin diagram associated to G. These diagrams appear in the classification of Lie groups, and each one corresponds to a unique compact, simple Lie group; they are the set of such diagrams containing no double or triple edges.

Each singularity $\mathbb{C}^2/G$ admits a unique crepant resolution (X, π). The preimage $\pi^{-1}(0)$ of the singular point is a union of a finite number of *rational curves* in X. These curves correspond naturally to the vertices of Γ. They all have self-intersection -2, and two curves intersect transversely at one point if and only if the corresponding vertices are joined by an edge in the diagram; otherwise the curves do not intersect.

These curves give a basis for the homology group $H_2(X, \mathbb{Z})$, which may be identified with the *root lattice* of the diagram, and the intersection form with respect to this basis is the negative of the Cartan matrix of Γ. Define Δ to be $\{\delta \in H_2(X, \mathbb{Z}) : \delta \cdot \delta = -2\}$. Then Δ is identified with the *set of roots* of the diagram. There are also 1-1 correspondences between the curves and the nonidentity conjugacy classes in G, and also the nontrivial representations of G; it makes sense to regard the nonidentity conjugacy classes as a basis for $H_2(X, \mathbb{Z})$, and the nontrivial representations as a basis for $H^2(X, \mathbb{Z})$.

This correspondence between the Kleinian singularities $\mathbb{C}^2/G$, Dynkin diagrams, and other areas of mathematics became known as the *McKay correspondence*, after John McKay, who pointed it out [158].

6.4.2 Crepant resolutions of $\mathbb{C}^m/G$ for $m \geqslant 3$

We start with a brief introduction to *toric geometry*. Let $\mathbb{C}^*$ be $\mathbb{C} \setminus \{0\}$, regarded as a complex Lie group, with multiplication as the group operation. A *toric variety* is a normal complex algebraic variety X of dimension m, equipped with a holomorphic action of $(\mathbb{C}^*)^m$, and with a dense open subset $T \subset X$ upon which $(\mathbb{C}^*)^m$ acts freely and transitively. For an introduction to toric varieties, see Fulton [78] or Oda [169]. Toric geometry is the geometry of toric varieties.

Each toric variety is the union of a finite number of orbits of $(\mathbb{C}^*)^m$. All of the information about these orbits, and the way they fit together, is represented in a finite collection of combinatorial data called a *fan* [78, §1.4], and the toric variety can be reconstructed from its fan. The importance of toric varieties is that they are very well understood, they are easy to work with and to compute invariants for, and they provide a large family of examples of varieties that can be used to test ideas on.

Now suppose $G \subset$ SL$(m, \mathbb{C})$ is a finite *abelian* group. Then we can choose a coordinate system $(z_1, \ldots, z_m)$ on $\mathbb{C}^m$ such that all elements of G are represented by *diagonal* matrices. The group of all invertible diagonal matrices is isomorphic to $(\mathbb{C}^*)^m$

and commutes with G, and so $(\mathbb{C}^*)^m$ acts on $\mathbb{C}^m/G$. This makes $\mathbb{C}^m/G$ into a toric variety, and its structure is described by a fan.

Any resolution of $\mathbb{C}^m/G$ that is also a toric variety is described by a *subdivision* of the fan of $\mathbb{C}^m/G$. There is a simple condition on this subdivision that determines whether or not the resolution is crepant. Moreover, one can show that any crepant resolution of $\mathbb{C}^m/G$ must be a toric variety. Thus, if $G \subset \mathrm{SL}(m, \mathbb{C})$ is a finite abelian group, then toric geometry gives a simple method for finding all the crepant resolutions of $\mathbb{C}^m/G$.

This method was described independently by Roan [181], and Markushevich [157, App.]. Both of them proved [181, p. 528], [157, p. 273] that if $m = 2$ or 3, then a toric crepant resolution of $\mathbb{C}^m/G$ always exists. For examples of such resolutions, see [157, p. 269–271]. When $m = 2$ the resolution is unique, but for $m \geqslant 3$ there can be finitely many different crepant resolutions of $\mathbb{C}^m/G$. If $G \subset \mathrm{SL}(m, \mathbb{C})$ is abelian and $m \geqslant 4$ then $\mathbb{C}^m/G$ may or may not admit a crepant resolution, depending on the fan of $\mathbb{C}^m/G$. For example, we will show below that $\mathbb{C}^4/\{\pm 1\}$ has no crepant resolution.

For more general subgroups of $\mathrm{SL}(3, \mathbb{C})$, Roan [183, Th. 1] proves

Theorem 6.4.1 *Let G be any finite subgroup of $\mathrm{SL}(3, \mathbb{C})$. Then the quotient singularity $\mathbb{C}^3/G$ admits a crepant resolution.*

Roan's proof is by explicit construction, using the classification of finite subgroups of $\mathrm{SU}(3)$, and it relies on previous work by Ito [112, 113] and Markushevich. In dimension four and above, singularities are less well understood. However, there are simple criteria to determine when a quotient singularity $\mathbb{C}^m/G$ is canonical or terminal, which we now give. To state them, we first define the *age grading* on G, following Reid [179, §3].

Definition 6.4.2 Let $G \subset \mathrm{SL}(m, \mathbb{C})$ be a finite group. Then each $\gamma \in G$ has m eigenvalues $e^{2\pi i a_1}, \ldots, e^{2\pi i a_m}$, where $a_1, \ldots, a_m \in [0, 1)$ are uniquely defined up to order. Define the *age* of γ to be $\mathrm{age}(\gamma) = a_1 + \cdots + a_m$. Then $\mathrm{age}(\gamma)$ is well-defined with $0 \leqslant \mathrm{age}(\gamma) < m$. Since $\det(\gamma) = 1$, we see that $e^{2\pi i\, \mathrm{age}(\gamma)} = 1$, so $\mathrm{age}(\gamma)$ is an integer. Thus $\mathrm{age}(\gamma) \in \{0, 1, \ldots, m-1\}$, and we have defined a mapping $\mathrm{age} : G \to \{0, 1, \ldots, m-1\}$.

The next result (see [167, Th. 2.3]) is due to Reid, Shephard-Barron and Tai.

Theorem 6.4.3 *Let $G \subset \mathrm{SL}(m, \mathbb{C})$ be a finite subgroup. Then $\mathbb{C}^m/G$ is a terminal singularity if and only if $\mathrm{age}(\gamma) \neq 1$ for all $\gamma \in G$.*

But terminal quotient singularities have no crepant resolutions:

Proposition 6.4.4 *Let G be a nontrivial subgroup of $\mathrm{SL}(m, \mathbb{C})$, and suppose $\mathbb{C}^m/G$ is a terminal singularity. Then $\mathbb{C}^m/G$ admits no crepant resolution.*

Proof From §6.3, any crepant resolution of a terminal singularity must be a *small resolution*. Thus it is enough to show that $\mathbb{C}^m/G$ admits no small resolutions. Suppose first that $\mathbb{C}^m/G$ has only an isolated singularity at 0, and that (X, π) is a small resolution of $\mathbb{C}^m/G$. Then $\pi^{-1}(0)$ is a finite union of compact algebraic varieties of dimension at least one. By constructing complex curves in $\pi^{-1}(0)$ one can show that $b_2(X) \geqslant 1$.

Regard X as a compact manifold with boundary S^{2m-1}/G. Poincaré duality for manifolds with boundary gives $b_2(X) = b_{2m-2}(X)$, and so $b_{2m-2}(X) \geq 1$. But X contracts onto $\pi^{-1}(0)$, and $\dim \pi^{-1}(0) < m-1$ since X is a small resolution, and together these imply that $b_{2m-2}(X) = 0$, a contradiction. If the singularities of $\mathbb{C}^m/G$ are not isolated, then the generic singular point looks locally like $(\mathbb{C}^k/H) \times \mathbb{C}^{m-k}$, where $H \subset \mathrm{SL}(k, \mathbb{C})$ and $\mathbb{C}^k/H$ has an isolated singularity at 0. Using this we reduce to the previous case. $\square$

By combining Theorem 6.4.3 and Proposition 6.4.4 we can show that some singularities $\mathbb{C}^m/G$ with $G \subset \mathrm{SL}(m, \mathbb{C})$ do not admit any crepant resolution, as in the next example.

Example 6.4.5 Let G be the group $\{\pm 1\} \subset \mathrm{SL}(4, \mathbb{C})$. Then $\mathrm{age}(1) = 0$ and $\mathrm{age}(-1) = 2$, so $\mathrm{age}(\gamma) \neq 1$ for all $\gamma \in G$. Thus $\mathbb{C}^4/\{\pm 1\}$ is a terminal singularity by Theorem 6.4.3, and does not admit any crepant resolution.

Here is a summary of the above. Let $G \subset \mathrm{SL}(m, \mathbb{C})$ be a nontrivial finite subgroup. If $m = 2$, there is a unique crepant resolution of $\mathbb{C}^2/G$. If $m = 3$ there is a crepant resolution of $\mathbb{C}^3/G$, but it may not be unique. If $m \geq 4$ then $\mathbb{C}^m/G$ may or may not have a crepant resolution, which then may or may not be unique. If G is abelian, one can calculate whether or not a crepant resolution exists using toric geometry. Also, Theorem 6.4.3 gives a criterion to determine whether $\mathbb{C}^m/G$ is terminal, and if it is then no crepant resolution exists.

6.4.3 The McKay correspondence

We now discuss some results and conjectures due to Reid [179], Ito and Reid [114], and Batyrev and Dais [25], which aim to describe the topology and geometry of crepant resolutions (X, π) of $\mathbb{C}^m/G$ in terms of the group theory of G. Their work was motivated in part by conjectures on the topology of resolutions of orbifolds made by Dixon et al. [62], who are physicists working in String Theory. We shall say more about the connection between String Theory and Calabi–Yau manifolds in §6.9.

The main idea is given in the following conjecture, from [114, p. 1–2]. Note that in Definition 6.4.2, the age grading $\mathrm{age}(\gamma)$ is unchanged under conjugation, and is therefore an invariant of the conjugacy class of γ.

Conjecture 6.4.6 Let G be a finite subgroup of $\mathrm{SL}(m, \mathbb{C})$, and (X, π) a crepant resolution of $\mathbb{C}^m/G$. Then there exists a basis of $H^*(X, \mathbb{Q})$ consisting of algebraic cycles in 1-1 correspondence with conjugacy classes of G, such that conjugacy classes with age k correspond to basis elements of $H^{2k}(X, \mathbb{Q})$. In particular, $b^{2k}(X)$ is the number of conjugacy classes of G with age k, and $b^{2k+1}(X) = 0$, so the Euler characteristic $\chi(X)$ is the number of conjugacy classes in G.

Reid calls this conjecture the *McKay correspondence*, because it generalizes the McKay correspondence for subgroups of $\mathrm{SL}(2, \mathbb{C})$ mentioned in §6.4.1 to higher dimensions. In the case $m = 2$ the conjecture is already known. A partial proof of Conjecture 6.4.6 for $m \geq 3$ is given by Ito and Reid [114, Cor. 1.5], in the following result.

Theorem 6.4.7 *Let G be a finite subgroup of $\mathrm{SL}(m, \mathbb{C})$, and (X, π) a crepant resolution of $\mathbb{C}^m/G$. Then there is a 1-1 correspondence between compact prime divisors in X, which form a basis for $H^2(X, \mathbb{Q})$, and elements of G with age 1.*

Ito and Reid then deduce that Conjecture 6.4.6 is true for $m = 3$, using Poincaré duality to relate $H^4(X, \mathbb{Q})$ and $H_c^2(X, \mathbb{Q})$. Batyrev and Dais [25, Th. 5.4] prove Conjecture 6.4.6 for arbitrary m when G is *abelian*, using toric geometry, and also give their own proof for the $m = 3$ case [25, Prop. 5.6].

One interesting feature of the McKay correspondence is that we can work out the Betti numbers of a crepant resolution of $\mathbb{C}^m/G$ without knowing anything about the resolution. Although singularities $\mathbb{C}^m/G$ for $m \geqslant 3$ often admit several different crepant resolutions, the correspondence shows that they must all have the same Betti numbers. In the case $m = 3$ this is true because all the different crepant resolutions of $\mathbb{C}^3/G$ are related by *flops*, as in §6.3, and a flop does not change the Betti numbers.

6.4.4 Deformations of $\mathbb{C}^m/G$

Section 4.9 defined two ways to desingularize a singular variety: *resolution* and *deformation*. Having discussed crepant resolutions of the quotient singularities $\mathbb{C}^m/G$, we now briefly consider their deformations. In particular we would like to understand the smoothings X_t of $\mathbb{C}^m/G$ with $c_1(X_t) = 0$ for $G \subset \mathrm{SL}(m, \mathbb{C})$. Such smoothings are analogues of crepant resolutions of $\mathbb{C}^m/G$.

The deformation theory of the Kleinian singularities $\mathbb{C}^2/G$ of §6.4.1 is very well understood (see for instance Slodowy [196]). In studying deformations of $\mathbb{C}^m/G$ for $m \geqslant 3$, it turns out that the *codimension* of the singularities of $\mathbb{C}^m/G$ is important. If $G \subset \mathrm{SL}(m, \mathbb{C})$ then $\mathbb{C}^m/G$ cannot have singularities of codimension one, since no nonidentity element of $\mathrm{SL}(m, \mathbb{C})$ can fix a subspace $\mathbb{C}^{m-1} \subset \mathbb{C}^m$. Thus the singularities of $\mathbb{C}^m/G$ are of codimension at least two. However, the *Schlessinger Rigidity Theorem* [191] shows that if the singularities are of codimension three or more, then $\mathbb{C}^m/G$ has no deformations.

Theorem 6.4.8. (Schlessinger) *Let G be a finite subgroup of $\mathrm{GL}(m, \mathbb{C})$ for some $m \geqslant 3$, that acts freely on $\mathbb{C}^m \setminus \{0\}$. Then the singularity $\mathbb{C}^m/G$ is rigid, that is, it admits no nontrivial deformations. More generally, if G is a finite subgroup of $\mathrm{GL}(m, \mathbb{C})$ and the singularities of $\mathbb{C}^m/G$ are all of codimension at least three, then $\mathbb{C}^m/G$ is rigid.*

Thus, if $G \subset \mathrm{SL}(m, \mathbb{C})$ then $\mathbb{C}^m/G$ can have nontrivial deformations X_t only if the singularities of $\mathbb{C}^m/G$ are of codimension two. It turns out that $c_1(X_t) = 0$ holds automatically in this case.

6.5 Complex orbifolds

Orbifolds are a special class of singular manifolds.

Definition 6.5.1 An *orbifold* is a singular real manifold X of dimension n whose singularities are locally isomorphic to quotient singularities $\mathbb{R}^n/G$ for finite subgroups $G \subset \mathrm{GL}(n, \mathbb{R})$, such that if $1 \neq \gamma \in G$ then the subspace V_γ of $\mathbb{R}^n$ fixed by γ has $\dim V_\gamma \leqslant n - 2$.

For each singular point $x \in X$ there is a finite subgroup $G_x \subset \mathrm{GL}(n, \mathbb{R})$, unique up to conjugation, such that open neighbourhoods of x in X and 0 in $\mathbb{R}^n/G_x$ are homeomorphic (and, in a suitable sense, diffeomorphic). We call x an *orbifold point* of X, and G_x the *orbifold group* or *isotropy group* of x.

Orbifolds are studied in detail by Satake [189], who calls them *V-manifolds*. The condition on γ means that the singularities of the orbifold have real codimension at least two. This makes orbifolds behave like manifolds in many respects. For instance, compact orbifolds satisfy Poincaré duality, but this fails if we allow singularities of codimension one.

There is a very easy method for constructing orbifolds, which will be used many times in the rest of the book. Suppose that M is a manifold, and G a finite group that acts smoothly on M, with nonidentity fixed point sets of codimension at least two. Then the quotient M/G is an orbifold. The following proposition describes the singular set of M/G. The proof is elementary, and we omit it.

Proposition 6.5.2 *Let M be an oriented manifold, and G a finite group acting smoothly and faithfully on M preserving orientation. Then M/G is an orbifold. For each $x \in M$, define the stabilizer subgroup of x to be $\mathrm{Stab}(x) = \{g \in G : g \cdot x = x\}$. If $\mathrm{Stab}(x) = \{1\}$ then xG is a nonsingular point of M/G. If $\mathrm{Stab}(x) \neq \{1\}$ then xG is a singular point of M/G, and has orbifold group $\mathrm{Stab}(x)$. Thus the singular set of M/G is*

$$S = \{xG \in M/G : x \in M \text{ and } g \cdot x = x \text{ for some } 1 \neq g \in G\}.$$

Here the condition that G preserves orientation eliminates the possibility that M/G could have singularities in codimension one, which is not allowed by Definition 6.5.1. In a similar way, we define complex orbifolds.

Definition 6.5.3 A *complex orbifold* is a singular complex manifold of dimension m whose singularities are all locally isomorphic to quotient singularities $\mathbb{C}^m/G$, for finite subgroups $G \subset \mathrm{GL}(m, \mathbb{C})$. *Orbifold points* x and *orbifold groups* $G_x \subset \mathrm{GL}(m, \mathbb{C})$ are defined as above.

Clearly, any complex orbifold is also a real orbifold. Notice that the singular points of orbifolds do not need to be isolated. For example, a complex orbifold of dimension m can have singularities locally modelled on $(\mathbb{C}^k/G) \times \mathbb{C}^{m-k}$, where G is a finite subgroup of $\mathrm{GL}(k, \mathbb{C})$. If G acts freely on $\mathbb{C}^k \setminus \{0\}$ then the singular set of $(\mathbb{C}^k/G) \times \mathbb{C}^{m-k}$ is a copy of $\mathbb{C}^{m-k}$.

The singular set of a complex orbifold is itself a locally finite union of complex orbifolds of lower dimension. If M is a complex manifold and G a finite group acting holomorphically on M then M/G is a complex orbifold, as in Proposition 6.5.2. The *weighted projective spaces* are a special class of complex orbifolds.

Definition 6.5.4 Let $m \geqslant 1$ be an integer, and let $a_0, a_1, \ldots, a_m$ be positive integers with highest common factor 1. Let $\mathbb{C}^{m+1}$ have complex coordinates $(z_0, \ldots, z_m)$, and define an action of the complex Lie group $\mathbb{C}^*$ on $\mathbb{C}^{m+1}$ by

$$(z_0, \ldots, z_m) \stackrel{u}{\longmapsto} (u^{a_0} z_0, \ldots, u^{a_m} z_m), \qquad \text{for } u \in \mathbb{C}^*. \tag{6.4}$$

Define the *weighted projective space* $\mathbb{CP}^m_{a_0,\ldots,a_m}$ to be $(\mathbb{C}^{m+1} \setminus \{0\})/\mathbb{C}^*$, where $\mathbb{C}^*$ acts on $\mathbb{C}^{m+1} \setminus \{0\}$ with the action (6.4). Then $\mathbb{CP}^m_{a_0,\ldots,a_m}$ is compact and Hausdorff, and has the structure of a *complex orbifold*.

Note that $\mathbb{CP}^m_{1,\ldots,1}$ is the usual complex projective space $\mathbb{CP}^m$. Thus the weighted projective spaces $\mathbb{CP}^m_{a_0,\ldots,a_m}$ are a large family of complex orbifolds that generalize the complex manifolds $\mathbb{CP}^m$. For each $(z_0,\ldots,z_m) \in \mathbb{C}^{m+1} \setminus \{0\}$, define $[z_0,\ldots,z_m] \in \mathbb{CP}^m_{a_0,\ldots,a_m}$ to be the orbit of $(z_0,\ldots,z_m)$ under $\mathbb{C}^*$.

Here is why $\mathbb{CP}^m_{a_0,\ldots,a_m}$ is a complex orbifold. Consider the point $[1, 0, \ldots, 0]$. Under the action of $u \in \mathbb{C}^*$ the point $(1, 0, \ldots, 0)$ is taken to $(u^{a_0}, 0, \ldots, 0)$. Therefore the stabilizer of $(1, 0, \ldots, 0)$ in $\mathbb{C}^*$ is $G = \{u \in \mathbb{C}^* : u^{a_0} = 1\}$, which is a finite group isomorphic to $\mathbb{Z}_{a_0}$. It can be shown that the open set $U_0 = \{[z_0,\ldots,z_m] \in \mathbb{CP}^m_{a_0,\ldots,a_m} : z_0 \ne 0\}$ is naturally isomorphic to $\mathbb{C}^m/G$, where $\mathbb{C}^m$ has complex coordinates $(z_1,\ldots,z_m)$ and $u \in G$ acts on $\mathbb{C}^m$ by

$$(z_1,\ldots,z_m) \xmapsto{u} (u^{a_1}z_1,\ldots,u^{a_m}z_m).$$

Thus, if $a_0 > 1$ then $[1, 0, \ldots, 0]$ is an orbifold point of $\mathbb{CP}^m_{a_0,\ldots,a_m}$ with orbifold group $\mathbb{Z}_{a_0}$. In the same way, one can prove the following. Let $[z_0,\ldots,z_m]$ be a point in $\mathbb{CP}^m_{a_0,\ldots,a_m}$, and let k be the highest common factor of the set of those a_j for which $z_j \ne 0$. If $k = 1$ then $[z_0,\ldots,z_m]$ is nonsingular, and if $k > 1$ then $[z_0,\ldots,z_m]$ is an orbifold point with orbifold group $\mathbb{Z}_k$.

Note that we cannot write $\mathbb{CP}^m_{a_0,\ldots,a_m}$ as M/G for M a complex manifold and G a *finite* group, and so not all orbifolds are of the form M/G. The construction of weighted projective spaces is an example of a more general phenomenon. Suppose a complex Lie group K acts holomorphically on a complex manifold M, such that the stabilizers of points in M are always finite subgroups of K. Then the quotient M/K is a complex orbifold, provided it is Hausdorff. Because of this, orbifold singularities occur naturally in moduli spaces and other geometrical problems.

6.5.1 *Kähler and Calabi–Yau orbifolds*

There is a natural notion of Kähler metric on complex orbifolds.

Definition 6.5.5 We say that g is a *Kähler metric* on a complex orbifold X if g is Kähler in the usual sense on the nonsingular part of X, and wherever X is locally isomorphic to $\mathbb{C}^m/G$, we can identify g with the quotient of a G-invariant Kähler metric defined near 0 in $\mathbb{C}^m$. A *Kähler orbifold* (X, J, g) is a complex orbifold (X, J) equipped with a Kähler metric g.

Examples of Kähler orbifolds are easy to find. For instance, all the weighted projective spaces $\mathbb{CP}^m_{a_0,\ldots,a_m}$ admit Kähler metrics, generalizing the Fubini-Study metric on $\mathbb{CP}^m$. Also, suppose (M, J, g) is a Kähler manifold, and G is a finite group acting holomorphically on M, not necessarily preserving the Kähler metric g. Then $g' = \frac{1}{|G|}\sum_{\alpha \in G} \alpha^*(g)$ is a G-invariant Kähler metric on M, and so $(M/G, J, g')$ is a Kähler orbifold. Thus, if M/G is a complex orbifold and M is Kähler, then M/G is Kähler too.

Because orbifold points are quite a mild form of singularity, orbifolds share many of the good properties of manifolds. Many definitions and results about manifolds can be very easily generalized to definitions and results about orbifolds, such as the definition of orbifold Kähler metrics above. In particular, the ideas of smooth k-forms and (p,q)-forms make sense on complex orbifolds. De Rham and Dolbeault cohomology are well-defined on orbifolds and have nearly all of their usual properties.

Another result of interest to us is that the *Calabi conjecture* holds for compact Kähler orbifolds. To interpret this, we first need to know what we mean by the first Chern class $c_1(X)$ when X is a complex orbifold. Now $c_1(X)$ is a characteristic class of the canonical bundle K_X of X. If all the orbifold groups of singular points in X lie in $\mathrm{SL}(m,\mathbb{C})$, then K_X is a genuine line bundle over X, and $c_1(X)$ is a well-defined element of $H^2(X,\mathbb{Z})$ in the usual way.

However, if the orbifold groups do not lie in $\mathrm{SL}(m,\mathbb{C})$ then K_X is a singular bundle with fibre $\mathbb{C}$ over nonsingular points of X, but with fibres $\mathbb{C}/\mathbb{Z}_k$ over orbifold points of X, for $k \geq 1$. It can be shown that in this case $c_1(X)$ is still well-defined, but exists in $H^2(X,\mathbb{Q})$ rather than $H^2(X,\mathbb{Z})$. So in both cases $c_1(X) \in H^2(X,\mathbb{R})$, and if X is Kähler with Ricci form ρ then $[\rho] = 2\pi\, c_1(X)$ in $H^2(X,\mathbb{R})$.

Thus the statement of the Calabi conjecture in the introduction to Chapter 5 does at least make sense in the category of complex orbifolds. Moreover, the proof of the conjecture also works for orbifolds, with only cosmetic changes. Since an orbifold is locally the quotient of a manifold by a finite group G, locally one can lift any problem on an orbifold up to a G-invariant problem on a manifold, and using this principle one can adapt many proofs in geometry and analysis to the orbifold case. So, by analogy with Theorem 6.2.1 we may prove:

Theorem 6.5.6 *Let X be a compact complex orbifold with $c_1(X) = 0$, admitting Kähler metrics. Then there is a unique Ricci-flat Kähler metric in every Kähler class on X.*

Let X be a real or complex orbifold with singular set S, and g a Riemannian or Kähler metric on X. We define the *holonomy group* $\mathrm{Hol}(g)$ to be the holonomy group (in the usual sense) of the restriction of g to $X \setminus S$. With this definition, holonomy groups on orbifolds have most of the good properties of the manifold case. However, $X \setminus S$ may not be simply-connected if X is simply-connected. Thus, if X is a simply-connected orbifold and g a metric on X then $\mathrm{Hol}(g)$ may not be connected, in contrast to the manifold case.

By analogy with Definition 6.1.2, we define a *Calabi–Yau orbifold* (X, J, g) to be a compact complex orbifold (X, J) of dimension m, such that $\mathrm{Hol}(g) = \mathrm{SU}(m)$. Since from §6.1 a Kähler metric g is Ricci-flat if and only if $\mathrm{Hol}^0(g) \subseteq \mathrm{SU}(m)$, we can use Theorem 6.5.6 to construct metrics with holonomy $\mathrm{SU}(m)$ on suitable compact complex orbifolds, making them into Calabi–Yau orbifolds.

Let (M, J, g) be a Calabi–Yau manifold, and G a finite group of holomorphic isometries of M. Since $\mathrm{Hol}(g) = \mathrm{SU}(m)$, from §6.1 there exists a holomorphic volume form θ on M, which is unique up to multiplication by $e^{i\phi}$. Clearly $(M/G, J, g)$ is a Calabi–Yau orbifold if and only if θ is G-invariant.

If (X, J, g) is a Kähler orbifold and an orbifold point x in X has orbifold group G, then there is a natural inclusion

$$G \subseteq \mathrm{Hol}(g) \subseteq \mathrm{U}(m) \subset \mathrm{GL}(m, \mathbb{C}). \tag{6.5}$$

In particular, G is a subgroup of $\mathrm{Hol}(g)$. So if (X, J, g) is a Calabi–Yau orbifold then $G \subset \mathrm{SU}(m) \subset \mathrm{SL}(m, \mathbb{C})$. This shows that if (X, J, g) is a Calabi–Yau orbifold then the orbifold groups of singular points of X all lie in $\mathrm{SL}(m, \mathbb{C})$.

Suppose G is a finite subgroup of $\mathrm{SL}(m, \mathbb{C})$. Choose $\alpha \in G$ with $\alpha \neq 1$, and let V_α be the vector subspace of $\mathbb{C}^m$ fixed by α. Then $\dim V_\alpha < m$ as $\alpha \neq 1$. If $\dim V_\alpha = m-1$ then α has exactly one eigenvalue that is not 1, contradicting $\det(\alpha) = 1$. So $\dim V_\alpha \leqslant m-2$. But the singular set of $\mathbb{C}^m/G$ is the image in $\mathbb{C}^m/G$ of the V_α for $\alpha \neq 1$. Thus the singular set of $\mathbb{C}^m/G$ has codimension at least two. Hence the singularities of a Calabi–Yau orbifold are of complex codimension at least two.

6.6 Crepant resolutions of orbifolds

Let X be a complex orbifold of dimension m. We shall consider *crepant resolutions* $(\tilde{X}, \pi)$ of X. Since the singularities of X are locally isomorphic to quotient singularities $\mathbb{C}^m/G$ for finite $G \subset \mathrm{GL}(m, \mathbb{C})$, any crepant resolution $(\tilde{X}, \pi)$ of X is locally isomorphic to crepant resolutions of $\mathbb{C}^m/G$. But from §6.4 we already understand the crepant resolutions of $\mathbb{C}^m/G$ quite well, especially when $m = 2$ or 3. Thus, we can use our knowledge of crepant resolutions of $\mathbb{C}^m/G$ to study crepant resolutions of general complex orbifolds.

From §6.4, a necessary condition for $\mathbb{C}^m/G$ to admit crepant resolutions is that $G \subset \mathrm{SL}(m, \mathbb{C})$. Thus, a complex orbifold X can admit a crepant resolution only if all its orbifold groups lie in $\mathrm{SL}(m, \mathbb{C})$. (From §6.5, this condition holds automatically for Calabi–Yau orbifolds.) So let X be a complex orbifold with all orbifold groups in $\mathrm{SL}(m, \mathbb{C})$. For each singular point there are a finite number of possible crepant resolutions (which may be zero if $m \geqslant 4$). If G is an orbifold group of X and $\mathbb{C}^m/G$ admits no crepant resolution, then clearly X has no crepant resolution either. So suppose that $\mathbb{C}^m/G$ has at least one crepant resolution for all orbifold groups G of X.

The obvious way to construct a crepant resolution of X is to choose a crepant resolution of $\mathbb{C}^m/G$ for each singular point of X with orbifold group G, and then to try to patch these resolutions together to form a crepant resolution $(\tilde{X}, \pi)$ of X. If the singularities of X are *isolated* then we can independently choose any crepant resolution of each singular point, and fit them together in a unique way to get a crepant resolution of X.

For nonisolated singularities things are more complicated, because choosing a crepant resolution for one singular point x then uniquely determines the choice of resolution of all other singular points in an open neighbourhood of x. The choice of resolution must vary continuously over the singular set, in an appropriate sense. However, in the case $m = 3$ this imposes no restrictions, as we show in the proof of the next result, due to Roan [183, p. 493].

Theorem 6.6.1 *Let X be a complex 3-orbifold with orbifold groups in $\mathrm{SL}(3, \mathbb{C})$. Then X admits a crepant resolution.*

Proof If X is a complex 3-orbifold with orbifold groups in SL(3, $\mathbb{C}$) then the singular points of X divide into two types:

(a) singular points modelled on $(\mathbb{C}^2/H) \times \mathbb{C}$, for H a finite subgroup of SU(2), and
(b) singular points not of type (a).

The singular points of type (a) form a complex 1-manifold in X, but the singular points of type (b) are a *discrete* set of isolated points.

Now singular points of type (a) have a *unique* crepant resolution, by §6.4.1. So we only have more than one possible choice of resolution at singular points of type (b). But these are isolated from one another, and so there are no compatibility conditions between the choices. By Theorem 6.4.1 there is at least one possible crepant resolution for each singular point of type (b). Making an arbitrary choice in each case, we can patch the resolutions together in a unique way to get a crepant resolution of X. □

It can happen that if X is a Kähler orbifold with nonisolated singularities, then some of the crepant resolutions $(\tilde{X}, \pi)$ of X are not Kähler. This is because, for reasons of global topology, it may not be possible to choose a class in $H^{1,1}(\tilde{X})$ which is simultaneously positive on homology classes of all the rational curves in $\tilde{X}$ introduced by the resolution, and this is a necessary condition for the existence of a Kähler class. However, if X is Kähler and admits crepant resolutions then at least one of them must be Kähler.

6.6.1 *Crepant resolutions of quotients of complex tori*

If X is a Calabi–Yau orbifold and $(\tilde{X}, \pi)$ is a Kähler crepant resolution of X, then $\tilde{X}$ has a family of Ricci-flat Kähler metrics which make it into a Calabi–Yau manifold. This gives a *method of constructing Calabi–Yau manifolds*.

We start with a compact Kähler orbifold (X, J, g) with $\text{Hol}(g) \subseteq \text{SU}(m)$. If $(\tilde{X}, \pi)$ is any Kähler crepant resolution of X, then $\tilde{X}$ has a family of Ricci-flat Kähler metrics $\tilde{g}$. It can be shown that $\text{Hol}(g) \subseteq \text{Hol}(\tilde{g}) \subseteq \text{SU}(m)$. Often we find that although $\text{Hol}(g)$ may be a proper subgroup of $\text{SU}(m)$, yet $\text{Hol}(\tilde{g}) = \text{SU}(m)$, and so $\tilde{X}$ is a Calabi–Yau manifold.

In particular we can take X to be T^{2m}/G, where T^{2m} is a flat Kähler torus and G a finite group. Let $\mathbb{C}^m$ have its standard complex structure J, Euclidean metric g, and holomorphic volume form θ, and let Λ be a *lattice* in $\mathbb{C}^m$. Then $\mathbb{C}^m/\Lambda$ is a compact torus T^{2m}, with a flat Kähler structure (J, g) and holomorphic volume form θ.

Let G be a finite group of automorphisms of T^{2m} preserving g, J and θ. Then $(T^{2m}/G, J, g)$ is a compact Kähler orbifold with orbifold groups in SL$(m, \mathbb{C})$. Suppose $(\tilde{X}, \pi)$ is a crepant resolution of T^{2m}/G. Under good conditions, $\tilde{X}$ turns out to be a Calabi–Yau manifold.

The simplest case of this is the *Kummer construction*, in which $T^4/\{\pm 1\}$ is resolved to give a $K3$ surface. It will be described in Examples 7.3.2 and 7.3.14. The method was studied in higher dimensions by Roan [181]. Here is a useful result in the 3-dimensional case.

Theorem 6.6.2 *Let (T^6, J, g) be a flat Kähler torus with a holomorphic volume form θ, and suppose G is a finite group of automorphisms of T^6 preserving J, g and θ. Then $X = T^6/G$ is a compact complex orbifold with at least one Kähler crepant* resolution $(\tilde{X}, \pi)$. There exist Ricci-flat Kähler metrics $\tilde{g}$ on $\tilde{X}$ with $\mathrm{Hol}(\tilde{g}) \subseteq \mathrm{SU}(3)$, and $\mathrm{Hol}(\tilde{g}) = \mathrm{SU}(3)$ if and only if $\pi_1(T^6/G)$ is finite.

Proof As G preserves θ the orbifold groups of T^6/G all lie in $\mathrm{SL}(3, \mathbb{C})$, so by Theorem 6.6.1 and the discussion after it, T^6/G has at least one Kähler crepant resolution $(\tilde{X}, \pi)$, which has trivial canonical bundle. This $\tilde{X}$ admits Ricci-flat Kähler metrics $\tilde{g}$ with $\mathrm{Hol}(\tilde{g}) \subseteq \mathrm{SU}(3)$ by Theorem 6.2.1 and Corollary 6.2.5.

By the classification of holonomy groups, $\mathrm{Hol}^0(\tilde{g})$ must be $\{1\}$, $\mathrm{SU}(2)$ or $\mathrm{SU}(3)$. Propositions 6.2.2 and 6.2.3 then show that $\mathrm{Hol}(\tilde{g}) = \mathrm{SU}(3)$ if and only if $\pi_1(\tilde{X})$ is finite. But $\pi_1(\tilde{X}) \cong \pi_1(T^6/G)$, as crepant resolutions of orbifolds replace each singular point by a simply-connected set. Therefore $\mathrm{Hol}(\tilde{g}) = \mathrm{SU}(3)$ if and only if $\pi_1(T^6/G)$ is finite, as we have to prove. □

Here is a simple example.

Example 6.6.3 Let $\zeta = -\frac{1}{2} + i\frac{\sqrt{3}}{2}$, so that $\zeta^3 = 1$, and define a lattice Λ in $\mathbb{C}^3$ by

$$\Lambda = \{(a_1 + b_1\zeta, a_2 + b_2\zeta, a_3 + b_3\zeta) : a_j, b_j \in \mathbb{Z}\}.$$

Then $T^6 = \mathbb{C}^3/\Lambda$ is a complex torus, with a natural metric g and holomorphic volume form θ. Define a map $\alpha : T^6 \to T^6$ by

$$\alpha : (z_1, z_2, z_3) + \Lambda \longmapsto (\zeta z_1, \zeta z_2, \zeta z_3) + \Lambda.$$

Then α is well-defined, preserves g and θ, and α^3 is the identity. Hence $G = \{1, \alpha, \alpha^2\}$ is a group of automorphisms of T^6 isomorphic to $\mathbb{Z}_3$, and T^6/G is a Kähler orbifold.

The fixed points of α on T^6 are the 27 points

$$\left\{(c_1, c_2, c_3) + \Lambda : c_1, c_2, c_3 \in \left\{0, \tfrac{i}{\sqrt{3}}, \tfrac{2i}{\sqrt{3}}\right\}\right\}.$$

Thus T^6/G has 27 isolated fixed points modelled on $\mathbb{C}^3/\mathbb{Z}_3$, where the action of $\mathbb{Z}_3$ on $\mathbb{C}^3$ is generated by $(z_1, z_2, z_3) \longmapsto (\zeta z_1, \zeta z_2, \zeta z_3)$. Now $\mathbb{C}^3/\mathbb{Z}_3$ has a unique crepant resolution, the blow-up of the singular point, in which the singular point is replaced by a copy of $\mathbb{CP}^2$. Therefore T^6/G has a unique crepant resolution Z, made by blowing up the 27 singular points.

Calculation shows that $\pi_1(T^6/G) = \{1\}$, so Z is a Calabi–Yau 3-fold by Theorem 6.6.2. Let us calculate the Hodge numbers $h^{p,q}$ of Z. Since

$$h^{p,q} = h^{q,p} = h^{3-p,3-q} = h^{3-q,3-p} \quad \text{and} \quad h^{0,0} = h^{3,0} = 1, \; h^{1,0} = h^{2,0} = 0$$

by (4.12) and Proposition 6.2.6, it is enough to find $h^{1,1}$ and $h^{2,1}$. Now the forms $dz_j \wedge d\bar{z}_k$ for $j, k = 1, 2, 3$ are a natural basis for $H^{1,1}(T^6)$, so that $h^{1,1}(T^6) = 9$. The action of $\alpha \in G$ multiplies dz_j by ζ and $d\bar{z}_k$ by $\bar{\zeta}$, so that $dz_j \wedge d\bar{z}_k$ is multiplied by $\zeta\bar{\zeta} = 1$. Thus G acts trivially on $H^{1,1}(T^6)$, and so $H^{1,1}(T^6/G) \cong H^{1,1}(T^6)$, and $h^{1,1}(T^6/G) = 9$. The resolution of each singular point adds 1 to $h^{1,1}$. Thus $h^{1,1}(Z) = 9 + 27 = 36$.

Similarly, a basis for $H^{2,1}(T^6)$ is $\mathrm{d}z_j \wedge \mathrm{d}z_k \wedge \mathrm{d}\bar{z}_l$ for $j, k, l = 1, 2, 3$ and $j < k$. The action of $\alpha \in G$ multiplies $\mathrm{d}z_j \wedge \mathrm{d}z_k \wedge \mathrm{d}\bar{z}_l$ by $\zeta^2 \bar{\zeta} = \zeta$. Thus the G-invariant part of $H^{2,1}(T^6)$ is $\{0\}$, and $h^{2,1}(T^6/G) = 0$. The resolution of the singular points does not change $h^{2,1}$, and so $h^{2,1}(Z) = 0$. Therefore the Hodge numbers of Z are $h^{1,1} = 36$ and $h^{2,1} = 0$. An interesting feature of this example is that $h^{2,1} = 0$. As we will see in §6.8, this implies that the complex 3-fold Z is *rigid*, and has no deformations.

Many other examples of Calabi–Yau manifolds can be constructed in this way by considering other finite groups G acting on T^{2m}, especially in the case $m = 3$. But one can show that the number of distinct manifolds arising in this way in any one dimension is finite.

6.7 Complete intersections

Let $\mathbb{CP}^m$ have homogeneous coordinates $[z_0, \ldots, z_m]$, let $f(z_0, \ldots, z_m)$ be a nonzero homogeneous polynomial of degree d, and define X to be

$$\{[z_0, \ldots, z_m] \in \mathbb{CP}^m : f(z_0, \ldots, z_m) = 0\}.$$

Then we call X a *hypersurface of degree d*. Suppose X is nonsingular, which is true for generic f, so that X is a compact complex manifold of dimension $m-1$. Now, under what conditions do we have $c_1(X) = 0$?

The *adjunction formula* for complex hypersurfaces [93, p. 147] gives that

$$K_X = (K_{\mathbb{CP}^m} \otimes L_X)|_X,$$

where L_X is the line bundle associated to the divisor $[X]$. In the notation of Example 4.10.1, any line bundle over $\mathbb{CP}^m$ is of the form $\mathcal{O}(k)$ for some integer k. It can be shown that $K_{\mathbb{CP}^m} = \mathcal{O}(-m-1)$ and $L_X = \mathcal{O}(d)$, and thus $K_X = \mathcal{O}(d-m-1)|_X$. Therefore K_X is trivial, so that $c_1(X) = 0$, if and only if $d = m+1$.

We have shown that any nonsingular hypersurface X in $\mathbb{CP}^m$ of degree $m + 1$ has $c_1(X) = 0$. But as X is also compact and Kähler, Theorem 6.2.1 shows that X has a family of Ricci-flat Kähler metrics. In fact these metrics have holonomy SU($m-1$), and X is a *Calabi–Yau manifold* for $m \geq 3$. This is perhaps the simplest known method of finding Calabi–Yau manifolds. Now all nonsingular hypersurfaces in $\mathbb{CP}^m$ of degree $m + 1$ are diffeomorphic, and thus this method yields only one smooth manifold admitting Calabi–Yau structures in each dimension. We shall now describe three ways of generalizing this idea that give Calabi–Yau structures on many more manifolds.

6.7.1 Complete intersections in $\mathbb{CP}^m$

An algebraic variety X in $\mathbb{CP}^m$ is a *complete intersection* if $X = H_1 \cap \cdots \cap H_k$, where $H_1, \ldots, H_k$ are hypersurfaces in $\mathbb{CP}^m$ which intersect transversely at X, so that $\dim X = m-k$. Suppose that X is a complete intersection of hypersurfaces $H_1, \ldots, H_k$, and let d_j be the degree of H_j. Finding an expression for $c_1(X)$ as above, one easily proves that $c_1(X) = 0$ if and only if $d_1 + \cdots + d_k = m + 1$, and in this case X is a Calabi–Yau manifold.

Since complete intersections with fixed m and $d_1, \ldots, d_k$ are all equivalent under deformation, X depends as a smooth manifold only on m and $d_1, \ldots, d_k$. By [87, Th. 1], if hypersurfaces H_j in $\mathbb{CP}^m$ of degree d_j are chosen generically then $X = H_1 \cap \cdots \cap H_k$ is a nonsingular complete intersection. Now if $d_j = 1$ for any j then X may be regarded as the intersection of $k-1$ hypersurfaces in $\mathbb{CP}^{m-1}$ of degrees $d_1, \ldots, d_k$, omitting d_j.

Thus Calabi–Yau manifolds of dimension $m - k$ which are complete intersections are classified, as smooth manifolds, by integers m and $d_1, \ldots, d_k$, where $d_j \geqslant 2$ for $j = 1, \ldots, k$ and $d_1 + \cdots + d_k = m + 1$. There are only a finite number of possibilities for m and $d_1, \ldots, d_k$ in each dimension. For example, using the notation $(m \mid d_1, \ldots, d_k)$, the five complete intersections giving Calabi–Yau 3-folds are

$$(4 \mid 5), \quad (5 \mid 2, 4), \quad (5 \mid 3, 3), \quad (6 \mid 2, 2, 3) \quad \text{and} \quad (7 \mid 2, 2, 2, 2).$$

By applying the Lefschetz Hyperplane Theorem, Theorem 4.10.4, we see that if X is a complete intersection of dimension $m - k$ in $\mathbb{CP}^m$ then $H^j(X, \mathbb{C}) \cong H^j(\mathbb{CP}^m, \mathbb{C})$ for $0 \leqslant j < m - k$. This gives the Betti numbers $b^j(X)$ of X, except in the middle dimension $m - k$. To determine $b^{m-k}(X)$ we calculate the Euler characteristic of X, using a formula for the Chern classes of X.

6.7.2 Hypersurfaces in weighted projective spaces and toric varieties

The construction above may be generalized by replacing $\mathbb{CP}^m$ by a *weighted projective space* $\mathbb{CP}^m_{a_0, \ldots, a_m}$, as in Definition 6.5.4. This was studied in depth by Candelas, Lynker and Schrimmrigk [55] in the case $m = 4$, and we now summarize their ideas. The weighted projective space $\mathbb{CP}^m_{a_0, \ldots, a_m}$ is the quotient of $\mathbb{C}^{m+1} \setminus \{0\}$ by the $\mathbb{C}^*$-action

$$(z_0, \ldots, z_m) \stackrel{u}{\longmapsto} (u^{a_0} z_0, \ldots, u^{a_m} z_m) \quad \text{for } u \in \mathbb{C}^*.$$

We call a nonzero polynomial $f(z_0, \ldots, z_m)$ *weighted homogeneous of degree d* if

$$f(u^{a_0} z_0, \ldots, u^{a_m} z_m) = u^d f(z_0, \ldots, z_m) \quad \text{for all } u, z_0, \ldots, z_m \in \mathbb{C}.$$

Let f be such a polynomial, and define a hypersurface X in $\mathbb{CP}^m_{a_0, \ldots, a_m}$ by

$$X = \{[z_0, \ldots, z_m] \in \mathbb{CP}^m_{a_0, \ldots, a_m} : f(z_0, \ldots, z_m) = 0\}.$$

Then we call X a *hypersurface of degree d* in $\mathbb{CP}^m_{a_0, \ldots, a_m}$.

Now $\mathbb{CP}^m_{a_0, \ldots, a_m}$ is an orbifold. Usually the hypersurface X intersects the singularities of $\mathbb{CP}^m_{a_0, \ldots, a_m}$, and at these points X itself is singular. We don't want to exclude all such X, and therefore we cannot restrict our attention to nonsingular X. Instead, we define the polynomial f to be *transverse* if $f(z_0, \ldots, z_m) = 0$ and $df(z_0, \ldots, z_m) = 0$ have no common solutions in $\mathbb{C}^{m+1} \setminus \{0\}$. If f is transverse then the only singular points of X are also singular points of $\mathbb{CP}^m_{a_0, \ldots, a_m}$, and in fact X is an *orbifold*, all of whose orbifold groups are cyclic.

So we restrict our attention to hypersurfaces X defined by transverse polynomials f. Note that for given weights $a_0, \ldots, a_m$ and degree d, there may not exist any transverse polynomials f. Any such f must be the sum of monomials $z_0^{b_0} \cdots z_m^{b_m}$, where $b_0, \ldots, b_m$ are nonnegative integers with $a_0 b_0 + \cdots + a_m b_m = d$. If there are not enough suitable solutions $\{b_j\}$ to this equation, then there are no transverse f. For example, by [55, p. 389] a necessary (but not sufficient) criterion for there to exist a transverse polynomial f of degree d is that for each $i = 0, \ldots, m$ there exists a j such that a_i divides $d - a_j$.

Let X be a hypersurface in $\mathbb{CP}^m_{a_0, \ldots, a_m}$ of degree d, defined by a transverse polynomial f. Then since X is an orbifold, the first Chern class $c_1(X)$ is well-defined. It can be shown that $c_1(X) = 0$ if and only if $d = a_0 + \cdots + a_m$. Moreover, in this case the canonical bundle K_X of X is trivial, and this implies that the orbifold groups of X lie in $SL(m-1, \mathbb{C})$. Therefore X is a Calabi–Yau orbifold, and if $(\tilde{X}, \pi)$ is a Kähler crepant resolution of X, then $\tilde{X}$ is a Calabi–Yau manifold.

In particular, when $m = 4$ the dimension of X is 3, and so by Theorem 6.6.1 and the discussion after it, X admits at least one Kähler crepant resolution $\tilde{X}$, which is then a Calabi–Yau 3-fold. This gives a method for constructing Calabi–Yau 3-folds:

- First choose a weighted projective space $\mathbb{CP}^4_{a_0, \ldots, a_4}$, where $a_0, \ldots, a_4$ are positive integers with highest common factor 1.
- If possible, find a hypersurface X in $\mathbb{CP}^4_{a_0, \ldots, a_4}$ defined by a transverse polynomial f of degree $a_0 + \cdots + a_4$.
- This X is a Calabi–Yau orbifold, whose orbifold groups are cyclic subgroups of $SL(3, \mathbb{C})$. Let $(\tilde{X}, \pi)$ be a Kähler crepant resolution of X. There is at least one such resolution. Then $\tilde{X}$ is a Calabi–Yau 3-fold.

The two big advantages of this method are that, firstly, there are many possibilities for $a_0, \ldots, a_4$ and so the construction yields many Calabi–Yau 3-folds, and secondly, the calculations involved are sufficiently mechanical that the construction can be implemented on a computer.

In particular, once $a_0, \ldots, a_4$ are fixed, all hypersurfaces X defined by transverse f of degree $a_0 + \cdots + a_4$ are deformation equivalent. Thus the topology of the orbifold X depends only on $a_0, \ldots, a_4$. Although X may admit several different crepant resolutions $\tilde{X}_1, \ldots, \tilde{X}_k$, the Hodge numbers of $\tilde{X}_j$ depend only on the topology of X, and hence only on $a_0, \ldots, a_4$.

Candelas et al. [55, §3] explained how to calculate the Hodge numbers of $\tilde{X}$ from $a_0, \ldots, a_4$. They then used a computer program to search for quintuples $(a_0, \ldots, a_4)$ for which a suitable transverse polynomial f exists, and to calculate the Hodge numbers of the corresponding Calabi–Yau 3-folds. In this way they constructed some 6000 examples of Calabi–Yau 3-folds, which realized 2339 distinct pairs of Hodge numbers $(h^{1,1}, h^{2,1})$. This was many more examples than were known at the time.

When Candelas et al. plotted the Hodge numbers $(h^{1,1}, h^{2,1})$ of their examples on a graph [55, Fig. 1, p. 384], they found that their graph had an approximate, but very persuasive symmetry: for nearly every Calabi–Yau 3-fold with $h^{1,1} = x$ and $h^{2,1} = y$ in their examples, there was another Calabi–Yau 3-fold with $h^{1,1} = y$ and $h^{2,1} = x$.

This was one of the first important pieces of experimental evidence supporting the idea of *mirror symmetry* for Calabi–Yau 3-folds, which is a curious conjecture arising from String Theory that will be discussed in §6.9.

Since the original paper [55], the subject has moved on. It now seems clear that rather than restricting to weighted projective spaces, it is better to consider hypersurfaces (or more general complete intersections) in *compact toric varieties*. As weighted projective spaces are toric varieties, this includes all the examples of Candelas et al. [55], and also some others.

This point of view was developed by Batyrev [23, 24]. He shows that using toric geometry, the 'mirror map' can be made very explicit, and that Calabi–Yau 3-folds arising from this construction automatically come in mirror pairs. Candelas et al. [52] discuss this, and give a graph of Hodge numbers of Calabi–Yau 3-folds which displays an exact (not just approximate) reflection symmetry.

6.8 Deformations of Calabi–Yau manifolds

If (X, J, g) is a Calabi–Yau manifold, then we can consider the *deformations* of (X, J) as a complex manifold. The Kodaira–Spencer–Kuranishi deformation theory for compact complex manifolds was explained in §4.9.2. It turns out that when X is a Calabi–Yau manifold, the deformation theory of X as a complex manifold is particularly simple. Our next result is due independently to Tian [202, Th. 1] and Todorov [209, Th. 1].

Theorem 6.8.1 *Let X be a compact Kähler manifold of dimension m with trivial canonical bundle. Then the local moduli space of deformations of the complex structure of X is a complex manifold of dimension $h^{m-1,1}(X)$. All the complex structures in this local moduli space are also Kähler with trivial canonical bundle.*

Here is why the moduli space has dimension $h^{m-1,1}(X)$. The deformation theory of a compact complex manifold X depends on the sheaf cohomology groups $H^*(X, \Theta_X)$, which are the cohomology of the complex

$$0 \to C^\infty(T^{1,0}X) \xrightarrow{\bar{\partial}} C^\infty(T^{1,0}X \otimes \Lambda^{0,1}X) \xrightarrow{\bar{\partial}} C^\infty(T^{1,0}X \otimes \Lambda^{0,2}X) \xrightarrow{\bar{\partial}} \cdots.$$

But since X has trivial canonical bundle there exists a holomorphic volume form θ on X, which is a section of $\Lambda^{m,0}$. Contraction with θ gives an isomorphism between $T^{1,0}X$ and $\Lambda^{m-1,0}X$. Thus

$$T^{1,0} \otimes \Lambda^{0,q}X \cong \Lambda^{m-1,0}X \otimes \Lambda^{0,q}X \cong \Lambda^{m-1,q}X.$$

Therefore the complex above is isomorphic to the complex

$$0 \to C^\infty(\Lambda^{m-1,0}X) \xrightarrow{\bar{\partial}} C^\infty(\Lambda^{m-1,1}X) \xrightarrow{\bar{\partial}} C^\infty(\Lambda^{m-1,2}X) \xrightarrow{\bar{\partial}} \cdots.$$

But this is part of the *Dolbeault complex* of X, as in §4.2.2 and §4.7.1. Therefore $H^q(X, \Theta_X)$ is isomorphic to $H^{m-1,q}(X)$.

Thus in the Kodaira–Spencer–Kuranishi deformation theory of §4.9.2, the space of *infinitesimal deformations* of X is $H^{m-1,1}(X)$, and the space of *obstructions* to deforming X is $H^{m-1,2}(X)$. Now Tian and Todorov show that even though $H^{m-1,2}(X)$ may be nonzero, the obstructions are ineffective, and every infinitesimal deformation of X lifts to an actual deformation. Hence the base space of the *Kuranishi family* of deformations of X is an open set in $H^{m-1,1}(X)$. They also show that the Kuranishi family is *universal*. So the local moduli space of deformations of the complex structure of X is isomorphic to an open set in $H^{m-1,1}(X)$, and is a complex manifold with dimension $h^{m-1,1}(X)$.

Corollary 6.8.2 *Let (X, J, g) be a Calabi–Yau manifold of dimension m. Then the local moduli space of deformations of the Calabi–Yau structure of X is a smooth real manifold of dimension $h^{1,1}(X) + 2h^{m-1,1}(X)$.*

Proof The geometric structure of a Calabi–Yau manifold X comes in two pieces, the complex structure J and the Kähler metric g. From Theorem 6.8.1, the local deformations of J form a manifold of complex dimension $h^{m-1,1}(X)$, and thus of real dimension $2h^{m-1,1}(X)$. But from §6.2, the family of Calabi–Yau metrics g on the fixed complex manifold (X, J) is a real manifold of dimension $h^{1,1}(X)$ isomorphic to the Kähler cone $\mathcal{K}$ of (X, J). This also holds on (X, J_t) for any small deformation J_t of J. So the local moduli space of Calabi–Yau structures on X is a real manifold of dimension $h^{1,1}(X) + 2h^{m-1,1}(X)$. $\square$

Now let X be a compact real manifold of dimension $2m$, let $\mathcal{A}_X$ be the set of integrable complex structures J on X for which there exists a Kähler metric g making (X, J, g) into a Calabi–Yau manifold, and let $\mathcal{D}_X$ be the group of diffeomorphisms of X isotopic to the identity map. Then $\mathcal{M}_X = \mathcal{A}_X / \mathcal{D}_X$ is the moduli space of *Calabi–Yau complex structures* on X.

Note that $\mathcal{A}_X$ and $\mathcal{D}_X$ are both infinite-dimensional. However, by Theorem 6.8.1 we see that $\mathcal{M}_X$ is a nonsingular complex manifold of dimension $h^{m-1,1}(X)$, where the Hodge number is a locally constant function of complex structure in $\mathcal{M}_X$. We would like to study the *global* geometry of $\mathcal{M}_X$, and hence to get some information about the entire collection of Calabi–Yau structures on a fixed real manifold. Here is a tool that helps us to do this.

For each $J \in \mathcal{A}_X$, write X_J for the complex manifold (X, J). Then $H^m(X, \mathbb{C})$ splits into a direct sum of Dolbeault groups $H^{p,q}(X_J)$ for $p+q = m$. In particular, $H^{m,0}(X_J)$ is a 1-dimensional subspace of $H^m(X, \mathbb{C})$, and so it defines a point in the complex projective space $P\big(H^m(X, \mathbb{C})\big)$. Clearly, this point depends only on the equivalence class of J in $\mathcal{M}_X$. Define a map $\Phi : \mathcal{M}_X \to P\big(H^m(X, \mathbb{C})\big)$ by $\Phi(J \cdot \mathcal{D}_X) = H^{m,0}(X_J)$. We call Φ the *period map* of X.

We showed above how to use a holomorphic volume form θ on X_J to define an isomorphism between $H^{m-1,1}(X_J)$ and the tangent space to $\mathcal{M}_X$ at the equivalence class of J. It is easy to see that in fact the tangent space is isomorphic to $H^{m,0}(X_J)^* \otimes H^{m-1,1}(X_J)$, without making any choice of θ. However, $H^{m,0}(X_J)^* \otimes H^{m-1,1}(X_J)$ is also isomorphic to a subspace of the tangent space to $P\big(H^m(X, \mathbb{C})\big)$ at the point $H^{m,0}(X_J)$.

Thus there is a natural isomorphism between the tangent space to $\mathcal{M}_X$ at $J \cdot \mathcal{D}_X$, and a subspace of the tangent space to $P(H^m(X, \mathbb{C}))$ at $\Phi(J \cdot \mathcal{D}_X)$. It can be shown that $d\Phi$ induces exactly this isomorphism. This has two important consequences: firstly, Φ is a holomorphic map of complex manifolds, and secondly, Φ is an immersion. Hence we have proved:

Theorem 6.8.3 *Let X be a compact real $2m$-manifold, $\mathcal{M}_X$ the moduli space of Calabi–Yau complex structures on X, and $\Phi : \mathcal{M}_X \to P(H^m(X, \mathbb{C}))$ the map defined above. Then $\mathcal{M}_X$ has the structure of a complex manifold of dimension $h^{m-1,1}(X)$, and Φ is a holomorphic immersion of complex manifolds.*

Now in good cases, we can hope that the map Φ is an *embedding* rather than just an immersion, and that its image $\text{Im}\,\Phi$ can be explicitly identified in $P(H^m(X, \mathbb{C}))$. We can then give a very precise description of the moduli space of Calabi–Yau complex structures $\mathcal{M}_X$ on X.

This happens for $K3$ surfaces, which are Calabi–Yau manifolds of dimension 2, as we will explain in §7.3. One can also give a good description of the moduli space $\mathcal{M}_X$ in the case that X is a Calabi–Yau m-fold for $m \geqslant 3$ defined as a complete intersection in some $\mathbb{CP}^n$ using this method, because all small deformations of X are also complete intersections in $\mathbb{CP}^n$.

6.9 String Theory and Mirror Symmetry

String Theory is a branch of high-energy theoretical physics in which particles are modelled not as points but as 1-dimensional objects – 'strings' – propagating in some background space-time M. String theorists aim to construct a *quantum theory* of the string's motion. The process of quantization is extremely complicated, and fraught with mathematical difficulties that are as yet still poorly understood.

One curious feature that emerges from the process of quantization is that the dimension of the background space-time M must take a particular value, which is not 4. Instead, $\dim M$ must be 10, 11 or 26, depending upon the details of the theory. The most popular theoretical framework is *Supersymmetric String Theory*, which requires $\dim M = 10$.

To account for the disparity between the 10 dimensions the theory requires, and the 4 space-time dimensions we observe, it is supposed that the space we live in looks locally like $M = \mathbb{R}^4 \times X$, where $\mathbb{R}^4$ is Minkowski space, and X is a compact Riemannian 6-manifold with radius of order 10^{-33} cm, the Planck length. Since the Planck length is so small, space then appears to macroscopic observers to be 4-dimensional.

It turns out that because of supersymmetry, X has to be a *Calabi–Yau 3-fold*. Moreover, the geometry and topology of X then determines the laws of the low-energy physics we observe in our 4-dimensional world. For instance, $|\chi(X)|$ should be twice the number of generations of fundamental particles, which according to experimental evidence is 3 (or possibly 4). Therefore the Euler characteristic $\chi(X)$ should be ± 6 (or possibly ± 8).

Because of this, String Theorists are interested in understanding and constructing examples of Calabi-Yau 3-folds. In particular, they hope to find examples of Calabi-Yau 3-folds with exactly the right geometric properties to reproduce the empirical observations of experimental particle physics, an idea referred to as *superstring phenomenology*.

String theorists believe that to each Calabi-Yau 3-fold X, equipped with a harmonic $(1, 1)$-form B, one can associate a *conformal field theory* (CFT), which is a Hilbert space $\mathcal{H}$ with a collection of operators satisfying some relations, to be interpreted as the quantum theory of strings moving in X. Although both Calabi-Yau 3-folds and conformal field theories are mathematically well-defined and fairly well understood, the relationship between them is based on physical ideas and has no good mathematical definition at present.

Now, String Theorists have been able to use this correspondence to translate their understanding of conformal field theories into conjectures about Calabi-Yau 3-folds, which have often turned out to be true, and mathematically very interesting. The most important of these conjectures is the idea of *mirror symmetry*.

If X is a Calabi-Yau 3-fold then $H^{1,1}(X)$ and $H^{2,1}(X)$ can be recovered from the associated CFT as eigenspaces of a certain operator. Furthermore, the only difference between the CFT representations of $H^{1,1}(X)$ and $H^{2,1}(X)$ is the sign of their eigenvalue under a particular U(1)-action, and the choice of sign is only a matter of convention. This led several physicists to conjecture that there should exist a second Calabi-Yau 3-fold Y which should have the same CFT but with the opposite sign, so that $H^{1,1}(X) \cong H^{2,1}(Y)$ and $H^{2,1}(X) \cong H^{1,1}(Y)$. We call X and Y a *mirror pair*.

This idea is very surprising from the mathematical point of view, since $H^{1,1}$ and $H^{2,1}$ are very different geometrical objects, and it seems odd that one could relate elements of $H^{1,1}$ on one manifold X to elements of $H^{2,1}$ on a different manifold Y. However, a lot of evidence has emerged in favour of mirror symmetry and the existence of mirror pairs.

For example, the large family of Calabi-Yau 3-folds constructed by Candelas et al. [52], described in §6.7.2, displays an exact reflection symmetry when $h^{1,1}$ and $h^{2,1}$ are exchanged. In another paper, Candelas et al. [53] study an explicit mirror pair in great detail. Using mirror symmetry they are able to conjecture values for the number of rational curves of each degree on their Calabi-Yau 3-folds. Some of these values have been checked using algebraic geometry, and have been found to be correct.

Another area in which String Theory produces interesting conjectures about Calabi-Yau 3-folds is that of *orbifolds*. If X is a Calabi-Yau 3-fold and G a finite group acting on X preserving the Calabi-Yau structure, then G acts on the CFT associated to X, and Dixon et al. [62, 63] show how to construct a new CFT associated to X/G. It turns out that even though X/G is singular, its CFT is non-singular.

Now crepant resolutions of X/G are supposed to correspond to smooth deformations of the CFT of X/G. By studying the CFT of X/G, physicists can predict the Hodge numbers of these crepant resolutions. Another aspect of the mirror symmetry picture explained by Aspinwall et al. [9, 10, 11] is that the Kähler cones of several topologically distinct Calabi-Yau 3-folds related by *flops* can fit together to make a single moduli space mirror to the deformations of a single Calabi–Yau 3-fold.

String Theory is a huge subject, and to discuss it properly would require several books the size of this one. Some introductory references will be given in the next section.

6.10 Further reading on Calabi–Yau manifolds

Here are some suggestions for further reading on the material of this chapter.

- **General references.** Beauville [27, 28] studies the differential geometry of Calabi–Yau manifolds, and Friedman [75] the algebraic geometry of Calabi–Yau 3-folds. The book by Hübsch [108] contains a lot of material on Calabi–Yau 3-folds, and is comprehensible to both mathematicians and physicists.
- **Calabi–Yau orbifolds and crepant resolutions.** Some references on crepant resolutions of quotient singularities and orbifolds, as in §6.4–§6.6, are Roan [182, 183], Ito [112, 113], Ito and Reid [114], Reid [179], and Bertin and Markushevich [32].

  References focussing more on the String Theory ideas of orbifold Euler numbers and Hodge numbers are (by physicists) Dixon et al. [62, 63], Vafa [212] and Zaslow [230], and (by mathematicians) Atiyah and Segal [14], Batyrev and Dais [25], and Hirzebruch and Höfer [103].
- **Constructions of Calabi–Yau 3-folds.** Papers on Calabi–Yau manifolds written by mathematicians and physicists use rather different language, and are not necessarily mutually comprehensible.

  For constructions of Calabi–Yau 3-folds written by mathematicians, see for instance Yau [227], Tian and Yau [205], Hirzebruch [102] and Roan [181].

  For constructions of Calabi–Yau 3-folds written by physicists, see for instance Strominger and Witten [198], Candelas et al. [52], [54] and [55], Green and Hübsch [87], Hübsch [107], Markushevich et al. [157], and Aspinwall et al. [12].
- **String Theory.** For a survey of String Theory, see Witten [222]. Some books on the basics of String Theory (not really getting as far as Calabi–Yau manifolds) are Lüst and Theisen [156] and Green, Schwarz and Witten [86]. Collections of papers on String Theory, including some introductory material, are given in Green and Gross [85] and Yau [228].
- **Mirror Symmetry.** Some survey papers on mirror symmetry are Greene and Plesser [91] and Vafa [213]. Collections of papers on Mirror Symmetry are given in Yau [229] and Greene and Yau [90].

Here are some papers on topics we have not been able to cover in this chapter.

- **Topology of higher-dimensional Calabi–Yau manifolds.** Salamon [187] finds relations on the Hodge numbers of Calabi–Yau m-folds for $m \geqslant 4$, using index theory.
- **Rational curves on Calabi–Yau 3-folds.** One can use mirror symmetry to count the number of rational curves with a given homology class on a Calabi–Yau 3-fold. Some papers on this are Morrison [166] and Katz [132].

- **Connecting moduli spaces of Calabi–Yau 3-folds.** Because there seem to be very many Calabi–Yau 3-folds, any realistic classification scheme will probably have to use a fairly coarse notion of equivalence. For instance, we could consider two Calabi–Yau 3-folds to be equivalent if we can continuously deform one into the other, and then try to classify Calabi–Yau 3-folds up to deformation.

  If we only allow deformations through smooth 3-folds, then two Calabi–Yau 3-folds will be equivalent only if they are diffeomorphic as real manifolds. However, if we also allow deformations through *singular* 3-folds, say with canonical singularities, then two Calabi–Yau 3-folds can be equivalent even though they are not diffeomorphic. For instance, if two 3-folds are related by a *flop* as in §6.3 then we can deform from one to the other through a singular 3-fold with one node.

  It has been conjectured by Reid that a large class of Calabi–Yau 3-folds may in fact be connected by deformations of this sort, and Green and Hübsch have proved this for complete intersection Calabi–Yau 3-folds. Some references on this subject are Reid [177], Green and Hübsch [89], and Aspinwall et al. [9, 10, 11].

- The **Kähler cone** of a Calabi–Yau 3-fold is studied by Wilson [220, 221].

- **Algebraic geometry of complex 3-folds and the Mori programme.** There is a rough classification scheme for complex algebraic 3-folds called the *Mori programme*, in which the understanding of 3-fold singularities and crepant resolutions play an important part. Some survey papers on this are Mori [164] and Kollar [137, 138].

7

HYPERKÄHLER MANIFOLDS

In this chapter we study Riemannian $4m$-manifolds (M, g) with holonomy $\mathrm{Sp}(m)$. Since $\mathrm{Sp}(m)$ is a subgroup of $\mathrm{SU}(2m)$, we see that g is Kähler and Ricci-flat, and such manifolds are closely related to the Calabi–Yau manifolds of Chapter 6. But as $\mathrm{Sp}(m)$ preserves three complex structures J_1, J_2, J_3 on $\mathbb{R}^{4m}$ with $J_1 J_2 = J_3$, there are corresponding constant complex structures J_1, J_2, J_3 on M. Furthermore, if $a_1, a_2, a_3 \in \mathbb{R}$ with $a_1^2 + a_2^2 + a_3^2 = 1$ then $a_1 J_1 + a_2 J_2 + a_3 J_3$ is a complex structure on M, and g is Kähler with respect to it.

Thus, a metric g on a $4m$-manifold M with $\mathrm{Hol}(g) \subseteq \mathrm{Sp}(m)$ is Kähler with respect to a whole 2-sphere $\mathcal{S}^2$ of complex structures. We call g a *hyperkähler metric*, (J_1, J_2, J_3, g) a *hyperkähler structure*, and (M, J_1, J_2, J_3, g) a *hyperkähler manifold*. Hyperkähler geometry should be understood in terms of the *quaternions* $\mathbb{H}$. The complex structures J_1, J_2, J_3 on M make each tangent space into a left $\mathbb{H}$-module isomorphic to $\mathbb{H}^m$. Hyperkähler manifolds can be thought of as a quaternionic analogue of Kähler manifolds.

Now there are actually *four* interesting families of geometrical structures on $4m$-manifolds based on the quaternions, called hyperkähler, hypercomplex, quaternionic Kähler and quaternionic structures. A *hypercomplex structure* is a triple (J_1, J_2, J_3) of complex structures J_j on M with $J_1 J_2 = J_3$, and can also be described as a torsion-free $\mathrm{GL}(m, \mathbb{H})$-structure. Hypercomplex manifolds are quaternionic analogues of complex manifolds, just as hyperkähler manifolds are quaternionic analogues of Kähler manifolds.

For $m \geqslant 2$, *quaternionic Kähler* and *quaternionic* structures are torsion-free $\mathrm{Sp}(m)\,\mathrm{Sp}(1)$- and $\mathrm{GL}(m, \mathbb{H})\,\mathrm{GL}(1, \mathbb{H})$-structures respectively. They are closely related to hyperkähler and hypercomplex structures. Note that hyperkähler and quaternionic Kähler structures include a Riemannian metric with holonomy contained in $\mathrm{Sp}(m)$ and $\mathrm{Sp}(m)\,\mathrm{Sp}(1)$, but hypercomplex and quaternionic structures do not include a metric.

The four quaternionic geometries are far too large a subject to cover properly in one chapter of a book. As the theme of the book is compact manifolds with special holonomy (and especially, Ricci-flat holonomy), we have chosen to focus mostly on compact hyperkähler manifolds. Many other interesting topics have been excluded, or relegated to a reading list at the end of the chapter.

We begin in §7.1 with an introduction to hyperkähler geometry. Section 7.2 describes some elementary examples of hyperkähler manifolds, the *hyperkähler ALE spaces*. These are noncompact hyperkähler 4-manifolds asymptotic at infinity to $\mathbb{H}/G$, for G a finite subgroup of $\mathrm{Sp}(1)$. They will be generalized in Chapter 8 to ALE manifolds with holonomy $\mathrm{SU}(m)$.

In §7.3 we consider $K3$ *surfaces*, which are compact hyperkähler 4-manifolds. These are very well understood, by treating them as complex surfaces and using complex algebraic geometry, and we give a very precise description of the moduli space of hyperkähler structures on $K3$. Then §7.4 explains the theory of compact hyperkähler manifolds in higher dimensions. In comparison to Calabi–Yau manifolds, rather few examples of compact hyperkähler manifolds are known.

Section 7.5 briefly discusses the other three quaternionic geometries, that is, hypercomplex, quaternionic Kähler and quaternionic manifolds. We close in §7.6 with a reading list on some topics in quaternionic geometry.

7.1 An introduction to hyperkähler geometry

We now introduce the holonomy groups $\mathrm{Sp}(m)$ in $4m$ dimensions. If (X, g) is a Riemannian $4m$-manifold and $\mathrm{Hol}(g) \subseteq \mathrm{Sp}(m)$, then g is Kähler with respect to complex structures J_1, J_2, J_3 on X with $J_1 J_2 = J_3$. If $a_1, a_2, a_3 \in \mathbb{R}$ and $a_1^2 + a_2^2 + a_3^2 = 1$ then $a_1 J_1 + a_2 J_2 + a_3 J_3$ is also a complex structure on X, which makes g Kähler. Thus g is Kähler with respect to a whole 2-sphere $\mathcal{S}^2$ of complex structures. We call (J_1, J_2, J_3, g) a *hyperkähler structure* on X.

It is convenient to define the Lie groups $\mathrm{Sp}(m)$ and hyperkähler manifolds using the *quaternions* $\mathbb{H}$. We shall also discuss *twistor spaces* of hyperkähler manifolds, a device for translating hyperkähler geometry in $4m$ real dimensions into holomorphic geometry in $2m + 1$ complex dimensions. Some references for this section are the foundational paper of Hitchin, Karlhede, Lindström and Roček [105], and Salamon [186, Ch. 8–9].

7.1.1 *The Lie groups* $\mathrm{Sp}(m)$

The quaternions $\mathbb{H}$ are the associative, nonabelian real algebra

$$\mathbb{H} = \{x_0 + x_1 i_1 + x_2 i_2 + x_3 i_3 : x_j \in \mathbb{R}\} \cong \mathbb{R}^4.$$

The imaginary quaternions are $\mathrm{Im}\,\mathbb{H} = \langle i_1, i_2, i_3 \rangle \cong \mathbb{R}^3$. Multiplication is given by

$$i_1 i_2 = -i_2 i_1 = i_3, \quad i_2 i_3 = -i_3 i_2 = i_1, \quad i_3 i_1 = -i_1 i_3 = i_2, \quad i_1^2 = i_2^2 = i_3^2 = -1.$$

When $x = x_0 + x_1 i_1 + x_2 i_2 + x_3 i_3$, we define $\bar{x}$ and $|x|$ by

$$\bar{x} = x_0 - x_1 i_1 - x_2 i_2 - x_3 i_3 \quad \text{and} \quad |x|^2 = x_0^2 + x_1^2 + x_2^2 + x_3^2.$$

These satisfy $\overline{(pq)} = \bar{q}\,\bar{p}$, $|p| = |\bar{p}|$ and $|pq| = |p||q|$ for all $p, q \in \mathbb{H}$.

Let $\mathbb{H}^m$ have coordinates $(q^1, \ldots, q^m)$, with $q^j = x_0^j + x_1^j i_1 + x_2^j i_2 + x_3^j i_3 \in \mathbb{H}$ and $x_k^j \in \mathbb{R}$. Define a metric g and 2-forms $\omega_1, \omega_2, \omega_3$ on $\mathbb{H}^m$ by

$$\begin{aligned}
g &= \sum_{j=1}^m \sum_{k=0}^3 (\mathrm{d}x_k^j)^2, & \omega_1 &= \sum_{j=1}^m \mathrm{d}x_0^j \wedge \mathrm{d}x_1^j + \mathrm{d}x_2^j \wedge \mathrm{d}x_3^j, \\
\omega_2 &= \sum_{j=1}^m \mathrm{d}x_0^j \wedge \mathrm{d}x_2^j - \mathrm{d}x_1^j \wedge \mathrm{d}x_3^j, & \omega_3 &= \sum_{j=1}^m \mathrm{d}x_0^j \wedge \mathrm{d}x_3^j + \mathrm{d}x_1^j \wedge \mathrm{d}x_2^j.
\end{aligned} \quad (7.1)$$

We can also express $g, \omega_1, \omega_2, \omega_3$ neatly as $g + \omega_1 i_1 + \omega_2 i_2 + \omega_3 i_3 = \sum_{j=1}^{m} d\bar{q}^j \otimes dq^j$, using multiplication in $\mathbb{H}$ to interpret $d\bar{q}^j \otimes dq^j$ as an $\mathbb{H}$-valued tensor.

Identify $\mathbb{H}^m$ with $\mathbb{R}^{4m}$. Then g is the Euclidean metric on $\mathbb{R}^{4m}$. Let J_1, J_2 and J_3 be the complex structures on $\mathbb{R}^{4m}$ corresponding to left multiplication by i_1, i_2 and i_3 in $\mathbb{H}^m$ respectively. Then g is Kähler with respect to each J_j, with Kähler form ω_j. Furthermore, if $a_1, a_2, a_3 \in \mathbb{R}$ with $a_1^2 + a_2^2 + a_3^2 = 1$ then $a_1 J_1 + a_2 J_2 + a_3 J_3$ is a complex structure on $\mathbb{R}^{4m}$, and g is Kähler with respect to it, with Kähler form $a_1 \omega_1 + a_2 \omega_2 + a_3 \omega_3$.

The subgroup of $GL(4m, \mathbb{R})$ preserving g, ω_1, ω_2 and ω_3 is $Sp(m)$. It is a compact, connected, simply-connected, semisimple Lie group of dimension $2m^2 + m$, a subgroup of $SO(4m)$. Since any of g, ω_j and J_j can be written in terms of the other two, $Sp(m)$ also preserves J_1, J_2 and J_3.

We can write $Sp(m)$ as a group of $m \times m$ matrices over $\mathbb{H}$ by

$$Sp(m) \cong \{A \in M_m(\mathbb{H}) : A\bar{A}^t = I\}.$$

To understand the action of $Sp(m)$ on $\mathbb{H}^m$ we think of $\mathbb{H}^m$ as row matrices over $\mathbb{H}$, and then $A \in Sp(m)$ acts by $(q^1 q^2 \cdots q^m) \mapsto (q^1 q^2 \cdots q^m)\bar{A}^t$. The point here is that J_1, J_2, J_3 are defined by left multiplication by i_2, i_2, i_3, so to commute with this A must act on the right; but as by convention mappings act on the left we right-multiply by $\bar{A}^t$, to preserve the order of multiplication.

We can also identify $\mathbb{H}^m = \mathbb{R}^{4m}$ with $\mathbb{C}^{2m}$. Define complex coordinates $(z_1, \ldots, z_{2m})$ on $\mathbb{R}^{4m}$ by $z_{2j-1} = x_0^j + i x_1^j$ and $z_{2j} = x_2^j + i x_3^j$ for $j = 1, \ldots, m$. Then g, ω_1, ω_2 and ω_3 satisfy

$$g = \sum_{j=1}^{2m} |dz_j|^2, \quad \omega_1 = \frac{i}{2} \sum_{j=1}^{2m} dz_j \wedge d\bar{z}_j \quad \text{and} \quad \omega_2 + i\omega_3 = \sum_{j=1}^{m} dz_{2j-1} \wedge dz_{2j}.$$

That is, g and ω_1 are the standard Hermitian metric and Hermitian form on $\mathbb{C}^{2m}$, and the (2,0)-form $\omega_2 + i\omega_3$ is a *complex symplectic form* on $\mathbb{C}^{2m}$.

Observe that $\frac{1}{m!}(\omega_2 + i\omega_3)^m = dz_1 \wedge \cdots \wedge dz_{2m}$, which is the usual holomorphic volume form θ on $\mathbb{C}^{2m}$. As $Sp(m)$ preserves ω_2 and ω_3 it preserves θ. So $Sp(m)$ preserves the metric g, Hermitian form ω_1 and complex volume form θ on $\mathbb{C}^{2m}$. From §6.1, this means that $Sp(m)$ is a subgroup of $SU(2m)$, the subgroup fixing the complex symplectic form $\omega_2 + i\omega_3$.

Now $\dim Sp(m) = 2m^2 + m$ and $\dim SU(2m) = 4m^2 - 1$. Thus, for all $m > 1$ we have $\dim Sp(m) < \dim SU(2m)$, and $Sp(m)$ is a proper subgroup of $SU(2m)$. However, when $m = 1$ we have $\dim Sp(1) = \dim SU(2) = 3$, and in fact $SU(2) = Sp(1)$, so the groups $SU(2)$ and $Sp(1)$ are the same.

7.1.2 The holonomy groups $Sp(m)$ and hyperkähler structures

Suppose (X, g) is a Riemannian $4m$-manifold with $Hol(g) \subseteq Sp(m)$. By Proposition 2.5.2, each $Sp(m)$-invariant tensor on $\mathbb{R}^{4m}$ corresponds to a tensor on X constant under the Levi-Civita connection ∇ of g. So from above there exist almost complex structures

J_1, J_2, J_3 and 2-forms $\omega_1, \omega_2, \omega_3$ on X, each constant under ∇, and isomorphic to the standard models on $\mathbb{R}^{4m}$ at each point of X.

As $\nabla J_j = 0$, each J_j is an integrable complex structure, and g is Kähler with respect to J_j, with Kähler form ω_j. Similarly, if $a_1, a_2, a_3 \in \mathbb{R}$ with $a_1^2 + a_2^2 + a_3^2 = 1$ then $a_1 J_1 + a_2 J_2 + a_3 J_3$ is a complex structure on X, and g is Kähler with respect to it, with Kähler form $a_1\omega_1 + a_2\omega_2 + a_3\omega_3$. Therefore g is Kähler in lots of different ways, with respect to a whole 2-sphere of complex structures. Because of this, we call g *hyperkähler*.

Definition 7.1.1 Let X be a $4m$-manifold. An *almost hyperkähler structure* on X is a quadruple (J_1, J_2, J_3, g), where J_j are almost complex structures on X with $J_1 J_2 = J_3$, and g is a Riemannian metric on X which is Hermitian with respect to J_1, J_2 and J_3.

We call (J_1, J_2, J_3, g) a *hyperkähler structure* on X if in addition $\nabla J_j = 0$ for $j = 1, 2, 3$, where ∇ is the Levi-Civita connection of g. Then (X, J_1, J_2, J_3, g) is a *hyperkähler manifold*, and g a *hyperkähler metric*. Each J_j is integrable, and g is Kähler with respect to J_j. We refer to the Kähler forms $\omega_1, \omega_2, \omega_3$ of J_1, J_2, J_3 as the *hyperkähler 2-forms* of X.

An almost hyperkähler structure is equivalent to an $\mathrm{Sp}(m)$-structure, and a hyperkähler structure to a torsion-free $\mathrm{Sp}(m)$-structure. The next proposition comes from [186, Lem. 8.4], except the last part, which follows from Proposition 6.1.1 using the inclusion $\mathrm{Sp}(m) \subseteq \mathrm{SU}(2m)$.

Proposition 7.1.2 *Suppose X is a $4m$-manifold and (J_1, J_2, J_3, g) an almost hyperkähler structure on X, and let $\omega_1, \omega_2, \omega_3$ be the Hermitian forms of J_1, J_2 and J_3. Then the following conditions are equivalent:*

(i) (J_1, J_2, J_3, g) *is a hyperkähler structure*,
(ii) $d\omega_1 = d\omega_2 = d\omega_3 = 0$,
(iii) $\nabla\omega_1 = \nabla\omega_2 = \nabla\omega_3 = 0$, *and*
(iv) $\mathrm{Hol}(g) \subseteq \mathrm{Sp}(m)$, *and J_1, J_2, J_3 are the induced complex structures.*

All hyperkähler metrics are Ricci-flat.

7.1.3 *Twistor spaces of hyperkähler manifolds*

Let (X, J_1, J_2, J_3, g) be a hyperkähler $4m$-manifold. Set $\mathbb{Z} = \mathbb{CP}^1 \times X$, where

$$\mathbb{CP}^1 \cong \mathcal{S}^2 = \{a_1 J_1 + a_2 J_2 + a_3 J_3 : a_j \in \mathbb{R}, \ a_1^2 + a_2^2 + a_3^2 = 1\}$$

is the natural 2-sphere of complex structures on X. Then $\mathbb{CP}^1$ has a natural complex structure J_0, say. Each point $z \in Z$ is of the form (J, x) for $J \in \mathbb{CP}^1$ and $x \in X$, and the tangent space $T_z Z$ is $T_z Z = T_J \mathbb{CP}^1 \oplus T_x X$.

Now J_0 is a complex structure on $T_J \mathbb{CP}^1$, and J is a complex structure on $T_x X$. Thus $J_Z = J_0 \oplus J$ is a complex structure on $T_z Z = T_J \mathbb{CP}^1 \oplus T_x X$. This defines an almost complex structure J_Z on Z, which turns out to be integrable. Hence (Z, J_Z) is a complex $(2m+1)$-manifold, called the *twistor space* of X.

Let $p : Z \to \mathbb{CP}^1$ and $\pi : Z \to X$ be the natural projections. Then p is holomorphic, and the hypersurface $p^{-1}(J)$ is isomorphic to X with complex structure J, for each $J \in \mathbb{CP}^1$. Define $\sigma : Z \to Z$ by $\sigma : (J, x) \mapsto (-J, x)$. Then σ is a free antiholomorphic involution of Z. For each $x \in X$, the fibre $\Sigma_x = \pi^{-1}(x)$ of π is a holomorphic curve in Z isomorphic to $\mathbb{CP}^1$, with normal bundle $2m\mathcal{O}(1)$, which is preserved by σ.

There is one other piece of holomorphic data on Z given by the hyperkähler structure on X. Let $\mathcal{D}$ be the kernel of $\mathrm{d}p : TZ \to T\mathbb{CP}^1$. Then $\mathcal{D}$ is a vector subbundle of TZ, which is holomorphic as p is holomorphic; $\mathcal{D}$ is the bundle of tangent spaces to the fibres of p. Then one can construct a nondegenerate holomorphic section ω of the holomorphic vector bundle $p^*(\mathcal{O}(2)) \otimes \Lambda^2 \mathcal{D}^*$ over Z.

Effectively this amounts to a choice of complex symplectic form on each of the fibres $p^{-1}(J) \cong (X, J)$ of p. For instance, the complex 2-form $\omega_2 + i\omega_3$ on X is holomorphic with respect to J_1, and on $p^{-1}(J_1)$ we take ω to be $\omega_2 + i\omega_3$, multiplied by some fixed element in the fibre of $\mathcal{O}(2)$ over J_1 in $\mathbb{CP}^1$. We summarize these ideas in the following theorem, taken from [105, §3(F)].

Theorem 7.1.3 *Let (X, J_1, J_2, J_3, g) be a hyperkähler $4m$-manifold. Then the twistor space (Z, J_Z) of X is a complex $(2m+1)$-manifold diffeomorphic to $\mathbb{CP}^1 \times X$. Let $p : Z \to \mathbb{CP}^1$ and $\pi : Z \to X$ be the projections. Then p is holomorphic, and there exists a holomorphic section ω of $p^*(\mathcal{O}(2)) \otimes \Lambda^2 \mathcal{D}^*$ which is symplectic on the fibres of the kernel $\mathcal{D}$ of $\mathrm{d}p : TZ \to T\mathbb{CP}^1$. There is also a natural, free antiholomorphic involution $\sigma : Z \to Z$ satisfying $\sigma^*(\omega) = \omega$, $\pi \circ \sigma = \pi$ and $p \circ \sigma = \sigma' \circ p$, where $\sigma' : \mathbb{CP}^1 \to \mathbb{CP}^1$ is the antipodal map.*

It turns out that the holomorphic data Z, p, ω and σ is sufficient to reconstruct (X, J_1, J_2, J_3, g). We express this in our next result, deduced from [105, Th. 3.3].

Theorem 7.1.4 *Suppose (Z, J_Z) is a complex $(2m+1)$-manifold equipped with*

(a) *a holomorphic projection $p : Z \to \mathbb{CP}^1$,*

(b) *a holomorphic section ω of $p^*(\mathcal{O}(2)) \otimes \Lambda^2 \mathcal{D}^*$ which is symplectic on the fibres of $\mathcal{D}$, where $\mathcal{D}$ is the kernel of $\mathrm{d}p : TZ \to T\mathbb{CP}^1$, and*

(c) *a free antiholomorphic involution $\sigma : Z \to Z$ satisfying $\sigma^*(\omega) = \omega$ and $p \circ \sigma = \sigma' \circ p$, where $\sigma' : \mathbb{CP}^1 \to \mathbb{CP}^1$ is the antipodal map.*

Define X' to be the set of holomorphic curves C in Z isomorphic to $\mathbb{CP}^1$, with normal bundle $2m\mathcal{O}(1)$ and $\sigma(C) = C$. Then X' is a hypercomplex $4m$-manifold, equipped with a natural pseudo-hyperkähler metric g. If g is positive definite, then X' is hyperkähler. Let Z' be the twistor space of X'. Then there is a natural, locally biholomorphic map $\iota : Z' \to Z$, which identifies p, ω and σ with their analogues on Z'.

Here a *pseudo-hyperkähler metric* is a pseudo-Riemannian metric g of type $(4k, 4m - 4k)$ for some $m/2 \leq k \leq m$, with $\mathrm{Hol}(g) \subseteq \mathrm{Sp}(k, m-k)$. It is positive definite if $m = k$. One moral of this theorem is that hyperkähler manifolds can be written solely in terms of *holomorphic* data, and so they can be studied and explicit examples found using *complex algebraic geometry*.

7.2 Hyperkähler ALE spaces

The *ALE spaces* are a special class of noncompact hyperkähler 4-manifolds.

Definition 7.2.1 Let G be a finite subgroup of $\mathrm{Sp}(1)$, and let $(\hat{J}_1, \hat{J}_2, \hat{J}_3, \hat{g})$ be the Euclidean hyperkähler structure and $r : \mathbb{H}/G \to [0, \infty)$ the radius function on $\mathbb{H}/G$. We say that a hyperkähler 4-manifold (X, J_1, J_2, J_3, g) is *Asymptotically Locally Euclidean*, or *ALE*, and asymptotic to $\mathbb{H}/G$, if there exists a compact subset $S \subset X$ and a map $\pi : X \setminus S \to \mathbb{H}/G$ that is a diffeomorphism between $X \setminus S$ and $\{x \in \mathbb{H}/G : r(x) > R\}$ for some $R > 0$, such that

$$\hat{\nabla}^k(\pi_*(g) - \hat{g}) = O(r^{-4-k}) \quad \text{and} \quad \hat{\nabla}^k(\pi_*(J_j) - \hat{J}_j) = O(r^{-4-k}) \tag{7.2}$$

as $r \to \infty$, for $j = 1, 2, 3$ and $k \geqslant 0$, where $\hat{\nabla}$ is the Levi-Civita connection of $\hat{g}$.

Hyperkähler ALE spaces are called *gravitational instantons* by physicists. What this definition means is that a hyperkähler ALE space is a noncompact hyperkähler 4-manifold X with one end which at infinity resembles $\mathbb{H}/G$, and the hyperkähler structure on X and its derivatives are required to approximate the Euclidean hyperkähler structure on $\mathbb{H}/G$ with a prescribed rate of decay.

One reason why hyperkähler ALE spaces are interesting is that they give a local model for how to desingularize hyperkähler 4-orbifolds to give hyperkähler 4-manifolds, as we will see when we discuss the Kummer construction of the $K3$ surface in §7.4. They are also the simplest examples of ALE manifolds with holonomy $\mathrm{SU}(m)$, which will be the subject of Chapter 8.

The first examples of hyperkähler ALE spaces, asymptotic to $\mathbb{H}/\{\pm 1\}$, were written down explicitly by Eguchi and Hanson [67] and are called *Eguchi–Hanson spaces*, which we define next.

Example 7.2.2 Consider $\mathbb{C}^2$ with complex coordinates (z_1, z_2), acted upon by the involution $-1 : (z_1, z_2) \mapsto (-z_1, -z_2)$. Let (X, π) be the blow-up of $\mathbb{C}^2/\{\pm 1\}$ at 0. Then X is a crepant resolution of $\mathbb{C}^2/\{\pm 1\}$. It is biholomorphic to $T^*\mathbb{CP}^1$, with $\pi_1(X) = \{1\}$ and $H^2(X, \mathbb{R}) = \mathbb{R}$. Define $f : X \setminus \pi^{-1}(0) \to \mathbb{R}$ by

$$f = \sqrt{r^4 + 1} + 2\log r - \log(\sqrt{r^4 + 1} + 1), \tag{7.3}$$

where $r = (|z_1|^2 + |z_2|^2)^{1/2}$ is the radius function on X.

Define a 2-form ω_1 on $X \setminus \pi^{-1}(0)$ by $\omega_1 = i\partial\bar{\partial} f$. Then ω_1 extends smoothly and uniquely to X. Furthermore, ω_1 is a closed, real, positive (1,1)-form, and is thus the Kähler form of a Kähler metric g on X. This is the *Eguchi–Hanson* metric on X, which has holonomy $\mathrm{SU}(2)$. It extends to a hyperkähler structure (J_1, J_2, J_3, g) on X, where J_1 is the natural complex structure on X, and the Kähler forms ω_2, ω_3 of J_2, J_3 satisfy $\omega_2 + i\omega_3 = \pi^*(dz_1 \wedge dz_2)$.

For large r we have $f = r^2 + O(r^{-2})$, so that $\omega = i\partial\bar{\partial}(r^2) + O(r^{-4})$. But $i\partial\bar{\partial}(r^2)$ is the Kähler form of the Euclidean metric $\hat{g}$ on $\mathbb{C}^2/\{\pm 1\}$, so that $g = \hat{g} + O(r^{-4})$ for large r, as in the first equation of (7.2). In the same way we can show that (X, J_1, J_2, J_3, g) is a *hyperkähler ALE space* asymptotic to $\mathbb{H}/\{\pm 1\}$.

There is a 3-dimensional family of hyperkähler ALE spaces $(X', J'_1, J'_2, J'_3, g')$ asymptotic to $\mathbb{H}/\{\pm 1\}$, which are called *Eguchi–Hanson spaces*, and are all isomorphic to the example (X, J_1, J_2, J_3, g) above under homotheties and rotations of the 2-sphere $\mathcal{S}^2$ of complex structures.

That is, if $(X', J'_1, J'_2, J'_3, g')$ is any hyperkähler ALE space asymptotic to $\mathbb{H}/\{\pm 1\}$ then X' is diffeomorphic to X, and we can choose a diffeomorphism $\phi : X \to X'$ such that $\phi^*(g') = t^2 g$ for some $t > 0$, and $\phi^*(J'_j) = \sum_{k=1}^{3} a_{jk} J_k$ for some 3×3 matrix (a_{jk}) in SO(3).

The Eguchi–Hanson spaces were soon generalized by Gibbons and Hawking [81], who gave explicit examples of hyperkähler ALE spaces asymptotic to $\mathbb{H}/\mathbb{Z}_k$ for all $k \geqslant 2$. Hitchin [104] constructed the same spaces using twistor methods.

Eventually, a complete construction and classification of all hyperkähler ALE spaces was achieved by Kronheimer [140, 141]. The construction makes use of the *McKay correspondence* of §6.4.1, which links the Kleinian singularities $\mathbb{C}^2/G$, their crepant resolutions, and the Dynkin diagrams of type A_r ($r \geqslant 1$), D_r ($r \geqslant 4$), E_6, E_7 and E_8. Here is a statement of Kronheimer's results, following from [140, 141], except part (d), which the author made up.

Theorem 7.2.3 *Let G be a nontrivial finite subgroup of* SU(2) *with Dynkin diagram Γ, and let $\mathfrak{h}_\mathbb{R}$ be the real vector space with basis the set of nontrivial irreducible representations of G. Let Δ be the set of roots and W the Weyl group of Γ. Then* $\text{Aut}(\Gamma) \ltimes W$ *acts naturally on $\mathfrak{h}_\mathbb{R}$, and Δ is a finite subset of $\mathfrak{h}_\mathbb{R}^*$ preserved by* $\text{Aut}(\Gamma) \ltimes W$. *Define $U \subset \mathfrak{h}_\mathbb{R} \otimes \mathbb{R}^3$ by*

$$U = \big\{ (\alpha_1, \alpha_2, \alpha_3) \in \mathfrak{h}_\mathbb{R} \otimes \mathbb{R}^3 : \text{for each } \delta \in \Delta, \\ \delta(\alpha_1), \delta(\alpha_2), \delta(\alpha_3) \text{ are not all zero} \big\}. \tag{7.4}$$

There exists a continuous family of noncompact hyperkähler 4-manifolds X_α parametrized by $\alpha = (\alpha_1, \alpha_2, \alpha_3)$ in $\mathfrak{h}_\mathbb{R} \otimes \mathbb{R}^3$, which can be written down explicitly using the hyperkähler quotient construction, satisfying the following conditions:

(a) *If $\alpha \in U$ then X_α is a hyperkähler ALE space asymptotic to $\mathbb{H}/G$, diffeomorphic to the crepant resolution X of $\mathbb{C}^2/G$. There is a natural isomorphism $H^2(X_\alpha, \mathbb{R}) \cong \mathfrak{h}_\mathbb{R}$.*

(b) *If $\alpha \notin U$ then X_α is a singular hyperkähler orbifold asymptotic to $\mathbb{H}/G$. Also $X_0 \cong \mathbb{H}/G$.*

(c) *Let $\alpha \in U$. Then the 2-forms $\omega_1, \omega_2, \omega_3$ have cohomology classes $[\omega_j]$ in $H^2(X_\alpha, \mathbb{R})$. The isomorphism $H^2(X_\alpha, \mathbb{R}) \cong \mathfrak{h}_\mathbb{R}$ identifies $[\omega_j]$ and α_j for $j = 1, 2, 3$.*

(d) *Let $\alpha, \beta \in \mathfrak{h}_\mathbb{R} \otimes \mathbb{R}^3$. Then X_α and X_β are isomorphic as hyperkähler manifolds if and only if $\alpha = (\gamma, w) \cdot \beta$ for some $(\gamma, w) \in \text{Aut}(\Gamma) \ltimes W$.*

(e) *Suppose X' is a hyperkähler ALE space asymptotic to $\mathbb{H}/G$. Then there exists $\alpha \in U$ such that $X' \cong X_\alpha$.*

Here the *hyperkähler quotient* is a method of producing hyperkähler manifolds due to Hitchin et al. [105, §3]. Given a hyperkähler $4m$-manifold (X, J_1, J_2, J_3, g) and a suitable k-dimensional Lie group G of automorphisms of it, the hyperkähler quotient of X by G is a new hyperkähler manifold of dimension $4(m-k)$.

Kronheimer's proof of this theorem comes in two parts. First, in [140], for each finite group $G \subset \mathrm{SU}(2)$ he uses the Dynkin diagram Γ of G to write down an explicit hyperkähler quotient of $\mathbb{H}^n$ by a product of unitary groups $\mathrm{U}(k)$, and shows that for suitable values of the moment map, this quotient is a hyperkähler ALE space asymptotic to $\mathbb{H}/G$.

Then in [141] he shows that every hyperkähler ALE space X asymptotic to $\mathbb{H}/G$ arises from this construction. The proof uses the twistor space of X, and facts about the deformations of $\mathbb{C}^2/G$ taken from Slodowy [196].

Here is the reason for the condition defining U in (7.4). Each element $\delta \in \Delta$ corresponds to a 2-sphere S^2 in X, with self-intersection -2. It turns out that the volume of the corresponding S^2 in X_α is $\left(\sum_{j=1}^{3} \delta(\alpha_j)^2\right)^{1/2}$. Thus if $\delta(\alpha_j) = 0$ for $j = 1, 2, 3$ then this S^2 collapses to a point, and X_α becomes singular.

We shall also explain the rôle of the group $\mathrm{Aut}(\Gamma) \ltimes W$ in part (d). Here W is the Weyl group, and $\mathrm{Aut}(\Gamma)$ the automorphism group of the graph Γ, given by

$$\mathrm{Aut}(\Gamma) = \begin{cases} \{1\} & \text{if } \Gamma = A_1, E_7 \text{ or } E_8, \\ \mathbb{Z}_2 & \text{if } \Gamma = A_k \ (k \geqslant 2), D_k \ (k \geqslant 5) \text{ or } E_6, \\ S_3 & \text{if } \Gamma = D_4. \end{cases}$$

There is a natural, surjective group homomorphism $\rho : \mathrm{Aut}(\mathbb{H}/G) \to \mathrm{Aut}(\Gamma)$, such that $\mathrm{Ker}\,\rho$ is the identity component of $\mathrm{Aut}(\mathbb{H}/G)$. Hence $\mathrm{Aut}(\Gamma)$ is the group of isotopy classes of automorphisms of $\mathbb{H}/G$.

Let X be the crepant resolution of $\mathbb{C}^2/G$, regarded as a real 4-manifold. Isotopy (continuous deformation) is an equivalence relation on the diffeomorphism group of X. It turns out that $\mathrm{Aut}(\Gamma) \ltimes W$ acts on X as a group of *isotopy classes of diffeomorphisms*. For each $(\gamma, w) \in \mathrm{Aut}(\Gamma) \ltimes W$ there is an isotopy class of diffeomorphisms $\Phi : X \to X$, such that $\Phi_* : H^2(X, \mathbb{C}) \to H^2(X, \mathbb{C})$ coincides with $(\gamma, w) : \mathfrak{h}_\mathbb{C} \to \mathfrak{h}_\mathbb{C}$ under the isomorphism $H^2(X, \mathbb{C}) \cong \mathfrak{h}_\mathbb{C}$.

If $\gamma = 1$, so that $(\gamma, w) \in W$, then we can choose Φ to be the identity outside a compact subset in X. More generally, we can choose Φ to be asymptotic to any $\tau \in \mathrm{Aut}(\mathbb{H}/G)$ with $\rho(\tau) = \gamma$. Thus $\mathrm{Aut}(\Gamma) \ltimes W$ is a kind of symmetry group of the topology of X.

The fact that ALE spaces admit automorphisms which act nontrivially on their topology will be important in Chapters 12 and 14, when we will use it to resolve a single orbifold T^7/Γ or T^8/Γ in many topologically distinct ways, and so get many different compact 7- or 8-manifolds with holonomy G_2 or $\mathrm{Spin}(7)$. The author also used this idea to find deformations of Calabi–Yau 3-orbifolds T^6/Γ with many different Hodge numbers in [126].

7.3 $K3$ surfaces

A $K3$ *surface* is defined to be a compact, complex surface (X, J) with $h^{1,0}(X) = 0$ and trivial canonical bundle. $K3$ surfaces occupy a special place in Kodaira's classification of complex surfaces. They are also important in Riemannian holonomy, as they carry Kähler metrics with holonomy $SU(2) = Sp(1)$, and are the only compact 4-manifold which does so.

Thus $K3$ surfaces are the lowest-dimensional examples of both Calabi–Yau manifolds (with holonomy $SU(m)$), and compact hyperkähler manifolds (with holonomy $Sp(m)$). But because their behaviour is more typical of the hyperkähler than the Calabi–Yau case, we discuss them here and not in Chapter 6.

We begin with some examples, and then discuss $K3$ surfaces first as complex surfaces, and then as hyperkähler 4-manifolds.

7.3.1 Examples of K3 surfaces

Example 7.3.1 Define S to be the *Fermat quartic*

$$S = \{[z_0, \ldots, z_3] \in \mathbb{CP}^3 : z_0^4 + z_1^4 + z_2^4 + z_3^4 = 0\}.$$

The *adjunction formula* [93, p. 147] shows that $K_S = (K_{\mathbb{CP}^3} \otimes L_S)|_S$, where L_S is the line bundle over $\mathbb{CP}^3$ associated to the divisor S. But $K_{\mathbb{CP}^3} = \mathcal{O}(-4)$ and $L_S = \mathcal{O}(4)$ as S is a quartic. So $K_{\mathbb{CP}^3} \otimes L_S = \mathcal{O}(0)$, and the canonical bundle K_S of S is trivial. Theorem 4.10.4 shows that $H^k(S, \mathbb{C}) \cong H^k(\mathbb{CP}^3, \mathbb{C})$ and $\pi_k(S) \cong \pi_k(\mathbb{CP}^3)$ for $k = 0, 1$, so S is connected and simply-connected. Thus $h^{1,0}(S) = 0$ and K_S is trivial, so that S is a $K3$ surface, by definition.

We shall work out the Hodge and Betti numbers $h^{p,q}$, b^k of S. The Riemann–Roch formula shows that $\chi(S) = 24$, and as $b^0 = 1$ and $b^1 = 0$ we have $b^2 = 22$. Also $h^{2,0} = 1$ as K_S is trivial, so $h^{0,2} = 1$ and $h^{1,1} = 20$. But the signature $\tau(S) = b_+^2 - b_-^2$ satisfies $\tau(S) = \sum_{p,q=0}^{2}(-1)^p h^{p,q} = -16$. Hence $b_+^2 = 3$ and $b_-^2 = 19$.

More generally, using §6.7 we find that the following are all $K3$ surfaces:

- Any nonsingular quartic in $\mathbb{CP}^3$.
- A complete intersection of a cubic and a quadric in $\mathbb{CP}^4$.
- A complete intersection of 3 quadrics in $\mathbb{CP}^5$.

But these are all projective varieties. Here are some *non-algebraic* $K3$ surfaces.

Example 7.3.2 Let Λ be a *lattice* in $\mathbb{C}^2$, so that $\Lambda \cong \mathbb{Z}^4$. Then $\mathbb{C}^2/\Lambda$ is a complex 4-torus T^4. Define a map $\sigma : T^4 \to T^4$ by $\sigma : (z_1, z_2) + \Lambda \mapsto (-z_1, -z_2) + \Lambda$. Then σ fixes the 16 points

$$\{(z_1, z_2) + \Lambda : (z_1, z_2) \in \tfrac{1}{2}\Lambda\}.$$

Thus $T^4/\langle\sigma\rangle$ is a *complex orbifold*, with 16 singular points modelled on $\mathbb{C}^2/\{\pm 1\}$.

Let S be the blow-up of $T^4/\langle\sigma\rangle$ at the 16 singular points. Then S is a *crepant resolution* of $T^4/\langle\sigma\rangle$, and is a $K3$ surface. We call this the *Kummer construction*, and S a *Kummer surface*. For a generic choice of lattice Λ, the torus T^4 is not an algebraic surface, and neither is the Kummer surface S. Thus there exist non-algebraic $K3$ surfaces.

Now $T^4/\langle\sigma\rangle$ is simply-connected, and $H^k(T^4/\langle\sigma\rangle, \mathbb{C})$ is the σ-invariant part of $H^k(T^4, \mathbb{C})$. Thus we find that $b_\pm^2(T^4/\langle\sigma\rangle) = 3$. The blow-up replaces each singular point with a copy of $\mathbb{CP}^1$, with self-intersection -2. This leaves π_1 and b_+^2 unchanged, and adds one to b_-^2 for each of the 16 points. Hence S is simply-connected and has $b_+^2 = 3$ and $b_-^2 = 19$, as in Example 7.3.1.

We shall return to the Kummer construction in Example 7.3.14.

7.3.2 *K3 surfaces as complex surfaces*

The theory of complex surfaces (including non-algebraic surfaces) is an old and very well understood branch of complex algebraic geometry. We now discuss what is known about $K3$ surfaces from this point of view. We begin with a result on the topology of $K3$ surfaces.

Theorem 7.3.3 *Let (X, J) be a $K3$ surface. Then X is simply-connected, with Betti numbers $b^2 = 22$, $b_+^2 = 3$, and $b_-^2 = 19$. Also X is Kähler, and has Hodge numbers $h^{2,0} = h^{0,2} = 1$ and $h^{1,1} = 20$. All $K3$ surfaces are diffeomorphic.*

Proof Kodaira [135, Th. 13] showed that every $K3$ surface is a deformation of a nonsingular quartic surface in $\mathbb{CP}^3$. Thus all $K3$ surfaces are diffeomorphic to the surface S of Example 7.3.1. Hence $\pi_1(X) = \{1\}$, $b^2 = 22$, $b_+^2 = 3$, and $b_-^2 = 19$, from above. Todorov [208, Th. 2] and Siu [195] prove that every $K3$ surface is Kähler. (This is not obvious.) As K_X is trivial we have $h^{2,0} = 1$, so $h^{0,2} = 1$ as X is Kähler, and this leaves $h^{1,1} = 20$ as $b^2 = 22$. $\square$

Work on $K3$ surfaces has focussed on two main areas: firstly, the study of *algebraic $K3$ surfaces*, and secondly, the description of the moduli space of *all $K3$ surfaces*, including the non-algebraic ones. We shall explain the principal results in this second area. Some good general references on the following material are Beauville et al. [29], Barth et al. [21, Ch. VIII] and Besse [33, p. 365–368].

Here are some important tools for studying the moduli space of $K3$ surfaces.

Definition 7.3.4 Let Λ be a lattice isomorphic to $\mathbb{Z}^{22}$, with an even, unimodular quadratic form $q_\Lambda : \Lambda \to \mathbb{Z}$ of signature $(3, 19)$. All such lattices are isomorphic. A *marked K3 surface* (X, J, ϕ) is a $K3$ surface (X, J) with an isomorphism $\phi : H^2(X, \mathbb{Z}) \to \Lambda$ identifying the intersection form q_X on $H^2(X, \mathbb{Z})$ with the quadratic form q_Λ on Λ. Let $\mathcal{M}_{K3}$ be the *moduli space of marked K3 surfaces*. It is locally a complex manifold of dimension 20, by Theorem 6.8.1. However, note that globally, $\mathcal{M}_{K3}$ is not Hausdorff.

Write $\Lambda_\mathbb{R} = \Lambda \otimes_\mathbb{Z} \mathbb{R}$ and $\Lambda_\mathbb{C} = \Lambda \otimes_\mathbb{Z} \mathbb{C}$. Then ϕ induces isomorphisms $\phi_\mathbb{R} : H^2(X, \mathbb{R}) \to \Lambda_\mathbb{R}$ and $\phi_\mathbb{C} : H^2(X, \mathbb{C}) \to \Lambda_\mathbb{C}$. Now $H^{2,0}(X) \cong \mathbb{C}$. We define the *period* of (X, J, ϕ) to be $\phi_\mathbb{C}(H^{2,0}(X))$, which we regard as a point in the complex projective space $P(\Lambda_\mathbb{C})$. Define the *period map* $\mathcal{P} : \mathcal{M}_{K3} \to P(\Lambda_\mathbb{C})$ by $\mathcal{P} : (X, J, \phi) \mapsto \phi_\mathbb{C}(H^{2,0}(X))$. Then $\mathcal{P}$ is holomorphic.

Now $\mathcal{M}_{K3}$ is a (non-Hausdorff) complex manifold of dimension 20, and $P(\Lambda_\mathbb{C}) \cong \mathbb{CP}^{21}$. Thus we expect $\mathcal{P}(\mathcal{M}_{K3})$ to be a complex hypersurface in $P(\Lambda_\mathbb{C})$. To identify which hypersurface, let (X, J, ϕ) be a marked $K3$ surface, and choose a holomorphic volume form $\omega_{\mathbb{C}}$ on X. Then $[\omega_{\mathbb{C}}] \in H^{2,0}(X)$. Define $\lambda_X = \phi_{\mathbb{C}}([\omega_{\mathbb{C}}]) \in \Lambda_{\mathbb{C}}$. As $\omega_{\mathbb{C}}$ is a (2,0)-form we have $\omega_{\mathbb{C}} \wedge \omega_{\mathbb{C}} = 0$. Thus

$$q_\Lambda(\lambda_X) = q_X([\omega_{\mathbb{C}}]) = \int_X \omega_{\mathbb{C}} \wedge \omega_{\mathbb{C}} = 0.$$

Near each point in X we can choose holomorphic coordinates (z_1, z_2) such that $\omega_{\mathbb{C}} = dz_1 \wedge dz_2$. Then

$$(\omega_{\mathbb{C}} + \bar{\omega}_{\mathbb{C}})^2 = (dz_1 \wedge dz_2 + d\bar{z}_1 \wedge d\bar{z}_2)^2 = 2 dz_1 \wedge dz_2 \wedge d\bar{z}_1 \wedge d\bar{z}_2,$$

which is a *positive* 4-form. Hence

$$q_\Lambda(\lambda_X + \bar{\lambda}_X) = q_X([\omega_{\mathbb{C}}] + [\bar{\omega}_{\mathbb{C}}]) = \int_X (\omega_{\mathbb{C}} + \bar{\omega}_{\mathbb{C}})^2 > 0.$$

We have shown that $q_\Lambda(\lambda_X) = 0$ and $q_\Lambda(\lambda_X + \bar{\lambda}_X) > 0$. But the period of X is $\langle \lambda_X \rangle \in P(\Lambda_{\mathbb{C}})$. Thus we have found two conditions on the period of (X, J, ϕ).

Definition 7.3.5 Define the *period domain* Q by

$$Q = \{[\lambda] \in P(\Lambda_{\mathbb{C}}) : \lambda \in \Lambda_{\mathbb{C}} \setminus \{0\},\ q_\Lambda(\lambda) = 0,\ q_\Lambda(\lambda + \bar{\lambda}) > 0\}. \tag{7.5}$$

From above, if (X, J, ϕ) is a marked $K3$ surface then the period of X lies in Q, and the period map $\mathcal{P}$ maps $\mathcal{M}_{K3} \to Q$.

We now state a series of important results on the period map. The first was published by Kodaira [135, Th. 17], who attributes it to Weil and Andreotti.

Theorem 7.3.6. (Local Torelli Theorem) *The period map* $\mathcal{P} : \mathcal{M}_{K3} \to Q$ *is a local isomorphism of complex manifolds.*

The second is due to Todorov [208, Th. 1] (see also Looijenga [155]).

Theorem 7.3.7 *The period map* $\mathcal{P} : \mathcal{M}_{K3} \to Q$ *is surjective.*

The third is due to Burns and Rapoport [48, Th. 1], and known as the Global Torelli Theorem. We state it in a weak form.

Theorem 7.3.8. (Weak Torelli Theorem) *Let* (X, J, ϕ) *and* (X', J', ϕ') *be marked $K3$ surfaces with the same period. Then* $(X, J) \cong (X', J')$.

Notice what is *not* said here: the theorem does not claim that the isomorphism $X \cong X'$ identifies ϕ and ϕ', and this is not in general true. To explain why, we must discuss the *Kähler cones* of $K3$ surfaces.

Definition 7.3.9 Let $\langle\, ,\, \rangle_{\mathbb{R}}$ be the indefinite inner product on $\Lambda_{\mathbb{R}}$ induced by q_Λ, and $\langle\, ,\, \rangle_{\mathbb{C}}$ the complex inner product on $\Lambda_{\mathbb{C}}$ induced by q_Λ. Let $\Pi \in P(\Lambda_{\mathbb{C}})$. Define the *root system* Δ_Π of Π by

$$\Delta_\Pi = \{\lambda \in \Lambda : q_\Lambda(\lambda) = -2,\ \langle \lambda, p \rangle_{\mathbb{C}} = 0 \text{ for all } p \in \Pi\}.$$

Define the *Kähler chambers* of Π to be the connected components of

$$\{\omega \in \Lambda_{\mathbb{R}} : q_\Lambda(\omega) > 0,\ \langle \omega, p \rangle_{\mathbb{C}} = 0 \text{ for all } p \in \Pi,\ \langle \omega, \lambda \rangle_{\mathbb{R}} \neq 0 \text{ for all } \lambda \in \Delta_\Pi\}.$$

It can be shown that the group G of automorphisms of Λ preserving q_Λ and Π acts transitively on the set of Kähler chambers of Π, so the Kähler chambers are really all isomorphic. Now Looijenga [155] proves

Theorem 7.3.10 *Let* (X, J, ϕ) *be a marked K3 surface with period* Π *and Kähler cone* $\mathcal{K}_X$. *Then* $\phi_{\mathbb{R}}(\mathcal{K}_X)$ *is one of the Kähler chambers of* Π.

Motivated by this, we define the *augmented period domain* and *map*:

Definition 7.3.11 Define the *augmented period domain* $\tilde{Q}$ by

$$\tilde{Q} = \{(\Pi, C) : \Pi \in Q \text{ and } C \subset \Lambda_{\mathbb{R}} \text{ is a Kähler chamber of } \Pi\},$$

where Q is given in (7.5). Define the *augmented period map* $\tilde{\mathcal{P}} : \mathcal{M}_{K3} \to \tilde{Q}$ by

$$\tilde{\mathcal{P}} : (X, J, \phi) \mapsto \big(\phi_{\mathbb{C}}(H^{2,0}(X)), \phi_{\mathbb{R}}(\mathcal{K}_X)\big),$$

where $\mathcal{K}_X$ is the Kähler cone of (X, J).

Observe that $\tilde{\mathcal{P}}\big((X, J, \phi)\big)$ lies in $\tilde{Q}$ by Theorem 7.3.7, so $\tilde{\mathcal{P}}$ does map $\mathcal{M}_{K3}$ to $\tilde{Q}$. Looijenga's main result [155] may then be written:

Theorem 7.3.12 *This map* $\tilde{\mathcal{P}} : \mathcal{M}_{K3} \to \tilde{Q}$ *is a 1-1 correspondence.*

Thus we have a very precise description of the moduli space $\mathcal{M}_{K3}$ of marked $K3$ surfaces. Note that $\mathcal{M}_{K3}$ is not Hausdorff. However, the moduli space of *unmarked* $K3$ surfaces is a Hausdorff complex orbifold isomorphic to $Q/\operatorname{Aut}(\Lambda)$.

7.3.3 *K3 surfaces as hyperkähler 4-manifolds*

We are primarily interested in $K3$ surfaces not as complex surfaces (X, J), but as Riemannian 4-manifolds (X, g) with holonomy $\operatorname{Sp}(1)$, or as hyperkähler 4-manifolds (X, J_1, J_2, J_3, g). By combining the above results with Yau's solution of the Calabi conjecture, we can deduce a great deal about metrics with holonomy $\operatorname{Sp}(1)$ and hyperkähler structures on the $K3$ 4-manifold.

Theorem 7.3.13 *Let* (X, J) *be a K3 surface. Then each Kähler class on X contains a unique metric with holonomy* $\operatorname{SU}(2)$. *Conversely, any compact Riemannian 4-manifold* (X, g) *with* $\operatorname{Hol}(g) = \operatorname{SU}(2) = \operatorname{Sp}(1)$ *admits a constant complex structure J such that* (X, J) *is a K3 surface.*

Proof As X is Kähler and $c_1(X) = 0$ there is a unique Ricci-flat Kähler metric g in every Kähler class on X by Theorem 6.2.1, and $\operatorname{Hol}^0(g) \subseteq \operatorname{SU}(2)$ by Proposition 6.1.1. From Berger's Theorem, the only possibilities for $\operatorname{Hol}^0(g)$ are $\operatorname{SU}(2)$ and $\{1\}$. But X is compact and simply-connected, so $\operatorname{Hol}^0(g) = \operatorname{Hol}(g)$, and $\operatorname{Hol}^0(g) = \{1\}$ is not possible. Hence $\operatorname{Hol}(g) = \operatorname{SU}(2)$.

Now let (X, g) be a compact Riemannian 4-manifold with $\operatorname{Hol}(g) = \operatorname{SU}(2)$. As $\operatorname{SU}(2)$ preserves the complex structure J and (2,0)-form $dz_1 \wedge dz_2$ on $\mathbb{C}^2$, there exists a constant complex structure J and a constant (2,0)-form θ on X. Proposition 6.2.3 shows that $\pi_1(X)$ is finite, so $h^{1,0}(X) = 0$, and K_X is trivial as θ is holomorphic. Thus (X, J) is a $K3$ surface. □

From §7.1, if (X, g) has $\mathrm{Hol}(g) = \mathrm{Sp}(1)$, then there are complex structures J_1, J_2, J_3 on X such that (X, J_1, J_2, J_3, g) is hyperkähler. In the context of the theorem, let (X, J_1, g) be a Kähler surface with $\mathrm{Hol}(g) = \mathrm{Sp}(1)$, and Kähler form ω_1. Choose a constant (2,0)-form $\omega_{\mathbb{C}}$ on X, scaled so that $\omega_{\mathbb{C}} \wedge \bar{\omega}_{\mathbb{C}} = 2\omega_1 \wedge \omega_1$. Then $\omega_2 = \mathrm{Re}(\omega_{\mathbb{C}})$ and $\omega_3 = \mathrm{Im}(\omega_{\mathbb{C}})$ are the Kähler forms of complex structures J_2, J_3 on X, and (J_1, J_2, J_3, g) is a hyperkähler structure.

Now this theorem tells us that holonomy SU(2) metrics exist on any $K3$ surface, but it tells us nothing about what these metrics actually look like. No explicit formulae are known for any holonomy SU(2) metric on a $K3$ surface, and it seems likely that no such formulae exist; that is, that these metrics are transcendental objects that admit no exact algebraic description.

However, there is one way of getting an *approximate* description of some of these hyperkähler $K3$ metrics, using the Kummer construction of Example 7.3.2. This idea was suggested by Page [172], and made rigorous by LeBrun and Singer [150] and Topiwala [210] using twistor theory, as an alternative proof of the existence of hyperkähler structures on $K3$. We explain it in our next example.

Example 7.3.14 Let the complex torus T^4, involution $\sigma : T^4 \to T^4$, and Kummer surface S resolving $T^4/\langle\sigma\rangle$ be as in Example 7.3.2. Then S is a $K3$ surface, and there is a 20-dimensional family of Kähler metrics on S with holonomy SU(2), isomorphic to the Kähler cone $\mathcal{K}$ of S, by Theorem 7.3.13.

Now $T^4 = \mathbb{C}^2/\Lambda$, where Λ acts on $\mathbb{C}^2$ by translations. Choose a Hermitian metric g_0 on $\mathbb{C}^2$. Then g_0 is invariant under Λ and σ, and so pushes down to give a flat Kähler orbifold metric g_0 on $T^4/\langle\sigma\rangle$. Let $\pi : S \to T^4/\langle\sigma\rangle$ be the blow-up map. Then $\pi^*(g_0)$ is a *singular* Kähler metric on S, which is degenerate at the 16 $\mathbb{CP}^1$ in S introduced by blowing up the 16 singular points of $T^4/\langle\sigma\rangle$.

We can think of $\pi^*(g_0)$ as a point in the boundary $\overline{\mathcal{K}}\setminus\mathcal{K}$ of the family $\mathcal{K}$ of holonomy SU(2) metrics on S, as there exists a smooth family $\{g_t : t \in (0, 1)\}$ of nonsingular holonomy SU(2) metrics on S such that $g_t \to \pi^*(g_0)$ as $t \to 0$, in a suitable sense. Thus $\pi^*(g_0)$ approximates g_t for small t.

However, $\pi^*(g_0)$ is not a good description of g_t near the 16 $\mathbb{CP}^1$ in S where $\pi^*(g_0)$ is singular. Instead, near each $\mathbb{CP}^1$ we can approximate g_t by an *Eguchi–Hanson metric*, as in Example 7.2.2. Near each $\mathbb{CP}^1$ we can naturally identify S with the blow-up X of $\mathbb{C}^2/\{\pm 1\}$ at 0, and when t is small g_t is close to one of the Eguchi–Hanson metrics on X.

Explicitly, we can choose g_t to be close to the Eguchi–Hanson metric on X with Kähler potential

$$f_t = \sqrt{r^4 + t^4} + 2t^2 \log r - t^2 \log\left(\sqrt{r^4 + t^4} + t^2\right),$$

which is a natural 1-parameter generalization of the function f of (7.3). Thus we regard the holonomy SU(2) metrics g_t on S as being the result of gluing a flat metric on $T^4/\langle\sigma\rangle$ together with 16 small Eguchi–Hanson spaces. The parameter $t > 0$ is a measure of the diameter of the central S^2 in each Eguchi–Hanson space, and the approximation is best when t is small.

This example is the model and motivation for more complicated 'Kummer constructions' of metrics with special holonomy in higher dimensions—in particular, those of Calabi–Yau 3-folds described in §6.6.1, and of compact manifolds with exceptional holonomy given in Chapters 11–14.

As we studied marked $K3$ surfaces (X, J, ϕ) above, so we can consider *marked hyperkähler $K3$ surfaces* $(X, J_1, J_2, J_3, g, \phi)$. Here are the appropriate notions of moduli space, period and period domain.

Definition 7.3.15 Let $\mathcal{M}_{K3}^{hk}$ be the moduli space of marked hyperkähler $K3$ surfaces. Define the *hyperkähler period map* $\mathcal{P}^{hk} : \mathcal{M}_{K3}^{hk} \to (\Lambda_\mathbb{R})^3$ by

$$\mathcal{P}^{hk} : (X, J_1, J_2, J_3, g, \phi) \mapsto \bigl(\phi_\mathbb{R}([\omega_1]), \phi_\mathbb{R}([\omega_2]), \phi_\mathbb{R}([\omega_3])\bigr),$$

where $\omega_1, \omega_2, \omega_3$ are the Kähler forms of J_1, J_2, J_3, and $\phi_\mathbb{R}$ maps $H^2(X, \mathbb{R}) \to \Lambda_\mathbb{R}$. Let $\langle\ ,\ \rangle_\mathbb{R}$ be as in Definition 7.3.5. Define the *hyperkähler period domain* Q^{hk} by

$$Q^{hk} = \bigl\{(\alpha_1, \alpha_2, \alpha_3) : \alpha_i \in \Lambda_\mathbb{R}, \langle \alpha_i, \alpha_j \rangle_\mathbb{R} = a\,\delta_{ij} \text{ for some } a > 0, \text{ and for each}$$
$$\lambda \in \Lambda \text{ with } q_\Lambda(\lambda) = -2, \text{ there exists } i = 1, 2 \text{ or } 3 \text{ with } \langle \alpha_i, \lambda \rangle_\mathbb{R} \neq 0 \bigr\}.$$

The reason for requiring $\langle \alpha_i, \alpha_j \rangle_\mathbb{R} = a\,\delta_{ij}$ here is that the 2-forms ω_i satisfy $\omega_i \wedge \omega_j = 2\delta_{ij} \mathrm{d}V_g$, where $\mathrm{d}V_g$ is the volume form on X. Hence

$$\langle \alpha_i, \alpha_j \rangle_\mathbb{R} = [\omega_i] \cup [\omega_j] = \int_X \omega_i \wedge \omega_j = 2\delta_{ij} \int_X \mathrm{d}V_g = 2\,\mathrm{vol}(X)\delta_{ij},$$

so that $\langle \alpha_i, \alpha_j \rangle_\mathbb{R} = a\,\delta_{ij}$ with $a = 2\,\mathrm{vol}(X) > 0$. As a marked $K3$ surface, (X, J_1, ϕ) has period $[\alpha_2 + i\alpha_3] \in P(\Lambda_\mathbb{C})$. So by Theorem 7.3.10, a necessary condition for $(\alpha_1, \alpha_2, \alpha_3)$ to be a hyperkähler period is that α_1 should lie in one of the Kähler chambers of $[\alpha_2 + i\alpha_3]$.

By Definition 7.3.9, this implies that $\langle \alpha_i, \lambda \rangle_\mathbb{R} \neq 0$ for some $i = 1, 2, 3$ whenever $\lambda \in \Lambda$ with $q_\Lambda(\lambda) = -2$. Thus we see that the period of $(X, J_1, J_2, J_3, g, \phi)$ lies in Q^{hk}, and $\mathcal{P}^{hk}$ maps $\mathcal{M}_{K3}^{hk} \to Q^{hk}$. From Theorems 7.3.10, 7.3.12 and 7.3.13 one can prove:

Theorem 7.3.16 *This map* $\mathcal{P}^{hk} : \mathcal{M}_{K3}^{hk} \to Q^{hk}$ *is a diffeomorphism.*

This description of the hyperkähler moduli space $\mathcal{M}_{K3}^{hk}$ is rather simpler than that of the complex structure moduli space $\mathcal{M}_{K3}$ in §7.3.2, and $\mathcal{M}_{K3}^{hk}$ is Hausdorff although $\mathcal{M}_{K3}$ is not. One can regard $\mathcal{M}_{K3}^{hk}$ as more fundamental, and $\mathcal{M}_{K3}$ as being derived from it.

Observe also that there is a strong similarity between Theorem 7.3.16, describing the moduli space of hyperkähler $K3$ surfaces, and Theorem 7.2.3, describing the moduli space of hyperkähler ALE spaces asymptotic to $\mathbb{H}/G$.

Here is a geometric interpretation of the condition that $\langle \alpha_i, \lambda \rangle_\mathbb{R} \neq 0$ for some $i = 1, 2, 3$ whenever $\lambda \in \Lambda$ with $q_\Lambda(\lambda) = -2$. Each such λ corresponds to a unique minimal 2-sphere $\mathcal{S}^2$ in X with self-intersection -2. The area A of this $\mathcal{S}^2$ is given by $A^2 = \sum_{j=1}^{3} \langle \alpha_i, \lambda \rangle_\mathbb{R}^2$. If $\langle \alpha_i, \lambda \rangle_\mathbb{R} = 0$ for $i = 1, 2, 3$ then $A = 0$, which is impossible.

That is, the periods for which $\langle \alpha_i, \lambda \rangle_{\mathbb{R}} = 0$ for $i = 1, 2, 3$ correspond to *singular* hyperkähler structures on X in which an $\mathcal{S}^2$ collapses down to a point, giving a hyperkähler orbifold. We can think of this process as blowing down one or more $\mathbb{CP}^1$'s in a complex surface X to get a singular complex surface X'.

7.4 Higher-dimensional compact hyperkähler manifolds

We now move on to discuss hyperkähler manifolds of dimension $4m$, for $m \geqslant 2$. There are many similarities with $K3$ surfaces. As in §7.3, we find it convenient to first explain the algebraic geometry of the underlying complex manifolds, which are called *complex symplectic manifolds*, and then translate this into results on hyperkähler manifolds. We finish with some examples. Our treatment is based on Huybrechts' excellent paper [109] and its sequel [110], though many of the important ideas are originally due to Beauville [28, §6–§9].

7.4.1 *Complex symplectic manifolds*

Complex symplectic manifolds are higher-dimensional analogues of $K3$ surfaces.

Definition 7.4.1 A *complex symplectic manifold* $(X, J, \omega_{\mathbb{C}})$ is a compact complex $2m$-manifold (X, J) admitting Kähler metrics, with a closed (2,0)-form $\omega_{\mathbb{C}}$, such that $\omega_{\mathbb{C}}^m$ is a nonvanishing $(2m, 0)$-form. We call $(X, J, \omega_{\mathbb{C}})$ *irreducible* if X is simply-connected and cannot be written as a product $X_1 \times X_2$ of lower-dimensional complex manifolds.

We shall discuss *deformations* of complex symplectic manifolds. If $(X, J, \omega_{\mathbb{C}})$ is an irreducible complex symplectic manifold, then the canonical bundle of X is trivial, so Theorem 6.8.1 shows that the local moduli space $\mathcal{M}$ of deformations of (X, J) is a complex manifold of dimension $h^{2m-1,1}(X)$. But we shall prove in (7.7) that $h^{2m-1,1}(X) = h^{1,1}(X)$, so $\dim \mathcal{M} = h^{1,1}(X)$. It can also be shown [28, §8], [109, §2.4] that small deformations of (X, J) are also irreducible complex symplectic manifolds. Thus we have:

Theorem 7.4.2 *Let* $(X, J, \omega_{\mathbb{C}})$ *be an irreducible complex symplectic manifold. Then the moduli space of deformations of (X, J) is locally a complex manifold of dimension* $h^{1,1}(X)$. *Every small deformation* (X', J') *of* (X, J) *has a* $(2, 0)$-*form* $\omega'_{\mathbb{C}}$ *such that* $(X', J', \omega'_{\mathbb{C}})$ *is an irreducible complex symplectic manifold.*

There is a natural quadratic form on the second cohomology of an irreducible complex symplectic manifold, due to Beauville ([28, Th. 5], see also [109, §1.9]).

Theorem 7.4.3 *Let* $(X, J, \omega_{\mathbb{C}})$ *be an irreducible complex symplectic $2m$-fold, scaled so that* $\int_X \omega_{\mathbb{C}}^m \wedge \bar{\omega}_{\mathbb{C}}^m = 1$. *Define a quadratic form f on $H^2(X, \mathbb{R})$ by*

$$f([\alpha]) = \frac{m}{2} \int_X \omega_{\mathbb{C}}^{m-1} \wedge \bar{\omega}_{\mathbb{C}}^{m-1} \wedge \alpha^2$$
$$+ (1-m) \left(\int_X \omega_{\mathbb{C}}^{m-1} \wedge \bar{\omega}_{\mathbb{C}}^m \wedge \alpha \right) \cdot \left(\int_X \omega_{\mathbb{C}}^m \wedge \bar{\omega}_{\mathbb{C}}^{m-1} \wedge \alpha \right).$$

Then there exists a unique constant $c > 0$ such that $q_X = c \cdot f$ is a primitive integral quadratic form on $H^2(X, \mathbb{Z})$, of index $\bigl(3, b^2(X) - 3\bigr)$.

When $m = 1$, so that X is a $K3$ surface, this q_X is just the intersection form on $H^2(X, \mathbb{Z})$. As q_X is integer-valued on $H^2(X, \mathbb{Z})$ it is invariant under continuous deformations of the complex symplectic structure on X, even though the definition appears to depend on the cohomology class $[\omega_{\mathbb{C}}] \in H^2(X, \mathbb{C})$ for $m > 1$.

Definition 7.4.4 Let Λ be a lattice, so that $\Lambda \cong \mathbb{Z}^d$, and let q_Λ be a primitive integral quadratic form on Λ of index $(3, d-3)$. Let $(X, J, \omega_{\mathbb{C}})$ be an irreducible complex symplectic manifold with $b^2(X) = d$. A *marking* of X is an isomorphism $\phi : H^2(X, \mathbb{Z}) \to \Lambda$ identifying the quadratic forms q_X on $H^2(X, \mathbb{Z})$ and q_Λ on Λ. We call $(X, J, \omega_{\mathbb{C}}, \phi)$ a *marked irreducible complex symplectic manifold*.

We would like to understand the moduli space of all marked irreducible complex symplectic manifolds $(X, J, \omega_{\mathbb{C}}, \phi)$. In fact it is more convenient to study the underlying marked complex manifolds (X, J, ϕ).

Definition 7.4.5 Let $(X, J, \omega_{\mathbb{C}}, \phi)$ be a marked irreducible complex symplectic manifold. Define $\mathcal{M}_X$ to be the moduli space of *marked complex deformations* (X', J', ϕ') of (X, J, ϕ). That is, (X', J') is a deformation of (X, J) admitting a $(2,0)$-form $\omega'_{\mathbb{C}}$ making $(X', J', \omega'_{\mathbb{C}})$ into an irreducible complex symplectic manifold, and ϕ' is a marking of $(X', J', \omega'_{\mathbb{C}})$. Then $\mathcal{M}_X$ is locally a complex manifold of dimension $h^{1,1}(X) = d - 2$, by Theorem 7.4.2. However, globally the topology of $\mathcal{M}_X$ may not be Hausdorff; that is, $\mathcal{M}_X$ is a *non-Hausdorff complex manifold*.

An important tool in studying the moduli space $\mathcal{M}_X$ is the *period map*.

Definition 7.4.6 Let $(X, J, \omega_{\mathbb{C}}, \phi)$ be a marked irreducible complex symplectic manifold, with lattice Λ. Write $\Lambda_\mathbb{R} = \Lambda \otimes_\mathbb{Z} \mathbb{R}$ and $\Lambda_\mathbb{C} = \Lambda \otimes_\mathbb{Z} \mathbb{C}$. Then ϕ induces isomorphisms $\phi_\mathbb{R} : H^2(X, \mathbb{R}) \to \Lambda_\mathbb{R}$ and $\phi_\mathbb{C} : H^2(X, \mathbb{C}) \to \Lambda_\mathbb{C}$. Now $H^{2,0}(X) \cong \mathbb{C}$. We define the *period* of $(X, J, \omega_{\mathbb{C}}, \phi)$ to be $\phi_\mathbb{C}(H^{2,0}(X))$, which we regard as a point in the complex projective space $P(\Lambda_\mathbb{C})$.

Observe that the period depends only on (X, J, ϕ), and not on $\omega_{\mathbb{C}}$. Let $\mathcal{M}_X$ be as above, and define the *period map* $\mathcal{P} : \mathcal{M}_X \to P(\Lambda_\mathbb{C})$ by $\mathcal{P} : (X', J', \phi') \mapsto \phi'_\mathbb{C}(H^{2,0}(X'))$. It is easy to show that $\mathcal{P}$ is holomorphic. From the definition of q_X in Theorem 7.4.3, we find that $q_X(\omega_{\mathbb{C}}) = 0$, and $q_X(\omega_{\mathbb{C}} + \bar{\omega}_{\mathbb{C}}) > 0$. Therefore, if as in (7.5) we define $Q \subset P(\Lambda_\mathbb{C})$ by

$$Q = \{[\lambda] \in P(\Lambda_\mathbb{C}) : \lambda \in \Lambda_\mathbb{C} \setminus \{0\}, \ q_\Lambda(\lambda) = 0, \ q_\Lambda(\lambda + \bar{\lambda}) > 0\}, \tag{7.6}$$

then $\mathcal{P}$ maps $\mathcal{M}_X \to Q$. We call Q the *period domain*.

Analogues of Theorems 7.3.6 and 7.3.7 (the Local Torelli Theorem and the Surjectivity Theorem for $K3$ surfaces) were proved by Beauville [28, Th. 5(b)] and Huybrechts [109, Th. 8.1], and we give them in our next result.

Theorem 7.4.7 *The period map* $\mathcal{P} : \mathcal{M}_X \to Q$ *is a local isomorphism of complex manifolds. Let* $\mathcal{M}_X^0$ *be a nonempty connected component of* $\mathcal{M}_X$. *Then* $\mathcal{P} : \mathcal{M}_X^0 \to Q$ *is surjective.*

However, at the time of writing no good analogue is known of the Global Torelli Theorem ([48, Th. 1], partially stated in Theorem 7.3.8). In fact Theorem 7.3.8 is false for complex symplectic manifolds in complex dimension 4 and above, as Debarre has found examples of complex symplectic manifolds with the same period, which are birational but not biholomorphic. For more information see Huybrechts [109, §2.5, §4, §10].

Also, in the $K3$ case Theorem 7.3.10 gives an exact description of the Kähler cone of (X, J), but for more general irreducible complex symplectic manifolds we do not yet have such a precise picture. For some results and conjectures on this, see Huybrechts [110].

7.4.2 General theory of compact hyperkähler manifolds

Here are three topological results on compact manifolds with holonomy $\mathrm{Sp}(m)$.

Proposition 7.4.8 *Let (X, g) be a compact $4m$-manifold with $\mathrm{Hol}(g) = \mathrm{Sp}(m)$. Then X is simply-connected and has $\hat{A}$-genus $\hat{A}(X) = m + 1$.*

Proof Theorem 3.6.5 shows that $\hat{A}(X) = m+1$, and Corollary 3.5.6 shows that $\pi_1(X)$ is finite. Let $(\tilde{X}, \tilde{g})$ be the universal cover of (X, g), and $d = |\pi_1(X)|$ the degree of the cover. Then $\tilde{X}$ is compact, as $\pi_1(X)$ is finite, and $\mathrm{Hol}(\tilde{g}) = \mathrm{Hol}^0(g) = \mathrm{Sp}(m)$. Thus $\hat{A}(\tilde{X}) = m + 1$ as above. But $\hat{A}(\tilde{X}) = d \cdot \hat{A}(X)$ by properties of characteristic classes. So $m + 1 = d(m + 1)$, giving $d = 1$, and X is simply-connected. $\square$

By analogy with Proposition 6.2.6 we have:

Proposition 7.4.9 *Let (X, J, g) be a compact Kähler manifold of dimension $2m$ with $\mathrm{Hol}(g) = \mathrm{Sp}(m)$, and let $h^{p,q}$ be the Hodge numbers of X. Then*

$$h^{2k,0} = 1 \text{ for } 0 \leqslant k \leqslant m, \ h^{2k+1,0} = 0 \text{ for all } k, \text{ and } h^{p,q} = h^{2m-p,q}. \tag{7.7}$$

Proof From §7.1.1, $\mathrm{Sp}(m)$ is the subgroup of $\mathrm{SU}(2m)$ fixing a complex symplectic form $\omega_2 + i\omega_3$ in $\Lambda^{2,0}\mathbb{C}^{2m}$. Thus $\mathrm{Sp}(m)$ also fixes the powers $(\omega_2 + i\omega_3)^k$ in $\Lambda^{2k,0}\mathbb{C}^{2m}$, for $k = 0, 1, \ldots, m$. Any form in $\Lambda^{p,0}\mathbb{C}^{2m}$ fixed by $\mathrm{Sp}(m)$ is proportional to some $(\omega_2 + i\omega_3)^k$. Therefore by Proposition 6.2.4, if $\mathrm{Hol}(g) = \mathrm{Sp}(m)$ then $H^{2k,0}(X)$ is $\mathbb{C}$ if $k = 0, \ldots, m$ and $H^{2k+1,0}(X) = 0$.

Suppose that $0 \leqslant p \leqslant m$, as otherwise we may replace p by $2m - p$. Define a linear map $\phi : \Lambda^{p,q}\mathbb{C}^{2m} \to \Lambda^{2m-p,q}\mathbb{C}^{2m}$ by $\phi(\alpha) = \alpha \wedge (\omega_2 + i\omega_3)^{m-p}$. Then ϕ is a vector space isomorphism, and as $\mathrm{Sp}(m)$ preserves $\omega_2 + i\omega_3$, it is also an isomorphism of $\mathrm{Sp}(m)$-representations. Thus Theorem 3.5.3 shows that $H^{p,q}(X) \cong H^{2m-p,q}(X)$, and so $h^{p,q} = h^{2m-p,q}$. $\square$

Combining the last part of the proposition with (4.12), we find that

$$h^{p,q} = h^{2m-p,q} = h^{p,2m-q} = h^{2m-p,2m-q} = h^{q,p} = h^{2m-q,p} = h^{q,2m-p} = h^{2m-q,2m-p}.$$

Next we discuss the relation between compact manifolds with holonomy $\mathrm{Sp}(m)$ and irreducible complex symplectic manifolds.

Theorem 7.4.10 *Let $(X, J, \omega_{\mathbb{C}})$ be an irreducible complex symplectic $2m$-fold. Then there is a unique metric with holonomy $\mathrm{Sp}(m)$ in each Kähler class on X. Conversely, if (X, g) is a compact Riemannian $4m$-manifold with holonomy $\mathrm{Sp}(m)$ then there exist a constant complex structure J and $(2, 0)$-form $\omega_{\mathbb{C}}$ on X such that $(X, J, \omega_{\mathbb{C}})$ is an irreducible complex symplectic $2m$-fold.*

Proof Observe that $\omega_{\mathbb{C}}^m$ is a nonvanishing holomorphic section of the canonical bundle K_X of X. Thus K_X is trivial, and $c_1(X) = 0$. So by Theorem 6.2.1, each Kähler class κ contains a unique Ricci-flat Kähler metric g. Let ∇ be the Levi-Civita connection of g. Then Proposition 6.2.4 shows that $\nabla \omega_{\mathbb{C}} = 0$, as $\omega_{\mathbb{C}}$ is a closed $(2,0)$-form.

Therefore $\nabla g = \nabla J = \nabla \omega_{\mathbb{C}} = 0$, and $\mathrm{Hol}(g)$ preserves a metric, complex structure and complex symplectic form on $\mathbb{C}^{2m}$. This forces $\mathrm{Hol}(g) \subseteq \mathrm{Sp}(m)$. But g is irreducible, as X is irreducible, and from the classification of holonomy groups we see that $\mathrm{Hol}(g) = \mathrm{Sp}(m)$. Any holonomy $\mathrm{Sp}(m)$ metric is Ricci-flat, so g is the only metric in κ with holonomy $\mathrm{Sp}(m)$, by Theorem 6.2.1.

Now let (X, g) be a compact Riemannian $4m$-manifold with holonomy $\mathrm{Sp}(m)$. Then from §7.1 there exist a constant complex structure $J = J_1$ and $(2,0)$-form $\omega_{\mathbb{C}} = \omega_2 + i\omega_3$ on X. Clearly $(X, J, \omega_{\mathbb{C}})$ is a complex symplectic manifold. Also $\pi_1(X) = \{1\}$ by Proposition 7.4.8, and X cannot be written as a product of lower-dimensional complex manifolds, as this would force g to be reducible. Hence $(X, J, \omega_{\mathbb{C}})$ is irreducible. $\square$

Now in §7.3.3 we applied results on the moduli of $K3$ surfaces to describe the moduli space of hyperkähler structures on $K3$. In a similar way, we can apply Theorem 7.4.7 and other results in [109, 110] to describe the moduli space of hyperkähler structures on X. But as we lack both a Global Torelli Theorem and a clear picture of the Kähler cone, our results will not be as strong.

Following Definition 7.3.15, we define:

Definition 7.4.11 Let $\mathcal{M}_X^{\mathrm{hk}}$ be the moduli space of marked hyperkähler structures on X. Define the *hyperkähler period map* $\mathcal{P}^{\mathrm{hk}} : \mathcal{M}_X^{\mathrm{hk}} \to (\Lambda_{\mathbb{R}})^3$ by

$$\mathcal{P}^{\mathrm{hk}} : (X, J_1, J_2, J_3, g, \phi) \mapsto \big(\phi_{\mathbb{R}}([\omega_1]), \phi_{\mathbb{R}}([\omega_2]), \phi_{\mathbb{R}}([\omega_3])\big),$$

where $\omega_1, \omega_2, \omega_3$ are the Kähler forms of J_1, J_2, J_3, and $\phi_{\mathbb{R}}$ maps $H^2(X, \mathbb{R}) \to \Lambda_{\mathbb{R}}$. Let $\langle \, , \, \rangle_{\mathbb{R}}$ be as in Definition 7.3.5. Define the *hyperkähler period domain* Q_X^{hk} by

$$Q_X^{\mathrm{hk}} = \big\{(\alpha_1, \alpha_2, \alpha_3) : \alpha_i \in \Lambda_{\mathbb{R}}, \langle \alpha_i, \alpha_j \rangle_{\mathbb{R}} = a\,\delta_{ij} \text{ for some } a > 0\big\}.$$

Here is a rough analogue of Theorem 7.3.16.

Theorem 7.4.12 *The image of $\mathcal{P}^{\mathrm{hk}} : \mathcal{M}_X^{\mathrm{hk}} \to Q_X^{\mathrm{hk}}$ is a dense open set in Q_X^{hk}.*

It is not yet known whether $\mathcal{P}^{\mathrm{hk}}$ is injective, nor what its image is in Q_X^{hk}.

7.4.3 Examples of compact hyperkähler manifolds

In comparison to Calabi–Yau manifolds, examples of compact hyperkähler manifolds are difficult to find, and only a few are known in each dimension. The first examples were two series of manifolds due to Beauville [28, §6–§7] and [27, §2], which generalize an example of Fujiki [77] in real dimension 8.

We now explain Beauville's examples. Let X be a compact complex surface. For $m > 1$, define X^m to be the product $X \times X \times \cdots \times X$ of m copies of X, and $X^{(m)}$ to be the mth symmetric product of X, that is, $X^{(m)} = X^m/S_m$, where S_m is the symmetric group acting on X^m by permutations. Then $X^{(m)}$ is a *complex orbifold*, of dimension $2m$.

Let $X^{[m]}$ be the *Hilbert scheme* or *Douady space* of zero-dimensional subspaces $(Z, \mathcal{O}_Z)$ of X of length $\dim_{\mathbb{C}} \mathcal{O}_Z = m$. Then $X^{[m]}$ is a compact, nonsingular complex $2m$-manifold, with a projection $\pi : X^{[m]} \to X^{(m)}$ which is a *crepant resolution*. By results of Varouchas [214], $X^{[m]}$ is Kähler whenever X is Kähler.

Now suppose that X is a $K3$ surface or a complex torus T^4, so that X is complex symplectic. Then X^m and $X^{(m)}$ both have complex symplectic forms. As $X^{[m]}$ is a crepant resolution of $X^{(m)}$, it also has a complex symplectic form, and it admits Kähler metrics from above, so $X^{[m]}$ is a complex symplectic manifold.

If X is a $K3$ surface then Beauville [28, Prop. 6] shows that $X^{[m]}$ is also irreducible, with $b^2 = 23$. Applying Theorems 7.4.10 and 7.4.12, we find:

Theorem 7.4.13 *Let X be a $K3$ surface and $m \geqslant 2$. Then the Hilbert scheme $X^{[m]}$ is an irreducible complex symplectic $2m$-manifold, with $b^2(X^{[m]}) = 23$. There exists a 61-dimensional family of metrics g on $X^{[m]}$ with holonomy $\mathrm{Sp}(m)$, and a 64-dimensional family of hyperkähler structures (J_1, J_2, J_3, g).*

These are the first known examples of compact Riemannian manifolds with holonomy $\mathrm{Sp}(m)$ for $m \geqslant 2$. Now let Y be a complex torus T^4. In this case $Y^{[m]}$ is complex symplectic, but it is not irreducible. Instead, $Y^{[m]}$ has a finite cover isomorphic to $K^{m-1}(Y) \times Y$, where $K^{m-1}(Y)$ is an irreducible complex symplectic $(2m-2)$-manifold.

To define $K^{m-1}(Y)$ explicitly, observe that we can regard T^4 as an abelian Lie group, and so there is a natural map $\Sigma : Y^{(m)} \to Y$ given by summing the m points. Let $K^{m-1}(Y)$ be the kernel of the composition $Y^{[m]} \xrightarrow{\pi} Y^{(m)} \xrightarrow{\Sigma} Y$. Then $K^{m-1}(Y)$ is a $(2m-2)$-dimensional complex submanifold of $Y^{[m]}$, and is an irreducible complex symplectic manifold.

Now $K^1(Y)$ is a $K3$ surface, got from Y by the Kummer construction as in Example 7.3.2. But for $m \geqslant 2$, Beauville [28, Prop. 8] shows that $K^m(Y)$ is an irreducible complex symplectic manifold with $b^2 = 7$. So as above we get:

Theorem 7.4.14 *Let Y be a complex torus T^4 and $m \geqslant 2$. Then $K^m(Y)$ is an irreducible complex symplectic $2m$-manifold, with $b^2(K^m(Y)) = 7$. There exists a 13-dimensional family of metrics g on $K^m(Y)$ with holonomy $\mathrm{Sp}(m)$, and a 16-dimensional family of hyperkähler structures (J_1, J_2, J_3, g).*

As compact real manifolds, $X^{[m]}$ and $K^m(Y)$ are independent of the choice of $K3$ surface X and complex torus Y. Thus Theorems 7.4.13 and 7.4.14 give only two distinct compact $4m$-manifolds with holonomy $\mathrm{Sp}(m)$, for each $m \geqslant 2$. Several other constructions of such manifolds are described in [109, §2], but they are all deformation equivalent (and hence diffeomorphic) to $X^{[m]}$ or $K^m(Y)$.

Until recently, all known examples of compact $4m$-manifolds with holonomy $\mathrm{Sp}(m)$ for $m \geqslant 2$ were diffeomorphic to $X^{[m]}$ or $K^m(Y)$. However, in 1998 Kieran O'Grady [170, 171] found a new example of an irreducible complex symplectic 10-manifold $\widetilde{\mathcal{M}}$, which he constructed as a crepant resolution of a moduli space $\mathcal{M}$ of a certain kind of sheaf on a $K3$ surface. Thus $\mathcal{M}$ carries metrics with holonomy $\mathrm{Sp}(5)$, by Theorem 7.4.10.

O'Grady was able to prove that $b^2(\widetilde{\mathcal{M}}) \geqslant 24$, so that $\widetilde{\mathcal{M}}$ is not diffeomorphic to either $X^{[5]}$ or $K^5(Y)$. It seems likely that a better understanding of O'Grady's example will lead to the discovery of other compact manifolds with holonomy $\mathrm{Sp}(m)$. A related, conjectural approach to finding new examples of compact hyperkähler manifolds is described by Verbitsky in [217].

7.5 The other quaternionic geometries

Apart from hyperkähler manifolds, there are three other interesting geometric structures on $4m$-manifolds based on the quaternions. These are *quaternionic Kähler structures* (metrics with holonomy $\mathrm{Sp}(m)\,\mathrm{Sp}(1)$), and two nonmetric geometries, *hypercomplex structures* (with holonomy $\mathrm{GL}(m, \mathbb{H})$) and *quaternionic structures* (with holonomy $\mathrm{GL}(m, \mathbb{H})\,\mathrm{GL}(1, \mathbb{H})$). We discuss each in turn. Some references for this section are Salamon [184, 185, 188] and [186, §9].

7.5.1 *Hypercomplex manifolds*

We call (X, J_1, J_2, J_3) a *hypercomplex manifold*, and (J_1, J_2, J_3) a *hypercomplex structure* on X, if X is a manifold of dimension $4m$ and J_1, J_2, J_3 are integrable complex structures on X such that $J_1 J_2 = J_3$. If (X, J_1, J_2, J_3, g) is hyperkähler then (X, J_1, J_2, J_3) is hypercomplex. However, not all hypercomplex manifolds admit hyperkähler metrics, even locally.

A hypercomplex manifold (X, J_1, J_2, J_3) carries a unique torsion-free connection ∇ on TM with $\nabla J_j = 0$ called the *Obata connection*, whose holonomy is a subgroup of $\mathrm{GL}(m, \mathbb{H})$. In the language of §2.6, a hypercomplex structure on X is equivalent to a *torsion-free* $\mathrm{GL}(m, \mathbb{H})$-*structure* on X. Note that if ∇ is not flat then (X, J_1, J_2, J_3) is not locally isomorphic to $\mathbb{H}^m$ with its flat hypercomplex structure. This is an important difference between complex and hypercomplex geometry, since all complex manifolds are locally trivial.

If (X, J_1, J_2, J_3) is hypercomplex and $a_1, a_2, a_3 \in \mathbb{R}$ with $a_1^2 + a_2^2 + a_3^2 = 1$, then $a_1 J_2 + a_2 J_2 + a_3 J_3$ is a complex structure on X. Thus a hypercomplex manifold has a 2-sphere $\mathcal{S}^2$ of complex structures. As for the hyperkähler case of §7.1.3, there is a *twistor construction* for hypercomplex manifolds. For examples of hypercomplex manifolds, including compact hypercomplex manifolds which are not locally hyperkähler, see Joyce [116, 117].

7.5.2 *Quaternionic Kähler manifolds*

A Riemannian $4m$-manifold (M, g) for $m \geqslant 2$ with $\mathrm{Hol}(g) \subseteq \mathrm{Sp}(m)\,\mathrm{Sp}(1)$ is called a *quaternionic Kähler manifold*. Some good references on quaternionic Kähler manifolds are Salamon [184, 188]. Such manifolds are automatically Einstein, with constant scalar curvature s, and are hyperkähler if $s = 0$. So we usually assume that $s \neq 0$, and split into the two cases $s > 0$ and $s < 0$, called *positive* and *negative quaternionic Kähler manifolds* respectively.

It should be stressed that quaternionic Kähler manifolds are *not* in general Kähler, as $\text{Sp}(m)\,\text{Sp}(1)$ is not a subgroup of $\text{U}(2m)$. In 4 dimensions we have $\text{Sp}(1)\,\text{Sp}(1) = \text{SO}(4)$, so that the holonomy group $\text{Sp}(1)\,\text{Sp}(1)$ is not interesting. Instead, we define a *quaternionic Kähler 4-manifold* to be an oriented Riemannian 4-manifold (M, g) which is Einstein, with self-dual Weyl curvature.

The geometric structures on a quaternionic Kähler manifold (M, g) are the metric g, a constant 4-form Ω, and a bundle P of almost complex structures on M with fibre S^2, such that over each $x \in M$ the fibre of P is $\{a_1 J_1 + a_2 J_2 + a_3 J_3 : a_j \in \mathbb{R},\ a_1^2 + a_2^2 + a_3^2 = 1\}$, where J_1, J_2, J_3 are almost complex structures at x compatible with g and satisfying $J_1 J_2 = J_3$. There are many local sections of P which are integrable complex structures.

There is a *twistor construction* for quaternionic Kähler manifolds, due to Salamon [184]. If (M, g) is a quaternionic Kähler $4m$-manifold with nonzero scalar curvature then the twistor space Z of M is a complex contact $(2m + 1)$-manifold, fibred by holomorphic curves $\mathbb{CP}^1$. It is the total space of the bundle P described above. If g has positive scalar curvature, then Z is also Kähler–Einstein with positive scalar curvature.

Now a compact Kähler–Einstein manifold Z with positive scalar curvature has ample anticanonical bundle K_Z^{-1}, and so is projective, and is by definition a *Fano manifold*. Thus, if (M, g) is a compact positive quaternionic Kähler manifold, then its twistor space Z is a *contact Fano manifold*. Such Z are very special objects in algebraic geometry, and a lot is known about them.

The most basic examples of quaternionic Kähler manifolds are the *Wolf spaces*, a family of compact quaternionic Kähler symmetric spaces described in 1965 by Wolf [223], who also discussed their twistor spaces. There is one Wolf space for each compact simple Lie group.

It is conjectured that the Wolf spaces are the only compact positive quaternionic Kähler manifolds, that is, that every compact positive quaternionic Kähler manifold is symmetric. In 8 dimensions this was proved by Poon and Salamon [175]. More generally, LeBrun and Salamon [149] show that up to homothety, there are only finitely many positive quaternionic Kähler manifolds of dimension $4m$, for any $m \geqslant 1$. Both these results are proved using complex algebraic geometry on the twistor space, and Fano manifolds.

There are many known examples of nonsymmetric quaternionic Kähler metrics on *noncompact* manifolds. Alekseevsky [7] found a large class of complete, nonsymmetric, negative quaternionic Kähler manifolds, which are homogeneous metrics on solvable Lie groups. Inhomogeneous quaternionic Kähler metrics are provided by the *quaternionic Kähler quotient construction* of Galicki and Lawson [80]. This produces examples of compact, positive quaternionic Kähler orbifolds.

7.5.3 *Quaternionic manifolds*

Underlying any hyperkähler structure (J_1, J_2, J_3, g) there is a hypercomplex structure (J_1, J_2, J_3), which is a weaker structure not involving a metric, and can be studied on its own. In just the same way, under any quaternionic Kähler structure is a weaker, metric-free geometry called a *quaternionic structure*, which is studied by Salamon [185].

In $4m$ dimensions for $m \geqslant 2$, a quaternionic structure on M is a *torsion-free* $\mathrm{GL}(m, \mathbb{H}) \mathrm{GL}(1, \mathbb{H})$-*structure* Q on M, where $\mathrm{GL}(m, \mathbb{H}) \mathrm{GL}(1, \mathbb{H})$ is a Lie subgroup of $\mathrm{GL}(4m, \mathbb{R})$. That is, Q is a $\mathrm{GL}(m, \mathbb{H}) \mathrm{GL}(1, \mathbb{H})$-structure on M, and there exists a torsion-free connection ∇ on TM preserving Q. However, this connection ∇ is not unique, and not part of the quaternionic structure.

We call (M, Q) a *quaternionic manifold*. A quaternionic Kähler structure is effectively a torsion-free $\mathrm{Sp}(m) \mathrm{Sp}(1)$-structure on M. As $\mathrm{Sp}(m) \mathrm{Sp}(1)$ is a subgroup of $\mathrm{GL}(m, \mathbb{H}) \mathrm{GL}(1, \mathbb{H})$, a quaternionic Kähler structure induces a quaternionic structure. In 4 dimensions, a quaternionic manifold is an oriented conformal 4-manifold $(M, [g])$ with self-dual Weyl curvature.

As in the quaternionic Kähler case, a quaternionic $4m$-manifold (M, Q) comes with a bundle P of almost complex structures on M with fibre S^2, such that at each $x \in M$ the fibre of P is $\{a_1 J_1 + a_2 J_2 + a_3 J_3 : a_j \in \mathbb{R}, a_1^2 + a_2^2 + a_3^2 = 1\}$, where J_1, J_2, J_3 are almost complex structures at x with $J_1 J_2 = J_3$. There are many local sections of P which are integrable complex structures.

The total space of P is the *twistor space* Z of M. As for the other three quaternionic geometries, Z is a complex $(2m+1)$-manifold, and (M, Q) can be reconstructed entirely from holomorphic data on Z, including a real structure.

One can argue that quaternionic manifolds are the weakest geometry for which the twistor transform works—in particular, quaternionic Kähler manifolds, hyperkähler manifolds and hypercomplex manifolds are all quaternionic manifolds with some extra structure. For examples of quaternionic manifolds, including compact quaternionic manifolds which are not locally quaternionic Kähler or hypercomplex, see Joyce [116, 117].

7.6 A reading list on quaternionic geometry

The four different kinds of quaternionic geometry form a rich and beautiful branch of mathematics. In this chapter we have concentrated on compact hyperkähler manifolds, and many other interesting topics have had to be left out, some of them dear to the author's heart. Here is a reading list on some of these. The choice is a personal one; things I would have liked to write about. We begin with subjects in hyperkähler geometry.

- **The hyperkähler quotient.** Suppose X is a hyperkähler $4m$-manifold, and G is a Lie group of dimension k which acts freely on X preserving the hyperkähler structure. Then under mild conditions on X and G, Hitchin et al. [105, §3] define a *moment map* $\mu : X \to \mathfrak{g}^* \otimes \mathbb{R}^3$, where $\mathfrak{g}$ is the Lie algebra of G, and show that $Y = \mu^{-1}(0)/G$ is a new hyperkähler manifold of dimension $4(m - k)$, called the *hyperkähler quotient* of X by G. Hyperkähler quotients of $\mathbb{H}^m$ by Lie subgroups of $\mathrm{Sp}(m)$ provide many explicit examples of hyperkähler manifolds, which are always noncompact.
- **Cohomology of compact hyperkähler manifolds.** Salamon [187] and Verbitsky [215, 216] prove some powerful results.

- **Instanton and monopole moduli spaces.** Let X be a hyperkähler 4-manifold such as T^4, $K3$ or $\mathbb{H}$, which should be either compact, or complete and well-behaved at infinity. Let $P \to X$ be a principal bundle over X. An *instanton* is a connection on P with anti-self-dual curvature.

  The moduli space $\mathcal{M}$ of (finite energy) instantons on P modulo gauge transformations will in general be a finite-dimensional hyperkähler manifold, which may be singular, and may be noncompact. Instanton moduli spaces are important in the study of 4-manifolds. For more information, see Donaldson and Kronheimer [65].

  A related problem is the study of *magnetic monopoles* on noncompact 3-manifolds. Moduli spaces of magnetic monopoles on $\mathbb{R}^3$ turn out to be hyperkähler. Atiyah and Hitchin [13] use these hyperkähler metrics to resolve questions about scattering of magnetic monopoles.

- **Coadjoint orbits.** Let G be a compact, semisimple Lie group, with complexification G^c, and complexified Lie algebra $\mathfrak{g}^c$. Then the coadjoint orbits of G^c in $(\mathfrak{g}^c)^*$ are all noncompact complex symplectic manifolds. It turns out that each such orbit admits G-invariant Kähler metrics making it into a hyperkähler manifold. These metrics were constructed by Kronheimer [142, 143] for the highest-dimensional orbits in $(\mathfrak{g}^c)^*$, and extended to general orbits by Biquard [34] and Kovalev [139].

  Kronheimer's construction was analytic: he obtained the coadjoint orbits as moduli spaces of SU(2)-invariant G-instantons on $\mathbb{R}^4 \setminus 0$, and used the fact that instanton moduli spaces are hyperkähler. The author considers the metrics from a more algebraic point of view in [125, §11–§12].

Here are some references about the other three quaternionic geometries.

- **Quotient constructions.** The hyperkähler quotient construction of [105] discussed above can be generalized to quaternionic Kähler, hypercomplex and quaternionic manifolds. The quaternionic Kähler case was done by Galicki and Lawson [79, 80], and the hypercomplex and quaternionic cases by the author [116]. The details are different for each case.

- **Compact hypercomplex and quaternionic 4-manifolds.** Compact hypercomplex 4-manifolds are classified by Boyer [40], and are either $K3$ surfaces or flat examples such as 4-tori and Hopf surfaces. Explicit examples of compact quaternionic 4-manifolds (that is, self-dual 4-manifolds) are given by LeBrun [146, 147, 148] and the author [118]. Analytic existence theorems for compact self-dual 4-manifolds are given by Donaldson and Friedman [64] and Taubes [199, 200].

- **Compact hypercomplex and quaternionic $4m$-manifolds for $m \geqslant 2$.** The first known compact hypercomplex manifolds which are not locally hyperkähler were hypercomplex structures on Lie groups, found by Spindel et al. [197]. They were generalized by the author [117] to hypercomplex and quaternionic structures on homogeneous spaces.

In [119] this is extended to inhomogeneous hypercomplex structures on *biquotients* $A\backslash B/C$ of Lie groups. The author also gave in [117] a second way of constructing compact, inhomogeneous hypercomplex and quaternionic manifolds, using the idea of 'twisting by an instanton'.

- **Other aspects of hypercomplex geometry.** The *deformations* of compact hypercomplex manifolds are studied by Pedersen and Poon [174]. The author has proposed a theory of *hypercomplex algebraic geometry* [125], a quaternionic analogue of complex algebraic geometry, in which one studies (noncompact) hypercomplex manifolds using an 'algebra' of $\mathbb{H}$-valued 'q-holomorphic functions' upon them.

8

ASYMPTOTICALLY LOCALLY EUCLIDEAN METRICS WITH HOLONOMY SU(m)

Let G be a finite subgroup of SO(n) acting freely on $\mathbb{R}^n \setminus \{0\}$. Then $\mathbb{R}^n/G$ has an *isolated quotient singularity* at 0. Suppose that X is a noncompact n-manifold with just one end which at infinity resembles $\mathbb{R}^n/G$, and that g is a metric on X which approximates the Euclidean metric h_0 on $\mathbb{R}^n/G$ by $g = h_0 + O(r^{-n})$, with appropriate decay in the derivatives of g. Then we call g *Asymptotically Locally Euclidean*, or *ALE* for short, and we call (X, g) an *ALE manifold asymptotic to* $\mathbb{R}^n/G$.

In this chapter we will study *ALE Kähler manifolds*, which are Kähler manifolds (X, J, g) that are also ALE manifolds asymptotic to $\mathbb{C}^m/G$. We are particularly interested in *Ricci-flat ALE Kähler manifolds*, or equivalently, *ALE manifolds with holonomy* SU(m). This is because ALE manifolds with holonomy SU(2), SU(3) or SU(4) play a very important rôle in the construction of compact 7- and 8-manifolds with holonomy G_2 and Spin(7), which will be described later in the book.

In fact we have already met ALE metrics with holonomy SU(2) in §7.2, and they are classified by Kronheimer. This chapter extends Kronheimer's results to the case $m \geqslant 3$. However, our proofs use analytic ideas and are very different to Kronheimer's algebraic methods. There are also some peculiarities in the case $m = 2$ that do not apply for $m \geqslant 3$, which will be discussed in §8.9.

The main result of the chapter is Theorem 8.2.3, which says that if G is a nontrivial finite subgroup of SU(m) acting freely on $\mathbb{C}^m \setminus \{0\}$ and X is a crepant resolution of $\mathbb{C}^m/G$, then every Kähler class of ALE Kähler metrics on X contains a unique Ricci-flat Kähler metric g, which has holonomy SU(m). This is an analogue for ALE manifolds of Theorem 6.2.1, and we can use it to construct many families of ALE metrics with holonomy SU(m).

To prove Theorem 8.2.3 we have to do quite a lot of work, which takes up sections 8.3–8.8. First we must develop some analytical tools for ALE manifolds (X, g), in particular weighted Sobolev and Hölder spaces of functions on X, and elliptic regularity theory for the Laplacian Δ on these spaces. Then we state a version of the Calabi conjecture for ALE manifolds, and prove it by following Chapter 5. Finally we apply this Calabi conjecture to construct a unique Ricci-flat Kähler metric on X.

It seems likely that many readers will not want to plough through all of this material, at least on a first reading. *Such people are advised to read §8.1 and §8.2 only*. These two sections cover all you need to know about ALE manifolds with holonomy SU(m) to understand the later chapters on exceptional holonomy. The main purpose of §8.3–§8.8 is to prove two theorems which are stated without proof in §8.2. Section 8.9 is a technical discussion about resolutions and deformations, which can also be omitted on a first reading.

The results of §8.2 and §8.5–§8.9 are original research by the author, announced in [127], and published here in full for the first time. However, a number of other people have already written papers on noncompact versions of the Calabi conjecture, and I should at once admit that there is significant overlap between their results and mine.

In particular, Tian and Yau [204, 206, 207] and independently Bando and Kobayashi [19, 20] prove the following result [207, Cor. 1.1], [20, Th. 1]:

Theorem *Let X be a compact Kähler manifold with $c_1(X) > 0$, and D a smooth reduced divisor on X such that $c_1(X) = \alpha[D]$ for some $\alpha > 1$. Suppose D admits a Kähler–Einstein metric with positive scalar curvature. Then $X \setminus D$ has a complete Ricci-flat Kähler metric.*

Also Tian and Yau allow X to be an orbifold, and they give estimates on the decay of the curvature of their Ricci-flat metric. It can be shown that the existence of the metrics of Theorem 8.2.3 follows from the theorem above. Tian and Yau show that the curvature of the metrics decays with order at least $O(r^{-3})$ for large r. Combining this with the results of Bando, Kasue and Nakajima [18] on the asymptotic behaviour of Ricci-flat ALE manifolds, one can deduce the asymptotic estimates on the metrics given in Theorem 8.2.3. So most of Theorem 8.2.3 is already known, but the proof we give is new.

In Chapter 9 we will extend the material of this chapter to construct a special class of Ricci-flat Kähler metrics on crepant resolutions of *non-isolated* singularities $\mathbb{C}^m/G$, which we will call *Quasi-ALE* metrics. The author believes that these metrics are not covered by the work of Tian and Yau or Bando and Kobayashi. One reason why we give a full proof of Theorem 8.2.3, although it is already known, is so that we will be able to generalize it in Chapter 9.

Many of the results of this chapter were announced in [127].

8.1 Introduction to ALE metrics

Suppose that G is a finite subgroup of $SO(n)$ that acts freely on $\mathbb{R}^n \setminus \{0\}$. Then $\mathbb{R}^n/G$ has an *isolated quotient singularity* at 0. Let h_0 be the standard Euclidean metric on $\mathbb{R}^n$. Then h_0 is preserved by G, as $G \subset SO(n)$, and so h_0 descends to $\mathbb{R}^n/G$. Let r be the radius function on $\mathbb{R}^n/G$, that is, $r(x)$ is the distance from 0 to x calculated using h_0. We will define a natural class of noncompact Riemannian manifolds (X, g) called *ALE manifolds*, that have one infinite end upon which the metric g asymptotically resembles the metric h_0 on $\mathbb{R}^n/G$ for large r.

Definition 8.1.1 Let X be a noncompact manifold of dimension n, and g a Riemannian metric on X. We say that (X, g) is an *Asymptotically Locally Euclidean manifold asymptotic to* $\mathbb{R}^n/G$, or an *ALE manifold* for short, and we say that g is an *ALE metric*, if the following conditions hold.

There should exist a compact subset $S \subset X$ and a map $\pi : X \setminus S \to \mathbb{R}^n/G$ that is a diffeomorphism between $X \setminus S$ and the subset $\{z \in \mathbb{R}^n/G : r(z) > R\}$ for some fixed $R > 0$. Under this diffeomorphism, the push-forward metric $\pi_*(g)$ should satisfy

$$\nabla^k\bigl(\pi_*(g) - h_0\bigr) = O(r^{-n-k}) \quad \text{on } \{z \in \mathbb{R}^n/G : r(z) > R\}, \tag{8.1}$$

for all $k \geqslant 0$. Here ∇ is the Levi-Civita connection of h_0, and $T = O(r^{-j})$ if $|T| \leqslant Kr^{-j}$ for some $K > 0$.

If $G = \{1\}$, so that (X, g) is asymptotic to $\mathbb{R}^n$, then we call (X, g) an *Asymptotically Euclidean manifold*, or *AE manifold*. We shall call the map $\pi : X \setminus S \to \mathbb{R}^n/G$ an *asymptotic coordinate system* for X. Note that an ALE manifold does not come equipped with one particular asymptotic coordinate system. Instead there are infinitely many different possible asymptotic coordinate systems upon X. Sometimes we find it convenient to choose one such coordinate system and work with it, but our results will be independent of this choice.

Equation (8.1) says that towards infinity the metric g on X (and its derivatives) must converge to the Euclidean metric on $\mathbb{R}^n/G$, with a given rate of decay. We will explain in §8.2.1 why we have chosen the powers r^{-n-k} here.

Definition 8.1.2 Let (X, g) be an ALE manifold asymptotic to $\mathbb{R}^n/G$. We say that a smooth function $\rho : X \to [1, \infty)$ is a *radius function* on X if, given any asymptotic coordinate system $\pi : X \setminus S \to \mathbb{R}^n/G$, we have

$$\nabla^k\bigl(\pi_*(\rho) - r\bigr) = O(r^{1-n-k}) \quad \text{on } \{z \in \mathbb{R}^n/G : r(z) > R\},$$

for all $k \geqslant 0$. This condition is independent of the choice of asymptotic coordinate system, and radius functions exist for every ALE manifold.

A radius function is a function ρ on X that approximates the function r on $\mathbb{R}^n/G$ near infinity. In doing analysis on ALE manifolds, we will find it useful to consider Banach spaces of functions in which the norms are weighted by powers ρ^β of a radius function. Note that by definition $\rho \geqslant 1$, so we do not have to worry about small values of ρ.

Here is one way to think about ALE metrics. The manifold X is noncompact, but it can be compactified in a natural way by adding the boundary S^{n-1}/G at infinity. So we can instead regard X as a *compact manifold with boundary*. Then ALE metrics are metrics on X satisfying a certain natural boundary condition.

It is a general principle in differential geometry that most results about compact manifolds can also be extended to results about compact manifolds with boundary, provided the right boundary conditions are imposed in the problem. ALE manifolds are an example of this principle, because many results about compact Riemannian manifolds have natural analogues for ALE manifolds.

Next we define *ALE Kähler metrics*. Suppose G is a finite subgroup of $U(m)$ acting freely on $\mathbb{C}^m \setminus \{0\}$. Then $\mathbb{C}^m/G$ has an isolated quotient singularity at 0, and the standard Hermitian metric h_0 on $\mathbb{C}^m$ descends to $\mathbb{C}^m/G$. Let r be the radius function on $\mathbb{C}^m/G$. Suppose (X, π) is a *resolution* of $\mathbb{C}^m/G$, in the sense of §4.9. Then we can consider metrics on X which are both Kähler, and ALE.

Definition 8.1.3 Let (X, π) be a resolution of $\mathbb{C}^m/G$, with complex structure J, and let g be a Kähler metric on X. We say that (X, J, g) is an *ALE Kähler manifold asymptotic to $\mathbb{C}^m/G$*, and that g is an *ALE Kähler metric*, if for some $R > 0$ we have

$$\nabla^k\bigl(\pi_*(g) - h_0\bigr) = O(r^{-2m-k}) \quad \text{on } \{z \in \mathbb{C}^m/G : r(z) > R\}, \tag{8.2}$$

for all $k \geqslant 0$. We say that a smooth function $\rho : X \to [1, \infty)$ is a *radius function* on X if $\rho = \pi^*(r)$ on the subset $\{x \in X : \pi^*(r) \geqslant 2\}$. A radius function exists for every ALE Kähler manifold.

Because X is a resolution of $\mathbb{C}^m/G$, it comes equipped with a resolving map $\pi : X \to \mathbb{C}^m/G$, which gives a natural asymptotic coordinate system for X. The consequence of using this preferred asymptotic coordinate system is that on an ALE Kähler manifold (X, J, g), both the metric g and the complex structure J are simultaneously asymptotic to the metric and complex structure on $\mathbb{C}^m/G$. We also use π to simplify the definition of radius function. Note that in dimension 2 we can also consider *deformations* of $\mathbb{C}^2/G$, which calls for a slightly different definition of ALE Kähler manifold. This will be discussed in §8.9.

8.2 Ricci-flat ALE Kähler manifolds

We now give a number of results on Ricci-flat ALE Kähler manifolds, and some examples. Most proofs will be deferred until §8.8. We begin by showing that Ricci-flat ALE Kähler metrics exist only on *crepant resolutions*.

Proposition 8.2.1 *Let G be a finite subgroup of $\mathrm{U}(m)$ acting freely on $\mathbb{C}^m \setminus \{0\}$, let (X, π) be a resolution of $\mathbb{C}^m/G$, and suppose g is a Ricci-flat ALE Kähler metric on X. Then X is a crepant resolution of $\mathbb{C}^m/G$ and $G \subset \mathrm{SU}(m)$.*

Proof Since g is Ricci-flat it has Ricci form zero. But the cohomology class of the Ricci form is $2\pi\, c_1(X) \in H^2(X, \mathbb{R})$. Thus $c_1(X) = 0$, and the proposition then follows from the material of §6.3 and §6.4. $\square$

Next we define *Kähler classes* and the *Kähler cone* for ALE manifolds.

Definition 8.2.2 Let (X, J, g) be an ALE Kähler manifold asymptotic to $\mathbb{C}^m/G$ for some $m > 1$, with Kähler form ω. Then ω defines a de Rham cohomology class $[\omega] \in H^2(X, \mathbb{R})$ called the *Kähler class* of g. Define the *Kähler cone* $\mathcal{K}$ of X to be the set of Kähler classes $[\omega] \in H^2(X, \mathbb{R})$ of ALE Kähler metrics on (X, J). It is not difficult to prove that $\mathcal{K}$ is an open convex cone in $H^2(X, \mathbb{R})$, which does not contain zero.

The following two theorems will be proved in §8.8.

Theorem 8.2.3 *Let G be a finite subgroup of $\mathrm{SU}(m)$ acting freely on $\mathbb{C}^m \setminus \{0\}$, and (X, π) a crepant resolution of $\mathbb{C}^m/G$. Then in each Kähler class of ALE Kähler metrics on X there is a unique Ricci-flat ALE Kähler metric g. The Kähler form ω of g satisfies*

$$\pi_*(\omega) = \omega_0 + A\, \mathrm{dd}^c(r^{2-2m}) + \mathrm{dd}^c \chi \tag{8.3}$$

on the set $\{z \in \mathbb{C}^m/G : r(z) > R\}$, where $A < 0$ and $R > 0$ are constants, ω_0 is the Kähler form of the Euclidean metric on $\mathbb{C}^m/G$, r the radius function on $\mathbb{C}^m/G$, and χ a smooth function on $\{z \in \mathbb{C}^m/G : r(z) > R\}$ such that $\nabla^k \chi = O(r^{\gamma-k})$ for each $k \geqslant 0$ and $\gamma \in (1 - 2m, 2 - 2m)$.

Theorem 8.2.4 *Let G be a nontrivial finite subgroup of $\mathrm{SU}(m)$ acting freely on $\mathbb{C}^m \setminus \{0\}$, let (X, π) be a crepant resolution of $\mathbb{C}^m/G$, and let g be a Ricci-flat ALE Kähler metric on X. Then g has holonomy $\mathrm{SU}(m)$.*

Theorem 8.2.3 is the main result of this chapter, and is an analogue of Theorem 6.2.1 for ALE Kähler manifolds. As we explained in the introduction to the chapter, nearly all of Theorem 8.2.3 can be deduced from the results of Tian and Yau [204, 206, 207], Bando and Kobayashi [19, 20] and Bando, Kasue and Nakajima [18]. The proof of Theorem 8.2.3 takes up most of §8.3–§8.8, and works by first proving a version of the Calabi conjecture for ALE manifolds, using the ideas of Chapter 5. Theorem 8.2.4 is quite easy to prove.

8.2.1 *A discussion of Theorem 8.2.3 and its proof*

Suppose (M, g) is a compact Riemannian manifold. Then the Hölder spaces $C^{k,\alpha}(M)$ have the important property that if $f \in C^{k,\alpha}(M)$ and $\int_M f \, dV_g = 0$, then there is a unique $u \in C^{k+2,\alpha}(M)$ with $\int_M u \, dV_g = 0$ and $\Delta u = f$, and $\|u\|_{C^{k+2,\alpha}} \leqslant C \|f\|_{C^{k,\alpha}}$ for $C > 0$ independent of u and f. This is called *elliptic regularity*, and plays an essential rôle in the proof of the Calabi conjecture.

However, if (X, g) is an ALE manifold then the spaces $C^{k,\alpha}(X)$ do not have this property, as X is noncompact. Instead, in §8.3 we define a family of *weighted Hölder spaces* $C^{k,\alpha}_\beta(X)$, in which the usual Hölder norm is weighted by a power ρ^β of a *radius function* ρ on X. If $f \in C^{k,\alpha}_\beta(X)$ then $f = O(\rho^\beta)$, and $\nabla^j f = O(\rho^{\beta-j})$ for $j \leqslant k$. We show that elliptic regularity results for Δ hold in these spaces for generic β.

Section 8.4 extends these ideas to exterior forms, and discusses the de Rham cohomology of X. Then §8.5 states a version of the Calabi conjecture for ALE manifolds, and proves it assuming Theorems A1–A4, which are analogous to Theorems C1–C4 of Chapter 5. After some preliminary estimates in §8.6 Theorems A1–A4 are proved in §8.7, completing the proof of the conjecture. We then apply this in §8.8 to construct Ricci-flat ALE Kähler metrics.

One way to write the Calabi conjecture is as an existence problem for a nonlinear elliptic p.d.e.: if f is a smooth function on X, then there should exist a smooth function ϕ on X with $(\omega + dd^c \phi)^m = e^f \omega^m$, where ω is the Kähler form on X. In the ALE case we must also impose decay conditions on f and ϕ, in terms of a radius function ρ on X. For instance, if $\beta \in (-2m, -2)$ and $\nabla^j f = O(\rho^{\beta-j})$, then ϕ satisfies $\nabla^j \phi = O(\rho^{\beta+2-j})$.

Suppose g is a Ricci-flat ALE Kähler metric on X with Kähler form ω, and let ω_0 be the Kähler form of the Euclidean metric on $\mathbb{C}^m/G$. Then on $\{z \in \mathbb{C}^m/G : r(z) > R\}$ we have $\pi_*(\omega) = \omega_0 + dd^c \phi$, where ϕ is a smooth function satisfying $(\omega_0 + dd^c \phi)^m = \omega_0^m$. This implies that $\Delta \phi = O(|dd^c \phi|^2)$. Thus $\Delta \phi \approx 0$ when $dd^c \phi$ is small, which is true for large r.

Now $\Delta(r^{2-2m}) = 0$ away from 0 in $\mathbb{C}^m/G$, and more generally if f is compactly supported on $\mathbb{C}^m/G$ and $\Delta \phi = f$ with $\phi \to 0$ as $r \to \infty$, then $\phi = Ar^{2-2m} + O(r^{1-2m})$ for large r, where A is a multiple of $\int f \, dV$. This explains why the term Ar^{2-2m} appears in (8.3). To construct a Ricci-flat ALE Kähler metric we must solve a nonlinear elliptic p.d.e. for a Kähler potential ϕ. Near infinity this p.d.e. approaches the equation $\Delta\phi = 0$ on $\mathbb{C}^m/G$. But if $\Delta\phi = 0$ near infinity then $\phi = Ar^{2-2m} + O(r^{1-2m})$, as we want.

Note that because $A < 0$ in Theorem 8.2.3, the term $A\,dd^c(r^{2-2m})$ in (8.3) is nonzero. Therefore $\pi_*(g) - h_0$ decays with order exactly $O(r^{-2m})$, and similarly $\nabla^k(\pi_*(g) - h_0)$ decays with order exactly $O(r^{-2m-k})$. Thus in Definition 8.1.1 the decay rates given in (8.1) are *sharp* for all Ricci-flat ALE Kähler metrics, and cannot be improved upon. This is why we chose the powers r^{-n-k} in our definition (8.1) of ALE metrics.

8.2.2 *Examples*

As we saw in §7.2, ALE Kähler manifolds with holonomy SU(2) are already very well understood. Thus, we can write down many explicit examples of ALE manifolds with holonomy SU(2). For $m \geqslant 3$, Calabi [51, p. 285] found an explicit ALE Kähler manifold with holonomy SU(m) asymptotic to $\mathbb{C}^m/\mathbb{Z}_m$, which we describe next. In the case $m = 2$, Calabi's example coincides with the Eguchi–Hanson metric of Example 7.2.2.

Example 8.2.5 Let $\mathbb{C}^m$ have complex coordinates $(z_1, \ldots, z_m)$, let $\zeta = e^{2\pi i/m}$, and let α act on $\mathbb{C}^m$ by $\alpha : (z_1, \ldots, z_m) \mapsto (\zeta z_1, \ldots, \zeta z_m)$. Then $\alpha^m = 1$, and the group $G = \langle \alpha \rangle$ generated by α is a subgroup of SU(m) isomorphic to $\mathbb{Z}_m$, which acts freely on $\mathbb{C}^m \setminus \{0\}$. Thus the quotient $\mathbb{C}^m/G$ has an isolated singular point at 0. Let (X, π) be the *blow-up* of $\mathbb{C}^m/G$ at 0, so that $\pi^{-1}(0) \cong \mathbb{CP}^{m-1}$. It is easy to show that X is in fact a *crepant resolution* of $\mathbb{C}^m/G$.

Let r be the radius function on $\mathbb{C}^m/G$, and define $f : \mathbb{C}^m/G \setminus \{0\} \to \mathbb{R}$ by

$$f = \sqrt[m]{r^{2m} + 1} + \frac{1}{m}\sum_{j=0}^{m-1} \zeta^j \log\left(\sqrt[m]{r^{2m} + 1} - \zeta^j\right). \tag{8.4}$$

To define the logarithm of the complex number $\sqrt[m]{r^{2m}+1} - \zeta^j$ we cut $\mathbb{C}$ along the negative real axis, and set $\log(Re^{i\theta}) = \log R + i\theta$ for $R > 0$ and $\theta \in (-\pi, \pi)$. Then f is well-defined, and it is a smooth *real* function on $\mathbb{C}^m/G \setminus \{0\}$, despite its complex definition.

Define a (1,1)-form ω on $X \setminus \pi^{-1}(0)$ by $\omega = dd^c \pi^*(f)$. It can be shown that ω extends to a smooth, closed, positive (1,1)-form on all of X. Let g be the Kähler metric on X with Kähler form ω. Then Calabi [51, §4] shows that g is complete and Ricci-flat, with Hol(g) = SU(m). Equation (8.4) is derived from [51, eqn (4.14), p. 285]. Note also that the action of U(m) on $\mathbb{C}^m$ pushes down to $\mathbb{C}^m/G$ and lifts through π to X, and g is invariant under this action of U(m) on X.

This metric g on X is an *ALE Kähler metric*. To prove this, we show using (8.4) that

$$f = r^2 - \frac{1}{m(m-1)}r^{2-2m} + O(r^{-2m}) \quad \text{on } \mathbb{C}^m/G \setminus \{0\}, \text{ for large } r.$$

Now the Kähler form of the Euclidean metric on $\mathbb{C}^m/G$ is $\omega_0 = \mathrm{dd}^c(r^2)$. Hence

$$\pi_*(\omega) = \omega_0 - \frac{1}{m(m-1)} \mathrm{dd}^c(r^{2-2m}) + \mathrm{dd}^c\chi \quad \text{on } \mathbb{C}^m/G \setminus \{0\}, \tag{8.5}$$

where $\chi = f - r^2 + \frac{1}{m(m-1)} r^{2-2m}$. It is easy to show that $\nabla^k \chi = O(r^{-k-2m})$ on $\mathbb{C}^m/G \setminus \{0\}$ for large r, and it quickly follows that g is an ALE Kähler metric on X, by Definition 8.1.3. Also, g is one of the Ricci-flat ALE Kähler metrics of Theorem 8.2.3, and comparing (8.5) with (8.3) we see that $A = -\frac{1}{m(m-1)}$, which verifies that $A < 0$.

For $m \geq 3$, the metrics of Example 8.2.5 are the *only* explicit examples of ALE metrics with holonomy $\mathrm{SU}(m)$ that are known, at least to the author. It is possible to find these metrics explicitly because they have a large symmetry group $\mathrm{U}(m)$, whose orbits are of real codimension 1 in X. Because of this, the problem can be reduced to a nonlinear, second-order o.d.e. in one real variable, which can then be explicitly solved.

It is a natural question whether we can find an explicit, algebraic form for any or all of the other ALE metrics with holonomy $\mathrm{SU}(m)$ for $m \geq 3$, that exist on crepant resolutions of $\mathbb{C}^m/G$ by Theorem 8.2.3. The author believes that general ALE metrics with holonomy $\mathrm{SU}(m)$ for $m \geq 3$ are essentially transcendental, nonalgebraic objects, and that one cannot write them down explicitly using simple functions. Furthermore, the author conjectures that for $m \geq 3$, the metrics of Example 8.2.5 are the only ALE metrics with holonomy $\mathrm{SU}(m)$ that can be written down explicitly in coordinates.

8.3 Analysis on ALE manifolds

In §1.2 we defined the Sobolev spaces $L_k^q(M)$, the C^k-spaces $C^k(M)$, and the Hölder spaces $C^{k,\alpha}(M)$. These are Banach spaces of functions upon a Riemannian manifold (M, g), and we saw in Theorem 1.4.1 that when M is *compact*, elliptic operators such as the Laplacian Δ have very good regularity properties on Sobolev spaces and Hölder spaces, so that they are useful tools in problems involving elliptic operators on compact manifolds.

Now suppose that (X, g) is an ALE manifold asymptotic to $\mathbb{R}^n/G$. Then X is noncompact, and the results of Theorem 1.4.1 are in fact false for X, even when P is the Laplacian Δ. What this tells us, is that $L_k^q(X)$ and $C^{k,\alpha}(X)$ are not good choices of Banach spaces of functions for studying elliptic operators on an ALE manifold (X, g). Instead, it turns out to be helpful to introduce *weighted Sobolev spaces* and *weighted Hölder spaces*, which we define next.

Definition 8.3.1 Let (X, g) be an ALE manifold asymptotic to $\mathbb{R}^n/G$, and let ρ be a radius function on X. For $q \geq 1$, $\beta \in \mathbb{R}$ and k a nonnegative integer, define the *weighted Sobolev space* $L_{k,\beta}^q(X)$ to be the set of functions f on X that are locally integrable and k times weakly differentiable, and for which the norm

$$\|f\|_{L_{k,\beta}^q} = \left(\sum_{j=0}^k \int_X |\rho^{j-\beta} \nabla^j f|^q \rho^{-n} \mathrm{d}V_g \right)^{1/q}$$

is finite. Then $L_{k,\beta}^q(X)$ is a Banach space, and $L_{k,\beta}^2(X)$ a Hilbert space.

Definition 8.3.2 Let (X, g) be an ALE manifold asymptotic to $\mathbb{R}^n/G$, and let ρ be a radius function on X. For $\beta \in \mathbb{R}$ and k a nonnegative integer, define $C^k_\beta(X)$ to be the space of continuous functions f on X with k continuous derivatives, such that $\rho^{j-\beta}|\nabla^j f|$ is bounded on X for $j = 0, \ldots, k$. Define the norm $\|.\|_{C^k_\beta}$ on $C^k_\beta(X)$ by

$$\|f\|_{C^k_\beta} = \sum_{j=0}^{k} \sup_X |\rho^{j-\beta} \nabla^j f|.$$

Let $\delta(g)$ be the injectivity radius of g, and write $d(x, y)$ for the distance between x, y in X. For T a tensor field on X and $\alpha, \gamma \in \mathbb{R}$, define

$$[T]_{\alpha,\gamma} = \sup_{\substack{x \neq y \in X \\ d(x,y) < \delta(g)}} \left[\min(\rho(x), \rho(y))^{-\gamma} \cdot \frac{|T(x) - T(y)|}{d(x, y)^\alpha} \right]. \tag{8.6}$$

Here we interpret $|T(x) - T(y)|$ using parallel translation along the unique geodesic of length $d(x, y)$ joining x and y, as in eqn (1.2) of §1.2.

For $\beta \in \mathbb{R}$, k a nonnegative integer, and $\alpha \in (0, 1)$, define the *weighted Hölder space* $C^{k,\alpha}_\beta(X)$ to be the set of $f \in C^k_\beta(X)$ for which the norm

$$\|f\|_{C^{k,\alpha}_\beta} = \|f\|_{C^k_\beta} + [\nabla^k f]_{\alpha, \beta-k-\alpha} \tag{8.7}$$

is finite. Define $C^\infty_\beta(X)$ to be the intersection of the $C^k_\beta(X)$ for all $k \geq 0$. Both $C^k_\beta(X)$ and $C^{k,\alpha}_\beta(X)$ are Banach spaces, but $C^\infty_\beta(X)$ is not a Banach space.

These definitions are adapted from Lee and Parker [151, §9]. What they really mean is that a function f in $L^q_{k,\beta}(X)$, $C^k_\beta(X)$ or $C^{k,\alpha}_\beta(X)$ grows at most like ρ^β as $\rho \to \infty$, and so the index β should be interpreted as an *order of growth*. Similarly, the derivatives $\nabla^j f$ grow at most like $\rho^{\beta-j}$ for $j = 1, \ldots, k$. As vector spaces of functions, $L^q_{k,\beta}(X)$, $C^k_\beta(X)$ and $C^{k,\alpha}_\beta(X)$ are independent of the choice of radius function ρ. The norms on these spaces do depend on the choice of ρ, but not in a significant way, as all choices of ρ give equivalent norms.

Suppose that V is a vector bundle on X with Euclidean metrics on its fibres, and ∇_V is a connection on V preserving these metrics. Then we can define $L^q_{k,\beta}(V)$, $C^k_\beta(V)$ and $C^{k,\alpha}_\beta(V)$ to be weighted Banach spaces of sections of V over X, by extending the definitions above in the obvious way. In particular, taking V to be $\Lambda^k T^*X$ gives $L^q_{k,\beta}(\Lambda^k T^*X)$, $C^k_\beta(\Lambda^k T^*X)$ and $C^{k,\alpha}_\beta(\Lambda^k T^*X)$, which are Banach spaces of k-forms on X.

Here is an analogue for ALE manifolds of the last part of Theorem 1.2.3, which follows from Chaljub-Simon and Choquet-Bruhat [57, Lem. 3, p. 11].

Theorem 8.3.3 *Suppose (X, g) is an ALE manifold. Let k be a nonnegative integer and $\alpha, \beta, \gamma \in \mathbb{R}$ with $0 < \alpha < 1$ and $\beta < \gamma$. Then the embedding $C^{k,\alpha}_\beta(X) \hookrightarrow C^k_\gamma(X)$ is compact.*

Next we discuss the analysis of the Laplacian Δ on ALE manifolds. Much work has been done on the behaviour of Δ on weighted Sobolev spaces and Hölder spaces on $\mathbb{R}^n$, and more generally on AE manifolds. Some papers on this subject are [22, 56, 57, 59, 154, 160, 168]. Many of their results apply immediately to ALE manifolds, with only very minor cosmetic changes to their proofs.

Proposition 8.3.4 *Let (X, g) be an ALE manifold of dimension n asymptotic to $\mathbb{R}^n/G$, let $\beta, \gamma \in \mathbb{R}$ satisfy $\beta + \gamma < 2 - n$, and suppose $u \in C^2_\beta(X)$ and $v \in C^2_\gamma(X)$. Then*

$$\int_X u \,\Delta v \,dV_g = \int_X v \,\Delta u \,dV_g. \tag{8.8}$$

Let ρ be a radius function on X. Then $\Delta(\rho^{2-n}) \in C^\infty_{-2n}(X)$ and

$$\int_X \Delta(\rho^{2-n}) \,dV_g = \frac{(n-2)\,\Omega_{n-1}}{|G|}, \tag{8.9}$$

where Ω_{n-1} is the volume of the unit sphere $\mathcal{S}^{n-1}$ in $\mathbb{R}^n$.

Proof Let S_R be the subset $\{x \in X : \rho(x) \leqslant R\}$ in X. Stokes' Theorem gives that

$$\int_{S_R} (u\Delta v - v\Delta u) \,dV_g = \int_{\partial S_R} \left[(u\nabla v - v\nabla u) \cdot \mathbf{n}\right] dV_g, \tag{8.10}$$

where $\mathbf{n}$ is the inward-pointing unit normal to ∂S_R. But for large R we have $\mathrm{vol}(\partial S_R) = O(R^{n-1})$ and $u\nabla v - v\nabla u = O(R^{\beta+\gamma-1})$ on ∂S_R, so that the right hand side of (8.10) is $O(R^{\beta+\gamma+n-2})$. Since $\beta + \gamma < 2 - n$ we see that the right hand side of (8.10) tends to zero as $R \to \infty$, and this proves (8.8).

The point about the power ρ^{2-n} is that $\Delta(r^{2-n}) = 0$ away from 0 in $\mathbb{R}^n/G$. Using the definitions of radius function and ALE manifold one can show that $\Delta(\rho^{2-n}) \in C^\infty_{-2n}(X)$, as we want. Using Stokes' Theorem again we find that

$$\int_{S_R} \Delta(\rho^{2-n}) \,dV_g = \int_{\partial S_R} \left[\nabla(\rho^{2-n}) \cdot \mathbf{n}\right] dV_g.$$

But for large R we have $\nabla(\rho^{2-n}) \cdot \mathbf{n} \approx (n-2)R^{1-n}$ and $\mathrm{vol}(S_R) \approx R^{n-1}\Omega_{n-1}/|G|$. Thus, letting $R \to \infty$ gives (8.9). $\square$

Theorem 8.3.5 *Let $n > 2$ and $k \geqslant 0$ be integers and $\alpha \in (0, 1)$, and let $\mathbb{R}^n$ have its Euclidean metric. Then*

(a) *Suppose $\beta \in (-n, -2)$. Then for each $f \in C^{k,\alpha}_\beta(\mathbb{R}^n)$ there is a unique $u \in C^{k+2,\alpha}_{\beta+2}(\mathbb{R}^n)$ with $\Delta u = f$.*

(b) *Suppose $\beta \in (-1-n, -n)$. Then for each $f \in C^{k,\alpha}_\beta(\mathbb{R}^n)$ there exists a unique $u \in C^{k+2,\alpha}_{\beta+2}(\mathbb{R}^n)$ with $\Delta u = f$ if and only if $\int_{\mathbb{R}^n} f \,dV = 0$.*

In each case $\|u\|_{C^{k+2,\alpha}_{\beta+2}} \leqslant C\|f\|_{C^{k,\alpha}_\beta}$ for some $C > 0$ depending only on n, k, α and β.

Proof This is a weighted version of the *Schauder estimates* of §1.4. We will explain how to prove that $u \in C^0_{\beta+2}(\mathbb{R}^n)$ and estimate $\|u\|_{C^0_{\beta+2}}$ in each case, and leave it to the reader to extend this to $C^{k+2,\alpha}_{\beta+2}(\mathbb{R}^n)$. From §1.4.1, if $u \in C^2_{\beta+2}(\mathbb{R}^n)$ for $\beta < -2$ and $\Delta u = f$, then by (1.10) we have

$$u(y) = \frac{1}{(n-2)\Omega_{n-1}} \int_{x \in \mathbb{R}^n} |x-y|^{2-n} f(x) dx, \tag{8.11}$$

where Ω_{n-1} is the volume of the unit sphere S^{n-1} in $\mathbb{R}^n$. This is *Green's representation* for u. Let ρ be a radius function on $\mathbb{R}^n$. Then $|f(x)| \le \|f\|_{C^0_\beta} \rho(x)^\beta$, so from (8.11) we see that

$$|u(y)| \le \frac{1}{(n-2)\Omega_{n-1}} \|f\|_{C^0_\beta} \cdot \int_{x \in \mathbb{R}^n} |x-y|^{2-n} \rho(x)^\beta dx.$$

We split this integral into integrals over the three regions $|x| \le \frac{1}{2}|y|$, $\frac{1}{2}|y| < |x| \le 2|y|$ and $|x| > 2|y|$ in $\mathbb{R}^n$. Estimating the integral on each region separately we prove

$$\int_{x \in \mathbb{R}^n} |x-y|^{2-n} \rho(x)^\beta dx \le \begin{cases} C' \rho(y)^{\beta+2} & \text{for } \beta \in (-n, -2), \\ C' \rho(y)^{2-n} & \text{for } \beta < -n. \end{cases}$$

Thus if $\beta \in (-n, -2)$ then $|u(y)| \le C \|f\|_{C^0_\beta} \rho(y)^{\beta+2}$ for some $C > 0$ depending only on n and β, and so $u \in C^0_{\beta+2}(\mathbb{R}^n)$ and $\|u\|_{C^0_{\beta+2}} \le C \|f\|_{C^0_\beta}$. This is part of case (a) of the theorem. The rest of case (a) follows from (8.11) by the usual Schauder estimate methods of [83, §6].

Next we prove case (b). Suppose $\beta \in (-1-n, -n)$, $u \in C^{k+2,\alpha}_{\beta+2}(\mathbb{R}^n)$ and $\Delta u = f$. Then

$$\int_{\mathbb{R}^n} f \, dV = \int_{\mathbb{R}^n} 1 \Delta u \, dV = \int_{\mathbb{R}^n} u \Delta(1) \, dV = 0$$

by Proposition 8.3.4, since $u \in C^2_{\beta+2}(\mathbb{R}^n)$ and $1 \in C^2_0(\mathbb{R}^n)$ and $\beta + 2 + 0 < 2 - n$. Thus, given $f \in C^{k,\alpha}_\beta(\mathbb{R}^n)$, there can only exist $u \in C^{k+2,\alpha}_{\beta+2}(\mathbb{R}^n)$ with $\Delta u = f$ if $\int_{\mathbb{R}^n} f \, dV = 0$. So suppose that $\int_{\mathbb{R}^n} f \, dV = 0$, and define u by

$$u(y) = \frac{1}{(n-2)\Omega_{n-1}} \int_{x \in \mathbb{R}^n} \Big[|x-y|^{2-n} - \rho(y)^{2-n} \Big] f(x) dx. \tag{8.12}$$

Since $\int_{\mathbb{R}^n} f \, dV = 0$ the term involving $\rho(y)^{2-n}$ in this integral vanishes, so the equation reduces to (8.11) and thus $\Delta u = f$. From (8.12) we see that

$$|u(y)| \le \frac{1}{(n-2)\Omega_{n-1}} \|f\|_{C^0_\beta} \cdot \int_{x \in \mathbb{R}^n} \Big| |x-y|^{2-n} - \rho(y)^{2-n} \Big| \rho(x)^\beta dx,$$

and estimating as before shows that $|u(y)| \le C \|f\|_{C^0_\beta} \rho(y)^{\beta+2}$ when $\beta \in (-1-n, -n)$. Thus $u \in C^0_{\beta+2}(\mathbb{R}^n)$ and $\|u\|_{C^0_{\beta+2}} \le C \|f\|_{C^0_\beta}$, which is part of case (b). The rest of case (b) follows as above. □

Now we extend Theorem 8.3.5 to ALE manifolds.

Theorem 8.3.6 *Suppose (X, g) is an ALE manifold asymptotic to $\mathbb{R}^n/G$ for $n > 2$, and ρ a radius function on X. Let $k \geqslant 0$ be an integer and $\alpha \in (0, 1)$. Then*
  (a) *Let $\beta \in (-n, -2)$. Then there exists $C > 0$ such that for each $f \in C_\beta^{k,\alpha}(X)$ there is a unique $u \in C_{\beta+2}^{k+2,\alpha}(X)$ with $\Delta u = f$, which satisfies $\|u\|_{C_{\beta+2}^{k+2,\alpha}} \leqslant C \|f\|_{C_\beta^{k,\alpha}}$.*
  (b) *Let $\beta \in (-1-n, -n)$. Then there exist $C_1, C_2 > 0$ such that for each $f \in C_\beta^{k,\alpha}(X)$ there is a unique $u \in C_{2-n}^{k+2,\alpha}(X)$ with $\Delta u = f$. Moreover $u = A\rho^{2-n} + v$, where*

$$A = \frac{|G|}{(n-2)\,\Omega_{n-1}} \cdot \int_X f \, dV_g \tag{8.13}$$

and $v \in C_{\beta+2}^{k+2,\alpha}(X)$ satisfy $|A| \leqslant C_1 \|f\|_{C_\beta^0}$ and $\|v\|_{C_{\beta+2}^{k+2,\alpha}} \leqslant C_2 \|f\|_{C_\beta^{k,\alpha}}$. Here Ω_{n-1} is the volume of the unit sphere $\mathcal{S}^{n-1}$ in $\mathbb{R}^n$.

Proof The theory of weighted Hölder spaces on AE manifolds and the Laplacian is developed by Chaljub-Simon and Choquet-Bruhat [57], who restrict their attention to the case $n = 3$. In particular, they prove part (a) of the theorem for the case $n = 3, k = 0$ and $G = \{1\}$, [57, p. 15–16]. Their proof uses a result equivalent to part (a) of Theorem 8.3.5 in the case $n = 3$ and $k = 0$. By using Theorem 8.3.5 together with the methods of [57] one can show that Theorem 8.3.5 applies not only to $\mathbb{R}^n$ with its Euclidean metric, but also to any ALE manifold (X, g) asymptotic to $\mathbb{R}^n/G$. This proves case (a) of the theorem immediately.

For case (b), let $f \in C_\beta^{k,\alpha}(X)$, and define A by (8.13). Then by eqn (8.9) we have $\int_X [f - \Delta(A\rho^{2-n})] dV_g = 0$. Also $\Delta(\rho^{2-n}) \in C_{-2n}^\infty(X)$ by Proposition 8.3.4, and so $f - \Delta(A\rho^{2-n})$ lies in $C_\beta^{k,\alpha}(X)$ and has integral zero on X. Since $|f| \leqslant \|f\|_{C_\beta^0} \rho^\beta$ we have $|A| \leqslant C_1 \|f\|_{C_\beta^0}$ for $C_1 = \int_X \rho^\beta \, dV_g$, as we have to prove.

Applying case (b) of Theorem 8.3.5 for X to $f - \Delta(A\rho^{2-n})$, we see that there is a unique $v \in C_{\beta+2}^{k+2,\alpha}(X)$ with $\Delta v = f - \Delta(A\rho^{2-n})$, which satisfies

$$\|v\|_{C_{\beta+2}^{k+2,\alpha}} \leqslant C\big(\|f\|_{C_\beta^{k,\alpha}} + |A| \cdot \|\Delta(\rho^{2-n})\|_{C_\beta^{k,\alpha}}\big). \tag{8.14}$$

Defining $u = A\rho^{2-n} + v$ gives $\Delta u = f$ as we want. Clearly $u \in C_{2-n}^{k+2,\alpha}(X)$, and the inequality $\|v\|_{C_{\beta+2}^{k+2,\alpha}} \leqslant C_2 \|f\|_{C_\beta^{k,\alpha}}$ then follows from (8.14) and the estimate on $|A|$ above. $\square$

8.4 Exterior forms and de Rham cohomology

Let (X, g) be an ALE manifold asymptotic to $\mathbb{R}^n/G$. Let $H^*(X, \mathbb{R})$ be the de Rham cohomology of X, and $H_c^*(X, \mathbb{R})$ the de Rham cohomology of X *with compact support*. That is,

$$H_c^k(X, \mathbb{R}) = \frac{\{\eta : \eta \text{ is a smooth, closed, compactly-supported } k\text{-form on } X\}}{\{d\zeta : \zeta \text{ is a smooth, compactly-supported } (k-1)\text{-form on } X\}}.$$

Both $H^k(X, \mathbb{R})$ and $H_c^k(X, \mathbb{R})$ are finite-dimensional vector spaces. Let us regard X as a compact manifold with boundary $\mathcal{S}^{n-1}/G$. Using the long exact sequence

$$\ldots \to H_c^k(X, \mathbb{R}) \to H^k(X, \mathbb{R}) \to H^k(\mathcal{S}^{n-1}/G, \mathbb{R}) \to H_c^{k+1}(X, \mathbb{R}) \to \ldots,$$

the cohomology of $\mathcal{S}^{n-1}/G$, and the fact that $H_c^k(X, \mathbb{R}) \cong [H^{n-k}(X, \mathbb{R})]^*$ by Poincaré duality for manifolds with boundary, one can show that

$$H^0(X, \mathbb{R}) = \mathbb{R}, \quad H_c^0(X, \mathbb{R}) = 0, \quad H^n(X, \mathbb{R}) = 0, \quad H_c^n(X, \mathbb{R}) = \mathbb{R}, \text{ and}$$
$$H^k(X, \mathbb{R}) \cong H_c^k(X, \mathbb{R}) \cong [H^{n-k}(X, \mathbb{R})]^* \cong [H_c^{n-k}(X, \mathbb{R})]^* \text{ for } k \neq 0, n.$$

Now the material on weighted Sobolev and Hölder spaces of functions in §8.3 generalizes naturally to weighted Sobolev spaces and Hölder spaces of k-forms on ALE manifolds (X, g), so we may define the spaces $L^q_{l,\beta}(\Lambda^k T^*X)$, $C^{l,\alpha}_\beta(\Lambda^k T^*X)$ and $C^\infty_\beta(\Lambda^k T^*X)$ in the obvious way. Similarly, the results of §8.3 on the Laplacian Δ on functions generalize to results on $\Delta = \mathrm{d}\mathrm{d}^* + \mathrm{d}^*\mathrm{d}$ on k-forms.

These tools can be used to generalize the ideas of Hodge theory, described in §1.1.3, to ALE manifolds. In particular, one can prove the following result.

Theorem 8.4.1 *Let (X, g) be an ALE manifold asymptotic to $\mathbb{R}^n/G$ for $n > 2$, and define*

$$\mathcal{H}^k(X) = \{\eta \in C^\infty_{1-n}(\Lambda^k T^*X) : \mathrm{d}\eta = \mathrm{d}^*\eta = 0\}.$$

Then $\mathcal{H}^0(X) = \mathcal{H}^n(X) = 0$, and the map $\mathcal{H}^k(X) \to H^k(X, \mathbb{R})$ given by $\eta \mapsto [\eta]$ induces natural isomorphisms $\mathcal{H}^k(X) \cong H^k(X, \mathbb{R}) \cong H_c^k(X, \mathbb{R})$ for $0 < k < n$. The Hodge star gives an isomorphism $* : \mathcal{H}^k(X) \to \mathcal{H}^{n-k}(X)$. Suppose $1 - n \leqslant \beta < -n/2$. Then

$$C^\infty_\beta(\Lambda^k T^*X) = \mathcal{H}^k(X) \oplus \mathrm{d}\left[C^\infty_{\beta+1}(\Lambda^{k-1}T^*X)\right] \oplus \mathrm{d}^*\left[C^\infty_{\beta+1}(\Lambda^{k+1}T^*X)\right],$$

where the summands are L^2-orthogonal.

This is an analogue of Theorems 1.1.3 and 1.1.4.

For the rest of the section we shall restrict our attention to ALE Kähler manifolds. If (X, J, g) is an ALE Kähler manifold then we can define the weighted Banach spaces of (p, q)-forms $L^r_{l,\beta}(\Lambda^{p,q}X)$ and $C^{l,\alpha}_\beta(\Lambda^{p,q}X)$ on X in the obvious way. The Laplacian Δ acts on these spaces, for instance by

$$\Delta : C^{l+2,\alpha}_{\beta+2}(\Lambda^{p,q}X) \to C^{l,\alpha}_\beta(\Lambda^{p,q}X).$$

These Banach spaces of (p, q)-forms have very similar analytic properties to the Banach spaces of functions on an ALE manifold discussed in §8.3.

We can use facts about the Laplacian on weighted Banach spaces of (p, q)-forms to develop an analogue for ALE Kähler manifolds of the Hodge theory for compact Kähler manifolds described in §4.7.1.

Theorem 8.4.2 *Let (X, J, g) be an ALE Kähler manifold asymptotic to $\mathbb{C}^m/G$. Define*

$$\mathcal{H}^{p,q}(X) = \{\eta \in C^\infty_{1-2m}(\Lambda^{p,q} X) : d\eta = d^*\eta = 0\}.$$

Then $\mathcal{H}^{p,q}(X)$ is a finite-dimensional vector space, and the map from $\mathcal{H}^{p,q}(X)$ to $H^{p+q}(X, \mathbb{C})$ defined by $\eta \mapsto [\eta]$ is injective. Define $H^{p,q}(X)$ to be the image of this map. Then

$$H^k(X, \mathbb{C}) = \bigoplus_{j=0}^{k} H^{j,k-j}(X) \quad \text{for } 0 < k < 2m.$$

However, $H^{p,q}(X) = 0$ for $p \neq q$ when X is a crepant resolution of $\mathbb{C}^m/G$.

Theorem 8.4.3 *Let (X, J, g) be an ALE Kähler manifold, where X is a resolution of $\mathbb{C}^m/G$. Then $H^{2,0}(X) = H^{0,2}(X) = 0$, and each element of $H^{1,1}(X)$ is represented by a closed, compactly-supported $(1, 1)$-form on X.*

Here is a sketch of the proof of this theorem. Since X is a resolution of $\mathbb{C}^m/G$, it can be shown that the homology group $H_{2m-2}(X, \mathbb{C})$ is generated by the homology classes of the exceptional divisors of the resolution. But $H_{2m-2}(X, \mathbb{C}) \cong H^2_c(X, \mathbb{C})$. Thus $H^2_c(X, \mathbb{C})$ is generated by cohomology classes dual to the homology classes $[D]$ of exceptional divisors D in X. If U is any open neighbourhood of D in X, then we can find a closed $(1, 1)$-form supported in U representing the cohomology class dual to $[D]$. Therefore $H^2_c(X, \mathbb{C})$ is generated by cohomology classes represented by closed, compactly-supported $(1, 1)$-forms. It easily follows that $H^{2,0}(X) = H^{0,2}(X) = 0$, and the proof is finished.

Next we prove a version of the Global dd^c-Lemma of §4.5 for ALE Kähler manifolds.

Theorem 8.4.4 *Let (X, J, g) be an ALE Kähler manifold asymptotic to $\mathbb{C}^m/G$ for some $m > 1$, and let $\beta < -m$. Suppose that $\eta \in C^\infty_\beta(\Lambda^{1,1}_\mathbb{R} X)$ is a closed real $(1,1)$-form and $[\eta] = 0$ in $H^2(X, \mathbb{R})$. Then there exists a unique real function $u \in C^\infty_{\beta+2}(X)$ with $\eta = dd^c u$.*

Proof Let ω be the Kähler form of g. If u is a smooth function on X we have

$$dd^c u \wedge \omega^{m-1} = -\tfrac{1}{m} \Delta u \, \omega^m. \tag{8.15}$$

Also, if ζ is a real $(1,1)$-form on X and $\zeta \wedge \omega^{m-1} = 0$ it can be shown that

$$\zeta \wedge \omega^{m-2} = -\tfrac{1}{2}(m-2)! * \zeta \quad \text{and} \quad \zeta \wedge \zeta \wedge \omega^{m-2} = -\tfrac{1}{2}(m-2)! |\zeta|^2 dV_g, \tag{8.16}$$

where $*$ is the Hodge star and dV_g the volume form of g. Equations (8.15) and (8.16) hold on any Kähler manifold of dimension m, with the conventions used in this book.

Define a function f on X by $\eta \wedge \omega^{m-1} = -\tfrac{1}{m} f \omega^m$. Since $\eta \in C^\infty_\beta(\Lambda^{1,1}_\mathbb{R} X)$, it follows that $f \in C^\infty_\beta(X)$. Now suppose for simplicity that $-2m < \beta < -m$. Then by part (a) of Theorem 8.3.6 there exists a unique function $u \in C^\infty_{\beta+2}(X)$ with $\Delta u = f$.

Set $\zeta = \eta - dd^c u$, which is an exact 2-form in $C^\infty_\beta(\Lambda^{1,1}_\mathbb{R} X)$. As $\beta < -m$ we can use the last part of Theorem 8.4.1 to prove that $\zeta = d\theta$, for some $\theta \in C^\infty_{\beta+1}(T^*X)$.

By (8.15) we have $\zeta \wedge \omega^{m-1} = -\frac{1}{m}(f - \Delta u)\omega^m = 0$, so (8.16) gives

$$d\big[\theta \wedge \zeta \wedge \omega^{m-2}\big] = \zeta \wedge \zeta \wedge \omega^{m-2} = -\frac{1}{2}(m-2)!\,|\zeta|^2 dV_g. \tag{8.17}$$

Let ρ be a radius function on X, and define $S_R = \{x \in X : \rho(x) \leq R\}$ for $R > 1$. Integrating (8.17) over S_R and using Stokes' Theorem gives that

$$-\frac{1}{2}(m-2)! \cdot \int_{S_R} |\zeta|^2 dV_g = \int_{\partial S_R} \theta \wedge \zeta \wedge \omega^{m-2}. \tag{8.18}$$

But for large R we have $\theta = O(R^{\beta+1})$, $\zeta = O(R^\beta)$ and $\omega = O(1)$ on ∂S_R, and $\mathrm{vol}(\partial S_R) = O(R^{2m-1})$. Thus the right hand side of (8.18) is $O(R^{2\beta+2m})$. As $\beta < -m$, taking the limit as $R \to \infty$ shows that $\int_X |\zeta|^2 dV_g = 0$, and so $\zeta = 0$ on X. Thus $\eta = dd^c u$, as we have to prove.

We have proved the theorem assuming that $-2m < \beta < -m$, but we wish to prove it for all $\beta < -m$. If $\beta \leq 2m$ and $\eta \in C^\infty_\beta(\Lambda^{1,1}_\mathbb{R} X)$ then $\eta \in C^\infty_\gamma(\Lambda^{1,1}_\mathbb{R} X)$ for any γ with $-2m < \gamma < -m$, and so from above we have $\eta = dd^c u$ for some unique u in $C^\infty_{\gamma+2}(X)$. However, if $u \in C^\infty_{\gamma+2}(X)$ and $dd^c u \in C^\infty_\beta(\Lambda^{1,1}_\mathbb{R} X)$, one can show that $u \in C^\infty_{\beta+2}(X)$ as we want. This is because $dd^c u$ is a stronger derivative of u than Δu is, and contains more information. $\square$

Finally, we show we can modify any ALE Kähler metric to be flat outside a compact set.

Proposition 8.4.5 *Let $\mathbb{C}^m/G$ have an isolated singularity at 0 for some $m > 1$, let (X, π) be a resolution of $\mathbb{C}^m/G$ that admits ALE Kähler metrics, and let ρ be a radius function on X. Then in each Kähler class there exists an ALE Kähler metric g' on X such that $g' = \pi^*(h_0)$ on the subset $\{x \in X : \rho(x) > R\}$, where h_0 is the Hermitian metric on $\mathbb{C}^m/G$ and $R > 0$ is a constant.*

Proof Let g be an ALE Kähler metric on X, with Kähler form ω. By Theorems 8.4.2 and 8.4.3 there exists a closed, compactly-supported, real (1,1)-form θ on X with $[\theta] = [\omega]$ in $H^2(X, \mathbb{R})$. Define $\eta = \omega - dd^c(\rho^2) - \theta$. Then η is an exact real (1,1)-form on X. Now the Kähler form of h_0 on $\mathbb{C}^m/G$ is $\omega_0 = dd^c(r^2)$. So from the definition of ALE Kähler metric we see that $\omega - dd^c(\rho^2) \in C^\infty_{-2m}(\Lambda^{1,1}_\mathbb{R} X)$, and therefore $\eta \in C^\infty_{-2m}(\Lambda^{1,1}_\mathbb{R} X)$, as θ has compact support. Thus by Theorem 8.4.4 there is a unique real function $u \in C^\infty_{2-2m}(X)$ with $\eta = dd^c u$, and we have $\omega = \theta + dd^c(\rho^2) + dd^c u$.

Let $\mu : \mathbb{R} \to [0,1]$ be a smooth function with $\mu(t) = 1$ for $t \leq -1$ and $\mu(t) = 0$ for $t \geq 0$. For each $R > 0$ define a closed (1,1)-form ω_R by

$$\omega_R = \theta + dd^c(\rho^2) + dd^c\big[\mu(\rho - R) \cdot u\big].$$

Then $\omega_R = \omega$ wherever $\rho < R - 1$, and $\omega_R = dd^c(\rho^2)$ wherever $\rho > R$ and outside the support of θ. It is easy to show that ω_R is a positive (1,1)-form for large R, which therefore defines a Kähler metric g_R on X. Define g' to be g_R for some R sufficiently large that ω_R is positive, $\rho \leqslant R$ on the support of θ and $R \geqslant 2$. Then g' is an ALE Kähler metric in the Kähler class of g, and where $\rho > R$ we have $g' = \pi^*(h_0)$, since the Kähler form of g' is $\mathrm{dd}^c(\rho^2)$, the Kähler form of h_0 is $\mathrm{dd}^c(r^2)$, and $\rho = \pi^*(r)$ as $\rho > R \geqslant 2$. □

8.5 The Calabi conjecture for ALE manifolds

We will now state and prove the following version of the Calabi conjecture for ALE Kähler manifolds, which is based on the third version of the Calabi conjecture in §5.1.

The Calabi conjecture for ALE manifolds *Suppose that (X, J, g) is an ALE Kähler manifold of dimension m asymptotic to $\mathbb{C}^m/G$ for some $m > 1$, with Kähler form ω, and that ρ is a radius function on X. Then*

(a) *Let $\beta \in (-2m, -2)$. Then for each $f \in C^\infty_\beta(X)$ there is a unique $\phi \in C^\infty_{\beta+2}(X)$ such that $\omega + \mathrm{dd}^c\phi$ is a positive $(1, 1)$-form and $(\omega + \mathrm{dd}^c\phi)^m = \mathrm{e}^f \omega^m$ on X.*

(b) *Let $\beta \in (-1-2m, -2m)$. Then for each $f \in C^\infty_\beta(X)$ there is a unique $\phi \in C^\infty_{2-2m}(X)$ such that $\omega+\mathrm{dd}^c\phi$ is a positive $(1, 1)$-form and $(\omega+\mathrm{dd}^c\phi)^m = \mathrm{e}^f \omega^m$ on X. Moreover we can write $\phi = A\rho^{2-2m} + \psi$, where $\psi \in C^\infty_{\beta+2}(X)$ and*

$$A = \frac{|G|}{(m-1)\Omega_{2m-1}} \cdot \int_X (1-\mathrm{e}^f)\mathrm{d}V_g. \tag{8.19}$$

Here Ω_{2m-1} is the volume of the unit sphere $\mathcal{S}^{2m-1}$ in $\mathbb{C}^m$.

It is easy to rewrite this in terms of the existence of ALE Kähler metrics with prescribed Ricci curvature, as in the original Calabi conjecture. The two cases (a) $\beta \in (-2m, -2)$ and (b) $\beta \in (-1-2m, -2m)$ come from Theorem 8.3.6. We will prove the conjecture using the following four theorems, which are based on Theorems C1–C4 of §5.2.

Theorem A1 *Let (X, J, g) be an ALE Kähler manifold asymptotic to $\mathbb{C}^m/G$ for some $m > 1$, with Kähler form ω. Let $Q_1 > 0$, $\alpha \in (0, 1)$ and $\beta + 2 \leqslant \gamma < 0$. Then there exists $Q_2 > 0$ such that if $f \in C^{3,\alpha}_\beta(X)$ and $\phi \in C^5_\gamma(X)$ satisfy $(\omega + \mathrm{dd}^c\phi)^m = \mathrm{e}^f \omega^m$ and $\|f\|_{C^{3,\alpha}_\beta} \leqslant Q_1$, then $\omega + \mathrm{dd}^c\phi$ is a positive $(1, 1)$-form, $\phi \in C^{5,\alpha}(X)$ and $\|\phi\|_{C^{5,\alpha}} \leqslant Q_2$. Also, if $f \in C^{k,\alpha}_\beta(X)$ for $k \geqslant 3$ then $\phi \in C^{k+2,\alpha}(X)$.*

Theorem A2 *Let (X, J, g) be an ALE Kähler manifold asymptotic to $\mathbb{C}^m/G$ for some $m > 1$, with Kähler form ω and radius function ρ. Let $Q_1, Q_2 > 0$, $\alpha \in (0, 1)$ and $\beta + 2 \leqslant \gamma < 0$. Then there exists $Q_3 > 0$ such that the following holds.*

Suppose $f \in C^{3,\alpha}_\beta(X)$ and $\phi \in C^5_\gamma(X) \cap C^{5,\alpha}(X)$ satisfy $(\omega + \mathrm{dd}^c\phi)^m = \mathrm{e}^f \omega^m$ and the inequalities $\|f\|_{C^{3,\alpha}_\beta} \leqslant Q_1$ and $\|\phi\|_{C^{5,\alpha}} \leqslant Q_2$. Then

(a) *Let $\beta \in (-2m, -2)$. Then $\phi \in C^{5,\alpha}_{\beta+2}(X)$ and $\|\phi\|_{C^{5,\alpha}_{\beta+2}} \leqslant Q_3$. If $f \in C^{k,\alpha}_\beta(X)$ for $k \geqslant 3$ then $\phi \in C^{k+2,\alpha}_{\beta+2}(X)$.*

(b) Let $\beta \in (-1-2m, -2m)$. Then $\phi = A\rho^{2-2m} + \psi$, where A is defined by (8.19) and $\psi \in C^{5,\alpha}_{\beta+2}(X)$ with $\|\psi\|_{C^{5,\alpha}_{\beta+2}} \leq Q_3$. If $f \in C^{k,\alpha}_{\beta}(X)$ for $k \geq 3$ then $\psi \in C^{k+2,\alpha}_{\beta+2}(X)$.

Theorem A3 *Let (X, J, g) be an ALE Kähler manifold asymptotic to $\mathbb{C}^m/G$ for some $m > 1$, with Kähler form ω. Let $\alpha \in (0, 1)$ and $\beta \in (-2m, -2)$. Suppose that $f' \in C^{3,\alpha}_{\beta}(X)$ and $\phi' \in C^{5,\alpha}_{\beta+2}(X)$ satisfy $(\omega + dd^c\phi')^m = e^{f'}\omega^m$. Then whenever $f \in C^{3,\alpha}_{\beta}(X)$ and $\|f - f'\|_{C^{3,\alpha}_{\beta}}$ is sufficiently small, there exists $\phi \in C^{5,\alpha}_{\beta+2}(X)$ such that $(\omega + dd^c\phi)^m = e^f\omega^m$.*

Theorem A4 *Let (X, J, g) be an ALE Kähler manifold asymptotic to $\mathbb{C}^m/G$ for some $m > 1$, with Kähler form ω, and let $\beta + 2 \leq \gamma < 0$. Then for each $f \in C^1_\beta(X)$ there is at most one function $\phi \in C^3_\gamma(X)$ such that $(\omega + dd^c\phi)^m = e^f\omega^m$ on X.*

These results will be proved in §8.7, using some a priori estimates of ϕ that will be found in §8.6. Assuming Theorems A1–A4, we will now prove the Calabi conjecture for ALE manifolds. The proof follows that in §5.2.1 closely, so we will pass quickly over the details.

Theorem 8.5.1 *The Calabi conjecture for ALE manifolds is true.*

Proof Let (X, J, g) be an ALE Kähler manifold asymptotic to $\mathbb{C}^m/G$ for $m > 1$ with Kähler form ω, and let $\alpha \in (0, 1)$. We begin by proving part (a) of the conjecture, so suppose that $\beta \in (-2m, -2)$. Pick some γ with $\beta + 2 < \gamma < 0$. Fix $f \in C^{3,\alpha}_\beta(X)$, and define S to be the set of all $t \in [0, 1]$ for which there exists $\phi \in C^{5,\alpha}_{\beta+2}(X)$ such that $(\omega + dd^c\phi)^m = e^{tf}\omega^m$ on X. Now, we will show using Theorems A1 and A2 that S is *closed*, and using Theorem A3 that S is *open*.

Proposition 8.5.2 *In the situation above, S is a closed subset of $[0, 1]$.*

Proof Let $\{t_j\}_{j=0}^\infty$ be a sequence in S, which converges to some $t' \in [0, 1]$. We will prove that $t' \in S$. Since $t_j \in S$, by definition there exists $\phi_j \in C^{5,\alpha}_{\beta+2}(X)$ such that $(\omega + dd^c\phi_j)^m = e^{t_j f}\omega^m$. Define Q_1 by $Q_1 = \|f\|_{C^{3,\alpha}_\beta}$. Let Q_2, Q_3 be the constants given by Theorems A1 and A2, which depend only on (X, J, g), Q_1, α, β and γ. Applying Theorem A1 with ϕ_j in place of ϕ and $t_j f$ in place of f, we see that $\|\phi_j\|_{C^{5,\alpha}} \leq Q_2$. Thus $\|\phi_j\|_{C^{5,\alpha}_{\beta+2}} \leq Q_3$ for all j by Theorem A2.

Since $\beta + 2 < \gamma$, Theorem 8.3.3 shows that the inclusion $C^{5,\alpha}_{\beta+2}(X) \to C^5_\gamma(X)$ is *compact*. Therefore, as $\{\phi_j\}_{j=0}^\infty$ is bounded in $C^{5,\alpha}_{\beta+2}(X)$, we can choose a subsequence $\{\phi_{i_j}\}_{j=0}^\infty$ which converges in $C^5_\gamma(X)$. Let $\phi' \in C^5_\gamma(X)$ be the limit of this subsequence. Taking the limit in the equation $(\omega + dd^c\phi_{i_j})^m = e^{t_{i_j} f}\omega^m$ shows that $(\omega + dd^c\phi')^m = e^{t' f}\omega^m$. Theorems A1 and A2 then prove that $\phi' \in C^{5,\alpha}_{\beta+2}(X)$, and so $t' \in S$. Thus S contains its limit points, and is closed. □

Proposition 8.5.3 *In the situation above, S is an open subset of $[0, 1]$.*

Proof Let $t' \in S$. Then there exists $\phi' \in C^{5,\alpha}_{\beta+2}(X)$ with $(\omega + dd^c \phi')^m = e^{t' f} \omega^m$ on X. Apply Theorem A3, with $t' f$ in place of f', and tf in place of f, for $t \in [0, 1]$. The theorem shows that whenever $|t - t'| \cdot \|f\|_{C^{3,\alpha}_\beta}$ is sufficiently small, there exists $\phi \in C^{5,\alpha}_{\beta+2}(X)$ with $(\omega + dd^c \phi)^m = e^{tf} \omega^m$, and so $t \in S$. Thus, if $t \in [0, 1]$ is sufficiently close to t' then $t \in S$, and S is open. $\square$

Now when $t = 0$ we can take $\phi = 0$ to show that $0 \in S$. So S is nonempty, and is open and closed in $[0, 1]$ by the last two propositions. Since $[0, 1]$ is connected, we have $S = [0, 1]$, and so $1 \in S$. Thus we have proved that if $f \in C^{3,\alpha}_\beta(X)$ then there exists $\phi \in C^{5,\alpha}_{\beta+2}(X)$ such that $(\omega + dd^c \phi)^m = e^f \omega^m$ on X. Suppose now that $f \in C^\infty_\beta(X)$. Then part (a) of Theorem A2 shows that $\phi \in C^{k+2,\alpha}_{\beta+2}(X)$ for all $k \geq 3$, and so $\phi \in C^\infty_{\beta+2}(X)$ as we want. Also $\omega + dd^c \phi$ is a positive (1,1)-form by Theorem A1, and ϕ is unique by Theorem A4. This proves part (a) of the conjecture.

Next we prove part (b), so suppose that $\beta \in (-1 - 2m, -2m)$ and $f \in C^\infty_\beta(X)$. Pick some $\beta' \in (-2m, -2)$. Then f lies in $C^\infty_{\beta'}(X)$ as $\beta' > \beta$, and so by part (a) of the conjecture there exists a unique $\phi \in C^\infty_{\beta'+2}(X)$ such that $\omega + dd^c \phi$ is a positive (1,1)-form and $(\omega + dd^c \phi)^m = e^f \omega^m$ on X. But applying part (b) of Theorem A2 then shows that $\phi = A\rho^{2-2m} + \psi$, where A is defined by (8.19) and $\psi \in C^\infty_{\beta+2}(X)$. Clearly $\phi \in C^\infty_{2-2m}(X)$, so part (b) of the conjecture holds. This completes the proof of Theorem 8.5.1. $\square$

8.6 A priori estimates of ϕ

We shall find a priori estimates for ϕ in $C^0(X)$, in $C^0_\delta(X)$ for $\delta < 0$ in a certain range, and finally in $C^{k,\alpha}_\delta(X)$. They are based on the proofs of Theorems C1 and C2 of §5.2, and will be used to prove Theorems A1 and A2 in §7.

8.6.1 Estimating ϕ in $C^0(X)$

First we construct an a priori bound for $\|\phi\|_{C^0}$ that will be needed to prove Theorem A1.

Theorem 8.6.1 *Let (X, J, g) be an ALE Kähler manifold asymptotic to $\mathbb{C}^m/G$ for $m > 1$, with Kähler form ω. Let $Q_1 > 0$ and $\beta + 2 \leq \gamma < 0$. Then there exists $P_1 > 0$ such that if $f \in C^0_\beta(X)$ and $\phi \in C^2_\gamma(X)$ satisfy $(\omega + dd^c \phi)^m = e^f \omega^m$ and $\|f\|_{C^0_\beta} \leq Q_1$, then $\omega + dd^c \phi$ is a positive (1, 1)-form, ϕ lies in $C^0(X)$ and $\|\phi\|_{C^0} \leq P_1$.*

Proof Let all notation be as in the theorem, and let ρ be a radius function on X. This result is an analogue of Corollary 5.4.5, and to prove it we will follow the method of §5.4.1. The next four results are analogues of Lemma 5.1.1, Proposition 5.4.1 and Lemmas 5.4.2 and 5.4.3, respectively.

Lemma 8.6.2 *The (1, 1)-form $\omega' = \omega + dd^c \phi$ is positive on X.*

Proof As in the proof of Lemma 5.1.1, if $\omega + dd^c \phi$ is positive at one point of X, then it is positive everywhere. But since $dd^c \phi = O(\rho^{\gamma-2})$ as $\phi \in C^2_\gamma(X)$, and $\gamma < 0$, we have $|dd^c \phi| < 1$ for large ρ, and so $\omega + dd^c \phi$ is positive wherever ρ is suffiently large. Thus $\omega + dd^c \phi$ is positive on all of X. $\square$

Proposition 8.6.3 *Let $p > (2 - 2m)/\gamma$. Then in the situation above,*

$$\int_X \left|\nabla |\phi|^{p/2}\right|^2_g \leqslant \frac{mp^2}{4(p-1)} \int_X (1 - e^f)\phi|\phi|^{p-2} dV_g. \tag{8.20}$$

Proof The argument follows the proof of Proposition 5.4.1 exactly, except that in proving (5.19) we cannot use Stokes' Theorem upon X because it is not compact. Instead, define $S_R = \{x \in X : \rho(x) \leqslant R\}$. Then by Stokes' Theorem

$$\int_{S_R} d\left[\phi|\phi|^{p-2} d^c\phi \wedge \left(\omega^{m-1} + \omega^{m-2} \wedge \omega' + \cdots + (\omega')^{m-1}\right)\right] = \int_{\partial S_R} \phi|\phi|^{p-2} d^c\phi \wedge \left(\omega^{m-1} + \omega^{m-2} \wedge \omega' + \cdots + (\omega')^{m-1}\right). \tag{8.21}$$

Since $\phi \in C^2_\gamma(X)$, on ∂S_R we have $\phi = O(R^\gamma)$, $d^c\phi = O(R^{\gamma-1})$ and $\omega, \omega' = O(1)$. Also $\text{vol}(\partial S_R) = O(R^{2m-1})$. Therefore the right hand side of (8.21) is $O(R^{p\gamma+2m-2})$, and $p\gamma + 2m - 2 < 0$ as $p > (2m-2)/\gamma$ and $\gamma < 0$. Taking the limit as $R \to \infty$ then shows that

$$\int_X d\left[\phi|\phi|^{p-2} d^c\phi \wedge \left(\omega^{m-1} + \omega^{m-2} \wedge \omega' + \cdots + (\omega')^{m-1}\right)\right] = 0,$$

as we want. The rest of the proof of Proposition 5.4.1 applies unchanged. □

Lemma 8.6.4 *Let $\epsilon = m/(m-1)$. There is a constant $C_1 > 0$ depending on X and g, such that if $\chi \in L^{2\epsilon}(X)$ is weakly differentiable and $\nabla \chi \in L^2(T^*X)$, then $\|\chi\|_{L^{2\epsilon}} \leqslant C_1 \|\nabla \chi\|_{L^2}$.*

Proof This is really a Sobolev embedding estimate for the embedding $L^2_1(X) \to L^{2\epsilon}(X)$. Three different proofs for the case $X = \mathbb{R}^n$ are given by Aubin [17, p. 37–44]. The result for $\mathbb{R}^n$ can be extended to any ALE manifold (X, g) asymptotic to $\mathbb{R}^n/G$ using the methods of Choquet-Bruhat and Christodoulu [59, §6]. □

Lemma 8.6.5 *There are constants $\tau > 1$ and $C_2 > 0$ depending on X, g and Q_1, such that if $p \in [2\tau, 2\tau\epsilon]$ then $\|\phi\|_{L^p} \leqslant C_2$.*

Proof Choose some $p \in \mathbb{R}$ satisfying $p > 1$ and $p > (2 - 2m)/\gamma$. Define $q = mp/(p+m+1)$ and $r = \epsilon p/(p-1)$. Then $q, r > 1$ and $\frac{1}{q} + \frac{1}{r} = 1$. Applying Lemma 8.6.4 with $\chi = |\phi|^{p/2}$ shows that $\|\phi\|^p_{L^{\epsilon p}} \leqslant C_1^2 \||\nabla|\phi|^{p/2}\|^2_{L^2}$, and substituting this into (8.20) gives

$$\|\phi\|^p_{L^{\epsilon p}} \leqslant \frac{C_1^2 mp^2}{4(p-1)} \int_X (1 - e^f)\phi|\phi|^{p-2} dV_g$$

$$\leqslant \frac{C_1^2 mp^2}{4(p-1)} \|1 - e^f\|_{L^q} \cdot \||\phi|^{p-1}\|_{L^r},$$

by Hölder's inequality. But $\||\phi|^{p-1}\|_{L^r} = \|\phi\|_{L^{r(p-1)}}^{p-1} = \|\phi\|_{L^{\epsilon p}}^{p-1}$, as $r(p-1) = \epsilon p$, and so

$$\|\phi\|_{L^{\epsilon p}} \leqslant \frac{C_1^2 m p^2}{4(p-1)} \|1 - e^f\|_{L^q}.$$

As $\beta + 2 \leqslant \gamma < 0$ and $p > (2 - 2m)/\gamma$ we have $p > (2 - 2m)/(\beta + 2)$, which implies that $q\beta < -2m$. Therefore $\|1 - e^f\|_{L^q}$ exists, and can be bounded solely in terms of X, g, Q_1 and q. Thus, given any p with $p > (2 - 2m)/\gamma$, we can find an a priori bound on $\|\phi\|_{L^{\epsilon p}}$ depending only on X, g, Q_1 and p. It is then easy to complete the lemma. □

Now we can follow the proofs of Proposition 5.4.4 and Corollary 5.4.5 to construct a constant $P_1 > 0$ depending on X, J, g and Q_1 such that $\phi \in C^0(X)$ and $\|\phi\|_{C^0} \leqslant P_1$. Note that one must start the 'induction' in the proof of Proposition 5.4.4 from $p = \tau$ rather than $p = 2$. Together with Lemma 8.6.2, this proves Theorem 8.6.1. □

8.6.2 Estimating ϕ in $C_\delta^0(X)$

Next we show that $|\phi| \leqslant P_2 \rho^\delta$ for suitable constants $P_2 > 0$ and $\delta < 0$. This will be needed to prove Theorem A2. Notice that the result depends upon the estimate $|dd^c \phi| \leqslant Q_2$, which will be proved in Theorem A1.

Theorem 8.6.6 *Let (X, J, g) be an ALE Kähler manifold asymptotic to $\mathbb{C}^m/G$ for $m > 1$, with Kähler form ω. Let $Q_1, Q_2 > 0$ and β, γ, δ satisfy $\beta + 2 \leqslant \gamma < \delta < 0$ and $\delta \geqslant 1 - m$. Then there exists $P_2 > 0$ such that if $f \in C_\beta^0(X)$ and $\phi \in C_\gamma^2(X)$ satisfy $(\omega + dd^c \phi)^m = e^f \omega^m$, $\|f\|_{C_\beta^0} \leqslant Q_1$ and $\|dd^c \phi\|_{C^0} \leqslant Q_2$, then $\phi \in C_\delta^0(X)$ and $\|\phi\|_{C_\delta^0} \leqslant P_2$.*

Proof Here is a 'weighted' version of Proposition 8.6.3.

Proposition 8.6.7 *Let $p > 1$ and $q \geqslant 0$ satisfy $p\gamma + q < 2 - 2m$. Then there exists $C_3 > 0$ depending on m, Q_2 and ρ but not on p, q, f and ϕ, such that*

$$\int_X \left|\nabla(|\phi|^{p/2} \rho^{q/2})\right|_g^2 dV_g \leqslant \frac{mp^2}{4(p-1)} \int_X (1 - e^f) \phi |\phi|^{p-2} \rho^q dV_g \qquad (8.22)$$
$$+ C_3 \frac{q(p+q)}{4(p-1)} \int_X |\phi|^p \rho^{q-2} dV_g.$$

Proof First observe that as $\phi \in C_\gamma^2(X)$, $1 - e^f \in C_\beta^0(X)$ and $\beta + 2 \leqslant \gamma$ all three integrands in (8.22) are $O(\rho^{p\gamma + q - 2})$, and since $p\gamma + q - 2 < -2m$ all three integrals do exist. Arguing using Stokes' Theorem as in Proposition 8.6.3, we prove that

$$\int_X d\left\{\left[p^2 \phi |\phi|^{p-2} \rho^q d^c \phi + (p-2) q |\phi|^p \rho^{q-1} d^c \rho\right] \wedge \left(\omega^{m-1} + \cdots + (\omega')^{m-1}\right)\right\} = 0,$$

provided $p\gamma + q < 2 - 2m$. Multiplying this by $m/4(p-1)$ and rearranging gives

$$m \int_X \mathrm{d}\bigl[|\phi|^{p/2}\rho^{q/2}\bigr] \wedge \mathrm{d}^c\bigl[|\phi|^{p/2}\rho^{q/2}\bigr] \wedge \bigl(\omega^{m-1} + \cdots + (\omega')^{m-1}\bigr) =$$

$$-\frac{mp^2}{4(p-1)} \int_X \phi|\phi|^{p-2}\rho^q \, \mathrm{dd}^c\phi \wedge \bigl(\omega^{m-1} + \cdots + (\omega')^{m-1}\bigr)$$

$$+\frac{mq}{4(p-1)} \int_X |\phi|^p \rho^{q-2}\bigl[(p+q-2)\mathrm{d}\rho \wedge \mathrm{d}^c\rho - (p-2)\rho \, \mathrm{dd}^c\rho\bigr]$$

$$\wedge \bigl(\omega^{m-1} + \cdots + (\omega')^{m-1}\bigr).$$

As in the proof of Proposition 5.4.1 one can show that

$$\int_X \bigl|\nabla(|\phi|^{p/2}\rho^{q/2})\bigr|_g^2 \omega^m \leqslant m \int_X \mathrm{d}\bigl[|\phi|^{p/2}\rho^{q/2}\bigr] \wedge \mathrm{d}^c\bigl[|\phi|^{p/2}\rho^{q/2}\bigr]$$
$$\wedge \bigl(\omega^{m-1} + \cdots + (\omega')^{m-1}\bigr),$$

$$\int_X \phi|\phi|^{p-2}\rho^q \, \mathrm{dd}^c\phi \wedge \bigl(\omega^{m-1} + \cdots + (\omega')^{m-1}\bigr) = -\int_X (1-\mathrm{e}^f)\phi|\phi|^{p-2}\rho^q\omega^m.$$

Also, since $|\nabla^j \rho| = O(\rho^{1-j})$ and $\|\mathrm{dd}^c\phi\|_{C^0} \leqslant Q_2$, there exists $C' > 0$ depending on ρ and Q_2 but independent of p, q, for which

$$\bigl|(p+q-2)\mathrm{d}\rho \wedge \mathrm{d}^c\rho - (p-2)\rho \, \mathrm{dd}^c\rho\bigr| \leqslant C'(p+q)$$
$$\text{and} \quad \bigl|\omega^{m-1} + \cdots + (\omega')^{m-1}\bigr| \leqslant C'. \tag{8.23}$$

Combining the last five equations and remembering that $\omega^m = m!\,\mathrm{d}V_g$, we prove (8.22). $\square$

Proposition 8.6.8 *Let p satisfy $p\delta \leqslant 2 - 2m$ and $p \geqslant 2$. Then there exists a constant $C_4 > 0$ depending on Q_1, Q_2, m, δ and ρ but independent of p, such that if $\phi \in L^p_{0,\delta}(X)$ then $\phi \in L^{p\epsilon}_{0,\delta}(X)$, where $\epsilon = m/(m-1)$, and*

$$\|\phi\|^p_{L^{p\epsilon}_{0,\delta}} \leqslant C_4 p\bigl(\|\phi\|^{p-1}_{L^{p-1}_{0,\delta}} + \|\phi\|^p_{L^p_{0,\delta}}\bigr). \tag{8.24}$$

Proof Define $q = 2 - 2m - p\delta$. Then there exist $C', C'', C''' > 0$ depending on m, Q_1 such that

$$\bigl|(1-\mathrm{e}^f)\rho^{2-\delta}\bigr| \leqslant C', \quad \frac{mp^2}{4(p-1)} \leqslant C''p, \quad \text{and} \quad \frac{q(p+q)}{4(p-1)} \leqslant C'''p. \tag{8.25}$$

The first inequality holds because $|1 - \mathrm{e}^f| = O(\rho^\beta)$, and thus $\bigl|(1-\mathrm{e}^f)\rho^{2-\delta}\bigr| = O(\rho^{\beta+2-\delta}) = O(1)$, as $\beta + 2 < \delta$. The second and third inequalities follow easily.

Now $p > 1, q \geqslant 0$ and $p\gamma + q < 2 - 2m$, so we may apply Proposition 8.6.7. From (8.22) and (8.25) we see that

$$\int_X \bigl|\nabla(|\phi|^{p/2}\rho^{1-m-p\delta/2})\bigr|_g^2 \mathrm{d}V_g \leqslant C'C''p\|\phi\|^{p-1}_{L^{p-1}_{0,\delta}} + C'''C_3 p\|\phi\|^p_{L^p_{0,\delta}}.$$

Applying Lemma 8.6.4 to the function $|\phi|^{p/2}\rho^{1-m-p\delta/2}$ gives

$$\|\phi\|_{L^{p\epsilon}_{0,\delta}}^p = \left[\int_X |\phi|^{p\epsilon}\rho^{-p\delta\epsilon-2m}dV_g\right]^{(m-1)/m} \leqslant C_1^2 \int_X \left|\nabla\bigl(|\phi|^{p/2}\rho^{1-m-p\delta/2}\bigr)\right|_g^2 dV_g.$$

The last two equations prove (8.24), with $C_4 = C_1^2 \max(C'C'', C'''C_3)$. □

Lemma 8.6.9 *In the situation above, there exists $C_5 > 0$ depending on Q_1, Q_2, m, δ and ρ such that $\phi \in L^p_{0,\delta}(X)$ and $\|\phi\|_{L^p_{0,\delta}} \leqslant C_5$ for all p with $1 \leqslant p \leqslant -2m/\delta$.*

Proof First we consider the case $p = 1$. Choose q such that

$$q < \frac{-2m}{\delta}, \quad q > \frac{-2m}{\gamma}, \quad q > \frac{-2m}{\beta+2} \quad \text{and} \quad q > \frac{m}{m-1}. \tag{8.26}$$

This is possible as $\beta + 2 \leqslant \gamma < \delta < 0$ and $\delta \geqslant 1 - m > 2 - 2m$. Define r, s by $\frac{1}{q} + \frac{1}{r} = 1$ and $s = -r(\delta + 2m)$. Then $r > 1$ and $s < -2m$ by (8.26). But

$$\|\phi\|_{L^1_{0,\delta}} = \int_X |\phi|\rho^{-\delta-2m}dV_g \leqslant \|\phi\|_{L^q} \cdot \left[\int_X \rho^s dV_g\right]^{1/r}, \tag{8.27}$$

using Hölder's inequality. Since $s < -2m$, the integral $\int_X \rho^s dV_g$ exists. Moreover, as $q > -2m/\gamma$, $q > -2m/(\beta+2)$ and $q > \epsilon$, the proof of Lemma 8.6.5 provides a bound for $\|\phi\|_{L^q}$. Thus (8.27) gives an a priori bound for $\|\phi\|_{L^1_{0,\delta}}$.

By definition of $L^p_{0,\delta}(X)$, when $p = -2m/\delta$ we have $L^p_{0,\delta}(X) = L^p(X)$ and $\|\phi\|_{L^p_{0,\delta}} = \|\phi\|_{L^p}$, and the proof of Lemma 8.6.5 therefore gives an a priori bound for $\|\phi\|_{L^p_{0,\delta}}$. Thus ϕ lies in both $L^1_{0,\delta}(X)$ and $L^{-2m/\delta}_{0,\delta}(X)$, and we can find a constant $C_5 > 0$ such that $\|\phi\|_{L^1_{0,\delta}} \leqslant C_5$ and $\|\phi\|_{L^{-2m/\delta}_{0,\delta}} \leqslant C_5$.

Now let $1 < p < -2m/\delta$. Then there are unique $q, r > 1$ such that $\frac{1}{q} + \frac{1}{r} = 1$ and $\frac{1}{q} + \frac{-\delta}{2mr} = \frac{1}{p}$. Using Hölder's inequality we can show that

$$\|\phi\|_{L^p_{0,\delta}} \leqslant \|\phi\|_{L^1_{0,\delta}}^{1/q} \cdot \|\phi\|_{L^{-2m/\delta}_{0,\delta}}^{1/r} \leqslant C_5^{1/q} \cdot C_5^{1/r} = C_5.$$

So $\|\phi\|_{L^p_{0,\delta}} \leqslant C_5$ for all p with $1 \leqslant p \leqslant -2m/\delta$, as we want. □

Now, by Lemma 8.6.9 and the method of Proposition 5.4.4 we can prove

Proposition 8.6.10 *There exist constants $P_2, C_6 > 0$ depending on Q_1, Q_2, m, δ and ρ such that ϕ lies in $L^p_{0,\delta}(X)$ for all $p \in [1, \infty)$ and satisfies*

$$\|\phi\|_{L^p_{0,\delta}} \leqslant P_2(C_6 p)^{-m/p}.$$

Finally, since $\phi \in C^2_\gamma(X)$ by assumption and $\gamma < \delta$ it follows that $\phi \in C^0_\delta(X)$, and one can show that $\|\phi\|_{C^0_\delta} = \lim_{p\to\infty}\|\phi\|_{L^p_{0,\delta}}$. Thus Proposition 8.6.10 gives

$$\|\phi\|_{C^0_\delta} \leqslant \lim_{p\to\infty} P_2(C_6 p)^{-m/p} = P_2,$$

and the proof of Theorem 8.6.6 is complete. □

8.6.3 Estimating ϕ in $C_\delta^{5,\alpha}(X)$

Now we prove an estimate of ϕ in $C_\delta^{5,\alpha}(X)$ that will be needed in Theorem A2. The result depends on a priori estimates of $\|\phi\|_{C^{5,\alpha}}$, which will be given by Theorem A1, and of $\|\phi\|_{C_\delta^0}$, which is provided by Theorem 8.6.6. Notice also that we drop the constant $\gamma < 0$ and the requirement that $\phi \in C_\gamma^2(X)$, and we relax the condition $\beta + 2 < \delta$ to $\beta + 2 \leqslant \delta$.

Theorem 8.6.11 *Let (X, J, g) be an ALE Kähler manifold asymptotic to $\mathbb{C}^m/G$ for $m > 1$, with Kähler form ω. Let $Q_1, Q_2, P_2 > 0$, $\alpha \in (0, 1)$ and $\beta + 2 \leqslant \delta < 0$. Then there exists $P_3 > 0$ such that if $f \in C_\beta^{3,\alpha}(X)$ and $\phi \in C_\delta^0(X) \cap C^{5,\alpha}(X)$ satisfy $\|f\|_{C_\beta^{3,\alpha}} \leqslant Q_1$, $\|\phi\|_{C^{5,\alpha}} \leqslant Q_2$, $\|\phi\|_{C_\delta^0} \leqslant P_2$ and $(\omega + dd^c\phi)^m = e^f\omega^m$, then $\phi \in C_\delta^{5,\alpha}(X)$ and $\|\phi\|_{C_\delta^{5,\alpha}} \leqslant P_3$. Furthermore, if $f \in C_\beta^{k,\alpha}(X)$ and $\phi \in C^{k+2,\alpha}(X)$ for some integer $k \geqslant 3$, then $\phi \in C_\delta^{k+2,\alpha}(X)$.*

Proof We begin with a proposition. Its proof is a little complicated, but the basic idea is that we will apply Theorem 1.4.3 of §1.4 to balls of radius $L\rho(x)^\lambda$ about x for each $x \in X$, where $L > 0$ is small and independent of x. It will probably help to look at Theorem 1.4.3 before reading the proof of the proposition.

Proposition 8.6.12 *Let $K_1, K_2 > 0$, $\lambda \in [0, 1]$ and $k \geqslant 3$ be an integer. Then there exists $K_3 > 0$ depending on (X, J, g), $\alpha, \beta, \delta, P_2$ and K_1, K_2, λ, k, such that the following is true.*

Suppose the hypotheses of Theorem 8.6.11 hold, that $\|f\|_{C_\beta^{k,\alpha}} \leqslant K_1$ and

$$\|\nabla^j dd^c\phi\|_{C_{-\lambda j}^0} \leqslant K_2 \quad \text{for } j = 0, \ldots, k,$$
$$\text{and} \quad [\nabla^k dd^c\phi]_{\alpha, -\lambda(k+\alpha)} \leqslant K_2, \qquad (8.28)$$

where all these norms exist. Then the following inequalities hold, where each norm exists:

$$\|\nabla^j \phi\|_{C_{\delta-\lambda j}^0} \leqslant K_3 \quad \text{for } j = 0, \ldots, k+2,$$
$$\text{and} \quad [\nabla^{k+2}\phi]_{\alpha, \delta-\lambda(k+2+\alpha)} \leqslant K_3. \qquad (8.29)$$

Proof Fix $x \in X$, and let $\delta_x(g)$ be the injectivity radius of g at x and $B_R(x)$ the geodesic ball of radius R about x in X. As g is complete, the exponential map $\exp_x : T_xX \to X$ is well-defined and when $0 < R < \delta_x(g)$ it induces a diffeomorphism between $B_R(x)$ and the ball of radius R about 0 in T_xX.

Let B_1 and B_2 be the balls of radius 1 and 2 about 0 in $\mathbb{C}^m$. Choose R such that $0 < 2R < \delta_x(g)$, and choose an isomorphism $T_xX \cong \mathbb{C}^m$ which identifies metrics and complex structures. Define a map $\Psi_{x,R} : B_2 \to B_{2R}(x)$ by $\Psi_{x,R}(y) = \exp_x(Ry)$, where we regard $Ry \in \mathbb{C}^m$ as an element of T_xX. Then $\Psi_{x,R}$ is a diffeomorphism, and $\Psi_{x,R}(B_1) = B_R(x)$.

Define an operator $P_{x,R} : C^{k+2,\alpha}(B_2) \to C^{k,\alpha}(B_2)$ by

$$(\Psi_{x,R})_* [P_{x,R}(\chi)] \cdot \omega^m = -R^2 \cdot dd^c [(\Psi_{x,R})_*(\chi)] \wedge \\ (\omega^{m-1} + \cdots + (\omega')^{m-1}) \tag{8.30}$$

on $B_{2R}(x) \subset X$. Then $P_{x,R}$ is a second-order partial differential operator on B_{2R}, and as ω, ω' are positive $(1,1)$-forms we can show that $P_{x,R}$ is *elliptic*. Moreover, from (5.18) we see that

$$P_{x,R}(\Psi_{x,R}^*(\phi)) = R^2(1 - \Psi_{x,R}^*(e^f)). \tag{8.31}$$

Now let $R = L\rho(x)^\lambda$, for some $L > 0$ satisfying the conditions

$$0 < 2R < \delta_x(g) \quad \text{for all } x \in X, \tag{8.32}$$

$$\tfrac{1}{2}\rho(x) \leqslant \rho(y) \leqslant 2\rho(x) \quad \text{for all } y \in B_{2R}(x) \text{ and } x \in X, \tag{8.33}$$

$$\|R^{-2}\Psi_{x,R}^*(g) - h\|_{C^{k,\alpha}} \leqslant \tfrac{1}{2} \quad \text{for all } x \in X, \text{ and} \tag{8.34}$$

$$\|R^{-2}\Psi_{x,R}^*(\omega) - \omega_0\|_{C^{k,\alpha}} \leqslant \tfrac{1}{2} \quad \text{for all } x \in X, \tag{8.35}$$

where h is the standard Hermitian metric on B_2, with Kähler form ω_0. It can be shown using the definition of ALE Kähler metric that these conditions hold for all sufficiently small $L > 0$, where 'sufficiently small' depends on $(X, J, g), \rho, k, \alpha$ and λ. We can also prove using (8.28), (8.33), (8.34) and (8.35) that

$$\|R^{-2}\Psi_{x,R}^*(\omega')\|_{C^{k,\alpha}} \leqslant C_7 \tag{8.36}$$

for some $C_7 > 0$ depending only on k, α and K_2. The choice of weights $-\lambda j$ and $-\lambda(k+\alpha)$ in (8.28) are exactly those needed to prove (8.36).

We shall apply Theorem 1.4.3 to the operator $P_{x,R}$ on B_2. As in (1.11), write

$$(P_{x,R}\chi)(z) = a^{ij}(z) \frac{\partial^2 \chi}{\partial x_i \partial x_j}(z) + b^i(z) \frac{\partial \chi}{\partial x_i}(z) + c(z)\chi(z)$$

for $z \in B_2$, where $(x_1, \ldots, x_{2m})$ is the standard real coordinate system on B_2 and a^{ij}, b^i and c are real functions on B_2.

From (8.30) we see that a^{ij}, b^i and c are nonlinear functions of $R^{-2}\Psi_{x,R}^*(g)$, $R^{-2}\Psi_{x,R}^*(\omega)$ and $R^{-2}\Psi_{x,R}^*(\omega')$. But (8.34)–(8.36) give $C^{k,\alpha}$ bounds on these, and so we can find constants $\lambda, \Lambda > 0$ depending only on k, α and K_2, such that $|a^{ij}(z)\xi_i\xi_j| \geqslant \lambda|\xi|^2$ for all $z \in B_2$ and $\xi \in \mathbb{C}^m$, and $\|a^{ij}\|_{C^{k,\alpha}} \leqslant \Lambda, \|b^i\|_{C^{k,\alpha}} \leqslant \Lambda$, and $\|c\|_{C^{k,\alpha}} \leqslant \Lambda$ on B_2 for all $i, j = 1, \ldots, 2m$. Therefore by eqn (1.14) of Theorem 1.4.3 there exists $C_8 > 0$ such that

$$\|\chi|_{B_1}\|_{C^{k+2,\alpha}} \leqslant C_8 \big(\|P_{x,R}(\chi)\|_{C^{k,\alpha}} + \|\chi\|_{C^0} \big) \tag{8.37}$$

for each $\chi \in C^{k+2,\alpha}(B_2)$. The important thing about this inequality is that it applies for all $x \in X$, but the constant C_8 is independent of x, and in fact depends only on k, α and K_2.

Putting $\chi = \Psi_{x,R}^*(\phi)$ in (8.37) and using (8.31) shows that

$$\|\Psi_{x,R}^*(\phi)|_{B_1}\|_{C^{k+2,\alpha}} \leq C_8\Big(R^2\|1-\Psi_{x,R}^*(e^f)\|_{C^{k,\alpha}} + \|\Psi_{x,R}^*(\phi)\|_{C^0}\Big).$$

All norms in this equation are taken with respect to the Euclidean metric h on B_2. But since the metric $R^{-2}\Psi_{x,R}^*(g)$ on B_2 is close to h in $C^{k,\alpha}$ by (8.34), we can take norms with respect to $R^{-2}\Psi_{x,R}^*(g)$ instead, at the cost of increasing C_8 to some $C_9 > 0$ depending only on k, α and K_2. Pushing down to X using $\Psi_{x,R}$, we see that

$$\sum_{j=0}^{k+2} R^j \|\nabla^j \phi|_{B_R(x)}\|_{C^0} + R^{k+2+\alpha}\big[\nabla^{k+2}\phi|_{B_R(x)}\big]_\alpha \leq \\ C_9\Bigg[\sum_{j=0}^{k} R^{j+2}\|\nabla^j f|_{B_{2R}(x)}\|_{C^0} + R^{k+2+\alpha}\big[\nabla^k f|_{B_{2R}(x)}\big]_\alpha + \|\phi|_{B_{2R}(x)}\|_{C^0}\Bigg]. \tag{8.38}$$

As $R = L\rho(x)^\lambda$, $\|f\|_{C_\beta^{k,\alpha}} \leq K_1$ and $\|\phi\|_{C_\delta^0} \leq P_2$ we can prove using (8.33) that

$$R^{j+2}\|\nabla^j f|_{B_{2R}(x)}\|_{C^0} \leq K_1 L^{j+2} 2^{j-\beta} \rho(x)^{\beta+2-(1-\lambda)(j+2)},$$
$$R^{k+2+\alpha}\big[\nabla^k f|_{B_{2R}(x)}\big]_\alpha \leq K_1 L^{k+2+\alpha} 2^{k+\alpha-\beta} \rho(x)^{\beta+2-(1-\lambda)(k+2+\alpha)},$$
$$\text{and}\quad \|\phi|_{B_{2R}(x)}\|_{C^0} \leq P_2 2^{-\delta} \rho(x)^\delta.$$

Since $\lambda \leq 1$ and $\beta + 2 \leq \delta$ this shows that each term on the second line of (8.38) is bounded by a multiple of $\rho(x)^\delta$, and from this it is easy to prove (8.29). $\square$

We can now prove Theorem 8.6.11 by repeatedly applying Proposition 8.6.12. Let $k \geq 3$ be an integer, and suppose that $f \in C_\beta^{k,\alpha}(X)$ with $\|f\|_{C_\beta^{k,\alpha}} \leq K_1$, and $\phi \in C^{k+2,\alpha}(X)$ with $\|\phi\|_{C^{k+2,\alpha}} \leq K_2$. In the case $k = 3$ dealt with by the first part of the theorem, we put $K_1 = Q_1$ and $K_2 = Q_2$. Let n be the smallest integer for which $-n\delta > k + \alpha$, and define a finite series $\lambda_0, \ldots, \lambda_n$ by $\lambda_i = -i\delta/(k+\alpha)$ for $i < n$ and $\lambda_n = 1$. Then $\lambda_j \in [0, 1]$ for all j. We shall show by induction on i that the hypotheses (8.28) of the proposition hold with λ replaced by λ_i and K_2 by $K_2(i)$. For the first step, observe that (8.28) holds with $\lambda = \lambda_0 = 0$ and $K_2(0) = K_2$, since $\|\phi\|_{C^{k+2,\alpha}} \leq K_2$.

For the inductive step, suppose that (8.28) holds with λ replaced by λ_i and K_2 by $K_2(i)$. Applying the Proposition then shows that (8.29) holds with $\lambda = \lambda_i$ and some constant $K_3(i)$. However, it easy to see that this implies (8.28) with $\lambda = \lambda_{i+1}$ and $K_2(i + 1) = K_3(i)$. Thus by induction on i, the hypotheses of Proposition 8.6.12 apply with $\lambda = \lambda_i$ and suitable constants $K_2(i)$ for all $i = 0, \ldots, n$. In particular they apply when $\lambda = \lambda_n = 1$. But in this case (8.29) shows that $\phi \in C_\delta^{k+2,\alpha}(X)$, and gives an a priori bound for $\|\phi\|_{C_\delta^{k+2,\alpha}}$. This completes the proof of Theorem 8.6.11. $\square$

Here is a variation on Theorem 8.6.11 in which we replace ϕ by $A\rho^{2-2m} + \psi$, and estimate ψ. It will be used to prove part (b) of Theorem A2.

Theorem 8.6.13 *Let (X, J, g) be an ALE Kähler manifold asymptotic to $\mathbb{C}^m/G$ for $m > 1$, with Kähler form ω. Let $Q_1, Q_2, P_4 > 0$, $\alpha \in (0, 1)$, $\beta < -2$ and $\delta < 0$ satisfy $\beta + 2 \leqslant \delta$ and $\beta > -4m$. Then there exists $P_5 > 0$ such that if $A \in \mathbb{R}$, $f \in C^{3,\alpha}_\beta(X)$, $\psi \in C^0_\delta(X) \cap C^{5,\alpha}(X)$ and $\phi = A\rho^{2-2m} + \psi$ satisfy $\|f\|_{C^{3,\alpha}_\beta} \leqslant Q_1$, $|A| \leqslant Q_2$, $\|\psi\|_{C^{5,\alpha}} \leqslant Q_2$, $\|\psi\|_{C^0_\delta} \leqslant P_4$ and $(\omega + \mathrm{dd}^c\phi)^m = \mathrm{e}^f \omega^m$, then $\psi \in C^{5,\alpha}_\delta(X)$ and $\|\psi\|_{C^{5,\alpha}_\delta} \leqslant P_5$. Furthermore, if $f \in C^{k,\alpha}_\beta(X)$ and $\psi \in C^{k+2,\alpha}(X)$ for some integer $k \geqslant 3$, then $\psi \in C^{k+2,\alpha}_\delta(X)$.*

The proof of this result is very similar to that of Theorem 8.6.11. The main differences are that ψ replaces ϕ throughout, and instead of (8.31) we have

$$P_{x,R}\bigl(\Psi^*_{x,R}(\psi)\bigr) = R^2\bigl(1 - \Psi^*_{x,R}(\mathrm{e}^f)\bigr) - A \cdot P_{x,R}\bigl(\Psi^*_{x,R}(\rho^{2-2m})\bigr). \tag{8.39}$$

It is not difficult to show that $P_{x,R}\bigl(\Psi^*_{x,R}(\rho^{2-2m})\bigr) = O\bigl(R^2\rho(x)^{-4m}\bigr)$, and to find analogous estimates for its derivatives. The condition $\beta > -4m$ in the theorem ensures that the terms used to estimate the right hand side of (8.31) are also sufficient to estimate the right hand side of (8.39).

8.7 The proofs of Theorems A1–A4

We now prove Theorem A1 of §8.5.

Theorem A1 *Let (X, J, g) be an ALE Kähler manifold asymptotic to $\mathbb{C}^m/G$ for some $m > 1$, with Kähler form ω. Let $Q_1 > 0$, $\alpha \in (0, 1)$ and $\beta + 2 \leqslant \gamma < 0$. Then there exists $Q_2 > 0$ such that if $f \in C^{3,\alpha}_\beta(X)$ and $\phi \in C^5_\gamma(X)$ satisfy $(\omega + \mathrm{dd}^c\phi)^m = \mathrm{e}^f \omega^m$ and $\|f\|_{C^{3,\alpha}_\beta} \leqslant Q_1$, then $\omega + \mathrm{dd}^c\phi$ is a positive $(1,1)$-form, $\phi \in C^{5,\alpha}(X)$ and $\|\phi\|_{C^{5,\alpha}} \leqslant Q_2$. Also, if $f \in C^{k,\alpha}_\beta(X)$ for $k \geqslant 3$ then $\phi \in C^{k+2,\alpha}(X)$.*

Proof This result is put together from Theorems C1 and C2 of §5.2, and we can prove it by following the proofs of Theorems C1 and C2 in §5.4 and §5.5. The only place where the proof for the ALE case differs significantly from that in the compact case is in estimating $\|\phi\|_{C^0}$ in §5.4.1. But we have already proved in Theorem 8.6.1 that $\omega + \mathrm{dd}^c\phi$ is a positive (1,1)-form, and that $\|\phi\|_{C^0} \leqslant P_1$ for some suitable constant P_1. We will now run briefly through the rest of the proofs of Theorems C1 and C2, explaining the differences in the ALE case.

It turns out that the material in §5.4.2 and §5.4.3 needs almost no changes at all. In §5.4.2 the only problem is that the function $F = \log(m - \Delta\phi) - \kappa\phi$ may not have a maximum on X, since X is noncompact. But clearly $F \to \log m$ as $\rho \to \infty$, and so if F has no maximum then $F \leqslant \log m$ on X, and this is enough to complete the proof. In §5.4.3 a similar argument applies to the function S. Thus by Theorem 8.6.1, Corollary 5.4.5 and Proposition 5.4.7 we can find a priori estimates for $\|\phi\|_{C^0}$, $\|\mathrm{dd}^c\phi\|_{C^0}$ and $\|\nabla\mathrm{dd}^c\phi\|_{C^0}$, and the conclusions of Theorem C1 apply in the ALE case.

Next we must extend Theorem C2 to the case of our ALE Kähler manifold (X, J, g). Now Theorem C2 is a *local* result. The principal tools in its proof are Lemmas 5.5.1–5.5.3, and these are proved by covering the compact manifold M with a finite number of small open balls, and then applying analysis results such as Theorem 1.4.3 on each open ball.

Since X has infinite volume, we need an infinite number of small open balls to cover it. But this does not matter, because outside a compact set the metric g on X is arbitrarily close to the Euclidean metric on $\mathbb{R}^n/G$. Thus Lemmas 5.5.1–5.5.3 apply on X. The rest of the proof of Theorem C2 then holds on X without change, except to observe that if $f \in C^{k,\alpha}_\beta(X)$ then $f \in C^{k,\alpha}(X)$ and $\|f\|_{C^{k,\alpha}} \leqslant \|f\|_{C^{k,\alpha}_\beta}$. So Theorem C2 applies to X, and Theorem A1 immediately follows. □

Theorem A2 *Let (X, J, g) be an ALE Kähler manifold asymptotic to $\mathbb{C}^m/G$ for some $m > 1$, with Kähler form ω and radius function ρ. Let $Q_1, Q_2 > 0$, $\alpha \in (0, 1)$ and $\beta + 2 \leqslant \gamma < 0$. Then there exists $Q_3 > 0$ such that the following holds.*

Suppose $f \in C^{3,\alpha}_\beta(X)$ and $\phi \in C^5_\gamma(X) \cap C^{5,\alpha}(X)$ satisfy $(\omega + dd^c\phi)^m = e^f \omega^m$ and the inequalities $\|f\|_{C^{3,\alpha}_\beta} \leqslant Q_1$ and $\|\phi\|_{C^{5,\alpha}} \leqslant Q_2$. Then

(a) *Let $\beta \in (-2m, -2)$. Then $\phi \in C^{5,\alpha}_{\beta+2}(X)$ and $\|\phi\|_{C^{5,\alpha}_{\beta+2}} \leqslant Q_3$. If $f \in C^{k,\alpha}_\beta(X)$ for $k \geqslant 3$ then $\phi \in C^{k+2,\alpha}_{\beta+2}(X)$.*

(b) *Let $\beta \in (-1-2m, -2m)$. Then $\phi = A\rho^{2-2m} + \psi$, where A is defined by (8.19) and $\psi \in C^{5,\alpha}_{\beta+2}(X)$ with $\|\psi\|_{C^{5,\alpha}_{\beta+2}} \leqslant Q_3$. If $f \in C^{k,\alpha}_\beta(X)$ for $k \geqslant 3$ then $\psi \in C^{k+2,\alpha}_{\beta+2}(X)$.*

Proof We begin by showing that $\Delta\phi = O(|f| + |dd^c\phi|^2)$.

Lemma 8.7.1 *There exists $D_1 > 0$ depending only on Q_2 such that $|\Delta\phi + e^f - 1| \leqslant D_1|dd^c\phi|^2$ and $|\nabla(\Delta\phi) + e^f \nabla f| \leqslant D_1|dd^c\phi| \cdot |\nabla dd^c\phi|$.*

Proof Let $x \in X$. Then as in §5.3 we can choose holomorphic coordinates $(z_1, \ldots, z_m)$ near x such that ω and $dd^c\phi$ are given at x by

$$\omega_x = i(dz_1 \wedge d\bar{z}_1 + \cdots + dz_m \wedge d\bar{z}_m),$$
$$dd^c\phi_x = i(\epsilon_1 dz_1 \wedge d\bar{z}_1 + \cdots + \epsilon_m dz_m \wedge d\bar{z}_m),$$

where $\epsilon_1, \ldots, \epsilon_m \in \mathbb{R}$. Writing $\Delta\phi$ and e^f in terms of $\epsilon_1, \ldots, \epsilon_m$ we have

$$\Delta\phi = -\sum_{j=1}^m \epsilon_j, \quad e^f = \prod_{j=1}^m (1 + \epsilon_j),$$
$$\text{so that} \quad \Delta\phi + e^f - 1 = \sum_{\substack{2 \leqslant j \leqslant m \\ i_1 < \cdots < i_j}} \epsilon_{i_1} \cdots \epsilon_{i_j}. \tag{8.40}$$

Also $|dd^c\phi|^2 = 2\sum_{j=1}^m \epsilon_j^2$, which implies that $|\epsilon_{i_1} \cdots \epsilon_{i_j}| \leqslant |dd^c\phi|^j$. But $|dd^c\phi| \leqslant Q_2$ since $\|\phi\|_{C^{5,\alpha}} \leqslant Q_2$, and thus $|\epsilon_{i_1} \cdots \epsilon_{i_j}| \leqslant Q_2^{j-2}|dd^c\phi|^2$ for $j \geqslant 2$. Using this to estimate the third equation of (8.7) gives the first equation of the lemma, with constant $D_1 = \sum_{j=2}^m \binom{m}{j} Q_2^{j-2}$. The second equation can be proved by a similar method. □

Next we show that if we have an estimate $|\phi| \le K\rho^\delta$ of the rate of decay of ϕ, where $\delta < 0$ is small, then we can get a stronger estimate of the same type.

Proposition 8.7.2 *Let $K > 0$ and $\delta < 0$ satisfy $\delta \ge \beta + 2$, and set $\delta' = \max(2\delta - 2, \beta + 2)$. Then there exists $K' > 0$ such that*
(a) *if $\delta' > 2 - 2m$ and $|\phi| \le K\rho^\delta$ then $|\phi| \le K'\rho^{\delta'}$,*
(b) *if $1 - 2m < \delta' < 2 - 2m$ and $|\phi| \le K\rho^\delta$ and $\phi = A\rho^{2-2m} + \psi$ where A is defined by (8.19), then $|\psi| \le K'\rho^{\delta'}$.*

Proof In both cases $\delta \ge \beta + 2$ and $\|\phi\|_{C_\delta^0} \le K$, so Theorem 8.6.11 proves that $\phi \in C_\delta^{5,\alpha}(X)$ and $\|\phi\|_{C_\delta^{5,\alpha}} \le P_3$ for some $P_3 > 0$ depending on (X, J, g), α, β, δ, Q_1, Q_2 and K. Thus $|dd^c\phi| \le P_3\rho^{\delta-2}$ and $|\nabla dd^c\phi| \le P_3\rho^{\delta-3}$ on X. But $|f| \le Q_1\rho^\beta$ and $|\nabla f| \le Q_1\rho^{\beta-1}$, and these imply that $|1 - e^f| \le e^{Q_1}\rho^\beta$ and $|e^f \nabla f| \le Q_1 e^{Q_1}\rho^{\beta-1}$. Hence by Lemma 8.7.1 we see that

$$|\Delta\phi| \le D_1 P_3^2 \rho^{2\delta-4} + e^{Q_1}\rho^\beta \quad \text{and} \quad |\nabla(\Delta\phi)| \le D_1 P_3^2 \rho^{2\delta-5} + Q_1 e^{Q_1}\rho^{\beta-1}.$$

From this we deduce that $\Delta\phi \in C_{\delta'-2}^{0,\alpha}(X)$ and $\|\Delta\phi\|_{C_{\delta'-2}^{0,\alpha}} \le D_2$ for some $D_2 > 0$ depending on D_1, P_3 and Q_1. If $\delta' > 2 - 2m$ then we apply part (a) of Theorem 8.3.6 to show that $\phi \in C_{\delta'}^{2,\alpha}(X)$ and $\|\phi\|_{C_{\delta'}^{2,\alpha}} \le K'$, for $K' = CD_2$. Hence $|\phi| \le K'\rho^{\delta'}$, and we have proved part (a) of the Proposition. If $1 - 2m < \delta' < 2 - 2m$ then part (b) of Theorem 8.3.6 shows that $\phi = A\rho^{2-2m} + \psi$, where A is given by

$$A = \frac{|G|}{(m-1)\Omega_{2m-1}} \int_X \Delta\phi\, dV_g, \tag{8.41}$$

and $\psi \in C_{\delta'}^{2,\alpha}(X)$ with $\|\psi\|_{C_{\delta'}^{2,\alpha}} \le K'$, where $K' = C_2 D_2$. Note that the constant in (8.41) differs from that in (8.13) by a factor of 2, because the Laplacian on Kähler manifolds is by convention half that on Riemannian manifolds.

Hence $|\psi| \le K'\rho^{\delta'}$, and it only remains to prove that (8.19) and (8.41) give the same values for A. Since $\omega^m = m!\cdot dV_g$, it is enough to show that $\int_X \Delta\phi\, \omega^m = \int_X (1-e^f)\,\omega^m$. Multiplying out $(\omega + dd^c\phi)^m = e^f \omega^m$ and substituting in $m\, dd^c\phi \wedge \omega^{m-1} = \Delta\phi\, \omega^m$, we find that

$$\Delta\phi\, \omega^m - (1 - e^f)\omega^m = \sum_{j=2}^m \binom{m}{j}(dd^c\phi)^j \wedge \omega^{m-j}. \tag{8.42}$$

But all the terms on the right hand side of this equation are exact $2m$-forms which decay at a rate of at least $O(\rho^{-4m})$, since $j \ge 2$. Thus by a Stokes' Theorem argument similar to that used in Proposition 8.6.3 we see that the integral of each side of (8.42) over X is zero, and the proposition is finished. □

Now we can prove Theorem A2. Pick some δ satisfying $\gamma < \delta < 0$ and $\delta \ge 1 - m$. This is possible as $\gamma < 0$ and $m > 1$. Then by Theorem 8.6.6 there exists $P_2 > 0$ such that whenever ϕ and f satisfy the hypotheses of Theorem A2, then $|\phi| \le P_2\rho^\delta$. Define a sequence $\delta_1, \delta_2, \ldots$ by $\delta_1 = \delta$ and $\delta_{i+1} = \max(2\delta_i - 2, \beta + 2)$. Clearly $\delta_n = \beta + 2$ for all large enough n.

Suppose $\beta \in (-2m, -2)$. Then $\delta_i > 2 - 2m$ for all i. Applying part (a) of Proposition 8.7.2 and using induction on i shows that there exist constants $K_1, K_2, \ldots$ with $K_1 = P_2$ such that $|\phi| \leqslant K_i \rho^{\delta_i}$ for all i. Hence $|\phi| \leqslant K_n \rho^{\beta+2}$ on X. Applying Theorem 8.6.11 with $\delta = \beta + 2$ then proves part (a) of Theorem A2. Suppose $\beta \in (-1 - 2m, -2m)$. Then using induction on i and both parts (a) and (b) of Proposition 8.7.2, we show in a similar way that $\phi = A\rho^{2-2m} + \psi$, where A is given by (8.19) and $|\psi| \leqslant K_n \rho^{\beta+2}$. Finally, applying Theorem 8.6.13 with $\delta = \beta + 2$ proves part (b) of Theorem A2. $\square$

It remains to prove Theorems A3 and A4. These are straightforward analogues of Theorems C3 and C4 of §5.2, and their proofs are simple generalizations of the proofs of Theorems C3 in §5.6 and C4 in §5.7, so we will leave them to the reader as an exercise. In the proof of Theorem A3, the main difference is that instead of using Theorem 1.5.4 we must use a version of Theorem 8.3.6 for metrics g' with coefficients in $C^{3,\alpha}$.

In the proof of Theorem A4, the problem is in applying Stokes' Theorem to prove an analogue of (5.35) on X. What we do is to pick $p \geqslant 2$ such that $p\gamma < 2 - 2m$, and replace (5.35) with the equation

$$\int_X d\Big[|\phi_1 - \phi_2|^{p-2}(\phi_1 - \phi_2) \, d^c(\phi_1 - \phi_2) \wedge \big(\omega_1^{m-1} + \cdots + \omega_2^{m-1}\big)\Big] = 0,$$

which can be proved by the method of Proposition 8.6.3. The argument in §5.7 generalizes to show that $d\big(|\phi_1 - \phi_2|^{p/2}\big) = 0$, forcing $\phi_1 = \phi_2$ since $\phi_1, \phi_2 \to 0$ as $\rho \to \infty$. This concludes our proof of the Calabi conjecture for ALE manifolds.

8.8 The proofs of Theorems 8.2.3 and 8.2.4

Using the material of §8.3–§8.7, we are now ready to prove the two theorems about Ricci-flat ALE Kähler manifolds that were stated in §8.2.

8.8.1 *The proof of Theorem 8.2.3*

Let X be a crepant resolution of $\mathbb{C}^m/G$, where G acts freely on $\mathbb{C}^m \setminus \{0\}$. By Proposition 8.4.5, in each Kähler class of ALE Kähler metrics on X we can choose a metric $\hat{g}$ with $\hat{g} = \pi^*(h_0)$ wherever $\rho > R \geqslant 2$, where h_0 is the Euclidean metric on $\mathbb{C}^m/G$. Let $\hat{\omega}$ be the Kähler form and η the Ricci form of $\hat{g}$. Then η is closed and $[\eta] = 2\pi \, c_1(X)$ in $H^2(X, \mathbb{R})$. But $c_1(X) = 0$ as X is a crepant resolution, so $[\eta] = 0$ in $H^2(X, \mathbb{R})$. Also, $\eta = 0$ wherever $\rho > R$, since g is flat there.

Thus η is a closed, compactly-supported $(1,1)$-form on X with $[\eta] = 0$ in $H^2(X, \mathbb{R})$, and by Theorem 8.4.4 there exists a unique function $f \in C^\infty_\beta(X)$ for each $\beta < 0$ with $\eta = \frac{1}{2} dd^c f$. In fact $f = 0$ wherever $\rho > R$, so f is compactly supported. The Calabi conjecture for ALE manifolds holds by Theorem 8.5.1. Part (b) of the conjecture shows that there exists a unique function $\phi = A\rho^{2-2m} + \psi$ where A is given by (8.19) and $\psi \in C^\infty_{\beta+2}(X)$ for $\beta \in (-1-2m, -2m)$, such that $\omega = \hat{\omega} + dd^c\phi$ is a positive $(1,1)$-form and $\omega^m = e^f \hat{\omega}^m$.

Let g be the Kähler metric on X with Kähler form ω. Then since the Ricci form of $\hat{g}$ is $\frac{1}{2}dd^c f$ it follows as in §5.1 that the Ricci form of g is zero, and so g is Ricci-flat. On $\{z \in \mathbb{C}^m/G : r(z) > R\}$ we have $\pi_*(\hat{g}) = h$, so that $\pi_*(\hat{\omega}) = \omega_0$, and $\pi_*(\rho) = r$. Thus defining $\chi = \pi_*(\psi)$ gives (8.3). Since $\psi \in C^\infty_{\beta+2}(X)$ for $\beta \in (-1-2m, -2m)$, putting $\gamma = \beta + 2$ we see that $\nabla^k \chi = O(r^{\gamma-k})$ for $k = 0, 1, 2, \ldots$ and $\gamma \in (1-2m, 2-2m)$, as we have to prove.

From (8.3) we see that $\nabla^k(\pi_*(g) - h_0) = O(r^{-2m-k})$ for $k \geq 0$, and thus g is an ALE Kähler metric by Definition 8.1.3. Also, g is unique in its Kähler class of ALE metrics because ϕ is unique. It only remains to prove that $A < 0$. We can do this by giving an explicit expression for A. Let ζ be the unique element of $\mathcal{H}^{1,1}(X)$ with $[\zeta] = [\omega]$. Then a calculation shows that

$$A = -\frac{|G|}{2m(m-1)^2 \Omega_{2m-1}} \int_X |\zeta|^2 dV_g, \qquad (8.43)$$

where Ω_{2m-1} is the volume of the unit sphere $\mathcal{S}^{2m-1}$ in $\mathbb{C}^m$. Now $[\omega] \neq 0$ as this is outside the Kähler cone, so $\zeta \neq 0$ and A is negative. This completes the proof of Theorem 8.2.3.

8.8.2 *The proof of Theorem* 8.2.4

First we show that $\mathbb{C}^m/G$ has no crepant resolutions when $m > 2$ and $G \subset \mathrm{Sp}(m/2)$.

Proposition 8.8.1 *Suppose that $m > 2$ is even and that G is a nontrivial finite subgroup of $\mathrm{Sp}(m/2)$ which acts freely on $\mathbb{C}^m \setminus \{0\}$. Then $\mathbb{C}^m/G$ is a terminal singularity, and admits no crepant resolutions.*

Proof Let $\gamma \neq 1$ in G. Then there exist coordinates $(z_1, \ldots, z_m)$ on $\mathbb{C}^m$ in which γ acts by

$$(z_1, \ldots, z_m) \overset{\gamma}{\mapsto} \left(e^{2\pi i a_1} z_1, \ldots, e^{2\pi i a_m} z_m\right). \qquad (8.44)$$

As γ acts freely on $\mathbb{C}^m \setminus \{0\}$ we can take $a_j \in (0, 1)$ for $j = 1, \ldots, m$. Since $G \subset \mathrm{Sp}(m/2)$ we know that γ preserves a complex symplectic form on $\mathbb{C}^m$, and we can choose $(z_1, \ldots, z_m)$ so that this form is $dz_1 \wedge dz_2 + \cdots + dz_{m-1} \wedge dz_m$. Thus (8.44) gives $e^{2\pi i a_{2j-1}} e^{2\pi i a_{2j}} = 1$ for $j = 1, \ldots, m/2$. But as $a_{2j-1}, a_{2j} \in (0, 1)$ this implies that $a_{2j-1} + a_{2j} = 1$ for $j = 1, \ldots, m/2$.

Now Definition 6.4.2 defines the *age* of γ to be $\mathrm{age}(\gamma) = a_1 + \cdots + a_m$. Substituting $a_{2j-1} + a_{2j} = 1$ gives $\mathrm{age}(\gamma) = m/2$. Thus $\mathrm{age}(\gamma) = m/2$ for all $\gamma \neq 1$ in G, and $\mathrm{age}(1) = 0$. Since $m > 2$ this shows that $\mathrm{age}(\gamma) \neq 1$ for all $\gamma \in G$. Therefore $\mathbb{C}^m/G$ is a terminal singularity by Theorem 6.4.3, and admits no crepant resolutions by Proposition 6.4.4. □

We now prove Theorem 8.2.4. Let X be a crepant resolution of $\mathbb{C}^m/G$, where G is nontrivial and acts freely on $\mathbb{C}^m \setminus \{0\}$, and let g be a Ricci-flat ALE Kähler metric on X. As X is a resolution of $\mathbb{C}^m/G$ it is simply-connected, and the metric g on X is complete. Also (X, g) is not a Riemannian product, because it is asymptotic to $\mathbb{C}^m/G$ and $\mathbb{C}^m/G$

is not a Riemannian product. Thus g is irreducible, by the ideas of §3.2. Since g is Ricci-flat it is nonsymmetric. Therefore $\mathrm{Hol}(g)$ lies on Berger's list by Theorem 3.4.1.

But g is Ricci-flat and Kähler, so $\mathrm{Hol}^0(g) \subseteq \mathrm{SU}(m)$ by Proposition 6.1.1. Thus either $\mathrm{Hol}(g) = \mathrm{SU}(m)$ or $\mathrm{Hol}(g) = \mathrm{Sp}(m/2)$. Now the holonomy of the Euclidean metric h_0 on $\mathbb{C}^m/G$ is $G \subset \mathrm{SU}(m)$. Since g is asymptotic to h_0 we can show that $G \subset \mathrm{Hol}(g) \subseteq \mathrm{SU}(m)$. Hence, if $\mathrm{Hol}(g) = \mathrm{Sp}(m/2)$ then $G \subset \mathrm{Sp}(m/2)$. But Proposition 8.8.1 then shows that $\mathbb{C}^m/G$ admits no crepant resolutions, a contradiction. So $\mathrm{Hol}(g) \neq \mathrm{Sp}(m/2)$, and thus $\mathrm{Hol}(g) = \mathrm{SU}(m)$, as we have to prove.

8.9 The case $m = 2$, and deformations

Some readers may have wondered why we restricted our attention to resolutions (X, π) of $\mathbb{C}^m/G$ in our study of ALE Kähler manifolds. In fact Definition 8.1.3 is not the only natural way to define ALE Kähler manifolds. Here is an alternative definition, which is closer in spirit to the definition of ALE manifold, and also to the definition of hyperkähler ALE space in §7.2.

Definition 8.9.1 (Alternative) Let G be a finite subgroup of $\mathrm{U}(m)$ acting freely on $\mathbb{C}^m \setminus \{0\}$, and let h_0 be the Euclidean metric and J_0 the complex structure on $\mathbb{C}^m/G$. Then we say that a Kähler manifold (X, J, g) is an *ALE Kähler manifold asymptotic to* $\mathbb{C}^m/G$ if there exists a compact subset $S \subset X$ and a map $\pi : X \setminus S \to \mathbb{C}^m/G$ that is a diffeomorphism between $X \setminus S$ and the subset $\{z \in \mathbb{C}^m/G : r(z) > R\}$ for some fixed $R \geqslant 0$, such that $\nabla^k \big(\pi_*(g) - h_0 \big) = O(r^{-2m-k})$ and $\nabla^k \big(\pi_*(J) - J_0 \big) = O(r^{-2m-k})$ for all $k \geqslant 0$.

Suppose (X, J, g) is an ALE Kähler manifold asymptotic to $\mathbb{C}^m/G$ in this sense. Then one can use results on the analysis of Δ on $C_\beta^\infty(X)$ to prove that there is a large collection of holomorphic functions of polynomial growth on X, which form an algebra, and make X into an affine complex algebraic variety. From this it is easy to show that X is birational to a deformation of $\mathbb{C}^m/G$.

However, if $m \geqslant 3$ then $\mathbb{C}^m/G$ has no nontrivial deformations by Theorem 6.4.8. Thus X is birational to $\mathbb{C}^m/G$, and so X is a resolution of $\mathbb{C}^m/G$. Therefore, when $m \geqslant 3$ Definition 8.9.1 is actually equivalent to the more restrictive Definition 8.1.3. But the Kleinian singularities $\mathbb{C}^2/G$ and their crepant resolutions have a large family of deformations, described by Slodowy [196]. In this case, (X, J) must be isomorphic to one of the deformations X_s of the crepant resolution X_0 of $\mathbb{C}^2/G$.

There remains the question of whether we can generalize the results of §8.4–§8.8 to the case in which (X, J, g) is an ALE Kähler manifold asymptotic to $\mathbb{C}^2/G$ in the sense of Definition 8.9.1, but (X, J) is not a resolution of $\mathbb{C}^2/G$. Now Kronheimer's classification of hyperkähler ALE spaces using explicit algebraic methods, described in §7.2, already gives us existence and uniqueness results for ALE metrics of holonomy $\mathrm{SU}(2)$ on surfaces (X, J) of this kind.

Clearly we have no real need of an analytic proof of the same result. However, in §9.9 we will study metrics of holonomy $\mathrm{SU}(m)$ on deformations of *nonisolated* quotient singularities $\mathbb{C}^m/G$. Technically this has much in common with the problem of finding metrics of holonomy $\mathrm{SU}(2)$ on (X, J), but it requires an analytic approach, as we have no algebraic construction of the relevant holonomy SU(m) metrics. Therefore we now explain how to modify the material of §8.4–§8.8 to the case when (X, J) is asymptotic to $\mathbb{C}^2/G$, but is not a resolution of $\mathbb{C}^2/G$.

In §8.4, Theorems 8.4.2 and 8.4.4 hold for (X, J). However, Theorem 8.4.3 and Proposition 8.4.5 *do not hold* for (X, J). The problem is this. Although every element of $H^2(X, \mathbb{C})$ can be represented both by a closed, compactly-supported 2-form on X and also by a closed (1,1)-form on X, in general there are classes in $H^2(X, \mathbb{C})$ which are not represented by a closed, compactly-supported (1,1)-form on X. This is a consequence of the following result.

Theorem 8.9.2 *Suppose $m \geqslant 3$ and η is a smooth, closed real (1, 1)-form on $\mathbb{C}^m \setminus \{0\}$. Then there exists a smooth real function f on $\mathbb{C}^m \setminus \{0\}$ such that $\eta = dd^c f$. But this is not true for $m = 2$. Let (z_1, z_2) be coordinates on $\mathbb{C}^2$, and define real (1, 1)-forms ζ_1, ζ_2 on $\mathbb{C}^2 \setminus \{0\}$ by*

$$\zeta_1 + i\zeta_2 = \frac{z_1^2 d\bar{z}_1 \wedge dz_2 + z_2^2 dz_1 \wedge d\bar{z}_2 + z_1 z_2 (dz_1 \wedge d\bar{z}_1 - dz_2 \wedge d\bar{z}_2)}{\left(|z_1|^2 + |z_2|^2\right)^3}. \tag{8.45}$$

Then ζ_1, ζ_2 are smooth, closed real (1, 1)-forms on $\mathbb{C}^2 \setminus \{0\}$, but neither of ζ_1, ζ_2 can be written in the form $dd^c f$ for any smooth real function f on $\mathbb{C}^2 \setminus \{0\}$. If η is a smooth, closed real (1, 1)-form on $\mathbb{C}^2 \setminus \{0\}$ then there exist unique $a_1, a_2 \in \mathbb{R}$ such that $\eta = a_1 \zeta_1 + a_2 \zeta_2 + dd^c f$ for some smooth function f on $\mathbb{C}^2 \setminus \{0\}$.

This result is closely related to Theorem 6.4.8, but I have not been able to find it in the literature. I have an outline proof (for analytic rather than smooth forms and functions), which I will not give here. The result also holds when $\mathbb{C}^m \setminus \{0\}$ is replaced by the annulus $\{z \in \mathbb{C}^m : R_1 < |z| < R_2\}$ for $0 \leqslant R_1 < R_2 \leqslant \infty$. The (1,1)-forms ζ_1, ζ_2 of (8.45) are invariant under the action of SU(2) on $\mathbb{C}^2 \setminus \{0\}$, and decay like r^{-4}. This is also true of the (1,1)-form $dd^c(r^{-2})$ on $\mathbb{C}^2 \setminus \{0\}$, which is linearly independent of ζ_1, ζ_2.

Using Theorem 8.9.2 we can show that every class in $H^2(X, \mathbb{C})$ can be represented by a closed (1,1)-form η on X asymptotic up to $O(\rho^{-6})$ to $a_1 \zeta_1 + a_2 \zeta_2$ for some $a_1, a_2 \in \mathbb{C}$. The class can be represented by a compactly-supported (1,1)-form if and only if $a_1 = a_2 = 0$.

Our results in §8.5–§8.7 on the Calabi conjecture for ALE manifolds extend without change to surfaces (X, J) asymptotic to $\mathbb{C}^2/G$ which are not resolutions of $\mathbb{C}^2/G$. But the material of §8.8 has to be substantially modified, because Theorem 8.2.3 makes use of Proposition 8.4.5, which does not hold in this case. In (8.3) we have to replace the leading term $A\, dd^c(r^{2-2m})$ with the expression $A_0\, dd^c(r^{-2}) + A_1 \zeta_1 + A_2 \zeta_2$, for real numbers A_0, A_1, A_2.

9

QUASI-ALE METRICS WITH HOLONOMY SU(m) AND Sp(m)

Let G be a finite subgroup of U(m) and suppose (X, π) is a resolution of $\mathbb{C}^m/G$. In the case in which G acts freely on $\mathbb{C}^m \setminus \{0\}$, so that $\mathbb{C}^m/G$ has an *isolated quotient singularity* at 0, Chapter 8 defined a special class of Kähler metrics on X called ALE Kähler metrics, and proved an existence result for Ricci-flat ALE Kähler metrics. We will now generalize these ideas to the case when G does not act freely on $\mathbb{C}^m \setminus \{0\}$, so that the singularities of $\mathbb{C}^m/G$ are not isolated.

The appropriate class of Kähler metrics on resolutions X of non-isolated quotient singularities $\mathbb{C}^m/G$ will be called *Quasi Asymptotically Locally Euclidean*, or *QALE* for short. Our particular interest is in *Ricci-flat* QALE Kähler manifolds, which usually have holonomy SU(m) or Sp($m/2$). The main result of the chapter is an existence theorem for Ricci-flat QALE Kähler metrics on crepant resolutions of $\mathbb{C}^m/G$.

The key to understanding the structure of nonisolated singularities $\mathbb{C}^m/G$ is to observe that if s is a singular point of $\mathbb{C}^m/G$ and $s \neq 0$, then an open neighbourhood of s in $\mathbb{C}^m/G$ is isomorphic to an open neighbourhood of $(0, 0)$ in $\mathbb{C}^k \times \mathbb{C}^{m-k}/H$, where $0 < k < m$ and H is a finite subgroup of U($m - k$), and also of G. Thus, away from zero the singularities of $\mathbb{C}^m/G$ look locally like *products* $\mathbb{C}^k \times \mathbb{C}^{m-k}/H$ for $k > 0$.

In §9.1 we will define a special class of resolutions X of $\mathbb{C}^m/G$ called *local product resolutions*, which have the property that if $\mathbb{C}^m/G$ is locally modelled on $\mathbb{C}^k \times \mathbb{C}^{m-k}/H$ then X is locally modelled on $\mathbb{C}^k \times Y$, where Y is a resolution of $\mathbb{C}^{m-k}/H$. Crepant resolutions are automatically of this form. If X is a local product resolution, we want to impose some suitable asymptotic conditions 'near infinity' on Kähler metrics g on X.

In §9.2 we define g_X to be a *QALE Kähler metric* on X if g_X is asymptotic to $h_{\mathbb{C}^k} \times g_Y$ on the part of X modelled on $\mathbb{C}^k \times Y$, where $h_{\mathbb{C}^k}$ is the Euclidean metric on $\mathbb{C}^k$, and g_Y is a metric on Y, the resolution of $\mathbb{C}^{m-k}/H$. So g_X converges to $h_{\mathbb{C}^k} \times g_Y$ at infinity. Now the points of $\mathbb{C}^m/G$ will be of several different kinds, modelled on $\mathbb{C}^k \times \mathbb{C}^{m-k}/H$ for different $0 \leq k \leq m$ and subgroups $H \subseteq G$. Thus we impose not one but many asymptotic conditions on g_X, which must all be satisfied for g_X to be QALE.

Section 9.3 discusses *Ricci-flat QALE Kähler manifolds*. We state the main result of the chapter, Theorem 9.3.3, an existence result for Ricci-flat QALE metrics on crepant resolutions X of $\mathbb{C}^m/G$. Its proof follows that of Theorem 8.2.3 in Chapter 8, and takes up most of sections 9.4–9.8. We first develop the appropriate ideas of weighted Sobolev and Hölder spaces and elliptic regularity on QALE manifolds. Then we prove two versions of the Calabi conjecture on QALE manifolds, and apply them to construct Ricci-flat QALE Kähler metrics on X.

As in Chapter 8, we expect that many readers will not want to work through the whole chapter, much of which is technical and difficult. *Such people are advised to read*

§9.1–§9.3 *only*, and perhaps also §9.9. Sections 9.1–9.3 contain the (rather complicated) definition of QALE Kähler manifolds and the important results on Ricci-flat QALE Kähler manifolds, and give a number of examples. Section 9.9 generalizes the idea of QALE manifold in two ways. These cover all you need to know about QALE manifolds to understand the rest of the book.

All of this chapter is original research by the author. Most of the results were announced in [130], which omitted the proofs in §9.6–§9.7 and the discussion in §9.9. Apart from this, the material is published here for the first time. The QALE Calabi conjecture ideas in §9.6 and §9.7 are naturally similar to previous papers on noncompact versions of the Calabi conjecture such as Tian and Yau [204, 206, 207] and Bando and Kobayashi [19, 20], and owe an intellectual debt to them. But as far as the author knows, neither the existence of the metrics of Theorem 9.3.3, nor their asymptotic properties, actually follow from previous published results.

Other than [130], the author is not aware of any papers in the literature on QALE manifolds, at the time of writing.

9.1 Local product resolutions

In this section we will study the structure of *nonisolated quotient singularities* $\mathbb{C}^m/G$ and their resolutions. We shall define a special kind of resolution of $\mathbb{C}^m/G$ called *local product resolutions*, which include all crepant resolutions. Our main goal is to set up a lot of notation describing resolutions of $\mathbb{C}^m/G$, that we will use in the rest of the chapter.

Let G be a finite subgroup of $U(m)$. If A is a subgroup of G and V a subspace of $\mathbb{C}^m$, define the *fixed point set* $\mathrm{Fix}(A)$, the *centralizer* $C(V)$ and the *normalizer* $N(V)$ by

$$\mathrm{Fix}(A) = \{x \in \mathbb{C}^m : a\,x = x \text{ for all } a \in A\},$$
$$C(V) = \{g \in G : g\,v = v \text{ for all } v \in V\}$$
$$\text{and} \quad N(V) = \{g \in G : g\,V = V\}.$$

Then $C(V)$ and $N(V)$ are subgroups of G, and $C(V)$ is a normal subgroup of $N(V)$.

Definition 9.1.1 Define a finite set $\mathcal{L}$ of linear subspaces of $\mathbb{C}^m$ by

$$\mathcal{L} = \big\{\mathrm{Fix}(A) : A \text{ is a subgroup of } G\big\}.$$

Let I be an *indexing set* for $\mathcal{L}$, so that we may write $\mathcal{L} = \{V_i : i \in I\}$. Let the indices of $\mathrm{Fix}(\{1\})$ and $\mathrm{Fix}(G)$ be $0, \infty$ respectively, so that $0, \infty \in I$ and $V_0 = \mathbb{C}^m$, $V_\infty = \mathrm{Fix}(G)$ by definition. Usually $V_\infty = \{0\}$.

Define a *partial order* $\succeq$ on I by $i \succeq j$ if $V_i \subseteq V_j$. Then $\infty \succeq i \succeq 0$ for all $i \in I$. Let W_i be the perpendicular subspace to V_i in $\mathbb{C}^m$, so that $\mathbb{C}^m = V_i \oplus W_i$. Define $A_i = C(V_i)$. Then $V_i = \mathrm{Fix}(A_i)$ and A_i acts on W_i, with $\mathbb{C}^m/A_i \cong V_i \times W_i/A_i$. If $i \succeq j$ then $W_i \supseteq W_j$ and $A_i \supseteq A_j$. Define B_i to be the quotient group $N(V_i)/C(V_i)$. Then B_i acts naturally on V_i and W_i/A_i, and $(V_i \times W_i/A_i)/B_i \cong \mathbb{C}^m/N(V_i)$. Hence, if $N(V_i) = G$ then $(V_i \times W_i/A_i)/B_i \cong \mathbb{C}^m/G$.

Let $V_i, V_j \in \mathcal{L}$, let A be the subgroup of G generated by A_i and A_j, and set $V = \text{Fix}(A)$. It is easy to show that $V = V_i \cap V_j$. Thus $V_i \cap V_j \in \mathcal{L}$, and $\mathcal{L}$ is closed under intersection of subspaces. Also, if $g \in G$ and $V_i \in \mathcal{L}$ then $gV_i \in \mathcal{L}$, since $gV_i = \text{Fix}(gA_ig^{-1})$. For each $g \in G$ and $i \in I$, let $g \cdot i$ be the unique element of I such that $V_{g \cdot i} = gV_i$. This defines an action of G on I, which satisfies $W_{g \cdot i} = gW_i$ and $A_{g \cdot i} = gA_ig^{-1}$.

Let $v \in \mathbb{C}^m$. Then vG is a singular point of $\mathbb{C}^m/G$ if and only if the subgroup of G fixing v is nontrivial, that is, if $v \in \text{Fix}(A)$ for some nontrivial subgroup $A \subset G$. Thus vG is a singular point if and only if $v \in V_i$ for some $i \in I$ with $i \neq 0$, and the *singular set* S of $\mathbb{C}^m/G$ is

$$S = \bigcup_{i \in I \setminus \{0\}} V_i \Big/ G.$$

For generic points $v \in V_i$, the subgroup of G fixing v is A_i, and the singularity of $\mathbb{C}^m/G$ at vG is locally modelled on the product $V_i \times W_i/A_i$.

Definition 9.1.2 Let (X, π) be a resolution of $\mathbb{C}^m/G$. We say that (X, π) is a *local product resolution* if for each $i \in I$ there exists a resolution (Y_i, π_i) of W_i/A_i such that the following conditions hold. Let $R > 0$, and define subsets S_i and T_i in $V_i \times W_i/A_i \cong \mathbb{C}^m/A_i$ by

$$S_i = \bigcup_{j \in I : i \not\leq j} V_j \Big/ A_i,$$
$$T_i = \{x \in \mathbb{C}^m/A_i : d(x, S_i) \leq R\}$$
$$= \{x \in \mathbb{C}^m : d(x, V_j) \leq R \text{ for some } j \in I \text{ with } i \not\leq j\} \Big/ A_i,$$

where $d(\,,\,)$ is the distance in $\mathbb{C}^m/A_i$ or $\mathbb{C}^m$. Let U_i be the pull-back of T_i to $V_i \times Y_i$ under $\text{id} \times \pi_i : V_i \times Y_i \to V_i \times W_i/A_i$. Let ϕ_i be the natural projection from $V_i \times W_i/A_i$ to $\mathbb{C}^m/G$. Then there should exist a map $\psi_i : V_i \times Y_i \setminus U_i \to X$ such that the following diagram commutes:

$$\begin{array}{ccc} V_i \times Y_i \setminus U_i & \xrightarrow{\psi_i} & X \\ \downarrow{\scriptstyle \text{id} \times \pi_i} & & \downarrow{\scriptstyle \pi} \\ V_i \times W_i/A_i \setminus T_i & \xrightarrow{\phi_i} & \mathbb{C}^m/G. \end{array} \qquad (9.1)$$

This ψ_i should be a local isomorphism of complex manifolds, and whenever $x \in X$ and $y \in V_i \times W_i/A_i \setminus T_i$ satisfy $\pi(x) = \phi_i(y)$, there should exist a unique $z \in V_i \times Y_i \setminus U_i$ with $x = \psi_i(z)$ and $y = (\text{id} \times \pi_i)(z)$.

Here is an equivalent way to write this. Define $\tilde{X}_i$ by

$$\tilde{X}_i = \left\{ (x, y) \in X \times \left(V_i \times W_i/A_i \setminus T_i \right) : \pi(x) = \phi_i(y) \right\}.$$

For any resolution X of $\mathbb{C}^m/G$, one can show that $\tilde{X}_i$ is a well-defined complex manifold, and the projection to X is a local isomorphism. We say that X is a local product resolution if $\tilde{X}_i$ is isomorphic to $V_i \times Y_i \backslash U_i$ in such a way that the natural projections to $V_i \times W_i/A_i \backslash T_i$ agree.

The idea of the definition is that every point in $\mathbb{C}^m/G$ has an open neighbourhood isomorphic to an open neighbourhood of 0 in $V_i \times W_i/A_i$ for some $i \in I$. (For nonsingular points we use $V_0 \times W_0/A_0 = \mathbb{C}^m$.) A resolution X of $\mathbb{C}^m/G$ is a local product resolution if and only if the resolution of each singular point modelled on $V_i \times W_i/A_i$ is locally isomorphic to $V_i \times Y_i$.

Proposition 9.1.3 *Let (X, π) be a local product resolution of $\mathbb{C}^m/G$. Then for each $g \in G$ and $i \in I$ there is a unique isomorphism $\chi_{g,i} : Y_i \to Y_{g \cdot i}$ making a commutative diagram*

$$\begin{array}{ccc} Y_i & \xrightarrow{\chi_{g,i}} & Y_{g \cdot i} \\ \downarrow \pi_i & & \downarrow \pi_{g \cdot i} \\ W_i/A_i & \xrightarrow{g} & W_{g \cdot i}/A_{g \cdot i}, \end{array} \quad (9.2)$$

where the map $W_i/A_i \xrightarrow{g} W_{g \cdot i}/A_{g \cdot i}$ is given by $g(w \, A_i) = (g \, w) A_{g \cdot i}$.

Proof If v is a generic point of V_i then $\mathbb{C}^m/G$ is locally isomorphic to $V_i \times W_i/A_i$ near vG, and X is locally isomorphic to $V_i \times Y_i$ near $\pi^{-1}(vG)$, as X is a local product resolution. But if $g \in G$ then $vG = (gv)G$, and gv is a generic point of $V_{g \cdot i}$, so that $\mathbb{C}^m/G$ is also locally isomorphic to $V_{g \cdot i} \times W_{g \cdot i}/A_{g \cdot i}$ near vG, and X is locally isomorphic to $V_{g \cdot i} \times Y_{g \cdot i}$ near $\pi^{-1}(vG)$. Hence $V_i \times Y_i$ and $V_{g \cdot i} \times Y_{g \cdot i}$ are locally isomorphic. In fact they are globally isomorphic, and this gives an isomorphism $\chi_{g,i} : Y_i \to Y_{g \cdot i}$, which makes (9.2) commutative. $\square$

If $g \in N(V_i)$ then $g \cdot i = i$, so that $\chi_{g,i}$ is an automorphism of Y_i. Also, if $g \in A_i$ then $\chi_{g,i}$ is the identity on Y_i. Thus $N(V_i)$ acts on Y_i, and the normal subgroup A_i of $N(V_i)$ acts trivially on Y_i. Therefore the action descends to an action of the quotient group $B_i = N(V_i)/A_i$ on Y_i. That is, the action of B_i on W_i/A_i must lift to an action of B_i on the resolution Y_i on W_i/A_i. So each resolution Y_i in a local product resolution must be B_i-equivariant.

We can use this to give a more thorough explanation of the idea of local product resolution, under a simplifying assumption. Suppose that $N(V_i) = G$ for each $i \in I$. This happens if G is abelian, and in other cases too. Then we have $(V_i \times W_i/A_i)/B_i \cong \mathbb{C}^m/G$ for $i \in I$, from Definition 9.1.1. Since B_i acts on V_i and Y_i we can take the quotient $(V_i \times Y_i)/B_i$, which is a complex orbifold with a natural projection to $(V_i \times W_i/A_i)/B_i \cong \mathbb{C}^m/G$.

Effectively, $(V_i \times Y_i)/B_i$ is a *partial resolution* of $\mathbb{C}^m/G$, which resolves the singularities of $\mathbb{C}^m/G$ due to fixed points of elements of A_i, but leaves unresolved those caused by fixed points of elements of $G \backslash A_i$. Now S_i is exactly the subset of $V_i \times W_i/A_i$ fixed by some $b \neq 1$ in B_i, and therefore S_i/B_i is the set of singularities of $\mathbb{C}^m/G$ due to fixed points of $G \setminus A_i$. Also T_i is the subset of $V_i \times W_i/A_i$ within distance R of S_i, so T_i/B_i is the subset of $\mathbb{C}^m/G$ within distance R of singularities due to fixed points of $G \setminus A_i$.

All the fixed points of the B_i-action on $V_i \times Y_i$ lie in $(\mathrm{id} \times \pi_i)^{-1}(S_i)$, which is in the interior of U_i. Thus B_i acts freely on $V_i \times Y_i \setminus U_i$, and $(V_i \times Y_i \setminus U_i)/B_i$ is nonsingular. In fact $\psi_i : (V_i \times Y_i \setminus U_i)/B_i \to X$ is an *isomorphism* between $(V_i \times Y_i \setminus U_i)/B_i$ and $X \setminus \pi^{-1}(T_i/B_i)$. Thus, the definition requires that the resolution X of $\mathbb{C}^m/G$ has to coincide with the partial resolution $(V_i \times Y_i)/B_i$ of $\mathbb{C}^m/G$ outside the set U_i/B_i which contains the singularities of $(V_i \times Y_i)/B_i$.

Local product resolutions are a very large class of resolutions of $\mathbb{C}^m/G$. There do exist (rather artificial) examples of resolutions which are not local product resolutions, but in practice they include all interesting resolutions of $\mathbb{C}^m/G$, in particular crepant resolutions.

Proposition 9.1.4 *Suppose X is a crepant resolution of $\mathbb{C}^m/G$, which implies that $G \subset SU(m)$. Then X is a local product resolution. Moreover, each resolution Y_i of W_i/A_i is also a crepant resolution.*

The proof of this proposition is fairly straightforward, and we omit it. Our next result shows that local product resolutions are built out of local product resolutions of smaller dimension.

Proposition 9.1.5 *Suppose X is a local product resolution of $\mathbb{C}^m/G$. Then each of the resolutions Y_i of W_i/A_i in Definition 9.1.2 is also a local product resolution.*

Proof Let $i \in I$ be fixed throughout the proof. Above we defined a lot of notation such as $\mathcal{L}, I, V_j, W_j$ and so on, associated to $\mathbb{C}^m/G$. The corresponding data associated to W_i/A_i will be written $\mathcal{L}', I', V_j', W_j'$, etc., in the obvious way. We will express the data for W_i/A_i in terms of that for $\mathbb{C}^m/G$. Define the indexing set I' by $I' = \{j \in I : i \succeq j\}$, and for each $j \in I'$ set $V_j' = V_j \cap W_i$. Then $\mathcal{L}' = \{V_j' : j \in I'\} = \{\mathrm{Fix}(A) : A \text{ is a subgroup of } A_i\}$ is a finite set of subspaces of W_i. The two special elements of I' are $0' = 0$ and $\infty' = i$. Also W_j', A_j' and Y_j' are the same as W_j, A_j and Y_j for each $j \in I'$.

Let ϕ_j', T_j' and U_j' be as in Definition 9.1.2. It can be shown that there is a unique map $\psi_j' : V_j' \times Y_j' \setminus U_j' \to Y_i$, such that the product $\mathrm{id} \times \psi_j'$ of ψ_j' with the identity on V_i makes the following picture into a commutative diagram:

$$\begin{array}{ccc} V_i \times V_j' \times Y_j' \setminus U_j & \xrightarrow{\mathrm{id} \times \psi_j'} & V_i \times Y_i \setminus U_i \\ \downarrow \cong & & \downarrow \psi_i \\ V_j \times Y_j \setminus U_j & \xrightarrow{\psi_j} & X. \end{array} \qquad (9.3)$$

It easily follows that Y_i is a local product resolution of W_i/A_i. □

9.2 Quasi-ALE Kähler metrics

We will now define a class of Kähler metrics on local product resolutions of $\mathbb{C}^m/G$ called *Quasi Asymptotically Locally Euclidean*, or *Quasi-ALE* or *QALE* for short, which generalize the ALE Kähler metrics considered in Chapter 8. Let G be a finite subgroup of $U(m)$, let X be a local product resolution of $\mathbb{C}^m/G$, and let all notation be as in §9.1.

Definition 9.2.1 For each pair $i, j \in I$, define $\mu_{i,j} : V_i \times Y_i \to [0, \infty)$ by $\mu_{i,j}(z) = d\big((\mathrm{id} \times \pi_i)(z), V_j A_i / A_i\big)$, where $V_j A_i / A_i = \{v A_i : v \in V_j\}$, as a subset of $\mathbb{C}^m/A_i$, and $d(y, T)$ is the shortest distance between the point y and the subset T in $\mathbb{C}^m/A_i$. For each $i \in I$, define $\nu_i : V_i \times Y_i \to [1, \infty)$ by $\nu_i(z) = 1 + \min\{\mu_{i,j}(z) : j \in I, j \neq 0\}$. Then $\mu_{i,j}$ and ν_i are both continuous functions on $V_i \times Y_i$. For each $i \in I$ define $d_i = 2 - 2 \dim W_i = 2 - 2m + 2 \dim V_i$, and let h_i be the Euclidean metric on V_i.

Let g be a Kähler metric on X. We say g is *Quasi-ALE* if the complex codimension n of the singular set S of $\mathbb{C}^m/G$, given by $n = \min\{\dim W_i : i \in I, i \neq 0\}$ satisfies $n \geqslant 2$, and for each $i \in I$ there is a Kähler metric g_i on Y_i, such that the metric $h_i \times g_i$ on $V_i \times Y_i$ satisfies

$$\nabla^l \big(\psi_i^*(g) - h_i \times g_i\big) = \sum_{j \in I : i \not\geq j} O\big(\mu_{i,j}^{d_j} \nu_i^{-2-l}\big) \tag{9.4}$$

on $V_i \times Y_i \setminus U_i$, for all $l \geqslant 0$. If X is a local product resolution of $\mathbb{C}^m/G$ with complex structure J, and g is a QALE metric on X, then we say that (X, J, g) is a *QALE Kähler manifold asymptotic to* $\mathbb{C}^m/G$.

We now discuss this definition. The pull-back to $V_i \times Y_i$ of the singular set S of $\mathbb{C}^m/G$ splits up into a number of pieces parametrized by $j \in I \setminus \{0\}$, and $\mu_{i,j}$ is a measure of the distance to piece j. Similarly, ν_i measures the distance in $V_i \times Y_i$ to the pull-back of S, but we add 1 to this to avoid problems when ν_i is small. Also, by definition U_i is the subset of $V_i \times Y_i$ on which $\mu_{i,j} \leqslant R$ for some $j \in I$ with $i \not\geq j$. Thus $\mu_{i,j} > R > 0$ in (9.4), so there are no problems when $\mu_{i,j}$ is small.

Thus eqn (9.4) says that at large distances from U_i, the pull-back $\psi_i^*(g)$ of g to $V_i \times Y_i$ must approximate the product metric $h_i \times g_i$ on $V_i \times Y_i$. As in §9.1, we can explain this more clearly if we assume that $N(V_i) = G$. In this case $h_i \times g_i$ pushes down to an orbifold metric on $(V_i \times Y_i)/B_i$, and X is isomorphic to $(V_i \times Y_i)/B_i$ outside U_i/B_i. So (9.4) says that the metrics g and $h_i \times g_i$ on the isomorphic subsets of X and $(V_i \times Y_i)/B_i$ must agree asymptotically at large distances from U_i/B_i.

We assume that the singular set S of $\mathbb{C}^m/G$ has codimension $n \geqslant 2$ because many of the results in the rest of the chapter are false when $n = 1$. One reason for this is that if $\dim W_j = 1$ then $d_j = 2 - 2 \dim W_j = 0$, and so the 'error term' $O(\mu_{i,j}^{d_j} \nu_i^{-2-l})$ in (9.4) may not be small when $\mu_{i,j}$ is large. But some of our results rely on the errors being small at large distances.

Another reason is that the equation $\Delta u = f$ on $\mathbb{C}^k$ behaves differently in the cases $k = 1$ and $k > 1$. When $k > 1$ and f is a smooth function on $\mathbb{C}^k$ that decays rapidly at infinity, there is a unique smooth function u on $\mathbb{C}^k$ with $\Delta u = f$ and $u = O(r^{2-2k})$

for large r. But when $k = 1$ this is false, and instead $u = O(\log r)$ and is not unique. Because of this, some of our results about the Laplacian on QALE manifolds are false when $n = 1$.

Consider what (9.4) means when $i = 0$ and $i = \infty$. Now $V_0 = \mathbb{C}^m$ and W_0 and Y_0 are both a single point, so $V_0 \times Y_0 \cong \mathbb{C}^m$. The metric $h_0 \times g_0$ on $V_0 \times Y_0$ is just the Euclidean metric h_0 on $\mathbb{C}^m$, and (9.4) says that the metrics g on X and h_0 on $\mathbb{C}^m/G$ must be asymptotic at large distances from the singular set of $\mathbb{C}^m/G$.

When $i = \infty$ we have $U_\infty = \emptyset$, the map $\psi_\infty : V_\infty \times Y_\infty \to X$ is an isomorphism, and the right hand side of (9.4) is zero, which forces $\psi_\infty^*(g) = h_\infty \times g_\infty$. Using ψ_∞ to identify X and $V_\infty \times Y_\infty$, we see that $g = h_\infty \times g_\infty$, the product of a Euclidean metric on V_∞ and a metric g_∞ on Y_∞. If $\text{Fix}(G) = \{0\}$ then $V_\infty = \{0\}$ and $Y_\infty = X$, and (9.4) holds trivially by taking $g_\infty = g$. It is often convenient to assume that $\text{Fix}(G) = \{0\}$.

In particular, suppose $\mathbb{C}^m/G$ has an *isolated* singularity at 0. Then $I = \{0, \infty\}$ and $\text{Fix}(G) = \{0\}$, so (9.4) is trivial for $i = \infty$. We have $d_\infty = 2 - 2m$ and $\mu_{0,\infty} = r$, the radius function on $\mathbb{C}^m$, and $v_0 = 1 + r$, so when $i = 0$ eqn (9.4) becomes

$$\nabla^l \big(\psi_0^*(g) - h_0 \big) = O\big(r^{2-2m}(1+r)^{-2-l}\big)$$

wherever $r > R$ on $\mathbb{C}^m$, for all $l \geq 0$. But this is equivalent to the equation (8.2) defining ALE Kähler metrics. So we have proved:

Lemma 9.2.2 *Suppose $\mathbb{C}^m/G$ has an isolated singularity at 0. Then QALE Kähler metrics on X are the same thing as ALE Kähler metrics on X, in the sense of Definition* 8.1.3.

The idea used to prove Proposition 9.1.3 also yields the following result.

Proposition 9.2.3 *In the situation of Definition* 9.2.1, *the metrics g_i satisfy $\chi_{\gamma,i}^*(g_{\gamma \cdot i}) = g_i$ for each $\gamma \in G$ and $i \in I$. Thus g_i is invariant under the natural action of B_i on Y_i.*

Next we show that QALE metrics are made out of other QALE metrics of lower dimension.

Proposition 9.2.4 *Let g_i be the Kähler metric on the resolution Y_i of W_i/A_i in Definition* 9.2.1. *Then g_i is also a QALE metric.*

Proof We shall use the notation defined in the proof of Proposition 9.1.5. In addition, let h'_j be the Euclidean metric on $V'_j = V_j \cap W_i$. Let $j \in I'$, so that $i \succeq j$ and $V_j = V_i \times V'_j$. Writing the Euclidean metric h_j on V_j as $h_i \times h'_j$, eqn (9.4) with i replaced by j becomes

$$\nabla^l \big(\psi_j^*(g) - h_i \times h'_j \times g_j \big) = \sum_{k \in I : j \not\succeq k} O\big(\mu_{j,k}^{d_k} v_j^{-2-l}\big) \tag{9.5}$$

on $V_i \times V'_j \times Y_j \setminus U_j$. Using $\text{id} \times \psi'_j$ to pull eqn (9.4) from $V_i \times Y_i \setminus U_i$ back to $V_i \times V'_j \times Y_j \setminus U_j$, and substituting $(\text{id} \times \psi'_j)^*(\psi_i^*(g)) = \psi_j^*(g)$ since $\psi_i \circ (\text{id} \times \psi'_j) = \psi_j$ by (9.3), gives

$$\nabla^l \big(\psi_j^*(g) - h_i \times (\psi_j')^*(g_i) \big) = \sum_{k \in I : i \not\leq k} O\big(\mu_{j,k}^{d_k} v_j^{-2-l} \big) \tag{9.6}$$

on $V_i \times V_j' \times Y_j \setminus U_j$. Hence, subtracting (9.5) and (9.6) gives

$$\nabla^l \big(h_i \times (\psi_j')^*(g_j) - h_i \times h_j' \times g_j \big) = \sum_{k \in I : j \not\leq k} O\big(\mu_{j,k}^{d_k} v_j^{-2-l} \big).$$

By restricting to $\{v\} \times V_j' \times Y_j$ for $v \in V_i$ and taking v to be large, we show that

$$\nabla^l \big((\psi_j')^*(g_i) - h_j' \times g_j \big) = \sum_{k \in I' : j \not\leq k} O\big((\mu_{j,k}')^{d_k} (v_j')^{-2-l} \big)$$

on $V_j' \times Y_j \setminus U_j'$, for all $l \geq 0$. Thus g_i is a QALE Kähler metric on Y_i. □

So local product resolutions and QALE Kähler metrics are built up by a kind of *induction on dimension*. If X is a local product resolution with a QALE Kähler metric, then the Y_i that appear as limits of X are also local product resolutions with QALE metrics, but in a lower dimension. This suggests a method of proving results about QALE metrics: we assume the result is true for all Y_i with $\dim Y_i < \dim X$, and prove it for X. Then the result holds for all QALE metrics, by induction on $\dim X$. We will use this idea several times later on.

9.3 Ricci-flat QALE Kähler manifolds

The following proposition is easily proved using the method of Proposition 8.2.1.

Proposition 9.3.1 *Let (X, J, g) be a Ricci-flat QALE Kähler manifold asymptotic to $\mathbb{C}^m / G$. Then $G \subset \mathrm{SU}(m)$ and X is a crepant resolution of $\mathbb{C}^m / G$.*

Next we define Kähler classes and the Kähler cone on QALE manifolds.

Definition 9.3.2 Let X be a local product resolution of $\mathbb{C}^m / G$, and g a QALE Kähler metric on X with Kähler form ω. The de Rham cohomology class $[\omega] \in H^2(X, \mathbb{R})$ is called the *Kähler class* of g. Define the *Kähler cone* $\mathcal{K}$ of X to be the set of Kähler classes $[\omega] \in H^2(X, \mathbb{R})$ of QALE Kähler metrics on X. Then $\mathcal{K}$ is an open convex cone in $H^2(X, \mathbb{R})$, not containing zero.

The following two theorems will be proved in §9.8.

Theorem 9.3.3 *Let G be a finite subgroup of $\mathrm{SU}(m)$, and X a crepant resolution of $\mathbb{C}^m / G$. Then each Kähler class of QALE metrics on X contains a unique Ricci-flat QALE Kähler metric.*

If X is a crepant resolution of $\mathbb{C}^m / G$ then it is a local product resolution, by Proposition 9.1.4. Also, since $G \subset \mathrm{SU}(m)$ the singular set of $\mathbb{C}^m / G$ has codimension $n \geq 2$, which was one of the conditions for X to admit QALE metrics in Definition 9.2.1. In practice this means that any Kähler crepant resolution of $\mathbb{C}^m / G$ also admits QALE Kähler metrics, and thus has a family of Ricci-flat QALE Kähler metrics.

We call a quotient singularity $\mathbb{C}^m/G$ *reducible* if we can write $\mathbb{C}^m/G = \mathbb{C}^{m_1}/G_1 \times \mathbb{C}^{m_2}/G_2$, where $G_j \subset \mathrm{U}(m_j)$ and $m_1, m_2 > 0$ satisfy $m_1 + m_2 = m$. Otherwise we say $\mathbb{C}^m/G$ is *irreducible*. If $\mathbb{C}^m/G$ is reducible and (X, J, g) is a Ricci-flat QALE Kähler manifold asymptotic to $\mathbb{C}^m/G$, then it is easy to see that (X, J, g) is a product of lower-dimensional Ricci-flat QALE Kähler manifolds. So we shall restrict our attention to irreducible $\mathbb{C}^m/G$. In this case we can prove:

Theorem 9.3.4 *Let (X, J, g) be a Ricci-flat QALE Kähler manifold asymptotic to $\mathbb{C}^m/G$, where $\mathbb{C}^m/G$ is irreducible. If $m \geqslant 4$ is even and G is conjugate to a subgroup of $\mathrm{Sp}(m/2)$ then $\mathrm{Hol}(g) = \mathrm{Sp}(m/2)$, and otherwise $\mathrm{Hol}(g) = \mathrm{SU}(m)$.*

Theorem 9.3.3 is the main result of this chapter, which generalizes Theorem 8.2.3 to the QALE case. Its proof takes up most of §9.4–§9.8, and works by first proving a version of the Calabi conjecture for QALE manifolds. Theorem 9.3.4 is quite easy to prove.

9.3.1 *Discussion*

Theorems 9.3.3 and 9.3.4 are analogues of Theorems 8.2.3 and 8.2.4 for ALE manifolds, and their proofs mostly follow those of Theorems 8.2.3 and 8.2.4 in Chapter 8. But there are many small differences, and we now describe two of the more important ones.

Firstly, on an ALE manifold X we defined weighted Sobolev and Hölder spaces $L^q_{k,\beta}(X)$, $C^{k,\alpha}_\beta(X)$, using powers ρ^β of a radius function ρ on X, which is an approximate measure of the distance to 0 in $\mathbb{C}^m/G$. For reasons explained in §9.5, on a QALE manifold X we instead define $L^q_{k,\beta,\gamma}(X)$ and $C^{k,\alpha}_{\beta,\gamma}(X)$ using powers $\rho^\beta \sigma^\gamma$ of two functions ρ, σ on X.

Here, roughly speaking, ρ measures the distance to 0 and σ the distance to the singular set S of $\mathbb{C}^m/G$. For (β, γ) in a certain open set in $\mathbb{R}^2$ we show that $\Delta : C^{k+2,\alpha}_{\beta,\gamma}(X) \to C^{k,\alpha}_{\beta,\gamma-2}(X)$ is an isomorphism, that is, it is bounded and invertible with bounded inverse. This is the main analytic property we need to prove a version of the Calabi conjecture for QALE manifolds.

Secondly, a number of the proofs in the QALE case work by *induction on dimension*, which is not a feature of the ALE case in Chapter 8. For example, to prove Theorem 9.3.3 we must find a Ricci-flat QALE Kähler metric g on X. To do this we first assume that suitable Ricci-flat QALE Kähler metrics g_i exist on Y_i for $i \neq 0$, so that $h_i \times g_i$ is Ricci-flat on $V_i \times Y_i$. These metrics $h_i \times g_i$ are then used to construct g, which is asymptotic to $h_i \times g_i$ at infinity. Since $\dim Y_i < \dim X$, the result then follows by induction on $\dim X$.

In §8.2 we saw that the estimate $\nabla^k(\pi_*(g) - h) = O(r^{-2m-k})$ in the definition of ALE Kähler manifolds is actually *sharp* for Ricci-flat ALE Kähler metrics, and cannot be improved upon. This was why we chose the powers r^{-n-k} in (8.1) and r^{-2m-k} in (8.2). So it is natural to ask whether the decay rates in the definition (9.4) of QALE Kähler manifolds are also sharp for Ricci-flat QALE Kähler manifolds, or not.

In fact (9.4) is not quite sharp in the Ricci-flat case. At large distances from U_i in $V_i \times Y_i$, where $\mu_{i,j}$ is of the same order as ν_i when $i \not\leq j$, eqn (9.4) accurately describes the decay of g. However, near U_i in $V_i \times Y_i$, where $\mu_{i,j} \gg \nu_i$ for some j with $i \not\leq j$, the error terms in (9.4) are probably too large. For instance, the author can show that a Ricci-flat QALE Kähler metric g actually satisfies

$$\psi_i^*(g) = h_i \times g_i + \sum_{j \in I : i \not\succeq j} O\big(\mu_{i,j}^{-2\dim W_j}\big) \quad \text{on } V_i \times Y_i \setminus U_i,$$

which is sharp, and slightly stronger than the case $l = 0$ of (9.4), provided that $\mathbb{C}^m/G$ has nonisolated singularities. We will not pursue this point, since (9.4) is sufficient for our needs, and this chapter is already too long.

9.3.2 *Examples of QALE manifolds with holonomy* SU(m)

Here are three examples of QALE manifolds with holonomy SU(m), for $m = 3$ and 4. We will return again to the first two examples in §9.9.2.

Example 9.3.5 Define $\alpha : \mathbb{C}^3 \to \mathbb{C}^3$ by

$$\alpha : (z_1, z_2, z_3) \mapsto (-z_1, iz_2, iz_3).$$

Then $\langle \alpha \rangle$ is a subgroup of SU(3) isomorphic to $\mathbb{Z}_4$. Let I be $\{0, 1, \infty\}$, and set

$$V_0 = \mathbb{C}^3, \quad V_1 = \{(z_1, 0, 0) : z_1 \in \mathbb{C}\} \quad \text{and} \quad V_\infty = \{0\}.$$

Then $\mathcal{L} = \{V_i : i \in I\}$, and the groups A_i are

$$A_0 = \{1\}, \quad A_1 = \{1, \alpha^2\}, \quad \text{and} \quad A_\infty = \{1, \alpha, \alpha^2, \alpha^3\} = \mathbb{Z}_4.$$

The quotient $\mathbb{C}^3/\mathbb{Z}_4$ has a unique crepant resolution X, with $b^2(X) = 2$. It can be constructed using toric geometry as in §6.4.2, and admits QALE Kähler metrics. Thus X has a 2-parameter family of Ricci-flat QALE Kähler metrics g by Theorem 9.3.3, which have holonomy SU(3) by Theorem 9.3.4.

We can describe the asymptotic behaviour of these Ricci-flat metrics g quite simply. The vector space W_1 is $\{(0, z_2, z_3) : z_2, z_3 \in \mathbb{C}\}$, and $\alpha^2 \in A_1$ acts as -1 on W_1. Thus $W_1/A_1 \cong \mathbb{C}^2/\{\pm 1\}$. Let Y_1 be the blow-up of $\mathbb{C}^2/\{\pm 1\}$ at 0. Then Y_1 carries a 1-parameter family of ALE metrics with holonomy SU(2), given explicitly by Eguchi and Hanson [67]. Let g_1 be an Eguchi–Hanson metric on Y_1, and h_1 the Euclidean metric on $V_1 \cong \mathbb{C}$. Then $h_1 \times g_1$ is a Ricci-flat metric on $V_1 \times Y_1$.

Now α acts on $V_1 \times Y_1$ with $\alpha^2 = 1$, and the fixed points of α are a copy of $\mathbb{CP}^1$ in $V_1 \times Y_1$. Thus $(V_1 \times Y_1)/\langle \alpha \rangle$ is a complex orbifold, and X is its unique crepant resolution. The metric $h_1 \times g_1$ is preserved by α and descends to $(V_1 \times Y_1)/\langle \alpha \rangle$. Identifying X with $(V_1 \times Y_1)/\langle \alpha \rangle$ outside $\pi^{-1}(0)$, the asymptotic conditions (9.4) on our QALE metric g become

$$\nabla^l(g - h_1 \times g_1) = O\big(\pi^*(r)^{-4}(1 + \pi^*(s))^{-2-l}\big),$$

where $r, s : \mathbb{C}^3/\mathbb{Z}_4 \to [0, \infty)$ are the distances in $\mathbb{C}^3/\mathbb{Z}_4$ to 0 and to the singular set $S = \{\pm(z_1, 0, 0) : z_1 \in \mathbb{C}\}$ respectively. Thus at infinity g is asymptotic to $h_1 \times g_1$, a metric we can write down explicitly in coordinates.

Example 9.3.6 Define $\alpha, \beta : \mathbb{C}^3 \to \mathbb{C}^3$ by

$$\alpha : (z_1, z_2, z_3) \mapsto (z_1, -z_2, -z_3), \quad \beta : (z_1, z_2, z_3) \mapsto (-z_1, z_2, -z_3).$$

Then $\langle \alpha, \beta \rangle$ is a subgroup of SU(3) isomorphic to $\mathbb{Z}_2^2$. Let I be $\{0, 1, 2, 3, \infty\}$, and set

$$V_0 = \mathbb{C}^3, \quad V_\infty = \{0\}, \quad V_1 = \{(z_1, 0, 0) : z_1 \in \mathbb{C}\},$$
$$V_2 = \{(0, z_2, 0) : z_2 \in \mathbb{C}\} \quad \text{and} \quad V_3 = \{(0, 0, z_3) : z_3 \in \mathbb{C}\}.$$

Then $\mathcal{L} = \{V_i : i \in I\}$, and the groups A_i are

$$A_0 = \{1\}, \quad A_1 = \{1, \alpha\}, \quad A_2 = \{1, \beta\}, \quad A_3 = \{1, \alpha\beta\}, \quad A_\infty = \mathbb{Z}_2^2.$$

The quotient $\mathbb{C}^3/\mathbb{Z}_2^2$ has four distinct crepant resolutions $X_1, \ldots, X_4$, with $b^2(X_j) = 3$. They can be constructed explicitly using toric geometry as in §6.4.2, and all admit QALE Kähler metrics. Thus $X_1, \ldots, X_4$ carry 3-parameter families of Ricci-flat QALE Kähler metrics by Theorem 9.3.3, which have holonomy SU(3) by Theorem 9.3.4.

Example 9.3.7 Define $\alpha, \beta, \gamma : \mathbb{C}^4 \to \mathbb{C}^4$ by

$$\alpha : (z_1, \ldots, z_4) \mapsto (-z_1, -z_2, z_3, z_4),$$
$$\beta : (z_1, \ldots, z_4) \mapsto (z_1, -z_2, -z_3, z_4)$$
$$\text{and} \quad \gamma : (z_1, \ldots, z_4) \mapsto (z_1, z_2, -z_3, -z_4).$$

Then $\langle \alpha, \beta, \gamma \rangle \cong \mathbb{Z}_2^3$ is a subgroup of SU(4). Define an 8-element indexing set I by $I = \{0, \infty\} \cup \{jk : 1 \leqslant j < k \leqslant 4\}$. Then $\mathcal{L} = \{V_i : i \in I\}$, where

$$V_0 = \mathbb{C}^4, \quad V_\infty = \{1\} \quad \text{and} \quad V_{jk} = \{(z_1, \ldots, z_4) \in \mathbb{C}^4 : z_j = z_k = 0\}.$$

There are 48 distinct crepant resolutions $X_1, \ldots, X_{48}$ of $\mathbb{C}^4/\mathbb{Z}_2^3$, with $b^2(X_j) = 6$. They can be constructed explicitly using toric geometry as in §6.4.2, and all admit QALE Kähler metrics. Thus $X_1, \ldots, X_{48}$ carry 6-parameter families of Ricci-flat QALE Kähler metrics by Theorem 9.3.3, which have holonomy SU(4) by Theorem 9.3.4.

In each of these examples we know by Theorem 9.3.3 that QALE metrics with holonomy SU(m) exist on the resolutions X of $\mathbb{C}^m/G$, but we are unable to write these metrics down explicitly in coordinates. As for the ALE case in §8.2.2, the author believes that QALE metrics with holonomy SU(m) for $m \geqslant 3$ are not algebraic objects, and it is not possible to give an explicit formula for these metrics in coordinates.

9.3.3 *Examples of QALE manifolds with holonomy* Sp(m)

We begin with a proposition.

Proposition 9.3.8 *Suppose* (X, J, g) *is a QALE manifold with holonomy* Sp(m) *for some* $m \geqslant 2$, *asymptotic to* $\mathbb{C}^{2m}/G$. *Then G is nonabelian.*

Proof Suppose for a contradiction that $G \subset \text{Sp}(m)$ is abelian, and (X, J, g) is a QALE Kähler manifold asymptotic to $\mathbb{C}^{2m}/G$ with $\text{Hol}(g) = \text{Sp}(m)$. Then there exists a coordinate system $(z_1, \ldots, z_{2m})$ on $\mathbb{C}^{2m}$ such that each $g \in G$ can be written

$$(z_1, \ldots, z_{2m}) \xmapsto{g} (e^{i\theta_1} z_1, e^{-i\theta_1} z_2, \ldots, e^{i\theta_m} z_{2m-1}, e^{-i\theta_m} z_{2m}),$$

where $\theta_1, \ldots, \theta_m \in [0, 2\pi)$. For $k = 1, \ldots, m$, let $H_k \subseteq G$ be the set of $g \in G$ for which $\theta_j = 0$ when $j \neq k$. Then H_k and $H_1 \times \cdots \times H_m$ are subgroups of G, and $\mathbb{C}^{2m}/H_1 \times \cdots \times H_m = \mathbb{C}^2/H_1 \times \cdots \times \mathbb{C}^2/H_m$.

Let Y_k be the crepant resolution of $\mathbb{C}^2/H_k$, and B the quotient group $G/(H_1 \times \cdots \times H_m)$. Then B acts naturally on $Y_1 \times \cdots \times Y_m$, and X must be a crepant resolution of $(Y_1 \times \cdots \times Y_m)/B$. Now if the fixed point set of $g \in G$ has codimension 2 in $\mathbb{C}^{2m}$, then $g \in H_k$ for some k. Thus, $H_1 \times \cdots \times H_m$ contains all $g \in G$ with fixed points of codimension 2. Using this one can prove that the singular set of $(Y_1 \times \cdots \times Y_m)/B$ has codimension at least 4.

If B is nontrivial, the idea of the proof of Proposition 8.8.1 shows that the singularities of $(Y_1 \times \cdots \times Y_m)/B$ are *terminal*, and admit no crepant resolutions. So B is trivial, and $G = H_1 \times \cdots \times H_m$. Therefore $\mathbb{C}^{2m}/G = \mathbb{C}^2/H_1 \times \cdots \times \mathbb{C}^2/H_m$ is *reducible*. It is then easy to see that $\text{Hol}(g) = \text{Sp}(1) \times \cdots \times \text{Sp}(1)$, which contradicts $\text{Hol}(g) = \text{Sp}(m)$ and $m \geq 2$. $\square$

The proposition shows that to find examples of QALE metrics with holonomy $\text{Sp}(m)$ for $m \geq 2$, we must consider nonabelian groups.

Example 9.3.9 Define $\alpha, \beta : \mathbb{C}^4 \to \mathbb{C}^4$ by

$$\alpha :(z_1, \ldots, z_4) \mapsto (e^{2\pi i/3} z_1, e^{4\pi i/3} z_2, e^{4\pi i/3} z_3, e^{2\pi i/3} z_4),$$
$$\beta :(z_1, \ldots, z_4) \mapsto (z_3, z_4, z_1, z_2).$$

Then $G = \langle \alpha, \beta \rangle$ is a nonabelian subgroup of $\text{Sp}(2)$ of order 6 isomorphic to the symmetric group S_3, that preserves the complex 2-form $dz_1 \wedge dz_2 + dz_3 \wedge dz_4$.

Let S be the singular set of $\mathbb{C}^4/G$, and define a map $\phi : \mathbb{C}^4/G \setminus S \to \mathbb{CP}^4$ by

$$\phi\big((z_1, \ldots, z_4)G\big) = \big[z_1 z_2 - z_3 z_4,\ z_1^3 - z_3^3,\ z_1^2 z_4 - z_2 z_3^2,\ z_1 z_4^2 - z_2^2 z_3,\ z_2^3 - z_4^3\big].$$

The five polynomials in $z_1, \ldots, z_4$ given here are invariant under α and change sign under β, and they are all zero if and only if $(z_1, z_2, z_3, z_4)G$ lies in S. Let X be the closure of the graph of ϕ in $\mathbb{C}^4/G \times \mathbb{CP}^4$, and $\pi : X \to \mathbb{C}^4/G$ the natural projection. Then a careful analysis shows that X is actually nonsingular, and (X, π) is a *crepant resolution* of $\mathbb{C}^4/G$.

Let $[x_0, x_1, x_2]$ be homogeneous coordinates on the *weighted projective space* $\mathbb{CP}^2_{3,1,1}$, as explained in Definition 6.5.4. Define a map $\psi : \mathbb{CP}^2_{3,1,1} \to \mathbb{CP}^4$ by

$$\psi\big([x_0, x_1, x_2]\big) = \big[x_0,\ x_1^3,\ x_1^2 x_2,\ x_1 x_2^2,\ x_2^3\big].$$

Then ψ is injective, and it can be shown that $\pi^{-1}(0)$ is the subset $\{0\} \times \text{Im}\,\psi$ in $\mathbb{C}^4/G \times \mathbb{CP}^4$. Thus $\pi^{-1}(0)$ is isomorphic to the (singular) weighted projective space $\mathbb{CP}^2_{3,1,1}$.

Since X retracts onto $\pi^{-1}(0)$, it follows that $b^2(X) = 1$, $b^4(X) = 1$ and $b^6(X) = 0$. Clearly X is a quasi-projective variety, and therefore Kähler. Theorem 9.3.3 shows that X has a 1-parameter family of Ricci-flat QALE Kähler metrics, which have holonomy Sp(2) by Theorem 9.3.4.

In fact Example 9.3.9 generalizes to give an action of the symmetric group S_{m+1} on $\mathbb{C}^{2m}$, and a crepant resolution X_m of $\mathbb{C}^{2m}/S_{m+1}$ carrying QALE Kähler metrics with holonomy Sp(m). To prove this we adapt an idea of Beauville, and regard the *Hilbert scheme* or *Douady space* $\text{Hilb}^{m+1}\mathbb{C}^2$ of $m+1$ points in $\mathbb{C}^2$ as a crepant resolution of $\mathbb{C}^{2m+2}/S_{m+1}$. Further details are given in [28, §6–§7].

Example 9.3.10 Let $\mathbb{C}^4$ have coordinates (z_1, z_2, z_3, z_4). Let H be a subgroup of SU(2). Then $H \times H$ acts on $\mathbb{C}^4$, with the first H acting only on the coordinates (z_1, z_2) and the second H acting only on (z_3, z_4). Define $\alpha : \mathbb{C}^4 \to \mathbb{C}^4$ by $\alpha : (z_1, \ldots, z_4) \mapsto (z_3, z_4, z_1, z_2)$. Let G be the subgroup of Sp(2) generated by $H \times H$ and α. Then G is a semidirect product $\mathbb{Z}_2 \ltimes (H \times H)$. We can define a crepant resolution X of $\mathbb{C}^4/G$ as follows.

Let Y be the unique crepant resolution of $\mathbb{C}^2/H$. Then $Y \times Y$ is a crepant resolution of $\mathbb{C}^4/(H \times H)$, and the action of α lifts to $Y \times Y$ by $\alpha : (y_1, y_2) \mapsto (y_2, y_1)$ in the obvious way. The singular set of the quotient $(Y \times Y)/\langle\alpha\rangle$ is the 'diagonal' $\Delta_Y = \{(y, y) : y \in Y\}$, and each singular point is modelled locally on $\mathbb{C}^2 \times (\mathbb{C}^2/\{\pm 1\})$. Let X be the blow-up of $(Y \times Y)/\langle\alpha\rangle$ along the diagonal Δ_Y. Then X is nonsingular, and is a crepant resolution of $\mathbb{C}^4/G$.

Since we understand Y very well, it is easy to compute the Betti numbers of X. For example, in the case $H = \mathbb{Z}_k$ we have $b^2(X) = k$, $b^4(X) = \frac{1}{2}(k+2)(k-1)$ and $b^6(X) = 0$. Again, each such resolution X has a family of Ricci-flat QALE Kähler metrics by Theorem 9.3.3, which have holonomy Sp(2) by Theorem 9.3.4.

We claimed in §9.3.2 that QALE metrics with holonomy SU(m) for $m \geqslant 3$ are nonalgebraic objects, and cannot be explicitly written down in coordinates. However, for QALE metrics with holonomy Sp(m), the reverse is true. The author has a proof that *every* QALE metric with holonomy Sp(m) has an algebraic description. This proof uses the theory of *hypercomplex algebraic geometry*, which is described in Joyce [125].

It seems likely that QALE manifolds with holonomy Sp(m) can be explicitly constructed using the hyperkähler quotient, as in Kronheimer's construction of ALE manifolds with holonomy Sp(1), [140]. However, at present the author has no proof of this, nor any explicit examples.

9.4 Kähler potentials on QALE Kähler manifolds

If g, g' are two QALE Kähler metrics in the same Kähler class on X, with Kähler forms ω, ω', then we expect that $\omega' = \omega + dd^c\phi$ for some function ϕ on X. Conversely, if g is a QALE Kähler metric on X with Kähler form ω, then we can try to define other QALE metrics g' on X with Kähler forms $\omega' = \omega + dd^c\phi$ for suitable functions ϕ on X. In this section we will study the properties of such functions ϕ in detail.

Definition 9.4.1 Let (X, J, g) be a QALE Kähler manifold asymptotic to $\mathbb{C}^m/G$, and use the notation of Definition 9.2.1. We say that a smooth real function ϕ on X is of *Kähler potential type* if for each $i \in I$ there exists a smooth real function ϕ_i on Y_i, with $\phi_0 = 0$, such that

$$\nabla^l(\psi_i^*(\phi) - \phi_i) = \sum_{j \in I: i \not\succeq j} O(\mu_{i,j}^{d_j} v_i^{-l}) \tag{9.7}$$

on $V_i \times Y_i \setminus U_i$, for all $l \geq 0$. Here we identify ϕ_i with its pull-back to $V_i \times Y_i$.

By comparing (9.7) with (9.4), we immediately deduce:

Proposition 9.4.2 Let (X, J, g) be a QALE Kähler manifold, ω the Kähler form of g, and ϕ a function of Kähler potential type on X. Suppose $\omega' = \omega + dd^c\phi$ is a positive $(1, 1)$-form on X. Then the Kähler metric g' on X with Kähler form ω' is QALE.

Here is a converse to this proposition. It can be proved by combining the method of Theorem 8.4.4 with the results on the Laplacian on QALE manifolds that we will prove in §9.5, and the ideas of §9.7.

Theorem 9.4.3 Let X be a local product resolution of $\mathbb{C}^m/G$, and suppose that g, g' are QALE metrics on X in the same Kähler class, with Kähler forms ω, ω'. Then there exists a unique function ϕ of Kähler potential type on X such that $\omega' = \omega + dd^c\phi$.

Proposition 9.4.2 and Theorem 9.4.3 show that functions of Kähler potential type are the natural class of functions to use as Kähler potentials on QALE Kähler manifolds. The next result can be proved by following the proofs of Propositions 9.2.3 and 9.2.4.

Proposition 9.4.4 Let (X, J, g) be a QALE Kähler manifold, and ϕ a function of Kähler potential type on X. Then the functions ϕ_i on Y_i introduced in Definition 9.4.1 satisfy the following conditions:

(i) For each $\gamma \in G$ and $i \in I$ we have $\chi_{\gamma,i}^*(\phi_{\gamma \cdot i}) = \phi_i$. Hence ϕ_i is invariant under the action of B_i on Y_i.

(ii) Whenever $i, j \in I$ with $i \succeq j$, then on $V_j' \times Y_j' \setminus U_j'$ we have

$$\nabla^l((\psi_j')^*(\phi_i) - \phi_j) = \sum_{\substack{k \in I: \\ i \succeq k, \, j \not\succeq k}} O((\mu_{j,k}')^{d_k}(v_j')^{-l})$$

for all $l \geq 0$, using the notation of Propositions 9.1.5 and 9.2.4. Thus the functions ϕ_i on Y_i are also of Kähler potential type.

The proposition shows that (i) and (ii) are *necessary* conditions for a set of functions ϕ_i to be associated to a function ϕ of Kähler potential type. In fact they are also *sufficient*.

Theorem 9.4.5 Let (X, J, g) be a QALE Kähler manifold, and suppose that for each $i \in I \setminus \{\infty\}$ there is a smooth function ϕ_i on Y_i, with $\phi_0 = 0$, such that the ϕ_i satisfy conditions (i) and (ii) of Proposition 9.4.4. Then there exists a function ϕ of Kähler potential type on X asymptotic to these ϕ_i for all $i \in I \setminus \{\infty\}$.

Proof Let $\eta : [0, \infty) \to [0, 1]$ be smooth with $\eta(x) = 0$ for $x \leqslant R$ and $\eta(x) = 1$ for $x \geqslant 2R$. For each $i \in I \setminus \{\infty\}$, define a smooth function Φ_i on $V_i \times Y_i$ by

$$\Phi_i(v, y) = \phi_i(y) \cdot \prod_{j \in I : i \not\succeq j} \eta\bigl(\mu_{i,j}(v, y)\bigr). \tag{9.8}$$

The idea here is that $\Phi_i = \phi_i$ at distance at least $2R$ from the pull-back of S_i in $V_i \times Y_i$, that $\Phi_i = 0$ at distance no more than R from the pull-back of S_i, and that between distances R and $2R$ we join the two possibilities $\Phi_i = \phi_i$ and $\Phi_i = 0$ smoothly together using a partition of unity. Note that $\Phi_i \equiv 0$ on U_i.

For $i \in I \setminus \{\infty\}$, let k_i be integers satisfying

$$\sum_{i \in I \setminus \{\infty\} : i \succeq j} k_i = 1 \qquad \text{for each } j \in I \setminus \{\infty\}. \tag{9.9}$$

It can be shown that these equations have a unique solution $\{k_i\}$. Now define

$$\phi(x) = \sum_{i \in I \setminus \{\infty\}} \frac{k_i |A_i|}{|G|} \sum_{\substack{(v,y) \in V_i \times Y_i \setminus U_i : \\ \psi_i(v, y) = x}} \Phi_i(v, y). \tag{9.10}$$

As $\Phi_i \equiv 0$ on U_i, we see that ϕ is smooth. It turns out that this ϕ is of Kähler potential type on X and asymptotic to ϕ_i for all $i \in I \setminus \{\infty\}$, so that it satisfies the conditions of the theorem. The proof of this is rather complicated, and we will not give it in full. Instead, we will explain the important points under the simplifying assumption that $N(V_i) = G$ for all $i \in I$.

In this case $B_i = G/A_i$, so $|B_i| = |G|/|A_i|$. Thus we may rewrite (9.10) as

$$\phi(x) = \sum_{i \in I \setminus \{\infty\}} k_i \Phi'_i(x), \quad \text{where} \quad \Phi'_i(x) = \frac{1}{|B_i|} \sum_{\substack{(v,y) \in V_i \times Y_i \setminus U_i : \\ \psi_i(v, y) = x}} \Phi_i(v, y). \tag{9.11}$$

Now B_i acts on Y_i and $V_i \times Y_i$, and ϕ_i is B_i-invariant by condition (i). Thus Φ_i is also B_i-invariant, and pushes down to a function on $(V_i \times Y_i)/B_i$. From §9.1, ψ_i induces an isomorphism between $(V_i \times Y_i \setminus U_i)/B_i$ and $X \setminus \pi^{-1}(T_i/B_i)$. In fact Φ'_i is the pushforward of Φ_i under this isomorphism. The factor $1/|B_i|$ in (9.11) ensures this, because each point of $X \setminus \pi^{-1}(T_i/B_i)$ pulls back to an orbit of B_i in $V_i \times Y_i$, consisting of $|B_i|$ points.

Thus, abusing notation a little, we can write $\phi = \sum_{i \neq \infty} k_i \Phi_i$, because the factor $|A_i|/|G|$ in (9.10) compensates for the fact that each generic point in X pulls back to $|G|/|A_i|$ points in $V_i \times Y_i$. But $\Phi_i = \phi_i$ away from U_i, and so away from $\pi^{-1}(S)$ in X we have $\phi = \sum_{i \neq \infty} k_i \phi_i$, again by an abuse of notation.

We must prove that ϕ satisfies (9.7). One way to interpret this is to say that X is divided roughly into overlapping regions corresponding to $j \in I$, where in the jth region X is locally isomorphic to $V_j \times Y_j$ and $\phi \approx \phi_j$. Now on the jth region we have

$\phi_i \approx \phi_j$ if $i \succeq j$ and $\phi_i \approx 0$ if $i \not\succeq j$, by condition (ii) of Proposition 9.4.2. Therefore on the jth region we have

$$\phi \approx \sum_{i \in I \setminus \{\infty\}: i \succeq j} k_i \phi_j = \phi_j,$$

by (9.9), as we want. This idea can be used to show that (9.7) holds away from $\pi^{-1}(S)$ in X.

It remains to consider the parts of X near $\pi^{-1}(S)$. Within distance $2R$ of $\pi^{-1}(S)$ in X, the functions $\eta(\mu_{i,j})$ in (9.8) are not all equal to 1, and so we do not have $\Phi_i = \phi_i$ for all i. The reason for introducing the $\eta(\mu_{i,j})$ is that the push-forward of ϕ_i to $X \setminus \pi^{-1}(S_i/B_i)$ may not extend smoothly to X. So we modify ϕ_i to get Φ_i, whose push-forward is zero near $\pi^{-1}(S_i/B_i)$ and does extend smoothly to X. But we then have to make sure that the 'errors' due to the $\eta(\mu_{i,j})$ are within the bounds allowed by (9.7). One can show using (9.9) that this is always so, and the proof is complete. □

The reason for excluding $i = \infty$ in this theorem is that in general $X = V_\infty \times Y_\infty$ and $\phi = \phi_\infty$, and so if we assumed that ϕ_∞ existed we could just take $\phi = \phi_\infty$ and there would be nothing to prove. Although the proof works by writing down ϕ explicitly, there are in fact many suitable ϕ, since if we add to ϕ any smooth function on Y_∞ with sufficiently fast decay at infinity, the resulting function also satisfies the theorem. The proof above will play an important part in the construction of Ricci-flat QALE metrics in §9.8, which is why we covered it in some detail.

Proposition 9.4.6 *In the situation of Theorem 9.4.5, let ω be the Kähler form of g and ω_i the Kähler form of g_i, and suppose $\omega_i + \mathrm{dd}^c\phi_i$ is a positive $(1, 1)$-form on Y_i for each $i \in I \setminus \{\infty\}$. Then we can choose the function ϕ such that $\omega' = \omega + \mathrm{dd}^c\phi$ is a positive $(1, 1)$-form on X. Thus the Kähler metric g' on X with Kähler form ω' is QALE, by Proposition 9.4.2.*

Proof Let ϕ be as in the proof of Theorem 9.4.5. Then $\omega + \mathrm{dd}^c\phi$ is positive outside a compact set in X, because at large distances from $\pi^{-1}(0)$ we have $\phi \approx \phi_i$ for some $i \neq \infty$, and so $\omega + \mathrm{dd}^c\phi$ is positive because $\omega_i + \mathrm{dd}^c\phi_i$ is positive. Let $\hat{R} > 0$, and choose a smooth function $\eta : X \to [0, 1]$ such that $\eta = 0$ at distance less than $\hat{R}$ from $\pi^{-1}(0)$ in X, and $\eta = 1$ at distance more than $2\hat{R}$ from $\pi^{-1}(0)$ in X, and $\nabla \eta = O(\hat{R}^{-1})$, $\nabla^2 \eta = O(\hat{R}^{-2})$. For large $\hat{R}$ it turns out that $\eta\phi$ satisfies the conditions of the proposition. □

One moral of this proposition is that QALE Kähler metrics on X are actually very abundant. That is, the asymptotic conditions on QALE metrics are not so restrictive that they admit few solutions. Rather, given any set of Kähler metrics g_i on Y_i for $i \in I \setminus \{\infty\}$ satisfying the obvious consistency conditions, we expect to find many QALE Kähler metrics g on X asymptotic to these g_i.

Finally, we give an analogue of Proposition 8.4.5 for QALE manifolds.

Theorem 9.4.7 *Let X be a local product resolution of $\mathbb{C}^m/G$ admitting QALE Kähler metrics. Then in each Kähler class there exists a QALE Kähler metric g' on X such that* *for each $i \in I$ we have $\psi_i^*(g') = h_i \times g_i'$ on the set $V_i \times Y_i \setminus U_i$, where g_i' is a Kähler metric on Y_i, and U_i is defined as in Definition 9.1.2 using a constant $R > 0$, which may depend on the Kähler class.*

Taking $i = 0$ gives $g' = \pi^*(h_0)$ on $\{x \in X : d(\pi(x), S) > R\}$, where h_0 is the Euclidean metric, S the singular set and $d(\,,\,)$ the distance on $\mathbb{C}^m/G$. Thus g' is flat outside a fixed distance from the exceptional set of the resolution X. Also, g' is a product metric $h_i \times g_i'$ on the parts of X modelled on $V_i \times Y_i$, except within a fixed distance of some other component of the exceptional set.

Here is a sketch of the proof of this theorem; we leave the details as an exercise for the reader. The basic idea is to start with a QALE Kähler manifold (X, J, g) with Kähler form ω, and to find a function ϕ of Kähler potential type on X such that $\omega' = \omega + \mathrm{dd}^c\phi$ is a positive (1,1)-form, and the QALE Kähler metric g' with Kähler form ω' satisfies the conditions of the theorem. We construct ϕ by induction on $m = \dim X$.

The inductive step of the proof works as follows. Assume by induction that such a ϕ exists when $\dim X < m$, and let (X, J, g) be a QALE Kähler manifold asymptotic to $\mathbb{C}^m/G$. The inductive hypothesis shows that for each $i \neq \infty$ in I there is a function ϕ_i on Y_i such that $\omega_i' = \omega_i + \mathrm{dd}^c\phi_i$ is positive on Y_i, and the corresponding Kähler metric g_i' on Y_i satisfies the theorem.

We can also arrange that these functions ϕ_i for $i \neq \infty$ satisfy conditions (i) and (ii) of Proposition 9.4.4. Then using the ideas of Theorem 9.4.5 and Proposition 9.4.6, we construct a function ϕ on X with the properties we need, that is asymptotic to ϕ_i on Y_i for each $i \neq \infty$. To prove such a ϕ exists requires the analytical results of §9.5 and also some discussion of the de Rham cohomology $H^2(X, \mathbb{R})$, along the lines of the proof of Proposition 8.4.5.

9.5 Analysis on QALE Kähler manifolds

In this section we shall define weighted Hölder spaces of functions on QALE Kähler manifolds X, and study the action of the Laplacian on them, as we did in §8.3 for the case of ALE manifolds. We begin by defining functions ρ, σ on X which are analogues of the *radius functions* used in Chapter 8.

Definition 9.5.1 Let G be a finite subgroup of $U(m)$, and S the singular set of $\mathbb{C}^m/G$. Define continuous functions $r : \mathbb{C}^m/G \to [0, \infty)$ and $s : \mathbb{C}^m/G \to [0, \infty)$ by $r(x) = d(x, 0)$ and $s(x) = d(x, S)$, where $d(\,,\,)$ is the distance on $\mathbb{C}^m/G$. Let (X, π) be a local product resolution of $\mathbb{C}^m/G$, and g a QALE Kähler metric on X. We say that (ρ, σ) is a *pair of radius functions on X* if $\rho, \sigma : X \to [1, \infty)$ are smooth functions such that $\rho \geqslant \sigma$ and for some $K > 0$ we have

$$\pi^*(r) + 1 \leqslant \rho \leqslant \pi^*(r) + 2, \quad |\nabla\rho| \leqslant K, \quad \text{and} \quad |\nabla^2\rho| \leqslant K\rho^{-1}, \tag{9.12}$$

$$\tfrac{1}{2}\pi^*(s) + 1 \leqslant \sigma \leqslant \pi^*(s) + 2, \quad |\nabla\sigma| \leqslant K, \quad \text{and} \quad |\nabla^2\sigma| \leqslant K\sigma^{-1}. \tag{9.13}$$

A pair of radius functions (ρ, σ) exists for every QALE Kähler manifold (X, J, g).

Here ρ and σ are smoothed versions of $\pi^*(r)$ and $\pi^*(s)$, adjusted to ensure that $\rho \geqslant 1$ and $\sigma \geqslant 1$. The reason why we have $\pi^*(r) + 1 \leqslant \rho$ but $\tfrac{1}{2}\pi^*(s) + 1 \leqslant \sigma$ is that s

is a rather less smooth function on $\mathbb{C}^m/G$ than r, and so to make a smoothed version σ with $\nabla^2\sigma$ small we must allow a greater difference between σ and $\pi^*(s)$ than between ρ and $\pi^*(r)$.

Section 8.3 defined weighted Sobolev and Hölder spaces $L^q_{k,\beta}(X)$ and $C^{k,\alpha}_\beta(X)$ on an ALE manifold X. These spaces were chosen because Δ behaves very well on them. In particular, for $\beta \in (-n, -2)$ the map $\Delta : C^{k+2,\alpha}_{\beta+2}(X) \to C^{k,\alpha}_\beta(X)$ is an isomorphism, by Theorem 8.3.6.

We wish to generalize these ideas to QALE manifolds. The most obvious way to do this is to apply Definitions 8.3.1 and 8.3.2 unchanged to QALE manifolds, using the function ρ defined above as a radius function. However, it turns out that Δ is not well-behaved on these spaces, as our next result shows.

Proposition 9.5.2 *Suppose $\mathbb{C}^m/G$ is a nonisolated singularity, (X, J, g) is a QALE Kähler manifold asymptotic to $\mathbb{C}^m/G$, and $\beta \in \mathbb{R}$. Then in general we can find smooth functions u, f on X such that $\Delta u = f$ and $u = O(\rho^{\beta+2})$ and $\nabla^k f = O(\rho^{\beta-k})$ for all $k \geqslant 0$, but $\nabla^k u \neq O(\rho^{\beta+2-k})$ for large k.*

Therefore, if $C^{k,\alpha}_\beta(X)$ is defined as in Definition 8.3.2, we can find functions u, f such that $\Delta u = f$ and $f \in C^{k,\alpha}_\beta(X)$, but $u \notin C^{k+2,\alpha}_{\beta+2}(X)$ even though $u = O(\rho^{\beta+2})$. If in addition $\beta < -2$, there does not exist any function $u' \in C^{k+2,\alpha}_{\beta+2}(X)$ with $\Delta u' = f$, and so $\Delta : C^{k+2,\alpha}_{\beta+2}(X) \to C^{k,\alpha}_\beta(X)$ is not surjective. In fact we can go further and say that $\Delta : C^{k+2,\alpha}_{\beta+2}(X) \to C^{k,\alpha}_\beta(X)$ is not Fredholm for any β.

We will not prove Proposition 9.5.2, but instead give a simple example of such behaviour, in which the function $f = 0$. The interested reader should be able to use the idea in the example to prove the general case. (The proof is quite messy, though.)

Example 9.5.3 Let $0 < k < m$, and let $\mathbb{C}^{m-k}/H$ be an isolated quotient singularity with resolution Y. Suppose g_Y is an ALE Kähler metric on Y, and $h_{\mathbb{C}^k}$ the Euclidean metric on $\mathbb{C}^k$. Then $\mathbb{C}^k \times Y$ is a local product resolution of $\mathbb{C}^k \times \mathbb{C}^{m-k}/H$, and $h_{\mathbb{C}^k} \times g_Y$ a QALE Kähler metric on $\mathbb{C}^k \times Y$.

Let v be a nonconstant harmonic function on Y with polynomial growth of order d. There is a well-understood theory of such functions (see for instance Bartnik [22, §1–§2]), which guarantees that such v exist. Let u be the pull-back of v from Y to $\mathbb{C}^k \times Y$. Then $\Delta u = 0$, and $\nabla^l u = O(\rho^{d-l})$ for $0 \leqslant l \leqslant d$. However, when $l > d$ it is not true that $\nabla^l u = O(\rho^{d-l})$, as $\nabla^l u$ is constant (and in general nonzero) on the slices $\mathbb{C}^k \times \{y\}$ for $y \in Y$, but $\rho^{d-l} \to 0$ approaching infinity in $\mathbb{C}^k \times \{y\}$.

So, to generalize the ideas of §8.3 to QALE manifolds, we need to find some other way to define weighted Sobolev and Hölder spaces, on which the Laplacian behaves well. A clue is provided by the function u of the example: although $\nabla^k u \neq O(\rho^{d-k})$, it is true that $\nabla^k u = O(\sigma^{d-k})$. So we could try replacing ρ by σ in the definition of $C^{k,\alpha}_\beta(X)$. It turns out that Δ is never Fredholm with this definition, either. However, if we use both ρ and σ at once, then we can define weighted Sobolev and Hölder spaces on which Δ behaves well.

Definition 9.5.4 Suppose (X, J, g) is a QALE Kähler manifold asymptotic to $\mathbb{C}^m/G$, and (ρ, σ) is a pair of radius functions on X. Let n be the complex codimension of the singular set of $\mathbb{C}^m/G$. For $\beta, \gamma \in \mathbb{R}$, $q \geq 1$ and k a nonnegative integer, define the *weighted Sobolev space* $L^q_{k,\beta,\gamma}(X)$ to be the set of functions f on X that are locally integrable and k times weakly differentiable, and for which the norm

$$\|f\|_{L^q_{k,\beta,\gamma}} = \left(\sum_{j=0}^{k} \int_X |\rho^{-\beta} \sigma^{j-\gamma} \nabla^j f|^q \rho^{2n-2m} \sigma^{-2n} dV_g \right)^{1/q} \tag{9.14}$$

is finite. Then $L^q_{k,\beta,\gamma}(X)$ is a Banach space, and $L^2_{k,\beta,\gamma}(X)$ a Hilbert space.

Definition 9.5.5 Let (X, J, g) be a QALE Kähler manifold and (ρ, σ) a pair of radius functions on X. For $\beta, \gamma \in \mathbb{R}$ and k a nonnegative integer, define $C^k_{\beta,\gamma}(X)$ to be the space of continuous functions f on X with k continuous derivatives, such that $\rho^{-\beta} \sigma^{j-\gamma} |\nabla^j f|$ is bounded on X for $j = 0, \ldots, k$. Define the norm $\|.\|_{C^k_{\beta,\gamma}}$ on $C^k_{\beta,\gamma}(X)$ by

$$\|f\|_{C^k_{\beta,\gamma}} = \sum_{j=0}^{k} \sup_X |\rho^{-\beta} \sigma^{j-\gamma} \nabla^j f|.$$

Let $\delta(g)$ be the injectivity radius of g, and write $d(x, y)$ for the distance between x, y in X. For T a tensor field on X and $\alpha, \beta, \gamma \in \mathbb{R}$, define

$$[T]_{\alpha,\beta,\gamma} = \sup_{\substack{x \neq y \in X \\ d(x,y) < \delta(g)}} \left[\min(\rho(x), \rho(y))^{-\beta} \cdot \min(\sigma(x), \sigma(y))^{-\gamma} \cdot \frac{|T(x) - T(y)|}{d(x,y)^\alpha} \right]. \tag{9.15}$$

Here we interpret $|T(x) - T(y)|$ using parallel translation along the unique geodesic of length $d(x, y)$ joining x and y. For $\beta, \gamma \in \mathbb{R}$, k a nonnegative integer, and $\alpha \in (0, 1)$, define the *weighted Hölder space* $C^{k,\alpha}_{\beta,\gamma}(X)$ to be the set of $f \in C^k_{\beta,\gamma}(X)$ for which the norm

$$\|f\|_{C^{k,\alpha}_{\beta,\gamma}} = \|f\|_{C^k_{\beta,\gamma}} + [\nabla^k f]_{\alpha,\beta,\gamma-k-\alpha} \tag{9.16}$$

is finite. Define $C^\infty_{\beta,\gamma}(X)$ to be the intersection of the $C^k_{\beta,\gamma}(X)$ for all $k \geq 0$. Both $C^k_{\beta,\gamma}(X)$ and $C^{k,\alpha}_{\beta,\gamma}(X)$ are Banach spaces, but $C^\infty_{\beta,\gamma}(X)$ is not a Banach space.

These definitions are adapted from Definitions 8.3.1 and 8.3.2. The basic idea is that if $f \in L^p_{k,\beta,\gamma}(X)$ or $C^k_{\beta,\gamma}(X)$ then $\nabla^j f = O(\rho^\beta \sigma^{\gamma-j})$ for $j \leq k$. The function $\rho^{2n-2m} \sigma^{-2n}$ in (9.14) seems to be the most natural choice, though it is difficult to explain why at this stage. If $\mathbb{C}^m/G$ has an isolated singularity, so that X is an ALE manifold, then $\rho = \sigma$ and $L^p_{k,\beta,\gamma}(X)$ and $C^{k,\alpha}_{\beta,\gamma}(X)$ agree with $L^p_{k,\beta+\gamma}(X)$ and $C^{k,\alpha}_{\beta+\gamma}(X)$, the usual weighted Sobolev and Hölder spaces on an ALE manifold.

In the rest of the section we will show that if (β, γ) lies in a certain open set $\mathcal{I}_X$ in $\mathbb{R}^2$, then $\Delta : C^{k+2,\alpha}_{\beta,\gamma}(X) \to C^{k,\alpha}_{\beta,\gamma-2}(X)$ is an isomorphism. We begin with an existence result for solutions of the equation $\Delta u = f$. It can be proved by adapting known methods for ALE manifolds.

Theorem 9.5.6 *Let X be a QALE Kähler manifold, $\alpha \in (0,1)$ and $\beta, \gamma < 0$. Then for each $f \in C^{0,\alpha}_{\beta,\gamma-2}(X)$ there exists a unique $u \in C^{2,\alpha}(X)$ such that $\Delta u = f$ and $u(x) \to 0$ as $x \to \infty$ in X.*

Here is a sufficient condition for $\Delta : C^{k+2,\alpha}_{\beta,\gamma}(X) \to C^{k,\alpha}_{\beta,\gamma-2}(X)$ to be an isomorphism.

Theorem 9.5.7 *Let (X, J, g) be a QALE Kähler manifold, (ρ, σ) a pair of radius functions on X, and $\beta, \gamma < 0$. Suppose there exists a smooth function $F : X \to (0, \infty)$ satisfying*

$$\Delta F \geq \rho^\beta \sigma^{\gamma-2} \quad \text{and} \quad K_1 \rho^\beta \sigma^\gamma \leq F \leq K_2 \rho^\beta \sigma^\gamma \qquad (9.17)$$

for some $K_1, K_2 > 0$. Then whenever $k \geq 0$ and $\alpha \in (0,1)$, there exists $C > 0$ such that for each $f \in C^{k,\alpha}_{\beta,\gamma-2}(X)$ there is a unique $u \in C^{k+2,\alpha}_{\beta,\gamma}(X)$ with $\Delta u = f$, which satisfies $\|u\|_{C^{k+2,\alpha}_{\beta,\gamma}} \leq C \|f\|_{C^{k,\alpha}_{\beta,\gamma-2}}$. In other words, $\Delta : C^{k+2,\alpha}_{\beta,\gamma}(X) \to C^{k,\alpha}_{\beta,\gamma-2}(X)$ is an isomorphism.

Proof Let $f \in C^{k,\alpha}_{\beta,\gamma-2}(X)$, and suppose for simplicity that $\|f\|_{C^{k,\alpha}_{\beta,\gamma-2}} \leq 1$. Then $f \in C^{0,\alpha}_{\beta,\gamma-2}(X)$, so by Theorem 9.5.6 there exists a unique $u \in C^{2,\alpha}(X)$ such that $\Delta u = f$ and $u(x) \to 0$ as $x \to \infty$ in X. Since $\Delta u = f \leq \rho^\beta \sigma^{\gamma-2} \leq \Delta F$, we see that $\Delta(u - F) \leq 0$ on X. Suppose that $u - F > 0$ at some point of X. Then $u - F$ is nonconstant and has a maximum in X, since $F > 0$ and $u(x) \to 0$ as $x \to \infty$ in X. But this contradicts the maximum principle, as $\Delta(u - F) \leq 0$. Therefore $u - F \leq 0$, and $u \leq F$. Similarly we show that $u \geq -F$, and so $|u| \leq F \leq K_2 \rho^\beta \sigma^\gamma$.

To complete the proof, we must prove that $u \in C^{k+2,\alpha}_{\beta,\gamma}(X)$ and $\|u\|_{C^{k+2,\alpha}_{\beta,\gamma}} \leq C$ for some $C > 0$ independent of f. We can do this by applying Theorem 1.4.3 to balls of radius $L\sigma(x)$ about x, for small $L \in (0, \frac{1}{2})$, using an argument very similar to that in the proof of Proposition 8.6.12. The basic idea is that on the ball $B_{2L\sigma(x)}(x)$ we know that $u = O(\rho(x)^\beta \sigma(x)^\gamma)$ from above, and $\nabla^j f = O(\rho(x)^\beta \sigma(x)^{\gamma-2-j})$ for $0 \leq j \leq k$, together with a Hölder estimate on $\nabla^k f$.

We use Theorem 1.4.3 to show that $\nabla^j u = O(\rho(x)^\beta \sigma(x)^{\gamma-j})$ on $B_{L\sigma(x)}(x)$ for $0 \leq j \leq k+2$, together with the appropriate Hölder estimate. The important point is that when we rescale distances on the ball $B_{2L\sigma(x)}(x)$ by a factor $(L\sigma(x))^{-1}$, the resulting metric is close to the Euclidean metric on the ball B_2 of radius 2 in $\mathbb{C}^m$, because of the definition of QALE metric. Therefore the Laplacian of the rescaled metric is close to the Euclidean Laplacian, Theorem 1.4.3 applies, and the proof is finished. □

Here when we say that Δ is an isomorphism between two Banach spaces of functions on X, we mean that it is an *isomorphism of topological vector spaces*. That is, Δ is a vector space isomorphism which is bounded with bounded inverse, but it does not necessarily identify the norms on the two Banach spaces. In fact (9.17) is also necessary for $\Delta : C^{k+2,\alpha}_{\beta,\gamma}(X) \to C^{k,\alpha}_{\beta,\gamma-2}(X)$ to be an isomorphism.

Proposition 9.5.8 *Let X be a QALE Kähler manifold, (ρ, σ) a pair of radius functions on X, and $\beta, \gamma < 0$. Suppose $\Delta : C^{2,\alpha}_{\beta,\gamma}(X) \to C^{0,\alpha}_{\beta,\gamma-2}(X)$ is an isomorphism. Then there exists a smooth function $F : X \to (0, \infty)$ satisfying (9.17).*

Proof From (9.12) and (9.13) we see that $\rho^\beta \sigma^{\gamma-2}$ lies in $C^1_{\beta,\gamma-2}(X)$, and hence in $C^{0,\alpha}_{\beta,\gamma-2}(X)$. So there exists a unique $F \in C^{2,\alpha}_{\beta,\gamma}(X)$ with $\Delta F = \rho^\beta \sigma^{\gamma-2}$. Clearly F is smooth, $\Delta F \geq \rho^\beta \sigma^{\gamma-2}$ and $F \leq K_2 \rho^\beta \sigma^\gamma$ for some $K_2 > 0$. Also, using (9.12) and (9.13) we can show that $\Delta(K_1 \rho^\beta \sigma^\gamma) \leq \rho^\beta \sigma^{\gamma-2}$ for some small $K_1 > 0$, and thus by the proof of the previous theorem we have $K_1 \rho^\beta \sigma^\gamma \leq F$. Thus F satisfies (9.17), as we want. □

Definition 9.5.9 Let (X, J, g) be a QALE Kähler manifold, and define $\mathcal{I}_X$ to be the set of pairs $(\beta, \gamma) \in \mathbb{R}^2$ such that $\beta < 0$, $\gamma < 0$ and $\Delta : C^{k+2,\alpha}_{\beta,\gamma}(X) \to C^{k,\alpha}_{\beta,\gamma-2}(X)$ is an isomorphism for $k \geq 0$ and $\alpha \in (0, 1)$. Theorem 9.5.7 and Proposition 9.5.8 prove that this condition is independent of k, α, and that $(\beta, \gamma) \in \mathcal{I}_X$ if and only if there exists a smooth function F on X satisfying (9.17). We can also show that $\mathcal{I}_X$ is an open set in $\mathbb{R}^2$.

Of course these ideas are of no use at all if $\mathcal{I}_X$ is empty. In the following three results we show that it is not. The proof of the next proposition is elementary, and we omit it.

Proposition 9.5.10 *Let $\mathbb{C}^{m-n}$ be a subspace of $\mathbb{C}^m$ for $n > 0$, and define $r, s : \mathbb{C}^m \to [0, \infty)$ by $r(x) = d(x, 0)$ and $s(x) = d(x, \mathbb{C}^{m-n})$. Let $\beta, \gamma \in \mathbb{R}$. Then on $\mathbb{C}^m \setminus \mathbb{C}^{m-n}$ we have*

$$\Delta(r^\beta s^\gamma) = -\tfrac{1}{2} r^{\beta-2} s^{\gamma-2} \left[\gamma(\gamma+2n-2)r^2 + \beta(2m-2+\beta+2\gamma)s^2\right]. \tag{9.18}$$

Since $0 \leq s \leq r$, there exists $C > 0$ such that $\Delta(C r^\beta s^\gamma) \geq r^\beta s^{\gamma-2}$ if and only if

$$\gamma(\gamma + 2n - 2) < 0 \text{ and } \gamma(\gamma + 2n - 2) + \beta(2m - 2 + \beta + 2\gamma) < 0. \tag{9.19}$$

The pair of inequalities (9.19) are equivalent to $2 - 2n < \gamma < 0$ and

$$|\beta + \gamma + m - 1| < \sqrt{(m-1)^2 + 2\gamma(m-n)}.$$

For these to have a solution we must have $n > 1$. Note also that

$$(m-1)^2 + 2\gamma(m-n) = (m+1-2n)^2 + 2(\gamma + 2n - 2)(m - n).$$

Thus if $2 - 2n < \gamma < 0$ the square root $\sqrt{(m-1)^2 + 2\gamma(m-n)}$ exists, and

$$|m + 1 - 2n| \leq \sqrt{(m-1)^2 + 2\gamma(m-n)} \leq m - 1.$$

Hence, any solutions β, γ to (9.19) must satisfy $2 - 2m < \beta + \gamma < 0$. Also, if $2 - 2n < \gamma < 0$ and $\beta + \gamma$ lies between $2 - 2n$ and $2(n-m)$ then β, γ satisfy (9.19). To get the factor $\tfrac{1}{2}$ in (9.18), remember that Δ on Kähler manifolds is half that on Riemannian manifolds.

Motivated by Proposition 9.5.10, we can prove:

Theorem 9.5.11 *Let X be a QALE Kähler manifold asymptotic to $\mathbb{C}^m/G$, and n the complex codimension of the singular set of $\mathbb{C}^m/G$. Suppose $\beta, \gamma \in \mathbb{R}$ satisfy (9.19). Then there exists a smooth function $F : X \to (0, \infty)$ satisfying (9.17).*

The basic idea of the proof is to model F on $\rho^\beta \sigma^\gamma$. The details are complicated, and we will not give them. But here is a sketch of the case when X is an ALE manifold. For $1 - m < \delta < 0$ we want to find a function F such that $\Delta F \geq \rho^{2\delta - 2}$ and $F \leq C \rho^{2\delta}$. On $\mathbb{C}^m / G$ we have $\Delta(r^2) = -2m$ and $|\nabla(r^2)|^2 = 4r^2$. We show that there exists a unique smooth function u on X such that $\Delta u = -2m$ and $u = \rho^2 + O(\rho^{2-2m})$ for large ρ, and $4u - |\nabla u|^2$ is bounded on X.

Choose $K \in \mathbb{R}$ such that $u + K \geq 1$ and $|\nabla u|^2 \leq 4(u + K)$ on X. Then

$$\Delta\big[(u + K)^\delta\big] = \delta(u + K)^{\delta - 1}\Delta u - \tfrac{1}{2}\delta(\delta - 1)(u + K)^{\delta - 2}|\nabla u|^2$$
$$= -2\delta(\delta + m - 1)(u + K)^{\delta - 1}$$
$$\quad + \tfrac{1}{2}\delta(\delta - 1)(u + K)^{\delta - 2}\big(4(u + K) - |\nabla u|^2\big)$$
$$\geq -2\delta(\delta + m - 1)(u + K)^{\delta - 1},$$

since $\delta(\delta - 1) > 0$ and $|\nabla u|^2 \leq 4(u + K)$. But there exist $C_1, C_2 > 0$ such that

$$-2\delta(\delta + m - 1)C_1(u + K)^{\delta - 1} \geq \rho^{2\delta - 2} \quad \text{and} \quad (u + K)^\delta \leq C_2 \rho^{2\delta}.$$

So putting $F = C_1(u + K)^\delta$ we see that $\Delta F \geq \rho^{2\delta - 2}$ and $F \leq C_1 C_2 \rho^{2\delta}$, and the proof is finished. To generalize this proof to the case that X is a QALE manifold, we use similar functions u on X and on each Y_i in the decomposition of X.

Corollary 9.5.12 *Let X be a QALE Kähler manifold asymptotic to $\mathbb{C}^m / G$, and n the complex codimension of the singular set of $\mathbb{C}^m / G$. Suppose $\beta, \gamma \in \mathbb{R}$ satisfy*

$$\beta < 0, \quad 2 - 2n < \gamma < 0, \quad |\beta + \gamma + m - 1| < \sqrt{(m-1)^2 + 2\gamma(m-n)}. \tag{9.20}$$

Then $(\beta, \gamma) \in \mathcal{I}_X$. Hence if $2 \leq n \leq m$ then $\mathcal{I}_X$ is nonempty.

To prove the last part, observe that if $2 \leq n \leq m$ then (9.20) holds for $\beta = 1 - m$ and $\gamma < 0$ small. Note that $n \geq 2$ by definition, since g is a QALE metric. If we allowed $n = 1$ then the equation $2 - 2n < \gamma < 0$ in (9.20) would have no solutions, which illustrates the problems when S has codimension one.

By definition, if $(\beta, \gamma) \in \mathcal{I}_X$ then $\beta < 0$ and $\gamma < 0$. However, (9.19) admits solutions with $\gamma < 0$ but $\beta \geq 0$. This suggests that $\Delta : C^{k+2,\alpha}_{\beta,\gamma}(X) \to C^{k,\alpha}_{\beta,\gamma-2}(X)$ can be an isomorphism when $\beta \geq 0$ as well. In fact this is the case, but we will not explore the idea because we shall need $\beta < 0$ in our applications anyway.

Using similar methods to those in the proof of Theorem 9.5.11, we can show:

Theorem 9.5.13 *Let X be a QALE Kähler manifold asymptotic to $\mathbb{C}^m / G$, and let $\gamma < 0$. Then there exists a smooth function $F : X \to (0, \infty)$ satisfying $\Delta F \geq \rho^{2-2m}\sigma^{\gamma-2}$ and $K_1 \rho^{2-2m} \leq F \leq K_2 \rho^{2-2m}$ for some $K_1, K_2 > 0$.*

From Theorem 9.5.13 and the proof of Theorem 9.5.7 we deduce

Corollary 9.5.14 *Let (X, J, g) be a QALE Kähler manifold of dimension m, and let $\gamma < 0$. Then whenever $k \geq 0$ and $\alpha \in (0, 1)$, there exists $C > 0$ such that for each $f \in C^{k,\alpha}_{2-2m, \gamma - 2}(X)$ there is a unique $u \in C^{k+2,\alpha}_{2-2m, 0}(X)$ with $\Delta u = f$, which satisfies $\|u\|_{C^{k+2,\alpha}_{2-2m,0}} \leq C \|f\|_{C^{k,\alpha}_{2-2m,\gamma-2}}$.*

The following result will be useful later.

Proposition 9.5.15 *Let (X, J, g) be a QALE Kähler manifold. If $(\beta, \gamma) \in \mathcal{I}_X$ then $(t\beta, t\gamma) \in \mathcal{I}_X$ for all $t \in (0, 1)$.*

Proof As $(\beta, \gamma) \in \mathcal{I}_X$ there is a smooth $F : X \to (0, \infty)$ satisfying (9.17). Let $t \in (0, 1)$. Then

$$\Delta(F^t) = tF^{t-1}\Delta F + \tfrac{1}{2}t(1-t)F^{t-2}|\nabla F|^2 \geq t\,K_2^{t-1}\rho^{t\beta}\sigma^{t\gamma-2},$$

using the inequalities $F \leq K_2 \rho^\beta \sigma^\gamma$, $\Delta F \geq \rho^\beta \sigma^{\gamma-2}$ and $t(1-t) > 0$. Hence, putting

$$F' = t^{-1}K_2^{1-t}F^t, \quad K_1' = t^{-1}K_1^t K_2^{1-t} \quad \text{and} \quad K_2' = t^{-1}K_2,$$

calculation shows that

$$\Delta F' \geq \rho^{t\beta}\sigma^{t\gamma-2} \quad \text{and} \quad K_1' \rho^{t\beta}\sigma^{t\gamma} \leq F' \leq K_2' \rho^{t\beta}\sigma^{t\gamma},$$

and so $(t\beta, t\gamma) \in \mathcal{I}_X$ by Theorem 9.5.7. □

The author believes that the following results are true.

Conjecture 9.5.16 *Let (X, J, g) be a QALE Kähler manifold asymptotic to $\mathbb{C}^m/G$, and let n be the complex codimension of the singular set of $\mathbb{C}^m/G$. Then*

$$\mathcal{I}_X = \{(\beta, \gamma) \in \mathbb{R}^2 : \beta < 0,\ 2 - 2n < \gamma < 0,\ \beta + \gamma > 2 - 2m\}. \tag{9.21}$$

Also, for generic $\beta, \gamma \in \mathbb{R}$ the map $\Delta : C^{k+2,\alpha}_{\beta,\gamma}(X) \to C^{k,\alpha}_{\beta,\gamma-2}(X)$ is Fredholm, with finite-dimensional kernel and cokernel.

9.6 The Calabi conjecture for QALE manifolds

We will now state and prove the following version of the Calabi conjecture for QALE manifolds, which is based on the Calabi conjecture for ALE manifolds given in §8.5.

The Calabi conjecture for QALE manifolds (first version) *Let (X, J, g) be a QALE Kähler manifold of dimension m, with Kähler form ω. Then*

(a) *Suppose $(\beta, \gamma) \in \mathcal{I}_X$, as in Definition 9.5.9. Then for each $f \in C^\infty_{\beta,\gamma-2}(X)$ there is a unique $\phi \in C^\infty_{\beta,\gamma}(X)$ such that $\omega + dd^c\phi$ is a positive $(1,1)$-form and $(\omega + dd^c\phi)^m = e^f \omega^m$ on X.*

(b) *Suppose $\gamma < 0$. Then for each $f \in C^\infty_{2-2m,\gamma-2}(X)$ there is a unique $\phi \in C^\infty_{2-2m,0}(X)$ such that $\omega + dd^c\phi$ is a positive $(1,1)$-form and $(\omega + dd^c\phi)^m = e^f \omega^m$ on X.*

Note that the assumption that $n \geq 2$ in Definition 9.2.1 forces $m \geq 2$. But if $m = 2$ then the singularities of $\mathbb{C}^2/G$ are isolated and X is an ALE manifold, which we have already dealt with in Chapter 8. Thus we may assume $m \geq 3$, and the first case in which the Calabi conjecture for QALE manifolds gives us new results is when $m = 3$ and $n = 2$.

Our goal is to prove:

Theorem 9.6.1 *The first version of the Calabi conjecture for QALE manifolds is true.*

Now the proof of this theorem follows the proof in Chapter 8 of the Calabi conjecture for ALE manifolds very closely. So to save space we will just indicate the important changes that must be made to convert the results and proofs in §8.5–§8.7 to the QALE case, and leave the reader to sort out the details.

9.6.1 Changes to the material in §8.5

Theorems A1–A4 should be replaced by the following four theorems:

Theorem Q1 *Let (X, J, g) be a QALE Kähler manifold of dimension m, with Kähler form ω. Let $Q_1 > 0$, $\alpha \in (0, 1)$ and $\beta, \gamma, \zeta < 0$. Then there exists $Q_2 > 0$ such that if $f \in C^{3,\alpha}_{\beta,\gamma-2}(X)$ and $\phi \in C^{5}_{\zeta,0}(X)$ satisfy $(\omega + \mathrm{dd}^c \phi)^m = \mathrm{e}^f \omega^m$ and $\|f\|_{C^{3,\alpha}_{\beta,\gamma-2}} \leqslant Q_1$, then $\omega + \mathrm{dd}^c \phi$ is a positive $(1,1)$-form, $\phi \in C^{5,\alpha}(X)$ and $\|\phi\|_{C^{5,\alpha}} \leqslant Q_2$. Also, if $f \in C^{k,\alpha}_{\beta,\gamma-2}(X)$ for $k \geqslant 3$ then $\phi \in C^{k+2,\alpha}(X)$.*

Theorem Q2 *Let (X, J, g) be a QALE Kähler manifold of dimension m, with Kähler form ω. Let $Q_1 > 0$, $Q_2 > 0$, $\alpha \in (0, 1)$ and $\beta, \gamma, \zeta < 0$. Then there exists $Q_3 > 0$ such that the following holds.*

Suppose $f \in C^{3,\alpha}_{\beta,\gamma-2}(X)$ and $\phi \in C^{5}_{\zeta,0}(X) \cap C^{5,\alpha}(X)$ satisfy $(\omega + \mathrm{dd}^c \phi)^m = \mathrm{e}^f \omega^m$ and the inequalities $\|f\|_{C^{3,\alpha}_{\beta,\gamma-2}} \leqslant Q_1$ and $\|\phi\|_{C^{5,\alpha}} \leqslant Q_2$. Then

(a) *Let $(\beta, \gamma) \in \mathcal{I}_X$. Then $\phi \in C^{5,\alpha}_{\beta,\gamma}(X)$ and $\|\phi\|_{C^{5,\alpha}_{\beta,\gamma}} \leqslant Q_3$.*
  If $f \in C^{k,\alpha}_{\beta,\gamma-2}(X)$ for $k \geqslant 3$ then $\phi \in C^{k+2,\alpha}_{\beta,\gamma}(X)$.
(b) *Let $\beta = 2 - 2m$. Then $\phi \in C^{5,\alpha}_{2-2m,0}(X)$ and $\|\phi\|_{C^{5,\alpha}_{2-2m,0}} \leqslant Q_3$.*
  If $f \in C^{k,\alpha}_{2-2m,\gamma-2}(X)$ for $k \geqslant 3$ then $\phi \in C^{k+2,\alpha}_{2-2m,0}(X)$.

Theorem Q3 *Let (X, J, g) be a QALE Kähler manifold of dimension m, with Kähler form ω. Let $\alpha \in (0, 1)$ and $(\beta, \gamma) \in \mathcal{I}_X$. Suppose that $f' \in C^{3,\alpha}_{\beta,\gamma-2}(X)$ and $\phi' \in C^{5,\alpha}_{\beta,\gamma}(X)$ satisfy $(\omega + \mathrm{dd}^c \phi')^m = \mathrm{e}^{f'} \omega^m$. Then whenever $f \in C^{3,\alpha}_{\beta,\gamma-2}(X)$ and $\|f - f'\|_{C^{3,\alpha}_{\beta,\gamma-2}}$ is sufficiently small, there exists $\phi \in C^{5,\alpha}_{\beta,\gamma}(X)$ such that $(\omega + \mathrm{dd}^c \phi)^m = \mathrm{e}^f \omega^m$.*

Theorem Q4 *Let (X, J, g) be a QALE Kähler manifold of dimension m, with Kähler form ω, and let $\beta, \gamma, \zeta < 0$. Then for each $f \in C^1_{\beta,\gamma-2}(X)$ there is at most one function $\phi \in C^3_{\zeta,0}(X)$ such that $(\omega + \mathrm{dd}^c \phi)^m = \mathrm{e}^f \omega^m$ on X.*

In the proof of Theorem 8.5.1, instead of picking γ with $\beta + 2 < \gamma < 0$ we choose ζ with $\beta < \zeta < 0$, and then in the proof of Proposition 8.5.2 we use the fact (following from [57, Lem. 3, p. 11]) that the inclusion $C^{5,\alpha}_{\beta,\gamma}(X) \to C^{5,0}_{\zeta,0}(X)$ is compact. The rest of the material of §8.5 generalizes in the obvious way.

9.6.2 Changes to the material in §8.6

In §8.6.1 the only significant change is to note that Lemma 8.6.4 also applies to QALE manifolds. It is somewhat more difficult to prove in the QALE case, but it can be done. In §8.6.2 we define ρ as in Definition 9.5.1. The only important property of ρ used in §8.6.2 is the first part of (8.23), which holds in the QALE case because $\nabla \rho$ and $\nabla^2 \rho$ are bounded on X by (9.12). So following the proofs we see that if $\beta < \delta$, $\zeta < \delta$ and $\delta \geqslant 1 - m$ then $|\phi| \leqslant P_2 \rho^\delta$, and hence that $\phi \in C^0_{\delta,0}(X)$ and $\|\phi\|_{C^0_{\delta,0}} \leqslant P_2$.

Section 8.6.3 needs a lot of small changes because the rôle of ρ in ALE manifolds is taken by ρ and σ in QALE manifolds, which leads to an abundance of indices. Here are new versions of Theorem 8.6.11 and Proposition 8.6.12:

Theorem 9.6.2 *Let (X, J, g) be a QALE Kähler manifold of dimension m, with Kähler form ω. Let $Q_1, Q_2, P_2 > 0$, $\alpha \in (0, 1)$, $\beta, \gamma, \delta < 0$ and $\epsilon \leqslant 0$ satisfy $\beta \leqslant \delta$ and $\gamma \leqslant \epsilon$. Then there exists $P_3 > 0$ such that if $f \in C^{3,\alpha}_{\beta,\gamma-2}(X)$ and $\phi \in C^0_{\delta,\epsilon}(X) \cap C^{5,\alpha}(X)$ satisfy $\|f\|_{C^{3,\alpha}_{\beta,\gamma-2}} \leqslant Q_1$, $\|\phi\|_{C^{5,\alpha}} \leqslant Q_2$, $\|\phi\|_{C^0_{\delta,\epsilon}} \leqslant P_2$ and $(\omega + dd^c\phi)^m = e^f \omega^m$, then $\phi \in C^{5,\alpha}_{\delta,\epsilon}(X)$ and $\|\phi\|_{C^{5,\alpha}_{\delta,\epsilon}} \leqslant P_3$. Furthermore, if $f \in C^{k,\alpha}_{\beta,\gamma-2}(X)$ and $\phi \in C^{k+2,\alpha}(X)$ for some integer $k \geqslant 3$, then $\phi \in C^{k+2,\alpha}_{\delta,\epsilon}(X)$.*

Proposition 9.6.3 *Let $K_1, K_2 > 0$, $\lambda \in [0, 1]$ and $k \geqslant 3$ be an integer. Then there exists $K_3 > 0$ depending on (X, J, g), $\alpha, \beta, \gamma, \delta, \epsilon, P_2$ and K_1, K_2, λ, k, such that the following is true.*

Suppose the hypotheses of Theorem 9.6.2 hold, and also that $\|f\|_{C^{k,\alpha}_{\beta,\gamma-2}} \leqslant K_1$ and $\|\nabla^j dd^c\phi\|_{C^0_{0,-\lambda j}} \leqslant K_2$ for $j = 0, \ldots, k$, and $\left[\nabla^k dd^c\phi\right]_{\alpha,0,-\lambda(k+\alpha)} \leqslant K_2$, where all these norms exist. Then $\|\nabla^j \phi\|_{C^0_{\delta,\epsilon-\lambda j}} \leqslant K_3$ for $j = 0, \ldots, k+2$, and $\left[\nabla^{k+2}\phi\right]_{\alpha,\delta,\epsilon-\lambda(k+2+\alpha)} \leqslant K_3$, where each norm exists.

To prove Proposition 9.6.3 we follow the proof of Proposition 8.6.12, but instead of choosing $R = L\rho(x)^\lambda$ we take $R = L\sigma(x)^\lambda$. The definition of QALE metrics then ensures that (8.32)–(8.35) hold for small L, and the remainder of the proofs of Proposition 8.6.12 and Theorem 8.6.11 generalize in a straightforward way. There is no need to extend Theorem 8.6.13 to the QALE case.

9.6.3 Changes to the material in §8.7

The proofs of Theorems A1–A4 must be modified to give proofs of Theorems Q1–Q4 above. The proof of Theorem A1 needs no significant changes, except to note that Lemmas 5.5.1–5.5.3 also apply in the QALE case.

The proof of Theorem Q2 needs more work. Lemma 8.7.1 holds in the QALE case with no changes. Here is a new version of Proposition 8.7.2:

Proposition 9.6.4 *Let $K > 0$, $\beta \leqslant \delta < 0$ and $\gamma \leqslant \epsilon \leqslant 0$, and suppose $\delta', \epsilon' < 0$ satisfy*

$$\beta \leqslant \delta', \quad 2\delta \leqslant \delta', \quad \beta + \gamma \leqslant \delta' + \epsilon', \quad \text{and } 2\delta + 2\epsilon - 2 \leqslant \delta' + \epsilon'. \tag{9.22}$$

Then there exists $K' > 0$ such that

(a) *if $(\delta', \epsilon') \in \mathcal{I}_X$ and $|\phi| \leqslant K\rho^\delta\sigma^\epsilon$ then $|\phi| \leqslant K'\rho^{\delta'}\sigma^{\epsilon'}$,*
(b) *if $\delta' = 2 - 2m$ and $|\phi| \leqslant K\rho^\delta\sigma^\epsilon$ then $|\phi| \leqslant K'\rho^{2-2m}$.*

Proof As $\|\phi\|_{C^0_{\delta,\epsilon}} \leqslant K$, Theorem 9.6.2 shows that $\phi \in C^{5,\alpha}_{\delta,\epsilon}(X)$ and $\|\phi\|_{C^{5,\alpha}_{\delta,\epsilon}} \leqslant P_3$ for some $P_3 > 0$ independent of ϕ, f. Thus $|dd^c\phi| \leqslant P_3\rho^\delta\sigma^{\epsilon-2}$ and $|\nabla dd^c\phi| \leqslant P_3\rho^\delta\sigma^{\epsilon-3}$ on X. But $|f| \leqslant Q_1\rho^\beta\sigma^{\gamma-2}$ and $|\nabla f| \leqslant Q_1\rho^\beta\sigma^{\gamma-3}$, so that $|1 - e^f| \leqslant e^{Q_1}\rho^\beta\sigma^{\gamma-2}$ and $|e^f \nabla f| \leqslant Q_1 e^{Q_1}\rho^\beta\sigma^{\gamma-3}$. Hence by Lemma 8.7.1 we see that

$$|\Delta\phi| \leq D_1 P_3^2 \rho^{2\delta} \sigma^{2\epsilon-4} + e^{Q_1} \rho^\beta \sigma^{\gamma-2}$$
and $$|\nabla(\Delta\phi)| \leq D_1 P_3^2 \rho^{2\delta} \sigma^{2\epsilon-5} + Q_1 e^{Q_1} \rho^\beta \sigma^{\gamma-3}.$$
(9.23)

From (9.22) and (9.23) we find that $\Delta\phi \in C^{0,\alpha}_{\delta',\epsilon'-2}(X)$ and $\|\Delta\phi\|_{C^{0,\alpha}_{\delta',\epsilon'-2}} \leq D_2$ for some $D_2 > 0$ depending on D_1, P_3 and Q_1.

To prove part (a), since $(\delta', \epsilon') \in \mathcal{I}_X$ we know $\Delta : C^{2,\alpha}_{\delta',\epsilon'}(X) \to C^{0,\alpha}_{\delta',\epsilon'-2}(X)$ is an isomorphism, and so $\phi \in C^{2,\alpha}_{\delta',\epsilon'}(X)$ with $\|\phi\|_{C^{2,\alpha}_{\delta',\epsilon'}} \leq K'$, where $K' = CD_2$ and C is a bound for $\Delta^{-1} : C^{0,\alpha}_{\delta',\epsilon'-2}(X) \to C^{2,\alpha}_{\delta',\epsilon'}(X)$. Thus $|\phi| \leq K' \rho^{\delta'} \sigma^{\epsilon'}$, as we want. To prove part (b) we use Corollary 9.5.14 in a similar way. $\square$

Now we shall prove Theorem Q2. We begin with part (a), so that $(\beta, \gamma) \in \mathcal{I}_X$. Pick some δ satisfying $\delta < 0$, $\beta < \delta$, $\zeta < \delta$ and $\delta \geq 1 - m$. Then by the QALE analogue of Theorem 8.6.6, there exists $P_2 > 0$ such that whenever ϕ and f satisfy the hypotheses of Theorem Q2, we have $|\phi| \leq P_2 \rho^\delta$.

Choose some integer $k \geq 1$ such that $2^{-k} \leq \delta/\beta$. Then we show by induction that there exist constants $K_1, \ldots, K_k$ such that $|\phi| \leq K_j \rho^{2^{j-k}\beta} \sigma^{2^{j-k}\gamma}$ for $j = 1, \ldots, k$. When $j = 1$, put $\epsilon = 0$, $\delta' = 2^{1-k}\beta$ and $\epsilon' = 2^{1-k}\gamma$. Then δ', ϵ' satisfy (9.22). Moreover $(\delta', \epsilon') \in \mathcal{I}_X$ by Proposition 9.5.15, since $(\beta, \gamma) \in \mathcal{I}_X$ and $2^{1-k} \in (0, 1]$. So applying part (a) of Proposition 9.6.4 we see that $|\phi| \leq K_1 \rho^{2^{1-k}\beta} \sigma^{2^{1-k}\gamma}$ for some $K_1 > 0$ independent of ϕ, f.

For the inductive step, suppose that $|\phi| \leq K_j \rho^{2^{j-k}\beta} \sigma^{2^{j-k}\gamma}$. Set $\delta = 2^{j-k}\beta$, $\epsilon = 2^{j-k}\gamma$, $\delta' = 2^{j+1-k}\beta$ and $\epsilon' = 2^{j+1-k}\gamma$. If $j + 1 \leq k$ then $(\delta', \epsilon') \in \mathcal{I}_X$ by Proposition 9.5.15, so part (a) of Proposition 9.6.4 shows that there is a constant K_{j+1} independent of ϕ, f, such that $|\phi| \leq K_{j+1} \rho^{2^{j+1-k}\beta} \sigma^{2^{j+1-k}\gamma}$. Therefore by induction the result holds, and putting $j = k$ we find that $|\phi| \leq K_k \rho^\beta \sigma^\gamma$. Applying Theorem 9.6.2 with $\delta = \beta$ and $\epsilon = \gamma$ proves part (a) of Theorem Q2.

Observe that if part (b) holds for some $\gamma < 0$, then it also holds for any $\gamma' \leq \gamma$. So we are free to increase γ if we want. If $\gamma < 0$ is sufficiently small then $(1 - m, \gamma/2) \in \mathcal{I}_X$ by Corollary 9.5.12. In this case the induction argument above shows that $|\phi| \leq K_k \rho^{1-m} \sigma^{\gamma/2}$ for some $K_k > 0$. Applying part (b) of Proposition 9.6.4 with $\delta = 1 - m$, $\epsilon = \gamma/2$, $\delta' = 2 - 2m$ and $\epsilon' = \gamma$ then gives $|\phi| \leq K' \rho^{2-2m}$. Finally, applying Theorem 9.6.2 with $\delta = 2 - 2m$ and $\epsilon = 0$ proves part (b) of Theorem Q2.

As in the ALE case, the proofs of Theorems Q3 and Q4 are simple generalizations of the proofs of Theorem C3 in §5.6 and C4 in §5.7, and we leave them to the reader as an exercise. The main point to note in the proof of Theorem Q3 is that $\Delta : C^{5,\alpha}_{\beta,\gamma}(X) \to C^{3,\alpha}_{\beta,\gamma-2}(X)$ is an isomorphism since $(\beta, \gamma) \in \mathcal{I}_X$, and we use this when we apply the Inverse Mapping Theorem, Theorem 1.2.4. This completes our explanation of the proof of Theorem 9.6.1.

9.7 A more complicated QALE Calabi conjecture

So far we have estimated the decay of functions ϕ and metrics g on X in two different ways. In §9.2–§9.4 we pulled g and ϕ back to $V_i \times Y_i$ and bounded them using powers of functions $\mu_{i,j}$ and ν_i on $V_i \times Y_i$. In §9.5 and §9.6 we estimated ϕ in terms of powers of functions ρ, σ on X. There are good reasons for using these two approaches in the way we have — attempting to define QALE metrics directly on X, and trying to solve the Calabi conjecture by pulling back to $V_i \times Y_i$, both seem to lead to disaster.

However, in order to construct QALE Kähler metrics with prescribed Ricci curvature on X we need to integrate these two approaches, because in solving the Calabi conjecture on X we want ϕ to be a function of *Kähler potential type*, as in Definition 9.4.1. Here is a version of the Calabi conjecture for QALE manifolds which achieves this.

The Calabi conjecture for QALE manifolds (second version) *Let (X, J, g) be a QALE Kähler manifold asymptotic to $\mathbb{C}^m/G$ with $\mathrm{Fix}(G) = \{0\}$, let ω be the Kähler form and ξ the Ricci form of g, and let $\epsilon < -2$. Suppose f is a smooth function on X such that*

$$\nabla^l \psi_i^*(f) = \sum_{j \in I : i \not\leq j} O\bigl(\mu_{i,j}^{d_j} \nu_i^{\epsilon - l}\bigr) \tag{9.24}$$

on $V_i \times Y_i \setminus U_i$, for all $i \in I \setminus \{\infty\}$ and $l \geq 0$. Then there is a unique smooth function ϕ on X such that $\omega' = \omega + \mathrm{dd}^c \phi$ is a positive $(1, 1)$-form and $(\omega')^m = \mathrm{e}^f \omega^m$ on X, and

$$\nabla^l \psi_i^*(\phi) = \sum_{j \in I : i \not\leq j} O\bigl(\mu_{i,j}^{d_j} \nu_i^{-l}\bigr) \tag{9.25}$$

on $V_i \times Y_i \setminus U_i$, for all $i \in I \setminus \{\infty\}$ and $l \geq 0$. Furthermore, ϕ is of Kähler potential type on X, and the Kähler metric g' on X with Kähler form ω' is QALE and has Ricci form $\xi' = \xi - \frac{1}{2} \mathrm{dd}^c f$.

The point of assuming that $\mathrm{Fix}(G) = V_\infty = \{0\}$ here is that taken over all $i \in I \setminus \{\infty\}$, eqns (9.24) and (9.25) prescribe the asymptotic behaviour of f and ϕ everywhere on X, except within a fixed distance of $\pi^{-1}(V_\infty)$. If $\mathrm{Fix}(G) = \{0\}$ this is all of X except a compact subset. But if $\mathrm{Fix}(G) \neq \{0\}$ then $\pi^{-1}(V_\infty)$ extends to infinity in X, and (9.24) and (9.25) only prescribe the behaviour on a part of the 'boundary' of X, so they are not good boundary conditions.

Note that the last sentence of this conjecture follows immediately from the existence of ϕ. For (9.25) implies that ϕ satisfies (9.7) for all $i \neq \infty$ in I, with $\phi_i = 0$. And as $\mathrm{Fix}(G) = \{0\}$ we can put $\phi_\infty = \phi$, and (9.7) holds for $i = \infty$. Thus ϕ is of Kähler potential type by definition. Proposition 9.4.2 then shows that g' is a QALE Kähler metric, since ω' is positive, and the equation $\xi' = \xi - \frac{1}{2} \mathrm{dd}^c f$ follows from §5.1.

We begin by defining some functions ρ_i on X, following Definition 9.5.1.

Definition 9.7.1 Let (X, J, g) be a QALE Kähler manifold asymptotic to $\mathbb{C}^m/G$. For each $i \in I$ define $r_i : \mathbb{C}^m/G \to [0, \infty)$ by $r_i(x) = d(x, V_i G/G)$, where $d(\,,\,)$ is the distance on $\mathbb{C}^m/G$ and $V_i G/G$ is the piece of the singular set of $\mathbb{C}^m/G$ coming from V_i. Let $\rho_i : X \to [1, \infty)$ be a smooth function such that $\sigma \leq \rho_i \leq \rho$ for $i \neq 0$ and

$$\tfrac{1}{2}\pi^*(r_i) + 1 \leq \rho_i \leq \pi^*(r_i) + 2, \quad |\nabla \rho_i| \leq K, \quad \text{and} \quad |\nabla^2 \rho_i| \leq K \rho_i^{-1} \tag{9.26}$$

for some $K > 0$. Such functions ρ_i exist for all $i \in I$. As $V_0 = \mathbb{C}^m$ and $r_0 \equiv 0$ we can choose $\rho_0 \equiv 1$, and if $\mathrm{Fix}(G) = \{0\}$ we can choose $\rho_\infty = \rho$.

Here the singular set S of $\mathbb{C}^m/G$ is the union of a number of pieces $V_i G/G$ for $i \neq 0$ in I, and the functions ρ_i are a rough measure of the distance in X to the subset $\pi^{-1}(V_i/G)$. They complement ρ and σ, which measure the distance in X to $\pi^{-1}(0)$ and $\pi^{-1}(S)$ respectively.

We shall prove the conjecture by rewriting (9.24) and (9.25) in terms of the functions ρ_i and σ on X, and then adapting the ideas of §9.5 and §9.6. It turns out that things are simpler when $N(V_i) = G$ for all $i \in I$, and so we will describe this case first.

9.7.1 Proof of the conjecture when $N(V_i) = G$ for all $i \in I$

Suppose (X, J, g) is a QALE Kähler manifold asymptotic to $\mathbb{C}^m/G$, and $N(V_i) = G$ for all $i \in I$. Then we can prove using Definitions 9.2.1, 9.5.1 and 9.7.1 that on $V_i \times Y_i \setminus U_i$ we have

$$\tfrac{1}{2}\mu_{i,j} + 1 \leqslant \psi_i^*(\rho_j) \leqslant \mu_{i,j} + 2 \quad \text{and} \quad \tfrac{1}{2}\nu_i + \tfrac{1}{2} \leqslant \psi_i^*(\sigma) \leqslant \nu_i + 1.$$

Thus $\mu_{i,j} \approx \psi_i^*(\rho_j)$ and $\nu_i \approx \psi_i^*(\sigma)$ on $V_i \times Y_i \setminus U_i$. So using ψ_i to compare estimates on X and on $V_i \times Y_i \setminus U_i$, we show:

Proposition 9.7.2 *Let (X, J, g) be a QALE Kähler manifold asymptotic to $\mathbb{C}^m/G$, where $\mathrm{Fix}(G) = \{0\}$ and $N(V_i) = G$ for all $i \in I$. Then (9.24) and (9.25) are equivalent to*

$$\nabla^l f = \sum_{j \in I : i \not\leq j} O(\rho_j^{d_j} \sigma^{\epsilon-l}) \quad \text{and} \quad \nabla^l \phi = \sum_{j \in I : i \not\leq j} O(\rho_j^{d_j} \sigma^{-l}) \quad (9.27)$$

for $i \in I \setminus \{\infty\}$ and $l \geqslant 0$.

Actually (9.24) and (9.25) do not prescribe the behaviour of $\psi_i^*(f)$ and $\psi_i^*(\phi)$ on U_i, but (9.27) requires that $\nabla^l f$ and $\nabla^l \phi$ should be $O(1)$ on the corresponding region of X. However, as $\mathrm{Fix}(G) = \{0\}$, taken over all $i \in I \setminus \{\infty\}$ equations (9.24) and (9.25) prescribe the behaviour of f and ϕ on all but a compact subset of X, and one can use this to show that the two pairs of equations really are equivalent.

The following result may be proved by the same method as Theorems 9.5.11 and 9.5.13.

Proposition 9.7.3 *Let (X, J, g) be a QALE Kähler manifold, let $j \neq 0$ lie in I, let $d_j \leqslant \beta_j < 0$ and $\epsilon < -2$. Then there exists a smooth function $F_j : X \to (0, \infty)$ such that*

$$\Delta F_j \geqslant \rho_j^{\beta_j} \sigma^\epsilon \quad \text{and} \quad F_j \leqslant K \rho_j^{\beta_j} \quad \text{for some } K > 0.$$

Using this we can show:

Proposition 9.7.4 *Suppose (X, J, g) is a QALE Kähler manifold asymptotic to $\mathbb{C}^m/G$ with $\mathrm{Fix}(G) = \{0\}$, and let $\epsilon < -2$. Then for each smooth function f on X satisfying*

$$\nabla^l f = \sum_{j \in I : i \not\leq j} O(\rho_j^{d_j} \sigma^{\epsilon-l}) \quad \text{for all } l \geqslant 0 \text{ and } i \in I \setminus \{\infty\}, \quad (9.28)$$

there exists a unique smooth function u on X satisfying $\Delta u = f$ and

$$\nabla^l u = \sum_{j \in I : i \not\trianglelefteq j} O(\rho_j^{d_j} \sigma^{-l}) \quad \text{for all } l \geqslant 0 \text{ and } i \in I \setminus \{\infty\}. \tag{9.29}$$

Proof Since Fix$(G) = \{0\}$ we see that (9.28) implies that $f \in C^\infty_{-2,\epsilon}(X)$. Thus by Theorem 9.5.6 there is a unique $u \in C^{2,\alpha}(X)$ with $\Delta u = f$ and $u(x) \to 0$ as $x \to \infty$ in X. Let F_j be the function on X given by Proposition 9.7.3, with $\beta_j = d_j$. Then for each $i \in I \setminus \{\infty\}$ we have

$$|f| \leqslant C \sum_{j \in I : i \not\trianglelefteq j} \rho_j^{d_j} \sigma^\epsilon \leqslant C \sum_{j \in I : i \not\trianglelefteq j} \Delta F_j.$$

Hence

$$|u| \leqslant C \sum_{j \in I : i \not\trianglelefteq j} F_j \leqslant CK \sum_{j \in I : i \not\trianglelefteq j} \rho_j^{d_j},$$

by the argument of Theorem 9.5.7. This proves (9.29) when $l = 0$. The case $l > 0$ also follows using the method of Theorem 9.5.7. $\square$

Since the Calabi conjecture is an existence problem for a nonlinear p.d.e. whose linearization is $\Delta u = f$, we can regard Proposition 9.7.4 as a linearized version of the Calabi conjecture for certain classes of functions u, f. Here is the corresponding, fully nonlinear existence result.

Theorem 9.7.5 *Suppose (X, J, g) is a QALE Kähler manifold asymptotic to $\mathbb{C}^m/G$ with Fix$(G) = \{0\}$, and let $\epsilon < -2$. Then for each smooth function f on X satisfying*

$$\nabla^l f = \sum_{j \in I : i \not\trianglelefteq j} O(\rho_j^{d_j} \sigma^{\epsilon-l}) \quad \text{for all } l \geqslant 0 \text{ and } i \in I \setminus \{\infty\},$$

there exists a unique smooth function ϕ on X such that $\omega + \mathrm{dd}^c \phi$ is a positive $(1,1)$-form and $(\omega + \mathrm{dd}^c \phi)^m = e^f \omega^m$ on X, and

$$\nabla^l \phi = \sum_{j \in I : i \not\trianglelefteq j} O(\rho_j^{d_j} \sigma^{-l}) \quad \text{for all } l \geqslant 0 \text{ and } i \in I \setminus \{\infty\}.$$

We can prove this by generalizing the material of §9.6, using the ideas of Proposition 9.7.4 as a guide. Proposition 9.7.2 and Theorem 9.7.5 give:

Corollary 9.7.6 *Suppose (X, J, g) is a QALE Kähler manifold asymptotic to $\mathbb{C}^m/G$, with Fix$(G) = \{0\}$ and $N(V_i) = G$ for all $i \in I$. Then the second version of the Calabi conjecture for QALE manifolds holds for X.*

Proposition 9.7.4 and Theorem 9.7.5 illustrate the general principle that if we can solve the equation $\Delta u = f$ uniquely on X for u, f smooth functions which decay in prescribed ways, then we expect the Calabi conjecture to hold under the same decay conditions on ϕ, f.

9.7.2 Proof of the conjecture in the general case

Suppose (X, J, g) is a QALE Kähler manifold asymptotic to $\mathbb{C}^m/G$, but that $N(V_j) \neq G$ for some $j \in I$. We can think of ρ_j as the distance to $V_j G/G$ in $\mathbb{C}^m/G$, and of $\mu_{i,j}$ as the distance to $V_j A_i/A_i$ in $\mathbb{C}^m/A_i$. The problem is that if $N(V_j) \neq G$, then the pull-back of $V_j G/G$ to $\mathbb{C}^m/A_i$ is $V_j G/A_i$, which need not coincide with $V_j A_i/A_i$. Because of this, it is no longer true that $\mu_{i,j} \approx \psi_i^*(\rho_j)$ on $V_i \times Y_i \setminus U_i$, although it is true that $\nu_i \approx \psi_i^*(\sigma)$.

Instead, for generic $x \in X$ there are $N = |G|/|A_i|$ points $y_1, \ldots, y_N$ in $V_i \times Y_i \setminus U_i$ with $\psi_i(y_k) = x$. Now $\psi_i^*(\rho_j)(y_k) = \rho_j(x)$ is the same for all k, but $\mu_{i,j}(y_k)$ can take very different values for different k. The relationship between ρ_j and $\mu_{i,j}$ is that $\rho_j(x) \approx \min_k \mu_{i,j}(y_k)$. However, if ϕ is a function on X and $|\psi_i^*(\phi)| \leqslant \mu_{i,j}^{d_j}$ then $|\phi(x)| \leqslant \mu_{i,j}(y_k)^{d_j}$ for all k, and so $|\phi(x)| \leqslant \left(\max_k \mu_{i,j}(y_k)\right)^{d_j}$, as $d_j < 0$.

Thus, since $\max_k \mu_{i,j}(y_k)$ may be significantly larger than $\min_k \mu_{i,j}(y_k)$, the statement $\psi_i^*(\phi) = O(\mu_{i,j}^{d_j})$ is rather stronger than the statement $\phi = O(\rho_j^{d_j})$. So in order to rewrite (9.24) and (9.25) in terms of ρ_i and σ we must take a different approach. We do this in our next result, whose proof we leave as a (difficult) exercise.

Theorem 9.7.7 *Suppose (X, J, g) is a QALE Kähler manifold asymptotic to $\mathbb{C}^m/G$, with $\text{Fix}(G) = \{0\}$. For $i \in I \setminus \{\infty\}$ let $I^i = \{j \in I : i \not\succeq g \cdot j \text{ for all } g \in G\}$, and for $j \in I^i$ define*

$$e_{i,j} = \min_{g \in G}\left(\max\{d_k : k \in I, \, g \cdot j \succeq k, \, i \not\succeq k\}\right). \tag{9.30}$$

Then (9.24) and (9.25) are equivalent to

$$\nabla^l f = \sum_{j \in I^i} O\left(\rho_j^{e_{i,j}} \sigma^{\epsilon - l}\right) \quad \text{and} \quad \nabla^l \phi = \sum_{j \in I^i} O\left(\rho_j^{e_{i,j}} \sigma^{-l}\right) \tag{9.31}$$

for $i \in I \setminus \{\infty\}$ and $l \geqslant 0$.

Observe in (9.30) that $k = g \cdot j$ satisfies $g \cdot j \succeq k$ and $i \not\succeq k$, and then $d_k = d_j$. Thus $\max\{d_k : k \in I, g \cdot j \succeq k, i \not\succeq k\}$ exists and is at least d_j, and hence $e_{i,j}$ is well-defined and $d_j \leqslant e_{i,j} < 0$. Therefore Proposition 9.7.3 applies with $\beta_j = e_{i,j}$. Following the method of §9.7.1 and making the obvious modifications, we soon prove:

Theorem 9.7.8 *The second version of the Calabi conjecture for QALE manifolds is true.*

9.8 The proofs of Theorems 9.3.3 and 9.3.4

Before proving the theorems, we give two preliminary propositions.

Proposition 9.8.1 *Suppose (X, J, g) is a Ricci-flat QALE Kähler manifold. Then g is the only Ricci-flat QALE Kähler metric in its Kähler class.*

Proof Suppose for a contradiction that g, g' are distinct Ricci-flat QALE Kähler metrics on X in the same Kähler class, and let X be of the smallest dimension in which this can happen. Clearly Fix$(G) = \{0\}$, since otherwise we can replace X by Y_∞, which has smaller dimension. Let g, g' be asymptotic to metrics g_i, g'_i on Y_i for $i \in I$, as in Definition 9.2.1. Then g_i, g'_i are Ricci-flat, and in the same Kähler class. Thus $g_i = g'_i$ for all $i \neq \infty$ in I, since dim $Y_i <$ dim X when $i \neq \infty$.

Let ω, ω' be the Kähler forms of g, g'. Then by Theorem 9.4.3 we have $\omega' = \omega + dd^c\phi$, where ϕ is a function of Kähler potential type on X. Since $g_i = g'_i$ for $i \neq \infty$, the functions ϕ_i of Definition 9.4.1 are zero for $i \neq \infty$, and ϕ satisfies (9.25). But g, g' have the same Ricci curvature, so that $(\omega')^m = \omega^m$. Thus ϕ and $f \equiv 0$ satisfy the second version of the Calabi conjecture for QALE manifolds. By uniqueness in the conjecture we see that $\phi \equiv 0$, so that $g = g'$, a contradiction. Thus g is unique in its Kähler class. $\square$

Proposition 9.8.2 *Let (X, π) be a crepant resolution of $\mathbb{C}^m/G$ with* Fix$(G) = \{0\}$, *and let g be a QALE Kähler metric on X asymptotic to metrics g_i on Y_i for $i \in I$, such that $\psi_i^*(g) = h_i \times g_i$ on $V_i \times Y_i \setminus U_i$. Suppose that for each $i \in I \setminus \{\infty\}$ there is a Ricci-flat Kähler metric g'_i on Y_i in the same Kähler class as g_i. Then there exists a QALE Kähler metric g' on X in the same Kähler class as g with Ricci form $\frac{1}{2}dd^c f$, where f is a smooth function on X satisfying*

$$\nabla^l \psi_i^*(f) = \sum_{\substack{j \neq k \in I \setminus \{\infty\}: \\ V_j \cap V_k = \{0\}}} O\left(\mu_{i,j}^{d_j} \mu_{i,k}^{d_k} v_i^{-4-l}\right) \quad \text{on } V_i \times Y_i \setminus U_i, \tag{9.32}$$

for $i \in I$ and $l \geq 0$. However, if there are no $j, k \in I \setminus \{\infty\}$ with $V_j \cap V_k = \{0\}$ then (9.32) does not hold, but instead f is compactly-supported on X.

Proof Let g_i, g'_i have Kähler forms ω_i, ω'_i. Then Theorem 9.4.3 shows that $\omega'_i = \omega_i + dd^c\phi_i$, where ϕ_i is a unique function of Kähler potential type on Y_i. As g'_i is the only Ricci-flat QALE metric in its Kähler class by Proposition 9.8.1, we see that $\chi^*_{\gamma,i}(g'_{\gamma \cdot i}) = g'_i$, and g'_i is asymptotic to g'_j when $i, j \neq \infty$ and $i \succeq j$. Using these facts one can show that the ϕ_i satisfy parts (i) and (ii) of Proposition 9.4.4.

Therefore we may apply Theorem 9.4.5 and Proposition 9.4.6 to find a function ϕ of Kähler potential type on X such that $\omega' = \omega + dd^c\phi$ is the Kähler form of a QALE Kähler metric g' on X, which is asymptotic to the Ricci-flat metrics g'_i on Y_i. Also, as Fix$(G) = \{0\}$, by construction ϕ satisfies (9.10) outside a compact subset T of X.

Let ω_0 be the Kähler form of the Euclidean metric h_0 on $\mathbb{C}^m/G$, and let $\theta_0 = dz_1 \wedge \cdots \wedge dz_m$ be the holomorphic volume form on $\mathbb{C}^m/G$, which is well-defined as $G \subset \mathrm{SU}(m)$, since X is a crepant resolution. Calculation shows that $\omega_0^m = C_m \theta_0 \wedge \bar\theta_0$ on $\mathbb{C}^m/G$, where $C_m = 2^{-m} i^m m! (-1)^{m(m-1)/2}$. Let $\theta = \pi^*(\theta_0)$. Then θ is a nonsingular holomorphic volume form on X. Define a smooth real function f on X by $e^f (\omega')^m = C_m \theta \wedge \bar\theta$. As in §5.1 we find that g' has Ricci form $\frac{1}{2} dd^c f$, as we want.

We must show that f satisfies (9.32). For simplicity we will restrict our attention to the case $N(V_i) = G$ for all $i \in I$. Then (9.11) gives

$$\phi = \sum_{i \in I \setminus \{\infty\}} k_i \Phi'_i \quad \text{on } X \setminus T. \tag{9.33}$$

Let $x \in X$ satisfy $d(\pi(x), S) > 2R$. Then putting $i = 0$ in $\psi_i^*(g) = h_i \times g_i$ on $V_i \times Y_i \setminus U_i$ shows that $g = \pi^*(h_0)$ near x. Thus $\omega = \pi^*(\omega_0)$ and $\theta = \pi^*(\theta)$, and so $\omega^m = C_m \theta \wedge \bar\theta$ near x.

If $i \neq \infty$ in I and $(v, y) \in V_i \times Y_i \setminus U_i$ with $\psi_i(v, y) = x$, then $\psi_i^*(g) = h_i \times g_i$ near (v, y). Also the function Φ_i defined in (9.8) satisfies $\Phi_i = \phi_i$ near (v, y), since $\mu_{i,j} > 2R$ near (v, y) for all $j \neq 0$. But $N(V_i) = G$, and so $\Phi_i = \psi_i^*(\Phi_i')$, where Φ_i' is defined by (9.11). Therefore the metric $h_i \times g_i'$ on $V_i \times Y_i$ has Kähler form $\psi_i^*(\omega + dd^c \Phi_i')$ near (v, y).

But $h_i \times g_i'$ is Ricci-flat. So $\omega + dd^c \Phi_i$ is the Kähler form of a Ricci-flat metric near x. It is then not difficult to show that $(\omega + dd^c \Phi_i')^m = C_m \theta \wedge \bar\theta$ near x. Thus $e^f (\omega + dd^c \phi)^m = \omega^m = (\omega + dd^c \Phi_i')^m$ wherever $d(\pi(x), S) > 2R$ on X, for all $i \in I \setminus \{\infty\}$. Multiplying out $(\omega + dd^c \Phi_i')^m = \omega^m$ and rearranging gives

$$m \, dd^c \Phi_i' \wedge \omega^{m-1} = -\tfrac{1}{2} m(m-1) (dd^c \Phi_i')^2 \wedge \omega^{m-2} + \ldots, \qquad (9.34)$$

where '$\ldots$' are terms of order at least 3 in $dd^c \Phi_i'$. Substituting (9.33) into the equation $e^f (\omega + dd^c \phi)^m = \omega^m$, rearranging and using (9.4) shows that

$$\frac{-2f}{m(m-1)} \omega^m = \sum_{i \neq j \in I \setminus \{\infty\}} k_i k_j \, dd^c \Phi_i' \wedge dd^c \Phi_j' \wedge \omega^{m-2}$$
$$+ \sum_{i \in I \setminus \{\infty\}} k_i (k_i - 1) (dd^c \Phi_i')^2 \wedge \omega^{m-2} + \ldots, \qquad (9.35)$$

where '$\ldots$' are terms of order at least 3 in the $dd^c \Phi_i'$ or at least 2 in f, and the equation holds for $x \in X \setminus T$ with $d(\pi(x), S) > 2R$. Thus f is roughly quadratic in the $dd^c \Phi_i'$, to highest order.

Using (9.35) and the special properties of the k_i and Φ_i', we can show that

$$\nabla^l f = \sum_{\substack{i \neq j \in I \setminus \{0\}: \\ V_i \cap V_j = \{0\}}} O\bigl(\rho_i^{d_i} \rho_j^{d_j} \sigma^{-4-l}\bigr) \quad \text{on } X, \qquad (9.36)$$

if there exist $i \neq j \in I \setminus \{0, \infty\}$ with $V_i \cap V_j = \{0\}$. Otherwise f is compactly-supported in X.

The proof of (9.36) is complicated, and we will not give it. The basic idea is that g' is made by combining the Ricci-flat metrics $h_i \times g_i'$. The dominant terms in $\mathrm{Ric}(g')$ result from interference between $h_i \times g_i'$ and $h_j \times g_j'$ for $i \neq j$, and contribute $O(\rho_i^{d_i} \sigma^{-2} \cdot \rho_j^{d_j} \sigma^{-2})$ to f. However, if $V_i \cap V_j = V_k$ with $k \neq \infty$ then $h_k \times g_k'$ is Ricci-flat and asymptotic to both $h_i \times g_i'$ and $h_j \times g_j'$. So we introduce no extra Ricci curvature by combining $h_i \times g_i'$ and $h_j \times g_j'$ in this case, which is why the sum in (9.36) is restricted to i, j with $V_i \cap V_j = \{0\}$.

As $N(V_i) = G$ for all $i \in I$ by assumption, the proof of Proposition 9.7.2 shows that (9.36) is equivalent to (9.32). So (9.32) holds, as we want. In the case when $N(V_i) \neq G$ for some i we can prove (9.32) by a similar method, using the ideas of §9.7.2. □

9.8.1 Proof of Theorem 9.3.3

We work by induction on $m = \dim X$. The result is trivial for $m = 0, 1$, giving the first step. For the inductive step, suppose that X is a crepant resolution of $\mathbb{C}^m/G$ for some $m \geqslant 2$, that κ is a Kähler class on X containing QALE Kähler metrics, and that the Theorem is true in dimensions $0, 1, \ldots, m-1$. We shall show that there exists a Ricci-flat QALE Kähler metric $\hat{g}$ on X in κ. This $\hat{g}$ is unique by Proposition 9.8.1, and so each Kähler class on X contains a unique Ricci-flat QALE Kähler metric. Thus by induction the Theorem is true for all m.

If $\text{Fix}(G) \neq \{0\}$ then $X = V_\infty \times Y_\infty$, and as $\dim Y_\infty < m$ there is a unique Ricci-flat QALE Kähler metric $\hat{g}_\infty$ on Y_∞ in $\kappa|_{Y_\infty}$. Then $\hat{g} = h_\infty \times \hat{g}_\infty$ is the metric on X that we seek. So suppose that $\text{Fix}(G) = \{0\}$.

By Theorem 9.4.7 we can choose QALE Kähler metrics g on X and g_i on Y_i such that g has Kähler class κ and $\psi_i^*(g) = h_i \times g_i$ on $V_i \times Y_i \setminus U_i$, where U_i is defined using some $R > 0$ depending on κ. Now Y_i is a crepant resolution of W_i/A_i, and $\dim Y_i < m = \dim X$ if $i \neq \infty$. Thus for each $i \neq \infty$ in I there is a unique Ricci-flat Kähler metric g_i' on Y_i in the Kähler class of g_i, by the inductive hypothesis.

Proposition 9.8.2 applies, and gives a QALE Kähler metric g' on X in the Kähler class κ with Ricci form $\frac{1}{2} dd^c f$, where either f satisfies (9.32) if there exist $j, k \in I \setminus \{\infty\}$ with $V_j \cap V_k = \{0\}$, or else f is compactly-supported on X. Now it is easy to show from (9.32) that f satisfies (9.24) with $\epsilon = -4$.

Therefore by the second version of the Calabi conjecture for QALE manifolds, which holds by Theorem 9.7.8, there is a unique function ϕ' on X such that $\hat{\omega} = \omega' + dd^c \phi'$ satisfies $\hat{\omega}^m = e^f (\omega')^m$ and (9.25) holds for ϕ'. The Kähler metric $\hat{g}$ on X with Kähler form $\hat{\omega}$ is QALE, as ϕ' is of Kähler potential type, and has Ricci form $\xi - \frac{1}{2} dd^c f$, where ξ is the Ricci form of g'. But $\xi = \frac{1}{2} dd^c f$ from above, and so $\hat{g}$ is Ricci-flat. Thus we have found a Ricci-flat QALE Kähler metric $\hat{g}$ in the Kähler class κ, and the proof is complete.

9.8.2 Proof of Theorem 9.3.4

Let (X, J, g) be a Ricci-flat QALE Kähler manifold asymptotic to $\mathbb{C}^m/G$, where $\mathbb{C}^m/G$ is irreducible. As X is simply-connected and g is Ricci-flat Kähler we have $\text{Hol}(g) \subseteq \text{SU}(m)$. But g is nonsymmetric as it is Ricci-flat, and since $\mathbb{C}^m/G$ is irreducible one can show that g is irreducible. So Theorem 3.4.1 shows that $\text{Hol}(g)$ is $\text{SU}(m)$ or $\text{Sp}(m/2)$. The Euclidean metric h_0 on $\mathbb{C}^m/G$ has $\text{Hol}(h_0) = G$, and as g is asymptotic to h_0 we see that $G \subset \text{Hol}(g)$. Thus, if G is not conjugate to a subgroup of $\text{Sp}(m/2)$ then $\text{Hol}(g) \neq \text{Sp}(m/2)$, which forces $\text{Hol}(g) = \text{SU}(m)$.

So suppose $m \geqslant 4$ is even and G is conjugate to a subgroup of $\text{Sp}(m/2)$. Then there exists a G-invariant, constant complex symplectic form $\omega_\mathbb{C}$ on $\mathbb{C}^m$, which pushes down to $\mathbb{C}^m/G$. The pull-back $\pi^*(\omega_\mathbb{C})$ is a nonsingular complex symplectic form on X, and using a Bochner argument we can prove that $\nabla \pi^*(\omega_\mathbb{C}) = 0$. Therefore $\text{Hol}(g) \subseteq \text{Sp}(m/2)$, and so $\text{Hol}(g) = \text{Sp}(m/2)$.

9.9 Generalized QALE manifolds

In §9.2 we defined QALE Kähler manifolds (X, J, g), where (X, J) is a local product resolution of $\mathbb{C}^m/G$. We now generalize this definition in two ways. The first way is to define *QALE manifolds* (X, g) asymptotic to $\mathbb{R}^n/G$. To do this we shall need a concept of real local product resolution of $\mathbb{R}^n/G$. The second way is to consider QALE Kähler manifolds (X, J, g) in which (X, J) is no longer required to be a local product resolution of $\mathbb{C}^m/G$, but can instead be a more general complex manifold asymptotic to $\mathbb{C}^m/G$, in a suitable sense.

Definition 9.9.1 Let G be a finite subgroup of $SO(n)$. Define a finite set $\mathcal{L}$ of linear subspaces of $\mathbb{R}^n$ by

$$\mathcal{L} = \{\text{Fix}(H) : H \text{ is a subgroup of } G\}.$$

Let I be an *indexing set* for $\mathcal{L}$, so that we may write $\mathcal{L} = \{V_i : i \in I\}$. Let the index of $\mathbb{R}^n \in \mathcal{L}$ be 0. Define a *partial order* $\succeq$ on I by $i \succeq j$ if $V_i \subseteq V_j$. For each $i \in I$, define W_i to be the perpendicular subspace to V_i in $\mathbb{R}^n$, so that $\mathbb{R}^n = V_i \oplus W_i$. Define $A_i = C(V_i)$. Then $V_i = \text{Fix}(A_i)$ and A_i acts on W_i, with $\mathbb{R}^n/A_i \cong V_i \times W_i/A_i$.

Let S be the singular set of $\mathbb{R}^n/G$. We say that (X, π) is a *real resolution* of $\mathbb{R}^n/G$ if X is a noncompact real manifold of dimension n, and $\pi : X \to \mathbb{R}^n/G$ is a continuous, proper, surjective map such that the restriction $\pi : X \setminus \pi^{-1}(S) \to \mathbb{R}^n/G \setminus S$ is a diffeomorphism.

Let (X, π) be a real resolution of $\mathbb{R}^n/G$. We call (X, π) a *real local product resolution* if for each $i \in I$ there is a real resolution (Y_i, π_i) of W_i/A_i such that the following conditions hold. Let $R > 0$, and define subsets S_i and T_i in $V_i \times W_i/A_i \cong \mathbb{R}^n/A_i$ by

$$S_i = \bigcup_{j \in I : i \not\succeq j} V_j \Big/ A_i, \quad T_i = \{x \in \mathbb{R}^n/A_i : d(x, S_i) \leqslant R\},$$

where $d(\,,\,)$ is the distance in $\mathbb{R}^n/A_i$. Let U_i be the pull-back of T_i to $V_i \times Y_i$ under $\text{id} \times \pi_i : V_i \times Y_i \to V_i \times W_i/A_i$. Let ϕ_i be the natural projection from $V_i \times W_i/A_i$ to $\mathbb{R}^n/G$. Then there should exist a local diffeomorphism $\psi_i : V_i \times Y_i \setminus U_i \to X$ such that the following diagram commutes:

$$\begin{array}{ccc} V_i \times Y_i \setminus U_i & \xrightarrow{\psi_i} & X \\ {\scriptstyle \text{id} \times \pi_i} \downarrow & & \downarrow {\scriptstyle \pi} \\ V_i \times W_i/A_i \setminus T_i & \xrightarrow{\phi_i} & \mathbb{R}^n/G. \end{array}$$

If (X, π) is a real resolution of $\mathbb{R}^n/G$ then π need not be differentiable at $\pi^{-1}(S)$, in any sense. Because of this, $\pi^{-1}(S)$ need not be a submanifold (or even a singular submanifold) of X. The most we can say is that $\pi^{-1}(S)$ is closed and connected, and $\pi^{-1}(s)$ is compact and connected for each $s \in S$. Next we define *QALE metrics*, following Definition 9.2.1.

Definition 9.9.2 Let (X, π) be a real local product resolution of $\mathbb{R}^n/G$. For each pair $i, j \in I$, define $\mu_{i,j} : V_i \times Y_i \to [0, \infty)$ by $\mu_{i,j}(z) = d\big((\text{id} \times \pi_i)(z), V_j A_i/A_i\big)$, where $d(y, T)$ is the shortest distance between the point y and the subset T in $\mathbb{R}^n/A_i$. Define

$v_i : V_i \times Y_i \to [0, \infty)$ by $1 + v_i(z) = \min\{\mu_{i,j}(z) : j \in I, j \neq 0\}$, let $d_i = 2 - \dim W_i$, and let h_i be the Euclidean metric on V_i for each $i \in I$.

We say that a Riemannian metric g on X is *Quasi-ALE* if the real codimension of the singular set S of $\mathbb{R}^n/G$ is at least four, and for each $i \in I$ there is a Riemannian metric g_i on Y_i, such that the metric $h_i \times g_i$ on $V_i \times Y_i$ satisfies

$$\nabla^l \big(\psi_i^*(g) - h_i \times g_i\big) = \sum_{j \in I : i \not\leq j} O\big(\mu_{i,j}^{d_j} v_i^{-2-l}\big) \qquad (9.37)$$

on $V_i \times Y_i \setminus U_i$, for all $l \geq 0$. If (X, π) is a real local product resolution of $\mathbb{R}^n/G$ and g a QALE metric on X, then we say that (X, g, π) is a *QALE manifold* asymptotic to $\mathbb{R}^n/G$. We will also call (X, g) a *QALE manifold* asymptotic to $\mathbb{R}^n/G$ if there exists a map $\pi : X \to \mathbb{R}^n/G$ such that (X, g, π) is a QALE manifold asymptotic to $\mathbb{R}^n/G$.

If (X, g) is a QALE manifold then the map $\pi : X \to \mathbb{R}^n/G$ will not be unique, because it is only the asymptotic behaviour of π at infinity that is important. It is natural to consider two such maps $\pi, \pi' : X \to \mathbb{R}^n/G$ *equivalent* if they are asymptotic at infinity in $\mathbb{R}^n/G$, in a suitable sense. It is sometimes helpful to implicitly assume that a QALE manifold (X, g) comes equipped with an equivalence class of such maps π, under this equivalence relation.

Although the definition of d_i above appears to differ from the expression given in Definition 9.2.1, the two are actually equivalent, since V_i, W_i are complex vector spaces in Definition 9.2.1, with real dimensions $2 \dim V_i$, $2 \dim W_i$. The same comment applies to the condition that S should have codimension at least four.

Having defined QALE manifolds, as opposed to QALE Kähler manifolds, we can go on to ask questions about them. For instance, do there exist QALE manifolds (X, g) with holonomy groups G_2 and $\mathrm{Spin}(7)$? The answer to this is yes, as we will discuss in §11.2 and §13.1. Note also that the analytic material of §9.5 holds for general QALE manifolds, and not just for QALE Kähler manifolds.

Definition 9.9.2 explains how to define QALE metrics on a real local product resolution. It is a simple matter to generalize this idea to other geometric structures. In particular, here is the natural idea of Quasi-ALE complex structure, which we will call a *local product desingularization*.

Definition 9.9.3 Let G be a finite subgroup of $\mathrm{U}(m)$. Then we may identify $\mathbb{C}^m/G$ with $\mathbb{R}^{2m}/G$, and so consider *real* local product resolutions X of $\mathbb{C}^m/G$. For each $i \in I$, we can regard V_i both as a real vector subspace in $\mathbb{R}^{2m}$, and as a complex vector subspace in $\mathbb{C}^m$. Thus V_i has a natural complex structure, which we will write as I_i.

We say that (X, J, π) is a *local product desingularization* of $\mathbb{C}^m/G$ if (X, π) is a real local product resolution of $\mathbb{C}^m/G$, and J is a complex structure on X, and for each $i \in I$ there is a complex structure J_i on Y_i such that the product complex structure $I_i \times J_i$ on $V_i \times Y_i$ satisfies

$$\nabla^l \big(\psi_i^*(J) - I_i \times J_i\big) = \sum_{j \in I : i \not\leq j} O\big(\mu_{i,j}^{d_j} v_i^{-2-l}\big) \qquad (9.38)$$

on $V_i \times Y_i \setminus U_i$, for all $l \geqslant 0$. We will also call (X, J) a *local product desingularization* of $\mathbb{C}^m/G$ if there exists a map $\pi : X \to \mathbb{C}^m/G$ such that (X, J, π) is a local product desingularization of $\mathbb{C}^m/G$.

A local product desingularization (X, J) of $\mathbb{C}^m/G$ is a noncompact, nonsingular complex manifold which is asymptotic at infinity to the singular complex manifold $\mathbb{C}^m/G$, in a well-defined way. *Local product resolutions* are examples of local product desingularizations, since they satisfy $\psi_i^*(J) = I_i \times J_i$ on $V_i \times Y_i \setminus U_i$, and so (9.38) holds automatically. Another way to construct local product desingularizations of $\mathbb{C}^m/G$ is by *deforming* the singular variety $\mathbb{C}^m/G$, or by a combination of deformations and resolutions.

It can be shown that *every* local product desingularization of $\mathbb{C}^m/G$ is a resolution of a deformation of $\mathbb{C}^m/G$. Deformations of $\mathbb{C}^m/G$ were discussed in §6.4.4, and Theorem 6.4.8 shows that if the singularities of $\mathbb{C}^m/G$ are of complex codimension at least three, then all deformations of $\mathbb{C}^m/G$ are trivial. In this case, every local product desingularization (X, J) of $\mathbb{C}^m/G$ is biholomorphic to a local product resolution.

However, if the singularities of $\mathbb{C}^m/G$ have codimension two, then there may exist local product desingularizations (X, J) of $\mathbb{C}^m/G$ which are not resolutions of $\mathbb{C}^m/G$. We will give some examples of this in §9.9.2.

9.9.1 *Generalized QALE Kähler metrics*

For simplicity, in Definition 9.2.1 we defined a *QALE Kähler manifold* to be a local product resolution X of $\mathbb{C}^m/G$, equipped with a Kähler metric g that is QALE in our usual sense. We can now generalize this definition by replacing X by a local product desingularization.

Definition 9.9.4 (Alternative) We say that (X, J, g) is a *QALE Kähler manifold asymptotic to* $\mathbb{C}^m/G$ if g is Kähler with respect to J, and there exists a map $\pi : X \to \mathbb{C}^m/G$ such that (X, J, π) is a local product desingularization of $\mathbb{C}^m/G$, and (X, g, π) a QALE manifold asymptotic to $\mathbb{C}^m/G$.

In the case that $\mathbb{C}^m/G$ has an isolated singularity at 0, this is equivalent to the alternative definition of ALE Kähler manifold given in Definition 8.9.1. Now sections 9.2–9.8 contain many results about QALE Kähler manifolds, in the sense of Definition 9.2.1. We shall now describe how these can be extended to the more general QALE Kähler manifolds of Definition 9.9.4.

The results of §9.2 hold with only the obvious changes. In §9.3, Proposition 9.3.1 does not hold in this context. Here are generalizations of Theorems 9.3.3 and 9.3.4:

Theorem 9.9.5 *Let G be a finite subgroup of* $\mathrm{U}(m)$, *and suppose (X, J) is a local product desingularization of $\mathbb{C}^m/G$ with $c_1(X) = 0$, that admits QALE Kähler metrics in the sense of Definition 9.9.4. Then in every Kähler class of QALE metrics on X, there exists a unique Ricci-flat QALE Kähler metric.*

Theorem 9.9.6 *Let (X, J, g) be a Ricci-flat QALE Kähler manifold asymptotic to $\mathbb{C}^m/G$, in the sense of Definition 9.9.4, and suppose $\mathbb{C}^m/G$ is irreducible and X is* simply-connected. If $m \geqslant 4$ is even and G is conjugate to a subgroup of $\text{Sp}(m/2)$ then $\text{Hol}(g) = \text{Sp}(m/2)$, and otherwise $\text{Hol}(g) = \text{SU}(m)$.

The proof of Theorem 9.9.5 will be discussed shortly. In Theorem 9.9.6 we assume that X is simply-connected. This condition is necessary, and we will give examples of Ricci-flat QALE Kähler manifolds (X, J, g) below in which X is not simply-connected, and $\text{Hol}(g)$ is neither $\text{Sp}(m/2)$ nor $\text{SU}(m)$.

The author believes that Proposition 9.3.8 also holds in the local product desingularization case, but does not have a proof of this. All the results of §9.4 extend to local product desingularizations, *except* Theorem 9.4.7, which only applies if X is a local product resolution. The results of §9.5–§9.7 apply unchanged to local product desingularizations. In §9.8, Proposition 9.8.1 is true for local product desingularizations.

To prove Theorem 9.9.6 we follow the proof of Theorem 9.3.4 in §9.8.2, except at one point: we can no longer pull back $\omega_{\mathbb{C}}$ from $\mathbb{C}^m/G$ to X. Instead, we use asymptotic analysis to construct a harmonic (2,0)-form $\tilde{\omega}_{\mathbb{C}}$ on X with the appropriate asymptotic behaviour, and show that $\nabla \tilde{\omega}_{\mathbb{C}} = 0$ by an integration by parts argument.

This leaves only the question of how to generalize the proof of Theorem 9.3.3 in §9.8.1 to a proof of Theorem 9.9.5 above, and the related problem of how to generalize Proposition 9.8.2. The difficulty is that Theorem 9.4.7 fails for local product desingularizations, but it is an essential step in §9.8.1. Also, Proposition 9.8.2 assumes the existence of a metric g on X with $\psi_i^*(g) = h_i \times g_i$ on $V_i \times Y_i \setminus U_i$, but no such g exists unless X is a local product resolution.

The solution is to prove an analogue of Proposition 9.8.2, in which one shows that every Kähler class of QALE metrics on a local product desingularization X with $c_1(X) = 0$ contains a QALE Kähler metric g' with Ricci form $\frac{1}{2}\text{dd}^c f$, where f satisfies an estimate similar to (9.32). This is not easy, but it can be done; the issues involved are similar to those discussed briefly at the end of §8.9. Having proved this analogue of Proposition 9.8.2, we use it to fill the gap in §9.8.1 left by the failure of Theorem 9.4.7, and Theorem 9.9.5 easily follows.

For QALE manifolds with holonomy $\text{Sp}(m)$, we can use the ideas of this chapter to prove the following result.

Theorem 9.9.7 *Suppose G is a finite subgroup of $\text{Sp}(m)$, and X is a crepant resolution of $\mathbb{C}^{2m}/G$ carrying QALE Kähler metrics with holonomy $\text{Sp}(m)$. Then X has a family of nontrivial complex deformations X_α parametrized by $\alpha \in H^2(X, \mathbb{C})$ with $X_0 \cong X$, such that each X_α is diffeomorphic to X, and is a local product desingularization of $\mathbb{C}^m/G$ admitting QALE metrics with holonomy $\text{Sp}(m)$.*

Thus the singularities $\mathbb{C}^{2m}/G$ given as examples in §9.3.3 have families of local product desingularizations, which also carry QALE metrics with holonomy $\text{Sp}(m)$.

9.9.2 *Examples*

Here are two examples, taken from Joyce [126, §4, §6].

Example 9.9.8 In Example 9.3.5 we defined a singularity $\mathbb{C}^3/\mathbb{Z}_4$, which has a unique crepant resolution carrying QALE metrics with holonomy $\text{SU}(3)$. We will now see that

$\mathbb{C}^3/\mathbb{Z}_4$ also has a family of local product desingularizations. Define a map $\phi : \mathbb{C}^3/\mathbb{Z}_4 \to \mathbb{C}^4/\{\pm 1\}$ by

$$\phi : (z_1, z_2, z_3)\mathbb{Z}_4 \longmapsto \pm(z_1, z_2^2, z_2 z_3, z_3^2).$$

Then ϕ is well-defined and injective, and has image

$$\operatorname{Im} \phi = \{\pm(w_1, w_2, w_3, w_4) \in \mathbb{C}^4/\{\pm 1\} : w_2 w_4 = w_3^2\}.$$

For each $\epsilon \in \mathbb{C}$, define a complex algebraic variety X_ϵ by

$$X_\epsilon = \{\pm(w_1, w_2, w_3, w_4) \in \mathbb{C}^4/\{\pm 1\} : w_2 w_4 = w_3^2 + \epsilon\}.$$

Then $X_0 \cong \mathbb{C}^3/\mathbb{Z}_4$, and $\{X_\epsilon : \epsilon \in \mathbb{C}\}$ is a family of deformations of $\mathbb{C}^3/\mathbb{Z}_4$, as in §4.9.

If $\epsilon \neq 0$ then X_ϵ is nonsingular, and has fundamental group $\mathbb{Z}_2$. The double cover of X_ϵ is $\mathbb{C} \times Y_\epsilon$, where Y_ϵ is a deformation of $\mathbb{C}^2/\{\pm 1\}$. Because of this, it is easy to show that X_ϵ is a local product desingularization of $\mathbb{C}^3/\mathbb{Z}_4$ with $c_1(X_\epsilon) = 0$, and has a family of Ricci-flat QALE Kähler metrics (in the sense of Definition 9.9.4), which have holonomy $\mathbb{Z}_2 \ltimes \mathrm{SU}(2)$ and can be written down explicitly in coordinates.

This example shows that we do need the assumption that X is simply-connected in Theorem 9.9.6, since here X is not simply-connected, and the metrics do not have holonomy $\mathrm{SU}(3)$. Note also that X_ϵ has Betti numbers $b^0 = 1$ and $b^1 = \cdots = b^6 = 0$, whereas the crepant resolution of $\mathbb{C}^3/\mathbb{Z}_4$ has $b^2 = 2$ and $b^4 = 1$. Thus the resolution and the deformations of $\mathbb{C}^3/\mathbb{Z}_4$ have rather different topology.

Example 9.9.9 In Example 9.3.6 we defined a singularity $\mathbb{C}^3/\mathbb{Z}_2^2$, which has 4 crepant resolutions, each admitting QALE metrics with holonomy $\mathrm{SU}(3)$. We shall show that $\mathbb{C}^3/\mathbb{Z}_2^2$ also has 5 other families of topologically distinct local product desingularizations, which carry QALE metrics with holonomy $\mathrm{SU}(3)$. These are produced by first deforming $\mathbb{C}^3/\mathbb{Z}_2^2$, and then resolving any remaining singularities in the deformation.

Define $\phi : \mathbb{C}^3/\mathbb{Z}_2^2 \to \mathbb{C}^4$ by $\phi : (z_1, z_2, z_3)\mathbb{Z}_2^2 \mapsto (z_1^2, z_2^2, z_3^2, z_1 z_2 z_3)$. Then ϕ is well-defined, and induces an isomorphism between $\mathbb{C}^3/\mathbb{Z}_2^2$ and the hypersurface

$$W_{0,0,0,0} = \{(x_1, x_2, x_3, x_4) \in \mathbb{C}^4 : x_1 x_2 x_3 - x_4^2 = 0\}$$

in $\mathbb{C}^4$. Let $\delta, \epsilon_1, \epsilon_2$ and ϵ_3 be complex numbers, and write

$$W_{\delta,\epsilon_1,\epsilon_2,\epsilon_3} = \{(x_1, x_2, x_3, x_4) \in \mathbb{C}^4 : x_1 x_2 x_3 - x_4^2 = \delta + \epsilon_1 x_1 + \epsilon_2 x_2 + \epsilon_3 x_3\}.$$

Then $W_{\delta,\epsilon_1,\epsilon_2,\epsilon_3}$ is a deformation of $\mathbb{C}^3/\mathbb{Z}_2^2$. For generic $\delta, \ldots, \epsilon_3$ the hypersurface $W_{\delta,\epsilon_1,\epsilon_2,\epsilon_3}$ is nonsingular, but for some special values of $\delta, \ldots, \epsilon_3$ it has singularities. If $W_{\delta,\epsilon_1,\epsilon_2,\epsilon_3}$ is singular, then it can be resolved with a crepant resolution to make it nonsingular.

Thus, each set of values of $\delta, \ldots, \epsilon_3$ give one or more ways to desingularize $\mathbb{C}^3/\mathbb{Z}_2^2$. Here are the different cases that arise in this way.

(i) $W_{0,0,0,0}$ is isomorphic to $\mathbb{C}^3/\mathbb{Z}_2^2$. It has 4 crepant resolutions, with Betti numbers $b^2 = 3$ and $b^3 = 0$.

(ii) $W_{\delta,0,0,0}$ for $\delta \neq 0$. This is nonsingular, with Betti numbers $b^2 = 2$ and $b^3 = 1$.
(iii) $W_{0,\epsilon_1,0,0}$ for $\epsilon_1 \neq 0$. This has singularities at the points $(0, x_2, x_3, 0)$ for $x_2 x_3 = \epsilon_1$. It has a unique crepant resolution, the blow-up of the singular set, which has Betti numbers $b^2 = 1$ and $b^3 = 1$.
(iv) $W_{\delta,\epsilon_1,0,0}$ for $\delta, \epsilon_1 \neq 0$. This is nonsingular, and is a smooth deformation of the resolution in (iii), with the same topology.
(v),(vi) As (iii) and (iv) but with ϵ_2 nonzero instead of ϵ_1.
(vii),(viii) As (iii) and (iv) but with ϵ_3 nonzero instead of ϵ_1.
(ix) $W_{\delta,\epsilon_1,\epsilon_2,\epsilon_3}$ with $\epsilon_1, \epsilon_2, \epsilon_3 \neq 0$ and $\delta^2 \neq 4\epsilon_1\epsilon_2\epsilon_3$. This is nonsingular, with Betti numbers $b^2 = 0$ and $b^3 = 2$.

Each of (i)–(ix) is a simply-connected local product desingularization with $c_1 = 0$, and admits QALE Kähler metrics. Thus by Theorems 9.9.5 and 9.9.6, each has a family of QALE Kähler metrics with holonomy SU(3). These realize four different sets of Betti numbers, and as (i) yields four different topologies but (iii)–(viii) are equivalent in pairs, they give nine distinct topologies.

In cases (ii), (iv), (vi) and (viii), not every Kähler class on X contains a QALE metric. Because of a curious phenomenon related to the *Weyl group* of the singularities, the Kähler classes containing QALE metrics are divided into several connected components — six in case (ii) and two in cases (iv), (vi) and (viii). The connected component of the Kähler class can be regarded as an extra topological choice in the desingularization. In case (ix) we have $b^2 = 0$, so that the QALE metric with holonomy SU(3) on $W_{\delta,\epsilon_1,\epsilon_2,\epsilon_3}$ is unique.

There are two remaining cases, which do not give QALE Kähler manifolds.

(x) If exactly one of $\epsilon_1, \epsilon_2, \epsilon_3$ is zero, say ϵ_1, then $W_{\delta,0,\epsilon_2,\epsilon_3}$ is nonsingular and is topologically equivalent to case (ix). However, although $W_{\delta,0,\epsilon_2,\epsilon_3}$ admits Kähler metrics, it does *not* admit QALE Kähler metrics.
(xi) $W_{\delta,\epsilon_1,\epsilon_2,\epsilon_3}$ with $\epsilon_1, \epsilon_2, \epsilon_3 \neq 0$ and $\delta^2 = 4\epsilon_1\epsilon_2\epsilon_3$ has a single node at $x_j = -\delta/2\epsilon_j$ for $j = 1, 2, 3$. We can resolve this by a small resolution in two different ways, as in Example 6.3.4. But neither resolution is Kähler.

This example shows that there may be several families of QALE Kähler manifolds (X, J, g) with holonomy SU(m) asymptotic to the same singularity $\mathbb{C}^m/G$, which have rather different topology and Betti numbers, and that understanding them all can be rather complicated.

10

INTRODUCTION TO THE EXCEPTIONAL HOLONOMY GROUPS

For the rest of the book we shall discuss the exceptional holonomy groups G_2 in 7 dimensions, and Spin(7) in 8 dimensions. This chapter introduces G_2 and Spin(7), and explains some basic topological and geometrical properties of compact manifolds with these holonomy groups. Later, Chapters 11 and 12 will construct examples of compact 7-manifolds with holonomy G_2, and Chapters 13, 14 and 15 examples of compact 8-manifolds with holonomy Spin(7).

Sections 10.1–10.4 define G_2, and study the topology of compact Riemannian 7-manifolds (M, g) with holonomy G_2. As g is Ricci-flat and irreducible, the Cheeger–Gromoll splitting theorem shows that M has finite fundamental group. We also prove that the moduli space of holonomy G_2 metrics on M is a smooth manifold with dimension $b^3(M)$.

Sections 10.5–10.7 give a similar treatment of Spin(7). If (M, g) is a compact Riemannian 8-manifold with holonomy Spin(7), then by considering the Dirac operator $D_+ : C^\infty(S_+) \to C^\infty(S_-)$ we show that M is simply-connected with $\hat{A}$-genus $\hat{A}(M) = 1$, and its Betti numbers satisfy $b^3 + b^4_+ = b^2 + 2b^4_- + 25$. Also, the moduli space of holonomy Spin(7) metrics on M is a manifold of dimension $1 + b^4_-$. In §10.8 we discuss *calibrated submanifolds* of manifolds with holonomy G_2 and Spin(7), and §10.9 gives a reading list on exceptional holonomy.

10.1 The holonomy group G_2

Here is a definition of G_2 as a subgroup of GL(7, $\mathbb{R}$).

Definition 10.1.1 Let $(x_1, \ldots, x_7)$ be coordinates on $\mathbb{R}^7$. Write $dx_{ij\ldots l}$ for the exterior form $dx_i \wedge dx_j \wedge \cdots \wedge dx_l$ on $\mathbb{R}^7$. Define a 3-form φ_0 on $\mathbb{R}^7$ by

$$\varphi_0 = dx_{123} + dx_{145} + dx_{167} + dx_{246} - dx_{257} - dx_{347} - dx_{356}. \tag{10.1}$$

The subgroup of GL(7, $\mathbb{R}$) preserving φ_0 is the exceptional Lie group G_2. It is compact, connected, simply-connected, semisimple and 14-dimensional, and it also fixes the 4-form

$$*\varphi_0 = dx_{4567} + dx_{2367} + dx_{2345} + dx_{1357} - dx_{1346} - dx_{1256} - dx_{1247}, \tag{10.2}$$

the Euclidean metric $g_0 = dx_1^2 + \cdots + dx_7^2$, and the orientation on $\mathbb{R}^7$. Note that φ_0 and $*\varphi_0$ are related by the Hodge star.

The forms φ_0 and $*\varphi_0$ above are those given by Bryant [43, p. 539]. They differ from those used in the author's papers [121, 122], but are equivalent to them under a permutation of coordinates $x_1, \ldots, x_7$.

Let M be an oriented 7-manifold. For each $p \in M$, define $\mathcal{P}_p^3 M$ to be the subset of 3-forms $\varphi \in \Lambda^3 T_p^* M$ for which there exists an oriented isomorphism between $T_p M$ and $\mathbb{R}^7$ identifying φ and the 3-form φ_0 of (10.1). Then $\mathcal{P}_p^3 M$ is isomorphic to $\mathrm{GL}_+(7, \mathbb{R})/G_2$, since φ_0 has symmetry group G_2.

Now $\dim \mathrm{GL}_+(7, \mathbb{R}) = 49$ and $\dim G_2 = 14$, so $\mathrm{GL}_+(7, \mathbb{R})/G_2$ has dimension $49 - 14 = 35$. But $\Lambda^3 T_p^* M$ also has dimension $\binom{7}{3} = 35$, and thus $\mathcal{P}_p^3 M$ is an *open subset* of $\Lambda^3 T_p^* M$. Let $\mathcal{P}^3 M$ be the bundle over M with fibre $\mathcal{P}_p^3 M$ at each $p \in M$. Then $\mathcal{P}^3 M$ is an open subbundle of $\Lambda^3 T^* M$ with fibre $\mathrm{GL}_+(7, \mathbb{R})/G_2$. Note that $\mathcal{P}^3 M$ is *not* a vector subbundle of $\Lambda^3 T^* M$. We say that a 3-form φ on M is *positive* if $\varphi|_p \in \mathcal{P}_p^3 M$ for each $p \in M$.

Similarly, define $\mathcal{P}_p^4 M$ to be the subset of 4-forms $\psi \in \Lambda^4 T_p^* M$ for which there exists an oriented isomorphism between $T_p M$ and $\mathbb{R}^7$ identifying ψ and the 4-form $*\varphi_0$ of (10.2), and let $\mathcal{P}^4 M$ have fibre $\mathcal{P}_p^4 M$ at each $p \in M$. Then $\mathcal{P}^4 M$ is an open subbundle of $\Lambda^4 T^* M$ with fibre $\mathrm{GL}_+(7, \mathbb{R})/G_2$, and sections of $\mathcal{P}^4 M$ are called *positive 4-forms*.

The *frame bundle* F of M is the bundle over M whose fibre at $p \in M$ is the set of isomorphisms between $T_p M$ and $\mathbb{R}^7$. Let φ be a positive 3-form on M, and let Q be the subset of F consisting of isomorphisms between $T_p M$ and $\mathbb{R}^7$ which identify $\varphi|_p$ and φ_0 of (10.1). It is easy to show that Q is a principal subbundle of F, with fibre G_2. That is, Q is a G_2-*structure*, as in Definition 2.6.1.

Conversely, if Q is a G_2-structure on M then, as φ_0, $*\varphi_0$ and g_0 are G_2-invariant, we can use Q to define a 3-form φ, a 4-form $*\varphi$ and a metric g on M corresponding to φ_0, $*\varphi_0$ and g_0. This 3-form φ will be positive if and only if Q is an *oriented G_2-structure*, that is, a G_2-structure which induces the given orientation on M.

Thus we have found a 1-1 correspondence between positive 3-forms φ and oriented G_2-structures Q on M. Furthermore, to any positive 3-form φ on M we can associate a unique positive 4-form $*\varphi$ and metric g, such that φ, $*\varphi$ and g are identified with φ_0, $*\varphi_0$ and g_0 under an isomorphism between $T_p M$ and $\mathbb{R}^7$, for each $p \in M$. We will call g and $*\varphi$ the *metric* and *4-form associated to φ*.

Definition 10.1.2 Let M be an oriented 7-manifold, φ a positive 3-form on M, and g the associated metric. For the rest of the book, we will adopt the following abuse of notation: we shall refer to the pair (φ, g) as a G_2-*structure*. Of course (φ, g) is not, exactly, a G_2-structure, but it does at least define a unique G_2-structure.

Define a map $\Theta : \mathcal{P}^3 M \to \mathcal{P}^4 M$ by $\Theta(\varphi) = *\varphi$. That is, if φ is a positive 3-form, then $\Theta(\varphi)$ is the associated 4-form $*\varphi$. It is important to note that Θ depends *solely* on M and its orientation, and also that Θ is a *nonlinear* map. Although we define $\Theta(\varphi) = *\varphi$ and the Hodge star $*$ is linear, actually $*$ depends on the metric g, which itself depends on φ, so $\Theta(\varphi)$ is not linear in φ.

Let M be a 7-manifold, (φ, g) a G_2-structure on M, and ∇ the Levi-Civita connection of g. We call $\nabla \varphi$ the *torsion* of (φ, g). If $\nabla \varphi = 0$ then (φ, g) is called *torsion-free*.

We define a G_2-*manifold* to be a triple (M, φ, g), where M is a 7-manifold, and (φ, g) a torsion-free G_2-structure on M.

This definition of G_2-manifold is brief and useful, but not widely known. The next proposition follows from [186, Lem. 11.5].

Proposition 10.1.3 *Let M be a 7-manifold and (φ, g) a G_2-structure on M. Then the following are equivalent:*

(i) (φ, g) *is torsion-free,*
(ii) $\mathrm{Hol}(g) \subseteq G_2$, *and φ is the induced 3-form,*
(iii) $\nabla \varphi = 0$ *on M, where ∇ is the Levi-Civita connection of g,*
(iv) $\mathrm{d}\varphi = \mathrm{d}^* \varphi = 0$ *on M, and*
(v) $\mathrm{d}\varphi = \mathrm{d}\,\Theta(\varphi) = 0$ *on M.*

Torsion-free G_2-structures will play an essential rôle in our construction of compact 7-manifolds with holonomy G_2 in Chapters 11 and 12. The basic idea is to find a torsion-free G_2-structure (φ, g) on M, and then show that $\mathrm{Hol}(g) = G_2$ provided $\pi_1(M)$ is finite.

The condition that (φ, g) be torsion-free is a *nonlinear p.d.e.* on the positive 3-form φ. This is most clearly seen in part (v) of Proposition 10.1.3, as Θ is a nonlinear map. Although in parts (iii) and (iv) the conditions $\nabla \varphi = 0$ and $\mathrm{d}\varphi = \mathrm{d}^* \varphi = 0$ appear linear in φ, in fact the operators ∇ and d^* depend on g, which depends on φ, so the equations $\nabla \varphi = 0$ and $\mathrm{d}^* \varphi = 0$ should be considered nonlinear in φ.

In §3.5.1 we explained that a G-structure on M induces a splitting of the bundles of tensors on M into irreducible components. Here is the decomposition of the exterior forms on a 7-manifold with a G_2-structure, which may be deduced from [186, Lem. 11.4].

Proposition 10.1.4 *Let M be a 7-manifold and (φ, g) a G_2-structure on M. Then $\Lambda^k T^* M$ splits orthogonally into components as follows, where Λ_l^k corresponds to an irreducible representation of G_2 of dimension l:*

(i) $\Lambda^1 T^* M = \Lambda_7^1$, (ii) $\Lambda^2 T^* M = \Lambda_7^2 \oplus \Lambda_{14}^2$,
(iii) $\Lambda^3 T^* M = \Lambda_1^3 \oplus \Lambda_7^3 \oplus \Lambda_{27}^3$, (iv) $\Lambda^4 T^* M = \Lambda_1^4 \oplus \Lambda_7^4 \oplus \Lambda_{27}^4$,
(v) $\Lambda^5 T^* M = \Lambda_7^5 \oplus \Lambda_{14}^5$, *and* (vi) $\Lambda^6 T^* M = \Lambda_7^6$.

The Hodge star $*$ of g gives an isometry between Λ_l^k and Λ_l^{7-k}. Note also that $\Lambda_1^3 = \langle \varphi \rangle$ and $\Lambda_1^4 = \langle *\varphi \rangle$, and that the spaces Λ_7^k for $k = 1, 2, \ldots, 6$ are canonically isomorphic.

Let the orthogonal projection from $\Lambda^k T^* M$ to Λ_l^k be denoted π_l. So, for instance, if $\xi \in C^\infty(\Lambda^2 T^* M)$, then $\xi = \pi_7(\xi) + \pi_{14}(\xi)$. We saw in Theorem 3.1.7 that the holonomy group of a Riemannian metric g constrains its Riemann curvature. Using this, Salamon [186, Lem. 11.8] shows:

Proposition 10.1.5 *Let (M, g) be a Riemannian 7-manifold. If $\mathrm{Hol}(g) \subseteq G_2$, then g is Ricci-flat.*

Since G_2 is a simply-connected subgroup of SO(7), any 7-manifold M with a G_2-structure must be a *spin manifold*, by Proposition 3.6.2. Furthermore, from Theorem 3.6.1 the natural representation of G_2 on the spinors on $\mathbb{R}^7$ fixes a nonzero spinor, and so a torsion-free G_2-structure has a corresponding *parallel spinor*. In fact if S is the spin bundle of M, then there is a natural isomorphism $S \cong \Lambda_1^0 \oplus \Lambda_7^1$. Thus we have:

Proposition 10.1.6 *Let (φ, g) be a G_2-structure on a 7-manifold M. Then M is spin, with a preferred spin structure. If (φ, g) is torsion-free, then (M, g) has a nonzero parallel spinor.*

10.2 The topology of compact G_2-manifolds

We now discuss the topology of *compact G_2-manifolds*, that is, compact manifolds M equipped with a torsion-free G_2-structure (φ, g). The material of this section can be found in Joyce [121, 122] and Bryant and Harvey [46]. Firstly, from the classification of Riemannian holonomy groups in §3.4 we deduce:

Theorem 10.2.1 *Write $\mathbb{R}^7 \cong \mathbb{R}^3 \oplus \mathbb{C}^2$, and let* SU(2) *act on $\mathbb{R}^7$ trivially on $\mathbb{R}^3$ and as usual on $\mathbb{C}^2$. Similarly, write $\mathbb{R}^7 \cong \mathbb{R} \oplus \mathbb{C}^3$, and let* SU(3) *act on $\mathbb{R}^7$ trivially on $\mathbb{R}$ and as usual on $\mathbb{C}^3$. Then* SU(2) $\subset$ SU(3) $\subset G_2 \subset$ SO(7).

The only connected Lie subgroups of G_2 which can be the holonomy group of a Riemannian metric on a 7-manifold are $\{1\}$, SU(2), SU(3) *and G_2, where the subgroups* SU(2) *and* SU(3) *are defined above. Hence, if (φ, g) is a torsion-free G_2-structure on a 7-manifold, then $\mathrm{Hol}^0(g)$ is one of $\{1\}$,* SU(2), SU(3) *or G_2.*

We use this to prove:

Proposition 10.2.2 *Let (M, φ, g) be a compact G_2-manifold. Then* $\mathrm{Hol}(g) = G_2$ *if and only if $\pi_1(M)$ is finite.*

Proof Since M is compact and g is Ricci-flat, Theorem 3.5.5 shows that M has a finite cover isometric to $N \times T^k$, where N is simply-connected. Thus $\pi_1(M) \cong F \ltimes \mathbb{Z}^k$, where F is a finite group. Clearly, $\pi_1(M)$ is finite if and only if $k = 0$.

Now $\mathrm{Hol}(g) \subseteq G_2$ as (φ, g) is torsion-free, so Theorem 10.2.1 shows that $\mathrm{Hol}^0(g)$ is $\{1\}$, SU(2), SU(3) or G_2. It is easy to show that $k = 7$ when $\mathrm{Hol}^0(g) = \{1\}$, $k = 3$ when $\mathrm{Hol}^0(g) = \mathrm{SU}(2)$, $k = 1$ when $\mathrm{Hol}^0(g) = \mathrm{SU}(3)$, and $k = 0$ when $\mathrm{Hol}^0(g) = G_2$. Thus $\mathrm{Hol}(g) = G_2$ if and only if $\pi_1(M)$ is finite. □

Next we apply the ideas of §3.5.2 to compact G_2-manifolds.

Definition 10.2.3 Let (M, φ, g) be a compact G_2-manifold. For each of the subbundles Λ_l^k of $\Lambda^k T^* M$ defined in Proposition 10.1.4, write

$$\mathcal{H}_l^k = \{\eta \in C^\infty(\Lambda_l^k) : \mathrm{d}\eta = \mathrm{d}^*\eta = 0\},$$

and let $H_l^k(M, \mathbb{R})$ be the vector subspace of $H^k(M, \mathbb{R})$ with representatives in $\mathcal{H}_l^k$. Define the *refined Betti numbers* b_l^k of M by $b_l^k = \dim H_l^k(M, \mathbb{R})$.

Combining Proposition 10.1.4 and Theorem 3.5.3 we find:

Theorem 10.2.4 *Let (M, φ, g) be a compact G_2-manifold. Then*

$$H^2(M, \mathbb{R}) = H_7^2(M, \mathbb{R}) \oplus H_{14}^2(M, \mathbb{R}),$$
$$H^3(M, \mathbb{R}) = H_1^3(M, \mathbb{R}) \oplus H_7^3(M, \mathbb{R}) \oplus H_{27}^3(M, \mathbb{R}),$$
$$H^4(M, \mathbb{R}) = H_1^4(M, \mathbb{R}) \oplus H_7^4(M, \mathbb{R}) \oplus H_{27}^4(M, \mathbb{R}),$$
and $\quad H^5(M, \mathbb{R}) = H_7^5(M, \mathbb{R}) \oplus H_{14}^5(M, \mathbb{R}).$

Here $H_1^3(M, \mathbb{R}) = \langle[\varphi]\rangle$, $H_1^4(M, \mathbb{R}) = \langle[*\varphi]\rangle$ and $H_l^k(M, \mathbb{R}) \cong H_l^{7-k}(M, \mathbb{R})$, so that $b_l^3 = b_l^4 = 1$ and $b_l^k = b_l^{7-k}$. Also, if $\mathrm{Hol}(g) = G_2$ then $H_7^k(M, \mathbb{R}) = \{0\}$ for $k = 1, \ldots, 6$.

This shows that if (M, g) is a compact Riemannian 7-manifold with holonomy G_2 then M has only two independent, nontrivial refined Betti numbers, b_{14}^2 and b_{27}^3, which satisfy $b^2 = b_{14}^2$ and $b^3 = b_{27}^3 + 1$. Thus we can calculate all the refined Betti numbers of M from b^2 and b^3. To prove the last part of the theorem, observe that $\pi_1(M)$ is finite by Proposition 10.2.2, and so $H^1(M, \mathbb{R}) = \{0\}$. But $H_7^k(M, \mathbb{R}) \cong H_7^1(M, \mathbb{R})$ for $k = 1, \ldots, 6$ by Theorem 3.5.3, as the G_2 representations associated to Λ_7^k and Λ_7^1 are isomorphic.

The following lemma is easily proved by calculating in coordinates.

Lemma 10.2.5 *Let (φ, g) be a G_2-structure on a 7-manifold M, and η a 2-form on M. Then $\eta \wedge \varphi = 2*\pi_7(\eta) - *\pi_{14}(\eta)$ and*

$$\eta \wedge \eta \wedge \varphi = \{2|\pi_7(\eta)|^2 - |\pi_{14}(\eta)|^2\} dV_g,$$

where dV_g is the volume form of g.

We use this to prove two results on the cohomology of M.

Proposition 10.2.6 *Let (M, φ, g) be a compact G_2-manifold, with $\mathrm{Hol}(g) = G_2$. Then $\langle \sigma \cup \sigma \cup [\varphi], [M] \rangle < 0$ for each nonzero $\sigma \in H^2(M, \mathbb{R})$.*

Proof Theorem 10.2.4 gives $H_7^2(M, \mathbb{R}) = \{0\}$, so that $H^2(M, \mathbb{R}) = H_{14}^2(M, \mathbb{R})$. Thus each $\sigma \in H^2(M, \mathbb{R})$ is represented by a unique $\eta \in \mathcal{H}_{14}^2$. So

$$\langle \sigma \cup \sigma \cup [\varphi], [M] \rangle = \int_M \eta \wedge \eta \wedge \varphi = -\int_M |\eta|^2 dV_g,$$

since $\eta \wedge \eta \wedge \varphi = -|\eta|^2 dV_g$ by Lemma 10.2.5. Therefore $\langle \sigma \cup \sigma \cup [\varphi], [M] \rangle < 0$ when $\sigma \neq 0$, as $\eta \neq 0$. □

Proposition 10.2.7 *Let (M, φ, g) be a compact G_2-manifold, R the Riemann curvature of g, and $p_1(M) \in H^4(M, \mathbb{Z})$ the first Pontryagin class. Then*

$$\langle p_1(M) \cup [\varphi], [M] \rangle = -\frac{1}{8\pi^2} \int_M |R|^2 dV_g.$$

If $\mathrm{Hol}(g) = G_2$ then g is not flat, so $\langle p_1(M) \cup [\varphi], [M] \rangle < 0$, and $p_1(M) \neq 0$.

Proof From Chern–Weil theory, $p_1(M)$ is represented by the closed 4-form $\frac{1}{8\pi^2} \text{Tr}(R \wedge R)$. But the 2-form part of R lies in Λ^2_{14}. Using this and the equation $\eta \wedge \eta \wedge \varphi = -|\eta|^2 dV_g$ for $\eta \in \Lambda^2_{14}$, we find that $\text{Tr}(R \wedge R) \wedge \varphi = -|R|^2 dV_g$. Integrating this over M, the result follows. □

We summarize Propositions 10.1.6, 10.2.2 and 10.2.7 in the following theorem, which gives topological restrictions on compact 7-manifolds M with holonomy G_2.

Theorem 10.2.8 *Suppose M is a compact 7-manifold admitting metrics with holonomy G_2. Then M is orientable and spin, $\pi_1(M)$ is finite and $p_1(M) \neq 0$.*

10.3 Exterior forms on G_2-manifolds

In this section we collect together some technical results on exterior forms on 7-manifolds with G_2-structures, that will be needed in §10.4 and Chapter 11. We begin with two lemmas.

Lemma 10.3.1 *Let (φ, g) be a G_2-structure on a 7-manifold M. Then there is a unique 1-form μ on M such that $\pi_7(d\varphi) = 3\mu \wedge \varphi$ and $\pi_7(d * \varphi) = 4\mu \wedge *\varphi$. In particular, if $d\varphi = 0$ then $\pi_7(d * \varphi) = 0$.*

Proof This follows from the identity $(*d\varphi) \wedge \varphi + (*d * \varphi) \wedge *\varphi = 0$ of Bryant [43, p. 553]. □

Lemma 10.3.2 *Let (M, φ, g) be a compact G_2-manifold, and η a 2-form on M. Then $\pi_7(d^*d\eta) = 0$ if and only if $d^*\pi_1(d\eta) = d^*\pi_7(d\eta) = 0$.*

Proof Suppose $\pi_7(d^*d\eta) = 0$. Let ξ be the unique d^*-exact 2-form on M with $d\xi = d\eta$. Then $\pi_7(dd^* + d^*d)\xi = 0$, as $d^*\xi = 0$. But π_7 commutes with $dd^* + d^*d$, so $(dd^* + d^*d)\pi_7(\xi) = 0$, and $\pi_7(\xi) \in \mathcal{H}^2_7$. Now ξ is d^*-exact, so ξ is L^2-orthogonal to $\mathcal{H}^2_7$, and $\pi_7(\xi) = 0$. Thus $\xi \wedge *\varphi = 0$, so $d\xi \wedge *\varphi = 0$ as $d * \varphi = 0$. But $d\xi = d\eta$ and $d\eta \wedge *\varphi \cong \pi_1(d\eta)$, so $\pi_1(d\eta) = 0$.

Lemma 10.2.5 shows that $\xi \wedge \varphi = -*\xi$, and $d * \xi = 0$ as ξ is d^*-exact, so $d\xi \wedge \varphi = 0$ as $d\varphi = 0$. But $d\eta = d\xi$ and $d\eta \wedge \varphi \cong \pi_7(d\eta)$, so $\pi_7(d\eta) = 0$. Thus $\pi_7(d^*d\eta) = 0$ implies $\pi_1(d\eta) = \pi_7(d\eta) = 0$, and hence $d^*\pi_1(d\eta) = d^*\pi_7(d\eta) = 0$. The converse follows from

$$\pi_7(d^*d\eta) = c_1\pi_7(d^*\pi_1(d\eta)) + c_2\pi_7(d^*\pi_7(d\eta)),$$

which holds for all 2-forms η, where c_1, c_2 are some constants. □

Definition 10.3.3 Let $\epsilon_1 > 0$ be a universal constant such that whenever (φ, g) is a G_2-structure on a 7-manifold M, then

  (i) If $\tilde\varphi \in C^\infty(\Lambda^3 T^*M)$ and $\|\tilde\varphi - \varphi\|_{C^0} \leq \epsilon_1$, then $\tilde\varphi \in C^\infty(\mathcal{P}^3 M)$. So $\tilde\varphi$ defines a G_2-structure $(\tilde\varphi, \tilde g)$. Let $\Lambda^5 T^*M = \tilde\Lambda^5_7 \oplus \tilde\Lambda^5_{14}$ be the corresponding splitting. Then each $\xi \in \tilde\Lambda^5_{14}$ satisfies $|\pi_7(\xi)| \leq |\pi_{14}(\xi)|$, where π_7, π_{14} and $|.|$ are taken with respect to the G_2-structure (φ, g).

(ii) If $\chi \in C^\infty(\Lambda^4 T^*M)$ and $\|\chi - *\varphi\|_{C^0} \leqslant \epsilon_1$ then $|\pi_{14}(\lambda \wedge \chi)| \leqslant \frac{1}{4}|\pi_7(\lambda \wedge \chi)|$ for all 1-forms λ on M.

Both conditions hold if ϵ_1 is sufficiently small. The idea in part (i) is that $\tilde{\Lambda}^2_{14}$ is close to Λ^2_{14} when ϵ_1 is small, so if $\xi \in \tilde{\Lambda}^2_{14}$ then $\pi_7(\xi)$ is small compared to $\pi_{14}(\xi)$. For part (ii), observe that $\lambda \wedge *\varphi$ lies in Λ^5_7, so $\lambda \wedge \chi$ is close to Λ^5_7 since χ is close to $*\varphi$. Thus $\pi_{14}(\lambda \wedge \chi)$ is small compared to $\pi_7(\lambda \wedge \chi)$.

The following rather curious result is based on [121, Lem. 3.1.5].

Proposition 10.3.4 *Let $\epsilon_1 > 0$ be as in Definition 10.3.3, and let M be a compact 7-manifold. Suppose (φ, g) is a G_2-structure, f a real function, α a 1-form, $\tilde{\varphi}$ a 3-form and χ a 4-form on M, satisfying $\|\tilde{\varphi} - \varphi\|_{C^0} \leqslant \epsilon_1$, $\|\chi - *\varphi\|_{C^0} \leqslant \epsilon_1$ and the equations*

$$\mathrm{d}\varphi = \mathrm{d}\tilde{\varphi} = \mathrm{d}\chi = 0 \quad \text{and} \quad \mathrm{d}\Theta(\tilde{\varphi}) = \mathrm{d}f \wedge \chi + \mathrm{d}\alpha \wedge \varphi. \tag{10.3}$$

Then $\mathrm{d}\Theta(\tilde{\varphi}) = 0$, $\mathrm{d}f = 0$ and $\mathrm{d}\alpha = 0$. Thus, $\tilde{\varphi}$ defines a torsion-free G_2-structure $(\tilde{\varphi}, \tilde{g})$ on M.

Proof Define $x_7, y_7, z_7 \in C^\infty(\Lambda^5_7)$ and $x_{14}, y_{14}, z_{14} \in C^\infty(\Lambda^5_{14})$ by

$$\begin{aligned} x_7 &= \pi_7\big(\mathrm{d}\Theta(\tilde{\varphi})\big), & y_7 &= \pi_7(\mathrm{d}f \wedge \chi), & z_7 &= \pi_7(\mathrm{d}\alpha \wedge \varphi), \\ x_{14} &= \pi_{14}\big(\mathrm{d}\Theta(\tilde{\varphi})\big), & y_{14} &= \pi_{14}(\mathrm{d}f \wedge \chi), & z_{14} &= \pi_{14}(\mathrm{d}\alpha \wedge \varphi). \end{aligned} \tag{10.4}$$

Taking π_7 and π_{14} of the last equation of (10.3) gives

$$x_7 = y_7 + z_7 \quad \text{and} \quad x_{14} = y_{14} + z_{14}. \tag{10.5}$$

As $\mathrm{d}\tilde{\varphi} = 0$, Lemma 10.3.1 shows that $\mathrm{d}\Theta(\tilde{\varphi}) \in C^\infty(\tilde{\Lambda}^5_{14})$, so (10.4) and part (i) of Definition 10.3.3 give that $|x_7| \leqslant |x_{14}|$. Also, putting $\lambda = \mathrm{d}f$ in part (ii) of Definition 10.3.3 and using (10.4) gives $|y_{14}| \leqslant \frac{1}{4}|y_7|$. Squaring these two inequalities and integrating over M gives

$$\|x_7\|_{L^2} \leqslant \|x_{14}\|_{L^2} \quad \text{and} \quad \|y_{14}\|_{L^2} \leqslant \frac{1}{4}\|y_7\|_{L^2}. \tag{10.6}$$

As φ is closed, $\mathrm{d}\alpha \wedge \mathrm{d}\alpha \wedge \varphi$ is exact, and $\int_M \mathrm{d}\alpha \wedge \mathrm{d}\alpha \wedge \varphi = 0$ by Stokes' Theorem. But

$$\begin{aligned} \mathrm{d}\alpha \wedge \mathrm{d}\alpha \wedge \varphi &= \{2|\pi_7(\mathrm{d}\alpha)|^2 - |\pi_{14}(\mathrm{d}\alpha)|^2\}\mathrm{d}V_g \\ &= \{\tfrac{1}{2}|z_7|^2 - |z_{14}|^2\}\mathrm{d}V_g, \end{aligned}$$

by Lemma 10.2.5 and the equations $\pi_7(\mathrm{d}\alpha) = \frac{1}{2}*z_7$ and $\pi_{14}(\mathrm{d}\alpha) = -*z_{14}$, which follow from (10.4) using Lemma 10.2.5. Integrating over M gives $\frac{1}{2}\|z_7\|^2_{L^2} - \|z_{14}\|^2_{L^2} = 0$, so that $\|z_7\|_{L^2} = \sqrt{2}\|z_{14}\|_{L^2}$. Similarly,

$$\int_M \mathrm{d}\Theta(\tilde{\varphi}) \wedge \mathrm{d}\alpha = \int_M (x_7 + x_{14}) \wedge (\tfrac{1}{2}*z_7 - *z_{14}) = 0.$$

Thus $\langle x_7, z_7 \rangle = 2\langle x_{14}, z_{14} \rangle$, where $\langle \, , \, \rangle$ is the L^2-inner product, and we have

$$\|z_7\|_{L^2} = \sqrt{2}\|z_{14}\|_{L^2} \quad \text{and} \quad \langle x_7, z_7 \rangle = 2\langle x_{14}, z_{14} \rangle. \tag{10.7}$$

From (10.5), (10.6) and (10.7) we have

$$\begin{aligned}\|x_{14} - z_{14}\|_{L^2} &= \|y_{14}\|_{L^2} \leqslant \tfrac{1}{4}\|y_7\|_{L^2} \leqslant \tfrac{1}{4}\bigl(\|x_7\|_{L^2} + \|z_7\|_{L^2}\bigr) \\ &\leqslant \tfrac{\sqrt{2}}{4}\bigl(\|x_{14}\|_{L^2} + \|z_{14}\|_{L^2}\bigr).\end{aligned} \tag{10.8}$$

Using this one can show that $\langle x_{14}, z_{14} \rangle \geqslant \tfrac{3}{4}\|x_{14}\|_{L^2}\|z_{14}\|_{L^2}$. Therefore

$$\begin{aligned}\|x_7\|_{L^2}\|z_7\|_{L^2} &\geqslant \langle x_7, z_7 \rangle = 2\langle x_{14}, z_{14} \rangle \\ &\geqslant \tfrac{3}{2}\|x_{14}\|_{L^2}\|z_{14}\|_{L^2} \geqslant \tfrac{3\sqrt{2}}{4}\|x_7\|_{L^2}\|z_7\|_{L^2},\end{aligned}$$

by (10.6) and (10.7). Thus $\|x_7\|_{L^2} = 0$ or $\|z_7\|_{L^2} = 0$, as $3\sqrt{2}/4 > 1$, and so $x_7 = 0$ or $z_7 = 0$. If $z_7 = 0$ then $z_{14} = 0$ as $\|z_7\|_{L^2} = \sqrt{2}\|z_{14}\|_{L^2}$, so $x_7 = y_7$ and $x_{14} = y_{14}$. But

$$\|x_7\|_{L^2} \leqslant \|x_{14}\|_{L^2} = \|y_{14}\|_{L^2} \leqslant \tfrac{1}{4}\|y_7\|_{L^2} = \tfrac{1}{4}\|x_7\|_{L^2},$$

so that $\|x_7\|_{L^2} = 0$, and hence $x_7 = x_{14} = y_7 = y_{14} = 0$.

If $x_7 = 0$ then $y_7 = -z_7$, and so $\|y_{14}\|_{L^2} \leqslant \tfrac{1}{4}\|y_7\|_{L^2} = \tfrac{1}{4}\|z_7\|_{L^2} = \tfrac{\sqrt{2}}{4}\|z_{14}\|_{L^2}$. Now if $z_{14} \neq 0$ then $\|y_{14}\|_{L^2} < \|z_{14}\|_{L^2}$ so that $\langle x_{14}, z_{14} \rangle > 0$, projecting the equation $x_{14} = y_{14} + z_{14}$ onto z_{14}. But this means that $\langle x_7, z_7 \rangle > 0$, contradicting $x_7 = 0$. Therefore $z_{14} = 0$, so $z_7 = 0$ and we have reduced to the previous case. Thus in both cases we have $x_7 = y_7 = z_7 = 0$ and $x_{14} = y_{14} = z_{14} = 0$, which gives $d\Theta(\tilde{\varphi}) = df = d\alpha = 0$, as we have to prove. □

Next we estimate the function Θ of Definition 10.1.2, as in [121, Lem. 3.1.1].

Proposition 10.3.5 *Let ϵ_1 be as in Definition 10.3.3. Then there exist constants $\epsilon_2, \epsilon_3 > 0$ such that whenever M is a 7-manifold and (φ, g) a G_2-structure on M with $d\varphi = 0$, then the following is true. Suppose $\chi \in C^\infty(\Lambda^3 T^*M)$ and $|\chi| \leqslant \epsilon_1$. Then $\varphi + \chi \in C^\infty(\mathcal{P}^3 M)$ and $\Theta(\varphi + \chi)$ is given by*

$$\begin{aligned}\Theta(\varphi + \chi) &= *\varphi + \tfrac{4}{3}*\pi_1(\chi) + *\pi_7(\chi) - *\pi_{27}(\chi) - F(\chi) \\ &= *\varphi + \tfrac{7}{3}*\pi_1(\chi) + 2*\pi_7(\chi) - *\chi - F(\chi),\end{aligned} \tag{10.9}$$

where F is a smooth function from the closed ball of radius ϵ_1 in $\Lambda^3 T^*M$ to $\Lambda^4 T^*M$ with $F(0) = 0$. Suppose $\chi, \xi \in C^\infty(\Lambda^3 T^*M)$ and $|\chi|, |\xi| \leqslant \epsilon_1$. Then

$$|F(\chi) - F(\xi)| \leqslant \epsilon_2 |\chi - \xi|(|\chi| + |\xi|) \quad \text{and} \tag{10.10}$$

$$\bigl|d\bigl(F(\chi) - F(\xi)\bigr)\bigr| \leqslant \epsilon_3 \Bigl\{|\chi - \xi|(|\chi| + |\xi|)|d^*\varphi| + \bigl|\nabla(\chi - \xi)\bigr|(|\chi| + |\xi|) \\ + |\chi - \xi|(|\nabla \chi| + |\nabla \xi|)\Bigr\}. \tag{10.11}$$

Proof By definition, the closed ball of radius ϵ_1 about φ in $\Lambda^3 T^*M$ is contained in $\mathcal{P}^3 M$. Let the function F be defined by (10.9). Then (10.9) holds, and F is a smooth function of ξ. By calculating in coordinates one can write $\Theta(\varphi+\chi)$ as an explicit function of χ up to first order in χ, and the answer is the first four terms of the right hand side of (10.9). So the function $F(\chi)$ has zero first derivative in χ at $\chi=0$. Thus the principal part of $F(\chi)$ is quadratic in χ.

It follows that (10.10) holds for some constant ϵ_2, and as this is a calculation at a point, ϵ_2 is independent of M, φ and g. By the chain rule, $dF(\chi)$ can be written $dF(\chi) = F_1(\varphi, \chi, \nabla\varphi) + F_2(\varphi, \chi, \nabla\chi)$, where F_1 and F_2 are linear in their third arguments. Since $d\varphi = 0$ by assumption, $\nabla\varphi$ is determined pointwise by $d^*\varphi$. Using this and estimates on F_1, F_2 it easy to show that dF satisfies an estimate of the form (10.11) for some $\epsilon_3 > 0$ independent of M, φ and g. $\square$

The following two theorems are the main results of this section.

Theorem 10.3.6 *Let (M, φ, g) be a compact G_2-manifold. Suppose $\xi \in \mathcal{H}^3$, and η is a 2-form on M such that $\|\xi + d\eta\|_{C^0} \leqslant \epsilon_1$. Let $\tilde\varphi = \varphi + \xi + d\eta$, so that $d\tilde\varphi = 0$. Then $(dd^* + d^*d)\eta = *dF(\xi + d\eta)$ if and only if $d^*\eta = 0$, $\pi_7(d^*\tilde\varphi) = 0$ and $d\Theta(\tilde\varphi) = 0$.*

Proof Suppose $(dd^* + d^*d)\eta = *dF(\xi + d\eta)$. As M is compact Im d and Im d^* are L^2-orthogonal, and so $d^*\eta = 0$ and $d^*d\eta = *dF(\xi+d\eta)$. This implies that $d*d\eta + dF(\xi+d\eta) = 0$, as $d^* = -*d*$ on $\Lambda^3 T^*M$. But (10.9) gives

$$d\Theta(\tilde\varphi) = \tfrac{7}{3} d*\pi_1(d\eta) + 2d*\pi_7(d\eta) - d*d\eta - dF(\xi+d\eta) \qquad (10.12)$$
$$= \tfrac{7}{3} d*\pi_1(d\eta) + 2d*\pi_7(d\eta),$$

using in the first line the fact that $*\pi_1(\xi)$, $*\pi_7(\xi)$ and $*\pi_{27}(\xi)$ are closed as $\xi \in \mathcal{H}^3$, and in the second the equation $d*d\eta + dF(\xi+d\eta) = 0$.

Applying Proposition 10.3.4 with $\psi = *\varphi$, $f\varphi = \tfrac{7}{3}\pi_1(d\eta)$ and $\alpha\wedge\varphi = 2\pi_7(d\eta)$, we see that

$$d\Theta(\tilde\varphi) = 0 \quad \text{and} \quad d*\pi_1(d\eta) = d*\pi_7(d\eta) = 0. \qquad (10.13)$$

Thus $d^*\pi_1(d\eta) = d^*\pi_7(d\eta) = 0$, and so $\pi_7(d^*d\eta) = 0$ by Lemma 10.3.2. But $d^*d\eta = d^*\tilde\varphi$. So we have shown that if $(dd^*+d^*d)\eta = *dF(\xi+d\eta)$ then $d^*\eta = 0$, $\pi_7(d^*\tilde\varphi) = 0$ and $d\Theta(\tilde\varphi) = 0$, which proves the 'only if' part of the theorem.

Now suppose that $d^*\eta = 0$, $\pi_7(d^*\tilde\varphi) = 0$ and $d\Theta(\tilde\varphi) = 0$. Then $d^*\pi_1(d\eta) = d^*\pi_7(d\eta) = 0$ by Lemma 10.3.2, which shows that

$$d\Theta(\tilde\varphi) = \tfrac{7}{3} d*\pi_1(d\eta) + 2d*\pi_7(d\eta) = 0,$$

and so $d*d\eta + dF(\xi+d\eta) = 0$ by the first line of (10.12). Since $d^*\eta = 0$, this gives $(dd^*+d^*d)\eta = *dF(\xi+d\eta)$, and the proof is complete. $\square$

This shows that the conditions $d\tilde\varphi = d\Theta(\tilde\varphi) = 0$ for $\tilde\varphi$ to represent a torsion-free G_2-structure, together with the 'gauge-fixing' condition $\pi_7(d^*\tilde\varphi) = 0$, are equivalent to the equation $(dd^*+d^*d)\eta = *dF(\xi+d\eta)$, which is a *nonlinear elliptic p.d.e.* upon the 2-form η. In §10.4 we will use Theorem 10.3.6 together with results on elliptic equations to study the family of torsion-free G_2-structures on a compact 7-manifold.

Theorem 10.3.7 *Let (φ, g) be a G_2-structure on a compact 7-manifold M with $d\varphi = 0$. Suppose η is a 2-form on M with $\|d\eta\|_{C^0} \leqslant \epsilon_1$, and ψ is a 3-form on M with $d^*\varphi = d^*\psi$ and $\|\psi\|_{C^0} \leqslant \epsilon_1$. Let $\tilde{\varphi} = \varphi + d\eta$, and define a real function f on M by $\frac{7}{3}\pi_1(d\eta) = f\varphi$. Then*

$$(dd^* + d^*d)\eta = d^*\psi + d^*(f\psi) + *dF(d\eta) \tag{10.14}$$

implies that $d\tilde{\varphi} = d\Theta(\tilde{\varphi}) = 0$, so $\tilde{\varphi}$ gives a torsion-free G_2-structure $(\tilde{\varphi}, \tilde{g})$ on M.

Proof As M is compact Im d and Im d* are L^2-orthogonal, so

$$d*d\eta = d*\psi + d*(f\psi) - dF(d\eta)$$

by (10.14). Combining this with (10.9) and the equation $d^*\varphi = d^*\psi$ gives

$$\begin{aligned}d\Theta(\tilde{\varphi}) &= d*\varphi + d(f*\varphi) + 2d*\pi_7(d\eta) - d*d\eta - dF(d\eta)\\ &= d*\varphi + d(f*\varphi) + 2d*\pi_7(d\eta) - d*\psi - d*(f\psi)\\ &= d(f\chi) + d(\alpha \wedge \varphi),\end{aligned}$$

using $\chi = *\varphi - *\psi$ and $\alpha \wedge \varphi = 2*\pi_7(d\eta)$. Clearly $d\tilde{\varphi} = 0$, and $d\chi = 0$ as $d^*\varphi = d^*\psi$. Also $\|\tilde{\varphi} - \varphi\|_{C^0} = \|d\eta\|_{C^0} \leqslant \epsilon_1$ and $\|\chi - *\varphi\|_{C^0} = \|\psi\|_{C^0} \leqslant \epsilon_1$ by assumption. So Proposition 10.3.4 applies, giving $d\Theta(\tilde{\varphi}) = 0$. □

Again, (10.14) is a *nonlinear elliptic p.d.e.* upon η. In Chapter 11 we will use results on elliptic equations to find a solution η to (10.14) for certain special M and (φ, g), and the theorem then gives us a torsion-free G_2-structure on M. Hence we shall construct metrics with holonomy G_2 on compact 7-manifolds.

10.4 The moduli space of holonomy G_2 metrics

Let M be a compact, oriented 7-manifold. Then as in §10.1, we can identify the set of all oriented G_2-structures on M with $C^\infty(\mathcal{P}^3 M)$, the set of positive 3-forms on M. Let $\mathcal{X}$ be the set of positive 3-forms corresponding to oriented, torsion-free G_2-structures. That is,

$$\mathcal{X} = \{\varphi \in C^\infty(\mathcal{P}^3 M) : d\varphi = d\Theta(\varphi) = 0\}. \tag{10.15}$$

Let $\mathcal{D}$ be the group of all diffeomorphisms Ψ of M isotopic to the identity. Then $\mathcal{D}$ acts naturally on $C^\infty(\mathcal{P}^3 M)$ and $\mathcal{X}$ by $\varphi \xmapsto{\Psi} \Psi_*(\varphi)$. We define the *moduli space of torsion-free G_2-structures* on M to be $\mathcal{M} = \mathcal{X}/\mathcal{D}$.

In this section we will show that $\mathcal{M}$ is a *nonsingular, smooth manifold*, with dimension $b^3(M)$. This is not obvious, as $\mathcal{X}$ is an infinite-dimensional manifold and $\mathcal{D}$ an infinite-dimensional Lie group, so their quotient $\mathcal{M} = \mathcal{X}/\mathcal{D}$ inherits the structure of a topological space, but little more than this.

The standard technique used in similar problems is to look for a 'slice' for the action of $\mathcal{D}$ on $\mathcal{X}$. This is a submanifold S of $\mathcal{X}$ which is (locally) transverse to the orbits of $\mathcal{D}$, so that each nearby orbit of $\mathcal{D}$ meets S in a single point. If such a slice can be found then $\mathcal{X}/\mathcal{D}$ is locally isomorphic to S, and thus has the structure of a manifold. Here is a 'slice' for the positive 3-forms on a 7-manifold.

252　INTRODUCTION TO THE EXCEPTIONAL HOLONOMY GROUPS

Theorem 10.4.1 *Suppose (M, φ, g) is a compact G_2-manifold, and let $\mathcal{D}$ be the group of diffeomorphisms of M isotopic to the identity. Then $\mathcal{D}$ acts on $C^\infty(\mathcal{P}^3 M)$. Let I_φ be the subgroup of $\mathcal{D}$ fixing (φ, g), and write*

$$L_\varphi = \{\tilde{\varphi} \in C^\infty(\mathcal{P}^3 M) : \pi_7(\mathrm{d}^* \tilde{\varphi}) = 0\}, \tag{10.16}$$

where π_7 and d^* come from the G_2-structure (φ, g). Then there is an open neighbourhood S_φ of φ in L_φ, invariant under I_φ, such that the natural projection from S_φ / I_φ to $C^\infty(\mathcal{P}^3 M)/\mathcal{D}$ induces a homeomorphism between S_φ/I_φ and a neighbourhood of $\varphi \mathcal{D}$ in $C^\infty(\mathcal{P}^3 M)/\mathcal{D}$.

This is based on Ebin's Slice Theorem [66] for the moduli space of Riemannian metrics, and can be proved in a very similar way. Note that I_φ is a compact, finite-dimensional Lie group (often trivial), and that L_φ and S_φ are open subsets of an infinite-dimensional vector space. Thus, S_φ/I_φ is a Hausdorff, singular, infinite-dimensional manifold.

Here is why the condition $\pi_7(\mathrm{d}^*\tilde{\varphi}) = 0$ occurs in (10.16). The tangent space at φ to the orbit $\varphi \mathcal{D}$ of $\mathcal{D}$ is the set of 3-forms $\mathcal{L}_v \varphi$, where $v \in C^\infty(TM)$ and $\mathcal{L}_v$ is the Lie derivative. Using the equations $\mathrm{d}\varphi = \mathrm{d}^*\varphi = 0$, one can show that (10.16) is equivalent to

$$L_\varphi = \{\tilde{\varphi} \in C^\infty(\mathcal{P}^3 M) : \langle \tilde{\varphi} - \varphi, \mathcal{L}_v \varphi \rangle_{L^2} = 0 \quad \text{for all } v \in C^\infty(TM)\}.$$

Thus L_φ is the L^2-orthogonal subspace to the orbit $\varphi \mathcal{D}$ of $\mathcal{D}$ at φ, and so it is the natural choice of slice at φ for the action of $\mathcal{D}$ on $C^\infty(\mathcal{P}^3 M)$.

Applying Theorem 10.4.1 to the set $\mathcal{X}$ of (10.15), we find:

Corollary 10.4.2 *Let M be a compact 7-manifold, and $\mathcal{M} = \mathcal{X}/\mathcal{D}$ the moduli space of torsion-free G_2-structures on M. Suppose $\varphi \mathcal{D} \in \mathcal{M}$, so that (φ, g) is a torsion-free G_2-structure on M, and let I_φ be the subgroup of $\mathcal{D}$ fixing (φ, g). Define*

$$L'_\varphi = \{\tilde{\varphi} \in C^\infty(\mathcal{P}^3 M) : \mathrm{d}\tilde{\varphi} = \mathrm{d}\,\Theta(\tilde{\varphi}) = 0 \text{ and } \pi_7(\mathrm{d}^*\tilde{\varphi}) = 0\}, \tag{10.17}$$

where π_7 and d^* come from (φ, g). Then there is an open neighbourhood S'_φ of φ in L'_φ, invariant under I_φ, such that the natural projection from S'_φ/I_φ to $\mathcal{M}$ induces a homeomorphism between S'_φ/I_φ and a neighbourhood of $\varphi \mathcal{D}$ in $\mathcal{M}$.

We shall use Theorem 10.3.6 to rewrite L'_φ in terms of a 2-form η.

Proposition 10.4.3 *Let (M, φ, g) be a compact G_2-manifold, and suppose $\tilde{\varphi}$ is a closed 3-form on M with $\|\tilde{\varphi} - \varphi\|_{C^0} \leq \epsilon_1$. Then we may write $\tilde{\varphi}$ uniquely as $\tilde{\varphi} = \varphi + \xi + \mathrm{d}\eta$, where $\xi \in \mathcal{H}^3$ and η is a d^*-exact 2-form. Moreover $\tilde{\varphi}$ lies in the set L'_φ of (10.17) if and only if $(\mathrm{d}\mathrm{d}^* + \mathrm{d}^*\mathrm{d})\eta = *\mathrm{d}F(\xi + \mathrm{d}\eta)$.*

Proof Let $[\varphi]$, $[\tilde{\varphi}]$ be the de Rham cohomology classes of $\varphi, \tilde{\varphi}$. As $\mathcal{H}^3 \cong H^3(M, \mathbb{R})$ there is a unique $\xi \in \mathcal{H}^3$ with $[\xi] = [\tilde{\varphi}] - [\varphi]$. Then $[\tilde{\varphi} - \varphi - \xi] = 0$ in $H^3(M, \mathbb{R})$, so $\tilde{\varphi} - \varphi - \xi$ is an exact 3-form. By Hodge theory, there is a unique d^*-exact 2-form η on M such that $\mathrm{d}\eta = \tilde{\varphi} - \varphi - \xi$. Thus $\tilde{\varphi} = \varphi + \xi + \mathrm{d}\eta$, as we want.

Now $\|\tilde{\varphi} - \varphi\|_{C^0} \leqslant \epsilon_1$ by assumption, and $d^*\eta = 0$ as η is d^*-exact. So Theorem 10.3.6 shows that $(dd^*+d^*d)\eta = *dF(\xi+d\eta)$ if and only if $\pi_7(d^*\tilde{\varphi}) = 0$ and $d\Theta(\tilde{\varphi}) = 0$. Since $d\tilde{\varphi} = 0$, from (10.17) we see that $\tilde{\varphi} \in L'_\varphi$ if and only if $(dd^*+d^*d)\eta = *dF(\xi+d\eta)$. □

Now we can prove our main result, based on [121, Th. C].

Theorem 10.4.4 *Let M be a compact 7-manifold, and $\mathcal{M} = \mathcal{X}/\mathcal{D}$ the moduli space of torsion-free G_2-structures on M defined above. Then $\mathcal{M}$ is a smooth manifold of dimension $b^3(M)$, and the natural projection $\pi : \mathcal{M} \to H^3(M, \mathbb{R})$ given by $\pi(\varphi\mathcal{D}) = [\varphi]$ is a local diffeomorphism.*

Proof Suppose (φ, g) is a torsion-free G_2-structure on M. For each $k \geqslant 0$ and $\alpha \in (0, 1)$, let $V^{k,\alpha}$ be the Banach space of 2-forms given by

$$V^{k,\alpha} = \{\eta \in C^{k,\alpha}(\Lambda^2 T^*M) : \eta \text{ is } L^2\text{-orthogonal to } \mathcal{H}^2\}.$$

Define an open set $U^{k+2,\alpha} \subset \mathcal{H}^3 \times V^{k+2,\alpha}$ and a map $\Phi : U^{k+2,\alpha} \to V^{k,\alpha}$ by

$$U^{k+2,\alpha} = \{(\xi, \eta) \in \mathcal{H}^3 \times V^{k+2,\alpha} : \|\xi + d\eta\|_{C^0} < \epsilon_1\}$$
and $\Phi(\xi, \eta) = (dd^*+d^*d)\eta - *dF(\xi + d\eta)$.

Then Φ is well-defined, and a smooth, nonlinear map of Banach spaces.

Since $F(\xi + d\eta)$ is at least quadratic in $\xi, d\eta$ by Proposition 10.3.5, the first derivative $d\Phi|_{(0,0)} : \mathcal{H}^3 \times V^{k+2,\alpha} \to V^{k,\alpha}$ is given by $d\Phi|_{(0,0)}(\xi, \eta) = (dd^*+d^*d)\eta$. But as $dd^* + d^*d$ is a self-adjoint elliptic operator on 2-forms, with kernel and cokernel $\mathcal{H}^2$, we can show using Theorem 1.5.3 that $dd^*+d^*d : V^{k+2,\alpha} \to V^{k,\alpha}$ is an isomorphism. Hence $d\Phi|_{(0,0)} : \mathcal{H}^3 \times V^{k+2,\alpha} \to V^{k,\alpha}$ is surjective, and has kernel $\mathcal{H}^3$.

By the Implicit Mapping Theorem, Theorem 1.2.5, we see that $\Phi^{-1}(0)$ is a manifold of dimension $b^3(M)$ in a neighbourhood of $(0, 0)$, and that the projection $(\xi, \eta) \mapsto \xi$ induces a diffeomorphism between neighbourhoods of $(0, 0)$ in $\Phi^{-1}(0)$ and 0 in $\mathcal{H}^3$. Also, for small $\|\xi + d\eta\|_{C^0}$ the equation $\Phi(\xi, \eta) = 0$ is a nonlinear elliptic equation in η, so by elliptic regularity any solution η must be smooth, rather than just $C^{k+2,\alpha}$.

Now by Proposition 10.4.3, the set L'_φ of (10.17) is locally isomorphic to

$$\{(\xi, \eta) \in \mathcal{H}^3 \times C^\infty(\Lambda^2 T^*M) : \eta \text{ is } L^2\text{-orthogonal to } \mathcal{H}^2,$$
$$\|\xi + d\eta\|_{C^0} < \epsilon_1 \text{ and } (dd^*+d^*d)\eta = *dF(\xi + d\eta)\}.$$

But we have just shown that near $(0, 0)$, this set is a manifold of dimension $b^3(M)$ and the projection to $\mathcal{H}^3$ is a diffeomorphism. Therefore L'_φ is a smooth manifold of dimension $b^3(M)$ in a neighbourhood of φ, and the projection from L'_φ to $H^3(M, \mathbb{R})$ mapping $\tilde{\varphi}$ to $[\tilde{\varphi}]$ induces a diffeomorphism between neighbourhoods of $\varphi \in L'_\varphi$ and $[\varphi] \in H^3(M, \mathbb{R})$.

By Corollary 10.4.2, the moduli space $\mathcal{M}$ is homeomorphic near $\varphi\mathcal{D}$ to a neighbourhood of φI_φ in L'_φ/I_φ, where I_φ is the subgroup of $\mathcal{D}$ fixing (φ, g). Now I_φ is a group of diffeomorphisms of M isotopic to the identity, so I_φ acts trivially on $H^3(M, \mathbb{R})$. But

L'_φ is isomorphic to $H^3(M, \mathbb{R})$ near φ, so I_φ acts trivially on L'_φ near φ, and L'_φ/I_φ is locally isomorphic to L'_φ.

Thus $\mathcal{M}$ is homeomorphic near $\varphi\mathcal{D}$ to a neighbourhood of φ in L'_φ. Therefore $\mathcal{M}$ is a smooth manifold of dimension $b^3(M)$ in a neighbourhood of $\varphi\mathcal{D}$, and the natural projection from $\pi : \mathcal{M} \to H^3(M, \mathbb{R})$ mapping $\tilde{\varphi}\mathcal{D}$ to $[\tilde{\varphi}]$ induces a diffeomorphism between neighbourhoods of $\varphi\mathcal{D} \in \mathcal{M}$ and $[\varphi] \in H^3(M, \mathbb{R})$. Since this holds for all $\varphi\mathcal{D} \in \mathcal{M}$, the proof is complete. □

Note that Theorem 10.4.4 is an entirely *local* result, and it gives little information about the global structure of $\mathcal{M}$. For instance, we do not know whether $\mathcal{M}$ is nonempty, or if it has one connected component or many, or whether the map $\pi : \mathcal{M} \to H^3(M, \mathbb{R})$ is injective, or what the image of π is.

Finally we show that the image of $\mathcal{M}$ in $H^3(M, \mathbb{R}) \times H^4(M, \mathbb{R})$ is a *Lagrangian submanifold*, as in [122, Lem. 1.1.3].

Proposition 10.4.5 *Let M be a compact, oriented 7-manifold. By Poincaré duality $H^4(M, \mathbb{R}) \cong H^3(M, \mathbb{R})^*$, so $H^3(M, \mathbb{R}) \times H^4(M, \mathbb{R})$ has a natural symplectic structure. Define a subset L in $H^3(M, \mathbb{R}) \times H^4(M, \mathbb{R})$ by*

$$L = \Big\{ ([\varphi], [*\varphi]) : (\varphi, g) \text{ is a torsion-free } G_2\text{-structure on } M \Big\}.$$

Then L is a Lagrangian submanifold of $H^3(M, \mathbb{R}) \times H^4(M, \mathbb{R})$.

Proof By Theorem 10.4.4, the moduli space $\mathcal{M}$ of torsion-free G_2-structures is a manifold locally isomorphic to $H^3(M, \mathbb{R})$. Thus L is a manifold of dimension $b^3(M)$, with nonsingular projection to $H^3(M, \mathbb{R})$. Identifying $H^4(M, \mathbb{R})$ with $H^3(M, \mathbb{R})^*$, we will show that L may be written locally in the form $(x, \mathrm{d}f(x))$, where f is a smooth real function on $H^3(M, \mathbb{R})$ defined by $f([\varphi]) = \frac{3}{7}[\varphi] \cup [\Theta(\varphi)]$.

Let $\{(\varphi_t, g_t) : t \in (-\epsilon, \epsilon)\}$ be a smooth family of torsion-free G_2-structures on M. Then

$$\begin{aligned}\frac{\partial f([\varphi_t])}{\partial t} &= \frac{3}{7} \int_M \Big\{ \frac{\partial \varphi_t}{\partial t} \wedge \Theta(\varphi_t) + \varphi_t \wedge \frac{\partial \Theta(\varphi_t)}{\partial t} \Big\} \\ &= \int_M \frac{\partial \varphi_t}{\partial t} \wedge \Theta(\varphi_t) = \frac{\partial [\varphi_t]}{\partial t} \cup [\Theta(\varphi_t)],\end{aligned} \quad (10.18)$$

as $\pi_1(\Theta(\varphi + \chi)) = *\varphi + \frac{4}{3} * \pi_1(\chi) + O(|\chi|^2)$ by Proposition 10.3.5. From (10.18) we see that $\mathrm{d}f([\varphi]) = [\Theta(\varphi)]$. Thus L is locally the graph of an exact 1-form on $H^3(M, \mathbb{R})$, so it is a Lagrangian submanifold. □

10.5 The holonomy group Spin(7)

We shall now discuss the holonomy group Spin(7), following the pattern of §10.1 for G_2. First we define Spin(7) as a subgroup of $\mathrm{GL}(8, \mathbb{R})$.

Definition 10.5.1 Let $\mathbb{R}^8$ have coordinates $(x_1, \ldots, x_8)$. Write $\mathrm{d}x_{ijkl}$ for the 4-form $\mathrm{d}x_i \wedge \mathrm{d}x_j \wedge \mathrm{d}x_k \wedge \mathrm{d}x_l$ on $\mathbb{R}^8$. Define a 4-form Ω_0 on $\mathbb{R}^8$ by

$$\begin{aligned}\Omega_0 = {} & \mathrm{d}x_{1234} + \mathrm{d}x_{1256} + \mathrm{d}x_{1278} + \mathrm{d}x_{1357} - \mathrm{d}x_{1368} \\ & - \mathrm{d}x_{1458} - \mathrm{d}x_{1467} - \mathrm{d}x_{2358} - \mathrm{d}x_{2367} - \mathrm{d}x_{2457} \\ & + \mathrm{d}x_{2468} + \mathrm{d}x_{3456} + \mathrm{d}x_{3478} + \mathrm{d}x_{5678}.\end{aligned} \tag{10.19}$$

The subgroup of $\mathrm{GL}(8, \mathbb{R})$ preserving Ω_0 is the holonomy group $\mathrm{Spin}(7)$. It is a compact, connected, simply-connected, semisimple, 21-dimensional Lie group, which is isomorphic as a Lie group to the double cover of $\mathrm{SO}(7)$. This group also preserves the orientation on $\mathbb{R}^8$ and the Euclidean metric $g_0 = \mathrm{d}x_1^2 + \cdots + \mathrm{d}x_8^2$ on $\mathbb{R}^8$. We have $*\Omega_0 = \Omega_0$, where $*$ is the Hodge star on $\mathbb{R}^8$, so that Ω_0 is a self-dual 4-form.

Let M be an oriented 8-manifold. For each $p \in M$, define $\mathcal{A}_p M$ to be the subset of 4-forms $\Omega \in \Lambda^4 T_p^* M$ for which there exists an oriented isomorphism between $T_p M$ and $\mathbb{R}^8$ identifying Ω and the 4-form Ω_0 of (10.19). Then $\mathcal{A}_p M$ is isomorphic to $\mathrm{GL}_+(8, \mathbb{R})/\mathrm{Spin}(7)$. Let $\mathcal{A}M$ be the bundle over M with fibre $\mathcal{A}_p M$ at each $p \in M$. Then $\mathcal{A}M$ is a subbundle of $\Lambda^4 T^* M$ with fibre $\mathrm{GL}_+(8, \mathbb{R})/\mathrm{Spin}(7)$. We say that a 4-form Ω on M is *admissible* if $\Omega|_p \in \mathcal{A}_p M$ for each $p \in M$.

Note that $\mathcal{A}M$ is *not* a vector subbundle of $\Lambda^4 T^* M$. As $\dim \mathrm{GL}_+(8, \mathbb{R}) = 64$ and $\dim \mathrm{Spin}(7) = 21$, we see that $\mathcal{A}_p M$ has dimension $64 - 21 = 43$. But $\Lambda^4 T_p^* M$ has dimension $\binom{8}{4} = 70$, so $\mathcal{A}_p M$ has codimension 27 in $\Lambda^4 T_p^* M$. This is rather different from the G_2 case of §10.1, in which $\mathcal{P}_p^3 M$ is *open* in $\Lambda^3 T_p^* M$. As in §10.1, there is a 1-1 correspondence between oriented Spin(7)-structures Q and admissible 4-forms $\Omega \in C^\infty(\mathcal{A}M)$ on M. Each Spin(7)-structure Q induces a 4-form Ω on M and a metric g on M, corresponding to Ω_0 and g_0 on $\mathbb{R}^8$.

Definition 10.5.2 Let M be an oriented 8-manifold, Ω an admissible 4-form on M, and g the associated metric. As for G_2, we shall abuse notation by referring to the pair (Ω, g) as a Spin(7)-*structure* on M. Let ∇ be the Levi-Civita connection of g. We call $\nabla \Omega$ the *torsion* of (Ω, g), and we say that (Ω, g) is *torsion-free* if $\nabla \Omega = 0$. A triple (M, Ω, g) is called a Spin(7)-*manifold* if M is an oriented 8-manifold, and (Ω, g) a torsion-free Spin(7)-structure on M.

This idea of Spin(7)-manifold is not in common use, but we will find it helpful. The next four results are analogues of Propositions 10.1.3–10.1.6. The first three follow from Salamon [186, Lem. 12.4], [186, Prop. 12.5] and [186, Cor. 12.6] respectively, and the fourth is proved in the same way as Proposition 10.1.6.

Proposition 10.5.3 *Let M be an 8-manifold and (Ω, g) a Spin(7)-structure on M. Then the following are equivalent:*

(i) *(Ω, g) is torsion-free,*
(ii) *$\mathrm{Hol}(g) \subseteq \mathrm{Spin}(7)$, and Ω is the induced 4-form,*
(iii) *$\nabla \Omega = 0$ on M, where ∇ is the Levi-Civita connection of g, and*
(iv) *$\mathrm{d}\Omega = 0$ on M.*

Here $d\Omega = 0$ is a linear equation on the 4-form Ω. However, the restriction that $\Omega \in C^\infty(\mathcal{A}M)$ is nonlinear. Thus, as in the G_2 case, the condition that (Ω, g) be a torsion-free Spin(7)-structure should be interpreted as a *nonlinear p.d.e.* upon Ω.

Proposition 10.5.4 *Let M be an oriented 8-manifold and (Ω, g) a Spin(7)-structure on M. Then $\Lambda^k T^*M$ splits orthogonally into components, where Λ_l^k corresponds to an irreducible representation of Spin(7) of dimension l:*

(i) $\Lambda^1 T^*M = \Lambda_8^1$, (ii) $\Lambda^2 T^*M = \Lambda_7^2 \oplus \Lambda_{21}^2$, (iii) $\Lambda^3 T^*M = \Lambda_8^3 \oplus \Lambda_{48}^3$,
(iv) $\Lambda^4 T^*M = \Lambda_+^4 T^*M \oplus \Lambda_-^4 T^*M$, $\Lambda_+^4 T^*M = \Lambda_1^4 \oplus \Lambda_7^4 \oplus \Lambda_{27}^4$, $\Lambda_-^4 T^*M = \Lambda_{35}^4$,
(v) $\Lambda^5 T^*M = \Lambda_8^5 \oplus \Lambda_{48}^5$, (vi) $\Lambda^6 T^*M = \Lambda_7^6 \oplus \Lambda_{21}^6$, (vii) $\Lambda^7 T^*M = \Lambda_8^7$.

The Hodge star $*$ gives an isometry between Λ_l^k and Λ_l^{8-k}. In part (iv), $\Lambda_\pm^4 T^*M$ are the ± 1-eigenspaces of $*$ on $\Lambda^4 T^*M$. Note also that $\Lambda_1^4 = \langle \Omega \rangle$, and that there are canonical isomorphisms $\Lambda_8^1 \cong \Lambda_8^3 \cong \Lambda_8^5 \cong \Lambda_8^7$ and $\Lambda_7^2 \cong \Lambda_7^4 \cong \Lambda_7^6$.

The orthogonal projection from $\Lambda^k T^*M$ to Λ_l^k will be written π_l.

Proposition 10.5.5 *Let (M, g) be a Riemannian 8-manifold. If Hol(g) is a subgroup of Spin(7), then g is Ricci-flat.*

Proposition 10.5.6 *Let (Ω, g) be a Spin(7)-structure on an 8-manifold M. Then M is orientable and spin, with a preferred spin structure and orientation. If (Ω, g) is torsion-free, then (M, g) has a nonzero parallel positive spinor.*

In fact, if $S = S_+ \oplus S_-$ is the spin bundle of M, then there are natural isomorphisms $S_+ \cong \Lambda_1^0 \oplus \Lambda_7^2$ and $S_- \cong \Lambda_8^1$. Here is the analogue of Theorem 10.2.1 in the Spin(7) case.

Theorem 10.5.7 *The only connected Lie subgroups of Spin(7) which can be the holonomy group of a Riemannian metric on an 8-manifold are:*
  (i) $\{1\}$,
  (ii) SU(2), *acting on* $\mathbb{R}^8 \cong \mathbb{R}^4 \oplus \mathbb{C}^2$ *trivially on* $\mathbb{R}^4$ *and as usual on* $\mathbb{C}^2$,
  (iii) SU(2) $\times$ SU(2), *acting on* $\mathbb{R}^8 \cong \mathbb{C}^2 \oplus \mathbb{C}^2$ *in the obvious way,*
  (iv) SU(3), *acting on* $\mathbb{R}^8 \cong \mathbb{R}^2 \oplus \mathbb{C}^3$ *trivially on* $\mathbb{R}^2$ *and as usual on* $\mathbb{C}^3$,
  (v) G_2, *acting on* $\mathbb{R}^8 \cong \mathbb{R} \oplus \mathbb{R}^7$ *trivially on* $\mathbb{R}$ *and as usual on* $\mathbb{R}^7$,
  (vi) Sp(2), (vii) SU(4), *and* (viii) Spin(7), *each acting as usual on* $\mathbb{R}^8$.
Thus, if (Ω, g) is a torsion-free Spin(7)-structure on an 8-manifold, then Hol$^0(g)$ *is one of* $\{1\}$, SU(2), SU(2) $\times$ SU(2), SU(3), G_2, Sp(2), SU(4) *or* Spin(7).

The inclusions '$\longrightarrow$' between these groups are shown below.

$$\begin{array}{ccccccc}
\mathrm{SU}(2) & = & \mathrm{SU}(2) & \longrightarrow & \mathrm{SU}(3) & \longrightarrow & G_2 \\
\downarrow & & \downarrow & & \downarrow & & \downarrow \\
\mathrm{SU}(2) \times \mathrm{SU}(2) & \longrightarrow & \mathrm{Sp}(2) & \longrightarrow & \mathrm{SU}(4) & \longrightarrow & \mathrm{Spin}(7).
\end{array}$$

Next we prove some results on the bundle of admissible 4-forms $\mathcal{A}M$. Let (Ω, g) be a Spin(7)-structure on an 8-manifold M, and let $p \in M$. From above, the fibre $\mathcal{A}_p M$ of $\mathcal{A}M$ at p is of codimension 27 in $\Lambda^4 T_p^* M$. Thus the tangent space $T_\Omega \mathcal{A}_p M$ of $\mathcal{A}_p M$ at Ω is a codimension 27 vector subspace of $\Lambda^4 T_p^* M$. By Proposition 10.5.4 we see that

$$T_\Omega \mathcal{A}_p M = (\Lambda_1^4 \oplus \Lambda_7^4 \oplus \Lambda_{35}^4)|_p,$$

since this is the only natural codimension 27 vector subspace of $\Lambda^4 T_p^* M$, and so the normal vector space to $\mathcal{A}_p M$ in $\Lambda^4 T_p^* M$ at Ω is Λ_{27}^4. This motivates the following definition.

Definition 10.5.8 Let M be an oriented 8-manifold. Let $\rho > 0$ be a small positive constant, and for each $p \in M$ define

$$\mathcal{T}_p M = \Big\{ \Omega_p + \phi_p : \Omega_p \in \mathcal{A}_p M, \text{ so that } \Omega_p \text{ induces a Spin(7)-structure}$$
$$(\Omega_p, g_p) \text{ at } p, \text{ and } \phi_p \text{ lies in } \Lambda_{27}^4|_p \text{ in the splitting induced}$$
$$\text{by } (\Omega_g, g_p) \text{ with } |\phi_p|_p < \rho, \text{ where } |\,.\,|_p \text{ is calculated using } g_p \Big\}.$$

Then $\mathcal{T}_p M$ is an open neighbourhood of $\mathcal{A}_p M$ in $\Lambda^4 T_p^* M$. Let $\mathcal{T}M$ be the subbundle of $\Lambda^4 T^* M$ with fibre $\mathcal{T}_p M$ at $p \in M$. Note that $\mathcal{T}M$ does *not* depend on a choice of Spin(7)-structure (Ω, g) on M, since the definition uses all possible Spin(7)-structures in an equal way. Instead, $\mathcal{T}M$ depends only on ρ and the structure of M as an oriented 8-manifold.

It is easy to see that if $\rho > 0$ is chosen small enough, then any 4-form $\chi_p \in \mathcal{T}_p M$ can be written *uniquely* as $\chi_p = \Omega_p + \phi_p$, where Ω_p and ϕ_p satisfy the properties above. Choose ρ this small, and define a map $\Theta_p : \mathcal{T}_p M \to \mathcal{A}_p M$ by $\Theta_p(\Omega_p + \phi_p) = \Omega_p$. Then Θ_p is smooth and surjective. Define $\Theta : \mathcal{T}M \to \mathcal{A}M$ to be Θ_p on $\mathcal{T}_p M$ for each $p \in M$. Then Θ is a smooth map of bundles. Again, Θ depends only on the structure of M as an oriented 8-manifold.

We think of $\mathcal{T}M$ as a *tubular neighbourhood* of $\mathcal{A}M$ in $\Lambda^4 T^* M$. If $\chi \in C^\infty(\mathcal{T}M)$, then $\Theta(\chi) \in C^\infty(\mathcal{A}M)$, and we regard $\Theta(\chi)$ as the section of $\mathcal{A}M$ closest to χ. In Chapter 13, we will use Θ in the following way. We will construct a compact 8-manifold M, and a closed 4-form χ on M near to $\mathcal{A}M$.

As χ is near to $\mathcal{A}M$ it lies in $C^\infty(\mathcal{T}M)$, and so $\Omega = \Theta(\chi)$ is a section of $\mathcal{A}M$, and defines a Spin(7)-structure (Ω, g) on M. Because $d\chi = 0$ and $\chi - \Omega$ is small we find that $d\Omega$ is small. Thus (Ω, g) has *small torsion*. We will then show that we can deform (Ω, g) to a nearby, *torsion-free* Spin(7)-structure $(\tilde{\Omega}, \tilde{g})$.

The following proposition is modelled on Proposition 10.3.5 in the G_2 case. The proof is very similar, so we omit it.

Proposition 10.5.9 *There exist constants $\epsilon_1, \epsilon_2, \epsilon_3 > 0$ such that whenever (Ω, g) is a Spin(7)-structure on an 8-manifold M, then the following is true. Suppose $\phi \in C^\infty(\Lambda^4 T^* M)$ and $|\phi| \leq \epsilon_1$. Then $\Omega + \phi \in C^\infty(\mathcal{T}M)$ and*

$$\Theta(\Omega + \phi) = \Omega + \pi_1(\phi) + \pi_7(\phi) + \pi_{35}(\phi) - F(\phi), \tag{10.20}$$

where F is a smooth function from the closed ball of radius ϵ_1 in $\Lambda^4 T^* M$ into $\Lambda^4 T^* M$ with $F(0) = 0$. Suppose $\phi, \psi \in C^\infty(\Lambda^4 T^* M)$ and $|\phi|, |\psi| \leqslant \epsilon_1$. Then

$$|F(\phi) - F(\psi)| \leqslant \epsilon_2 |\phi - \psi|(|\phi| + |\psi|) \quad \text{and} \tag{10.21}$$

$$|\nabla(F(\phi) - F(\psi))| \leqslant \epsilon_3 \Big\{ |\phi - \psi|(|\phi| + |\psi|)|d\Omega| + |\nabla(\phi - \psi)|(|\phi| + |\psi|) \\ + |\phi - \psi|(|\nabla \phi| + |\nabla \psi|) \Big\}. \tag{10.22}$$

Our next result will be used in §13.3 to construct Spin(7)-structures with small torsion. Note that Ω' and ϕ' depend only on the closed 4-form $\Omega + \phi$.

Proposition 10.5.10 *There exist $\epsilon_4, \epsilon_5 > 0$ such that the following holds. Let (Ω, g) be a Spin(7)-structure on an 8-manifold M, and suppose $\phi \in C^\infty(\Lambda^4 T^* M)$ satisfies $|\phi| \leqslant \epsilon_1$ and $d\Omega + d\phi = 0$. Then $\Omega + \phi \in C^\infty(\mathcal{T} M)$, so $\Omega' = \Theta(\Omega + \phi)$ lies in $C^\infty(\mathcal{A} M)$. Define (Ω', g') to be the induced Spin(7)-structure, and set $\phi' = \Omega + \phi - \Omega'$. Then $d\Omega' + d\phi' = 0$, and*

$$|\phi'|_{g'} \leqslant \epsilon_4 \big(|\pi_{27}(\phi)|_g + |\phi|_g^2 \big) \quad \text{and} \quad |\nabla' \phi'|_{g'} \leqslant \epsilon_5 |\nabla \phi|_g. \tag{10.23}$$

Here $| \cdot |_g$, ∇ and π_{27} are calculated using g, Ω, and $| \cdot |_{g'}$, ∇' using g'.

Proof As $d\Omega + d\phi = 0$ and $\Omega' + \phi' = \Omega + \phi$, we have $d\Omega' + d\phi' = 0$. Now $\phi' = \Omega + \phi - \Theta(\Omega + \phi)$, so (10.20) gives $\phi' = \pi_{27}(\phi) + F(\phi)$. Therefore

$$|\phi'|_g \leqslant |\pi_{27}(\phi)|_g + |F(\phi)|_g \leqslant |\pi_{27}(\phi)|_g + \epsilon_2 |\phi|_g^2,$$

using (10.21) with $\psi = 0$ and $F(\psi) = 0$ to estimate $|F(\phi)|_g$. But since g, g' are close $|\phi'|_g$ and $|\phi'|_{g'}$ differ by no more than a fixed factor, and so there exists $\epsilon_4 > 0$ such that the first equation of (10.23) holds.

As $\phi' = \pi_{27}(\phi) + F(\phi)$ we have $|\nabla' \phi|_g \leqslant |\nabla \pi_{27}(\phi)|_g + |\nabla F(\phi)|_g$. Now there exists $C > 0$ independent of M, Ω, ϕ such that

$$|\nabla \pi_{27}(\phi)|_g \leqslant |\nabla \phi|_g + C|\phi|_g |d\Omega|_g \leqslant |\nabla \phi|_g + C|\phi|_g |\nabla \phi|_g,$$

where $|d\Omega|_g = |d\phi|_g \leqslant |\nabla \phi|_g$ as $d\Omega = -d\phi$. Putting $\psi = 0$ in (10.22) gives

$$|\nabla F(\phi)|_g \leqslant \epsilon_3 \big(|\phi|_g^2 |d\Omega|_g + 2|\phi|_g |\nabla \phi|_g \big) \leqslant \epsilon_3 \big(|\phi|_g^2 |\nabla \phi|_g + 2|\phi|_g |\nabla \phi|_g \big).$$

Combining the last few equations and $|\phi|_g \leqslant \epsilon_1$, we see that

$$|\nabla \phi'|_g \leqslant \big\{ 1 + C\epsilon_1 + \epsilon_3(\epsilon_1^2 + 2\epsilon_1) \big\} |\nabla \phi|_g.$$

By bounding $|\nabla' \phi' - \nabla \phi'|_g$ and as g, g' are close, we show that the second equation of (10.23) holds for some $\epsilon_5 > 0$ independent of M, Ω, ϕ. □

10.6 The topology of compact Spin(7)-manifolds

Next we study the topology of *compact* Spin(7)*-manifolds* (M, Ω, g), that is, compact 8-manifolds M equipped with a torsion-free Spin(7)-structure (Ω, g). Because the dimension is divisible by four, we are able to use the *spin geometry* discussed in §3.6 to deduce some important topological restrictions on M, which have no parallel in the G_2 case. Since M is oriented and spin by Proposition 10.5.6, we may consider the spin bundle $S = S_+ \oplus S_-$ and the positive Dirac operator $D_+ : C^\infty(S_+) \to C^\infty(S_-)$ on M.

As in §3.6.2, the *index* ind(D_+) of D_+ is determined by Hol(g), since g is Ricci-flat. But in §3.6.3 we explained that by the Atiyah–Singer Index Theorem, ind(D_+) is equal to $\hat{A}(M)$, a characteristic class of M. Thus the *geometric* invariant Hol(g) determines the *topological* invariant $\hat{A}(M)$. We use this idea in the following theorem, taken from [120, Th. C].

Theorem 10.6.1 *Let* (M, Ω, g) *be a compact* Spin(7)*-manifold. Then the* $\hat{A}$*-genus* $\hat{A}(M)$ *of M satisfies*

$$24\hat{A}(M) = -1 + b^1 - b^2 + b^3 + b_+^4 - 2b_-^4, \qquad (10.24)$$

where b^i are the Betti numbers of M and $b_\pm^4$ the dimensions of $H_\pm^4(M, \mathbb{R})$. Moreover, if M is simply-connected then $\hat{A}(M)$ is 1, 2, 3 *or* 4, *and the holonomy group* Hol(g) *of g is determined by $\hat{A}(M)$ as follows:*

(i) Hol(g) = Spin(7) *if and only if* $\hat{A}(M) = 1$,
(ii) Hol(g) = SU(4) *if and only if* $\hat{A}(M) = 2$,
(iii) Hol(g) = Sp(2) *if and only if* $\hat{A}(M) = 3$, *and*
(iv) Hol(g) = SU(2) × SU(2) *if and only if* $\hat{A}(M) = 4$.

Proof To prove (10.24) we follow the reasoning of [184, §7]. For an 8-manifold, $\hat{A}(M)$ is given in terms of the Pontryagin classes $p_1(M)$ and $p_2(M)$ by

$$45.2^7 \hat{A}(M) = 7 p_1(M)^2 - 4 p_2(M).$$

From [184, eqn (7.1)], the signature $b_+^4 - b_-^4$ of M is given by

$$7 p_2(M) - p_1(M)^2 = 45(b_+^4 - b_-^4),$$

and by [184, eqn (7.3)], which applies to manifolds with structure group Spin(7) by the remark on [184, p. 166], the Euler characteristic of M satisfies

$$4 p_2(M) - p_1(M)^2 = 8(2 - 2b^1 + 2b^2 - 2b^3 + b^4).$$

Combining the last three equations gives (10.24), as we want.

Now suppose M is simply-connected. Theorem 10.5.7 shows that Hol(g) must be Spin(7), SU(4), Sp(2) or SU(2) × SU(2). Then Theorem 3.6.5 gives the value of $\hat{A}(M)$ for each holonomy group, which proves the 'only if' part of (i)–(iv). But as $\hat{A}(M)$ takes different values in the four cases, $\hat{A}(M)$ determines Hol(g). This gives the 'if' part of (i)–(iv), and completes the proof. □

Next we show that a compact 8-manifold with holonomy Spin(7) is simply-connected.

Proposition 10.6.2 *Let (M, g) be a compact Riemannian 8-manifold, and suppose $\mathrm{Hol}(g)$ is one of Spin(7), SU(4), Sp(2) and SU(2) × SU(2). Then M is simply-connected.*

Proof As g is Ricci-flat, $\pi_1(M)$ is finite by the argument used to prove Proposition 10.2.2. Let $\tilde{M}$ be the universal cover of M, and d the degree of the covering. Then $\tilde{M}$ is compact, and g lifts to a metric $\tilde{g}$ on $\tilde{M}$ with $\mathrm{Hol}(\tilde{g}) = \mathrm{Hol}^0(g) = \mathrm{Hol}(g)$. Thus $\hat{A}(\tilde{M}) = \hat{A}(M)$ by Theorem 10.6.1, and $\hat{A}(M) = 1, 2, 3$ or 4, depending on $\mathrm{Hol}(g)$. But $\hat{A}(\tilde{M}) = d \cdot \hat{A}(M)$, as the $\hat{A}$-genus is a characteristic class. Since $\hat{A}(M) \neq 0$, we see that $d = 1$, and so M is simply-connected. $\square$

By analogy with Definition 10.2.3 and Theorem 10.2.4, we have:

Definition 10.6.3 Let (M, Ω, g) be a compact Spin(7)-manifold. For each of the sub-bundles Λ^k_l of $\Lambda^k T^* M$ defined in Proposition 10.5.4, write

$$\mathcal{H}^k_l = \{\xi \in C^\infty(\Lambda^k_l) : \mathrm{d}\xi = \mathrm{d}^*\xi = 0\},$$

and let $H^k_l(M, \mathbb{R})$ be the vector subspace of $H^k(M, \mathbb{R})$ with representatives in $\mathcal{H}^k_l$. Define the *refined Betti numbers* b^k_l of M by $b^k_l = \dim H^k_l(M, \mathbb{R})$.

Theorem 10.6.4 *Let (M, Ω, g) be a compact Spin(7)-manifold. Then*

$$H^2(M, \mathbb{R}) = H^2_7(M, \mathbb{R}) \oplus H^2_{21}(M, \mathbb{R}), \qquad H^3(M, \mathbb{R}) = H^3_8(M, \mathbb{R}) \oplus H^3_{48}(M, \mathbb{R}),$$

$$H^4_+(M, \mathbb{R}) = H^4_1(M, \mathbb{R}) \oplus H^4_7(M, \mathbb{R}) \oplus H^4_{27}(M, \mathbb{R}), \qquad H^4_-(M, \mathbb{R}) = H^4_{35}(M, \mathbb{R}),$$

$$H^5(M, \mathbb{R}) = H^5_8(M, \mathbb{R}) \oplus H^5_{48}(M, \mathbb{R}), \qquad H^6(M, \mathbb{R}) = H^6_7(M, \mathbb{R}) \oplus H^6_{21}(M, \mathbb{R}).$$

Here $H^4_1(M, \mathbb{R}) = \langle [\Omega] \rangle$ and $H^k_l(M, \mathbb{R}) \cong H^{8-k}_l(M, \mathbb{R})$, so $b^4_1 = 1$ and $b^k_l = b^{8-k}_l$.

If (M, Ω, g) is compact Spin(7)-manifold, then

$$\hat{A}(M) = b^4_1 + b^4_7 - b^5_8 = 1 + b^2_7 - b^1. \tag{10.25}$$

To prove this we identify the Dirac operator $D_+ : C^\infty(S_+) \to C^\infty(S_-)$ with $\pi_8 \circ \mathrm{d} : C^\infty(\Lambda^4_1 \oplus \Lambda^4_7) \to C^\infty(\Lambda^5_8)$. This also suggests (correctly) that $\mathcal{H}^4_1$, $\mathcal{H}^4_7$ and $\mathcal{H}^5_8$ should consist of constant k-forms.

Proposition 10.6.5 *Let (M, Ω, g) be a compact Spin(7)-manifold. Then the spaces $H^k_8(M, \mathbb{R})$ for $k = 1, 3, 5, 7$ and $H^k_7(M, \mathbb{R})$ for $k = 2, 4, 6$ are represented by constant k-forms, and so are determined solely by $\mathrm{Hol}(g)$. In particular, if $\mathrm{Hol}(g) = \mathrm{Spin}(7)$ then these spaces are all zero, so that $b^1_8 = b^3_8 = b^5_8 = b^7_8 = 0$ and $b^2_7 = b^4_7 = b^6_7 = 0$.*

Proof We shall use the ideas of §3.5.2. Let $\xi \in \mathcal{H}^k_l$, so that $\xi \in C^\infty(\Lambda^k_l)$ and $(\mathrm{dd}^* + \mathrm{d}^*\mathrm{d})\xi = 0$. But $(\mathrm{dd}^* + \mathrm{d}^*\mathrm{d})\xi = \nabla^*\nabla\xi - 2\tilde{R}(\xi)$, by (3.18). By considering the representations of Spin(7), one can show that if ξ lies in $\mathcal{H}_8^k$ for $k = 1, 3, 5, 7$ or $\mathcal{H}_7^k$ for $k = 2, 4, 6$, then $\tilde{R}(\xi) = 0$. Thus $\nabla^*\nabla\xi = 0$, and integrating by parts we see that $\nabla\xi = 0$.

Thus $\mathcal{H}_8^k$ for $k = 1, 3, 5, 7$ and $\mathcal{H}_7^k$ for $k = 2, 4, 6$ are vector spaces of constant k-forms. But the constant k-forms on M are determined by $\mathrm{Hol}(g)$, and $\mathcal{H}_l^k \cong H_l^k(M, \mathbb{R})$. Thus $H_8^k(M, \mathbb{R})$ and $H_7^k(M, \mathbb{R})$ are determined by $\mathrm{Hol}(g)$, as we want. When $\mathrm{Hol}(g) = \mathrm{Spin}(7)$ these spaces are zero, as the corresponding Spin(7)-representations are nontrivial. □

The last two results show that if (M, g) is a compact Riemannian 8-manifold with holonomy Spin(7) then M has only four nontrivial refined Betti numbers, $b_{21}^2, b_{48}^3, b_{27}^4$ and b_{35}^4. However, these are not independent, since the equation $\hat{A}(M) = 1$ implies that

$$b_{48}^3 + b_{27}^4 = b_{21}^2 + 2b_{35}^4 + 24.$$

Thus we see that there are only three independent Betti-type invariants of a compact 8-manifold with holonomy Spin(7), and we can calculate all the refined Betti numbers from b^2, b^3 and b^4.

Let (Ω, g) be a Spin(7)-structure on an 8-manifold M. As in the G_2 case, if $\xi \in C^\infty(\Lambda_{21}^2)$ then $\xi \wedge \xi \wedge \Omega = -|\xi|^2 dV_g$. So, by analogy with Propositions 10.2.6 and 10.2.7, we prove:

Proposition 10.6.6 *Let (M, Ω, g) be a compact Spin(7)-manifold, and suppose $\mathrm{Hol}(g) = \mathrm{Spin}(7)$. Then $\langle \sigma \cup \sigma \cup [\Omega], [M] \rangle < 0$ for each nonzero $\sigma \in H^2(M, \mathbb{R})$.*

Proposition 10.6.7 *Let (M, Ω, g) be a compact Spin(7)-manifold, R the Riemann curvature of g, and $p_1(M) \in H^4(M, \mathbb{Z})$ the first Pontryagin class. Then*

$$\langle p_1(M) \cup [\Omega], [M] \rangle = -\frac{1}{8\pi^2} \int_M |R|^2 dV_g.$$

If $\mathrm{Hol}(g) = \mathrm{Spin}(7)$ then g is not flat, so $\langle p_1(M) \cup [\Omega], [M] \rangle < 0$, and $p_1(M) \neq 0$.

We summarize the results of this section as follows.

Theorem 10.6.8 *Suppose (M, Ω, g) is a compact Spin(7)-manifold. Then g has holonomy Spin(7) if and only if M is simply-connected and the Betti numbers of M satisfy $b^3 + b_+^4 = b^2 + 2b_-^4 + 25$. Also, if $\mathrm{Hol}(g) = \mathrm{Spin}(7)$ then M is spin and $p_1(M) \neq 0$.*

10.7 The moduli space of holonomy Spin(7) metrics

Let M be a compact, oriented 8-manifold. Then as in §10.5, we can identify the set of all oriented Spin(7)-structures on M with $C^\infty(\mathcal{A}M)$. Let $\mathcal{X} = \{\Omega \in C^\infty(\mathcal{A}M) : d\Omega = 0\}$ be the set of admissible 4-forms corresponding to torsion-free Spin(7)-structures, and let $\mathcal{D}$ be the group of all diffeomorphisms of M isotopic to the identity. Then $\mathcal{D}$ acts naturally on $C^\infty(\mathcal{A}M)$ and $\mathcal{X}$. We define the *moduli space of torsion-free Spin(7)-structures* on M to be $\mathcal{M} = \mathcal{X}/\mathcal{D}$. We shall show that $\mathcal{M}$ is a smooth manifold, with prescribed dimension.

Here is our main result, which is adapted from [120, Th. D]. Note that if M admits metrics with holonomy Spin(7) then $\hat{A}(M) = 1$ and $b^1(M) = 0$, so dim $\mathcal{M} = 1 + b_-^4$.

Theorem 10.7.1 *Let M be a compact, oriented 8-manifold, and $\mathcal{M} = \mathcal{X}/\mathcal{D}$ the moduli space of torsion-free* Spin(7)*-structures on M. Then $\mathcal{M}$ is a smooth manifold with*

$$\dim \mathcal{M} = b_1^4 + b_7^4 + b_{35}^4 = \hat{A}(M) + b^1(M) + b_-^4(M), \tag{10.26}$$

and the natural projection $\pi : \mathcal{M} \to H^4(M, \mathbb{R})$ given by $\pi(\Omega \mathcal{D}) = [\Omega]$ is an immersion.

Proof The theorem is analogous to Theorem 10.4.4, and its proof follows §10.4 fairly closely, so we will omit some of the details. Let M be a compact, oriented 8-manifold and (Ω, g) a torsion-free Spin(7)-structure on M, and let $\mathcal{M}, \mathcal{X}$ and $\mathcal{D}$ be as above. We begin with the following result, which is an analogue of Corollary 10.4.2, and is proved in the same way.

Proposition 10.7.2 *Let I_Ω be the subgroup of $\mathcal{D}$ fixing (Ω, g). Define*

$$L_\Omega = \{\tilde{\Omega} \in C^\infty(\mathcal{A}M) : d\tilde{\Omega} = 0 \text{ and } \pi_8(d^*\tilde{\Omega}) = 0\}, \tag{10.27}$$

where π_8 and d^* come from (Ω, g). Then there is an open neighbourhood S_Ω of Ω in L_Ω, invariant under I_Ω, such that the natural projection from S_Ω/I_Ω to $\mathcal{M}$ induces a homeomorphism between S_Ω/I_Ω and a neighbourhood of $\Omega \mathcal{D}$ in $\mathcal{M}$.

We shall show that L_Ω is a manifold with dimension $b_1^4 + b_7^4 + b_{35}^4$ near Ω, and that the projection $\tilde{\Omega} \mapsto [\tilde{\Omega}]$ from L_Ω to $H^4(M, \mathbb{R})$ is an immersion near Ω. Let $k \geq 0$ be an integer and $\alpha \in (0, 1)$. Define Banach spaces $V^{k,\alpha}$, $W^{k,\alpha}$ by

$$V^{k,\alpha} = \{\xi \in C^{k,\alpha}(\Lambda^5 T^*M) : \xi \text{ is exact}\},$$
$$W^{k,\alpha} = \{\xi \in C^{k,\alpha}(\Lambda_8^3) : \xi \text{ is } L^2\text{-orthogonal to } \mathcal{H}_8^3\}.$$

Define a map $\Phi : C^{k+1,\alpha}(\mathcal{A}M) \to V^{k,\alpha} \oplus W^{k,\alpha}$ by $\Phi(\tilde{\Omega}) = (d\tilde{\Omega}, \pi_8(d^*\tilde{\Omega}))$. The projection from $C^{k+1,\alpha}(\mathcal{A}M)$ to $C^{k+1,\alpha}(\Lambda_1^4 \oplus \Lambda_7^4 \oplus \Lambda_{35}^4)$ is a homeomorphism near Ω, as $T_\Omega \mathcal{A}M = \Lambda_1^4 \oplus \Lambda_7^4 \oplus \Lambda_{35}^4$. Thus near Ω we can identify $C^{k+1,\alpha}(\mathcal{A}M)$ with the Banach space $C^{k+1,\alpha}(\Lambda_1^4 \oplus \Lambda_7^4 \oplus \Lambda_{35}^4)$.

Proposition 10.7.3 *The derivative $d\Phi|_\Omega : C^{k+1,\alpha}(\Lambda_1^4 \oplus \Lambda_7^4 \oplus \Lambda_{35}^4) \to V^{k,\alpha} \oplus W^{k,\alpha}$ is given by $d\Phi|_\Omega(\xi) = (d\xi, \pi_8(d^*\xi))$. It is surjective, with kernel $\mathcal{H}_1^4 \oplus \mathcal{H}_7^4 \oplus \mathcal{H}_{35}^4$.*

Proof The expression for $d\Phi|_\Omega$ follows from the definition of Φ. Choose $\xi = \xi_1 + \xi_7 + \xi_{35}$ in $C^{k+1,\alpha}(\Lambda_1^4 \oplus \Lambda_7^4 \oplus \Lambda_{35}^4)$, where ξ_1, ξ_7, ξ_{35} are the components of ξ in Λ_1^4, Λ_7^4 and Λ_{35}^4. Let $d\Phi|_\Omega = (v, w)$, so that $v = d\xi$ lies in $V^{k,\alpha}$ and $w = \pi_8(d^*\xi)$ in $W^{k,\alpha}$. By part (iv) of Proposition 10.5.4 we have $*\xi = \xi_1 + \xi_7 - \xi_{35}$. Thus

$$\pi_8(*v) = \pi_8(*d\xi) = -\pi_8 \circ d^*(*\xi) = -\pi_8 \circ d^*(\xi_1 + \xi_7 - \xi_{35}).$$

Combining this with the equation $w = \pi_8 \circ d^*(\xi_1 + \xi_7 + \xi_{35})$ gives

$$\pi_8 \circ d^*(\xi_1 + \xi_7) = \tfrac{1}{2}\big(w - *\pi_8(v)\big). \tag{10.28}$$

Let $S_\pm$ be the positive and negative spin bundles of M. Then there are natural isomorphisms $S_+ \cong \Lambda_1^4 \oplus \Lambda_7^4$ and $S_- \cong \Lambda_8^3$ such that the Dirac operator $D_+ : C^{k+1,\alpha}(S_+) \to C^{k,\alpha}(S_-)$ discussed in §10.6 is identified with

$$\pi_8 \circ d^* : C^{k+1,\alpha}(\Lambda_1^4 \oplus \Lambda_7^4) \to C^{k,\alpha}(\Lambda_8^3). \tag{10.29}$$

By properties of the Dirac operator (in particular, ellipticity) we find that (10.29) has kernel $\mathcal{H}_1^4 \oplus \mathcal{H}_7^4$ and cokernel $\mathcal{H}_8^3$. But $W^{k,\alpha}$ is the subspace of $C^{k,\alpha}(\Lambda_8^3)$ orthogonal to $\mathcal{H}_8^3$, so $W^{k,\alpha}$ is the image of the operator $\pi_8 \circ d^*$ of (10.29). Also one can show that $d : C^{k+1,\alpha}(\Lambda_{35}^4) \to V^{k,\alpha}$ is surjective, with kernel $\mathcal{H}_{35}^4$.

Let $v \in V^{k,\alpha}$ and $w \in W^{k,\alpha}$. Then $\tfrac{1}{2}\big(w - *\pi_8(v)\big) \in W^{k,\alpha}$, so there exists $\xi_1 + \xi_7 \in C^{k+1,\alpha}(\Lambda_1^4 \oplus \Lambda_7^4)$ with $\pi_8 \circ d^*(\xi_1 + \xi_7) = \tfrac{1}{2}\big(w - *\pi_8(v)\big)$, since (10.29) has image $W^{k,\alpha}$. Also $v - d(\xi_1 + \xi_7) \in V^{k,\alpha}$. As $d : C^{k+1,\alpha}(\Lambda_{35}^4) \to V^{k,\alpha}$ is surjective, there exists $\xi_{35} \in C^{k+1,\alpha}(\Lambda_{35}^4)$ with $d\xi_{35} = v - d(\xi_1 + \xi_7)$. But then $d\Phi|_\Omega(\xi_1 + \xi_7 + \xi_{35}) = (v,w)$, and so $d\Phi|_\Omega$ is surjective, as we have to prove.

Suppose $\xi = \xi_1 + \xi_7 + \xi_{35}$ satisfies $d\Phi|_\Omega(\xi) = 0$. Then $\pi_8 \circ d^*(\xi_1 + \xi_7) = 0$ by (10.28), so $\xi_1 + \xi_7$ lies in $\mathcal{H}_1^4 \oplus \mathcal{H}_7^4$, as this is the kernel of (10.29). But then $d\xi_1 + d\xi_7 = 0$, so $d\xi_{35} = 0$, and $\xi_{35} \in \mathcal{H}_{35}^4$, as this is the kernel of $d : C^{k+1,\alpha}(\Lambda_{35}^4 \to V^{k,\alpha}$. Thus $\xi \in \mathcal{H}_1^4 \oplus \mathcal{H}_7^4 \oplus \mathcal{H}_{35}^4$. Conversely, if $\xi \in \mathcal{H}_1^4 \oplus \mathcal{H}_7^4 \oplus \mathcal{H}_{35}^4$ then $d\Phi|_\Omega(\xi) = 0$, so $\operatorname{Ker} d\Phi|_\Omega = \mathcal{H}_1^4 \oplus \mathcal{H}_7^4 \oplus \mathcal{H}_{35}^4$, and the proof is complete. $\square$

Applying the Implicit Mapping Theorem, Theorem 1.2.5, to Φ and using the proposition, we find that $\Phi^{-1}(0)$ is a manifold of dimension $b_1^4 + b_7^4 + b_{35}^4$ in a neighbourhood of Ω. Also, as the tangent space to $\Phi^{-1}(0)$ at Ω is $\mathcal{H}_1^4 \oplus \mathcal{H}_7^4 \oplus \mathcal{H}_{35}^4$, the natural map $\Phi^{-1}(\Omega) \to H^4(M, \mathbb{R})$ has injective first derivative at Ω, since the projection from $\mathcal{H}_1^4 \oplus \mathcal{H}_7^4 \oplus \mathcal{H}_{35}^4$ to $H^4(M, \mathbb{R})$ is injective. Thus the map $\tilde{\Omega} \mapsto [\tilde{\Omega}]$ from $\Phi^{-1}(0)$ to $H^4(M, \mathbb{R})$ is an *embedding* near Ω.

For small $\|\tilde{\Omega} - \Omega\|_{C^0}$ the equations $\tilde{\Omega} \in C^{k+1,\alpha}(\mathcal{A}M)$ and $\Phi(\tilde{\Omega}) = 0$ are a system of (overdetermined) nonlinear elliptic equations on $\tilde{\Omega}$. So by elliptic regularity, if $\tilde{\Omega} \in \Phi^{-1}(0)$ is close to Ω then $\tilde{\Omega}$ is smooth, rather than just $C^{k+1,\alpha}$. Thus $\Phi^{-1}(0)$ coincides with the set L_Ω of (10.27) near Ω. We have proved that near Ω, the set L_Ω is a submanifold of $C^\infty(\mathcal{A}M)$ of dimension $b_1^4 + b_7^4 + b_{35}^4$ and the projection $\tilde{\Omega} \to [\tilde{\Omega}]$ from L_Ω to $H^4(M, \mathbb{R})$ is an embedding.

By Proposition 10.7.2, the moduli space $\mathcal{M}$ is homeomorphic near $\Omega\mathcal{D}$ to a neighbourhood of ΩI_Ω in L_Ω/I_Ω. But I_Ω is a group of diffeomorphisms of M isotopic to the identity, so I_Ω acts trivially on $H^4(M, \mathbb{R})$. As L_Ω is embedded in $H^4(M, \mathbb{R})$ near Ω by the natural projection, we see that I_Ω acts trivially on L_Ω near Ω, and L_Ω/I_Ω is locally isomorphic to L_Ω. Thus $\mathcal{M}$ is homeomorphic near $\Omega\mathcal{D}$ to a neighbourhood of Ω in L_Ω.

Therefore $\mathcal{M}$ is a manifold of dimension $b_1^4 + b_7^4 + b_{35}^4$ near $\Omega\mathcal{D}$, and $\pi : \mathcal{M} \to H^4(M, \mathbb{R})$ is an embedding near $\Omega\mathcal{D}$. As this holds for all $\Omega\mathcal{D} \in \mathcal{M}$, we see that $\mathcal{M}$ is a smooth manifold with dimension given by the first part of (10.26). The second part of (10.26) then follows from (10.25). Finally, since $\pi : \mathcal{M} \to H^4(M, \mathbb{R})$ is an embedding near each point, it is an immersion, and the proof of Theorem 10.7.1 is complete. $\square$

10.8 Exceptional holonomy and calibrated geometry

In §3.7 we defined calibrations and calibrated submanifolds. We now apply these ideas to manifolds with exceptional holonomy. There are two types of calibrated submanifolds in 7-manifolds with holonomy G_2, called *associative* 3-*folds* and *coassociative* 4-*folds*, and one in 8-manifolds with holonomy Spin(7), called *Cayley* 4-*folds*. We discuss each in turn.

10.8.1 Associative 3-folds in 7-manifolds with holonomy G_2

Let φ_0 be the 3-form on $\mathbb{R}^7$ defined by (10.1). Then φ_0 is a *calibration* on $\mathbb{R}^7$ by [99, Th. IV.1.4]. Define

$$U = \{(x_1, x_2, x_3, 0, 0, 0, 0) : x_1, x_2, x_3 \in \mathbb{R}\} \subset \mathbb{R}^7. \tag{10.30}$$

Then U is a 3-plane in $\mathbb{R}^7$, and $\varphi_0|_U = dx_1 \wedge dx_2 \wedge dx_3$, so $\varphi_0|_U = \text{vol}_U$ with the appropriate choice of orientation on U. By [99, Th. IV.1.8] the subgroup of G_2 preserving U is SO(4), and an oriented 3-plane V in $\mathbb{R}^7$ satisfies $\varphi_0|_V = \text{vol}_V$ if and only if $V = \gamma \cdot U$ for some $\gamma \in G_2$.

We define an *associative* 3-*plane* in $\mathbb{R}^7$ to be an oriented 3-plane V in $\mathbb{R}^7$ with $\varphi_0|_V = \text{vol}_V$. Then the set of all associative 3-planes is isomorphic to $G_2/\text{SO}(4)$, which has dimension $14 - 6 = 8$. But the Grassmannian of all oriented 3-planes in $\mathbb{R}^7$ has dimension 12. Hence the associative 3-planes are of codimension 4 in the set of all 3-planes.

Let (φ, g) be a torsion-free G_2-structure on a 7-manifold M. Then the 3-form φ is a calibration on M. We refer to φ-submanifolds as *associative* 3-*folds*. That is, an associative 3-fold is an oriented 3-dimensional submanifold N of M such that $\text{vol}_N = \varphi|_N$, where vol_N is the volume form of N, depending on the metric $g|_N$ and the orientation of N.

McLean [159, §5] studied the deformations of a compact, associative 3-fold N in a 7-manifold M with torsion-free G_2-structure (φ, g). He found the deformation problem to be elliptic, with index zero. Thus, when (φ, g) is *generic* in a suitable sense, N will admit no deformations, and will persist under small deformations of (φ, g). But when (φ, g) is not generic, compact associative 3-folds may occur in moduli spaces of positive dimension, and may vanish under small deformations of (φ, g).

For example, the 7-torus T^7 has a 35-dimensional family of flat G_2-structures. Writing T^7 as a product $T^3 \times T^4$, it turns out that for a 31-dimensional subfamily of these G_2-structures, $T^3 \times \{p\}$ is an associative 3-fold for each $p \in T^4$. These associative T^3's deform in a moduli space of dimension 4. However, for generic flat G_2-structures there exist no associative 3-folds in T^7 at all. Thus the family of associative T^3's vanishes under small deformations of the underlying G_2-structure.

One might be tempted by this example to look for compact 7-manifolds with holonomy G_2 which are fibred by compact associative 3-folds diffeomorphic to T^3, with some singular fibres. Indeed, physicists have used this to explain some string dualities. However, for geometrical reasons this is probably not a workable idea.

The author believes that associative 3-folds do not occur in moduli spaces of positive dimension in compact 7-manifolds with holonomy G_2. The fact that associative fibrations are possible in T^7, $K3 \times T^3$ and $N \times \mathcal{S}^1$ for N a Calabi–Yau 3-fold, is to do with the special geometry of these spaces, and should not be expected to extend to general compact 7-manifolds with holonomy G_2.

Our next result will be used in Chapter 12 to find examples of compact associative 3-folds in compact 7-manifolds with holonomy G_2. Here a *nontrivial isometric involution* of (M, g) is a diffeomorphism $\sigma : M \to M$ such that $\sigma^*(g) = g$, and $\sigma \neq \text{id}$ but $\sigma^2 = \text{id}$, where id is the identity on M.

Proposition 10.8.1 *Let (φ, g) be a torsion-free G_2-structure on a 7-manifold M, and let $\sigma : M \to M$ be a nontrivial isometric involution with $\sigma^*(\varphi) = \varphi$. Then $N = \{p \in M : \sigma(p) = p\}$ is an associative 3-fold in M.*

Proof Clearly N is a closed submanifold of M. If $p \in N$ then $d\sigma : T_pM \to T_pM$ satisfies $(d\sigma)^2 = 1$, and T_pN is the subspace of T_pM fixed by $d\sigma$. If $d\sigma = 1$ then $T_pN = T_pM$ and $\dim N = 7$, so $N = M$ as M is connected. But this contradicts $\sigma \neq \text{id}$, and so $d\sigma \neq 1$. Also, $d\sigma$ preserves $\varphi|_p$. Thus, identifying T_pM with $\mathbb{R}^7$, we can regard $d\sigma : T_pM \to T_pM$ as an element of G_2.

It can be shown that if $\gamma \in G_2$ and $\gamma \neq 1$ but $\gamma^2 = 1$, then γ is conjugate in G_2 to the map

$$(x_1, \ldots, x_7) \longmapsto (x_1, x_2, x_3, -x_4, -x_5, -x_6, -x_7).$$

But the fixed set of this is the subspace U of (10.30). Thus, if $\gamma \in G_2$ and $\gamma \neq 1$ but $\gamma^2 = 1$, then the fixed set of γ in $\mathbb{R}^7$ is an associative 3-plane, from above. So T_pN is an associative 3-plane in T_pM. As this holds for all $p \in N$, we see that N is an associative 3-fold. $\square$

10.8.2 *Coassociative 4-folds in 7-manifolds with holonomy G_2*

Let $*\varphi_0$ be the 4-form on $\mathbb{R}^7$ defined by (10.2). Then $*\varphi_0$ is a *calibration* on $\mathbb{R}^7$ by [99, Th. IV.1.16]. Define

$$U = \{(0, 0, 0, x_4, x_5, x_6, x_7) : x_4, \ldots, x_7 \in \mathbb{R}\} \subset \mathbb{R}^7. \tag{10.31}$$

Then U is a 4-plane in $\mathbb{R}^7$, and $*\varphi_0|_U = dx_4 \wedge dx_5 \wedge dx_6 \wedge dx_7$, so $*\varphi_0|_U = \text{vol}_U$ with the appropriate orientation on U. As above the subgroup of G_2 preserving U is $SO(4)$, and an oriented 4-plane V in $\mathbb{R}^7$ satisfies $*\varphi_0|_V = \text{vol}_V$ if and only if $V = \gamma \cdot U$ for some $\gamma \in G_2$.

We define a *coassociative 4-plane* in $\mathbb{R}^7$ to be an oriented 4-plane V in $\mathbb{R}^7$ with $*\varphi_0|_V = \text{vol}_V$. Then the set of all coassociative 4-planes is isomorphic to $G_2/SO(4)$, which has dimension 8. But the Grassmannian of all oriented 4-planes in $\mathbb{R}^7$ has dimension 12. Hence the coassociative 4-planes are of codimension 4 in the set of all 4-planes.

From (10.1) and (10.31) we see that $\varphi_0|_U = 0$. Therefore $\varphi_0|_V = 0$ for every coassociative 4-plane V in $\mathbb{R}^7$, since $V = \gamma \cdot U$ for some $\gamma \in G_2$. In fact the converse is also true: if V is a 4-plane in $\mathbb{R}^7$ and $\varphi_0|_V = 0$, then there is a unique orientation on V making V into a coassociative 4-plane.

Let (φ, g) be a torsion-free G_2-structure on a 7-manifold M. Then the 4-form $*\varphi$ is a calibration on M. We refer to $*\varphi$-submanifolds as *coassociative 4-folds*. Thus, a coassociative 4-fold is an oriented 4-submanifold N of M with $\text{vol}_N = *\varphi|_N$, where vol_N is the volume form of N. As a 4-plane V in $\mathbb{R}^7$ is coassociative if and only if $\varphi_0|_V = 0$, we deduce

Lemma 10.8.2 *Let (φ, g) be a torsion-free G_2-structure on a 7-manifold M, and N a 4-dimensional submanifold of M. Then N is coassociative if and only if $\varphi|_N \equiv 0$.*

This gives a topological restriction on coassociative 4-folds.

Corollary 10.8.3 *Let (φ, g) be a torsion-free G_2-structure on a 7-manifold M, and N a coassociative 4-fold in M. Then $[\varphi|_N] = 0$ in $H^3(N, \mathbb{R})$.*

Next we discuss the *deformations* of coassociative 4-folds, following McLean [159, §4]. Let (φ, g) be a torsion-free G_2-structure on a 7-manifold M, and N a coassociative 4-fold in M. Then there is a natural isomorphism $TM|_N \cong TN \oplus \Lambda^2_+ N$, where $\Lambda^2_+ N$ is the bundle of self-dual 2-forms on N. So the normal bundle ν of N in M is isomorphic to $\Lambda^2_+ N$. Note that McLean uses the opposite sign for φ_0 and $*\varphi_0$, and so has $\Lambda^2_- N$ instead of $\Lambda^2_+ N$.

Sections of ν parametrize infinitesimal deformations of N as a submanifold of M. It turns out that a section of ν corresponds to an infinitesimal deformation of N as a coassociative 4-fold if and only if the associated self-dual 2-form on N is *closed*. In this way McLean [159, Th. 4.5] proves

Theorem 10.8.4 *Let (φ, g) be a torsion-free G_2-structure on a 7-manifold M, and N a compact coassociative 4-fold in M. Then the moduli space of coassociative 4-folds isotopic to N in M is a smooth manifold of dimension $b^2_+(N)$.*

It is well-known that some $K3$ surfaces admit *elliptic fibrations*, in which the surface is fibred by elliptic curves T^2, with some singular fibres. Similar phenomena also occur for higher-dimensional Calabi–Yau manifolds. It is natural to wonder whether compact 7-manifolds with holonomy G_2 can be fibred by coassociative 4-manifolds, with some singular fibres. Such fibrations are of interest to physicists [1, 3], in connection with string dualities. The existence of fibrations of this kind seems plausible, at least locally.

To fibre M by deformations of a compact coassociative 4-fold N we must have $b^2_+(N) = 3$, so that the moduli space is 3-dimensional by Theorem 10.8.4, and the normal bundle ν must be trivial, so $\Lambda^2_+ N$ is trivial. There are two obvious 4-manifolds N with these properties: the torus T^4, and the $K3$ surface. In §12.6 we will give examples of local fibrations of compact 7-manifolds with holonomy G_2 by both T^4 and $K3$.

If $\alpha : \mathbb{R}^7 \to \mathbb{R}^7$ is linear with $\alpha^2 = 1$ and $\alpha^*(\varphi_0) = -\varphi_0$, then either $\alpha = -1$, or α is conjugate under an element of G_2 to the map

$$(x_1, \ldots, x_7) \longmapsto (-x_1, -x_2, -x_3, x_4, x_5, x_6, x_7).$$

The fixed set of this map is the coassociative 4-plane U of (10.31). Thus, the fixed point set of α is either $\{0\}$, or a coassociative 4-plane in $\mathbb{R}^7$. Using this and the proof of Proposition 10.8.1 gives

Proposition 10.8.5 *Let (φ, g) be a torsion-free G_2-structure on a 7-manifold M, and $\sigma : M \to M$ a nontrivial isometric involution with $\sigma^*(\varphi) = -\varphi$. Then each connected component of the fixed point set $\{p \in M : \sigma(p) = p\}$ of σ is either a coassociative 4-fold or a single point.*

10.8.3 Cayley 4-folds in 8-manifolds with holonomy Spin(7)

Let Ω_0 be the 4-form on $\mathbb{R}^8$ defined by (10.19). Then Ω_0 is a *calibration* on $\mathbb{R}^8$ by [99, Th. IV.1.24]. Define

$$U = \{(x_1, x_2, x_3, x_4, 0, 0, 0, 0) : x_1, \ldots, x_4 \in \mathbb{R}\} \subset \mathbb{R}^8.$$

Then U is a 4-plane in $\mathbb{R}^8$, and $\Omega_0|_U = dx_1 \wedge \cdots \wedge dx_4$, so $\Omega_0|_U = \text{vol}_U$ with the appropriate choice of orientation on U. By [99, Th. IV.1.38] the subgroup of Spin(7) preserving U is $K = (SU(2) \times SU(2) \times SU(2))/\mathbb{Z}_2$, and an oriented 4-plane V in $\mathbb{R}^8$ satisfies $\Omega_0|_V = \text{vol}_V$ if and only if $V = \gamma \cdot U$ for some $\gamma \in \text{Spin}(7)$.

We define a *Cayley 4-plane* in $\mathbb{R}^8$ to be an oriented 4-plane V in $\mathbb{R}^8$ with $\Omega_0|_V = \text{vol}_V$. Then the set of all Cayley 4-planes is isomorphic to Spin(7)/K, which has dimension $21 - 9 = 12$. But the Grassmannian of all oriented 4-planes in $\mathbb{R}^8$ has dimension 16. Hence the Cayley 4-planes are of codimension 4 in the set of all 4-planes.

Now let (Ω, g) be a torsion-free Spin(7)-structure on an 8-manifold M. Then the 4-form Ω is a calibration on M. We refer to Ω-submanifolds as *Cayley 4-folds*. That is, a Cayley 4-fold is an oriented 4-dimensional submanifold N of M such that $\text{vol}_N = \varphi|_N$, where vol_N is the volume form of N.

McLean [159, §6] studied the deformations of a compact, Cayley 4-fold N in an 8-manifold M with torsion-free Spin(7)-structure (Ω, g). He found that the deformation problem depends upon a certain elliptic *twisted Dirac operator* D^F over N. A calculation by Christopher Lewis and the author shows that the index of D^F is

$$\text{ind}(D^F) = \tau(N) - \tfrac{1}{2}\chi(N) - \tfrac{1}{2}[N] \cdot [N], \tag{10.32}$$

where $\tau(N)$ is the signature, $\chi(N)$ the Euler characteristic and $[N] \cdot [N]$ the self-intersection of N.

We expect that in the *generic* case, the deformations of N as a Cayley 4-fold will locally form a smooth moduli space with dimension $\text{ind}(D^F)$, provided that $\text{ind}(D^F) \geq 0$. If $\text{ind}(D^F) < 0$ then N is nongeneric, and should vanish under small perturbations of (Ω, g). However, if (Ω, g) and N are not generic then we do not know very much about the deformations of N.

As with coassociative 4-folds above, it is natural to ask whether a compact 8-manifold with holonomy Spin(7) can be fibred by Cayley 4-folds N with some singular fibres, a question which also interests physicists [3]. If N has trivial normal bundle then the structure group of N reduces to SU(2), suggesting that N should be T^4 or $K3$, and $[N] \cdot [N] = 0$.

From (10.32) we have $\text{ind}(D^F) = 0$ when $N = T^4$ and $\text{ind}(D^F) = 4$ when $N = K3$. Thus T^4 is not suitable, as generically it forms moduli spaces of dimension 0, but

$K3$ generically forms moduli spaces of dimension 4, which is what we want. In Chapter 14 we will give examples of compact 8-manifolds with holonomy Spin(7) which are locally fibred by Cayley $K3$'s.

Finally, by the method of Propositions 10.8.1 and 10.8.5 one can prove

Proposition 10.8.6 *Let M be an 8-manifold, (Ω, g) a torsion-free Spin(7)-structure on M, and $\sigma : M \to M$ a nontrivial isometric involution with $\sigma^*(\Omega) = \Omega$. Then each connected component of the fixed point set $\{p \in M : \sigma(p) = p\}$ is either a Cayley 4-fold or a single point.*

10.9 A reading list on the exceptional holonomy groups

We begin by suggesting some books and survey papers about exceptional holonomy. The book by Salamon [186, §11–§12] includes a useful introduction to G_2 and Spin(7). Bryant [44] surveys the exceptional holonomy groups up to 1986, and by the author, two short papers [123], [124] and the book chapter [131] discuss compact manifolds with exceptional holonomy.

Next we list some of the papers on G_2 and Spin(7) in the literature. Three important landmarks in the history of the exceptional holonomy groups are Berger's classification of holonomy groups in 1955, Bryant's proof of the local existence of metrics with exceptional holonomy in 1984, and the author's construction of compact manifolds with exceptional holonomy in 1994. We use these to divide our list into three periods.

- **Early papers, 1955–1984.** Bonan [36] wrote down the G_2-invariant forms φ_0, $*\varphi_0$ of (10.1) and (10.2) and the Spin(7)-invariant 4-form Ω_0 of (10.19), and showed that metrics with holonomy G_2 and Spin(7) are Ricci-flat.

  Fernández and Gray [71] took a G_2-structure (φ, g) on a 7-manifold, and decomposed $\nabla\varphi$ into irreducible pieces. Similarly, Fernández [68] took a Spin(7)-structure (Ω, g) on an 8-manifold, and decomposed $\nabla\Omega$ into irreducible pieces.

- **Existence of exceptional holonomy metrics, 1984–1994.** In a very significant paper, Bryant [43] used the theory of exterior differential systems to prove that there exist many metrics with holonomy G_2 and Spin(7) on small balls in $\mathbb{R}^7$ and $\mathbb{R}^8$. He also gave some explicit, noncomplete examples of such metrics. His results were announced in advance in [42].

  Later, Bryant and Salamon [47] wrote down explicit, complete metrics with holonomy G_2 and Spin(7) on noncompact manifolds, which are the total spaces of vector bundles over manifolds of dimension 3 and 4, and have large symmetry groups. The same metrics were also found by Gibbons et al. [82].

  In 1986–7 Fernández and others [60, 69, 70, 72] gave examples of compact 7-manifolds M with G_2-structures (φ, g) such that either $d\varphi = 0$ or $d^*\varphi = 0$, but not both. Simple examples with $d^*\varphi = 0$ are also provided by real hypersurfaces in $\mathbb{O} = \mathbb{R}^8$.

- **Compact manifolds with exceptional holonomy, from 1994.** In 1994–5 the author constructed examples of compact 7-manifolds with holonomy G_2 [121, 122], and of compact 8-manifolds with holonomy Spin(7) [120]. These constructions will be explained at length in Chapters 11–14.

At about the same time, physicists working in String Theory became interested in using compact manifolds with holonomy G_2 and Spin(7) as vacua for string theories, in the same way that they had been using Calabi–Yau 3-folds for a number of years. Some papers on exceptional holonomy, written by physicists in the language of String Theory, are Papadopoulos and Townsend [173], Shatashvili and Vafa [192, 193], Acharya [1, 2, 3], Figueroa-O'Farrill [73] and Vafa [211].

Here are some papers on other topics related to exceptional holonomy:

- **Calibrated submanifolds.** Some good mathematical references on calibrated geometry, discussed in §3.7 and §10.8, are Harvey and Lawson [98, 99], Harvey [97] and McLean [159]. Physicists working in String Theory are also interested in calibrated geometry. Some physics papers on calibrated geometry and exceptional holonomy (not necessarily comprehensible for mathematicians) are Becker at al. [30] and Acharya [1, 2, 3].
- **Gauge theory over compact 8-manifolds with holonomy** Spin(7). Let (Ω, g) be a Spin(7)-structure on a compact 8-manifold M, let E be a vector bundle over M, and A a connection on E with curvature F_A. We call A a Spin(7) *instanton* if $\pi_7(F_A) = 0$. Such connections occur in finite-dimensional moduli spaces, and have many properties in common with instantons on 4-manifolds, which are the subject of Donaldson theory.

  Spin(7) instantons have been studied from the mathematical point of view by Thomas [201], Lewis [152] (who proves an existence theorem for nontrivial examples of Spin(7) instantons over compact 8-manifolds with holonomy Spin(7)), Reyes Carrión [180] and Tian [203], and from the String Theory point of view by Acharya et al. [4] and Baulieu et al. [26].
- **Nearly parallel G_2-structures.** These are G_2-manifolds (M, φ, g) with a *Killing spinor*, rather than a constant spinor. Such manifolds satisfy $d\varphi = -8\lambda * \varphi$ and $d * \varphi = 0$, for some $\lambda \in \mathbb{R}$, and are automatically Einstein with nonnegative scalar curvature $168\lambda^2$. They include as special cases 3-Sasakian 7-manifolds, Einstein–Sasakian 7-manifolds and holonomy G_2 7-manifolds. For more details see Friedrich et al. [76].

11

CONSTRUCTION OF COMPACT 7-MANIFOLDS WITH HOLONOMY G_2

In this chapter we explain how to construct examples of compact 7-manifolds with holonomy G_2. Here is a sketch of the method. We begin with a torus T^7 equipped with a flat G_2-structure (φ_0, g_0), and a finite group Γ of automorphisms of T^7 preserving (φ_0, g_0). Then T^7/Γ is an orbifold with a flat G_2-structure (φ_0, g_0).

We resolve the singularities of T^7/Γ to get a compact 7-manifold M, and define a 1-parameter family of G_2-structures (φ^t, g^t) on M depending on $t \in (0, \epsilon)$, such that the torsion of (φ^t, g^t) is $O(t^4)$. Then we show using analysis that when t is sufficiently small, we can deform (φ^t, g^t) to a nearby torsion-free G_2-structure $(\tilde{\varphi}, \tilde{g})$ on M. If $\pi_1(M)$ is finite then $\text{Hol}(\tilde{g}) = G_2$, so M admits metrics with holonomy G_2, as we want.

To carry this out we need to understand how to resolve the singularities of T^7/Γ, and how to put a G_2-structure with small torsion on the resolution. We introduce the idea of *QALE G_2-manifold*, which is a 7-manifold with a torsion-free G_2-structure that is asymptotic to the flat G_2-structure on $\mathbb{R}^7/G$ in the Quasi-ALE sense defined in Chapter 9.

The information we need to define M and the G_2-structures upon it is a choice of QALE G_2-manifold for each stratum of the singular set of T^7/Γ satisfying some compatibility conditions; we call this information *R-data*. The main result of the chapter is Theorem 11.6.2. It says that given an orbifold T^7/Γ with a flat G_2-structure (φ_0, g_0) and a set of R-data, there exists a resolution M of T^7/Γ which admits torsion-free G_2-structures.

Compact 7-manifolds with holonomy G_2 were first constructed by the author in [121, 122], using the above method. But the construction we shall describe here is more elaborate and quite a lot more powerful than that used in [121, 122]. The difference is in the kinds of singularities which the construction allows you to resolve.

In [121, 122] the author essentially considered only singularities that could be resolved using $\mathbb{R}^3 \times Y$ or $\mathbb{R} \times Y$, for Y an ALE manifold with holonomy SU(2) or SU(3). This placed strong restrictions on the choice of group Γ acting on T^7, and meant that the construction yielded a comparatively small number of 7-manifolds — for example, [122] found only 29 simply-connected and 39 non-simply-connected topologically distinct 7-manifolds with holonomy G_2.

However, the introduction of QALE G_2-manifolds and the existence results of Chapter 9 for QALE manifolds with holonomy SU(3) enable us to resolve many more kinds of singularity in T^7/Γ. This means we now have lots more possibilities for the group Γ. Also, it turns out that there are often many topological choices involved in resolving non-isolated singularities, and this means that a single orbifold T^7/Γ may admit a hundred or more topologically distinct resolutions with holonomy G_2.

So, we present a more sophisticated construction of compact G_2-manifolds than that previously known. It builds on the work of Chapters 8, 9 and 10, and is perhaps the central point of the book. Unfortunately, there are disadvantages in this for the casual reader, who may find the chapter tough going. In our search for greater generality, and to describe some subtle asymptotic behaviour precisely, we introduce a great deal of notation, symbols and indices. Also, all examples are banished to Chapter 12.

Readers who get into difficulty are advised to first have a look at one of the author's short survey papers [123, 124], the book chapter [131], or the paper [121]. All of these explain the main issues involved, shorn of the extra layer of complexity needed to include QALE G_2-manifolds in the construction.

We begin in §11.1 by considering the subgroups SU(2) and SU(3) of G_2, and showing that if Y has holonomy SU(2) (or SU(3)) then $\mathbb{R}^3 \times Y$ (or $\mathbb{R} \times Y$) has a torsion-free G_2-structure. Section 11.2 defines QALE G_2-manifolds, and §11.3 gives some notation to describe T^7/Γ. Sections 11.4 and 11.5 define R-data, the resolution M of T^7/Γ, and the G_2-structures (φ^t, g^t) on M.

The main result of the chapter, the existence of torsion-free G_2-structures on M, is given in §11.6. The proof depends on two analytic results, Theorems G1 and G2, which are proved in §11.7 and §11.8. Finally, in §11.9 we sketch some other possible methods of constructing compact 7-manifolds with holonomy G_2, which have not yet been proved.

Our presentation is based loosely on the author's papers [121, 122], but has been altered in many places. Theorem G2 is a strengthened version of [121, Th. A], and the proof we give of it in §11.8 was sketched in [131]. Apart from this, the results of this chapter are published here for the first time.

11.1 Resolving G_2-singularities with holonomy SU(2), SU(3)

We begin this chapter with some preparatory work on the subgroups SU(2) and SU(3) of G_2, which will be used in defining QALE G_2-manifolds in §11.2. Theorem 10.2.1 showed that if (φ, g) is a torsion-free G_2-structure on a 7-manifold, then $\text{Hol}^0(g)$ is one of $\{1\}$, SU(2), SU(3) and G_2. We start by considering torsion-free G_2-structures with holonomy SU(2) and SU(3).

Proposition 11.1.1 *Suppose (Y, g_Y) is a Riemannian 4-manifold with holonomy SU(2). Then Y admits a complex structure J, a Kähler form ω and a holomorphic volume form θ with $d\omega = d\theta = 0$.*

Let $\mathbb{R}^3$ have coordinates (x_1, x_2, x_3), and Euclidean metric $h = dx_1^2 + dx_2^2 + dx_3^2$. Define a metric g and a 3-form φ on $\mathbb{R}^3 \times Y$ by $g = h \times g_Y$ and

$$\varphi = dx_1 \wedge dx_2 \wedge dx_3 + dx_1 \wedge \omega + dx_2 \wedge \text{Re}\,\theta - dx_3 \wedge \text{Im}\,\theta. \tag{11.1}$$

Then (φ, g) is a torsion-free G_2-structure on $\mathbb{R}^3 \times Y$, and

$$*\varphi = \tfrac{1}{2}\omega \wedge \omega + dx_2 \wedge dx_3 \wedge \omega \\ - dx_1 \wedge dx_3 \wedge \text{Re}\,\theta - dx_1 \wedge dx_2 \wedge \text{Im}\,\theta. \tag{11.2}$$

Proof The existence of J, ω and θ and the equation $d\omega = d\theta = 0$ follow from the ideas of §6.1. For each $p \in N$ there exist complex coordinates (z_1, z_2) near p, such that

$$g_Y = |dz_1|^2 + |dz_2|^2, \quad \omega = \frac{i}{2}(dz_1 \wedge d\bar{z}_1 + dz_2 \wedge d\bar{z}_2)$$

$$\text{and} \quad \theta = dz_1 \wedge dz_2 \quad \text{at } p,$$

as in (6.1). Setting $z_1 = x_4 + ix_5$ and $z_2 = x_6 + ix_7$, we find that

$$g_Y = dx_4^2 + \cdots + dx_7^2, \qquad \omega = dx_4 \wedge dx_5 + dx_6 \wedge dx_7,$$
$$\operatorname{Re}\theta = dx_4 \wedge dx_6 - dx_5 \wedge dx_7, \quad \operatorname{Im}\theta = dx_4 \wedge dx_7 + dx_5 \wedge dx_6$$

at p. Substituting these into the equations $g = h \times g_Y$ and (11.1) gives $g = dx_1^2 + \cdots + dx_7^2$ and

$$\varphi = dx_1 \wedge dx_2 \wedge dx_3 + dx_1 \wedge (dx_4 \wedge dx_5 + dx_6 \wedge dx_7)$$
$$+ dx_2 \wedge (dx_4 \wedge dx_6 - dx_5 \wedge dx_7) - dx_3 \wedge (dx_4 \wedge dx_7 + dx_5 \wedge dx_6)$$

at p. But these agree with the standard metric g_0 and 3-form φ_0 on $\mathbb{R}^7$ given in Definition 10.1.1. Therefore (φ, g) is a G_2-structure on $\mathbb{R}^3 \times Y$. The expression (11.2) for $*\varphi$ follows in the same way. As dx_1, dx_2, dx_3, ω and θ are all closed, (11.1) and (11.2) show that $d\varphi = d(*\varphi) = 0$. Thus (φ, g) is torsion-free, by Proposition 10.1.3. $\square$

As in Theorem 10.2.1, we may write $\mathbb{R}^7 \cong \mathbb{R}^3 \oplus \mathbb{C}^2$, and if $SU(2)$ acts trivially on $\mathbb{R}^3$ and as usual on $\mathbb{C}^2$, then $SU(2) \subset G_2$. Let G be a finite subgroup of $SU(2)$. Then $G \subset G_2$, and $\mathbb{R}^7/G$ is an orbifold isomorphic to $\mathbb{R}^3 \times \mathbb{C}^2/G$. Moreover, the flat G_2-structure (φ_0, g_0) on $\mathbb{R}^7$ is invariant under G, and descends to $\mathbb{R}^7/G$.

Now suppose (Y, g_Y) is an *ALE manifold* with holonomy $SU(2)$ asymptotic to $\mathbb{C}^2/G$, as discussed in §7.2 and Chapter 8. Then by Proposition 11.1.1 there is a torsion-free G_2-structure (φ, g) on $\mathbb{R}^3 \times Y$. Since Y is asymptotic to $\mathbb{C}^2/G$, we can regard $\mathbb{R}^3 \times Y$ as asymptotic to $\mathbb{R}^3 \times \mathbb{C}^2/G$.

Furthermore, the G_2-structure (φ, g) on $\mathbb{R}^3 \times Y$ is asymptotic to the flat G_2-structure (φ_0, g_0) on $\mathbb{R}^7/G$ in a suitable sense. In fact, we should interpret $\mathbb{R}^3 \times Y$ as an example of a *QALE G_2-manifold* asymptotic to $\mathbb{R}^7/G$, in a similar way to Chapter 9. We will define this idea shortly.

Here is an analogous result for $SU(3) \subset G_2$.

Proposition 11.1.2 *Suppose (Y, g_Y) is a Riemannian 6-manifold with holonomy $SU(3)$. Then Y admits a complex structure J, a Kähler form ω and a holomorphic volume form θ with $d\omega = d\theta = 0$.*

Let $\mathbb{R}$ have coordinate x. Define a metric g and a 3-form φ on $\mathbb{R} \times Y$ by

$$g = dx^2 \times g_Y \quad \text{and} \quad \varphi = dx \wedge \omega + \operatorname{Re}\theta. \tag{11.3}$$

Then (φ, g) is a torsion-free G_2-structure on $\mathbb{R} \times Y$, and

$$*\varphi = \tfrac{1}{2}\omega \wedge \omega - dx \wedge \operatorname{Im}\theta. \tag{11.4}$$

Proof We choose complex coordinates (z_1, z_2, z_3) near a point $p \in Y$ such that

$$g_Y = |dz_1|^2 + \cdots + |dz_3|^2, \quad \omega = \frac{i}{2}(dz_1 \wedge d\bar{z}_1 + \cdots + dz_3 \wedge d\bar{z}_3)$$
$$\text{and} \quad \theta = dz_1 \wedge dz_2 \wedge dz_3 \quad \text{at } p,$$

and define real coordinates $(x_1, \ldots, x_7)$ on $\mathbb{R} \times Y$ by

$$x = x_1, \quad z_1 = x_2 + ix_3, \quad z_2 = x_4 + ix_5 \quad \text{and} \quad z_3 = x_6 + ix_7.$$

The rest of the proof follows that of Proposition 11.1.1. □

As in Theorem 10.2.1, we may write $\mathbb{R}^7 \cong \mathbb{R} \oplus \mathbb{C}^3$, and if SU(3) acts trivially on $\mathbb{R}$ and as usual on $\mathbb{C}^3$, then SU(3) $\subset G_2$. Let G be a finite subgroup of SU(3). Then $G \subset G_2$, and $\mathbb{R}^7/G$ is an orbifold isomorphic to $\mathbb{R} \times \mathbb{C}^3/G$, and the flat G_2-structure (φ_0, g_0) on $\mathbb{R}^7$ descends to $\mathbb{R}^7/G$.

Now suppose (Y, g_Y) is an *ALE manifold* or *QALE manifold* with holonomy SU(3) asymptotic to $\mathbb{C}^3/G$, as in Chapters 8 and 9. Then by Proposition 11.1.2 there is a torsion-free G_2-structure (φ, g) on $\mathbb{R} \times Y$. This is another example of a *QALE G_2-manifold* asymptotic to $\mathbb{R}^7/G$.

We have seen that if G is a finite subgroup of G_2, then $\mathbb{R}^7/G$ may be a product $\mathbb{R}^3 \times \mathbb{C}^2/G$ or $\mathbb{R} \times \mathbb{C}^3/G$, where G is a subgroup of SU(2) or SU(3). We will now show that these are the only nontrivial ways in which $\mathbb{R}^7/G$ can be isomorphic to a product.

Proposition 11.1.3 *Let G be a finite subgroup of G_2, and let*

$$V = \{v \in \mathbb{R}^7 : \gamma \cdot v = v \text{ for all } \gamma \in G\}.$$

Then either
 (i) $V = \mathbb{R}^7$;
 (ii) V *is an associative 3-plane (as in §10.8.1), G is conjugate to a subgroup of* SU(2), *and* $\mathbb{R}^7/G \cong \mathbb{R}^3 \times \mathbb{C}^2/G$;
 (iii) $V \cong \mathbb{R}$, *and G is conjugate in G_2 to a subgroup of* SU(3), *and* $\mathbb{R}^7/G \cong \mathbb{R} \times \mathbb{C}^3/G$; *or*
 (iv) $V = \{0\}$.

Proof Let (φ_0, g_0) be the flat G_2-structure on $\mathbb{R}^7$. We can use φ_0 and g_0 to construct a G_2-invariant vector product $\wedge$ on $\mathbb{R}^7$, given in index notation by

$$(u \wedge v)^a = (g_0)^{ab}(\varphi_0)_{bcd} u^c v^d.$$

As $G \subset G_2$, if $u, v \in \mathbb{R}^7$ are G-invariant, then $u \wedge v$ is G-invariant. Thus V is closed under the vector product $\wedge$. It is not difficult to show that any vector subspace $V \subseteq \mathbb{R}^7$ closed under $\wedge$ is $\mathbb{R}^7$ or $\{0\}$, or isomorphic to $\mathbb{R}^3$ and an associative 3-plane, or isomorphic to $\mathbb{R}$.

From §10.8.1, if V is an associative 3-plane, then V is conjugate to
$$U_3 = \{(x_1, x_2, x_3, 0, 0, 0, 0) : x_j \in \mathbb{R}\},$$
so G is conjugate to a subgroup of SU(2), the subgroup of G_2 fixing U_3. If $V \cong \mathbb{R}$, then V is conjugate under G_2 to
$$U_1 = \{(x_1, 0, 0, 0, 0, 0, 0) : x_1 \in \mathbb{R}\}.$$
Thus G is conjugate to a subgroup of SU(3), as this is the subgroup of G_2 fixing U_1. $\square$

11.2 Quasi-ALE G_2-manifolds

In this section we consider *QALE G_2-manifolds*, a natural class of resolutions of quotient singularities $\mathbb{R}^7/G$ for G a finite subgroup of G_2, equipped with torsion-free G_2-structures (φ, g) that are asymptotic to the flat G_2-structure (φ_0, g_0) on $\mathbb{R}^7/G$ in a certain way. Note that the idea of *ALE G_2-manifold* does not make sense, since the singularities of $\mathbb{R}^7/G$ are never isolated. This is because every element of G_2 fixes a nonzero vector subspace in $\mathbb{R}^7$, as $\mathbb{R}^7$ is of odd dimension.

Our treatment is modelled on §9.9, and the reader is advised to look at the beginning of §9.9 before continuing. We begin with the following definition, motivated by Proposition 11.1.3.

Definition 11.2.1 We say that a torsion-free G_2-structure (φ, g) on $V \times Y$ is a *product G_2-structure* if V is a Euclidean vector space and either

(i) $V \cong \mathbb{R}^7$, and Y is a point, and (φ, g) is the flat G_2-structure (φ_0, g_0) on $\mathbb{R}^7$ given in Definition 10.1.1;

(ii) $V \cong \mathbb{R}^3$, and Y is a 4-manifold, and (φ, g) is isomorphic to one of the G_2-structures on $\mathbb{R}^3 \times Y$ constructed in Proposition 11.1.1;

(iii) $V \cong \mathbb{R}$, and Y is a 6-manifold, and (φ, g) is isomorphic to one of the G_2-structures on $\mathbb{R} \times Y$ constructed in Proposition 11.1.2; or

(iv) $V = \{0\}$ and Y is a 7-manifold.

In each case g is a product metric $h_V \times g_Y$ on $V \times Y$, where h_V is Euclidean.

Here is the definition of QALE G_2-manifolds.

Definition 11.2.2 Let G be a finite subgroup of G_2, and let (X, π) be a *real local product resolution* of $\mathbb{R}^7/G$, as in Definition 9.9.1. Let the notation $\mathcal{L}, I, V_i, W_i, \ldots$ be as in Definitions 9.9.1 and 9.9.2.

Let (φ, g) be a torsion-free G_2-structure on X. We say that (X, φ, g) is a *QALE G_2-manifold* if for each $i \in I$, there exists a product G_2-structure (φ_i, g_i) on $V_i \times Y_i$, in the sense of Definition 11.2.1, such that

$$\nabla^l \big(\psi_i^*(\varphi) - \varphi_i\big) = \sum_{j \in I : i \not\trianglelefteq j} O\big(\mu_{i,j}^{d_j} v_i^{-2-l}\big) \tag{11.5}$$

on $V_i \times Y_i \setminus U_i$, for all $l \geq 0$. For each $i \in I$ we say that (X, φ, g) is *asymptotic to* $(V_i \times Y_i, \varphi_i, g_i)$.

This is based on Definitions 9.2.1 and 9.9.2, which define QALE Kähler manifolds and QALE manifolds. Those definitions also include the condition that the real codimension of the singularities of $\mathbb{C}^m/G$ or $\mathbb{R}^n/G$ should be at least 4. Using Proposition 11.1.3 we can see that for nontrivial $G \subset G_2$, the singularities of $\mathbb{R}^7/G$ always have codimension 4 or 6. Thus the condition holds automatically, and there is no need to assume it.

Now φ determines g, and in fact (11.5) implies that

$$\nabla^l\big(\psi_i^*(g) - g_i\big) = \sum_{j \in I: i \not\succeq j} O\big(\mu_{i,j}^{d_j} v_i^{-2-l}\big). \tag{11.6}$$

Since each g_i is the product of the Euclidean metric on V_i and some metric on Y_i, it follows from Definition 9.9.2 that g is a *QALE metric* on X. Putting $i = 0$, eqns (11.5) and (11.6) say that (φ, g) is asymptotic to the flat G_2-structure (φ_0, g_0) on $\mathbb{R}^7/G$.

The following result will be needed in §11.5.

Proposition 11.2.3 *Let (X, φ, g) be a QALE G_2-manifold. Then for each $i \in I$ there is a smooth 2-form σ_i and a smooth 3-form τ_i on $V_i \times Y_i \setminus U_i$, satisfying*

$$\psi_i^*(\varphi) - \varphi_i = d\sigma_i \quad \text{and} \quad \psi_i^*\big(\Theta(\varphi)\big) - \Theta(\varphi_i) = d\tau_i,$$

such that on $V_i \times Y_i \setminus U_i$, for all $l \geq 0$ we have

$$\nabla^l \sigma_i = \sum_{j \in I: i \not\succeq j} O\big(\mu_{i,j}^{d_j} v_i^{-1-l}\big) \quad \text{and} \quad \nabla^l \tau_i = \sum_{j \in I: i \not\succeq j} O\big(\mu_{i,j}^{d_j} v_i^{-1-l}\big). \tag{11.7}$$

We give only a sketch of the proof, leaving the details to the reader. Clearly $\psi_i^*(\varphi) - \varphi_i$ is closed, as $d\varphi = d\varphi_i = 0$. Now if α is a closed 3-form representing a nonzero de Rham cohomology class on $V_i \times Y_i \setminus U_i$, then α decays with order at best $O(v_i^{-3})$. As $d_j \leq -2$ for $j \neq 0$ we see that $\psi_i^*(\varphi) - \varphi_i$ is $O(v_i^{-4})$ by (11.5), so it must be zero in de Rham cohomology, and thus $\psi_i^*(\varphi) - \varphi_i = d\sigma_i$ for some 2-form σ_i. Similarly $\psi_i^*\big(\Theta(\varphi)\big) - \Theta(\varphi_i) = d\tau_i$ for some τ_i.

We can use the results of Chapters 8 and 9 to prove the proposition in the case that X is a product G_2-manifold $V \times Y$, where dim $V > 0$. But in the general case, (φ, g) is modelled near infinity in X on product G_2-manifolds $(V_j \times Y_j, \varphi_j, g_j)$ with dim $V_j > 0$, for which the proposition holds.

From this we can build up suitable asymptotic models σ_i', τ_i' for σ_i and τ_i near infinity, by adding up contributions from $V_j \times Y_j$ with $j \not\succeq i$ and dim $V_j > 0$, such that $\psi_i^*(\varphi) - \varphi_i - d\sigma_i'$ and $\psi_i^*\big(\Theta(\varphi)\big) - \Theta(\varphi_i) - d\tau_i'$ decay quickly near infinity on $V_i \times Y_i$. Finally, we use a version of Theorem 8.4.1 for QALE manifolds to find forms σ_i'', τ_i'' with suitable decay satisfying

$$\psi_i^*(\varphi) - \varphi_i - d\sigma_i' = d\sigma_i'' \quad \text{and} \quad \psi_i^*\big(\Theta(\varphi)\big) - \Theta(\varphi_i) - d\tau_i' = d\tau_i'',$$

and then $\sigma_i = \sigma_i' + \sigma_i''$ and $\tau_i = \tau_i' + \tau_i''$ satisfy the proposition.

11.2.1 Examples of QALE G_2-manifolds

It is easy to see that if $G \subset \mathrm{SU}(2)$ is a finite group and Y an ALE manifold with holonomy $\mathrm{SU}(2)$, then the G_2-manifold $(\mathbb{R}^3 \times Y, \varphi, g)$ given by Proposition 11.1.1 is a QALE G_2-manifold. Similarly, if Y is an ALE or QALE manifold with holonomy $\mathrm{SU}(3)$ then the G_2-manifold $(\mathbb{R} \times Y, \varphi, g)$ given by Proposition 11.1.2 is a QALE G_2-manifold. Thus we prove:

Theorem 11.2.4 *Let $G \subset G_2$ be a finite group, and suppose G is conjugate to a subgroup of $\mathrm{SU}(2)$ or $\mathrm{SU}(3)$, where the inclusions $\mathrm{SU}(2) \subset \mathrm{SU}(3) \subset G_2$ are defined in Theorem 10.2.1. Then there exists a QALE G_2-manifold (X, φ, g) asymptotic to $\mathbb{R}^7/G$.*

Proof If G is conjugate to a subgroup of $\mathrm{SU}(2)$ then there is a splitting $\mathbb{R}^7 \cong \mathbb{R}^3 \oplus \mathbb{C}^2$ such that $\mathbb{R}^7/G \cong \mathbb{R}^3 \times \mathbb{C}^2/G$, where G acts on $\mathbb{C}^2$ as a finite subgroup of $\mathrm{SU}(2)$. Then from §7.2 there exists a family of ALE manifolds (Y, g_Y) with holonomy $\mathrm{SU}(2)$ asymptotic to $\mathbb{C}^2/G$. Proposition 11.1.1 constructs a G_2-manifold $(\mathbb{R}^3 \times Y, \varphi, g)$, and this is a QALE G_2-manifold asymptotic to $\mathbb{R}^7/G$.

Similarly, if G is conjugate to a subgroup of $\mathrm{SU}(3)$ then there is a splitting $\mathbb{R}^7 \cong \mathbb{R} \oplus \mathbb{C}^3$ such that $\mathbb{R}^7/G \cong \mathbb{R} \times \mathbb{C}^3/G$, where G acts on $\mathbb{C}^3$ as a finite subgroup of $\mathrm{SU}(3)$. By Theorem 6.4.1 there exists a crepant resolution Y of $\mathbb{C}^3/G$, and Theorem 9.3.3 gives a family of Ricci-flat QALE Kähler metrics g_Y on Y. If G does not lie in any $\mathrm{SU}(2)$ in $\mathrm{SU}(3)$ then g_Y has holonomy $\mathrm{SU}(3)$ by Theorem 9.3.4. Proposition 11.1.2 constructs a G_2-manifold $(\mathbb{R} \times Y, \varphi, g)$, and this is a QALE G_2-manifold asymptotic to $\mathbb{R}^7/G$. $\square$

The theorem gives us many examples of QALE G_2-manifolds, but they are all products $\mathbb{R}^3 \times Y$ or $\mathbb{R} \times Y$. Here are some examples of QALE G_2-manifolds which are not products, which will be used in §12.8.2.

Example 11.2.5 Identify $\mathbb{R}^7$ with $\mathbb{R} \oplus \mathbb{C}^3$ with coordinates (x, z_1, z_2, z_3), where $x \in \mathbb{R}$ and $z_j \in \mathbb{C}$. Define $\alpha, \beta : \mathbb{R}^7 \to \mathbb{R}^7$ by

$$\alpha : (x, z_1, z_2, z_3) \mapsto (x, -z_1, iz_2, iz_3), \quad \beta : (x, z_1, z_2, z_3) \mapsto (-x, \bar{z}_1, \bar{z}_3, -\bar{z}_2).$$

Then α, β satisfy $\alpha^4 = \beta^4 = 1$, $\alpha^2 = \beta^2$ and $\beta\alpha\beta^{-1} = \alpha^3$, and so $G = \langle \alpha, \beta \rangle$ is a finite nonabelian group of order 8.

As in the proof of Proposition 11.1.2, define a flat G_2-structure (φ_0, g_0) on $\mathbb{R}^7$ by $\varphi_0 = dx \wedge \omega_0 + \mathrm{Re}\,\theta_0$, where $\omega_0 = \frac{i}{2}\sum_{j=1}^3 dz_j \wedge d\bar{z}_j$ and $\theta_0 = dz_1 \wedge dz_2 \wedge dz_3$. Then α and β preserve φ_0, so G is a subgroup of G_2. We will explain how to construct QALE G_2-manifolds asymptotic to $\mathbb{R}^7/G$.

Observe that $\mathbb{R}^7/\langle\alpha\rangle = \mathbb{R} \times \mathbb{C}^3/\mathbb{Z}_4$, and this complex singularity $\mathbb{C}^3/\mathbb{Z}_4$ has a unique crepant resolution (Y, π), which can be constructed using toric geometry. By Theorems 9.3.3 and 9.3.4, Y admits a family of QALE metrics g_Y with holonomy $\mathrm{SU}(3)$. Proposition 11.1.2 then gives a torsion-free G_2-structure (φ, g) on $\mathbb{R} \times Y$, and this makes $\mathbb{R} \times Y$ into a QALE G_2-manifold asymptotic to $\mathbb{R}^7/\langle\alpha\rangle$.

Now β acts on $\mathbb{R}^7/\langle\alpha\rangle$ with $\beta^2 = 1$, and $(\mathbb{R}^7/\langle\alpha\rangle)/\langle\beta\rangle = \mathbb{R}^7/G$. This action of β on $\mathbb{R}^7/\langle\alpha\rangle$ lifts through π to an action on $\mathbb{R} \times Y$, which preserves (φ, g). Furthermore, this action of $\langle\beta\rangle = \mathbb{Z}_2$ on $\mathbb{R} \times Y$ is *free*. Therefore $X = (\mathbb{R} \times Y)/\langle\beta\rangle$ is a nonsingular 7-manifold, and (φ, g) pushes down to X, making it into a QALE G_2-manifold asymptotic to $\mathbb{R}^7/G$. Note that X is not a product, and that $\pi_1(X) \cong \mathbb{Z}_2$.

In fact there is not just one way of resolving $\mathbb{R}^7/G$ with a QALE G_2-manifold, but *three*. Our method above was to write $\mathbb{R}^7 \cong \mathbb{R} \oplus \mathbb{C}^3$ such that α acts trivially on $\mathbb{R}$ and as an element of SU(3) on $\mathbb{C}^3$, to take a crepant resolution Y of $\mathbb{C}^3/\langle\alpha\rangle$, and lift the action of $G/\langle\alpha\rangle$ up to Y. In the same way, there exist two other splittings $\mathbb{R}^7 \cong \mathbb{R} \oplus \mathbb{C}^3$ in which β (in the first) and $\alpha\beta$ (in the second) act trivially on $\mathbb{R}$ and as an element of SU(3) on $\mathbb{C}^3$, and we can apply the same method to these splittings. We will use these QALE G_2-manifolds to desingularize an orbifold T^7/Γ in §12.8.2.

The G_2-structures in this example have holonomy $\mathbb{Z}_2 \ltimes \text{SU}(3)$. Clearly, it would be interesting to find QALE G_2-manifolds (X, φ, g) for which $\text{Hol}(g) = G_2$, and not some proper subgroup. In fact the author can construct examples of such G_2-manifolds (X, φ, g). The proof will not be given here, as it is rather long and not very interesting. But here is a sketch of how it goes.

Suppose G is a finite subgroup of G_2, and H is a normal subgroup of G which lies in SU(2) $\subset G_2$. Let K be the quotient group G/H. Then $\mathbb{R}^7/H \cong \mathbb{R}^3 \times \mathbb{C}^2/H$, and K acts on $\mathbb{R}^3 \times \mathbb{C}^2/H$ with $(\mathbb{R}^3 \times \mathbb{C}^2/H) = \mathbb{R}^7/G$. Choose an ALE manifold Y with holonomy SU(2) asymptotic to $\mathbb{C}^2/H$. Then $\mathbb{R}^3 \times Y$ is a QALE G_2-manifold asymptotic to $\mathbb{R}^7/H$. Suppose the K-action on $\mathbb{R}^7/H$ lifts to a K-action on $\mathbb{R}^3 \times Y$ preserving the G_2-structure on $\mathbb{R}^3 \times Y$. Then $(\mathbb{R}^3 \times Y)/K$ is a 7-orbifold equipped with a torsion-free G_2-structure.

The idea is to construct a 7-manifold X which resolves the singularities of $(\mathbb{R}^3 \times Y)/K$, and a family of metrics with holonomy G_2 on X making X into a QALE manifold asymptotic to $\mathbb{R}^7/G$. This is done following the method used to construct compact 7-manifolds with holonomy G_2 by desingularizing orbifolds T^7/Γ described later in the chapter, but instead of compact 7-manifolds we work with QALE 7-manifolds, using the analytic ideas of Chapter 9 to choose suitable Banach spaces of k-forms in which the elliptic operators are well-behaved.

The trick in doing this is to choose G, H and Y such that all the singularities of $(\mathbb{R}^3 \times Y)/K$ are locally modelled on SU(3) singularities, which can be desingularized as QALE manifolds with holonomy SU(3) by the results of Chapter 9. We then glue these local models together as in §11.5 to get a QALE G_2-structure with small torsion on X, and deform it using analysis as in §11.6–§11.8 to get a torsion-free QALE G_2-structure on X.

Here is an example of a suitable group G.

Example 11.2.6 Identify $\mathbb{R}^7$ with $\mathbb{R}^3 \oplus \mathbb{C}^2$ with coordinates $(x_1, x_2, x_3, z_1, z_2)$, where $x_j \in \mathbb{R}$ and $z_j \in \mathbb{C}$. Define $\alpha, \beta, \gamma : \mathbb{R}^7 \to \mathbb{R}^7$ by

$$\alpha : (x_1, x_2, x_3, z_1, z_2) \mapsto (x_1, x_2, x_3, -z_1, -z_2),$$
$$\beta : (x_1, x_2, x_3, z_1, z_2) \mapsto (x_1, -x_2, -x_3, iz_2, -iz_1)$$
$$\text{and} \quad \gamma : (x_1, x_2, x_3, z_1, z_2) \mapsto (-x_1, x_2, -x_3, \bar{z}_1, \bar{z}_2).$$

Let $G = \langle \alpha, \beta, \gamma \rangle$ and $H = \langle \alpha \rangle$. Then G is a nonabelian subgroup of G_2 of order 8, and $H \cong \mathbb{Z}_2$ is a normal subgroup of G lying in $\mathrm{SU}(2) \subset G_2$.

Choose Y to be the blow-up of $\mathbb{C}^2/\{\pm 1\}$ at 0, equipped with an ALE metric with holonomy $\mathrm{SU}(2)$. Then $K = G/H$ acts naturally on $\mathbb{R}^3 \times Y$. The three non-identity elements of K are βH, γH and $\beta \gamma H$. Of these βH and γH have fixed points which do not intersect, and $\beta \gamma H$ has no fixed points in $\mathbb{R}^3 \times Y$. So the fixed point set of $(\mathbb{R}^3 \times Y)/K$ splits into two connected components, one each from the fixed points of βH and γH.

As the two pieces of the singular set do not intersect, we can resolve each piece separately. But because the subgroups $\langle \alpha, \beta \rangle$ and $\langle \alpha, \gamma \rangle$ of G both lie in a copy of $\mathrm{SU}(3)$ in G_2, each piece can be resolved with holonomy $\mathrm{SU}(3)$ using the ideas of Chapter 9, and so by the discussion above we can construct a QALE manifold X with holonomy G_2 asymptotic to $\mathbb{R}^7/G$.

11.3 Some notation to describe the orbifolds T^7/Γ

Let Γ be a finite group acting on T^7 preserving the flat G_2-structure (φ_0, g_0). Then T^7/Γ is an *orbifold*. In this section we introduce some notation to describe the structure of the singularities of T^7/Γ, which is based on the ideas of Chapter 9 and §11.2.

Let Λ be a *lattice* in $\mathbb{R}^7$, that is, a discrete additive subgroup isomorphic to $\mathbb{Z}^7$. Then $\mathbb{R}^7/\Lambda$ is the 7-torus T^7. It is a compact 7-manifold. Each point $x \in T^7$ may be written $v + \Lambda$ for some $v \in \mathbb{R}^7$. Every tangent space $T_x T^7$ is naturally isomorphic to $\mathbb{R}^7$. The Euclidean G_2-structure on $\mathbb{R}^7$ given in Definition 10.1.1 pushes down to a flat G_2-structure (φ_0, g_0) on T^7.

Let Γ be a finite group of automorphisms of T^7 which preserve (φ_0, g_0). It is easy to show that every element $\gamma \in \Gamma$ acts by

$$\gamma(v + \Lambda) = \alpha(v) + \beta + \Lambda \quad \text{for } v \in \mathbb{R}^7,$$

where $\alpha \in G_2$ and $\alpha(\Lambda) = \Lambda$, and $\beta \in \mathbb{R}^7$, so that $\beta + \Lambda \in T^7$. That is, γ is the combination of a 'rotation' α and a 'translation' $\beta + \Lambda$ of T^7. Note that Γ also has a natural, linear action on $\mathbb{R}^7$ given by $\gamma : v \mapsto \alpha(v)$.

Define $\mathrm{SL}(\Lambda)$ to be the subgroup of $\alpha \in \mathrm{SL}(7, \mathbb{R})$ such that $\alpha(\Lambda) = \Lambda$. Then since $\Lambda \cong \mathbb{Z}^7$, we see that $\mathrm{SL}(\Lambda) \cong \mathrm{SL}(7, \mathbb{Z})$. The full group of automorphisms of T^7 preserving (φ_0, g_0) is $(G_2 \cap \mathrm{SL}(\Lambda)) \ltimes T^7$, and Γ is a finite subgroup of this. The quotient T^7/Γ is in general a *compact orbifold*, as in Definition 6.5.1.

If A is a subgroup of Γ and F a subset of T^7, define the *fixed point set* $\mathrm{Fix}(A)$, the *centralizer* $C(F)$ and the *normalizer* $N(F)$ by

$$\mathrm{Fix}(A) = \{x \in T^7 : ax = x \text{ for all } a \in A\},$$
$$C(F) = \{\gamma \in \Gamma : \gamma x = x \text{ for all } x \in F\},$$
$$\text{and} \quad N(F) = \{\gamma \in \Gamma : \gamma F = F\}.$$

Then $C(F)$ and $N(F)$ are subgroups of Γ, and $C(F)$ is normal in $N(F)$.

Definition 11.3.1 Define a finite set $\mathcal{L}$ of subsets of T^7 by

$$\mathcal{L} = \{F : F \text{ is a connected component of Fix}(A), \text{where } A \text{ is a subgroup of } \Gamma\}.$$

Let I be an *indexing set* for $\mathcal{L}$, so that we may write $\mathcal{L} = \{F_i : i \in I\}$. Since $T^7 = \text{Fix}(\{1\})$ is connected, we see that $T^7 \in \mathcal{L}$. Let the index of T^7 be 0, so that $F_0 = T^7$. Define a *partial order* $\succeq$ on I by $i \succeq j$ if $F_i \subseteq F_j$. Then $i \succeq 0$ for all $i \in I$. We shall also write $i \succ j$ if $i \succeq j$ and $i \neq j$.

Define $A_i = C(F_i)$ for each $i \in I$. Then A_i is a subgroup of Γ with $A_0 = \{1\}$, and F_i is a connected component of $\text{Fix}(A_i)$. Thus A_i acts on $\mathbb{R}^7$, as Γ does. Define V_i to be the vector subspace of $\mathbb{R}^7$ fixed by A_i, and let $n_i = \dim V_i$. It is easy to show that $\Lambda_i = V_i \cap \Lambda$ is a *lattice* in V_i, and V_i/Λ_i is a torus T^{n_i}. Choose a point $\beta + \Lambda$ in F_i. Then the map $V_i/\Lambda_i \to F_i$ given by $v + \Lambda_i \mapsto v + \beta + \Lambda$ is a 1-1 correspondence. So F_i is isomorphic to V_i/Λ_i, and F_i is a torus T^{n_i} in T^7. Moreover, each tangent space $T_p F_i$ is naturally isomorphic to V_i.

For each $i \in I$, define W_i to be the perpendicular subspace to V_i in $\mathbb{R}^7$, using the Euclidean metric. Then $\mathbb{R}^7 = V_i \oplus W_i$, and A_i acts on W_i, with $\mathbb{R}^7/A_i \cong V_i \times W_i/A_i$. Note also that W_i is the normal vector space to F_i in T^7 at every point of F_i. Define B_i to be the quotient group $N(F_i)/C(F_i)$. Then B_i acts naturally on F_i, V_i and W_i/A_i.

Let $F_i, F_j \in \mathcal{L}$. Then $F_i \cap F_j$ is a finite (possibly empty) union of isomorphic tori T^k. Let A be the subgroup of Γ generated by A_i and A_j. Then each connected component of $F_i \cap F_j$ is also a connected component of $\text{Fix}(A)$, and so lies in $\mathcal{L}$. If $\gamma \in \Gamma$ and $F_i \in \mathcal{L}$ then $\gamma F_i \in \mathcal{L}$, as γF_i is a component of $\text{Fix}(\gamma A_i \gamma^{-1})$. For each $\gamma \in \Gamma$ and $i \in I$, let $\gamma \cdot i$ be the unique element of I such that $F_{\gamma \cdot i} = \gamma F_i$. This defines an action of Γ on I, which also satisfies $V_{\gamma \cdot i} = \gamma V_i$, $W_{\gamma \cdot i} = \gamma W_i$, and $A_{\gamma \cdot i} = \gamma A_i \gamma^{-1}$.

Let $x \in T^7$. Then there exists a unique $i \in I$ such that $x \in F_i$ and $C(\{x\}) = A_i$. Furthermore, T^7/Γ is locally isomorphic to $\mathbb{R}^7/A_i = V_i \times W_i/A_i$ near $x\Gamma$. Thus T^7/Γ is an orbifold, and $x\Gamma$ is a singular point if and only if $i \neq 0$. Equivalently, $x\Gamma$ is a singular point of T^7/Γ if and only if $x \in F_i$ for some $i \in I$ with $i \neq 0$, and the *singular set* S of T^7/Γ is

$$S = \bigcup_{i \in I \setminus \{0\}} F_i \bigg/ \Gamma.$$

For generic $x \in F_i$ the subgroup of Γ fixing x is A_i, and the singularity of T^7/Γ at $x\Gamma$ is locally modelled on the product $V_i \times W_i/A_i$.

Next we consider what T^7/Γ looks like near the part of S coming from F_i, for $i \neq 0$ in I. This is the subset

$$F_i \Gamma / \Gamma = \bigcup_{\gamma \in \Gamma} F_{\gamma \cdot i} \bigg/ \Gamma.$$

We distinguish four cases, in increasing order of complexity.

- Suppose $A_i = \Gamma$. Then T^7/Γ is locally isomorphic near $F_i\Gamma/\Gamma$ to $F_i \times W_i/A_i$.
- Suppose $N(F_i) = \Gamma$, so that $\gamma \cdot i = i$ for all $i \in I$. Then T^7/Γ is locally isomorphic near $F_i\Gamma/\Gamma$ to

$$\big(F_i \times W_i/A_i\big)/B_i. \tag{11.8}$$

If B_i acts freely on F_i then the quotient by B_i introduces no extra singularities near F_i, but otherwise the B_i-action creates new singularities.
- If $F_i \cap F_{\gamma \cdot i} = \emptyset$ whenever $\gamma \cdot i \neq i$, then (11.8) is again a local model for T^7/Γ near $F_i\Gamma/\Gamma$.
- However, if $\gamma \cdot i \neq i$ and $F_i \cap F_{\gamma \cdot i} \neq \emptyset$, then things are more complicated. The subset of S coming from F_i effectively *intersects itself*, and we cannot write down a local model in such a neat way.

11.4 R-data and resolutions of T^7/Γ

Let T^7/Γ be an orbifold with a flat G_2-structure (φ_0, g_0), and let all notation be as in §11.3. We will now define a *resolution* M of T^7/Γ, a compact 7-manifold with a projection $\pi : M \to T^7/\Gamma$ resolving the singularities of T^7/Γ. Then in §11.5 we shall write down a family of G_2-structures (φ^t, g^t) on M with small torsion, that in some sense converge to the flat G_2-structure (φ_0, g_0) on T^7/Γ.

To do this, we need to know how to resolve each of the singularities of T^7/Γ, and how to put a torsion-free G_2-structure on its resolution. The information we use to do this will be called *R-data*, which we define next. Section 11.4.2 then constructs an open cover of T^7/Γ, which is used in §11.4.3 to define the resolution M.

11.4.1 R-data

Here is the definition of R-data, short for *Resolution data*.

Definition 11.4.1 We define a *set of R-data for T^7/Γ* to be

(a) a real local product resolution (Y_i, π_i) of W_i/A_i for each $i \in I$;
(b) a torsion-free G_2-structure (φ_i, g_i) on $V_i \times Y_i$ for each $i \in I$; and
(c) a diffeomorphism $\chi_{\gamma, i} : Y_i \to Y_{\gamma \cdot i}$ for each $i \in I$ and $\gamma \in \Gamma$.

These must satisfy the conditions:

(i) $(V_i \times Y_i, \varphi_i, g_i)$ is both a *product G_2-manifold* and a *QALE G_2-manifold* asymptotic to $\mathbb{R}^7/A_i$ in the sense of Definitions 11.2.1 and 11.2.2, with projection $\mathrm{id} \times \pi_i : V_i \times Y_i \to V_i \times W_i/A_i \cong \mathbb{R}^7/A_i$;
(ii) If $i \succeq j$ then $(V_i \times Y_i, \varphi_i, g_i)$ is asymptotic to $(V_j \times Y_j, \varphi_j, g_j)$ in the sense of Definition 11.2.2; and
(iii) For each $i \in I$ and $\gamma \in \Gamma$, the following diagram commutes:

$$\begin{array}{ccc} Y_i & \xrightarrow{\chi_{\gamma,i}} & Y_{\gamma \cdot i} \\ \downarrow{\pi_i} & & \downarrow{\pi_{\gamma \cdot i}} \\ W_i/A_i & \xrightarrow{\gamma} & W_{\gamma \cdot i}/A_{\gamma \cdot i}, \end{array}$$

and the map $\gamma \times \chi_{\gamma,i} : V_i \times Y_i \to V_{\gamma \cdot i} \times Y_{\gamma \cdot i}$ satisfies $(\gamma \times \chi_{\gamma,i})^*(\varphi_{\gamma \cdot i}) = \varphi_i$ and $(\gamma \times \chi_{\gamma,i})^*(g_{\gamma \cdot i}) = g_i$.

The G_2-structures (φ_i, g_i) will not be used until §11.5. Roughly speaking, (ii) says that the resolutions we choose for intersecting strata of the singular set of T^7/Γ must fit together, and (iii) says that our choice of resolutions must be Γ-equivariant. Part (iii) is based on Propositions 9.1.3 and 9.2.3.

In fact, by following the proofs of these propositions we can show that if F_j is a connected component of $F_i \cap F_{\gamma \cdot i}$ and $(V_j \times Y_j, \varphi_j, g_j)$ is asymptotic to both $(V_i \times Y_i, \varphi_i, g_i)$ and $(V_{\gamma \cdot i} \times Y_{\gamma \cdot i}, \varphi_{\gamma \cdot i}, g_{\gamma \cdot i})$, then there is a unique diffeomorphism $\chi_{\gamma,i} : Y_i \to Y_{\gamma \cdot i}$ satisfying (iii).

So in this case (ii) implies (iii). However, if $F_i \cap F_{\gamma \cdot i} = \emptyset$ then (ii) does not imply (iii), and this is why we have included part (iii) as an assumption. There was no need for a comparable assumption in Chapter 9, since there $V_i \cap V_{\gamma \cdot i}$ is never empty. It is also easy to show that the maps $\chi_{\gamma,i}$ satisfy

$$\chi_{\gamma_1 \gamma_2, i} = \chi_{\gamma_1, \gamma_2 \cdot i} \circ \chi_{\gamma_2, i} \quad \text{for all } \gamma_1, \gamma_2 \in \Gamma \text{ and } i \in I,$$

which means that the $\chi_{\gamma,i}$ define a Γ-action on the disjoint union of the Y_i.

11.4.2 Open covers of T^7 and T^7/Γ

We define *tubular neighbourhoods* O_i, P_i of F_i in T^7.

Definition 11.4.2 Choose constants $\zeta_0, \zeta_1, \zeta_3, \zeta_7 > 0$. For each $i \in I$ we have $n_i = \dim V_i$, so $n_i = 0, 1, 3$ or 7 by Proposition 11.1.3. Define $O_i \subset P_i \subset T^7$ by

$$O_i = \{x \in T^7 : d(x, F_i) \leqslant \zeta_{n_i}\}, \quad P_i = \{x \in T^7 : d(x, F_i) < 2\zeta_{n_i}\},$$

where $d(,)$ is the distance in T^7 induced by g_0. The constants ζ_j must satisfy:

(i) There are natural isomorphisms

$$O_i \cong F_i \times \overline{B}_i(\zeta_{n_i}) \quad \text{and} \quad P_i \cong F_i \times B_i(2\zeta_{n_i}), \tag{11.9}$$

where $\overline{B}_i(\zeta_{n_i})$ is the closed ball of radius ζ_{n_i} and $B_i(2\zeta_{n_i})$ the open ball of radius $2\zeta_{n_i}$ about 0 in W_i; and

(ii) For each $i, j \in I$ with $i \not\succ j$ and $j \not\succ i$, we have

$$P_i \cap P_j \subseteq \bigcup_{\substack{k \in I: \\ k \succ i, k \succ j}} O_k.$$

When $F_i \cap F_j = \emptyset$ there are no $k \in I$ with $k \succ i, k \succ j$, so $P_i \cap P_j = \emptyset$.

We can always choose the constants ζ_i to satisfy (i) and (ii). Part (i) holds when the ζ_i are sufficiently small, and part (ii) when ζ_{n_i} is small compared to ζ_{n_k}, for $k \succ i$. So (i), (ii) are both true if $1 \gg \zeta_0 \gg \zeta_1 \gg \zeta_3 \gg \zeta_7 > 0$. We use the O_i, P_i to define a special *open cover* $\{Q_i : i \in I\}$ for T^7.

Proposition 11.4.3 *For each $i \in I$, define an open set Q_i in T^7 by*

$$Q_i = P_i \setminus \bigcup_{j \in I : j \succ i} O_j. \tag{11.10}$$

Then $T^7 = \bigcup_{i \in I} Q_i$, and if $i, j \in I$ and $Q_i \cap Q_j \neq \emptyset$ then $i \succeq j$ or $j \succeq i$.

The group A_i acts on Q_i, and Q_i/A_i is naturally isomorphic to an open subset in $F_i \times W_i/A_i$. Also, B_i acts freely on Q_i/A_i, and the image of Q_i in T^7/Γ is isomorphic to $Q_i/N(F_i) = (Q_i/A_i)/B_i$.

Proof Let $x \in T^7$, and let i be a maximal element of the set

$$\{i \in I : d(x, F_i) \leq \zeta_{n_i}\}$$

with respect to the partial order $\succeq$. This set is finite and contains 0, so i certainly exists, and it is easy to see that $x \in Q_i$. Thus $T^7 = \bigcup_{i \in I} Q_i$.

Suppose $i, j \in I$ and $i \not\succeq j$, $j \not\succeq i$. Then by part (ii) above, $P_i \cap P_j$ is contained in the union of O_k with $k \succ i$, $k \succ j$. But each of these O_k is excluded from Q_i, Q_j, and so $Q_i \cap Q_j = \emptyset$. Conversely, if $Q_i \cap Q_j \neq \emptyset$ then $i \succeq j$ or $j \succeq i$.

Clearly A_i acts on Q_i. Now Q_i is an open subset of P_i, which is isomorphic to a subset of $F_i \times W_i$ by (11.9). Here A_i acts trivially on F_i and as usual on W_i. Thus Q_i/A_i is isomorphic to an open subset of $F_i \times W_i/A_i$.

It is also clear that B_i acts on Q_i/A_i. Suppose some $b \neq 1$ in B_i has fixed points in Q_i/A_i. These fixed points must come from some F_j in T^7, where $i \not\succeq j$. But if $j \succ i$ then F_j is excluded from Q_i by (11.10), and if $j \not\succeq i$ then F_j is excluded from Q_i by part (ii) above. Thus, no $b \neq 1$ in B_i can have fixed points, and B_i acts freely on Q_i/A_i.

The image of Q_i in T^7/Γ is $\bigcup_{\gamma \in \Gamma} Q_{\gamma \cdot i}/\Gamma$. Now if $\gamma \in \Gamma$ and $\gamma \cdot i \neq i$, then $i \not\succeq \gamma \cdot i$ and $\gamma \cdot i \not\succeq i$, and so $Q_i \cap Q_{\gamma \cdot i} = \emptyset$, from above. It easily follows that the image of Q_i in T^7/Γ is isomorphic to $Q_i/N(F_i) = (Q_i/A_i)/B_i$. □

The proposition shows that the Q_i form an open cover for T^7, with simple rules for when the overlaps $Q_i \cap Q_j$ can be nonempty. Similarly, T^7/Γ is covered by open sets $(Q_i/A_i)/B_i$. Now Q_i/A_i is an open set in $F_i \times W_i/A_i$, so its singularities are those of W_i/A_i, and B_i acts freely on Q_i/A_i, so the quotient by B_i introduces no new singularities.

Thus the singularities of $(Q_i/A_i)/B_i$ are essentially only product singularities. We have found an open cover for T^7/Γ in which both the singularities of each open set, and the overlaps between open sets, are simple and easily understood.

11.4.3 *Resolutions of T^7/Γ*

Now suppose we are given a set of R-data (Y_i, π_i), (φ_i, g_i), $\chi_{\gamma, i}$ for T^7/Γ, as in Definition 11.4.1. We will use it to define a *resolution* M of T^7/Γ, a compact 7-manifold with a projection $\pi : M \to T^7/\Gamma$ resolving the singularities of T^7/Γ.

Definition 11.4.4 Regard Q_i/A_i as an open subset of $F_i \times W_i/A_i$, as in Proposition 11.4.3. For each $i \in I$, define M_i to be the inverse image of Q_i/A_i under the map $\mathrm{id} \times \pi_i : F_i \times Y_i \to F_i \times W_i/A_i$. Then M_i is an open subset of $F_i \times Y_i$.

The free action of B_i on Q_i/A_i lifts through $\mathrm{id}\times\pi_i$ to a free action of B_i on M_i. Thus, M_i/B_i is a manifold, which resolves the singularities of $(Q_i/A_i)/B_i$. Now from above, the $(Q_i/A_i)/B_i$ are an open cover of T^7/Γ. The M_i/B_i form a corresponding open cover of M, and we make M by gluing together the overlapping patches M_i/B_i in a natural way.

This gluing process is simple in conception, but complicated to write down. Our method is to construct an equivalence relation '$\sim$' on the disjoint union $\coprod_{i\in I} M_i$. Then we will define M to be the quotient $\coprod_{i\in I} M_i \big/ \sim$.

Definition 11.4.5 Let $i \in I$, so (Y_i, π_i) is a *real local product resolution* of W_i/A_i. Explicitly, let $I' = \{j \in I : i \succeq j\}$ and for each $j \in I'$ define $V'_j = V_j \cap W_i$,

$$S'_j = \bigcup_{k\in I': j \not\succeq k} V'_k \Big/ A_j, \quad T'_j = \{x \in W_i/A_j : d(x, S'_j) \leqslant R_i\},$$

and $U'_j = (\mathrm{id}\times\pi_j)^{-1}(T'_j)$, where $\mathrm{id}\times\pi_j$ maps $V'_j \times Y_j \to V'_j \times W_j/A_j$, and $R_i > 0$. Then we have $\psi'_j : V'_j \times Y_j \setminus U'_j \to Y_i$. So $\mathrm{id}\times\psi'_j$ maps

$$\mathrm{id}\times\psi'_j : F_i \times V'_j \times Y_j \setminus F_i \times U'_j \to F_i \times Y_i. \tag{11.11}$$

As in Definition 11.4.2, there is a natural isomorphism

$$F_j \cap P_i \cong F_i \times B'_j(2\zeta_{n_i}) \subset F_i \times V'_j,$$

where $B'_j(2\zeta_{n_i})$ is the ball of radius $2\zeta_{n_i}$ about 0 in V'_j. Thus, we can identify $M_j \cap (P_i \times Y_j)$ with an open set in $F_i \times V'_j \times Y_j$. For simplicity we require that $R_i \leqslant \zeta_{n_i}$ for all i. (We shall see in §11.5 that this assumption can be removed using a rescaling argument.) It then follows by definition of M_j that

$$M_j \cap (P_i \times Y_j) \subset F_i \times V'_j \times Y_j \setminus F_i \times U'_j. \tag{11.12}$$

Therefore the map $\mathrm{id}\times\psi'_j$ of (11.11) sends $M_j \cap (P_i \times Y_j) \to F_i \times Y_i$, and in fact its image lies in M_i. So $\mathrm{id}\times\psi'_j$ maps $M_j \cap (P_i \times Y_j) \to M_i$.

Define an *equivalence relation* '$\sim$' on the disjoint union $\coprod_{i\in I} M_i$ as follows. Let $j, k \in I$ and $x \in M_j$, $y \in M_k$. We say that $x \sim y$ if either

(a) there exists $\gamma \in \Gamma$ such that $i = \gamma \cdot k$ satisfies $i \succeq j$, $x \in M_j \cap (P_i \times Y_j)$ and $\mathrm{id}\times\psi'_j(x) = \gamma \times \chi_{\gamma,k}(y)$, where $\gamma \times \chi_{\gamma,k} : F_k \times Y_k \to F_i \times Y_i$, or
(b) there exists $\gamma \in \Gamma$ such that $i = \gamma \cdot j$ satisfies $i \succeq k$, $y \in M_k \cap (P_i \times Y_k)$ and $\gamma \times \chi_{\gamma,j}(x) = \mathrm{id}\times\psi'_k(y)$, where $\gamma \times \chi_{\gamma,j} : F_j \times Y_j \to F_i \times Y_i$.

Define the *resolution* M of T^7/Γ to be $\coprod_{i\in I} M_i \big/ \sim$. It is easy to show that if $x, y \in M_i$ then $x \sim y$ if and only if $x = b \cdot y$ for some $b \in B_i$. Thus, the image of M_i in M is M_i/B_i, and M is covered by patches M_i/B_i for $i \in I$. But from above M_i/B_i is a manifold, and in fact M is a *compact 7-manifold*.

If $x \in M_i$ then $\mathrm{id}\times\pi_i(x) \in Q_i/A_i$. But $Q_i \subset T^7$, so $Q_i/A_i \subset T^7/A_i$. Let $\phi_i : T^7/A_i \to T^7/\Gamma$ be the natural projection. Then $\phi_i\big(\mathrm{id}\times\pi_i(x)\big) \in T^7/\Gamma$. Let $[x]$

be the equivalence class of x under $\sim$, and define the *projection* $\pi : M \to T^7/\Gamma$ by $\pi([x]) = \phi_i(\text{id} \times \pi_i(x))$. Then π is well-defined. Moreover, π is continuous and surjective, and $\pi : M \setminus \pi^{-1}(S) \to T^7/\Gamma \setminus S$ is a diffeomorphism, where S is the singular set of T^7/Γ. We shall also call (M, π) a *resolution* of T^7/Γ.

Here is some motivation for the definition of '$\sim$' above. By Proposition 11.4.3, the sets Q_i and Q_j in T^7 overlap if and only if $i \succeq j$ or $j \succeq i$. Passing to the quotient T^7/Γ, each Q_i becomes $(Q_i/A_i)/B_i$, and if $i = \gamma \cdot k$ for some $\gamma \in \Gamma$ then $(Q_k/A_k)/B_k = (Q_i/A_i)/B_i$ as subsets of T^7/Γ. Therefore $(Q_j/A_j)/B_j$ and $(Q_k/A_k)/B_k$ overlap in T^7/Γ if and only if either

(a) there exists $\gamma \in \Gamma$ such that $i = \gamma \cdot k$ satisfies $i \succeq j$, or
(b) there exists $\gamma \in \Gamma$ such that $i = \gamma \cdot j$ satisfies $i \succeq k$.

As the patches M_i/B_i are resolutions of $(Q_i/A_i)/B_i$, we therefore expect M_j/B_j and M_k/B_k to overlap in M if and only if (a) or (b) holds, and when they overlap we need to 'glue them together' with $\sim$. This is where parts (a), (b) of the definition come from.

When $i = 0$ we have $O_0 = P_0 = T^7$ and $Q_0 = T^7 \setminus \bigcup_{j \in I \setminus \{0\}} O_j = M_0$. Also $B_0 = \Gamma$, so M_0/B_0 is a nonsingular open set in T^7/Γ, the complement of a small open neighbourhood of the singular set S of T^7/Γ. So we can think of M as T^7/Γ with a neighbourhood of the singular set cut out, to give M_0/B_0, and with patches M_i/B_i for $i \neq 0$ glued in, which are resolutions of the various pieces of the singular set.

11.5 G_2-structures on M with small torsion

In this section we use the R-data of §11.4 to define a 1-parameter family of G_2-structures (φ^t, g^t) on the resolution M of T^7/Γ depending on $t \in (0, \epsilon]$. The idea is that (φ^t, g^t) converges in some sense to the flat, singular G_2-structure (φ_0, g_0) on T^7/Γ as $t \to 0$, and that t is a rough measure of the diameter of the blow-up of each singular point.

We will also estimate several geometrical properties of (φ^t, g^t). In particular, the torsion of (φ^t, g^t) has size somewhere between $O(t^4)$ and $O(t^2)$, depending on how you measure it. Thus, (φ^t, g^t) has *small torsion* when t is small. We will exploit this fact in §11.6–§11.8, by showing that if t is small enough then we can deform (φ^t, g^t) to a nearby, *torsion-free* G_2-structure $(\tilde{\varphi}^t, \tilde{g}^t)$.

The basic idea is that we shrink the G_2-structure (φ_i, g_i) on $V_i \times Y_i$ homothetically by a factor t, and make this into a G_2-structure (φ_i^t, g_i^t) on the open set M_i/B_i in M. We then join these φ_i^t with a partition of unity to get a smooth positive 3-form φ^t on M, which induces the G_2-structure (φ^t, g^t).

Unfortunately, to write this down explicitly is an arduous process involving many steps and a lot of notation, and this section is rather long as a result. Our main result is Theorem 11.5.7 in §11.5.4. *Some readers may wish just to read and understand the statement of Theorem 11.5.7, and the discussion after it, and then move on to §11.6.*

11.5.1   *The G_2-structures (φ_i, g_i) on M_i/B_i in M*

Suppose we are given an orbifold T^7/Γ equipped with a flat G_2-structure (φ_0, g_0) and a set of R-data (Y_i, π_i), (φ_i, g_i), $\chi_{\gamma \cdot i}$. Let all notation be as in §11.4, and let M be the resolution of T^7/Γ defined there.

Then (φ_i, g_i) is a product G_2-structure on $V_i \times Y_i$ for each $i \in I$. From §11.3 we can identify F_i with V_i/Λ_i, where $\Lambda_i = \Lambda \cap V_i$. Thus $F_i \times Y_i \cong (V_i \times Y_i)/\Lambda_i$, where Λ_i acts by translation on V_i and trivially on Y_i. Now (φ_i, g_i) is Λ_i-invariant, and so descends to a G_2-structure (φ_i, g_i) on $F_i \times Y_i$.

As M_i is an open set in $F_i \times Y_i$, we can interpret (φ_i, g_i) as a G_2-structure on M_i. Part (iii) of Definition 11.4.1 implies that (φ_i, g_i) is invariant under the action of B_i on $F_i \times Y_i$, so (φ_i, g_i) also pushes down to M_i/B_i. But from §11.4.3, the resolution M of T^7/Γ is covered by open sets M_i/B_i. Thus, on each open set in the cover $\{M_i/B_i : i \in I\}$ of M we have a torsion-free G_2-structure (φ_i, g_i).

We will need to compare the G_2-structures (φ_i, g_i) and (φ_j, g_j) on the overlaps $(M_i/B_i) \cap (M_j/B_j)$ in M, and the calculations in the next two pages will eventually help us to do this. By part (ii) of Definition 11.4.1, if $i, j \in I$ and $i \succ j$ then (φ_i, g_i) is asymptotic to (φ_j, g_j). Here is what this means explicitly.

As in Definition 11.4.5 we put $V'_j = V_j \cap W_i$, and we have a map $\psi'_j : V'_j \times Y_j \setminus U'_j \to Y_i$. Thus $\text{id} \times \psi'_j$ sends

$$\text{id} \times \psi'_j : V_i \times V'_j \times Y_j \setminus V_i \times U'_j \to V_i \times Y_i,$$

where $V_i \times V'_j \times Y_j$ is an open set in $V_j \times Y_j$. Then by Definition 11.2.2 we have

$$\nabla^l\big((\text{id} \times \psi'_j)^*(\varphi_i) - \varphi_j\big) = \sum_{k \in I' : j \not\succeq k} O\big(\mu_{j,k}^{d_k}(v'_j)^{-2-l}\big) \tag{11.13}$$

on $V_j \times Y_j \setminus V_i \times U'_j$, for all $l \geq 0$, where

$$I' = \{k \in I : i \succeq k\} \quad \text{and} \quad v'_j(z) = 1 + \min\{\mu_{j,k}(z) : i \succ k\}.$$

Everything in (11.13) is invariant under translation by Λ_i, and so pushes down to $F_i \times V'_j \times Y_j \setminus F_i \times U'_j$.

By Proposition 11.2.3 there are 2- and 3-forms $\sigma_{i,j}, \tau_{i,j}$ on $V_j \times Y_j \setminus V_i \times U'_j$, such that $(\text{id} \times \psi'_j)^*(\varphi_i) - \varphi_j = d\sigma_{i,j}$ and $(\text{id} \times \psi'_j)^*\big(\Theta(\varphi_i)\big) - \Theta(\varphi_j) = d\tau_{i,j}$, and

$$\nabla^l \sigma_{i,j} = \sum_{k \in I' : j \not\succeq k} O\big(\mu_{j,k}^{d_k}(v'_j)^{-1-l}\big) = \nabla^l \tau_{i,j} \tag{11.14}$$

on $V_j \times Y_j \setminus V_i \times U'_j$, for all $l \geq 0$.

We choose the forms $\sigma_{i,j}, \tau_{i,j}$ to be Γ-invariant, in the sense that

$$(\gamma \times \chi_{\gamma,j})^*(\sigma_{\gamma \cdot i, \gamma \cdot j}) = \sigma_{i,j} \quad \text{and} \quad (\gamma \times \chi_{\gamma,j})^*(\tau_{\gamma \cdot i, \gamma \cdot j}) = \tau_{i,j}, \tag{11.15}$$

under the natural map $\gamma \times \chi_{\gamma,j} : V_j \times Y_j \to V_{\gamma \cdot j} \times Y_{\gamma \cdot j}$. Also, we choose $\sigma_{i,j}, \tau_{i,j}$ to be invariant under the action of V_i on $V_i \times V'_j \times Y_j \setminus V_i \times U'_j$ by translations. Taking the quotient by $\Lambda_i \subset V_i$ shows that $\sigma_{i,j}, \tau_{i,j}$ push down to $F_i \times V'_j \times Y_j \setminus F_i \times U'_j$. Everything in (11.14) is Λ_i-invariant, so (11.14) holds on $F_i \times V'_j \times Y_j \setminus F_i \times U'_j$.

If $i \succ j$ then M_i/B_i and M_j/B_j overlap in M. Their overlap is isomorphic to $(M_j \cap P_i \times Y_j)/B'_j$, where $B'_j = \bigl(N(F_i) \cap N(F_j)\bigr)/A_j$. Actually, in general this is only one connected component of the overlap, as if $\gamma \in \Gamma$ satisfies $i \succ \gamma \cdot j$ then $(M_{\gamma \cdot j} \cap P_i \times Y_{\gamma \cdot j})/B'_{\gamma \cdot j}$ is another component which may be distinct. For simplicity we will ignore this possibility, as it makes very little difference.

Now $M_j \cap P_i \times Y_j$ is an open subset of $F_i \times V'_j \times Y_j$ by (11.12), and $\sigma_{i,j}, \tau_{i,j}$ are forms on $F_i \times V'_j \times Y_j$, so they restrict to $M_j \cap P_i \times Y_j$. Also $\sigma_{i,j}, \tau_{i,j}$ are B'_j-invariant by (11.15), so they push down to $(M_j \cap P_i \times Y_j)/B'_j$, the intersection of M_i/B_i and M_j/B_j in M.

Thus, when $i \succ j$ we have G_2-structures (φ_i, g_i) on M_i/B_i in M and (φ_j, g_j) on M_j/B_j in M, and on the overlap $(M_i/B_i) \cap (M_j/B_j)$ we have 2- and 3-forms $\sigma_{i,j}, \tau_{i,j}$ which satisfy

$$\varphi_i - \varphi_j = d\sigma_{i,j} \quad \text{and} \quad \Theta(\varphi_i) - \Theta(\varphi_j) = d\tau_{i,j},$$

since $\text{id} \times \psi'_j$ is the identity map between M_j/B_j and M_i/B_i on the overlap.

Now suppose $i, j, k \in I$ and $i \succ j \succ k$. Then on the triple overlap $(M_i/B_i) \cap (M_j/B_j) \cap (M_k/B_k)$ we have 2-forms $\sigma_{i,j}, \sigma_{j,k}$ and $\sigma_{i,k}$ such that

$$d(\sigma_{i,j} + \sigma_{j,k} - \sigma_{i,k}) = (\varphi_i - \varphi_j) + (\varphi_j - \varphi_k) - (\varphi_i - \varphi_k) = 0.$$

So $\sigma_{i,j} + \sigma_{j,k} - \sigma_{i,k}$ is a closed 2-form. But $\sigma_{i,j}, \sigma_{j,k}$ and $\sigma_{i,k}$ are themselves only defined up to addition of a closed 2-form. We shall assume that the $\sigma_{i,j}$ have been chosen such that $\sigma_{i,j} + \sigma_{j,k} - \sigma_{i,k} = 0$ on $(M_i/B_i) \cap (M_j/B_j) \cap (M_k/B_k)$ whenever $i \succ j \succ k$. It can be shown that this is possible.

11.5.2 G_2-structures and forms depending on $t > 0$

Our goal is to join the φ_i on M_i/B_i using a partition of unity to get a positive 3-form φ on M, inducing a G_2-structure (φ, g). We want (φ, g) to have *small torsion*. This will be true if $\varphi_i - \varphi_j$ is small on $(M_i/B_i) \cap (M_j/B_j)$, and hence if $\sigma_{i,j}, \tau_{i,j}$ and their derivatives are small on $(M_i/B_i) \cap (M_j/B_j)$.

However, the estimates (11.14) tell us only that $\nabla^l \sigma_{i,j}$ and $\nabla^l \tau_{i,j}$ are $O(1)$ on $(M_i/B_i) \cap (M_j/B_j)$, not that they are small. To get round this, we introduce an important new idea: we rescale $(V_i \times Y_i, \varphi_i, g_i)$ by a factor $t > 0$ to get a new QALE G_2-manifold $(V_i \times Y_i^t, \varphi_i^t, g_i^t)$.

Definition 11.5.1 For each $i \in I$ and $t > 0$, define a resolution (Y_i^t, π_i^t) of W_i/A_i by $Y_i^t = Y_i$ and $\pi_i^t = t\pi_i$. Define a G_2-structure (φ_i^t, g_i^t) on $V_i \times Y_i^t$ by

$$\varphi_i^t = t^3 (t^{-1} \times \text{id})^*(\varphi_i) \quad \text{and} \quad g_i^t = t^2 (t^{-1} \times \text{id})^*(g_i), \tag{11.16}$$

where $t^{-1} \times \text{id} : V_i \times Y_i^t \to V_i \times Y_i$ maps (v, y) to $(t^{-1}v, y)$ in the obvious way. It is easily shown that (φ_i^t, g_i^t) is a product G_2-structure on $V_i \times Y_i^t$, and that $(V_i \times Y_i^t, \varphi_i^t, g_i^t)$ is a QALE G_2-manifold asymptotic to $\mathbb{R}^7/A_i$. For $i \in I$ and $\gamma \in \Gamma$, we also define $\chi_{\gamma,i}^t : Y_i^t \to Y_{\gamma \cdot i}^t$ by $\chi_{\gamma,i}^t = \chi_{\gamma,i}$.

Let us explain this definition. If we apply a *homothety* to the G_2-structure (φ_i, g_i) which multiplies distances by a factor t, we get a new G_2-structure $(t^3\varphi_i, t^2 g_i)$, which is isomorphic to (φ_i^t, g_i^t). Here g_i, φ_i are multiplied by t^2, t^3 as they are tensors $(g_i)_{ab}$, $(\varphi_i)_{abc}$ with 2 and 3 covariant indices respectively.

We wish to regard $(t^3\varphi_i, t^2 g_i)$ as a QALE G_2-structure. However, it is asymptotic to $(t^3\varphi_0, t^2 g_0)$ on $\mathbb{R}^7/A_i$, rather than (φ_0, g_0). We therefore conjugate the V_i factor by the map $t^{-1} : V_i \to V_i$, and replace π_i by $\pi_i^t = t\pi_i$. This gives (φ_i^t, g_i^t) on $V_i \times Y_i^t$, which is asymptotic to (φ_0, g_0), as we want. When $t = 1$ we have $(Y_i^1, \pi_i^1) = (Y_i, \pi_i)$ and $(\varphi_i^1, g_i^1) = (\varphi_i, g_i)$.

For each $t > 0$, we see that the (Y_i^t, π_i^t), (φ_i^t, g_i^t) and $\chi_{\gamma,i}^t$ are a set of *R-data* for T^7/Γ. Thus, the definition of the resolution M of T^7/Γ in §11.4.3, and all of the discussion in §11.5.1 above, can be carried out using $Y_i^t, \pi_i^t, \varphi_i^t, \ldots$ instead of $Y_i, \pi_i, \varphi_i \ldots$, and adding superscripts t to all our notation in the obvious way.

Since the resolutions (M^t, π^t) of T^7/Γ that we get in this way form a smooth connected family, they are all diffeomorphic. So we will identify them all with M, and write M and π instead of M^t and π^t. Also, the inequality $R_i \leqslant \zeta_{n_i}$ in Definition 11.4.5 is replaced by $tR_i \leqslant \zeta_{n_i}$. Thus, by taking t small we can avoid assuming that $R_i \leqslant \zeta_{n_i}$, as we claimed in Definition 11.4.5.

When we replace $Y_i, \pi_i, \varphi_i, \ldots$ by $Y_i^t, \pi_i^t, \varphi_i^t, \ldots$ in (11.13), we get

$$\nabla^l \big((\mathrm{id} \times \psi_j^{t\prime})^*(\varphi_i^t) - \varphi_j^t\big) = \sum_{k \in I': j \not\succ k} t^{\dim W_k} O\big((\mu_{j,k}^t)^{d_k}(\nu_j^{t\prime})^{-2-l}\big), \tag{11.17}$$

where $\mu_{j,k}^t(z) = d\big(\mathrm{id} \times \pi_j^t(z), V_k A_j/A_j\big)$ and $\nu_j^{t\prime}(z) = t + \min\{\mu_{j,k}^t(z) : i \succ k\}$.

It takes some care to get the correct power $t^{\dim W_k}$ in (11.17). Here is how to do it. In (11.13), we must replace $\mu_{j,k}$ by $t^{-1}\mu_{j,k}^t$, and ν_j' by $t^{-1}\nu_j^{t\prime}$. Thus, the term $O\big(\mu_{j,k}^{d_k}(\nu_j')^{-2-l}\big)$ becomes $t^{2+l-d_k} \cdot O\big((\mu_{j,k}^t)^{d_k}(\nu_j^{t\prime})^{-2-l}\big)$. However, the left hand side of (11.13) is a tensor with $3 + l$ covariant indices, and because we multiply distances by t, we must multiply the right hand side of (11.13) by t^{-3-l}. Also, φ_i^t has an extra factor of t^3 by (11.16). So the total power of t is

$$t^{2+l-d_k} \cdot t^{-3-l} \cdot t^3 = t^{2-d_k} = t^{\dim W_k}, \quad \text{as } d_k = 2 - \dim W_k.$$

Similarly, there are 2- and 3-forms $\sigma_{i,j}^t, \tau_{i,j}^t$ on $V_j \times Y_j^t \setminus V_i \times U_j^{t\prime}$ such that $(\mathrm{id} \times \psi_j^{t\prime})^*(\varphi_i^t) - \varphi_j^t = d\sigma_{i,j}^t$ and $(\mathrm{id} \times \psi_j^{t\prime})^*\big(\Theta(\varphi_i^t)\big) - \Theta(\varphi_j^t) = d\tau_{i,j}^t$, and

$$\nabla^l \sigma_{i,j}^t = \sum_{k \in I': j \not\succ k} t^{\dim W_k} O\big((\mu_{j,k}^t)^{d_k}(\nu_j^{t\prime})^{-1-l}\big) = \nabla^l \tau_{i,j}^t. \tag{11.18}$$

As above, we can regard M_i^t/B_i as an open set in M, and (φ_i^t, g_i^t) as a torsion-free G_2-structure on M_i^t/B_i, and $\sigma_{i,j}^t, \tau_{i,j}^t$ as forms on the intersection $(M_i^t/B_i) \cap (M_j^t/B_j)$ in M. We summarize the important properties of the $\varphi_i^t, \sigma_{i,j}^t$ and $\tau_{i,j}^t$ in the following result.

Proposition 11.5.2 *For each $t \in (0, 1]$ the compact 7-manifold $M = M^t$ is covered by open sets M_i^t/B_i for $i \in I$. On each M_i^t/B_i there is a torsion-free G_2-structure (φ_i^t, g_i^t). If $i, j \in I$ with $i \succ j$, then there is a 2-form $\sigma_{i,j}^t$ and a 3-form $\tau_{i,j}^t$ on $M_i^t/B_i \cap M_j^t/B_j$ such that*

$$\varphi_i^t - \varphi_j^t = d\sigma_{i,j}^t \quad \text{and} \quad \Theta(\varphi_i^t) - \Theta(\varphi_j^t) = d\tau_{i,j}^t. \tag{11.19}$$

These satisfy $\sigma_{i,j}^t + \sigma_{j,k}^t - \sigma_{i,k}^t = 0$ and $\tau_{i,j}^t + \tau_{j,k}^t - \tau_{i,k}^t = 0$ on $(M_i/B_i) \cap (M_j/B_j) \cap (M_k/B_k)$ when $i \succ j \succ k$. Define a function $\rho_t : M^t \to (0, t^{-1}]$ by $\rho_t(z) = [t + d(\pi^t(z), S)]^{-1}$, where S is the singular set of T^7/Γ and $d(,)$ the distance in T^7/Γ. Then

$$\nabla^l \sigma_{i,j}^t = t^4 \, O(\rho_t^{1+l}) \quad \text{and} \quad \nabla^l \tau_{i,j}^t = t^4 \, O(\rho_t^{1+l}) \tag{11.20}$$

in $(M_i^t/B_i) \cap (M_j^t/B_j)$, for all $l \geq 0$.

Proof The first parts of the proposition all follow immediately from the discussion of φ_i, $\sigma_{i,j}$ and $\tau_{i,j}$ above. It remains only to prove (11.20), which we deduce from (11.18). Suppose $i, j \in I$ with $i \succ j$. Clearly $(\nu_j^{t'})^{-1} \leq \rho_t$ in M_j^t/B_j. Also, if $k \in I'$ and $j \not\leq k$ then $\zeta_{n_k} < \mu_{j,k}^t < 2\zeta_{n_i}$ in $(M_i^t/B_i) \cap (M_j^t/B_j)$, so $\mu_{j,k}^t = O(1)$.

Thus, in $(M_i^t/B_i) \cap (M_j^t/B_j)$ we may replace the term $O\big((\mu_{j,k}^t)^{d_k}(\nu_j^{t'})^{-1-l}\big)$ in (11.18) by $O(\rho_t^{1+l})$. Moreover $\dim W_k \geq 4$ as $k \neq 0$, so as $t \leq 1$ we can replace $t^{\dim W_k}$ by t^4. Therefore (11.18) implies (11.20), as we want. $\square$

Since $\rho_t \leq t^{-1}$ on M, eqn (11.20) shows that $\nabla^l \sigma_{i,j}^t = O(t^{3-l})$. In particular, $\sigma_{i,j}^t$, $\nabla \sigma_{i,j}^t$ and $\nabla^2 \sigma_{i,j}^t$ are small when t is small. As $\varphi_i^t - \varphi_j^t = d\sigma_{i,j}^t$ this means that φ_i^t is close to φ_j^t on $(M_i^t/B_i) \cap (M_j^t/B_j)$ when t is small.

11.5.3 The definition of (φ^t, g^t)

We shall now define 1-parameter families of 3-forms φ^t and 4-forms υ^t on M, for $t \in (0, \epsilon)$. We do this by gluing together the 3-forms φ_i^t and 4-forms $\Theta(\varphi_i^t)$ on the patches M_i^t/B_i using a partition of unity. The $\sigma_{i,j}^t$ and $\tau_{i,j}^t$ are used to make φ^t and υ^t closed.

Later we will show that φ^t induces a G_2-structure (φ^t, g^t) on M, and that $\Theta(\varphi^t) - \upsilon^t$ is small when t is small. So $d\varphi^t = 0$ and $d\Theta(\varphi^t)$ is small, as $\Theta(\varphi^t)$ is close to υ^t and $d\upsilon^t = 0$. Therefore (φ^t, g^t) has *small torsion* when t is small.

Definition 11.5.3 Let I/Γ be the set of orbits of Γ in I, and for each $i \in I$ define $[i] \in I/\Gamma$ by $[i] = \{\gamma \cdot i : \gamma \in \Gamma\}$. Then as $M_{\gamma \cdot i}^t/B_{\gamma \cdot i} = M_i^t/B_i$ for all $\gamma \in \Gamma$, the open sets M_i^t/B_i really depend only on $[i] \in I/\Gamma$ and not on $i \in I$.

For each $t \in (0, 1]$ choose a *partition of unity* $\{\eta_{[i]}^t : [i] \in I/\Gamma\}$ on M^t with

(a) $\eta_{[i]}^t : M^t \to [0, 1]$ is smooth, and zero outside M_i^t/B_i for each $i \in I$;

(b) $\sum_{[i] \in I/\Gamma} \eta_{[i]}^t \equiv 1$; and

(c) $|d\eta^t_{[i]}| = O(1)$ and $|\nabla d\eta^t_{[i]}| = O(1)$ on M^t_j/B_j for all $j \in I$, where $|.|$ and ∇ are defined using the metric g^t_j.

We can always find functions $\eta^t_{[i]}$ satisfying (a)–(c). The $\eta^t_{[i]}$ can be regarded as essentially independent of t, and this is why part (c) holds. Some similar functions $\eta^t_{[i]}$ will be explained in more detail in Definitions 13.2.6 and 13.5.1.

Define a 3-form φ^t and a 4-form υ^t on M^t by

$$\varphi^t = \varphi^t_i + \sum_{\substack{[j]\in I/\Gamma: \\ j\succ i}} d(\eta^t_{[j]}\sigma^t_{j,i}) - \sum_{\substack{[j]\in I/\Gamma: \\ i\succ j}} d(\eta^t_{[j]}\sigma^t_{i,j}) \qquad (11.21)$$

and $$\upsilon^t = \Theta(\varphi^t_i) + \sum_{\substack{[j]\in I/\Gamma: \\ j\succ i}} d(\eta^t_{[j]}\tau^t_{j,i}) - \sum_{\substack{[j]\in I/\Gamma: \\ i\succ j}} d(\eta^t_{[j]}\tau^t_{i,j}) \qquad (11.22)$$

on M^t_i/B_i, for each $i \in I$. Here we take $\eta^t_{[j]}\sigma^t_{j,i} \equiv 0$ outside M^t_j/B_j, and so on.

We shall derive alternative expressions for φ^t and υ^t. Since $d\sigma^t_{j,i} = \varphi^t_j - \varphi^t_i$ we have $d(\eta^t_{[j]}\sigma^t_{j,i}) = \eta^t_{[j]}(\varphi^t_j - \varphi^t_i) + d\eta^t_{[j]} \wedge \sigma^t_{j,i}$. Substituting this and a similar equation into (11.21) gives

$$\begin{aligned}\varphi^t = \varphi^t_i &+ \sum_{\substack{[j]\in I/\Gamma: \\ j\succ i}} \eta^t_{[j]}(\varphi^t_j - \varphi^t_i) - \sum_{\substack{[j]\in I/\Gamma: \\ i\succ j}} \eta^t_{[j]}(\varphi^t_i - \varphi^t_j) \\ &+ \sum_{\substack{[j]\in I/\Gamma: \\ j\succ i}} d\eta^t_{[j]} \wedge \sigma^t_{j,i} - \sum_{\substack{[j]\in I/\Gamma: \\ i\succ j}} d\eta^t_{[j]} \wedge \sigma^t_{i,j}\end{aligned} \qquad (11.23)$$

on M^t_i/B_i. But it is easy to see that

$$\varphi^t_i + \sum_{\substack{[j]\in I/\Gamma: \\ j\succ i}} \eta^t_{[j]}(\varphi^t_j - \varphi^t_i) - \sum_{\substack{[j]\in I/\Gamma: \\ i\succ j}} \eta^t_{[j]}(\varphi^t_i - \varphi^t_j) = \sum_{[j]\in I/\Gamma} \eta^t_{[j]}\varphi^t_j$$

on M^t_i/B_i, where we take $\eta^t_{[j]}\varphi^t_j \equiv 0$ outside M^t_j/B_j. Thus in M^t_i/B_i we have

$$\varphi^t = \sum_{[j]\in I/\Gamma} \eta^t_{[j]}\varphi^t_j + \sum_{\substack{[j]\in I/\Gamma: \\ j\succ i}} d\eta^t_{[j]} \wedge \sigma^t_{j,i} - \sum_{\substack{[j]\in I/\Gamma: \\ i\succ j}} d\eta^t_{[j]} \wedge \sigma^t_{i,j}, \qquad (11.24)$$

and similarly

$$\upsilon^t = \sum_{[j]\in I/\Gamma} \eta^t_{[j]}\Theta(\varphi^t_j) + \sum_{\substack{[j]\in I/\Gamma: \\ j\succ i}} d\eta^t_{[j]} \wedge \tau^t_{j,i} - \sum_{\substack{[j]\in I/\Gamma: \\ i\succ j}} d\eta^t_{[j]} \wedge \tau^t_{i,j}. \qquad (11.25)$$

Proposition 11.5.4 *The forms φ^t and υ^t given above are well-defined, smooth and closed, and satisfy*

$$|\varphi^t - \varphi^t_i| = t^4\, O(\rho^2_t) \quad \text{and} \quad |\upsilon^t - \Theta(\varphi^t_i)| = t^4\, O(\rho^2_t) \qquad (11.26)$$

on M^t_i/B_i for each $i \in I$, where $|.|$ is defined using the metric g^t_i.

Proof Distinct M_i^t/B_i and M_j^t/B_j have nonempty intersection only when $i \succ j$ or $j \succ i$, up to the action of Γ. Thus, to prove that φ^t is well-defined, we need to show that the definitions (11.21) of φ^t on M_i^t/B_i and M_j^t/B_j agree on $(M_i^t/B_i) \cap (M_j^t/B_j)$ when $i, j \in I$ with $i \succ j$. That is, we must prove that

$$\varphi_i^t + \sum_{\substack{[k] \in I/\Gamma: \\ k \succ i}} d(\eta_{[k]}^t \sigma_{k,i}^t) - \sum_{\substack{[k] \in I/\Gamma: \\ i \succ k}} d(\eta_{[k]}^t \sigma_{i,k}^t) = \varphi_j^t + \sum_{\substack{[k] \in I/\Gamma: \\ k \succ j}} d(\eta_{[k]}^t \sigma_{k,j}^t) - \sum_{\substack{[k] \in I/\Gamma: \\ j \succ k}} d(\eta_{[k]}^t \sigma_{j,k}^t)$$

on $(M_i^t/B_i) \cap (M_j^t/B_j)$. We can split the nonzero terms in this equation into five cases: $k = i$, $k = j$, $k \succ i$, $j \succ k$, and $i \succ k \succ j$. If k does not satisfy one of these then $\eta_k^t \equiv 0$ on $(M_i^t/B_i) \cap (M_j^t/B_j)$.

Dividing into sums over each case, the equation above becomes

$$\varphi_i^t - \varphi_j^t - d(\eta_{[i]}^t \sigma_{i,j}^t) - d(\eta_{[j]}^t \sigma_{i,j}^t) - \sum_{\substack{[k] \in I/\Gamma: \\ k \succ i}} d(\eta_{[k]}^t (\sigma_{k,j}^t - \sigma_{k,i}^t))$$

$$- \sum_{\substack{[k] \in I/\Gamma: \\ j \succ k}} d(\eta_{[k]}^t (\sigma_{i,k}^t - \sigma_{j,k}^t)) - \sum_{\substack{[k] \in I/\Gamma: \\ i \succ k \succ j}} d(\eta_{[k]}^t (\sigma_{i,k}^t + \sigma_{k,j}^t)) = 0.$$

However, each of $\sigma_{k,j}^t - \sigma_{k,i}^t$, $\sigma_{i,k}^t - \sigma_{j,k}^t$ and $\sigma_{i,k}^t + \sigma_{k,j}^t$ here is $\sigma_{i,j}^t$ by the equation $\sigma_{i,j}^t + \sigma_{j,k}^t - \sigma_{i,k}^t = 0$ in Proposition 11.5.2. Thus φ^t is well-defined if

$$\varphi_i^t - \varphi_j^t - \sum_{[k] \in I/\Gamma} d(\eta_{[k]}^t \sigma_{i,j}^t) = 0 \quad \text{on } (M_i^t/B_i) \cap (M_j^t/B_j).$$

But $\sum_{[k] \in I/\Gamma} \eta_k^t \equiv 1$ by part (b) of Definition 11.5.3, so this equation is $\varphi_i^t - \varphi_j^t - d\sigma_{i,j}^t = 0$, which holds by (11.19). Therefore φ^t is well-defined on M^t. It is clear that φ^t is smooth, and φ^t is closed as φ_i^t is closed. By the same argument, υ^t is well-defined, smooth and closed. The inequalities (11.26) follow easily from equations (11.20)–(11.22). □

Here is the definition of the G_2-structure (φ^t, g^t) and the 3-form ψ^t.

Definition 11.5.5 Let ϵ_1 be the constant of Definition 10.3.3. As $\rho_t \leqslant t^{-1}$ on M, eqn (11.26) shows that $|\varphi^t - \varphi_i^t| = O(t^2)$ on M_i^t/B_i for all $i \in I$, where $|\,.\,|$ is defined using g_i^t. Thus $|\varphi^t - \varphi_i^t| \leqslant \epsilon_1$ on M_i^t/B_i when t is sufficiently small. Choose $\epsilon \in (0, 1]$ such that $|\varphi^t - \varphi_i^t| \leqslant \epsilon_1$ on M_i^t/B_i for all $i \in I$ when $t \in (0, \epsilon]$.

Thus φ^t is a *positive 3-form* on M_i^t/B_i when $t \in (0, \epsilon]$, by definition of ϵ_1. As this holds for all $i \in I$, we see that φ^t is positive on M, and $\varphi^t \in C^\infty(\mathcal{P}^3 M)$. Therefore φ^t induces a unique G_2-structure (φ^t, g^t) on M. Define a smooth 3-form ψ^t on M by $\psi^t = \varphi^t - *\upsilon^t$, where '$*$' is the Hodge star of g^t.

This 3-form ψ^t has the following important properties.

Proposition 11.5.6 *For each $t \in (0, \epsilon]$ the 3-form ψ^t satisfies*

$$d^*\psi^t = d^*\varphi^t, \quad |\psi^t| \leqslant Ct^4 \rho_t \quad \text{and} \quad |d^*\psi^t| \leqslant Ct^4 \rho_t^2 \tag{11.27}$$

on M^t, where d^* and $|\,.\,|$ are defined using g^t and $C > 0$ is independent of t.

Proof We have $d^*\psi^t = d^*\varphi^t - d^*(*\upsilon^t) = d^*\varphi^t + *d\upsilon^t = d^*\varphi^t$, proving the first equation, since $d^** = -*d$ on 3-forms and $d\upsilon^t = 0$ by Proposition 11.5.4. To prove the second equation of (11.27) we write

$$|\psi^t| = |*\psi^t| = |\Theta(\varphi^t) - \upsilon^t| \leqslant \left|\Theta(\varphi^t) - \Theta\left(\sum_{[j]\in I/\Gamma} \eta^t_{[j]}\varphi^t_j\right)\right|$$

$$+ \left|\Theta\left(\sum_{[j]\in I/\Gamma} \eta^t_{[j]}\varphi^t_j\right) - \sum_{[j]\in I/\Gamma} \eta^t_{[j]}\Theta(\varphi^t_j)\right| + \left|\sum_{[j]\in I/\Gamma} \eta^t_{[j]}\Theta(\varphi^t_j) - \upsilon^t\right|.$$

We shall estimate the three terms on the right hand side of this equation. By Proposition 10.3.5 we find that $\left|\Theta(\varphi^t) - \Theta\left(\sum_{[j]\in I/\Gamma} \eta^t_{[j]}\varphi^t_j\right)\right|$ is of the same order as $\left|\varphi^t - \sum_{[j]\in I/\Gamma} \eta^t_{[j]}\varphi^t_j\right|$. But (11.24) shows that

$$\varphi^t - \sum_{[j]\in I/\Gamma} \eta^t_{[j]}\varphi^t_j = \sum_{[j]\in I/\Gamma: j\succ i} d\eta^t_{[j]} \wedge \sigma^t_{j,i} - \sum_{[j]\in I/\Gamma: i\succ j} d\eta^t_{[j]} \wedge \sigma^t_{i,j},$$

and as $|d\eta^t_{[j]}| = O(1)$ by part (c) of Definition 11.5.3 and $|\sigma^t_{j,i}|, |\sigma^t_{i,j}| = t^4 O(\rho_t)$ by (11.20), we have

$$\left|\Theta(\varphi^t) - \Theta\left(\sum_{[j]\in I/\Gamma} \eta^t_{[j]}\varphi^t_j\right)\right| = t^4 O(\rho_t).$$

Now $\Theta\left(\sum_{[j]\in I/\Gamma} \eta^t_j\varphi^t_j\right)$ is $\sum_{[j]\in I/\Gamma} \eta^t_j\Theta(\varphi^t_j)$ plus terms at least quadratic in $\varphi^t_i - \varphi^t_j$. But $\varphi^t_i - \varphi^t_j = t^4 O(\rho_t^2)$ in $(M^t_i/B_i) \cap (M^t_j/B_j)$ by (11.20). Therefore

$$\left|\Theta\left(\sum_{[j]\in I/\Gamma} \eta^t_{[j]}\varphi^t_j\right) - \sum_{[j]\in I/\Gamma} \eta^t_{[j]}\Theta(\varphi^t_j)\right| = t^8 O(\rho_t^4).$$

Using (11.25) to get an expression for $\sum_{[j]\in I/\Gamma} \eta^t_{[j]}\Theta(\varphi^t_j) - \upsilon^t$ and estimating it with (11.20) and part (c) of Definition 11.5.3, we find that

$$\left|\sum_{[j]\in I/\Gamma} \eta^t_{[j]}\Theta(\varphi^t_j) - \upsilon^t\right| = t^4 O(\rho_t).$$

Combining the last few equations shows that

$$\psi^t = t^4 O(\rho_t) + t^8 O(\rho_t^4) + t^4 O(\rho_t). \tag{11.28}$$

In a similar way we prove that

$$d^*\psi^t = t^4 O(\rho_t^2) + t^8 O(\rho_t^5) + t^4 O(\rho_t^2), \tag{11.29}$$

the difference being that $d^*\psi^t$ involves one more derivative of $\sigma^t_{i,j}$ and $\tau^t_{i,j}$ than ψ^t, and so by (11.20) our estimate of $d^*\psi^t$ has one more factor of ρ_t. It then follows from (11.28), (11.29) and $\rho_t \leqslant t^{-1}$ that the last two equations of (11.27) hold for some $C > 0$ independent of t. □

11.5.4 Estimates of φ^t, g^t and ψ^t

We can now now prove the main result of this section, the following theorem. In it we bound, by powers of t, various norms of the 3-form ψ^t and analytic properties of the metric g^t defined above. The things we estimate are chosen to meet the needs of the analytic proofs in §11.6–§11.8.

Theorem 11.5.7 *Let T^7/Γ be an orbifold of T^7 with flat G_2-structure (φ_0, g_0). Suppose we are given a set of R-data for T^7/Γ, and (M, π) is the corresponding resolution. Then we can write down the following data on M explicitly:*

- *Constants $\epsilon \in (0, 1]$ and $\lambda, \mu, \nu > 0$,*
- *A G_2-structure (φ^t, g^t) on M with $d\varphi^t = 0$ for each $t \in (0, \epsilon]$, and*
- *A smooth 3-form ψ^t on M with $d^*\psi^t = d^*\varphi^t$ for each $t \in (0, \epsilon]$.*

These satisfy the three conditions

(i) $\|\psi^t\|_{L^2} \leq \lambda t^4$, $\|\psi^t\|_{C^0} \leq \lambda t^3$ and $\|d^*\psi^t\|_{L^{14}} \leq \lambda t^{16/7}$;

(ii) *the injectivity radius $\delta(g^t)$ satisfies $\delta(g^t) \geq \mu t$; and*

(iii) *the Riemann curvature $R(g^t)$ satisfies $\|R(g^t)\|_{C^0} \leq \nu t^{-2}$.*

Here the operator d^* and all norms are calculated using the metric g^t.

Proof The G_2-structure (φ^t, g^t), 3-form ψ^t and constant $\epsilon > 0$ are given in Definition 11.5.5, and Propositions 11.5.4 and 11.5.6 show that $d\varphi^t = 0$ and $d^*\psi^t = d^*\varphi^t$. It remains only to find constants $\lambda, \mu, \nu > 0$ such that (i)–(iii) hold. Since we can make ϵ smaller if necessary, it is enough to prove (i)–(iii) when t is sufficiently small.

From (11.27) we see that $\|\psi^t\|_{L^2} \leq Ct^4 \|\rho_t\|_{L^2}$, $\|\psi^t\|_{C^0} \leq Ct^4 \|\rho_t\|_{C^0}$ and $\|d^*\psi^t\|_{L^{14}} \leq Ct^4 \|\rho_t^2\|_{L^{14}}$. But $\|\rho_t\|_{C^0} \leq t^{-1}$ as $\rho_t \leq t^{-1}$ on M, so $\|\psi^t\|_{C^0} \leq Ct^3$, and thus $\|\psi^t\|_{C^0} \leq \lambda t^3$ provided $\lambda \geq C$. To prove the other two inequalities in (i) we need to estimate $\|\rho_t\|_{L^2}$ and $\|\rho_t^2\|_{L^{14}}$.

Now $\rho_t : M \to (0, t^{-1}]$ is given by $\rho_t(z) = [t + d(\pi^t(z), S)]^{-1}$, where S is the singular set of T^7/Γ. Thus ρ_t is $O(t^{-1})$ close to $\pi_t^{-1}(S)$ in M, and $O(1)$ further away from $\pi_t^{-1}(S)$. Clearly, norms of ρ_t will be largest when $\pi_t^{-1}(S)$ is 'as big as possible' in M — that is, when S has maximal dimension. But $\dim S \leq 3$ by Proposition 11.1.3. So we need only study the worst case, in which $\dim S = 3$.

Consider a simple situation, in which one of the connected components of S is a copy F_i of T^3, and T^7/Γ is isomorphic to $T^3 \times \mathbb{C}^2/A_i$ near F_i, where A_i is a finite subgroup of $SU(2)$. Let Y_i^t be the resolution of $\mathbb{C}^2/A_i$, with projection $\pi_i^t : Y_i^t \to \mathbb{C}^2/A_i$, so that M is modelled on $T^3 \times Y_i^t$ near $(\pi_i^t)^{-1}(F_i)$. Write r for both the radius function on $\mathbb{C}^2/A_i$, and its pull-back to Y_i^t under π_i^t.

We can think of ρ_t near $\pi^{-1}(F_i)$ in the following way. Define two regions W^t and X^t in Y_i^t by $W^t = \{y \in Y_i^t : r \leq t\}$ and $X^t = \{y \in Y_i^t : t \leq r \leq \zeta_3\}$. Here W^t is the region in the centre of Y_i^t containing all the interesting topology, with diameter $O(t)$ and volume $O(t^4)$, and X^t is an *annulus* in $\mathbb{C}^2/A_i$, with inner radius t and outer radius ζ_3.

Then M_i^t/B_i is the union of $T^3 \times W^t$ and $T^3 \times X^t$. Since $\rho_t = (t+r)^{-1}$ we have $\rho_t \leqslant t^{-1}$ and $\rho_t \leqslant r^{-1}$. We approximate ρ_t by t^{-1} on $T^3 \times W^t$ and by r^{-1} on $T^3 \times X^t$. Thus if $\alpha > 0$ we have

$$\int_{M_i/B_i} \rho_t^\alpha dV_{g_t} \leqslant \int_{T^3 \times W^t} t^{-\alpha} dV_{g_t} + \int_{T^3 \times X^t} r^{-\alpha} dV_{g_t} \\ \approx t^{-\alpha} O(t^4) + \text{vol}(T^3) \cdot \int_t^{\zeta_3} r^{-\alpha} \frac{\Omega_3 r^3}{|A_i|} dr, \tag{11.30}$$

where Ω_3 is the volume of the unit 3-sphere $\mathcal{S}^3$. In the second line we approximate the ALE metric on Y_i^t by the flat metric on $\mathbb{C}^2/A_i$ in X^t, so that we can evaluate the integral explicitly. By properties of ALE metrics, this changes the volume form by no more than a constant factor independent of t.

Equation (11.30) estimates the integral of ρ_t^α on the open set M_i/B_i in M, in the special case we are considering. In the general case we get exactly the same behaviour, or better if $\dim S < 3$. Therefore, from (11.30) we see that

$$\int_M \rho_t^\alpha dV_{g_t} = \begin{cases} O(1) & \text{when } 0 < \alpha < 4, \\ O(t^{4-\alpha}) & \text{when } \alpha > 4. \end{cases}$$

Putting $\alpha = 2$ and $\alpha = 28$ and taking roots gives $\|\rho_t\|_{L^2} = O(1)$ and $\|\rho_t^2\|_{L^{14}} = O(t^{-12/7})$. As $\|\psi^t\|_{L^2} \leqslant Ct^4 \|\rho_t\|_{L^2}$ and $\|d^*\psi^t\|_{L^{14}} \leqslant Ct^4 \|\rho_t^2\|_{L^{14}}$, this proves the first and third equations of (i), for some $\lambda > 0$.

Parts (ii) and (iii) are in fact elementary. The metric g_i^t is made by scaling g_i by a factor t. Thus $\delta(g_i^t) = t\delta(g_i)$ and $\|R(g_i^t)\|_{C^0} = t^{-2} \|R(g_i)\|_{C^0}$. We make g^t by gluing together the g_i^t on the patches M_i^t/B_i. It is clear that for small t, the dominant contributions to $\delta(g^t)$ and $\|R(g^t)\|_{C^0}$ come from $\delta(g_i^t)$ and $\|R(g_i^t)\|_{C^0}$ for some i, and these are proportional to t and t^{-2}. This proves (ii) and (iii) for some $\mu, \nu > 0$, and the theorem is complete. □

11.5.5 *Discussion of Theorem* 11.5.7

The 7-manifold M may be thought of as a kind of 'blow-up' of the orbifold T^7/Γ where each point s in the singular set S of T^7/Γ is replaced by a compact subspace $\pi^{-1}(s)$ of M, often a submanifold of M, or a union of several submanifolds. In the 1-parameter family of G_2-structures (φ^t, g^t) that we have constructed on M, we should think of t as a measure of the *diameter* of $\pi^{-1}(s)$ for each singular point s in T^7/Γ.

For example, if T^7/Γ has a singular T^3 locally isomorphic to $T^3 \times \mathbb{C}^2/\{\pm 1\}$, then M is locally diffeomorphic to $T^3 \times Y$, where Y is the blow-up of $\mathbb{C}^2/\{\pm 1\}$ at 0. Thus, each s in T^3 is replaced by a copy of $\mathbb{CP}^1 = \mathcal{S}^2$ in M, and T^3 is replaced by $T^3 \times \mathcal{S}^2$. The restriction of g^t to $T^3 \times \mathcal{S}^2$ is the product of a fixed metric on T^3, and the round metric on $\mathcal{S}^2$ with diameter πt.

The G_2-structure (φ^t, g^t) on M tends to the singular G_2-structure (φ_0, g_0) on T^7/Γ as $t \to 0$. Here are two ways to say this rigorously. Firstly, if U is any open subset in T^7/Γ with $\overline{U} \cap S = \emptyset$, then π is a diffeomorphism between $\pi^{-1}(U)$ and U, and $(\varphi^t, g^t)|_{\pi^{-1}(U)} \to \pi^{-1}((\varphi_0, g_0)|_U)$ as $t \to 0$. Secondly, the *Gromov–Hausdorff distance* between the metric spaces (M, g^t) and $(T^7/\Gamma, g_0)$ is $O(t)$, and therefore (M, g^t) tends to $(T^7/\Gamma, g_0)$ in the Gromov–Hausdorff sense as $t \to 0$.

Thus g^t approaches a singular metric as $t \to 0$, so for small t we expect g^t to be 'badly behaved' near $\pi^{-1}(S)$. Parts (ii) and (iii) of Theorem 11.5.7 measure this. The injectivity radius $\delta(g^t)$ is $O(t)$ and the Riemannian curvature $R(g^t)$ is $O(t^{-2})$ near $\pi^{-1}(S)$, so g^t has *short injectivity radius* and *large curvature* for small t. These both make analysis proofs on (M, g^t) more difficult.

It may help to picture T^7/Γ as *the surface of a cube*, with sharp edges and corners, which are the singular set. Then (M, g^t) is like a cube with its edges and corners slightly rounded off, so that it is a nonsingular manifold. The metric g^t is quite flat on the interior of the faces of the cube, which is nearly all of M, but is highly curved on the rounded edges and corners.

This is also a good way to imagine the singular set of T^7/Γ, as it shows how the singular set can have a number of pieces of varying dimension – the edges and corners – which can meet each other in complicated ways, as three edges meet at a corner in a cube.

Now (φ^t, g^t) satisfies $d\varphi^t = 0$ and $d^*\varphi^t = d^*\psi^t$. As the *torsion* $\nabla\varphi^t$ depends only on $d\varphi^t$ and $d^*\varphi^t$, we see that $\nabla\varphi^t$ is determined by $d^*\varphi^t$, and hence by $d^*\psi^t$. Thus we may interpret ψ^t as a *first integral of the torsion* of (φ^t, g^t), since $\nabla\varphi^t$ depends on the first derivative of ψ^t. We will need to estimate ψ^t as well as $d^*\psi^t$ because of a trick in §11.8 involving integration by parts.

Therefore part (i) of Theorem 11.5.7 gives estimates on the torsion of (φ^t, g^t), and its first integral. Morally, the powers t^4, t^3 and $t^{16/7}$ in (i) ought all to be t^4, and we should think of the torsion as $O(t^4)$. Here the power 4 is really the *codimension of the singular set* S of T^7/Γ, in the worst case. If S has codimension 6, then the torsion is $O(t^6)$. The codimension enters as it determines the asymptotic rate at which QALE metrics approach the Euclidean metric.

The reason why we instead have the weaker estimates t^3 and $t^{16/7}$ in part (i) is that the equation (11.5) defining QALE G_2-manifolds involves the term $O\left(\mu_{i,j}^{d_j} v_i^{-2-l}\right)$, but this should probably be replaced by $O\left(\mu_{i,j}^{-\dim W_j} v_i^{-l}\right)$. We chose the weaker estimate $O\left(\mu_{i,j}^{d_j} v_i^{-2-l}\right)$ to make the proofs easier.

11.6 An existence result for compact G_2-manifolds

In this section and §11.7–§11.8 we will use analysis to derive the following existence result for torsion-free G_2-structures. Its proof will be given in §11.6.1.

Theorem 11.6.1 *Let λ, μ, ν be positive constants. Then there exist positive constants κ, K such that whenever $0 < t \leqslant \kappa$, the following is true.*

Let M be a compact 7-manifold, and (φ, g) a G_2-structure on M with $d\varphi = 0$. Suppose ψ is a smooth 3-form on M with $d^*\psi = d^*\varphi$, and

(i) $\|\psi\|_{L^2} \leqslant \lambda t^4$, $\|\psi\|_{C^0} \leqslant \lambda t^{1/2}$ and $\|d^*\psi\|_{L^{14}} \leqslant \lambda$,
(ii) *the injectivity radius $\delta(g)$ satisfies $\delta(g) \geqslant \mu t$, and*
(iii) *the Riemann curvature $R(g)$ satisfies $\|R(g)\|_{C^0} \leqslant \nu t^{-2}$.*

Then there exists a smooth, torsion-free G_2-structure $(\tilde{\varphi}, \tilde{g})$ on M with $\|\tilde{\varphi} - \varphi\|_{C^0} \leqslant Kt^{1/2}$.

Combining this with Theorem 11.5.7, we get the main result of the chapter:

Theorem 11.6.2 *Suppose T^7/Γ has a flat G_2-structure (φ_0, g_0) and a set of R-data (Y_i, π_i), (φ_i, g_i) and $\chi_{y,i}$, and let M be the resolution of T^7/Γ defined in §11.4.3. Then M is a compact 7-manifold with a nonempty, smooth family of torsion-free G_2-structures. That is, M is a compact G_2-manifold.*

Proof Let $\epsilon \in (0, 1]$ and $\lambda, \mu, \nu > 0$ be the constants given by Theorem 11.5.7. Then Theorem 11.6.1 gives constants $\kappa, K > 0$ depending on λ, μ, ν. Choose $t > 0$ with $t \leqslant \epsilon$ and $t \leqslant \kappa$, and let (φ, g) be the G_2-structure (φ^t, g^t) of Theorem 11.5.7, and ψ the 3-form ψ^t.

Then Theorem 11.5.7 shows that $d\varphi = 0$ and $d^*\psi = d^*\varphi$, and (i)–(iii) of Theorem 11.5.7 give (i)–(iii) of Theorem 11.6.1. Note that $\|\psi\|_{C^0} \leqslant \lambda t^3$ and $\|d^*\psi\|_{L^{14}} \leqslant \lambda t^{16/7}$ imply that $\|\psi\|_{C^0} \leqslant \lambda t^{1/2}$ and $\|d^*\psi\|_{L^{14}} \leqslant \lambda$, as $t \leqslant \epsilon \leqslant 1$.

Therefore Theorem 11.6.1 gives a torsion-free G_2-structure $(\tilde{\varphi}, \tilde{g})$ on M. By Theorem 10.4.4, the moduli space of torsion-free G_2-structures on M is a smooth manifold. This completes the proof. □

Together with Proposition 10.2.2, this gives:

Corollary 11.6.3 *Suppose T^7/Γ has a flat G_2-structure (φ_0, g_0) and a set of R-data, let M be the corresponding resolution of T^7/Γ, and suppose $\pi_1(M)$ is finite. Then M admits metrics with holonomy G_2.*

In this corollary, we finally achieve our goal: *we now have a method to construct compact 7-manifolds with holonomy G_2*. We will use this construction in Chapter 12 to write down hundreds of examples of such 7-manifolds. The first examples of compact 7-manifolds with holonomy G_2 were found by the author [121, 122].

This chapter expands and improves upon the methods of [121, 122]. In particular, by introducing the idea of QALE G_2-manifolds and using the results of Chapter 9 we will be able to resolve orbifolds T^7/Γ with singularities of much greater complexity than those considered in [121, 122], and this leads to a more powerful construction producing many more examples.

11.6.1 *The proof of Theorem 11.6.1*

The proof depends upon the following two results, to be proved in §11.7 and §11.8.

Theorem G1 *Let μ, ν and t be positive constants, and suppose (M, g) is a complete Riemannian 7-manifold, whose injectivity radius $\delta(g)$ and Riemann curvature $R(g)$ satisfy $\delta(g) \geqslant \mu t$ and $\|R(g)\|_{C^0} \leqslant \nu t^{-2}$. Then there exist $C_1, C_2 > 0$ depending only on μ and ν, such that if $\chi \in L_1^{14}(\Lambda^3 T^*M) \cap L^2(\Lambda^3 T^*M)$ then*

$$\|\nabla\chi\|_{L^{14}} \leqslant C_1\big(\|d\chi\|_{L^{14}} + \|d^*\chi\|_{L^{14}} + t^{-4}\|\chi\|_{L^2}\big) \quad (11.31)$$

$$\text{and} \quad \|\chi\|_{C^0} \leqslant C_2\big(t^{1/2}\|\nabla\chi\|_{L^{14}} + t^{-7/2}\|\chi\|_{L^2}\big). \quad (11.32)$$

We shall prove Theorem G1 by considering small balls in M of radius $O(t)$, and comparing them to balls of the same radius in $\mathbb{R}^7$. It is a purely *local* result, in that we use no global geometric information about M, but only information about small balls in M. We don't even need to assume that M is compact, so we suppose instead that M is complete.

Theorem G2 *Let λ, C_1 and C_2 be positive constants. Then there exist positive constants κ, K such that whenever $0 < t \leqslant \kappa$, the following is true.*

Let M be a compact 7-manifold, and (φ, g) a G_2-structure on M with $\mathrm{d}\varphi = 0$. Suppose ψ is a smooth 3-form on M with $\mathrm{d}^*\psi = \mathrm{d}^*\varphi$, and

(i) $\|\psi\|_{L^2} \leqslant \lambda t^4$, $\|\psi\|_{C^0} \leqslant \lambda t^{1/2}$ and $\|\mathrm{d}^*\psi\|_{L^{14}} \leqslant \lambda$,
(ii) *if* $\chi \in L_1^{14}(\Lambda^3 T^*M)$ *then* $\|\nabla \chi\|_{L^{14}} \leqslant C_1 \bigl(\|\mathrm{d}\chi\|_{L^{14}} + \|\mathrm{d}^*\chi\|_{L^{14}} + t^{-4}\|\chi\|_{L^2}\bigr)$,
(iii) *if* $\chi \in L_1^{14}(\Lambda^3 T^*M)$ *then* $\|\chi\|_{C^0} \leqslant C_2 \bigl(t^{1/2}\|\nabla \chi\|_{L^{14}} + t^{-7/2}\|\chi\|_{L^2}\bigr)$.

Let ϵ_1 be as in Definition 10.3.3, and F as in Proposition 10.3.5. Then there exist $\eta \in C^\infty(\Lambda^2 T^*M)$ with $\|\mathrm{d}\eta\|_{C^0} \leqslant K t^{1/2} \leqslant \epsilon_1$ and $f \in C^\infty(M)$, such that

$$(\mathrm{d}\mathrm{d}^* + \mathrm{d}^*\mathrm{d})\eta = \mathrm{d}^*\psi + \mathrm{d}^*(f\psi) + *\mathrm{d}F(\mathrm{d}\eta) \quad \text{and} \quad f\varphi = \tfrac{7}{3}\pi_1(\mathrm{d}\eta). \tag{11.33}$$

This equation is a *nonlinear elliptic p.d.e.* in η. On the left hand side we have $\Delta \eta$, where Δ is the Laplacian $\mathrm{d}\mathrm{d}^* + \mathrm{d}^*\mathrm{d}$, a linear elliptic operator of order 2. On the right hand side we have terms in ψ, and the nonlinear term $*\mathrm{d}F(\mathrm{d}\eta)$, which is at least quadratic in $\mathrm{d}\eta$ by Proposition 10.3.5. Thus, provided $\mathrm{d}\eta$ remains small $*\mathrm{d}F(\mathrm{d}\eta)$ should be negligible, and (11.33) approximates a linear elliptic p.d.e.

We solve (11.33) by inductively constructing sequences of 2-forms $\{\eta_j\}_{j=0}^\infty$ and functions $\{f_j\}_{j=0}^\infty$ with $\eta_0 = f_0 = 0$, satisfying

$$\Delta \eta_j = \mathrm{d}^*\psi + \mathrm{d}^*(f_{j-1}\psi) + *\mathrm{d}F(\mathrm{d}\eta_{j-1}) \quad \text{and} \quad f_j\varphi = \tfrac{7}{3}\pi_1(\mathrm{d}\eta_j).$$

Thus at each stage we have only to solve a *linear* equation $\Delta \eta_j = \ldots$, rather than a nonlinear one. We then show that $\{\eta_j\}_{j=0}^\infty$ and $\{f_j\}_{j=0}^\infty$ converge in the Sobolev spaces $L_2^{14}(\Lambda^2 T^*M)$ and $L_1^{14}(M)$ to limits η, f, which turn out to be smooth and to satisfy (11.33).

Most of the work is in deriving inductive a priori estimates on the η_j and f_j using inequalities (i)–(iii) of Theorem G2, so that we can show that the sequences converge. To do this we first use an integration by parts argument to get an a priori bound for $\|\mathrm{d}\eta_j\|_{L^2}$ in terms of $\|\psi\|_{L^2}$, and then use elliptic regularity of Δ to improve this to a bound for $\|\mathrm{d}\eta_j\|_{L_1^{14}}$.

In part (i) of Theorem G2 the powers of t are to a certain extent arbitrary, and are chosen for their simplicity. If we assume instead that $\|\psi\|_{L^2} \leqslant \lambda t^\alpha$, $\|\psi\|_{C^0} \leqslant \lambda t^\beta$ and $\|\mathrm{d}^*\psi\|_{L^{14}} \leqslant \lambda t^\gamma$, then the proof of Theorem G2 in §11.8 will still work (with appropriate changes to all the powers of t) provided that $\alpha > 7/2$, $\beta > 0$ and $\gamma > -\tfrac{1}{2}$.

Here is the proof of Theorem 11.6.1, assuming Theorems G1 and G2.

Proof of Theorem 11.6.1 By Theorem G1 and parts (ii), (iii) of Theorem 11.6.1, there exist constants C_1, C_2 depending only on μ, ν, such that if $\chi \in L_1^{14}(\Lambda^3 T^*M)$ then

(11.31) and (11.32) hold. This proves parts (ii) and (iii) of Theorem G2, and part (i) of Theorem G2 agrees with that of Theorem 11.6.1.

Thus Theorem G2 gives constants κ, K depending only on λ, C_1, C_2, and hence on λ, μ, ν, such that if $0 < t \leqslant \kappa$ then there exist $\eta \in C^\infty(\Lambda^2 T^*M)$ and $f \in C^\infty(M)$, satisfying $\|d\eta\|_{C^0} \leqslant Kt^{1/2} \leqslant \epsilon_1$ and

$$(dd^* + d^*d)\eta = d^*\psi + d^*(f\psi) + *dF(d\eta) \text{ and } f\varphi = \tfrac{7}{3}\pi_1(d\eta).$$

Theorem 10.3.7 then shows that $\tilde{\varphi} = \varphi + d\eta \in C^\infty(\mathcal{P}^3 M)$ and $d\tilde{\varphi} = d\,\Theta(\tilde{\varphi}) = 0$, so $\tilde{\varphi}$ defines a smooth, torsion-free G_2-structure $(\tilde{\varphi}, \tilde{g})$ on M. Moreover, $\|\tilde{\varphi} - \varphi\|_{C^0} = \|d\eta\|_{C^0} \leqslant Kt^{1/2}$, as we want. $\square$

11.7 The proof of Theorem G1

We begin by proving inequalities on balls in $\mathbb{R}^7$ with nearly Euclidean metrics.

Proposition 11.7.1 *Let B_2, B_3 be the balls of radii $2, 3$ about 0 in $\mathbb{R}^7$, and h the Euclidean metric on B_3. Then there exist F_1, F_2, $F_3 > 0$ such that if g is a Riemannian metric on B_3 and $\|g - h\|_{C^{1,1/2}} \leqslant F_1$, then the following is true.*

Let $\chi \in L_1^{14}(\Lambda^3 T^*B_3)$ on B_3. Then $\|\chi|_{B_2}\|_{C^0} \leqslant F_2(\|\nabla\chi\|_{L^{14}} + \|\chi\|_{L^2})$, and

$$\int_{B_2} |\nabla\chi|^{14} dV_g \leqslant F_3 \int_{B_3} (|d\chi|^{14} + |d^*\chi|^{14}) dV_g + F_3 \left[\int_{B_3} |\chi|^2 dV_g\right]^7. \tag{11.34}$$

Here the operators ∇ and d^* and all norms are with respect to the metric g on B_3.

Proof By the Sobolev Embedding Theorem, Theorem 1.2.1, L_1^{14} embeds in C^0 in 7 dimensions. We can use this to show that $\|\chi|_{B_2}\|_{C^0} \leqslant F_2(\|\nabla\chi\|_{L^{14}} + \|\chi\|_{L^2})$, following the proof of [17, Lem. 2.22]. Let V be the vector bundle $\bigoplus_{k=0}^{7} \Lambda^k T^*B_3$ over B_3. Then $d + d^* : L_1^{14}(V) \to L^{14}(V)$ is an elliptic operator, and so we can apply *elliptic regularity*, as in §1.4. In particular, by modifying the proof of [5, Th. 10.3] we can prove that if $\xi \in L_1^{14}(V)$ then

$$\|\nabla\xi|_{B_2}\|_{L^{14}} \leqslant C(\|d\xi\|_{L^{14}} + \|d^*\xi\|_{L^{14}} + \|\xi\|_{L^2}), \tag{11.35}$$

for some $C > 0$ independent of ξ, g. Equation (11.34) then follows by putting $\xi = \chi$ and raising (11.35) to the power 14, with constant $F_3 = 3^{13} C^{14}$. $\square$

In this proof the inequality $\|g - h\|_{C^{1,1/2}} \leqslant F_1$ serves two purposes. Firstly, since g is close to h in C^0, norms taken with respect to g and with respect to h differ by a bounded factor, and we can replace norms with respect to g by norms with respect to h. Secondly, we get an a priori $C^{0,1/2}$ bound on the coefficients of the operator $d + d^*$ in coordinates on B_3, and we need this to apply the elliptic regularity results in [5].

Next we show that if (M, g) is a complete 7-manifold with bounds on $\delta(g)$ and $\|R(g)\|_{C^0}$, then all sufficiently small balls in M are nearly Euclidean, in the $C^{1,1/2}$ sense required by the previous proposition.

Proposition 11.7.2 *Let μ, ν and F_1 be positive constants. Then there exist positive constants L, F_4, F_5 such that the following is true.*

Let $t > 0$ be given, and let $r = Lt$. Suppose (M, g) is a complete Riemannian 7-manifold with $\delta(g) \geq \mu t$ and $\|R(g)\|_{C^0} \leq \nu t^{-2}$. Then for each $m \in M$ we have $F_4 t^7 \leq \text{vol}(B_r(m)) \leq \text{vol}(B_{4r}(m)) \leq F_5 t^7$, where $B_r(m)$ is the geodesic ball of radius r about m, and there is a smooth, injective map $\Psi_m : B_3 \to M$ satisfying $\|r^{-2}\Psi_m^*(g) - h\|_{C^{1,1/2}} \leq F_1$ and $B_r(m) \subset \Psi_m(B_2) \subset \Psi_m(B_3) \subset B_{4r}(m)$.

Proof For simplicity, first suppose that $t = 1$. We require systems of coordinates on open balls in M, in which the metric g appears close to the Euclidean metric in the $C^{1,1/2}$ norm. These are provided by Jost and Karcher's theory of *harmonic coordinates* ([115], [92, p. 124]). Jost and Karcher show that if the injectivity radius is bounded below and the sectional curvature is bounded above, then there exist coordinate systems on all balls of a given radius, in which the $C^{1,\alpha}$ norm of the metric is bounded in terms of α for each $\alpha \in (0, 1)$.

Therefore, if $\delta(g) \geq \mu$ and $\|R(g)\|_{C^0} \leq \nu$ then for $L > 0$ depending only on μ, ν and F_1, for each $m \in M$ there exists a coordinate system Ψ_m about m, which we may write as a map $\Psi_m : B_3 \to M$ with $\Psi_m(0) = m$, such that $\|L^{-2}\Psi_m^*(g) - h\|_{C^{1,1/2}} \leq F_1$, as we have to prove. Now the radius and volume of balls are controlled by the C^0 norm of the metric on the balls. Thus if F_1 is small, the balls $\Psi_m(B_2)$, $\Psi_m(B_3)$ in M must have volume and radius close to those of the balls of radius $2L$ and $3L$ in $\mathbb{R}^7$.

By making F_1 and L smaller if necessary, we can ensure that $F_4 \leq \text{vol}(B_L(m))$ and $\text{vol}(B_{4L}(m)) \leq F_5$ for some F_4, $F_5 > 0$ depending only on μ, ν and F_1, and that $B_L(m) \subset \Psi_m(B_2)$ and $\Psi_m(B_3) \subset B_{4L}(m)$, for all $m \in M$. This completes the proof when $t = 1$. To prove the proposition for general $t > 0$, apply the case $t = 1$ to the rescaled metric $t^{-2}g$. □

We can now prove Theorem G1 of §11.6.

Theorem G1 *Let μ, ν and t be positive constants, and suppose (M, g) is a complete Riemannian 7-manifold, whose injectivity radius $\delta(g)$ and Riemann curvature $R(g)$ satisfy $\delta(g) \geq \mu t$ and $\|R(g)\|_{C^0} \leq \nu t^{-2}$. Then there exist $C_1, C_2 > 0$ depending only on μ and ν, such that if $\chi \in L_1^{14}(\Lambda^3 T^*M) \cap L^2(\Lambda^3 T^*M)$ then*

$$\|\nabla \chi\|_{L^{14}} \leq C_1 \left(\|d\chi\|_{L^{14}} + \|d^*\chi\|_{L^{14}} + t^{-4}\|\chi\|_{L^2} \right) \qquad (11.36)$$

and $\quad \|\chi\|_{C^0} \leq C_2 \left(t^{1/2} \|\nabla \chi\|_{L^{14}} + t^{-7/2} \|\chi\|_{L^2} \right). \qquad (11.37)$

Proof Let $F_1 > 0$ be the constant defined by Proposition 11.7.1, and apply Proposition 11.7.2. This constructs a constant $L > 0$ depending on μ, ν and F_1, such that for each $m \in M$ there exists a smooth, injective map $\Psi_m : B_3 \to M$ with $\|r^{-2}\Psi_m^*(g) - h\|_{C^{1,1/2}} \leq F_1$, where $r = Lt$. Therefore Proposition 11.7.1 applies to the metric $r^{-2}\Psi_m^*(g)$ on B_3.

Thus (11.34) shows that if $\chi \in L_1^{14}(\Lambda^3 T^*M)$ and $d\chi = 0$, then

$$\int_{\Psi_m(B_2)} |\nabla \chi|^{14} dV_g \leq F_3 \int_{\Psi_m(B_3)} \left(|d\chi|^{14} + |d^*\chi|^{14} \right) dV_g + F_3 r^{-56} \left[\int_{\Psi_m(B_3)} |\chi|^2 dV_g \right]^7.$$

Here r^{-56} is inserted because the norms are with respect to g and not $r^{-2}g$. But $B_r(m) \subset \Psi_m(B_2)$ and $\Psi_m(B_3) \subset B_{4r}(m)$ by Proposition 11.7.2, so

$$\int_{B_r(m)} |\nabla\chi|^{14} dV_g \leq F_3 \int_{B_{4r}(m)} \left(|d\chi|^{14} + |d^*\chi|^{14}\right) dV_g + F_3 r^{-56} \|\chi\|_{L^2}^{12} \cdot \int_{B_{4r}(m)} |\chi|^2 dV_g.$$

Integrating this over M and reversing the order of integration, we find that

$$\int_{x \in M} \text{vol}(B_r(x)) |\nabla\chi(x)|^{14} dV_g$$
$$\leq F_3 \int_{x \in M} \text{vol}(B_{4r}(x)) \left(|d\chi(x)|^{14} + |d^*\chi(x)|^{14}\right) dV_g$$
$$+ F_3 r^{-56} \|\chi\|_{L^2}^{12} \cdot \int_{x \in M} \text{vol}(B_{4r}(x)) |\chi(x)|^2 dV_g.$$

But $r = Lt$ and $F_4 t^7 \leq \text{vol}(B_r(x))$, $\text{vol}(B_{4r}(x)) \leq F_5 t^7$ by Proposition 11.7.2, so

$$F_4 t^7 \|\nabla\chi\|_{L^{14}}^{14} \leq F_3 F_5 t^7 \left(\|d\chi\|_{L^{14}}^{14} + \|d^*\chi\|_{L^{14}}^{14}\right) + F_3 F_5 t^7 (Lt)^{-56} \|\chi\|_{L^2}^{14}.$$

Raising to the power $1/14$ gives (11.36), with $C_1 = (F_3 F_4^{-1} F_5)^{1/14} \max(1, L^{-4})$.

Similarly, the inequality $\|\chi|_{B_2}\|_{C^0} \leq F_2(\|\nabla\chi\|_{L^{14}} + \|\chi\|_{L^2})$ of Proposition 11.7.1 shows that if $\chi \in L_1^{14}(\Lambda^3 T^*M)$, then

$$\|\chi|_{\Psi_m(B_2)}\|_{C^0} \leq F_2 \left(r^{1/2} \|\nabla\chi|_{\Psi_m(B_3)}\|_{L^{14}} + r^{-7/2} \|\chi|_{\Psi_m(B_3)}\|_{L^2}\right).$$

Again, the powers of r are inserted because the norms are with respect to g and not $r^{-2}g$. Taking the supremum of this equation over $m \in M$ and putting $r = Lt$ shows that

$$\|\chi\|_{C^0} \leq F_2\left((Lt)^{1/2} \|\nabla\chi\|_{L^{14}} + (Lt)^{-7/2} \|\chi\|_{L^2}\right),$$

so (11.37) holds with $C_2 = F_2 \max(L^{1/2}, L^{-7/2})$. $\square$

11.8 The proof of Theorem G2

Next we prove Theorem G2 of §11.6.

Theorem G2 *Let λ, C_1 and C_2 be positive constants. Then there exist positive constants κ, K such that whenever $0 < t \leq \kappa$, the following is true.*

Let M be a compact 7-manifold, and (φ, g) a G_2-structure on M with $d\varphi = 0$. Suppose ψ is a smooth 3-form on M with $d^*\psi = d^*\varphi$, and

(i) $\|\psi\|_{L^2} \leq \lambda t^4$, $\|\psi\|_{C^0} \leq \lambda t^{1/2}$ and $\|d^*\psi\|_{L^{14}} \leq \lambda$,
(ii) *if $\chi \in L_1^{14}(\Lambda^3 T^*M)$ then $\|\nabla\chi\|_{L^{14}} \leq C_1(\|d\chi\|_{L^{14}} + \|d^*\chi\|_{L^{14}} + t^{-4} \|\chi\|_{L^2})$,*
(iii) *if $\chi \in L_1^{14}(\Lambda^3 T^*M)$ then $\|\chi\|_{C^0} \leq C_2(t^{1/2} \|\nabla\chi\|_{L^{14}} + t^{-7/2} \|\chi\|_{L^2})$.*

Let ϵ_1 be as in Definition 10.3.3, and F as in Proposition 10.3.5. Then there exist $\eta \in C^\infty(\Lambda^2 T^*M)$ with $\|d\eta\|_{C^0} \leqslant Kt^{1/2} \leqslant \epsilon_1$ and $f \in C^\infty(M)$, such that

$$(dd^* + d^*d)\eta = d^*\psi + d^*(f\psi) + *dF(d\eta) \quad \text{and} \quad f\varphi = \tfrac{7}{3}\pi_1(d\eta). \tag{11.38}$$

Proof We begin with the following proposition.

Proposition 11.8.1 *In the situation above, there exist positive constants κ, C_3 and K depending only on λ, C_1, C_2, such that if $0 < t \leqslant \kappa$ then there are sequences $\{\eta_j\}_{j=0}^\infty$ in $L_2^{14}(\Lambda^2 T^*M)$ and $\{f_j\}_{j=0}^\infty$ in $L_1^{14}(M)$ with $\eta_0 = f_0 = 0$, satisfying the equations*

$$(dd^*+d^*d)\eta_j = d^*\psi + d^*(f_{j-1}\psi) + *dF(d\eta_{j-1}) \quad \text{and} \quad f_j\varphi = \tfrac{7}{3}\pi_1(d\eta_j) \tag{11.39}$$

for each $j > 0$, and the inequalities

(a) $\|d\eta_j\|_{L^2} \leqslant 2\lambda t^4$,
(b) $\|\nabla d\eta_j\|_{L^{14}} \leqslant C_3$,
(c) $\|d\eta_j\|_{C^0} \leqslant Kt^{1/2} \leqslant \epsilon_1$ and

(d) $\|d\eta_j - d\eta_{j-1}\|_{L^2} \leqslant 2\lambda 2^{-j} t^4$,
(e) $\|\nabla(d\eta_j - d\eta_{j-1})\|_{L^{14}} \leqslant C_3 2^{-j}$,
(f) $\|d\eta_j - d\eta_{j-1}\|_{C^0} \leqslant K 2^{-j} t^{1/2}$.

Proof Define $C_3 = 4C_1\lambda$ and $K = C_2(C_3 + 2\lambda)$. We shall not give an explicit value for κ. Instead, at a number of points in the proof we shall need t to be smaller than some positive constant defined in terms of λ, C_1, C_2 and the constants ϵ_1, ϵ_2, ϵ_3 of Proposition 10.3.5. As a shorthand we shall simply say that this holds since $t \leqslant \kappa$, and suppose without remark that κ has been chosen such that the relevant restriction holds.

The proof works by induction in j. We begin with $j = 1$. Since $\eta_0 = f_0 = 0$ and $F(0) = 0$, we have $(dd^*+d^*d)\eta_1 = d^*\psi$ by (11.39). Now $dd^*+d^*d : C^\infty(\Lambda^2 T^*M) \to C^\infty(\Lambda^2 T^*M)$ has image $\operatorname{Im} d \oplus \operatorname{Im} d^*$ and kernel $\mathcal{H}^2$. As $d^*\psi$ lies in $\operatorname{Im} d^*$, there is a unique $\eta_1 \in C^\infty(\Lambda^2 T^*M)$ satisfying $(dd^*+d^*d)\eta_1 = d^*\psi$ which is L^2-orthogonal to $\mathcal{H}^2$. The equation $f_1\varphi = \tfrac{7}{3}\pi_1(d\eta_1)$ then defines f_1.

Clearly $d^*\eta_1 = 0$, as $\operatorname{Im} d$ and $\operatorname{Im} d^*$ are L^2-orthogonal, so $d^*d\eta_1 = d^*\psi$. Taking the inner product with η_1 and integrating by parts we find that

$$\|d\eta_1\|_{L^2}^2 = \langle d\eta_1, \psi\rangle_{L^2} \leqslant \|d\eta_1\|_{L^2}\|\psi\|_{L^2},$$

so cancelling $\|d\eta_1\|_{L^2}$ gives $\|d\eta_1\|_{L^2} \leqslant \|\psi\|_{L^2}$. Thus $\|d\eta_1\|_{L^2} \leqslant \lambda t^4$ by part (i) of Theorem G2, and this proves parts (a) and (d) for $j = 1$.

Applying part (ii) of Theorem G2 to $\chi = d\eta_1$ gives

$$\|\nabla d\eta_1\|_{L^{14}} \leqslant C_1\big(\|d^*\psi\|_{L^{14}} + t^{-4}\|d\eta_1\|_{L^2}\big) \leqslant C_1(\lambda + t^{-4}\lambda t^4) = \tfrac{1}{2}C_3,$$

since $d^*d\eta_1 = d^*\psi$, $\|d^*\psi\|_{L^{14}} \leqslant \lambda$, $\|d\eta_1\|_{L^2} \leqslant \lambda t^4$ and $C_3 = 4C_1\lambda$. Thus parts (b) and (e) hold when $j = 1$. So by part (iii) of Theorem G2 we have

$$\|d\eta_1\|_{C^0} \leqslant C_2\big(t^{1/2}\|\nabla d\eta_1\|_{L^{14}} + t^{-7/2}\|d\eta_1\|_{L^2}\big) \leqslant C_2 t^{1/2}(\tfrac{1}{2}C_3 + \lambda) = \tfrac{1}{2}Kt^{1/2},$$

as $K = C_2(C_3 + 2\lambda)$. Also $Kt^{1/2} \leqslant \epsilon_1$ since $t \leqslant \kappa$. This gives parts (c) and (f) when $j = 1$.

Next we prove the inductive step. Suppose smooth $\eta_0, \ldots, \eta_k$ and $f_0, \ldots, f_k$ exist and satisfy the conditions of the proposition for $j \leqslant k$. Then $\|d\eta_k\|_{C^0} \leqslant \epsilon_1$ by (c), so $F(d\eta_k)$ is well-defined. Now $d^*\psi + d^*(f_k\psi) + *dF(d\eta_k)$ lies in Im d^*, so by the argument above there exist $\eta_{k+1} \in C^\infty(\Lambda^2 T^*M)$ and $f_{k+1} \in C^\infty(M)$ satisfying (11.39) for $j = k+1$, and if η_{k+1} is L^2-orthogonal to $\mathcal{H}^2$ then η_{k+1}, f_{k+1} are unique.

Taking the difference between (11.39) for $j = k, k+1$, we find

$$d^*d\eta_{k+1} - d^*d\eta_k = d^*\bigl((f_k - f_{k-1})\psi\bigr) - d^*\bigl(*F(d\eta_k) - *F(d\eta_{k-1})\bigr), \tag{11.40}$$

as $d^*\eta_k = d^*\eta_{k+1} = 0$ and $*d = -d^**$ on 4-forms. Integrating the inner product of this with $\eta_{k+1} - \eta_k$ by parts shows that

$$\|d\eta_{k+1} - d\eta_k\|_{L^2}^2 = \langle d\eta_{k+1} - d\eta_k, (f_k - f_{k-1})\psi\rangle_{L^2} - \langle d\eta_{k+1} - d\eta_k, *F(d\eta_k) - *F(d\eta_{k-1})\rangle_{L^2},$$

so using $\langle u, v\rangle_{L^2} \leqslant \|u\|_{L^2}\|v\|_{L^2}$ and cancelling $\|d\eta_{k+1} - d\eta_k\|_{L^2}$ gives

$$\|d\eta_{k+1} - d\eta_k\|_{L^2} \leqslant \|(f_k - f_{k-1})\psi\|_{L^2} + \|F(d\eta_k) - F(d\eta_{k-1})\|_{L^2}.$$

By (11.39), part (i) of Theorem G2 and part (f) of the proposition, we have

$$\|(f_k - f_{k-1})\psi\|_{L^2} \leqslant \|f_k - f_{k-1}\|_{C^0}\|\psi\|_{L^2} \leqslant \tfrac{7}{3}\|d\eta_k - d\eta_{k-1}\|_{C^0}\lambda t^4 \leqslant \tfrac{7}{3}K\lambda 2^{-k}t^{9/2}.$$

Also, eqn (10.10) of Proposition 10.3.5 and parts (b), (e) show that

$$\|F(d\eta_k) - F(d\eta_{k-1})\|_{L^2} \leqslant \epsilon_2\|d\eta_k - d\eta_{k-1}\|_{L^2}\bigl(\|d\eta_k\|_{C^0} + \|d\eta_{k-1}\|_{C^0}\bigr)$$
$$\leqslant \epsilon_2 2\lambda 2^{-k}t^4(Kt^{1/2} + Kt^{1/2}) = 4\epsilon_2\lambda K 2^{-k}t^{9/2}.$$

Combining the last three equations, we see that

$$\|d\eta_{k+1} - d\eta_k\|_{L^2} \leqslant K(\tfrac{7}{3}\lambda + 4\epsilon_2)2^{-k}t^{9/2} \leqslant 2\lambda 2^{-k-1}t^4,$$

as $t \leqslant \kappa$. Thus part (b) holds for $j = k+1$.

Applying part (ii) of Theorem G2 to $\chi = d\eta_{k+1} - d\eta_k$ and using (11.40) gives

$$\|\nabla(d\eta_{k+1} - d\eta_k)\|_{L^{14}} \leqslant C_1\Bigl(\|d^*(f_k - f_{k-1})\psi\|_{L^{14}} + \|dF(d\eta_k) - dF(d\eta_{k-1})\|_{L^{14}} + t^{-4}\|d\eta_{k+1} - d\eta_k\|_{L^2}\Bigr).$$

Now (11.39), part (i) of Theorem G2 and parts (e), (f) show that

$$\|d^*(f_k - f_{k-1})\psi\|_{L^{14}} \leqslant \|\nabla(f_k - f_{k-1})\|_{L^{14}}\|\psi\|_{C^0} + \|f_k - f_{k-1}\|_{C^0}\|d^*\psi\|_{L^{14}}$$
$$\leqslant \tfrac{7}{3}\|\nabla(\eta_k - \eta_{k-1})\|_{L^{14}}\lambda t^{1/2} + \tfrac{7}{3}\|\eta_k - \eta_{k-1}\|_{C^0}\lambda$$
$$\leqslant \tfrac{7}{3}C_3\lambda 2^{-k}t^{1/2} + \tfrac{7}{3}K\lambda 2^{-k}t^{1/2}.$$

Also, eqn (10.11) of Proposition 10.3.5 and parts (b), (c), (e) and (f) give

$$\|dF(d\eta_k) - dF(d\eta_{k-1})\|_{L^{14}}$$
$$\leq \epsilon_3 \Big\{ \|d\eta_k - d\eta_{k-1}\|_{C^0} \big(\|d\eta_k\|_{C^0} + \|d\eta_{k-1}\|_{C^0}\big) \|d^*\psi\|_{L^{14}}$$
$$+ \|\nabla(d\eta_k - d\eta_{k-1})\|_{L^{14}} \big(\|d\eta_k\|_{C^0} + \|d\eta_{k-1}\|_{C^0}\big)$$
$$+ \|d\eta_k - d\eta_{k-1}\|_{C^0} \big(\|\nabla d\eta_k\|_{L^{14}} + \|\nabla d\eta_{k-1}\|_{L^{14}}\big) \Big\}$$
$$\leq 2^{-k} t^{1/2} \cdot 2K\epsilon_3 (K\lambda t^{1/2} + 2C_3).$$

Combining the last three equations and part (d) of the proposition shows that

$$\|\nabla(d\eta_{k+1} - d\eta_k)\|_{L^{14}} \leq C_1 \Big(\tfrac{7}{3} C_3 \lambda 2^{-k} t^{1/2} + \tfrac{7}{3} K \lambda 2^{-k} t^{1/2}$$
$$+ 2^{-k} t^{1/2} \cdot 2K\epsilon_3 (K\lambda t^{1/2} + 2C_3) + 2\lambda 2^{-1-k} \Big)$$
$$\leq C_3 2^{-1-k},$$

as $2\lambda C_1 < C_3$ and $t \leq \kappa$. This proves part (e) for $j = k+1$.

Part (f) for $j = k+1$ follows easily from parts (d) and (e), part (ii) of Theorem G2, and $K = C_2(C_3 + 2\lambda)$. Thus parts (d)–(f) hold for $j = k+1$. But parts (a)–(c) for $j = k+1$ follow by induction from parts (d)–(f) for $j = 1, \ldots, k+1$ and $\eta_0 = 0$. Thus parts (a)–(f) of the proposition hold for $j = k+1$. This concludes the inductive step, and the proof of Proposition 11.8.1. □

Consider the sequence $\{d\eta_j\}_{j=0}^\infty$ in $C^0(\Lambda^3 T^*M)$. By part (f), if $j < k$ then

$$\|d\eta_j - d\eta_k\|_{C^0} \leq \|d\eta_j - d\eta_{j+1}\|_{C^0} + \cdots + \|d\eta_{k-1} - d\eta_k\|_{C^0}$$
$$\leq Kt^{1/2}(2^{-1-j} + \cdots + 2^{-k}) \leq Kt^{1/2} 2^{-j}.$$

Hence, $\{d\eta_j\}_{j=0}^\infty$ is a *Cauchy sequence* in $C^0(\Lambda^3 T^*M)$, and must converge as $C^0(\Lambda^3 T^*M)$ is a Banach space. Similarly, parts (d), (e) of the proposition show that $\{d\eta_j\}_{j=0}^\infty$ converges in $L^2(\Lambda^3 T^*M)$ and $L_1^{14}(\Lambda^3 T^*M)$.

However, we don't yet know that $\{\eta_j\}_{j=0}^\infty$ converges in any Banach space of 2-forms, since we have no estimates on the η_j. Here is a way to do this. Using Proposition 1.5.2 and the ellipticity of $d + d^*$, one can show that there exists a constant $C_g > 0$ depending on M and g, such that if $\zeta \in L_2^{14}(\Lambda^2 T^*M)$ is L^2-orthogonal to $\mathcal{H}^2$ and $d^*\zeta = 0$, then

$$\|\zeta\|_{L_2^{14}} \leq C_g \big(\|\nabla d\zeta\|_{L^{14}} + \|d\zeta\|_{C^0}\big).$$

Since the η_j are L^2-orthogonal to $\mathcal{H}^2$ and $d^*\eta_j = 0$, parts (e) and (f) give

$$\|\eta_j - \eta_{j-1}\|_{L_2^{14}} \leq C_g (C_3 + Kt^{1/2}) 2^{-j}.$$

So $\{\eta_j\}_{j=0}^\infty$ is Cauchy and converges in the Banach space $L_2^{14}(\Lambda^2 T^*M)$, as above.

Let η be the limit of $\{\eta_j\}_{j=0}^\infty$ in $L_2^{14}(\Lambda^2 T^*M)$, and define $f \in L_1^{14}(M)$ by $f\varphi = \frac{7}{3}\pi_1(d\eta)$. Taking the limit in (11.39) then shows that η and f satisfy (11.38), and part (c) of the proposition shows that $\|d\eta\|_{C^0} \leqslant Kt^{1/2} \leqslant \epsilon_1$, as we have to prove.

It remains only to show that η is smooth. We shall do this using elliptic regularity and the 'bootstrap method'. We can rewrite (11.38) as

$$(dd^* + d^*d)\eta + G(\psi, d\eta, \nabla d\eta) = H(d^*\psi, d\eta), \qquad (11.41)$$

where $-G$ is the sum of the term in ∇f from $d^*(f\psi)$ and the term in $\nabla d\eta$ from $*dF(d\eta)$. Here G and H are smooth functions of their arguments, and $G(x, y, z)$ is linear in z. Also, as $F(d\eta)$ is at least quadratic in $d\eta$ by Proposition 10.3.5, we find that $G(0, 0, z) = 0$.

Define a second-order linear partial differential operator $P : L_2^{14}(\Lambda^2 T^*M) \to L^{14}(\Lambda^2 T^*M)$ by $P(\zeta) = (dd^* + d^*d)\zeta + G(\psi, d\eta, \nabla d\zeta)$. Since $dd^* + d^*d$ is elliptic and ellipticity is an open condition, we see that P is elliptic if the perturbation $G(\psi, d\eta, \nabla d\zeta)$ is small enough. As $G(0, 0, z) = 0$, this is true whenever ψ and $d\eta$ are sufficiently small in C^0. But $\|\psi\|_{C^0} \leqslant \lambda t^{1/2}$ and $\|d\eta\|_{C^0} \leqslant Kt^{1/2}$ from above, and $t \leqslant \kappa$. So by making κ smaller if necessary, we can assume that P is elliptic.

Thus η satisfies $P(\eta) = H(d^*\psi, d\eta)$ by (11.41), where P is a second-order linear elliptic operator. Now $\eta \in L_2^{14}(\Lambda^2 T^*M)$, and so Theorem 1.2.1 shows that $\eta \in C^{1,1/2}(\Lambda^2 T^*M)$, as L_2^{14} embeds in $C^{1,1/2}$. Suppose by induction that $\eta \in C^{k,1/2}(\Lambda^2 T^*M)$ for some $k \geqslant 1$. Then $d\eta \in C^{k-1,1/2}(\Lambda^3 T^*M)$, so the coefficients of P and the right hand side of (11.41) are both $C^{k-1,1/2}$. Therefore $\eta \in C^{k+1,1/2}(\Lambda^2 T^*M)$ by Theorem 1.4.2. Hence by induction $\eta \in C^{k,1/2}(\Lambda^2 T^*M)$ for all $k \geqslant 1$, and η is smooth. This completes the proof of Theorem G2. $\square$

11.9 Other constructions of compact 7-manifolds with holonomy G_2

Here are sketches of three methods which could be used to construct compact 7-manifolds with holonomy G_2, but which have not yet been proved. The first two are due to the author, and the third to Simon Donaldson.

Method 1. Choose a flat metric on T^3, and a metric with holonomy SU(2) on the $K3$ surface. Define a torsion-free G_2-structure (φ, g) on $T^3 \times K3$ as in Proposition 11.1.1. Let Γ be a finite group of automorphisms of $T^3 \times K3$ preserving (φ, g). Then $(T^3 \times K3)/\Gamma$ is an orbifold with torsion-free G_2-structure (φ, g). We aim to resolve $(T^3 \times K3)/\Gamma$ to get a compact 7-manifold M with a torsion-free G_2-structure, in a similar way to the resolutions of T^7/Γ described in this chapter. If $\pi_1(M)$ is finite, then M has holonomy G_2 by Proposition 10.2.2.

Here is a condition that makes $(T^3 \times K3)/\Gamma$ particularly easy to resolve. Suppose that the stabilizer in Γ of every point in $T^3 \times K3$ is *cyclic*. Then the singularities of $(T^3 \times K3)/\Gamma$ divide up into connected components, each of which can locally be resolved within holonomy SU(3). To make this into a formal proof would require some Calabi conjecture type analysis to model Calabi–Yau metrics on crepant resolutions of orbifolds. We will see in Chapter 12 that the examples of §12.5 and §12.7 can be thought of as resolutions of $(T^3 \times K3)/\mathbb{Z}_2^2$.

The second method was described in [122, §4.3].

Method 2. Let (Y, J, g) be a Calabi–Yau 3-fold, as in Chapter 6, with Kähler form ω and holomorphic volume form θ. Suppose $\sigma : Y \to Y$ is an involution, satisfying $\sigma^*(g) = g, \sigma^*(J) = -J$ and $\sigma^*(\theta) = \bar{\theta}$. We call σ a *real structure* on Y. Let N be the fixed point set of σ in Y. Then N is a real 3-dimensional submanifold of Y, and is in fact a special Lagrangian 3-fold.

Let $\mathcal{S}^1 = \mathbb{R}/\mathbb{Z}$, and define a torsion-free G_2-structure (φ, g) on $\mathcal{S}^1 \times Y$ as in Proposition 11.1.2. Then $\varphi = \mathrm{d}x \wedge \omega + \mathrm{Re}\,\theta$, where $x \in \mathbb{R}/\mathbb{Z}$ is the coordinate on $\mathcal{S}^1$. Define $\hat{\sigma} : \mathcal{S}^1 \times Y \to \mathcal{S}^1 \times Y$ by $\hat{\sigma}\big((x, y)\big) = \big(-x, \sigma(y)\big)$. Then $\hat{\sigma}$ preserves (φ, g) and $(\hat{\sigma})^2 = 1$. The fixed points of $\hat{\sigma}$ in $\mathcal{S}^1 \times Y$ are $\{\mathbb{Z}, \frac{1}{2} + \mathbb{Z}\} \times N$. Thus $(\mathcal{S}^1 \times Y)/\langle\hat{\sigma}\rangle$ is an orbifold. Its singular set is two copies of N, and each singular point is modelled on $\mathbb{R}^3 \times \mathbb{R}^4/\{\pm 1\}$.

We aim to resolve $(\mathcal{S}^1 \times Y)/\langle\hat{\sigma}\rangle$ to get a compact 7-manifold M with holonomy G_2. Locally, each singular point should be resolved like $\mathbb{R}^3 \times Y$, where Y is an ALE 4-manifold with holonomy SU(2) asymptotic to $\mathbb{R}^4/\{\pm 1\}$. There is a 3-dimensional family of such Y, and we need to choose one member of this family for each singular point in the singular set.

Calculations by the author indicate that the data needed to do this in one way is a 1-form α on N that is nonzero at every point of N and satisfies $\mathrm{d}\alpha = \mathrm{d}^*\alpha = 0$. (There is a second way involving double covers of N, which we will not discuss.) Clearly $[\alpha] \neq 0$ in $H^1(N, \mathbb{R})$, so we need $b^1(N) \geqslant 1$. But this is not a sufficient condition, as all the harmonic 1-forms on N could still have zeroes. The existence of a suitable 1-form α depends on the metric on N, which is the restriction of the metric g on Y.

For general Y we know very little about g, as it is constructed using the Calabi conjecture, and we cannot guarantee that α exists and has no zeroes. However, it may be possible to get round this in some cases by taking Y close to the 'large complex structure limit', and arranging that N is $k\,T^3$ with a metric that is nearly flat, and therefore must have harmonic 1-forms without zeroes. This would probably involve a lot of rather difficult analysis.

The third method was explained to me by Simon Donaldson, and I am grateful to him for permission to include it here.

Method 3. Let X be a projective complex 3-manifold with canonical bundle K_X, and s a holomorphic section of K_X^{-1} which vanishes to order 1 on a smooth divisor D in X. The adjunction formula [93, p. 147] shows that D has trivial canonical bundle, and so D is T^4 or a $K3$ surface. Suppose D is a $K3$ surface, with trivial normal bundle in X. Define $Y = X \setminus D$, and suppose Y is simply-connected.

Then Y is a noncompact complex 3-fold with trivial canonical bundle, which has one infinite end modelled on $D \times \mathcal{S}^1 \times [0, \infty)$. There are good reasons for believing that there should exist a complete metric with holonomy SU(3) on Y, which is asymptotic to the product on $D \times \mathcal{S}^1 \times [0, \infty)$ of a metric with holonomy SU(2) on D, and Euclidean metrics on $\mathcal{S}^1$ and $[0, \infty)$. To prove this we would need a version for $\alpha = 1$ of the theorem for $\alpha > 1$ (due to Tian and Yau, and Bando and Kobayashi) stated in the introduction to Chapter 8.

Suppose we have such a metric on Y. Define a torsion-free G_2-structure (φ, g) on $\mathcal{S}^1 \times Y$ as in Proposition 11.1.2. Then $\mathcal{S}^1 \times Y$ is a noncompact G_2-manifold with one end modelled on $D \times T^2 \times [0, \infty)$, whose metric is asymptotic to the product on $D \times T^2 \times [0, \infty)$ of a metric with holonomy SU(2) on D, and Euclidean metrics on T^2 and $[0, \infty)$.

Donaldson's idea is that we should take two such products $\mathcal{S}^1 \times Y_1$ and $\mathcal{S}^1 \times Y_2$ whose infinite ends are isomorphic in a suitable way, and glue them together to get a compact 7-manifold M with holonomy G_2. The gluing process would actually be quite easy, using the results of this chapter — the hard work is in constructing the metrics on Y_1 and Y_2 in the first place. Note that the gluing process should swap round the $\mathcal{S}^1$ factors. That is, the $\mathcal{S}^1$ factor in $\mathcal{S}^1 \times Y_1$ is identified with the asymptotic $\mathcal{S}^1$ factor in $Y_2 \sim D_2 \times \mathcal{S}^1 \times [0, \infty)$, and vice versa.

12

EXAMPLES OF COMPACT 7-MANIFOLDS WITH HOLONOMY G_2

We shall now use the results of Chapter 11 to construct examples of compact 7-manifolds with holonomy G_2. Our general plan is to start with carefully explained, simple examples, and gradually introduce more complex behaviour. Later on we leave out a lot of the details, to keep the chapter to a manageable length.

The main focus of the chapter is the construction of many examples of compact, simply-connected 7-manifolds M with holonomy G_2, and the calculation of their Betti numbers $b^2(M)$ and $b^3(M)$. We find that one orbifold can admit a surprisingly large number of topologically distinct resolutions (M, π) with holonomy G_2, realizing many different pairs of Betti numbers $(b^2(M), b^3(M))$.

We begin in §12.1 by explaining how to compute the fundamental group $\pi_1(M)$ and Betti numbers $b^k(M)$ of a resolution M of T^7/Γ. Then sections 12.2, 12.3, 12.5, 12.7 and 12.8 study eight orbifolds T^7/Γ and their resolutions. In all, our examples give 252 different sets of Betti numbers of compact, simply-connected 7-manifolds with holonomy G_2, which are displayed on a graph in §12.9.

In §12.5 we study an orbifold T^7/Γ which has at least $3^{16} = 43\,046\,721$ topologically distinct resolutions (M, π) with holonomy G_2. Distinct resolutions may be diffeomorphic as 7-manifolds, so we have not found this many distinct 7-manifolds with holonomy G_2, but we do show that this one example yields at least 120 different sets of Betti numbers $(b^2(M), b^3(M))$, and thus at least 120 distinct 7-manifolds.

We also consider two other topics. Firstly, §12.4 constructs some families of compact 7-manifolds M with holonomy G_2 and nontrivial fundamental group $\pi_1(M)$. Many more examples could be constructed using the methods of this chapter, but we have chosen to concentrate on simply-connected manifolds.

Secondly, §12.6 gives examples of compact associative 3-folds and coassociative 4-folds with a variety of different topological types, in a compact 7-manifold M with holonomy G_2. They are constructed as fixed point sets of isometric involutions. For simplicity we take M to be our first example from §12.2, but we could give many more examples in different 7-manifolds.

Sections 12.1–12.4 and 12.6 are based on the author's second paper [122] on G_2-manifolds, with some alterations and improvements. The rest of the chapter is original research by the author, and is published here for the first time.

12.1 Calculating topological invariants of resolutions of T^7/Γ

Let Γ be a finite group of automorphisms of T^7 preserving a flat G_2-structure (φ_0, g_0) on T^7, and suppose (Y_i, π_i), (φ_i, g_i) and $\chi_{\gamma,i}$ is a set of R-data for T^7/Γ, as in §11.4.1. Then in Definition 11.4.5 we defined a compact 7-manifold M and a map $\pi : M \to$

T^7/Γ, and in Theorem 11.6.2 we proved that there exist torsion-free G_2-structures on M.

In the rest of the chapter we will give many examples of such compact 7-manifolds M. For each example we will want to know the most basic topological invariants—the fundamental group $\pi_1(M)$, and the Betti numbers $b^2(M)$ and $b^3(M)$. The fundamental group is important because it determines the holonomy group of the metrics, by Proposition 10.2.2, and $b^3(M)$ determines the dimension of the family of torsion-free G_2-structures on M, by Theorem 10.4.4. All three invariants will be used to distinguish different 7-manifolds M.

In this section we explain how to compute $\pi_1(M)$ and $b^k(M)$. Later we will leave out most these calculations, giving only the results and commenting only on difficult points.

12.1.1 How to find the fundamental group $\pi_1(M)$ of M

Let T^7/Γ, M and π be as above. We begin by showing how to calculate $\pi_1(T^7/\Gamma)$, and then discuss how to use this to work out $\pi_1(M)$. Now $T^7 = \mathbb{R}^7/\mathbb{Z}^7$, where $\mathbb{Z}^7$ acts on $\mathbb{R}^7$ by translations. Therefore $T^7/\Gamma = \mathbb{R}^7/(\Gamma \ltimes \mathbb{Z}^7)$, where $\Gamma \ltimes \mathbb{Z}^7$ is a natural semidirect product of Γ and $\mathbb{Z}^7$ which acts on $\mathbb{R}^7$ by affine transformations—roughly speaking, as a combination of rotations and translations.

Let $\Pi : \mathbb{R}^7 \to T^7/\Gamma$ be the natural projection, let x_0 be a base-point in $\mathbb{R}^7$, and let $y_0 = \Pi(x_0)$. Choose $\gamma \in \Gamma \ltimes \mathbb{Z}^7$, and let $f_\gamma : [0,1] \to \mathbb{R}^7$ be a continuous path with $f_\gamma(0) = x_0$ and $f_\gamma(1) = \gamma \cdot x_0$. Then $\Pi \circ f_\gamma : [0,1] \to T^7/\Gamma$ is a loop in T^7/Γ based at y_0, and so defines a class in the fundamental group $\pi_1(T^7/\Gamma)$ calculated using the base-point y_0. As $\mathbb{R}^7$ is simply-connected, this class is independent of the choice of f_γ, and depends only on γ.

Thus we have defined a map $\rho : \Gamma \ltimes \mathbb{Z}^7 \to \pi_1(T^7/\Gamma)$. It is easy to show that ρ is surjective, and a group homomorphism. Therefore $\operatorname{Ker} \rho$ is a normal subgroup of $\Gamma \ltimes \mathbb{Z}^7$, and $\pi_1(T^7/\Gamma) \cong (\Gamma \ltimes \mathbb{Z}^7)/\operatorname{Ker} \rho$. So, if we can determine $\operatorname{Ker} \rho$ then we know $\pi_1(T^7/\Gamma)$. The following lemma characterizes $\operatorname{Ker} \rho$.

Lemma 12.1.1 *In the situation above, $\operatorname{Ker} \rho$ is the subgroup of $\Gamma \ltimes \mathbb{Z}^7$ generated by the elements of $\Gamma \ltimes \mathbb{Z}^7$ which have fixed points in $\mathbb{R}^7$.*

Proof We shall show that if $\gamma \in \Gamma \ltimes \mathbb{Z}^7$ has fixed points then $\gamma \in \operatorname{Ker} \rho$, and then leave the rest of the proof of as an exercise in algebraic topology. Suppose γ fixes x in $\mathbb{R}^7$. Define the path $f_\gamma : [0,1] \to \mathbb{R}^7$ to be the straight line from x_0 to x followed by the straight line from x to $\gamma \cdot x_0$. These two line segments are identified by Π, so that $\Pi \circ f_\gamma$ goes along a line segment and then back along the same line segment. Thus $\Pi \circ f_\gamma$ is a contractible loop, and $\rho(\gamma) = 1$. □

Our next result is trivial to prove.

Lemma 12.1.2 *In the situation above, let Γ' be the subgroup of Γ generated by the elements of Γ with fixed points in T^7. Then Γ' is normal in Γ, and $\rho(\mathbb{Z}^7)$ is normal in $\pi_1(T^7/\Gamma)$, and the quotient groups Γ/Γ' and $\pi_1(T^7/\Gamma)/\rho(\mathbb{Z}^7)$ are isomorphic. Hence, if $\Gamma' = \Gamma$ then $\pi_1(T^7/\Gamma) = \rho(\mathbb{Z}^7)$. In this case $\pi_1(T^7/\Gamma)$ is abelian, and isomorphic to $H_1(T^7/\Gamma, \mathbb{Z})$.*

In investigating an orbifold T^7/Γ, we have to work out which elements of Γ have fixed points to find the singularities of T^7/Γ. As Γ is a finite group and usually fairly small, working out the group Γ' in the lemma is very little extra work. Usually we find that $\Gamma = \Gamma'$.

Then $\pi_1(T^7/\Gamma) \cong \mathbb{Z}^7/\operatorname{Ker}(\rho|_{\mathbb{Z}^7})$, and so calculating $\pi_1(T^7/\Gamma)$ reduces to working out which elements of $\mathbb{Z}^7$ are sent to zero in $\pi_1(T^7/\Gamma)$, or equivalently $H_1(T^7/\Gamma, \mathbb{Z})$. This may be a lengthy calculation to do explicitly. If we can show that a basis of loops in T^7 are sent to contractible loops in T^7/Γ, then T^7/Γ is simply-connected.

Next we consider how to determine $\pi_1(M)$, knowing $\pi_1(T^7/\Gamma)$. The following proposition, which is elementary to prove, is very useful for doing this.

Proposition 12.1.3 *In the situation above, the projection $\pi : M \to T^7/\Gamma$ induces a surjective group homomorphism $\pi_* : \pi_1(M) \to \pi_1(T^7/\Gamma)$. If the resolutions (Y_i, π_i) in the R-data are all simply-connected, then π_* is an isomorphism.*

Note that Y_i is always simply-connected if either Y_i is an ALE space with holonomy SU(2), or Y_i is a crepant resolution of $\mathbb{C}^3/A_i$. Thus, when all the (Y_i, π_i) are of one of these two sorts, then $\pi_1(M) \cong \pi_1(T^7/\Gamma)$.

Resolutions (Y_i, π_i) with nontrivial fundamental group were given in Examples 9.9.8 and 11.2.5. If we have included such resolutions in the R-data then we must think carefully about whether they cause π_* to have a nonzero kernel. This will happen very rarely in the examples we give, so we will not discuss it here.

12.1.2 *How to find the Betti numbers $b^k(M)$ of M*

The Betti numbers $b^k(M)$ are the dimensions of the de Rham cohomology groups $H^k(M, \mathbb{R})$. They satisfy $b^0(M) = 1$ as M is connected and $b^k(M) = b^{7-k}(M)$ as M is compact and orientable, and if $\pi_1(M)$ is finite (as it will be in all our examples) then $b^1(M) = 0$. Thus, it is enough to calculate $b^2(M)$ and $b^3(M)$. Our general method is to work out $b^k(T^7/\Gamma)$, and then add on contributions from the resolution of singularities in T^7/Γ to get $b^k(M)$.

The Betti numbers $b^k(T^7/\Gamma)$ are very easy to work out. It can be shown that $H^k(T^7/\Gamma, \mathbb{R})$ is the Γ-invariant part of $H^k(T^7, \mathbb{R})$. But by Hodge theory each class in $H^k(T^7, \mathbb{R})$ is represented by a unique constant k-form. Thus $H^k(T^7, \mathbb{R})$ is isomorphic to $\Lambda^k(\mathbb{R}^7)^*$, where $T^7 = \mathbb{R}^7/\mathbb{Z}^7$, and

$$b^k(T^7/\Gamma) = \dim\{\alpha \in \Lambda^k(\mathbb{R}^7)^* : \alpha \text{ is } \Gamma\text{-invariant}\}.$$

So to find $b^k(T^7/\Gamma)$ we calculate the Γ-invariant k-forms on $\mathbb{R}^7$.

Now $\pi : M \to T^7/\Gamma$ induces a map $\pi^* : H^k(T^7/\Gamma, \mathbb{R}) \to H^k(M, \mathbb{R})$. It is easy to see that π^* is injective, and therefore $b^k(M) \geq b^k(T^7/\Gamma)$. We can think of $H^k(M, \mathbb{R})$ as being $H^k(T^7/\Gamma)$ together with some extra cohomology classes, which are introduced by resolving the singularities of T^7/Γ, and are supported in a small neighbourhood of $\pi^{-1}(S)$, where S is the singular set of T^7/Γ.

Here is a method to calculate $b^k(M)$. First find $H^k(T^7/\Gamma, \mathbb{R})$ as above, work out the singular set S of T^7/Γ and calculate $H^{k-1}(S, \mathbb{R})$ and $H^k(S, \mathbb{R})$. Then, using the exact sequence

$$H^{k-1}(S,\mathbb{R}) \to H^k(T^7/\Gamma; S, \mathbb{R}) \to H^k(T^7/\Gamma, \mathbb{R}) \to H^k(S, \mathbb{R})$$

find $H^k(T^7/\Gamma; S, \mathbb{R})$. But $H^k(T^7/\Gamma; S, \mathbb{R}) \cong H^k(M; \pi^{-1}(S), \mathbb{R})$ by excision, so this gives us $H^k(M; \pi^{-1}(S), \mathbb{R})$.

Then calculate the cohomology of $\pi^{-1}(S)$, and use the exact sequence

$$H^{k-1}(\pi^{-1}(S), \mathbb{R}) \to H^k(M; \pi^{-1}(S), \mathbb{R}) \to H^k(M, \mathbb{R}) \to H^k(\pi^{-1}(S), \mathbb{R})$$

to get $H^k(M, \mathbb{R})$. The most difficult step in this process is understanding $\pi^{-1}(S)$ and finding its cohomology.

To write this method out in full for every example (or even for one complicated example) would be very long and tedious, and we will not do it. When performing the calculations at home the author actually uses a series of short-cuts to speed things up, which are based on experience, intuition and visualization of the singular set and its resolution, aided by drawing pictures of the pieces of the singular set and how they intersect.

Most of these short-cuts will be explained in passing later in the chapter. Any reader who carries out a lot of these calculations will soon find short-cuts of their own. Here are two useful principles for performing such calculations.

- Each connected component of S can be treated separately. That is, in calculating the additions to $b^k(M)$ from the resolution of the singularities of S there is no interaction between different components, so we can calculate the contribution from each component in isolation, and then add them up.
- If Y_i is an ALE space with holonomy SU(2) asymptotic to $\mathbb{C}^2/A_i$ for $A_i \subset \mathrm{SU}(2)$ a finite group, or if Y_i is a crepant resolution of $\mathbb{C}^3/A_i$ for $A_i \subset \mathrm{SU}(3)$ a finite group, then we can calculate the Betti numbers of Y_i very easily using the *McKay correspondence* of §6.4. We use this to work out the contributions to $b^k(M)$ from resolving singularities locally modelled on $\mathbb{R}^3 \times \mathbb{C}^2/A_i$ or $\mathbb{R} \times \mathbb{C}^3/A_i$ using Y_i.

12.2 A simple example of a compact 7-manifold with holonomy G_2

We shall now explain in detail the simplest example of a compact 7-manifold with holonomy G_2 that the author knows. It is based on the 7-manifold studied in [121]. Let $(x_1, \ldots, x_7)$ be coordinates on $\mathbb{R}^7$, and let (φ_0, g_0) be the flat G_2-structure on $\mathbb{R}^7$ given in Definition 10.1.1. Then

$$\varphi_0 = \mathrm{d}\mathbf{x}_{123} + \mathrm{d}\mathbf{x}_{145} + \mathrm{d}\mathbf{x}_{167} + \mathrm{d}\mathbf{x}_{246} - \mathrm{d}\mathbf{x}_{257} - \mathrm{d}\mathbf{x}_{347} - \mathrm{d}\mathbf{x}_{356}$$

by (10.1), where $\mathrm{d}\mathbf{x}_{ijk} = \mathrm{d}x_i \wedge \mathrm{d}x_j \wedge \mathrm{d}x_k$. Let $\mathbb{Z}^7$ act on $\mathbb{R}^7$ by translations in the obvious way, and let $T^7 = \mathbb{R}^7/\mathbb{Z}^7$. Then (φ_0, g_0) descends to T^7. We shall regard $(x_1, \ldots, x_7)$ as coordinates on T^7, where each x_j lies in $\mathbb{R}/\mathbb{Z}$ rather than $\mathbb{R}$.

Let α, β and γ be the involutions of T^7 defined by

$$\alpha : (x_1, \ldots, x_7) \mapsto (x_1, x_2, x_3, -x_4, -x_5, -x_6, -x_7), \qquad (12.1)$$

$$\beta : (x_1, \ldots, x_7) \mapsto (x_1, -x_2, -x_3, x_4, x_5, \tfrac{1}{2} - x_6, -x_7), \qquad (12.2)$$

$$\gamma : (x_1, \ldots, x_7) \mapsto (-x_1, x_2, -x_3, x_4, \tfrac{1}{2} - x_5, x_6, \tfrac{1}{2} - x_7). \qquad (12.3)$$

By inspection α, β and γ preserve φ_0, because of the careful choice of exactly which signs to change. Thus they generate a group $\Gamma = \langle \alpha, \beta, \gamma \rangle$ of isometries of T^7 preserving the flat G_2-structure (φ_0, g_0).

It is easy to verify that $\alpha^2 = \beta^2 = \gamma^2 = 1$, and α, β and γ commute. Thus Γ is isomorphic to $\mathbb{Z}_2^3$. Note that the values $\frac{1}{2}$ in (12.2) and (12.3) are required to make α, β and γ commute. For example, $\beta\alpha$ acts by

$$\beta\alpha : (x_1, \ldots, x_7) \mapsto (x_1, -x_2, -x_3, -x_4, -x_5, \tfrac{1}{2} + x_6, x_7),$$

so that $(\beta\alpha)^2$ acts by

$$(\beta\alpha)^2 : (x_1, \ldots, x_7) \mapsto (x_1, x_2, x_3, x_4, x_5, 1 + x_6, x_7).$$

But $x_6 = 1 + x_6$ as x_j takes values in $\mathbb{R}/\mathbb{Z}$, so $(\beta\alpha)^2$ is the identity, implying that α and β commute. If we had written say $\frac{1}{3} - x_6$ instead of $\frac{1}{2} - x_6$ in (12.2) then α and β would not commute, and $\langle \alpha, \beta, \gamma \rangle$ would be larger, nonabelian, and more complicated to analyse.

In the next two lemmas, we shall describe the singular set S of T^7/Γ.

Lemma 12.2.1 *The elements $\beta\gamma$, $\gamma\alpha$, $\alpha\beta$ and $\alpha\beta\gamma$ of Γ have no fixed points on T^7. The fixed points of α in T^7 are 16 copies of T^3, and the group $\langle \beta, \gamma \rangle$ acts freely on the set of 16 T^3 fixed by α. Similarly, the fixed points of β, γ in T^7 are each 16 copies of T^3, and the groups $\langle \alpha, \gamma \rangle$ and $\langle \alpha, \beta \rangle$ act freely on the sets of 16 T^3 fixed by β, γ respectively.*

Proof Observe that $\beta\gamma$ acts on x_7 by $x_7 \mapsto x_7 - \frac{1}{2}$. Thus $\beta\gamma$ can have no fixed point y, because the x_7-coordinates of y and $\beta\gamma(y)$ are different. Similarly, $\gamma\alpha$ changes x_5 and x_7, $\alpha\beta$ changes x_6, and $\alpha\beta\gamma$ changes x_5. So none of these elements have fixed points.

By inspection, the fixed points of α are $x_4, x_5, x_6, x_7 \in \{\mathbb{Z}, \frac{1}{2} + \mathbb{Z}\}$, which clearly divide into 16 disjoint copies of T^3. The action of β on these 16 T^3 fixes x_4, x_5 and x_7, and takes x_6 to $x_6 + \frac{1}{2}$. The action of γ on the 16 T^3 fixes x_4 and x_6 and takes x_5 to $x_5 + \frac{1}{2}$ and x_7 to $x_7 + \frac{1}{2}$. Therefore the group $\langle \beta, \gamma \rangle$ does act freely on the set of 16 fixed 3-tori of α. The rest of the lemma uses the same argument. □

Lemma 12.2.2 *The singular set S of T^7/Γ is a disjoint union of 12 copies of T^3, and the singularity at each T^3 is locally modelled on $T^3 \times \mathbb{C}^2/\{\pm 1\}$.*

Proof Let S' be the set of points in T^7 fixed by some nonidentity element of Γ. Then S is the image of S' in T^7/Γ. By Lemma 12.2.1, S' is the union of three sets of 16 T^3 fixed by α, β and γ. Suppose that two T^3 fixed by different elements, say α and β, intersect. Then the intersection is fixed by both α and β, so $\alpha\beta$ has fixed points, contradicting Lemma 12.2.1.

Thus S' is a disjoint union of 48 T^3 in T^7, and $S = S'/\Gamma$. As $\langle \beta, \gamma \rangle$ acts freely on the 16 α T^3, these 16 T^3 contribute 4 T^3 to S. Similarly, the 16 β T^3 and 16 γ T^3 each contribute 4 T^3 to S. Therefore S is a disjoint union of 12 T^3. Examining the action of α, β and γ near their fixed sets, we see that the singularity near each T^3 in S is modelled on $T^3 \times \mathbb{C}^2/\{\pm 1\}$. □

Following the method of §12.1.1 we find that T^7/Γ is simply-connected. As in §12.1.2, the Betti numbers $b^k(T^7/\Gamma)$ are the dimensions of the Γ-invariant subspaces of $\Lambda^k(\mathbb{R}^7)^*$. Calculation shows that there are no nonzero Γ-invariant 1-forms or 2-forms, and that the Γ-invariant 3-forms are

$$\langle \mathrm{d}\mathbf{x}_{123}, \mathrm{d}\mathbf{x}_{145}, \mathrm{d}\mathbf{x}_{167}, \mathrm{d}\mathbf{x}_{246}, \mathrm{d}\mathbf{x}_{257}, \mathrm{d}\mathbf{x}_{347}, \mathrm{d}\mathbf{x}_{356} \rangle,$$

where $\mathrm{d}x_{ijk} = \mathrm{d}x_i \wedge \mathrm{d}x_j \wedge \mathrm{d}x_k$. Therefore the Betti numbers of T^7/Γ are $b^1(T^7/\Gamma) = b^2(T^7/\Gamma) = 0$ and $b^3(T^7/\Gamma) = 7$.

Next we shall write down a set of *R-data* for T^7/Γ, as in §11.4. In the notation of §11.3, the elements of $\mathcal{L}$ are the 48 T^3 in the set S' of the previous lemma, together with T^7. So we choose I to be $\{0, 1, \ldots, 48\}$, where 0 is the index of T^7 and $1, \ldots, 48$ the indices of the 48 T^3.

Then $V_0 = \mathbb{R}^7$, $W_0 = \{0\}$ and $A_0 = \{1\}$, and $V_j \cong \mathbb{R}^3$, $W_j \cong \mathbb{C}^2$ and $A_j \cong \{\pm 1\}$ for $j = 1, \ldots, 48$. By Definition 11.4.1, a set of R-data for T^7/Γ is a choice of resolution (Y_i, π_i) for W_i/A_i and a torsion-free product G_2-structure (φ_i, g_i) on $V_i \times Y_i$ for each i, together with isomorphisms $\chi_{\gamma,i} : Y_i \to Y_{\gamma \cdot i}$ for each $\gamma \in \Gamma$ and $i \in I$, satisfying certain conditions.

As W_0/A_0 is a point, Y_0 is just a point. For $i = 1, \ldots, 48$, for the conditions to be satisfied Y_i must be an ALE 4-manifold with holonomy SU(2), asymptotic to $\mathbb{C}^2/\{\pm 1\}$. From §7.2, this implies that Y_i is one of the 3-dimensional family of *Eguchi–Hanson spaces*, described in Example 7.2.2. As a 4-manifold, Y_i is diffeomorphic to the blow-up at 0 of $\mathbb{C}^2/\{\pm 1\}$, and has Betti numbers $b^0 = b^2 = 1$ and $b^1 = b^3 = b^4 = 0$.

We saw in Lemma 12.2.2 that the 48 T^3 in S' are identified in fours by the action of Γ to give the 12 T^3 in S. In the R-data, the maps $\chi_{\gamma,i}$ must identify the corresponding Y_i's in fours. So we are not free to choose the metric on each Y_i independently, but in each set of four Y_i's identified by the $\chi_{\gamma,i}$ action the metrics must be the same. Effectively this means that we can choose one Eguchi–Hanson space Y_i for each of the 12 T^3 in S.

Thus we can choose a set of R-data (Y_i, π_i), (φ_i, g_i) and $\chi_{\gamma,i}$ for T^7/Γ satisfying the required conditions. So by the results of Chapter 11, T^7/Γ has a resolution M which carries torsion-free G_2-structures. Let us work out the topological invariants of M. By Proposition 12.1.3 we see that $\pi_1(M) \cong \pi_1(T^7/\Gamma)$. But T^7/Γ is simply-connected, from above. Thus M is simply-connected, and thus the torsion-free G_2-structures on M have holonomy G_2 by Proposition 10.2.2.

We make M by desingularizing the 12 T^3 in S. Near each T^3, T^7/Γ is modelled on $T^3 \times \mathbb{C}^2/\{\pm 1\}$, and M is modelled on $T^3 \times Y_i$. Effectively we cut out a copy of $T^3 \times \mathbb{C}^2/\{\pm 1\}$ and glue in a copy of $T^3 \times Y_i$. Intuitively we would expect this process to contribute $b^k(T^3 \times Y_i) - b^k(T^3 \times \mathbb{C}^2/\{\pm 1\})$ to $b^k(M)$. This is in fact true, and can be justified by the exact sequence argument in §12.1.2.

By the Künneth formula $b^2(T^3 \times Y_i) = b^3(T^3 \times Y_i) = 4$, and $T^3 \times \mathbb{C}^2/\{\pm 1\}$ retracts onto T^3, giving $b^2(T^3 \times \mathbb{C}^2/\{\pm 1\}) = 3$ and $b^3(T^3 \times \mathbb{C}^2/\{\pm 1\}) = 1$. So each T^3 in S contributes $4 - 3 = 1$ to $b^2(M)$ and $4 - 1 = 3$ to $b^3(M)$. Thus

$$b^2(M) = b^2(T^7/\Gamma) + 12 \cdot 1 = 12 \quad \text{and} \quad b^3(M) = b^3(T^7/\Gamma) + 12 \cdot 3 = 43.$$

Therefore we have constructed a compact, simply-connected 7-manifold M with $b^2(M) = 12$ and $b^3(M) = 43$, which admits metrics with holonomy G_2.

12.3 A modification of the example of §12.2

We now study a variation of the previous example, based on [122, Ex. 4]. In it we first meet a phenomenon that will crop up again and again in what follows: there can be topological choices in desingularizing T^7/Γ, so that one orbifold T^7/Γ gives rise to many 7-manifolds M. Let T^7 and (φ_0, g_0) be as in §12.2, but this time define $\alpha, \beta, \gamma : T^7 \to T^7$ by

$$\alpha : (x_1, \ldots, x_7) \mapsto (x_1, x_2, x_3, -x_4, -x_5, -x_6, -x_7), \tag{12.4}$$

$$\beta : (x_1, \ldots, x_7) \mapsto (x_1, -x_2, -x_3, x_4, x_5, \tfrac{1}{2} - x_6, -x_7), \tag{12.5}$$

$$\gamma : (x_1, \ldots, x_7) \mapsto (-x_1, x_2, -x_3, x_4, -x_5, x_6, \tfrac{1}{2} - x_7). \tag{12.6}$$

The only difference between (12.1)–(12.3) and (12.4)–(12.6) is that (12.6) has $-x_5$ where (12.3) has $\tfrac{1}{2} - x_5$. As before α, β, γ commute, satisfy $\alpha^2 = \beta^2 = \gamma^2 = 1$ and preserve (φ_0, g_0), so $\Gamma = \langle \alpha, \beta, \gamma \rangle$ is a group of automorphisms of T^7 isomorphic to $\mathbb{Z}_2^3$ preserving (φ_0, g_0).

Here are the analogues of Lemmas 12.2.1 and 12.2.2 for this group Γ.

Lemma 12.3.1 *The elements $\beta\gamma$, $\gamma\alpha$, $\alpha\beta$ and $\alpha\beta\gamma$ of Γ have no fixed points on T^7. The fixed points of α, β and γ in T^7 are each 16 copies of T^3, where $\langle \beta, \gamma \rangle$ acts freely on the 16 α T^3, and $\langle \alpha, \gamma \rangle$ acts freely on the 16 β T^3. However, $\langle \alpha, \beta \rangle$ does not act freely on the set of 16 γ T^3, since $\alpha\beta$ acts trivially.*

Lemma 12.3.2 *The singular set S of T^7/Γ is a disjoint union of 8 copies of T^3 and 8 copies of $T^3/\mathbb{Z}_2$. The singularity at each T^3 in S is locally modelled on $T^3 \times \mathbb{C}^2/\{\pm 1\}$. Choose coordinates (x_2, x_4, x_6) on T^3 and (z_1, z_2) on $\mathbb{C}^2$, where $x_j \in \mathbb{R}/\mathbb{Z}$ and $z_j \in \mathbb{C}$. Then the singularity at each $T^3/\mathbb{Z}_2$ in S is locally modelled on $(T^3 \times \mathbb{C}^2/\{\pm 1\})/\langle \alpha\beta \rangle$, where $\alpha\beta$ acts on $T^3 \times \mathbb{C}^2/\{\pm 1\}$ by*

$$\alpha\beta : \big((x_2, x_4, x_6), \pm(z_1, z_2)\big) \mapsto \big((x_2, x_4, \tfrac{1}{2} + x_6), \pm(z_1, -z_2)\big). \tag{12.7}$$

Here is what the lemmas mean. The fixed set of γ in T^7 is

$$\{(x_1, \ldots, x_7) \in T^7 : x_1, x_3, x_5 \in \{\mathbb{Z}, \tfrac{1}{2} + \mathbb{Z}\} \text{ and } x_7 \in \{\tfrac{1}{4} + \mathbb{Z}, \tfrac{3}{4} + \mathbb{Z}\}\},$$

which is 16 copies of T^3, where each T^3 is defined by a choice of values for x_1, x_3, x_5 and x_7, and the coordinates on T^3 are x_2, x_4, x_6 in $\mathbb{R}/\mathbb{Z}$. The corresponding part of the singular set S of T^7/Γ is the quotient of these 16 T^3 by the subgroup $\langle \alpha, \beta \rangle$ in Γ.

In §12.2 we found that $\langle \alpha, \beta \rangle$ acted freely on the set of 16 T^3, identifying them in fours, and contributing 4 T^3 to S. But in this case $\alpha\beta$ takes each T^3 to itself, and thus the quotient of the 16 T^3 by $\langle \alpha, \beta \rangle$ is 8 copies of $T^3/\mathbb{Z}_2$ rather than 4 copies of T^3. So the γ fixed points contribute 8 $T^3/\mathbb{Z}_2$ to S.

A MODIFICATION OF THE EXAMPLE OF §12.2

Let us pick one of the 16 T^3 fixed by γ, say that with $x_1 = x_3 = x_5 = \mathbb{Z}$ and $x_7 = \frac{1}{4} + \mathbb{Z}$, and investigate what T^7/Γ looks like near the image of this T^3. A point $(x_1, \ldots, x_7)$ in T^7 near this T^3 may be written

$$(y_1 + \mathbb{Z}, x_2, y_3 + \mathbb{Z}, x_4, y_5 + \mathbb{Z}, x_6, y_7 + \tfrac{1}{4} + \mathbb{Z}),$$

where $x_2, x_4, x_6 \in \mathbb{R}/\mathbb{Z}$ and $y_1, y_3, y_5, y_7 \in \mathbb{R}$ with y_j *small*. Identify this point with $\big((x_2, x_4, x_6), (z_1, z_2)\big)$ in $T^3 \times \mathbb{C}^2$, where $z_1 = y_1 + iy_7$ and $z_2 = y_3 + iy_5$. Then γ fixes (x_2, x_4, x_6) and takes (z_1, z_2) to $-(z_1, z_2)$, so converting $\mathbb{C}^2$ to $\mathbb{C}^2/\{\pm 1\}$, and $\alpha\beta$ acts as in (12.7). Thus, T^7/Γ looks like $(T^3 \times \mathbb{C}^2/\{\pm 1\})/\langle\alpha\beta\rangle$ near this $T^3/\mathbb{Z}_2$.

Here is how to choose a set of R-data for T^7/Γ, and so construct a compact 7-manifold M with holonomy G_2 by the results of Chapter 11. Each of the 8 T^3 in S must be resolved using an Eguchi–Hanson space Y_i, as in §11.2, which contributes 1 to $b^2(M)$ and 3 to $b^3(M)$. However, there are actually two *topologically distinct* ways of resolving each $T^3/\mathbb{Z}_2$ in S.

Near each $T^3/\mathbb{Z}_2$ the local model for T^7/Γ is $(T^3 \times \mathbb{C}^2/\{\pm 1\})/\langle\alpha\beta\rangle$, and so the local model for M should be $(T^3 \times Y_i)/\langle\alpha\beta\rangle$, where Y_i is an ALE space with holonomy SU(2) asymptotic to $\mathbb{C}^2/\{\pm 1\}$, and $\alpha\beta$ acts on Y_i with action asymptotic to $\pm(z_1, z_2) \mapsto \pm(z_1, -z_2)$, as in (12.7). Here are the two possible choices for Y_i.

(a) Let Y_1 be the blow-up of $\mathbb{C}^2/\{\pm 1\}$ at 0. Then Y_1 is the unique crepant resolution of $\mathbb{C}^2/\{\pm 1\}$ and carries ALE metrics with holonomy SU(2). The exceptional divisor in Y_1 is $\Sigma_1 = \mathbb{CP}^1$ with homogeneous coordinates $[z_1, z_2]$, and its homology class generates $H_2(Y_1, \mathbb{R}) = \mathbb{R}$.

The action of $\alpha\beta$ lifts to Y_1, and on Σ_1 it acts by $[z_1, z_2] \mapsto [z_1, -z_2]$, which preserves the orientation on Σ_1. Hence, the induced map $(\alpha\beta)_*$ on $H_2(Y_1, \mathbb{R})$ is the identity.

(b) The map $\sigma : \mathbb{C}^2/\{\pm 1\} \to \mathbb{C}^3$ given by

$$\sigma : \pm(z_1, z_2) \to (z_1^2 - z_2^2, iz_1^2 + iz_2^2, 2z_1 z_2) \qquad (12.8)$$

identifies $\mathbb{C}^2/\{\pm 1\}$ with $\{(w_1, w_2, w_3) \in \mathbb{C}^3 : w_1^2 + w_2^2 + w_3^2 = 0\}$. Let $\epsilon \in \mathbb{C}$ be small and nonzero, and define Y_2 to be

$$Y_2 = \{(w_1, w_2, w_3) \in \mathbb{C}^3 : w_1^2 + w_2^2 + w_3^2 = \epsilon\}.$$

Then Y_2 is a *smoothing* of $\mathbb{C}^2/\{\pm 1\}$, which is diffeomorphic to Y_1. Let $\epsilon = re^{2i\theta}$ for $r > 0$ and $\theta \in [0, \pi)$, and define

$$\Sigma_2 = \{(e^{i\theta}x_1, e^{i\theta}x_2, e^{i\theta}x_3) : x_j \in \mathbb{R}, \quad x_1^2 + x_2^2 + x_3^2 = r\}.$$

Then Σ_2 is an $\mathcal{S}^2$ in Y_2, and $H_2(Y_2, \mathbb{R}) = \langle[\Sigma_2]\rangle$.

As $\alpha\beta$ takes $\pm(z_1, z_2)$ to $\pm(z_1, -z_2)$ we see from (12.8) that the appropriate action of $\alpha\beta$ on Y_2 is

$$\alpha\beta : (w_1, w_2, w_3) \mapsto (w_1, w_2, -w_3).$$

There is a unique ALE Kähler metric with holonomy SU(2) on Y_2 that is invariant under this action of $\alpha\beta$. Also $\alpha\beta$ preserves Σ_2 but reverses its orientation. Thus

$(\alpha\beta)_*[\Sigma_2] = -[\Sigma_2]$ in $H_2(Y_2, \mathbb{R})$, and the induced map $(\alpha\beta)_*$ on $H_2(Y_2, \mathbb{R})$ is multiplication by -1.

The important idea here is that although Y_1 and Y_2 are diffeomorphic, the actions of $\alpha\beta$ on Y_1 and Y_2 are *not* topologically equivalent, since one acts as 1 and one as -1 on $H_2(Y_j, \mathbb{R})$. So, for each of the 8 $T^3/\mathbb{Z}_2$ in S, we can choose to resolve using either Y_1 as in (a) or Y_2 as in (b). In the R-data notation of §11.3 the action of $\alpha\beta$ on Y_i is $\chi_{\alpha\beta,i}$, so the point is that $\alpha\beta \cdot i = i$ and $\chi_{\alpha\beta,i} : Y_i \to Y_i$ is an automorphism of Y_i, which can be one of two topological possibilities.

Let k be an integer with $0 \leqslant k \leqslant 8$, and consider a 7-manifold M_k made by desingularizing T^7/Γ using Y_1 as in (a) for k of the 8 $T^3/\mathbb{Z}_2$'s in S, and Y_2 as in (b) for the remaining $8-k$ $T^3/\mathbb{Z}_2$'s. Then $\pi_1(M_k) \cong \pi_1(T^7/\Gamma)$ by Proposition 12.1.3. But $\pi_1(T^7/\Gamma) = \{1\}$ as in §12.2, so M_k is simply-connected.

To compute the Betti numbers of M_k, we begin with $b^2(T^7/\Gamma) = 0$ and $b^3(T^7/\Gamma) = 7$, as in §12.2. Each of the 8 T^3 in S adds 1 to b^2 and 3 to b^3. It can also be shown that each of the k $T^3/\mathbb{Z}_2$ resolved using part (a) above adds 1 to b^2 and 1 to b^3, and each of the $8-k$ $T^3/\mathbb{Z}_2$ resolved using part (b) above adds 0 to b^2 and 2 to b^3. Therefore the Betti numbers of M_k are

$$b^2(M_k) = 8+k, \quad b^3(M_k) = 47-k, \quad k = 0, 1, \ldots, 8.$$

In this example there are $2^8 = 256$ possible topological choices in desingularizing T^7/Γ to get a compact 7-manifold with holonomy G_2. Not all of these 256 7-manifolds will be topologically distinct, but as they realize 9 sets of Betti numbers we have found at least 9 topologically distinct compact, simply-connected 7-manifolds with holonomy G_2.

12.4 Examples with nontrivial fundamental group

By Proposition 10.2.2, if M is compact with holonomy G_2 then $\pi_1(M)$ is finite. So the universal cover $\tilde{M}$ of M is a compact, simply-connected 7-manifold with holonomy G_2. Thus every 7-manifold M with holonomy G_2 and $\pi_1(M) = A$ is of the form $\tilde{M}/A$, where $(\tilde{M}, \varphi, g)$ is a compact, simply-connected G_2-manifold with holonomy G_2, and A a finite group acting freely on $\tilde{M}$ preserving (φ, g).

We shall use this to construct examples of compact 7-manifolds M with holonomy G_2 and nontrivial fundamental group. Let (φ_0, g_0) be a flat G_2-structure on T^7, and Γ a finite group of automorphisms of T^7 preserving (φ_0, g_0), such that T^7/Γ is simply-connected. Let A be a finite group which acts *freely* on T^7/Γ preserving (φ_0, g_0). Then $(T^7/\Gamma)/A$ has fundamental group A, and there is a finite group $\tilde{\Gamma} = \Gamma \rtimes A$ of automorphisms of T^7 preserving (φ_0, g_0), such that $T^7/\tilde{\Gamma} = (T^7/\Gamma)/A$.

Our goal is to resolve $T^7/\tilde{\Gamma}$ to get a compact 7-manifold M with holonomy G_2 and fundamental group A. We shall describe a family of examples modelled on [122, Ex. 5], in which Γ is the group defined in §12.3, and A is a finite group of translations of T^7 which commute with Γ, so that $\tilde{\Gamma} = \Gamma \times A$. Let Γ be the group of §12.3, and define translations $\sigma_1, \sigma_2, \sigma_3$ on T^7 by

$$\sigma_1((x_1,\ldots,x_7)) = (x_1, x_2, \tfrac{1}{2}+x_3, \tfrac{1}{2}+x_4, \tfrac{1}{2}+x_5, x_6, x_7),$$
$$\sigma_2((x_1,\ldots,x_7)) = (x_1, \tfrac{1}{2}+x_2, x_3, \tfrac{1}{2}+x_4, x_5, x_6, x_7),$$
$$\sigma_3((x_1,\ldots,x_7)) = (\tfrac{1}{2}+x_1, x_2, x_3, x_4, \tfrac{1}{2}+x_5, \tfrac{1}{2}+x_6, x_7).$$

Then σ_1, σ_2 and σ_3 commute with Γ. We claim that $\langle\sigma_1,\sigma_2,\sigma_3\rangle$ acts freely on T^7/Γ. To prove this claim, it is enough to show that the only nonidentity elements of $\Gamma \times \langle\sigma_1,\sigma_2,\sigma_3\rangle$ that have fixed points in T^7 are α, β and γ. For instance, $\alpha\sigma_1$ has no fixed points on T^7 because it acts on x_3 as $x_3 \mapsto \tfrac{1}{2}+x_3$. Reasoning in this way, the claim is easily proved.

Table 12.1 *Fundamental group and Betti numbers for the examples of §12.4*

Group A	Singular set of $T^7/\tilde{\Gamma}$	$\pi_1(M)$	$b^2(M)$	$b^3(M)$	Range of k
$\{1\}$	$8T^3$ and $8T^3/\mathbb{Z}_2$	$\{1\}$	$8+k$	$47-k$	$k=0,\ldots,8$
$\langle\sigma_2\rangle$	$4T^3$ and $8T^3/\mathbb{Z}_2$	$\mathbb{Z}_2$	$4+k$	$35-k$	$k=0,\ldots,8$
$\langle\sigma_3\rangle$	$2T^3$ and $8T^3/\mathbb{Z}_2$	$\mathbb{Z}_2$	$2+k$	$29-k$	$k=0,\ldots,8$
$\langle\sigma_2,\sigma_3\rangle$	T^3 and $6T^3/\mathbb{Z}_2$	$\mathbb{Z}_2^2$	$1+k$	$22-k$	$k=0,\ldots,6$
$\langle\sigma_1\sigma_2,\sigma_3\rangle$	$6T^3/\mathbb{Z}_2$	$\mathbb{Z}_2^2$	k	$19-k$	$k=0,\ldots,6$
$\langle\sigma_1,\sigma_2,\sigma_3\rangle$	$4T^3/\mathbb{Z}_2$	$\mathbb{Z}_2^3$	k	$15-k$	$k=0,\ldots,4$

Let A be a subgroup of $\langle\sigma_1,\sigma_2,\sigma_3\rangle$, and define $\tilde{\Gamma} = \Gamma \times A$. Then $T^7/\tilde{\Gamma}$ is a singular 7-manifold with fundamental group A. It can be shown that the singular set S of $T^7/\tilde{\Gamma}$ breaks up into disjoint components of the form T^3 and $T^3/\mathbb{Z}_2$, as §12.3. Therefore we may desingularize $T^7/\tilde{\Gamma}$ as in §12.3, to get a compact 7-manifold M admitting metrics with holonomy G_2, with $\pi_1(M) = A$. Each of the $T^3/\mathbb{Z}_2$ components in the singular set may be resolved in two distinct ways, as in parts (a) and (b) of §12.3, giving a number of different topological types for M.

We present in Table 12.1 the form of the singular set of $T^7/\tilde{\Gamma}$ for six choices of the group A, and $\pi_1(M)$, $b^2(M)$ and $b^3(M)$ for the resulting 7-manifold M. As in §12.3 these Betti numbers depend on an integer k, which is the number of $T^3/\mathbb{Z}_2$ components of S that are resolved using part (a) of §12.3. The range of k is also given in the table. The calculations to determine the singular set of $T^7/\tilde{\Gamma}$ are elementary but long. When $A = \{1\}$ we retrieve the case of §12.3. The other cases yield at least 37 distinct non-simply-connected 7-manifolds admitting metrics with holonomy G_2.

12.5 A more complicated example

We now consider an orbifold T^7/Γ very similar to those in §12.2 and §12.3, but with a more complicated singular set that must be resolved using QALE manifolds with holonomy SU(3). It turns out that this leads to a striking increase both in the complexity of resolving T^7/Γ, and in the number of topologically distinct 7-manifolds that result—just from this one example we can find compact 7-manifolds with at least 120 different sets of Betti numbers, and the number of distinct 7-manifolds may be many more than this.

Let T^7 and (φ_0, g_0) be as in §12.2, and define $\alpha, \beta, \gamma : T^7 \to T^7$ by

$$\alpha : (x_1, \ldots, x_7) \mapsto (x_1, x_2, x_3, -x_4, -x_5, -x_6, -x_7), \tag{12.9}$$

$$\beta : (x_1, \ldots, x_7) \mapsto (x_1, -x_2, -x_3, x_4, x_5, -x_6, -x_7), \tag{12.10}$$

$$\gamma : (x_1, \ldots, x_7) \mapsto (-x_1, x_2, \tfrac{1}{2} - x_3, x_4, -x_5, x_6, -x_7). \tag{12.11}$$

These differ from (12.1)–(12.3) and (12.4)–(12.6) only in where $\frac{1}{2}$ appears. As before $\Gamma = \langle \alpha, \beta, \gamma \rangle$ is a group isomorphic to $\mathbb{Z}_2^3$, which acts on T^7 preserving (φ_0, g_0). Here are the analogues of Lemmas 12.2.1 and 12.2.2 for this group Γ.

Lemma 12.5.1 *The elements $\beta\gamma$ and $\alpha\beta\gamma$ of Γ have no fixed points on T^7. The fixed points of $\alpha, \beta, \alpha\beta, \gamma$ and $\alpha\gamma$ in T^7 are each 16 copies of T^3. Moreover $\langle \beta, \gamma \rangle$ acts trivially on the set of 16 T^3 fixed by α, and α acts trivially on the sets of 16 T^3 fixed by $\beta, \alpha\beta, \gamma$ and $\alpha\gamma$. The fixed points of α, β and $\alpha\beta$ intersect in 64 S^1 in T^7, which is the fixed point set of $\langle \alpha, \beta \rangle$. Similarly, the fixed points of α, γ and $\alpha\gamma$ intersect in 64 S^1 in T^7, the fixed point set of $\langle \alpha, \gamma \rangle$.*

Lemma 12.5.2 *The singular set S of T^7/Γ is the union of*

(i) *16 $T^3/\mathbb{Z}_2^2$ from the α fixed points,*

(ii) *8 $T^3/\mathbb{Z}_2$ from the β fixed points,*

(iii) *8 $T^3/\mathbb{Z}_2$ from the $\alpha\beta$ fixed points,*

(iv) *8 $T^3/\mathbb{Z}_2$ from the γ fixed points,*

(v) *8 $T^3/\mathbb{Z}_2$ from the $\alpha\gamma$ fixed points.*

This union is not disjoint. Instead, the sets (i), (ii) and (iii) intersect in 32 S^1 in T^7/Γ from the fixed points of $\langle \alpha, \beta \rangle$, and the sets (i), (iv) and (v) intersect in 32 S^1 in T^7/Γ from the fixed points of $\langle \alpha, \gamma \rangle$.

The fixed points of $\langle \alpha, \beta \rangle$ are $\{(x_1, \ldots, x_7) \in T^7 : x_2, \ldots, x_7 \in \{\mathbb{Z}, \tfrac{1}{2} + \mathbb{Z}\}\}$. These are 64 copies of S^1 in T^7. As γ acts freely upon them, their image in T^7/Γ is 32 S^1. We shall describe the singularities of T^7/Γ close to one of these S^1's, say that with $x_2 = \cdots = x_7 = \mathbb{Z}$. Near this S^1, identify T^7 with $S^1 \times \mathbb{C}^3$ by equating

$$(x_1, y_2 + \mathbb{Z}, \ldots, y_7 + \mathbb{Z}) \quad \text{with} \quad \big(x_1, (y_2 + iy_3, y_4 + iy_5, y_6 + iy_7)\big),$$

where $x_1 \in \mathbb{R}/\mathbb{Z}$, and $y_2, \ldots, y_7 \in \mathbb{R}$ are *small*.

Then by (12.9) and (12.10) the actions of α and β on $S^1 \times \mathbb{C}^3$ are

$$\alpha : \big(x_1, (z_1, z_2, z_3)\big) \mapsto \big(x_1, (z_1, -z_2, -z_3)\big),$$

$$\beta : \big(x_1, (z_1, z_2, z_3)\big) \mapsto \big(x_1, (-z_1, z_2, -z_3)\big).$$

Thus T^7/Γ is locally isomorphic to $S^1 \times \mathbb{C}^3/\mathbb{Z}_2^2$ near the image S^1. Now we have already studied this complex singularity $\mathbb{C}^3/\mathbb{Z}_2^2$ in Chapter 9, in Examples 9.3.6 and 9.9.9. We found that $\mathbb{C}^3/\mathbb{Z}_2^2$ has a number of topologically distinct resolutions and deformations, labelled (i)–(ix) in Example 9.9.9, which all carry QALE Kähler metrics with holonomy SU(3).

Our basic plan is that for each of the 32 $\mathcal{S}^1$ in T^7/Γ coming from the $\langle \alpha, \beta \rangle$ fixed points, we will choose one of cases (i)–(ix) in Example 9.9.9 to be the corresponding resolution Y_i in the R-data for T^7/Γ. Similarly, near each of the 32 $\mathcal{S}^1$ in T^7/Γ coming from the $\langle \alpha, \gamma \rangle$ fixed points, T^7/Γ is also locally isomorphic to $\mathcal{S}^1 \times \mathbb{C}^3/\mathbb{Z}_2^2$, and we again resolve using one of (i)–(ix) in Example 9.9.9.

This gives us a huge number of choices for the R-data of T^7/Γ. But these choices are also subject to many consistency conditions, which rule out almost all the possibilities, but still leave a very large number. Then for each of the allowed possibilities for the R-data, we want to calculate the fundamental group and Betti numbers of the associated resolution M of T^7/Γ. This is a task too big to attempt without the aid of a powerful computer.

So instead we will simplify the problem by splitting it up into two stages, and limiting the kinds of resolution we consider. In the first stage we resolve $T^7/\langle \alpha \rangle$ to get $T^3 \times K3$, such that the action of $\langle \beta, \gamma \rangle$ on $T^7/\langle \alpha \rangle$ lifts to $T^3 \times K3$. Then $(T^3 \times K3)/\langle \beta, \gamma \rangle$ is an orbifold. In the second stage we resolve $(T^3 \times K3)/\langle \beta, \gamma \rangle$ to get M. It turns out there are 3^{16} topologically distinct ways of doing the first stage of this process, and all of them can be carried through to the second stage.

Note that this idea of first resolving the α singularities and then resolving the $\langle \beta, \gamma \rangle$ singularities is just a way of simplifying the choice of R-data for T^7/Γ, and understanding the topology of the resulting resolution M. We do the actual construction of metrics with holonomy G_2 on M in one stage, and not two. If we were to do it in two stages as above then we would need analytic results on how to resolve $(T^3 \times K3)/\langle \beta, \gamma \rangle$, which we have not included in this book.

12.5.1 *Resolving $T^7/\langle \alpha \rangle$ in a $\langle \beta, \gamma \rangle$-equivariant way*

Consider one of the 16 T^3 fixed by α, say $x_4 = \cdots = x_7 = \mathbb{Z}$. Near this T^3 we can identify T^7 with $T^3 \times \mathbb{C}^2$ by equating

$$(x_1, x_2, x_3, y_4 + \mathbb{Z}, \ldots, y_7 + \mathbb{Z}) \quad \text{with} \quad \big((x_1, x_2, x_3), (y_4 + iy_5, y_6 + iy_7)\big),$$

for $x_j \in \mathbb{R}/\mathbb{Z}$ and small $y_j \in \mathbb{R}$. By (12.9)–(12.11), α, β, γ act on $T^3 \times \mathbb{C}^2$ by

$$\alpha : \big((x_1, x_2, x_3), (z_1, z_2)\big) \mapsto \big((x_1, x_2, x_3), (-z_1, -z_2)\big),$$
$$\beta : \big((x_1, x_2, x_3), (z_1, z_2)\big) \mapsto \big((x_1, -x_2, -x_3), (z_1, -z_2)\big),$$
$$\gamma : \big((x_1, x_2, x_3), (z_1, z_2)\big) \mapsto \big((-x_1, x_2, \tfrac{1}{2} - x_3), (\bar{z}_1, \bar{z}_2)\big).$$

To resolve the singularities at this T^3 in $T^7/\langle \alpha \rangle$ we need a 4-manifold Y_i with holonomy SU(2) asymptotic to $\mathbb{C}^2/\{\pm 1\}$, and the resolution will be modelled on $T^3 \times Y_i$. To lift the action of $\langle \beta, \gamma \rangle$ from $T^3 \times \mathbb{C}^2/\{\pm 1\}$ to $T^3 \times Y_i$, we need an action of $\langle \beta, \gamma \rangle$ on Y_i such that β, γ are asymptotic to

$$\beta : \pm(z_1, z_2) \mapsto \pm(z_1, -z_2) \quad \text{and} \quad \gamma : \pm(z_1, z_2) \mapsto \pm(\bar{z}_1, \bar{z}_2) \quad (12.12)$$

on $\mathbb{C}^2/\{\pm 1\}$, by the equations above. An investigation similar to (a) and (b) in §12.3 shows that there are in fact three topologically distinct possibilities for Y_i, which are as follows.

(a) Let Y_1 be the blow-up of $\mathbb{C}^2/\{\pm 1\}$ at 0. The exceptional divisor in Y_1 is $\Sigma_1 = \mathbb{CP}^1$ with homogeneous coordinates $[z_1, z_2]$, and its homology class generates $H_2(Y_1, \mathbb{R}) = \mathbb{R}$. The actions of β and γ lift to Y_1, and on Σ_1 they act by $\beta : [z_1, z_2] \mapsto [z_1, -z_2]$ (which is orientation-preserving) and $\gamma : [z_1, z_2] \mapsto [\bar{z}_1, \bar{z}_2]$ (which is orientation-reversing). Hence, β_* is the identity on $H_2(Y_1, \mathbb{R})$ and γ_* multiplies by -1 on $H_2(Y_1, \mathbb{R})$.

(b) Let $\epsilon > 0$ be real and positive, and define Y_2 by

$$Y_2 = \{(w_1, w_2, w_3) \in \mathbb{C}^3 : w_1^2 + w_2^2 + w_3^2 = \epsilon\}$$

as in part (b) of §12.3. Then Y_2 is a smoothing of $\mathbb{C}^2/\{\pm 1\}$, diffeomorphic to Y_1. Define

$$\Sigma_2 = \{(x_1, x_2, x_3) : x_j \in \mathbb{R}, \quad x_1^2 + x_2^2 + x_3^2 = \epsilon\}.$$

Then Σ_2 is an S^2 in Y_2, and $H_2(Y_2, \mathbb{R}) = \langle[\Sigma_2]\rangle$.

We see from (12.8) and (12.12) that the actions of β and γ on Y_2 are

$$\beta : (w_1, w_2, w_3) \mapsto (w_1, w_2, -w_3), \quad \gamma : (w_1, w_2, w_3) \mapsto (\bar{w}_1, -\bar{w}_2, \bar{w}_3).$$

There is a unique ALE Kähler metric with holonomy SU(2) on Y_2 that is invariant under these actions of β and γ. Also β and γ both preserve Σ_2 but reverse its orientation. Thus β_* and γ_* both multiply by -1 on $H_2(Y_2, \mathbb{R})$.

(c) Let $\epsilon < 0$ be real and negative, and define Y_3 by

$$Y_3 = \{(w_1, w_2, w_3) \in \mathbb{C}^3 : w_1^2 + w_2^2 + w_3^2 = \epsilon\}.$$

Again, Y_3 is a smoothing of $\mathbb{C}^2/\{\pm 1\}$, and diffeomorphic to Y_1. Define

$$\Sigma_3 = \{(ix_1, ix_2, ix_3) : x_j \in \mathbb{R}, \quad x_1^2 + x_2^2 + x_3^2 = -\epsilon\}.$$

Then Σ_3 is an S^2 in Y_3, and $H_2(Y_3, \mathbb{R}) = \langle[\Sigma_3]\rangle$.

The actions of β and γ on Y_3 are the same as in part (b). There is a unique ALE Kähler metric with holonomy SU(2) on Y_3 that is invariant under these actions of β and γ. Also β, γ preserve Σ_3, but β reverses its orientation while γ preserves it. Thus β_* multiplies by -1 on $H_2(Y_3, \mathbb{R})$, and γ_* is the identity on $H_2(Y_3, \mathbb{R})$.

As a shorthand we shall write $\beta_* = 1$ on $H_2(Y_i, \mathbb{R})$ if β_* is the identity on $H_2(Y_i, \mathbb{R})$, and $\beta_* = -1$ if β_* multiplies by -1 on $H_2(Y_i, \mathbb{R})$, and similarly for γ_*. Possibilities (a)–(c) are topologically distinct because in (a) $\beta_* = 1$ and $\gamma_* = -1$, in (b) $\beta_* = \gamma_* = -1$, and in (c) $\beta_* = -1$ and $\gamma_* = 1$ on $H_2(Y_i, \mathbb{R})$. Thus in each case one of β_*, γ_* and $(\beta\gamma)_*$ are 1, and two are -1.

To resolve $T^7/\langle\alpha\rangle$ in a $\langle\beta, \gamma\rangle$-equivariant way, we must choose one of (a)–(c) for each of the 16 T^3 in T^7 fixed by α. As $T^7/\langle\alpha\rangle$ is $T^3 \times T^4/\{\pm 1\}$, and the resolution of $T^4/\{\pm 1\}$ is the $K3$ surface, the resolution of $T^7/\langle\alpha\rangle$ is $T^3 \times K3$. Choose one of (a)–(c) for each of the 16 α T^3 in T^7. This choice gives us a resolution $T^3 \times K3$ of $T^7/\langle\alpha\rangle$

and an action of $\langle \beta, \gamma \rangle$ on $T^3 \times K3$, so that $(T^3 \times K3)/\langle \beta, \gamma \rangle$ is an orbifold. We shall describe the singularities of $(T^3 \times K3)/\langle \beta, \gamma \rangle$, and how to resolve it with holonomy G_2.

For each of the 16 α T^3 we know that $\beta_* = \pm 1$ and $\gamma_* = \pm 1$ on the corresponding $H_2(Y_i, \mathbb{R})$. Let $\beta_* = 1$ for k of the 16 α T^3, and $\gamma_* = 1$ for l of the 16 α T^3. Then k of the 16 α T^3 are resolved using (a) above, and l of the 16 β T^3 using (c) above. Hence $16 - k - l$ are resolved using (b) above. Thus k, l satisfy $0 \leqslant k \leqslant 16$, $0 \leqslant l \leqslant 16$ and $k + l \leqslant 16$.

Let Σ_β be the subset of $K3$ fixed by β, and Σ_γ the subset of $K3$ fixed by γ. Then $\Sigma_\beta, \Sigma_\gamma$ are compact real submanifolds of $K3$ of dimension 2, which are oriented but not necessarily connected. Calculation shows that the Euler characteristics of $\Sigma_\beta, \Sigma_\gamma$ are given by

$$\chi(\Sigma_\beta) = 2b^0(\Sigma_\beta) - b^1(\Sigma_\beta) = 2k - 16$$
$$\text{and} \quad \chi(\Sigma_\gamma) = 2b^0(\Sigma_\gamma) - b^1(\Sigma_\gamma) = 2l - 16. \tag{12.13}$$

The fixed points of β in T^3 are $\{(x_1, x_2, x_3) \in T^3 : x_2, x_2 \in \{\mathbb{Z}, \frac{1}{2} + \mathbb{Z}\}\}$, which is 4 copies of $\mathcal{S}^1$. Hence, the fixed points of β in $T^3 \times K3$ are 4 copies of $\mathcal{S}^1 \times \Sigma_\beta$. Now γ acts freely on the 4 $\mathcal{S}^1$ and preserves Σ_β, so the image of the fixed points of β in $(T^3 \times K3)/\langle \beta, \gamma \rangle$ is 2 copies of $\mathcal{S}^1 \times \Sigma_\beta$, which is part of the singular set of $(T^3 \times K3)/\langle \beta, \gamma \rangle$.

Similarly, the fixed points of γ in $T^3 \times K3$ are 4 copies of $\mathcal{S}^1 \times \Sigma_\gamma$, and their image in $(T^3 \times K3)/\langle \beta, \gamma \rangle$ is 2 copies of $\mathcal{S}^1 \times \Sigma_\gamma$, which are also part of the singular set. As $\beta\gamma$ acts freely on T^3 it has no fixed points in $T^3 \times K3$, and this also means that the fixed points of β and γ do not intersect. Thus we have proved:

Lemma 12.5.3 *The singular set S of $(T^3 \times K3)/\langle \beta, \gamma \rangle$ is the disjoint union of two copies of $\mathcal{S}^1 \times \Sigma_\beta$ and two copies of $\mathcal{S}^1 \times \Sigma_\gamma$. At each point of S the singularity is locally modelled on $\mathbb{R}^3 \times \mathbb{R}^4/\{\pm 1\}$.*

We wish to resolve $(T^3 \times K3)/\langle \beta, \gamma \rangle$ to get a compact 7-manifold M with holonomy G_2. In general there will be several ways of doing this, but to simplify matters we will study only two. For the β singularities, the first way can be roughly described as taking a 'crepant resolution' of the $\mathcal{S}^1 \times \Sigma_\beta$ singularities, and the second way as fixing $k = 0$ and taking a 'smoothing' of the $\mathcal{S}^1 \times \Sigma_\beta$ singularities. We will use these to construct 3 families of resolutions M of T^7/Γ with holonomy G_2, in §12.5.3–§12.5.5. But first, we explain how to find $b^0(\Sigma_\beta)$ and $b^0(\Sigma_\gamma)$.

12.5.2 *How to work out $b^0(\Sigma_\beta)$ and $b^0(\Sigma_\gamma)$*

Later we will find that we need to know $b^0(\Sigma_\beta)$ and $b^0(\Sigma_\gamma)$, in order to compute the Betti numbers $b^2(M), b^3(M)$ of the resolutions M of $(T^3 \times K3)/\langle \beta, \gamma \rangle$ that we construct. Here is a method for calculating $b^0(\Sigma_\beta)$ by drawing diagrams. First draw a grid of 4 vertical and 4 horizontal lines, intersecting in 16 points. Label the 4 vertical lines with the possible values of the variables x_4, x_5 in the set $\{\mathbb{Z}, \frac{1}{2} + \mathbb{Z}\}$. Similarly, label the 4 horizontal lines with the possible values of the variables x_6, x_7 in the set $\{\mathbb{Z}, \frac{1}{2} + \mathbb{Z}\}$.

Thus each of the 16 vertices of the grid corresponds to a set of values of the variables x_4, x_5, x_6, x_7 in the set $\{\mathbb{Z}, \frac{1}{2} + \mathbb{Z}\}$. But the 16 T^3 fixed by α in T^7 are defined by x_4, x_5, x_6, x_7 taking constant values in $\{\mathbb{Z}, \frac{1}{2} + \mathbb{Z}\}$. This gives a 1-1 correspondence between the 16 vertices of the grid and the 16 α T^3.

Next, at each of the vertices of the grid draw either an empty circle 'o' if $\beta_* = 1$ for the corresponding α T^3, or a solid circle '•' if $\beta_* = -1$ for the corresponding α T^3. As $\beta_* = 1$ for k of the 16 α T^3, there will be k empty circles 'o' and $16 - k$ solid circles '•'. An example of what your diagram should now look like is given in Figure 12.1.

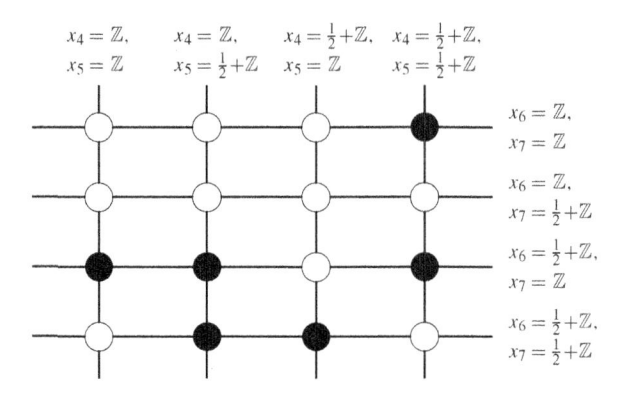

Figure 12.1. Diagrammatic notation for finding $b^0(\Sigma_\beta)$

We then work out $b^0(\Sigma_\beta)$ from the diagram in the following way. When two lines cross at a solid circle '•', we consider that they are joined together at that point. But when they cross at an empty circle 'o' we consider that they are not joined together. With this convention, $b^0(\Sigma_\beta)$ is the *number of connected components* in the diagram. Once you have the hang of this idea, you can count the components by just looking at the picture; there is no need for any calculation.

The example in Figure 12.1 has $k = 10$ and $b^0(\Sigma_\beta) = 2$. That is, there are 10 empty circles 'o' and two connected components in the diagram. One component is the second horizontal line from the top, and the other component is the union of the other 3 horizontal lines and the 4 vertical lines. Another six examples are given in Figure 12.2.

Here is why this method works. The $K3$ surface is the resolution of $T^4/\langle\alpha\rangle$, where T^4 has coordinates (x_4, x_5, x_6, x_7). Now β acts on $T^4/\langle\alpha\rangle$, and its fixed points are the union of 8 copies of $T^2/\{\pm 1\}$, of which 4 copies are defined by x_4, x_5 taking values in $\{0, \frac{1}{2}\}$, and 4 copies by x_6, x_7 taking values in $\{0, \frac{1}{2}\}$. Each of the x_4, x_5 copies of $T^2/\{\pm 1\}$ intersects each of the x_6, x_7 copies of $T^2/\{\pm 1\}$ in one point, which is one of the 16 singular points of $T^4/\langle\alpha\rangle$.

When we resolve $T^4/\langle\alpha\rangle$ to get $K3$, the fixed points Σ_β of β in $K3$ are derived from the fixed points of β in $T^4/\langle\alpha\rangle$ in the following way. When we resolve a singular point of $T^4/\langle\alpha\rangle$ with $\beta_* = 1$, the two copies of $T^4/\{\pm 1\}$ meeting at that point in $T^4/\langle\alpha\rangle$ are

A MORE COMPLICATED EXAMPLE   321

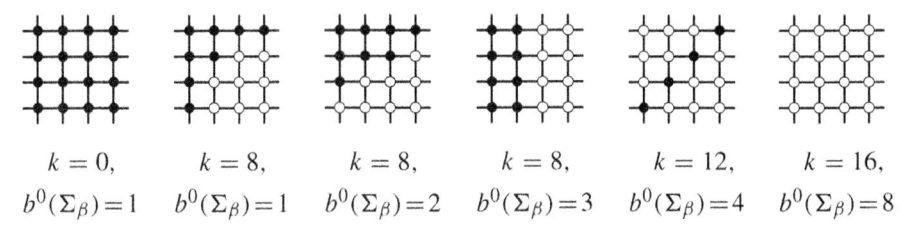

| $k=0$, | $k=8$, | $k=8$, | $k=8$, | $k=12$, | $k=16$, |
| $b^0(\Sigma_\beta)=1$ | $b^0(\Sigma_\beta)=1$ | $b^0(\Sigma_\beta)=2$ | $b^0(\Sigma_\beta)=3$ | $b^0(\Sigma_\beta)=4$ | $b^0(\Sigma_\beta)=8$ |

Figure 12.2. Examples of calculations of $b^0(\Sigma_\beta)$

separated in $K3$. But when we resolve with $\beta_* = -1$, these two copies of $T^4/\{\pm 1\}$ are joined together in $K3$. So our diagrams are just a way of visualizing the fixed points of β in $K3$.

To calculate $b^0(\Sigma_\gamma)$ we follow the same method, except that the 4 vertical lines should be labelled with the possible values of the variables x_4, x_6 in the set $\{\mathbb{Z}, \frac{1}{2} + \mathbb{Z}\}$ and the 4 horizontal lines with the possible values of the variables x_5, x_7 in the set $\{\mathbb{Z}, \frac{1}{2} + \mathbb{Z}\}$, and we label vertices 'o' or '•' according to whether $\gamma_* = 1$ or $\gamma_* = -1$.

12.5.3 A family of 7-manifolds resolving T^7/Γ

Let $(T^3 \times K3)/\langle \beta, \gamma \rangle$ be as above, and consider the β singularities $2 S^1 \times \Sigma_\beta$. Near these singularities we can identify $(T^3 \times K3)/\langle \beta, \gamma \rangle$ with $S^1 \times (T^2 \times K3)/\langle \beta \rangle$, where $(T^2 \times K3)/\langle \beta \rangle$ is a singular *complex* 3-fold with singularities locally modelled on $\mathbb{C} \times \mathbb{C}^2/\{\pm 1\}$. This complex 3-fold has a unique crepant resolution X, obtained by blowing up the singular set. We can use $S^1 \times X$ as a local model for how to resolve the $2 S^1 \times \Sigma_\beta$ singularities in $(T^3 \times K3)/\langle \beta, \gamma \rangle$.

Similarly, the γ singularities $2 S^1 \times \Sigma_\gamma$ are also locally modelled on $S^1 \times (T^2 \times K3)/\langle \gamma \rangle$, but using a different splitting $T^3 \times K3 \cong S^1 \times T^2 \times K3$ and a different complex structure on $T^2 \times K3$, and crepant resolution of the complex singularities of $(T^2 \times K3)/\langle \gamma \rangle$ gives a local model for how to resolve the $2 S^1 \times \Sigma_\gamma$ singularities. Using these two local models we can resolve $(T^3 \times K3)/\langle \beta, \gamma \rangle$ to get a compact 7-manifold M, which is also a resolution of T^7/Γ.

Theorem 12.5.4 *The method of resolving T^7/Γ described above gives a compact 7-manifold M, which is simply-connected and has Betti numbers*

$$b^2(M) = 2\bigl(b^0(\Sigma_\beta) + b^0(\Sigma_\gamma)\bigr)$$
$$\text{and}\quad b^3(M) = 87 - 4(k+l) + 6\bigl(b^0(\Sigma_\beta) + b^0(\Sigma_\gamma)\bigr). \tag{12.14}$$

There exists a set of R-data for T^7/Γ yielding this 7-manifold M as the corresponding resolution of T^7/Γ. Thus M admits metrics with holonomy G_2, by the results of Chapter 11.

Proof As in §12.2 T^7/Γ is simply-connected, and resolving to $T^3 \times K3/\langle \beta, \gamma \rangle$ and then to M does not change the fundamental group by the argument of Proposition 12.1.3. Thus M is simply-connected. Initially T^7/Γ has $b^2 = 0$ and $b^3 = 7$, as in

§12.2. Resolving the α singularities to get $(T^3 \times K3)/\langle \beta, \gamma \rangle$ fixes b^2 and adds 1 to b^3 for each of the 16 α T^3, so $(T^3 \times K3)/\langle \beta, \gamma \rangle$ has $b^2 = 0$ and $b^3 = 23$.

It can be shown that resolving the 2 $\mathcal{S}^1 \times \Sigma_\beta$ singularities as above adds $2b^0(\Sigma_\beta)$ to b^2 and $2b^0(\Sigma_\beta) + 2b^1(\Sigma_\beta)$ to b^3. Similarly, resolving the 2 $\mathcal{S}^1 \times \Sigma_\gamma$ singularities adds $2b^0(\Sigma_\gamma)$ to b^2 and $2b^0(\Sigma_\gamma) + 2b^1(\Sigma_\gamma)$ to b^3. Therefore we have

$$b^2(M) = 2\bigl(b^0(\Sigma_\beta) + b^0(\Sigma_\gamma)\bigr)$$
$$\text{and} \quad b^3(M) = 23 + 2\bigl(b^0(\Sigma_\beta) + b^1(\Sigma_\beta) + b^0(\Sigma_\gamma) + b^1(\Sigma_\gamma)\bigr).$$

Using (12.13) to eliminate $b^1(\Sigma_\beta)$ and $b^1(\Sigma_\gamma)$ we get (12.14).

One can prove that there is a set of R-data for T^7/Γ yielding this 7-manifold M as its resolution by following the construction of M above, and making appropriate R-data choices at each stage. The R-data includes QALE manifolds with holonomy SU(3) for each $\mathcal{S}^1$ fixed by $\langle \alpha, \beta \rangle$ and $\langle \alpha, \gamma \rangle$, and we use the results of Chapter 9 to show that these exist.

As a suitable choice of R-data exists, M admits torsion-free G_2-structures by Theorem 11.6.2. But from above M is simply-connected, so Proposition 10.2.2 shows that these structures have holonomy G_2. This completes the proof. $\square$

Above we made $T^3 \times K3$ with its $\langle \beta, \gamma \rangle$ action by choosing one of (a)–(c) for each of the 16 α T^3. There are 3^{16} ways of making these choices, and the theorem shows that every one of these 3^{16} possibilities leads to a resolution M of T^7/Γ with holonomy G_2. Thus we have:

Corollary 12.5.5 *There are at least* $3^{16} = 43\,046\,721$ *topologically distinct ways* (M, π) *of resolving the orbifold* T^7/Γ *to get a compact 7-manifold M with holonomy* G_2.

Here we mean that the pairs (M, π) are topologically distinct, where $\pi : M \to T^7/\Gamma$ is the resolution map. Equivalently, there are at least 3^{16} topologically distinct ways to choose R-data for T^7/Γ. We do *not* mean that all these 7-manifolds M are topologically distinct, and probably there are a large number of diffeomorphisms between them, making the number of distinct smooth 7-manifolds M arising from this construction rather smaller than 3^{16}.

Nonetheless, it seems surprising that there are *so many* different ways of resolving T^7/Γ. One lesson to be learnt from this is that for orbifolds T^7/Γ with complicated singular sets, the business of resolving T^7/Γ can be very complex with many possible choices and many conditions on these choices. Often it is not really feasible to classify the resolutions using only pencil and paper; one would need a powerful computer to do it.

We would like to work out the set of pairs of Betti numbers $\bigl(b^2(M), b^3(M)\bigr)$ realized by these 3^{16} resolutions M of T^7/Γ. By (12.14) these depend only on $k + l$ and $b^0(\Sigma_\beta) + b^0(\Sigma_\gamma)$. Now $0 \leqslant k + l \leqslant 16$, and calculations by the author indicate that $2 \leqslant b^0(\Sigma_\beta) + b^0(\Sigma_\gamma) \leqslant 9$. So there are 17 possibilities for $k + l$ and 8 possibilities for $b^0(\Sigma_\beta) + b^0(\Sigma_\gamma)$, giving at most $17 \times 8 = 136$ possibilities for $\bigl(b^2(M), b^3(M)\bigr)$. In fact not all of these possibilities occur.

Table 12.2 *Betti numbers $\bigl(b^2(M), b^3(M)\bigr)$ from the construction of §12.5.3*

Value of $k+l$	\multicolumn{8}{c}{Value of $b^0(\Sigma_\beta) + b^0(\Sigma_\gamma)$}							
	2	3	4	5	6	7	8	9
0	(4,99)							
1	(4,95)							
2	(4,91)							
3	(4,87)							
4	(4,83)	(6,89)						
5	(4,79)	(6,85)						
6	(4,75)	(6,81)						
7	(4,71)	(6,77)	(8,83)					
8	(4,67)	(6,73)	(8,79)					
9	(4,63)	(6,69)	(8,75)					
10	(4,59)	(6,65)	(8,71)	(10,77)				
11	(4,55)	(6,61)	(8,67)	(10,73)				
12	(4,51)	(6,57)	(8,63)	(10,69)	(12,75)			
13	(4,47)	(6,53)	(8,59)	(10,65)	(12,71)			
14	(4,43)	(6,49)	(8,55)	(10,61)	(12,67)	(14,73)		
15	(4,39)	(6,45)	(8,51)	(10,57)	(12,63)	(14,69)	(16,75)	
16	(4,35)	(6,41)	(8,47)	(10,53)	(12,59)	(14,65)	(16,71)	(18,77)

The author has worked out, by hand, some sets of Betti numbers which do occur, and the results are presented in Table 12.2. The method used was simply to try out around a hundred of the 3^{16} possible choices of (a)–(c) for the 16 α T^3, and to calculate $k, l, b^0(\Sigma_\beta)$ and $b^0(\Sigma_\gamma)$ for each choice, using the diagrammatic method explained in §12.5.2. Having done one such calculation you can deduce others from it by changing β_* or γ_* from 1 to -1 on one α T^3 and leaving everything else fixed; this cuts down the amount of work considerably. The values of $b^2(M)$ and $b^3(M)$ in the table are calculated from $k + l$ and $b^0(\Sigma_\beta) + b^0(\Sigma_\gamma)$ using (12.14).

Each of the 58 pairs of Betti numbers in Table 12.2 are the Betti numbers of a 7-manifold with holonomy G_2, so we have found at least 58 topologically distinct 7-manifolds M with holonomy G_2 in just this one example. The author is confident that the Betti numbers given in Table 12.2 are nearly all of the possible Betti numbers arising from this construction, but a few sets of Betti numbers may have slipped through the net.

12.5.4 *A second family of 7-manifolds resolving T^7/Γ*

We now describe another family of 7-manifolds M resolving $(T^3 \times K3)/\langle \beta, \gamma \rangle$, in which the β singularities $2\, S^1 \times \Sigma_\beta$ may be resolved using either a 'crepant resolution' or a 'smoothing'. Let $(T^3 \times K3)/\langle \beta, \gamma \rangle$ be as in §12.5.1, and suppose $k = 0$. This means that $\beta_* = -1$ on all 16 T^3 fixed by α, which corresponds to choosing either (b) or (c) for each T^3, but not (a). Since there are now only 2 options for each of the 16 T^3, there are $2^{16} = 65\,536$ possible choices.

This requirement fixes the action of β on $K3$ up to isomorphism. It is easy to see that Σ_β is connected, so $b^0(\Sigma_\beta) = 1$, and $b^1(\Sigma_\beta) = 18$ by (12.13). Thus the β singularities in $(T^3 \times K3)/\langle\beta, \gamma\rangle$ are two connected components $\mathcal{S}^1 \times \Sigma_\beta$, and near each component we can identify $(T^3 \times K3)/\langle\beta, \gamma\rangle$ with $\mathcal{S}^1 \times (T^2 \times K3)/\langle\beta\rangle$, where $(T^2 \times K3)/\langle\beta\rangle$ is a singular complex 3-fold.

As in §12.5.3 we can take a crepant resolution of $(T^2 \times K3)/\langle\beta\rangle$, and this gives a local model for how to resolve $(T^3 \times K3)/\langle\beta, \gamma\rangle$ at each $\mathcal{S}^1 \times \Sigma_\beta$. But we can also make a Calabi-Yau 3-fold by *smoothing* $(T^2 \times K3)/\langle\beta\rangle$, and this provides a second local model for how to resolve $(T^3 \times K3)/\langle\beta, \gamma\rangle$ at each $\mathcal{S}^1 \times \Sigma_\beta$.

We assume $k = 0$ to simplify the calculations. It is very likely that $(T^2 \times K3)/\langle\beta\rangle$ will admit smoothings or deformations followed by crepant resolutions in other cases as well, and these would give further local models for how to resolve the β singularities in $(T^3 \times K3)/\langle\beta, \gamma\rangle$. But the author does not have time or patience to work these cases out.

When $k = 0$, the smoothing of $(T^2 \times K3)/\langle\beta\rangle$ can be thought of as the smoothing of $T^6/\langle\alpha, \beta\rangle$ in which all of the 64 singular points modelled on $\mathbb{C}^3/\mathbb{Z}_2^2$ are desingularized using part (ix) of Example 9.9.9. By studying the topology of the 3-fold in (ix) we can work out the contribution to the Betti numbers of M when we resolve $(T^3 \times K3)/\langle\beta, \gamma\rangle$ in this way.

So, let $(T^3 \times K3)/\langle\beta, \gamma\rangle$ be as above, with $k = 0$ and $0 \leqslant l \leqslant 16$, and singular set S. Let $m = 0$, 1 or 2. Make a compact 7-manifold M by resolving $(T^3 \times K3)/\langle\beta, \gamma\rangle$ using a 'crepant resolution' for m of the $\mathcal{S}^1 \times \Sigma_\beta$ in S and a 'smoothing' for the remaining $2 - m$ $\mathcal{S}^1 \times \Sigma_\beta$ in S, and using a 'crepant resolution' for the 2 $\mathcal{S}^1 \times \Sigma_\gamma$ in S. As with Theorem 12.5.4, we can prove:

Theorem 12.5.6 *The method of resolving T^7/Γ described above gives a compact 7-manifold M, which is simply-connected and has Betti numbers*

$$b^2(M) = m + 2b^0(\Sigma_\gamma), \quad b^3(M) = 151 - 29m - 4l + 6b^0(\Sigma_\gamma). \tag{12.15}$$

There exists a set of R-data for T^7/Γ yielding this 7-manifold M as the corresponding resolution of T^7/Γ. Thus M admits metrics with holonomy G_2.

Proof As above M is simply-connected, and $(T^3 \times K3)/\langle\beta, \gamma\rangle$ has $b^2 = 0$ and $b^3 = 23$. It can be shown that resolving m $\mathcal{S}^1 \times \Sigma_\beta$ singularities adds m to b^2 and $19m$ to b^3, and smoothing $2 - m$ $\mathcal{S}^1 \times \Sigma_\beta$ singularities fixes b^2 and adds $48(2 - m)$ to b^3. Also, resolving the 2 $\mathcal{S}^1 \times \Sigma_\gamma$ singularities adds $2b^0(\Sigma_\gamma)$ to b^2 and $2b^0(\Sigma_\gamma) + 2b^1(\Sigma_\gamma)$ to b^3 as before. Therefore we have

$$b^2(M) = m + 2b^0(\Sigma_\gamma), \quad b^3(M) = 23 + 19m + 48(2 - m) + 2b^0(\Sigma_\gamma) + 2b^1(\Sigma_\gamma).$$

Using (12.13) to eliminate $b^1(\Sigma_\gamma)$ we get (12.15). The remainder of the proof follows as in Theorem 12.5.4. □

In this construction we have 3 possibilities 0, 1, 2 for m, and 2^{16} possibilities for the choice of either (b) or (c) for each of the 16 α T^3. Thus there are 3×2^{16} topologically distinct resolutions (M,π) of T^7/Γ arising from this construction. Of these, the cases with $m=2$ were already considered in §12.5.3, but those with $m=0$ or 1 are new.

We wish to work out the set of pairs of Betti numbers $\bigl(b^2(M), b^3(M)\bigr)$ realized by these 3×2^{16} resolutions M of T^7/Γ. By (12.15) these depend only on l, m and $b^0(\Sigma_\gamma)$. Now $0\leqslant l \leqslant 16$, $0\leqslant m \leqslant 2$ and $1\leqslant b^0(\Sigma_\gamma)\leqslant 8$, so there are 17 possibilities for l, 3 for m and 8 for $b^0(\Sigma_\gamma)$, giving at most $17\times 3\times 8=408$ sets of Betti numbers.

However, l and $b^0(\Sigma_\gamma)$ are not really independent. For instance, (12.13) implies that $b^0(\Sigma_\gamma) \geqslant l-8$, as $b^1(\Sigma_\gamma)\geqslant 0$. The author has worked out, by hand, all the possibilities for the pair $l, b^0(\Sigma_\gamma)$, using the diagrammatic notation of §12.5.2. Knowing the possibilities for l and $b^0(\Sigma_\gamma)$ we then calculate the values of $\bigl(b^2(M), b^3(M)\bigr)$ for $m=0,1,2$ using (12.15).

The results are presented in Table 12.3. They give 90 sets of Betti numbers of compact, simply-connected 7-manifolds with holonomy G_2, of which 30 (those with $m=2$) already occur in Table 12.2, and 60 are new.

12.5.5 A third family of 7-manifolds resolving T^7/Γ

In §12.5.4 we fixed $k=0$, which allowed us to resolve the β singularities either by 'crepant resolution' or by 'smoothing'. We now apply this method to both the β and the γ singularities. Construct $(T^3\times K3)/\langle\beta,\gamma\rangle$ as in §12.5.1, using resolution (b) for all 16 T^3 fixed by α. Then $k=l=0$, and Σ_β, Σ_γ are both connected with $b^0=1$ and $b^1=18$, as in §12.5.4. So the singular set S of $(T^3\times K3)/\langle\beta,\gamma\rangle$ has 4 connected components, 2 copies of $S^1\times\Sigma_\beta$ and 2 copies of $S^1\times\Sigma_\gamma$.

Table 12.3 Betti numbers $\bigl(b^2(M), b^3(M)\bigr)$ from the construction of §12.5.4

l	$b^0(\Sigma_\gamma)$	$m=0$	$m=1$	$m=2$
0	1	(2,157)	(3,128)	(4,99)
1	1	(2,153)	(3,124)	(4,95)
2	1	(2,149)	(3,120)	(4,91)
3	1	(2,145)	(3,116)	(4,87)
4	1,2	(2,141), (4,147)	(3,112), (5,118)	(4,83), (6,89)
5	1,2	(2,137), (4,143)	(3,108), (5,114)	(4,79), (6,85)
6	1,2	(2,133), (4,139)	(3,104), (5,110)	(4,75), (6,81)
7	1,2,3	(2,129),(4,135),(6,141)	(3,100),(5,106),(7,112)	(4,71),(6,77),(8,83)
8	1,2,3	(2,125),(4,131),(6,137)	(3,96),(5,102),(7,108)	(4,67),(6,73),(8,79)
9	1,2,3	(2,121),(4,127),(6,133)	(3,92),(5,98),(7,104)	(4,63),(6,69),(8,75)
10	2,3,4	(4,123),(6,129),(8,135)	(5,94),(7,100),(9,106)	(6,65),(8,71),(10,77)
11	3,4	(6,125), (8,131)	(7,96), (9,102)	(8,67), (10,73)
12	4,5	(8,127), (10,133)	(9,98), (11,104)	(10,69), (12,75)
13	5	(10,129)	(11,100)	(12,71)
14	6	(12,131)	(13,102)	(14,73)
15	7	(14,133)	(15,104)	(16,75)
16	8	(16,135)	(17,106)	(18,77)

As in §12.5.4, each component of S can be resolved using either a 'crepant resolution' or a 'smoothing'. Let $m = 0, 1, 2, 3$ or 4, and define a compact 7-manifold M by resolving m of the components of S with a 'crepant resolution', and the remaining $4 - m$ components with a 'smoothing'. Following the proofs of Theorems 12.5.4 and 12.5.6, we find:

Theorem 12.5.7 *The method of resolving T^7/Γ described above gives a compact 7-manifold M, which is simply-connected and has Betti numbers*

$$b^2(M) = m \quad \text{and} \quad b^3(M) = 215 - 29m. \tag{12.16}$$

There exists a set of R-data for T^7/Γ yielding this 7-manifold M as the corresponding resolution of T^7/Γ. Thus M admits metrics with holonomy G_2.

Table 12.4 Betti numbers $(b^2(M), b^3(M))$ from the construction of §12.5.5

$m=0$	$m=1$	$m=2$	$m=3$	$m=4$
(0,215)	(1,186)	(2,157)	(3,128)	(4,99)

Table 12.4 gives the resulting Betti numbers, calculated from (12.16). Here the Betti numbers for $m = 2, 3$ and 4 already appear in Table 12.3, but those for $m = 0$ and 1 are new. Table 12.2 gives 58 sets of Betti numbers from resolutions of T^7/Γ, Table 12.3 a further 60, and Table 12.4 a further 2, excluding repeated Betti numbers. Thus, we have found $58 + 60 + 2 = 120$ different sets of Betti numbers of compact, simply-connected 7-manifolds with holonomy G_2 that are resolutions of the one orbifold T^7/Γ.

12.6 Examples of associative and coassociative submanifolds

Calibrated geometry was discussed in general terms in §3.7, and for the exceptional holonomy groups in §10.8. We saw in §10.8.1 and §10.8.2 that there are two interesting classes of calibrated submanifolds in G_2-manifolds (M, φ, g), called *associative* 3-*folds* and *coassociative* 4-*folds*.

We also showed in Propositions 10.8.1 and 10.8.5 that the fixed point set of an isometric involution $\sigma : M \to M$ with $\sigma^*(\varphi) = \pm \varphi$ is automatically associative or coassociative, provided it has the right dimension. We now use this idea to find examples of associative and coassociative submanifolds in the compact 7-manifold M with holonomy G_2 constructed in §12.2. Throughout this section, let T^7, Γ, M and all other notation be as in §12.2.

Example 12.6.1 Define an isometry $\sigma : T^7 \to T^7$ by

$$\sigma : (x_1, \ldots, x_7) \mapsto (x_1, x_2, x_3, \tfrac{1}{2} - x_4, -x_5, -x_6, -x_7).$$

Then σ commutes with Γ, and so its action pushes down to T^7/Γ. The fixed points of σ on T^7 are 16 copies of T^3, and $\sigma\delta$ has no fixed points in T^7 for all $\delta \neq 1$ in Γ. Thus the fixed points of σ in T^7/Γ are the image of the 16 T^3 fixed by σ in T^7.

But calculation shows that these 16 T^3 do not intersect the fixed points of α, β or γ, and that Γ acts freely on the set of 16 T^3 fixed by σ. So the image of the 16 T^3 in T^7 is 2 T^3 in T^7/Γ, which do not intersect the singular set of T^7/Γ. When we resolve T^7/Γ in a σ-equivariant way to get M with holonomy G_2, the fixed points of σ in M are again 2 copies of T^3, which are associative 3-folds by Proposition 10.8.1.

Example 12.6.2 Define an isometry $\sigma : T^7 \to T^7$ by

$$\sigma : (x_1, \ldots, x_7) \mapsto (-x_1, x_2, \tfrac{1}{2} - x_3, -x_4, x_5, \tfrac{1}{2} - x_6, x_7).$$

Everything works as in the previous example, except that Γ does not act freely on the set of 16 σ T^3 in T^7, as $\beta\gamma$ acts trivially. Therefore the fixed points of σ in T^7/Γ are 4 copies of $T^3/\mathbb{Z}_2$, which do not intersect the singular set of T^7/Γ. Here the $\mathbb{Z}_2$ action on T^3 is free, and generated by $\beta\gamma$. A σ-equivariant resolution of T^7/Γ gives a 7-manifold M with holonomy G_2, and the fixed points of σ in M are again 4 copies of $T^3/\mathbb{Z}_2$, which are associative by Proposition 10.8.1.

Example 12.6.3 Here is a more complex example. Define $\sigma : T^7 \to T^7$ by

$$\sigma : (x_1, \ldots, x_7) \mapsto (x_1, \tfrac{1}{2} - x_2, \tfrac{1}{2} - x_3, x_4, x_5, -x_6, -x_7).$$

This time, σ and $\sigma\alpha$ both fix 16 T^3, which intersect in 64 $\mathcal{S}^1$, part of the fixed points of α. Also, if $\delta \in \Gamma$ then $\sigma\delta$ has fixed points only if $\delta = 1, \alpha$. Thus the fixed points of σ in T^7/Γ are the image in T^7/Γ of the 16 σ T^3 and 16 $\sigma\alpha$ T^3, and they intersect the 4 singular T^3 in T^7/Γ coming from the fixed points of α.

Calculation shows that the fixed points of σ in T^7/Γ are $\mathcal{S}^1 \times \Sigma_0$, where $\mathcal{S}^1$ has coordinate x_1, and Σ_0 is a connected, singular surface, the union of two sets of 4 $T^2/\{\pm 1\}$, which intersect in 16 points. It makes sense to regard Σ_0 as the fixed points of σ in $T^4/\langle\alpha\rangle$, where T^4 has coordinates (x_4, x_5, x_6, x_7).

To resolve T^7/Γ in a σ-equivariant way, for each of the 4 singular α T^3 in T^7/Γ we must choose whether σ acts on the corresponding resolution as in part (a) or part (b) of §12.3. This choice determines a σ-action on the resolution $K3$ of $T^4/\langle\alpha\rangle$. The fixed points of σ in M are then isomorphic to $\mathcal{S}^1 \times \Sigma$, where Σ is the nonsingular surface in $K3$ fixed by σ. By Proposition 10.8.1, this is an associative 3-fold.

We can work out $b^0(\Sigma)$ and $b^1(\Sigma)$ using the method of §12.5.2. Choosing part (a) of §12.3 for all 4 α T^3 in T^7/Γ gives $b^0(\Sigma) = 8$ and $b^1(\Sigma) = 0$, so Σ is 8 $\mathcal{S}^2$, and M contains 8 associative $\mathcal{S}^1 \times \mathcal{S}^2$'s. Choosing part (b) for all 4 α T^3 gives $b^0(\Sigma) = 1$ and $b^1(\Sigma) = 18$, so Σ is connected with genus 9.

Here are some examples of coassociative 4-folds.

Example 12.6.4 Define an orientation-reversing isometry $\sigma : T^7 \to T^7$ by

$$\sigma : (x_1, \ldots, x_7) \mapsto (\tfrac{1}{2} - x_1, x_2, x_3, x_4, x_5, \tfrac{1}{2} - x_6, \tfrac{1}{2} - x_7).$$

Then σ commutes with Γ, preserves g_0 and takes φ_0 to $-\varphi_0$. The fixed points of σ in T^7 are 8 copies of T^4, and the fixed points of $\sigma\alpha\beta$ in T^7 are 128 points. If $\delta \in \Gamma$ then

$\sigma\delta$ has no fixed points unless $\delta = 1, \alpha\beta$. Thus the fixed points of σ in T^7/Γ are the image of the fixed points of σ and $\sigma\alpha\beta$ in T^7.

Now Γ acts freely on the sets of 8 σ T^4 and 128 $\sigma\alpha\beta$ points. So the fixed point set of σ in T^7/Γ is the union of T^4 and 16 isolated points, none of which intersect the singular set of T^7/Γ. When we resolve T^7/Γ to get M with holonomy G_2 in a σ-equivariant way, the action of σ on M again fixes T^4 and 16 points. By Proposition 10.8.5 this T^4 is coassociative. As $b_+^2(T^4) = 3$, there is a 3-parameter family of coassociative 4-tori in M near N, by Theorem 10.8.4.

Example 12.6.5 Define an orientation-reversing isometry $\sigma : T^7 \to T^7$ by

$$\sigma : (x_1, \ldots, x_7) \mapsto (\tfrac{1}{2} - x_1, \tfrac{1}{2} - x_2, \tfrac{1}{2} - x_3, x_4, x_5, x_6, x_7).$$

Then σ commutes with Γ, preserves g_0 and takes φ_0 to $-\varphi_0$. The fixed points of σ are 8 T^4, and the fixed points of $\sigma\alpha$ are 128 points. But this time α acts as -1 on each T^4 fixed by σ, with 16 fixed points, which are also fixed points of $\sigma\alpha$. Thus the fixed set of σ in T^7/Γ is 2 copies of $T^4/\mathbb{Z}_2$.

Resolving T^7/Γ in a σ-equivariant way to get M with holonomy G_2, the fixed points of σ in M are 2 copies of the $K3$ surface, which are coassociative by Proposition 10.8.5. As $b_+^2(K3) = 3$, each of these admits a 3-parameter family of coassociative deformations by Theorem 10.8.4.

In the previous two examples the moduli space of coassociative 4-folds is of dimension 3, and so we expect M to be locally fibred by coassociative 4-folds. It is an interesting question whether this extends to a global fibration of M by coassociative 4-folds, allowing some singular fibres. This idea was discussed briefly in §10.8.2.

Example 12.6.6 Here is a more complex example. Define $\sigma : T^7 \to T^7$ by

$$\sigma : (x_1, \ldots, x_7) \mapsto (x_1, x_2, \tfrac{1}{2} - x_3, -x_4, x_5, x_6, -x_7).$$

Then σ and $\sigma\alpha$ both fix 8 T^4, which intersect in 32 T^2, part of the fixed points of α. Also, if $\delta \in \Gamma$ then $\sigma\delta$ has fixed points only if $\delta = 1, \alpha$. Thus the fixed points of σ in T^7/Γ are the image in T^7/Γ of the 8 σ T^4 and 8 $\sigma\alpha$ T^4, and they intersect the 4 singular T^3 in T^7/Γ coming from the fixed points of α.

Calculation shows that the fixed points of σ in T^7/Γ are $(T^2 \times \Sigma_0)/\mathbb{Z}_2$, where T^2 has coordinates (x_1, x_2), and Σ_0 is as in Example 12.6.3, and the $\mathbb{Z}_2$-action on $T^2 \times \Sigma_0$ is free and generated by $\beta\gamma$. It makes sense to regard Σ_0 as the fixed points of σ in $T^4/\langle\alpha\rangle$, where T^4 has coordinates (x_4, x_5, x_6, x_7).

As in Example 12.6.3, to resolve T^7/Γ to get N in a σ-equivariant way we must choose (a) or (b) for each of the 4 singular α T^3 in T^7/Γ. This choice determines a σ-action on the resolution $K3$ of $T^4/\langle\alpha\rangle$, with fixed points Σ. It then turns out that the fixed points of σ in M are isomorphic to $(T^2 \times \Sigma)/\mathbb{Z}_2$, where the $\mathbb{Z}_2$ action is free on $T^2 \times \Sigma$. By Proposition 10.8.5, this $(T^2 \times \Sigma)/\mathbb{Z}_2$ is a coassociative 4-fold in M.

When we choose part (a) for all 4 α T^3 in T^7/Γ, we find that Σ is 8 $\mathcal{S}^2$. Then the $\mathbb{Z}_2$-action on the 8 $T^2 \times \mathcal{S}^2$ is free, so the fixed set of σ in M is 4 copies of $T^2 \times \mathcal{S}^2$. These are coassociative, and each has a 1-parameter family of coassociative deformations by Theorem 10.8.4.

12.7 An example of a resolution of T^7/Γ with Γ nonabelian

Sections 12.2, 12.3 and 12.5 considered orbifolds T^7/Γ where $\Gamma \cong \mathbb{Z}_2^3$. We now consider an orbifold T^7/Γ in which Γ is the dihedral group D_4 of order 8. Let T^7 and (φ_0, g_0) be as in §12.2, and define $\alpha, \beta, \gamma : T^7 \to T^7$ by

$$\alpha : (x_1, \ldots, x_7) \mapsto (x_1, x_2, x_3, -x_4, -x_5, -x_6, -x_7), \quad (12.17)$$

$$\beta : (x_1, \ldots, x_7) \mapsto (x_1, -x_2, -x_3, x_4, x_5, -x_6, -x_7), \quad (12.18)$$

$$\gamma : (x_1, \ldots, x_7) \mapsto (-x_1, x_2, \tfrac{1}{2} - x_3, x_7, x_6, x_5, x_4). \quad (12.19)$$

Then $\alpha^2 = \beta^2 = \gamma^2 = 1$, $\alpha\beta = \beta\alpha$, $\alpha\gamma = \gamma\alpha$ and $\gamma\beta = \alpha\beta\gamma$, so that $(\beta\gamma)^2 = \alpha$. Thus $\Gamma = \langle \alpha, \beta, \gamma \rangle$ is a nonabelian group of order 8, which acts on T^7 preserving (φ_0, g_0). Here are the analogues of Lemmas 12.2.1 and 12.2.2 for this group Γ.

Lemma 12.7.1 *The elements $\beta\gamma$ and $\alpha\beta\gamma$ of Γ have no fixed points on T^7. The fixed points of α, β and $\alpha\beta$ in T^7 are each 16 copies of T^3. The fixed points of γ and $\alpha\gamma$ in T^7 are each 4 copies of T^3. Moreover β swaps the 16 γ T^3 and the 16 $\alpha\gamma$ T^3, and γ swaps the 16 β T^3 and the 16 β T^3, and γ fixes 4 of the 16 α T^3 and swaps the remaining 12 in pairs. The fixed points of α, β and $\alpha\beta$ intersect in 64 $\mathcal{S}^1$ in T^7. The fixed points of α, γ and $\alpha\gamma$ intersect in 16 $\mathcal{S}^1$ in T^7.*

Lemma 12.7.2 *The singular set S of T^7/Γ is the union of*

  (i) *4 $T^3/\mathbb{Z}_2^2$ from the α fixed points,*
  (ii) *6 $T^3/\mathbb{Z}_2$ from the α fixed points,*
  (iii) *16 $T^3/\mathbb{Z}_2$ from the β (equivalently $\alpha\beta$) fixed points,*
  (iv) *4 $T^3/\mathbb{Z}_2$ from the γ (equivalently $\alpha\gamma$) fixed points.*

Here (i) and (iii) intersect in 8 $\mathcal{S}^1$, and (ii) and (iii) intersect in 24 $\mathcal{S}^1$, and (i) and (iv) intersect in 8 $\mathcal{S}^1$.

Since Γ is nonabelian, distinct elements in Γ can be conjugate, and this affects the form of the singular set. Now $\gamma\beta\gamma^{-1} = \alpha\beta$, so β and $\alpha\beta$ are conjugate in Γ. This also means that γ takes the fixed points of β in T^7 to the fixed points of $\alpha\beta$ in T^7. Therefore the β fixed points and the $\alpha\beta$ fixed points yield the same piece (iii) of the singular set of T^7/Γ. Similarly, the γ and $\alpha\gamma$ fixed points yield the same piece (iv) of the singular set.

This phenomenon can create difficulties in resolving orbifolds T^7/Γ with Γ nonabelian, for the following reason. Let δ be a nonidentity element of Γ with fixed points in T^7, and F a connected component of the fixed point set of δ. Let ϵ be conjugate to δ in Γ. Then there is a component F' of the fixed points of ϵ which is mapped to F by some element of Γ.

Suppose F and F' are distinct, but intersect. As the images of F and F' in T^7/Γ are the same, this means that the corresponding piece of the singular set of T^7/Γ *intersects itself*. One must be careful not to make mistakes in resolving such orbifolds. Fortunately, this does not occur in the case we are considering.

As in §12.5, we simplify the problem of resolving T^7/Γ by dividing it into two stages. First we resolve $T^7/\langle\alpha\rangle$ to get $T^3 \times K3$, in such a way that the action of $\langle\beta, \gamma\rangle$

on $T^7/\langle\alpha\rangle$ lifts to $T^3 \times K3$, and then we resolve $(T^3 \times K3)/\langle\beta,\gamma\rangle$ to get a 7-manifold M. Note that the group acting on $T^7/\langle\alpha\rangle$ and $T^3 \times K3$ is really the quotient group $\Gamma/\langle\alpha\rangle$, which is isomorphic to $\mathbb{Z}_2^2$, but we will write it as $\langle\beta,\gamma\rangle$.

12.7.1 Resolving $T^7/\langle\alpha\rangle$ in a $\langle\beta,\gamma\rangle$-equivariant way

To resolve $T^7/\langle\alpha\rangle$ we must assign to each of the 16 α T^3 an ALE 4-manifold Y_i with holonomy SU(2) asymptotic to $\mathbb{C}^2/\{\pm 1\}$, and to do this in a $\langle\beta,\gamma\rangle$-equivariant way we must ensure that $\langle\beta,\gamma\rangle$ acts appropriately on the Y_i. Now 4 of the α T^3 are fixed by γ and 12 are swapped in pairs, by Lemma 12.7.1. We must treat the 4 T^3 and the 12 T^3 differently.

For each of the 4 α T^3 fixed by γ there are 3 topologically distinct choices for Y_i, very similar to parts (a)–(c) of §12.5. They are distinguished by the actions of β_* and γ_* on $H_2(Y_i, \mathbb{R})$; as in §12.5, we have 3 cases (a)–(c) where in (a) $\beta_* = 1$, $\gamma_* = -1$, in (b) $\beta_* = \gamma_* = -1$, and in (c) $\beta_* = -1$, $\gamma_* = 1$ on $H_2(Y_i, \mathbb{R})$. For the 12 α T^3 swapped in pairs by γ there are two topologically distinct possibilities for each pair, very similar to parts (a) and (b) of §12.3. They are distinguished by the action of β_* on $H_2(Y_i, \mathbb{R})$; in (a) $\beta_* = 1$ and in (b) $\beta_* = -1$ on $H_2(Y_i, \mathbb{R})$.

Thus there are 3 choices for each of the 4 α T^3 fixed by γ and 2 choices for each of the 6 pairs of α T^3 swapped by γ, so there are in total $3^4 \times 2^6 = 5184$ possible choices, each of which leads to a resolution $T^3 \times K3$ of $T^7/\langle\alpha\rangle$ with an action of $\langle\beta,\gamma\rangle$ upon it. Fix one such choice, and define integers k, l, m by

- $\beta_* = 1$ on k pairs of Y_i assigned to the 6 pairs of α T^3 swapped by γ,
- $\beta_* = 1$ on l of the 4 Y_i assigned to the 4 α T^3 fixed by γ, and
- $\gamma_* = 1$ on m of the 4 Y_i assigned to the 4 α T^3 fixed by γ.

Then $0 \leq k \leq 6$, $0 \leq l \leq 4$ and $0 \leq m \leq 4$. Also, we must choose resolution (a) for l and resolution (c) for m of the 4 α T^3 fixed by γ, and therefore $l + m \leq 4$.

Let Σ_β be the subset of $K3$ fixed by β, and Σ_γ the subset of $K3$ fixed by γ. Then Σ_β, Σ_γ are compact real submanifolds of $K3$ of dimension 2, which are oriented but not necessarily connected. Then as in Lemma 12.5.3 we have:

Lemma 12.7.3 *The singular set S of $(T^3 \times K3)/\langle\beta,\gamma\rangle$ is the disjoint union of two copies of $\mathcal{S}^1 \times \Sigma_\beta$ and two copies of $\mathcal{S}^1 \times \Sigma_\gamma$. At each point of S the singularity is locally modelled on $\mathbb{R}^3 \times \mathbb{R}^4/\{\pm 1\}$.*

Calculation shows that the Euler characteristics of Σ_β, Σ_γ are given by

$$\chi(\Sigma_\beta) = 2b^0(\Sigma_\beta) - b^1(\Sigma_\beta) = 4k + 2l - 16$$
$$\text{and} \quad \chi(\Sigma_\gamma) = 2b^0(\Sigma_\gamma) - b^1(\Sigma_\gamma) = 2m - 4. \tag{12.20}$$

Furthermore, the topology of Σ_γ is determined by m, as Σ_γ is 2 copies of $\mathcal{S}^2$ when $m = 4$, and Σ_γ is connected when $0 \leq m \leq 3$. Thus by (12.20) we have

$$\begin{aligned} b^0(\Sigma_\gamma) &= 1 \quad \text{and} \quad b^1(\Sigma_\gamma) = 6 - 2m \quad \text{when } 0 \leq m \leq 3, \\ b^0(\Sigma_\gamma) &= 2 \quad \text{and} \quad b^1(\Sigma_\gamma) = 0 \quad \text{when } m = 4. \end{aligned} \tag{12.21}$$

As in §12.5 we will consider 3 families of 7-manifolds M resolving T^7/Γ, which will differ according to whether we resolve the β and γ singularities by 'crepant resolution' or 'smoothing'.

12.7.2 A family of 7-manifolds resolving T^7/Γ

Let $(T^3 \times K3)/\langle \beta, \gamma \rangle$ be as above. Then as in §12.5.3 we can resolve the singularities using 'crepant resolution' to get a unique 7-manifold M. Here is the analogue of Theorem 12.5.4 for this resolution, which is proved by a similar method.

Theorem 12.7.4 *The method of resolving T^7/Γ described above gives a compact 7-manifold M, which is simply-connected and has Betti numbers*

$$\begin{aligned} b^2(M) &= 1 + k + 2\big(b^0(\Sigma_\beta) + b^0(\Sigma_\gamma)\big) \quad \text{and} \\ b^3(M) &= 62 - 9k - 4l - 4m + 6\big(b^0(\Sigma_\beta) + b^0(\Sigma_\gamma)\big). \end{aligned} \quad (12.22)$$

There exists a set of R-data for T^7/Γ yielding this 7-manifold M as the corresponding resolution of T^7/Γ. Thus M admits metrics with holonomy G_2.

We would like to work out the set of pairs of Betti numbers $\big(b^2(M), b^3(M)\big)$ realized by these 5184 resolutions M of T^7/Γ. By (12.22) these depend only on the 5 variables $k, l, m, b^0(\Sigma_\beta)$ and $b^0(\Sigma_\gamma)$. However, $b^0(\Sigma_\gamma)$ depends only on m by (12.21). Furthermore, k, l and $b^0(\Sigma_\beta)$ depend only on whether $\beta_* = \pm 1$ for each of the 16 α T^3, and not on the values of γ_* for the 4 α T^3.

This means that we can proceed as follows. We first determine all the possible values of the three variables $k, l, b^0(\Sigma_\beta)$, noting that we only need to consider the values of β_* to do this. For each set of values of the triple $k, l, b^0(\Sigma_\beta)$, the possible values of m are $0 \leqslant m \leqslant 4 - l$. Knowing m then tells us $b^0(\Sigma_\gamma)$. So by working out the possible values of $k, l, b^0(\Sigma_\beta)$ we immediately deduce the possibilities for the quintuple $k, l, m, b^0(\Sigma_\beta), b^0(\Sigma_\gamma)$, and these give us $b^2(M)$ and $b^3(M)$ by (12.22).

The author has worked out the possibilities for the triple $k, l, b^0(\Sigma_\beta)$ using the diagrammatic notation of §12.5.2, and deduced from this the set of pairs of Betti numbers $\big(b^2(M), b^3(M)\big)$ realized by this construction. The results are given in Table 12.5. There are 69 sets of Betti numbers, excluding repeats.

By (12.21) and (12.22), when $m \leqslant 3$ we have $b^2(M) = 3 + k + 2b^0(\Sigma_\beta)$ and $b^3(M) = 68 - 9k - 4(l + m) + 6b^0(\Sigma_\beta)$. Thus when $m \leqslant 3$ the Betti numbers depend only on the variables $k, l + m$ and $b^0(\Sigma_\beta)$, and not on l, m individually. To avoid having to repeat pairs of Betti numbers up to four times, in Table 12.5 we give the range of possible values of l for each set of values of k and $b^0(\Sigma_\beta)$, and then we list Betti numbers under the corresponding possible values of $l + m$, separating the cases $l + m = 4, m \leqslant 3$ and $l = 0, m = 4$.

12.7.3 A second family of 7-manifolds resolving T^7/Γ

Now fix $m = 0$. Then the γ singularities in $(T^3 \times K3)/\langle \beta, \gamma \rangle$ are 2 copies of $\mathcal{S}^1 \times \Sigma_\gamma$, where Σ_γ is connected with $b^1(\Sigma_\gamma) = 6$. As in §12.5.4, it turns out that we are free to resolve each copy of $\mathcal{S}^1 \times \Sigma_\gamma$ by either 'crepant resolution' or 'smoothing'. Choose

Table 12.5 Betti numbers $(b^2(M), b^3(M))$ from the construction of §12.7.2

k	$b^0(\Sigma_\beta)$	l	$l+m=0$	$l+m=1$	$l+m=2$	$l+m=3$	$l+m=4,$ $m\leqslant 3$	$l=0,$ $m=4$
0	1	0,1,2,3,4	(5,74)	(5,70)	(5,66)	(5,62)	(5,58)	(7,64)
1	1	0,1,2,3,4	(6,65)	(6,61)	(6,57)	(6,53)	(6,49)	(8,55)
2	1	0,1,2,3,4	(7,56)	(7,52)	(7,48)	(7,44)	(7,40)	(9,46)
2	2	4					(9,46)	
3	1	0,1,2,3	(8,47)	(8,43)	(8,39)	(8,35)	(8,31)	(10,37)
3	2	0,1,2,3,4	(10,53)	(10,49)	(10,45)	(10,41)	(10,37)	(12,43)
3	3	1,2,3,4		(12,55)	(12,51)	(12,47)	(12,43)	
4	2	0,1,2	(11,44)	(11,40)	(11,36)	(11,32)	(11,28)	(13,34)
4	3	1,2,3		(13,46)	(13,42)	(13,38)	(13,34)	
4	4	4					(15,40)	
5	3	0,1	(14,41)	(14,37)	(14,33)	(14,29)	(14,25)	(16,31)
5	4	1,2		(16,43)	(16,39)	(16,35)	(16,31)	
5	5	2,3			(18,45)	(18,41)	(18,37)	
5	6	4					(20,43)	
6	4	0	(17,38)	(17,34)	(17,30)	(17,26)		(19,28)
6	5	1		(19,40)	(19,36)	(19,32)	(19,28)	
6	6	2			(21,42)	(21,38)	(21,34)	
6	7	3				(23,44)	(23,40)	
6	8	4					(25,46)	

$n = 0, 1$ or 2, and make a compact 7-manifold M by resolving n of the 2 $S^1 \times \Sigma_\gamma$ singularities with a 'crepant resolution' and the remaining $2 - n$ with a 'smoothing', and by resolving the β singularities with a 'crepant resolution'. Then as in Theorem 12.5.6 we can prove:

Theorem 12.7.5 *The method of resolving T^7/Γ described above gives a compact 7-manifold M, which is simply-connected and has Betti numbers*

$$b^2(M) = 1 + k + n + 2b^0(\Sigma_\beta)$$
$$\text{and} \quad b^3(M) = 78 - 9k - 4l - 5n + 6b^0(\Sigma_\beta). \quad (12.23)$$

There exists a set of R-data for T^7/Γ yielding this 7-manifold M as the corresponding resolution of T^7/Γ. Thus M admits metrics with holonomy G_2.

The sets of possible values for k, $b^0(\Sigma_\beta)$ and l are given in the 3 left hand columns of Table 12.5, and we use these and (12.23) to write down all the sets of Betti numbers $(b^2(M), b^3(M))$ arising from this construction. The results are given in Table 12.6. We have only included the Betti numbers for $n = 0, 1$, since those with $n = 2$ are already part of Table 12.5.

12.7.4 A third family of 7-manifolds resolving T^7/Γ

Return to the situation of §12.7.1, but this time fix $k = l = 0$. That is, $\beta_* = -1$ on all of the 16 α T^3 in T^7. Then Σ_β is connected, and $b^1(\Sigma_\beta) = 18$ by (12.20). So

AN EXAMPLE OF A RESOLUTION OF T^7/Γ WITH Γ NONABELIAN 333

Table 12.6 Betti numbers $(b^2(M), b^3(M))$ from the construction of §12.7.3

k	$b^0(\Sigma_\beta)$	l	$n=0$	$n=1$
0	1	0,1,2,3,4	(3,84),(3,80),(3,76),(3,72),(3,68)	(4,79),(4,75),(4,71),(4,67),(4,63)
1	1	0,1,2,3,4	(4,75),(4,71),(4,67),(4,63),(4,59)	(5,70),(5,66),(5,62),(5,58),(5,54)
2	1	0,1,2,3,4	(5,66),(5,62),(5,58),(5,54),(5,50)	(6,61),(6,57),(6,53),(6,49),(6,45)
2	2	4	(7,56)	(8,51)
3	1	0,1,2,3	(6,57), (6,53), (6,49), (6,45)	(7,52), (7,48), (7,44), (7,40)
3	2	0,1,2,3,4	(8,63),(8,59),(8,55),(8,51),(8,47)	(9,58),(9,54),(9,50),(9,46),(9,42)
3	3	1,2,3,4	(10,65), (10,61), (10,57), (10,53)	(11,60), (11,56), (11,52), (11,48)
4	2	0,1,2	(9,54), (9,50), (9,46)	(10,49), (10,45), (10,41)
4	3	1,2,3	(11,56), (11,52), (11,48)	(12,51), (12,47), (12,43)
4	4	4	(13,50)	(14,45)
5	3	0,1	(12,51), (12,47)	(13,46), (13,42)
5	4	1,2	(14,53), (14,49)	(15,48), (15,44)
5	5	2,3	(16,55), (16,51)	(17,50), (17,46)
5	6	4	(18,53)	(19,48)
6	4	0	(15,48)	(16,43)
6	5	1	(17,50)	(18,45)
6	6	2	(19,52)	(20,47)
6	7	3	(21,54)	(22,49)
6	8	4	(23,56)	(24,51)

the β singularities in $(T^3 \times K3)/\langle \beta, \gamma \rangle$ have 2 connected components, each a copy of $S^1 \times \Sigma_\beta$. As in §12.5.5 we are free to resolve each component with either a 'crepant resolution' or a 'smoothing'.

As $l = 0$, the possibilities for m are 0, 1, 2, 3, 4. We can resolve the γ singularities by 'crepant resolution'. But when $m = 0$, we also have the choice of resolving the γ singularities by 'smoothing'. Let us make a compact 7-manifold M as follows. Choose $j = 0$, 1 or 2, and resolve j of the $S^1 \times \Sigma_\beta$ singularities with a 'crepant resolution', and the remaining $2 - j$ with a 'smoothing'. If $1 \leqslant m \leqslant 4$, then resolve the γ singularities with a 'crepant resolution', as in §12.7.2. But if $m = 0$, choose $n = 0$, 1 or 2, and resolve n of the 2 $S^1 \times \Sigma_\gamma$ singularities with a 'crepant resolution' and the remaining $2 - n$ with a 'smoothing', as in §12.7.3. Following the proofs of Theorems 12.7.4 and 12.7.5, we find:

Theorem 12.7.6 *The method of resolving T^7/Γ described above gives a compact 7-manifold M, which is simply-connected and has Betti numbers*

$$b^2(M) = \begin{cases} 1+j+n, & m=0, \\ 3+j, & 1 \leqslant m \leqslant 3, \\ 5+j, & m=4, \end{cases} \quad b^3(M) = \begin{cases} 142-29j-5n, & m=0, \\ 132-29j-4m, & 1 \leqslant m \leqslant 3, \\ 122-29j, & m=4. \end{cases}$$

There exists a set of R-data for T^7/Γ yielding this 7-manifold M as the corresponding resolution of T^7/Γ. Thus M admits metrics with holonomy G_2.

Table 12.7 *Betti numbers $(b^2(M), b^3(M))$ from §12.7.4*

	$\begin{array}{c}m=0\\n=0\end{array}$	$\begin{array}{c}m=0\\n=1\end{array}$	$\begin{array}{c}m=0\\n=2\end{array}$	$m=1$	$m=2$	$m=3$	$m=4$
$j=0$	(1,142)	(2,137)	(3,132)	(3,128)	(3,124)	(3,120)	(5,122)
$j=1$	(2,113)	(3,108)	(4,103)	(4,99)	(4,95)	(4,91)	(6,93)
$j=2$	(3,84)	(4,79)	(5,74)	(5,70)	(5,66)	(5,62)	(7,64)

Table 12.7 gives the resulting Betti numbers. Tables 12.5, 12.6 and 12.7 contain 191 pairs of Betti numbers $(b^2(M), b^3(M))$, but because of repeats there are only 141 distinct pairs. Thus we have shown that there are at least 141 topologically distinct compact 7-manifolds M with holonomy G_2 arising as resolutions of this one orbifold T^7/Γ.

12.8 More examples

We finish with four more examples of orbifolds T^7/Γ and their resolutions.

12.8.1 *A variation on the examples of §12.2, §12.3 and §12.5*

Let T^7 and (φ_0, g_0) be as in §12.2, and define $\alpha, \beta, \gamma : T^7 \to T^7$ by

$$\alpha : (x_1, \ldots, x_7) \mapsto (x_1, x_2, x_3, -x_4, -x_5, -x_6, -x_7),$$
$$\beta : (x_1, \ldots, x_7) \mapsto (x_1, -x_2, -x_3, x_4, x_5, -x_6, -x_7),$$
$$\gamma : (x_1, \ldots, x_7) \mapsto (-x_1, x_2, -x_3, x_4, \tfrac{1}{2} - x_5, x_6, \tfrac{1}{2} - x_7).$$

These differ from (12.1)–(12.3) only in where $\frac{1}{2}$ appears. As before $\Gamma = \langle \alpha, \beta, \gamma \rangle$ is isomorphic to $\mathbb{Z}_2^3$ and preserves (φ_0, g_0). Here is the analogue of Lemma 12.2.2.

Lemma 12.8.1 *The singular set S of T^7/Γ is the union of*

(i) *$8\ T^3/\mathbb{Z}_2$ from the α fixed points,*
(ii) *$8\ T^3/\mathbb{Z}_2$ from the β fixed points,*
(iii) *$8\ T^3/\mathbb{Z}_2$ from the $\alpha\beta$ fixed points, and*
(iv) *$4\ T^3$ from the γ fixed points.*

The sets (i), (ii) and (iii) intersect in $32\ S^1$ in T^7/Γ.

We can apply the method of sections 12.2, 12.3 and 12.5 to find a number of different resolutions M of T^7/Γ with holonomy G_2. Let us do this using 'crepant resolutions' for all of the 24 α, β and $\alpha\beta$ $T^3/\mathbb{Z}_2$'s. Initially $b^2 = 0$ and $b^3 = 7$. Resolving each of the 24 $T^3/\mathbb{Z}_2$ in S adds 1 to b^2 and 1 to b^3. Resolving each of the 4 γ T^3 adds 1 to b^2 and 3 to b^3.

Therefore we get a compact, simply-connected 7-manifold M with holonomy G_2 and Betti numbers $b^2(M) = 28$ and $b^3(M) = 43$. This has the largest value of b^2 we have met so far. By using a combination of 'crepant resolution' and 'smoothing' we can also find a number of other 7-manifolds M resolving T^7/Γ, but it turns out that we have met nearly all the resulting Betti numbers before in §12.5 and §12.7, so we will not list them.

12.8.2 Another group Γ of order 8

Let T^7 and (φ_0, g_0) be as in §12.2, and define $\alpha, \beta : T^7 \to T^7$ by

$$\alpha : (x_1, \ldots, x_7) \mapsto (x_1, -x_2, -x_3, x_5, -x_4, x_7, -x_6),$$
$$\beta : (x_1, \ldots, x_7) \mapsto (-x_1, x_2, -x_3, x_6, -x_7, -x_4, x_5).$$

Then $\alpha^4 = \beta^4 = 1$, $\alpha^2 = \beta^2$ and $\alpha\beta = \beta\alpha^3$. Thus $\Gamma = \langle \alpha, \beta \rangle$ is a nonabelian group of order 8, which preserves (φ_0, g_0), and is not isomorphic to the group D_4 of §12.7. It has centre $\langle \alpha^2 \rangle$, where α^2 acts by

$$\alpha^2 : (x_1, \ldots, x_7) \mapsto (x_1, x_2, x_3, -x_4, -x_5, -x_6, -x_7).$$

Now Γ consists of $1, \alpha^2$ and 6 other elements whose square is α^2. Thus, any fixed point of a nonidentity element of Γ is also a fixed point of α^2, and the singular set S of T^7/Γ is the image in T^7/Γ of the 16 T^3 fixed by α^2, which are given by constant values of $x_4, \ldots, x_7$ in $\{\mathbb{Z}, \frac{1}{2} + \mathbb{Z}\}$. Using this we prove:

Lemma 12.8.2 *The singular set S of T^7/Γ is the disjoint union of*

(i) $2\ T^3/\mathbb{Z}_2^2$ locally modelled on $(T^3 \times \mathbb{R}^4)/\langle \alpha, \beta \rangle$, where α, β act on $T^3 \times \mathbb{R}^4$ by

$$\alpha : (x_1, x_2, x_3, y_4, y_5, y_6, y_7) \mapsto (x_1, -x_2, -x_3, y_5, -y_4, y_7, -y_6),$$
$$\beta : (x_1, x_2, x_3, y_4, y_5, y_6, y_7) \mapsto (-x_1, x_2, -x_3, y_6, -y_7, -y_4, y_5).$$

Here x_j are coordinates in $\mathbb{R}/\mathbb{Z}$ and y_j coordinates in $\mathbb{R}$.

(ii) $3\ T^3/\mathbb{Z}_2$ locally modelled on $(T^3 \times \mathbb{R}^4)/\langle \alpha \rangle$, where α acts on $T^3 \times \mathbb{R}^4$ by

$$\alpha : (x_1, x_2, x_3, y_4, y_5, y_6, y_7) \mapsto (x_1, -x_2, -x_3, y_5, -y_4, y_7, -y_6).$$

(iii) $2\ T^3$ locally modelled on $T^3 \times \mathbb{R}^4/\{\pm 1\}$.

Observe that each $T^3/\mathbb{Z}_2^2$ in part (i) contains 8 points $x_j \in \{\mathbb{Z}, \frac{1}{2} + \mathbb{Z}\}$, $y_j = 0$ modelled on $\mathbb{R}^7/\Gamma$. To resolve these singular points we cannot use a product $\mathbb{R}^3 \times Y$ or $\mathbb{R} \times Y$ where Y has holonomy SU(2) or SU(3)—we need a QALE G_2-manifold which is not a product. Fortunately, we have already constructed 3 topologically distinct families of QALE G_2-manifolds asymptotic to $\mathbb{R}^7/\Gamma$ in Example 11.2.5, and we shall use these.

Here is how we desingularize T^7/Γ to get a compact 7-manifold M with holonomy G_2. For each of the 2 $T^3/\mathbb{Z}_2^2$ in (i) we let δ be one of α, β or $\alpha\beta$. Then we can write $(T^3 \times \mathbb{R}^4)/\langle\delta\rangle = \mathcal{S}^1 \times (T^2 \times \mathbb{R}^4)/\langle\delta\rangle$, where we regard $(T^2 \times \mathbb{R}^4)/\langle\delta\rangle$ as a complex singularity. This singularity has a unique crepant resolution Y, upon which the quotient group $\Gamma/\langle\delta\rangle$ acts freely. Then $(\mathcal{S}^1 \times Y)/(\Gamma/\langle\delta\rangle)$ is a local model for how to resolve $T^3/\mathbb{Z}_2^2$.

The 3 $T^3/\mathbb{Z}_2$ in (ii) are locally modelled on $S^1 \times (T^2 \times \mathbb{R}^4)/\langle\alpha\rangle$, where we regard $(T^2 \times \mathbb{R}^4)/\langle\alpha\rangle$ as a complex singularity. There are two topologically distinct ways to resolve this. The singularity has a unique crepant resolution. But it also has a smoothing, constructed by first smoothing $T^2 \times (\mathbb{R}^4/\langle\alpha^2\rangle)$ to get $T^2 \times Y$, and then observing that $\langle\alpha\rangle = \mathbb{Z}_2$ acts freely on $T^2 \times Y$, and so $(T^2 \times Y)/\langle\alpha\rangle$ is a smoothing of $(T^2 \times \mathbb{R}^4)/\langle\alpha\rangle$ with fundamental group $\mathbb{Z}_2$. The 2 T^3 in (iii) have a unique resolution, as in §12.2.

Let $k = 0, 1, 2$ or 3, and let M be a compact 7-manifold constructed by resolving T^7/Γ using a 'crepant resolution' for k of the $T^3/\mathbb{Z}_2$ in (i) and a 'smoothing' for the remaining $3 - k$ $T^3/\mathbb{Z}_2$. Initially T^7/Γ has $b^2 = 3$ and $b^3 = 4$. Resolving each of the 2 $T^3/\mathbb{Z}_2^2$ in (i) fixes b^2 and adds 5 to b^3. Each of the k $T^3/\mathbb{Z}_2$ in (ii) resolved with a 'crepant resolution' adds 5 to b^2 and 5 to b^3, and each of the $3 - k$ $T^3/\mathbb{Z}_2$ resolved with a 'smoothing' fixes b^2 and adds 2 to b^3. Resolving each of the 2 T^3 in (iii) adds 1 to b^2 and 3 to b^3.

Adding up all the contributions we get

$$b^2(M) = 5 + 5k, \quad b^3(M) = 26 + 3k, \quad k = 0, 1, 2, 3.$$

Thus we have the four sets of Betti numbers

$$\bigl(b^2(M), b^3(M)\bigr) = (5, 26), \quad (10, 29), \quad (15, 32) \quad \text{and} \quad (20, 35).$$

Now T^7/Γ is simply-connected, but in calculating $\pi_1(M)$ we have to be a little careful because some of the QALE G_2-manifolds we are using to resolve T^7/Γ are not simply-connected, but have fundamental group $\mathbb{Z}_2$. It turns out that if $k > 0$, or if $k = 0$ and we choose δ differently for the 2 $T^3/\mathbb{Z}_2^2$ in (i), then M is simply-connected and has holonomy G_2. But if $k = 0$ and we choose δ the same for both $T^3/\mathbb{Z}_2^2$ in (i) then M has fundamental group $\mathbb{Z}_2 \ltimes \mathbb{Z}$ and holonomy $\mathbb{Z}_2 \ltimes \mathrm{SU}(3)$.

12.8.3 A group Γ of order 24

Let T^7 and (φ_0, g_0) be as in §12.2, but define $\alpha, \beta, \gamma : T^7 \to T^7$ by

$$\alpha : (x_1, \ldots, x_7) \mapsto (x_2, x_3, x_1, -x_6, x_4, -x_5, x_7),$$
$$\beta : (x_1, \ldots, x_7) \mapsto (-x_1, -x_2, x_3, \tfrac{1}{2} + x_4, \tfrac{1}{2} - x_5, -x_6, x_7),$$
$$\gamma : (x_1, \ldots, x_7) \mapsto (x_1, x_2, x_3, -x_4, -x_5, -x_6, -x_7).$$

Then $\alpha^3 = \beta^2 = \gamma^2 = 1$, and γ commutes with α, β. The group $\langle\alpha, \beta\rangle$ is isomorphic to A_4, of order 12, and thus $\Gamma = \langle\alpha, \beta, \gamma\rangle$ is isomorphic to $A_4 \times \mathbb{Z}_2$, of order 24. Also Γ preserves (φ_0, g_0). Our next lemma describes the singular set of T^7/Γ.

Lemma 12.8.3 *The singular set S of T^7/Γ is the union of*

(i) $T^3/\mathbb{Z}_2$ *from the α fixed points, locally modelled on* $(T^3 \times \mathbb{C}^2)/\langle\alpha, \gamma\rangle$, *where α, γ act on $T^3 \times \mathbb{C}^2$ by*

$$\alpha : (w_1, w_2, x_7, z_1, z_2) \mapsto (w_1, w_2, x_7, e^{2\pi i/3} z_1, e^{-2\pi i/3} z_2),$$
$$\gamma : (w_1, w_2, x_7, z_1, z_2) \mapsto (w_1, -w_2, -x_7, z_1, -z_2).$$

Here w_1, w_2 and x_7 are coordinates in $\mathbb{R}/\mathbb{Z}$ with $w_1 = x_1 + x_2 + x_3$ and $w_2 = x_4 + x_5 - x_6$, and z_1, z_2 are coordinates in $\mathbb{C}$.

(ii) 4 $T^3/\mathbb{Z}_3$ *from the γ fixed points, modelled on* $(T^3 \times \mathbb{R}^4/\{\pm 1\})/\langle \alpha \rangle$, *where α acts on* $T^3 \times \mathbb{R}^4/\{\pm 1\}$ *by*

$$\alpha : \big((x_1, x_2, x_3), \pm(y_4, y_5, y_6, y_7)\big) \mapsto \big((x_2, x_3, x_1), \pm(-y_6, y_4, -y_5, y_7)\big).$$

Here x_1, x_2, x_3 are coordinates in $\mathbb{R}/\mathbb{Z}$ and y_4, y_5, y_6, y_7 coordinates in $\mathbb{R}$. Each $T^3/\mathbb{Z}_3$ in (ii) *intersects the $T^3/\mathbb{Z}_2$ in* (i) *in $\mathcal{S}^1$.*

The singular set has only these two pieces because if δ is a nonidentity element of Γ with fixed points then either δ is conjugate to α or α^2, so that the fixed points of δ give the set (i), or δ is conjugate to $\alpha\gamma$ or $\alpha^2\gamma$, giving the intersection of (i) and (ii), or $\delta = \gamma$ giving the set (ii).

We can write $T^7/\langle \alpha, \beta \rangle$ as $\mathcal{S}^1 \times T^6/A_4$, where T^6/A_4 is a complex orbifold with singular set T^2, modelled on $T^2 \times \mathbb{C}^2/\mathbb{Z}_3$. We resolve the $\mathbb{C}^2/\mathbb{Z}_3$ singularity using an ALE space Y with holonomy $SU(2)$, and $H_2(Y, \mathbb{R}) = \mathbb{R}^2$, to get $\mathcal{S}^1 \times N$, where N is a Calabi–Yau 3-fold. There are two topologically distinct ways to choose Y such that the action of γ on $T^7/\langle \alpha, \beta \rangle$ lifts to $\mathcal{S}^1 \times N$, in which γ_* either acts as the identity on $H_2(Y, \mathbb{R})$, or with eigenvalues ± 1.

Calculation shows that in the first way, resolving $(\mathcal{S}^1 \times N)/\langle \gamma \rangle$ gives a compact, simply-connected 7-manifold M with holonomy G_2 and Betti numbers $b^2(M) = 10$ and $b^3(M) = 13$. In the second way, resolving $(\mathcal{S}^1 \times N)/\langle \gamma \rangle$ gives a compact, simply-connected 7-manifold M with holonomy G_2 and Betti numbers $b^2(M) = 5$ and $b^3(M) = 10$.

12.8.4 *A group Γ of order* 18

Write $\mathbb{R}^7 = \mathbb{R} \oplus \mathbb{C}^3$ with coordinates (x, z_1, z_2, z_3) with $x \in \mathbb{R}$ and $z_j \in \mathbb{C}$, and let (φ_0, g_0) be the flat G_2-structure on $\mathbb{R} \oplus \mathbb{C}^3$ defined as in Proposition 11.1.2. Define a lattice Λ in $\mathbb{R} \oplus \mathbb{C}^3$ by

$$\Lambda = \big\{(a, a_1 + e^{2\pi i/3}b_1, a_2 + e^{2\pi i/3}b_2, a_3 + e^{2\pi i/3}b_3) : a, a_j, b_j \in \mathbb{Z}\big\}.$$

Let $T^7 = (\mathbb{R} \oplus \mathbb{C}^3)/\Lambda$, and define $\alpha, \beta, \gamma : T^7 \to T^7$ by

$$\alpha : (x, z_1, z_2, z_3) + \Lambda \mapsto (x, e^{2\pi i/3}z_1, e^{-2\pi i/3}z_2, z_3) + \Lambda,$$
$$\beta : (x, z_1, z_2, z_3) + \Lambda \mapsto (\tfrac{1}{3} + x, z_1, e^{2\pi i/3}z_2, e^{-2\pi i/3}z_3) + \Lambda,$$
$$\gamma : (x, z_1, z_2, z_3) + \Lambda \mapsto (-x, \bar{z}_1, \bar{z}_2, \bar{z}_3) + \Lambda.$$

Then $\alpha^3 = \beta^3 = \gamma^2 = 1$, and $\Gamma = \langle \alpha, \beta, \gamma \rangle$ is a nonabelian group of order 18 which preserves (φ_0, g_0) on T^7. Note that $\langle \alpha, \beta \rangle \cong \mathbb{Z}_3 \times \mathbb{Z}_3$, and the 9 elements of $\Gamma \setminus \langle \alpha, \beta \rangle$ are all conjugate to γ.

Using $e^{2\pi i/3} = -\tfrac{1}{2} + i\tfrac{\sqrt{3}}{2}$, we find that the fixed points of α in T^7 are

$$\big\{(x, z_1, z_2, z_3) + \Lambda : z_1, z_2 \in \{0, i/\sqrt{3}, -i/\sqrt{3}\}\big\},$$

which is 9 copies of T^3. The action of β on the set of these 9 T^3 is trivial, and γ fixes one and switches the other 8 in pairs. The fixed points of γ in T^7 are 2 T^3, and the same is true for the other 8 conjugates of γ. So we can prove:

Lemma 12.8.4 *The singular set S of T^7/Γ is the union of*
(i) *$T^3/\langle\beta,\gamma\rangle$ from the T^3 fixed by α with $z_1 = z_2 = 0$, where $\langle\beta,\gamma\rangle \cong S_3$,*
(ii) *$4\ T^3/\langle\beta\rangle$ from the other 8 $\alpha\ T^3$, where $\langle\beta\rangle \cong \mathbb{Z}_3$ acts freely on T^3, and*
(iii) *$2\ T^3$ from the γ fixed points.*
The sets (i), (iii) *intersect in* $2\ \mathcal{S}^1$.

It seems odd that the γ singularities are $2\ T^3$ even though they intersect other parts of the singular set, but it is true. Each of the $2\ \mathcal{S}^1$ in the intersection of (i) and (iii) is locally modelled on $\mathcal{S}^1 \times \mathbb{C}^3/\langle\alpha,\gamma\rangle$, where α, γ act on $\mathbb{C}^3$ by

$$\alpha : (w_1, w_2, w_3) \mapsto (e^{2\pi i/3}w_1, e^{-2\pi i/3}w_2, w_3),$$
$$\gamma : (w_1, w_2, w_3) \mapsto (w_2, w_1, -w_3).$$

These complex coordinates w_1, w_2, w_3 are different to z_1, z_2, z_3.

We can resolve T^7/Γ to get a compact 7-manifold M with holonomy G_2. Initially T^7/Γ has $b^2 = 0$ and $b^3 = 4$. There are two topological choices for the resolution of the set (i), of which the first adds 1 to b^2 and 1 to b^3, and the second fixes b^2 and adds 1 to b^3. For each of the 4 $T^3/\langle\beta\rangle$ there are also two topological choices, the first adding 2 to b^2 and 2 to b^3, and the second fixing b^2 and adding 2 to b^3. In both cases, the choices arise because the α singularities look locally like $\mathbb{R}^3 \times \mathbb{C}^2/\mathbb{Z}_3$, so their resolution looks like $\mathbb{R}^3 \times Y$ for an ALE 4-manifold Y with holonomy $SU(2)$, and there are two possible topologically distinct actions of β on $H_2(Y,\mathbb{R})$.

Having made these choices, there is only one way to resolve each of the $\gamma\ T^3$, which adds 1 to b^2 and 3 to b^3. Summing up all the contributions gives

$$b^2(M) = 2 + k \quad \text{and} \quad b^3(M) = 19 \quad \text{for } k = 0, 1, \ldots, 9.$$

The resulting 7-manifold M is simply-connected, with holonomy G_2. This gives us 10 sets of Betti numbers. Note that they are the first examples we have found which do not satisfy $b^2 + b^3 \equiv 3 \bmod 4$.

12.9 Conclusions

In Figure 12.3 we give a graph of the Betti numbers $b^2(M)$ and $b^3(M)$ of known examples of compact, simply-connected 7-manifolds with holonomy G_2. The data is taken from the examples of this chapter, and those of [122]. There are 252 different sets of Betti numbers on the graph. The range of b^2 is between 0 and 28, and the range of b^3 between 4 and 215.

One striking thing about the graph is that the data points seem to group themselves on diagonal lines. This is because nearly all (238 out of 252 pairs) of the Betti numbers on the graph satisfy $b^2(M) + b^3(M) \equiv 3 \bmod 4$. The only ones which don't are the examples of §12.8.3 and some of the manifolds of [122]. The author doesn't know whether this is significant; but we note that $b^2 + b^3$ is the dimension of the Conformal Field Theory moduli space in String Theory, and that 'mirror' G_2-manifolds are expected to have the same value of $b^2 + b^3$.

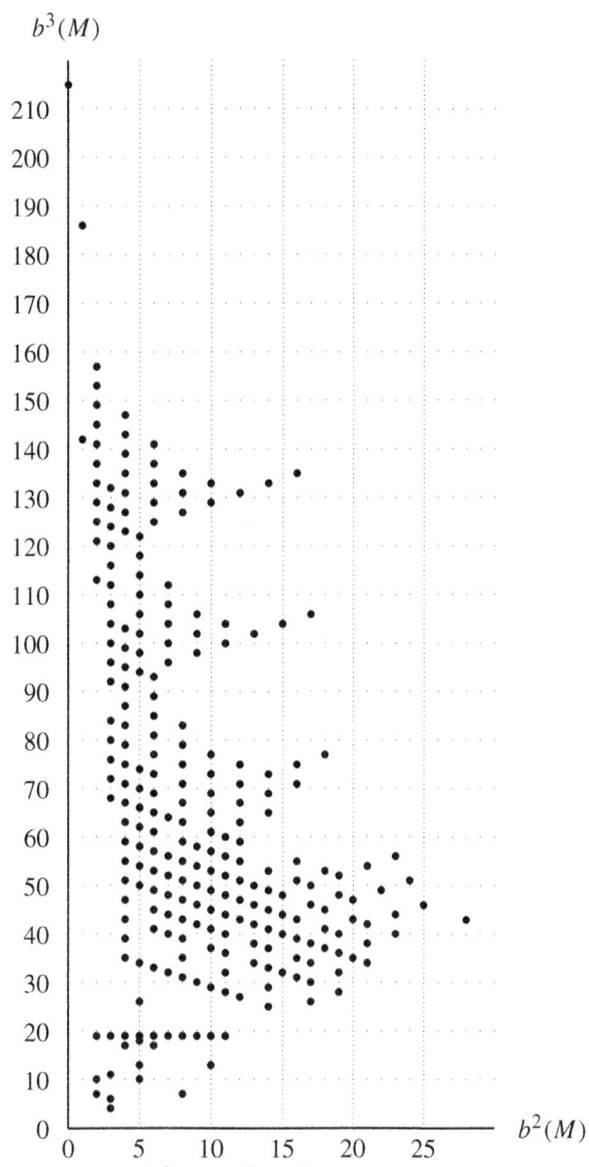

Figure 12.3. Betti numbers $(b^2(M), b^3(M))$ of compact, simply-connected 7-manifolds with holonomy G_2

A possible explanation is that the large majority of our examples can in fact be regarded as resolutions of $(T^3 \times K3)/\Gamma$, as in Method 1 of §11.9, where Γ is usually $\mathbb{Z}_2^2$. If for some reason resolutions M of $(T^3 \times K3)/\Gamma$ tend to have $b^2(M) + b^3(M) \equiv 3 \bmod 4$, then this would explain why it holds for so many of our examples.

We also note that the construction of Method 1 is likely to be quite powerful, producing many examples; one reason for this is that as a smooth 4-manifold the $K3$ surface has a large group of isotopy classes of diffeomorphisms acting nontrivially upon its cohomology, and so there are a lot of choices for Γ acting on $T^3 \times K3$.

Another obvious feature of the graph is that $b^3(M)$ is usually bigger than $b^2(M)$, sometimes by a lot. The only example we know of with $b^2(M) > b^3(M)$ is [122, Ex. 17], with $b^2(M) = 8$ and $b^3(M) = 7$. Here is one reason why. It can be shown that when we desingularize T^7/Γ using metrics with holonomy SU(2) and SU(3), we always add at least as much to b^3 as we do to b^2.

This is because all contributions to b^2 represent degrees of freedom in varying the Kähler form of the resolution. But changing the Kähler form changes the G_2-structure on M, so these degrees of freedom must also appear in $b^3(M)$, the dimension of the moduli space of G_2-structures.

However, there are two ways in which a resolution can increase b^3 but not b^2: deformations of the complex structure of the holonomy SU(2) and SU(3) metrics add to b^3 but fix b^2, and dividing an SU(2) or SU(3) metric by an antiholomorphic involution can kill contributions to b^2, but leave the contributions to b^3.

We usually find that using 'crepant resolutions' adds about the same amount to b^2 and b^3, whereas using 'smoothings' adds less to b^2 and more to b^3. In orbifolds which admit many different resolutions we can choose to resolve different parts of the singular set by 'crepant resolutions' or 'smoothings', resulting in a spectrum of different Betti numbers ranging from b^3 a bit bigger than b^2, to b^2 small and b^3 large.

In the author's experience, it is a general principle that the orbifolds T^7/Γ which have the greatest number of topologically distinct resolutions, and whose resolutions have the largest Betti numbers, tend to come from groups Γ with a small number of elements (say 8) such that the singular set of T^7/Γ is fairly simple.

There seem to be two reasons for this. The first is that the size and variety of the Betti numbers you get is roughly proportional to the number of 'pieces' in the singular set. But adding more elements to Γ does not necessarily increase the number of pieces in the singular set, because although the new elements create new pieces with their fixed points, they also identify pieces which were previously distinct.

The second reason is that topological choices in the resolution of singularities modelled on $\mathbb{C}^3/G$ work best if G is something small such as $\mathbb{Z}_2^2$, $\mathbb{Z}_3^2$ or $\mathbb{Z}_2 \times \mathbb{Z}_3$. If G is larger then $\mathbb{C}^3/G$ tends to be rigid, and can only be desingularized with a crepant resolution.

13

CONSTRUCTION OF COMPACT 8-MANIFOLDS WITH HOLONOMY Spin(7)

In this chapter we explain how to construct compact 8-manifolds with holonomy Spin(7), following the G_2 case in Chapter 11 closely. We begin with a torus T^8 equipped with a flat Spin(7)-structure (Ω_0, g_0), and a finite group Γ of automorphisms of T^8 preserving (Ω_0, g_0). Then T^8/Γ is an orbifold with a flat Spin(7)-structure (Ω_0, g_0).

We resolve the singularities of T^8/Γ to get a compact 8-manifold M, and define a 1-parameter family of Spin(7)-structures (Ω^t, g^t) on M depending on $t \in (0, \epsilon]$, such that the torsion of (Ω^t, g^t) is $O(t^{13/3})$. Then we show that for small t we can deform (Ω^t, g^t) to a nearby torsion-free Spin(7)-structure $(\tilde{\Omega}, \tilde{g})$ on M. If M is simply-connected and $\hat{A}(M) = 1$ then $\text{Hol}(\tilde{g}) = \text{Spin}(7)$.

In outline this proof is very similar to the G_2 case of Chapter 11. However, there are important technical differences between the two. In the G_2 case we wanted a 3-form $\varphi \in C^\infty(\mathcal{P}^3 M)$ satisfying $d\varphi = d\Theta(\varphi) = 0$, and in the Spin(7) case we want a 4-form $\Omega \in C^\infty(\mathcal{A} M)$ satisfying $d\Omega = 0$. These equations are superficially rather dissimilar, which creates differences in the proofs.

But there is also a deeper reason why the Spin(7) case is more difficult than the G_2 case, which is solely to do with the increase in dimension from 7 to 8. If we define (Ω^t, g^t) by the method used in Chapter 11, then its torsion is $O(t^4)$, where the power 4 is the expected codimension of the singularities of T^8/Γ. Now it turns out that in 8 dimensions this $O(t^4)$ estimate is not good enough for us to prove that for small t we can deform (Ω^t, g^t) to a nearby torsion-free Spin(7)-structure $(\tilde{\Omega}, \tilde{g})$.

To get round this we give a more complicated definition of (Ω^t, g^t), in which we use a devious technique to cancel out the $O(t^4)$ contribution to the torsion. In the end we show that (Ω^t, g^t) has torsion of order $O(t^{13/3})$, which is good enough for our purposes. Unfortunately, this does mean that we spend a lot of time defining and estimating exterior forms.

Most readers probably do not need to read the whole of this chapter. I suggest that on a first reading you should look at §13.1 and §13.2 in order to understand the notation, and then jump straight to the statement of Theorem 13.5.8, and the discussion of it in §13.5.2. The proof of Theorem 13.5.8, which basically takes the whole of §13.3–§13.5, can be missed out, or returned to later.

Section 13.6 completes our construction of compact 8-manifolds with holonomy Spin(7), which is the main result of the chapter. The proof of the analytic part of the construction, Theorem S2, is postponed to §13.7, and is similar to the proof of Theorem G2 in §11.8. As with the G_2 case, we shall dedicate the whole of Chapter 14 to examples of this construction, and so we give no examples in this chapter.

This chapter is original research by the author. It is based on the author's paper [120], but has been made more general and powerful by incorporating the ideas of Chapters 8 and 9. We will also explain a second way of constructing compact 8-manifolds with holonomy Spin(7) in Chapter 15.

13.1 Resolving Spin(7)-singularities and QALE Spin(7)-manifolds

Sections 11.1 and 11.2 explained how to resolve G_2-singularities $\mathbb{R}^7/G$ with holonomy SU(2) and SU(3), and defined and discussed QALE G_2-manifolds. We shall now carry out the same programme for Spin(7). Since we have seen most of the ideas before in Chapter 11 we will be brief, and leave out some of the proofs. We begin with an analogue of Proposition 11.1.1.

Proposition 13.1.1 *Suppose (Y, g_Y) is a Riemannian 4-manifold with holonomy* SU(2). *Then Y has a complex structure J, a Kähler form ω and a holomorphic volume form θ. Let $\mathbb{R}^4$ have coordinates $(x_1, \ldots, x_4)$ and metric $h = dx_1^2 + \cdots + dx_4^2$, and write $d\mathbf{x}_{ij\ldots l}$ for $dx_i \wedge dx_j \wedge \cdots \wedge dx_l$. Define a metric g and a 4-form Ω on $\mathbb{R}^4 \times Y$ by $g = h \times g_Y$ and*

$$\Omega = d\mathbf{x}_{1234} + (d\mathbf{x}_{12} + d\mathbf{x}_{34}) \wedge \omega + (d\mathbf{x}_{13} - d\mathbf{x}_{24}) \wedge \operatorname{Re}\theta \\ - (d\mathbf{x}_{14} + d\mathbf{x}_{23}) \wedge \operatorname{Im}\theta + \tfrac{1}{2}\omega \wedge \omega. \tag{13.1}$$

Then (Ω, g) is a torsion-free Spin(7)-*structure on $\mathbb{R}^4 \times Y$.*

Proof Near each $p \in N$ there exist complex coordinates (z_1, z_2), such that

$$g_Y = |dz_1|^2 + |dz_2|^2, \quad \omega = \frac{i}{2}(dz_1 \wedge d\bar{z}_1 + dz_2 \wedge d\bar{z}_2)$$
$$\text{and} \quad \theta = dz_1 \wedge dz_2 \quad \text{at } p,$$

as in (6.1). Setting $z_1 = x_5 + ix_6$ and $z_2 = x_7 + ix_8$, we find that

$$g_Y = dx_5^2 + \cdots + dx_8^2, \qquad \omega = d\mathbf{x}_{56} + d\mathbf{x}_{78},$$
$$\operatorname{Re}\theta = d\mathbf{x}_{57} - d\mathbf{x}_{68} \quad \text{and} \quad \operatorname{Im}\theta = d\mathbf{x}_{58} + d\mathbf{x}_{67} \quad \text{at } p.$$

Substituting these into $g = h \times g_Y$ and (13.1) gives $g = dx_1^2 + \cdots + dx_8^2$ and

$$\Omega = d\mathbf{x}_{1234} + (d\mathbf{x}_{12} + d\mathbf{x}_{34}) \wedge (d\mathbf{x}_{56} + d\mathbf{x}_{78}) + (d\mathbf{x}_{13} - d\mathbf{x}_{24}) \wedge (d\mathbf{x}_{57} - d\mathbf{x}_{68}) \\ - (d\mathbf{x}_{14} + d\mathbf{x}_{23}) \wedge (d\mathbf{x}_{58} + d\mathbf{x}_{67}) + d\mathbf{x}_{5678}.$$

at p. But these agree with the standard metric g_0 and 4-form Ω_0 on $\mathbb{R}^8$ given in Definition 10.5.1. Therefore (Ω, g) is a Spin(7)-structure on $\mathbb{R}^4 \times Y$. As $d\mathbf{x}_{ij\ldots l}$, ω and θ are all closed, (13.1) shows that $d\Omega = 0$. Thus (Ω, g) is torsion-free, by Proposition 10.5.3. □

Here are versions of Proposition 13.1.1 for SU(3), G_2 and SU(4).

Proposition 13.1.2 *Suppose (Y, g_Y) is a Riemannian 6-manifold with holonomy $SU(3)$, and associated Kähler form ω and holomorphic volume form θ. Let $\mathbb{R}^2$ have coordinates (x_1, x_2). Define a metric g and a 4-form Ω on $\mathbb{R}^2 \times Y$ by $g = (dx_1^2 + dx_2^2) \times g_Y$ and*

$$\Omega = dx_1 \wedge dx_2 \wedge \omega + dx_1 \wedge \operatorname{Re}\theta - dx_2 \wedge \operatorname{Im}\theta + \tfrac{1}{2}\omega \wedge \omega. \tag{13.2}$$

Then (Ω, g) is a torsion-free $Spin(7)$-structure on $\mathbb{R}^2 \times Y$.

Proposition 13.1.3 *Let (Y, g_Y) be a Riemannian 7-manifold with holonomy G_2, and associated 3-form φ and 4-form $*\varphi$. Define a metric g and a 4-form Ω on $\mathbb{R} \times Y$ by $g = dx^2 \times g_Y$ and $\Omega = dx \wedge \varphi + *\varphi$, where x is the coordinate on $\mathbb{R}$. Then (Ω, g) is a torsion-free $Spin(7)$-structure on $\mathbb{R} \times Y$.*

Proposition 13.1.4 *Suppose (Y, g) is a Riemannian 8-manifold with holonomy $SU(4)$, $Sp(2)$ or $SU(2) \times SU(2)$, and associated Kähler form ω and holomorphic volume form θ. Define a 4-form Ω on Y by $\Omega = \tfrac{1}{2}\omega \wedge \omega + \operatorname{Re}\theta$. Then (Ω, g) is a torsion-free $Spin(7)$-structure on Y.*

Note that in Proposition 13.1.4 we deal with the holonomy groups $SU(4)$, $Sp(2)$ and $SU(2) \times SU(2)$ together. Here is the analogue for $Spin(7)$ of Proposition 11.1.3. It is proved in the same way.

Proposition 13.1.5 *Let G be a finite subgroup of $Spin(7)$, and let*

$$V = \{v \in \mathbb{R}^8 : \gamma \cdot v = v \text{ for all } \gamma \in G\}.$$

Then either
 (i) $V = \mathbb{R}^8$;
 (ii) *V is a Cayley 4-plane (as in §10.8.3), G is conjugate to a subgroup of $SU(2)$, and $\mathbb{R}^8/G \cong \mathbb{R}^4 \times \mathbb{C}^2/G$;*
 (iii) *$V \cong \mathbb{R}^2$, and G is conjugate in $Spin(7)$ to a subgroup of $SU(3)$, and $\mathbb{R}^8/G \cong \mathbb{R}^2 \times \mathbb{C}^3/G$;*
 (iv) *$V \cong \mathbb{R}$, and G is conjugate in $Spin(7)$ to a subgroup of G_2, and $\mathbb{R}^8/G \cong \mathbb{R} \times \mathbb{R}^7/G$; or*
 (v) $V = \{0\}$.

Motivated by this we define *product* $Spin(7)$-*manifolds*.

Definition 13.1.6 We say that a torsion-free $Spin(7)$-structure (Ω, g) on $V \times Y$ is a *product $Spin(7)$-structure* if V is a Euclidean vector space and either
 (i) $V = \mathbb{R}^8$, and Y is a point, and (Ω, g) is the flat $Spin(7)$-structure (Ω_0, g_0) on $\mathbb{R}^8$ given in Definition 10.5.1;
 (ii) $V \cong \mathbb{R}^4$, and Y is a 4-manifold, and (Ω, g) is isomorphic to one of the $Spin(7)$-structures on $\mathbb{R}^4 \times Y$ constructed in Proposition 13.1.1;
 (iii) $V \cong \mathbb{R}^2$, and Y is a 6-manifold, and (Ω, g) is isomorphic to one of the $Spin(7)$-structures on $\mathbb{R}^2 \times Y$ constructed in Proposition 13.1.2;

(iv) $V \cong \mathbb{R}$, and Y is a 7-manifold, and (Ω, g) is isomorphic to one of the Spin(7)-structures on $\mathbb{R} \times Y$ constructed in Proposition 13.1.3; or

(v) $V = \{0\}$ and Y is an 8-manifold.

In each case g is a product metric $h_V \times g_Y$ on $V \times Y$, where h_V is Euclidean.

If (Y_1, g_1) and (Y_2, g_2) are Riemannian 4-manifolds with holonomy SU(2), then $(Y_1 \times Y_2, g_1 \times g_2)$ is an 8-manifold with holonomy SU(2) × SU(2), and so carries a torsion-free Spin(7)-structure (Ω, g) by Proposition 13.1.4. However, we have chosen not to define $Y_1 \times Y_2$ to be a product Spin(7)-structure. This is because we want the first factor V in $V \times Y$ to be a Euclidean vector space.

13.1.1   Quasi-ALE Spin(7)-manifolds

Next we define *ALE* and *QALE* Spin(7)-*manifolds*, two natural classes of resolutions of quotient singularities $\mathbb{R}^8/G$ for G a finite subgroup of Spin(7), equipped with torsion-free Spin(7)-structures (Ω, g) that are asymptotic to the flat Spin(7)-structure (Ω_0, g_0) on $\mathbb{R}^8/G$ in a certain way.

Definition 13.1.7 Let G be a finite subgroup of Spin(7) which acts freely on $\mathbb{R}^8 \setminus \{0\}$. Let (X, π) be a *real resolution* of $\mathbb{R}^8/G$, as in Definition 9.9.1, and (Ω, g) a torsion-free Spin(7)-structure on X. We say that (X, Ω, g) is an *ALE* Spin(7)-*manifold* if

$$\nabla^l(\pi_*(\Omega) - \Omega_0) = O(r^{-8-l}) \quad \text{on } \{x \in \mathbb{R}^8/G : r(x) > R\}, \text{ for all } l \geq 0.$$

Here Ω_0 is the usual Spin(7) 4-form on $\mathbb{R}^8/G$ given in Definition 10.5.1, r is the radius function on $\mathbb{R}^8/G$, and $R > 0$ is a constant.

Definition 13.1.8 Let G be a finite subgroup of Spin(7), and let (X, π) be a *real local product resolution* of $\mathbb{R}^8/G$, as in Definition 9.9.1. Let the notation $\mathcal{L}, I, V_i, W_i, \ldots$ be as in Definitions 9.9.1 and 9.9.2.

Let (Ω, g) be a torsion-free Spin(7)-structure on X. We say that (X, Ω, g) is a *QALE* Spin(7)-*manifold* if for each $i \in I$, there exists a product Spin(7)-structure (Ω_i, g_i) on $V_i \times Y_i$, such that on $V_i \times Y_i \setminus U_i$ we have

$$\nabla^l(\psi_i^*(\Omega) - \Omega_i) = \sum_{j \in I : i \not\leq j} O\bigl(\mu_{i,j}^{d_j} v_i^{-2-l}\bigr) \quad \text{for all } l \geq 0. \tag{13.3}$$

We say that (X, Ω, g) is *asymptotic to* $(V_i \times Y_i, \Omega_i, g_i)$. Now (13.3) implies that

$$\nabla^l(\psi_i^*(g) - g_i) = \sum_{j \in I : i \not\leq j} O\bigl(\mu_{i,j}^{d_j} v_i^{-2-l}\bigr) \tag{13.4}$$

as Ω determines g, and so g is a *QALE metric* on X by Definition 9.9.2.

Here is the analogue of Proposition 11.2.3, proved in the same way.

Proposition 13.1.9 *Let (X, Ω, g) be a QALE Spin(7)-manifold. Then for each $i \in I$ there is a smooth 3-form σ_i on $V_i \times Y_i \setminus U_i$, satisfying $\psi_i^*(\Omega) - \Omega_i = \mathrm{d}\sigma_i$, such that on $V_i \times Y_i \setminus U_i$ we have*

$$\nabla^l \sigma_i = \sum_{j \in I: i \not\leq j} O\bigl(\mu_{i,j}^{d_j} v_i^{-1-l}\bigr) \quad \textit{for all } l \geqslant 0. \tag{13.5}$$

We can use the results of Chapters 8 and 9 to construct ALE and QALE Spin(7)-manifolds, as in Theorem 11.2.4.

Theorem 13.1.10 *Let $G \subset \mathrm{Spin}(7)$ be a finite group.*
  (a) *Suppose G is conjugate to a subgroup of $\mathrm{SU}(2)$ or $\mathrm{SU}(3)$. Then there exists a QALE Spin(7)-manifold (X, Ω, g) asymptotic to $\mathbb{R}^8/G$.*
  (b) *Suppose G is conjugate to $G' \subset \mathrm{SU}(4)$ and acts freely on $\mathbb{R}^8 \setminus \{0\}$, and that $\mathbb{C}^4/G'$ admits a crepant resolution. Then there exists an ALE Spin(7)-manifold (X, Ω, g) asymptotic to $\mathbb{R}^8/G$.*
  (c) *Suppose G is conjugate to a subgroup G' of $\mathrm{SU}(4)$ or $\mathrm{Sp}(2)$, and that $\mathbb{C}^4/G'$ admits a crepant resolution. Then there exists a QALE Spin(7)-manifold (X, Ω, g) asymptotic to $\mathbb{R}^8/G$.*

Here the inclusions $\mathrm{SU}(m), \mathrm{Sp}(2) \subset \mathrm{Spin}(7)$ are defined in Theorem 10.5.7.

Proof If G is conjugate to a subgroup of $\mathrm{SU}(2)$ then there is a splitting $\mathbb{R}^8 \cong \mathbb{R}^4 \oplus \mathbb{C}^2$ such that $\mathbb{R}^8/G \cong \mathbb{R}^4 \times \mathbb{C}^2/G$, where G acts on $\mathbb{C}^2$ as a finite subgroup of $\mathrm{SU}(2)$. Then from §7.2 there exists a family of ALE manifolds (Y, g_Y) with holonomy $\mathrm{SU}(2)$ asymptotic to $\mathbb{C}^2/Y$. Proposition 13.1.1 constructs a Spin(7)-manifold $(\mathbb{R}^4 \times Y, \Omega, g)$, and this is a QALE Spin(7)-manifold asymptotic to $\mathbb{R}^8/G$.

Similarly, if G is conjugate to a subgroup of $\mathrm{SU}(3)$ then we can write $\mathbb{R}^8/G \cong \mathbb{R}^2 \times \mathbb{C}^3/G$. By Theorem 6.4.1 there exists a crepant resolution Y of $\mathbb{C}^3/G$, and Theorem 9.3.3 gives a family of Ricci-flat QALE Kähler metrics g_Y on Y. Proposition 13.1.2 then constructs a Spin(7)-manifold $(\mathbb{R}^2 \times Y, \Omega, g)$, and this is a QALE Spin(7)-manifold asymptotic to $\mathbb{R}^8/G$. This proves part (a).

To prove (b) we take a crepant resolution X of $\mathbb{C}^4/G'$ and use Theorems 8.2.3 and 8.2.4 to construct an ALE Kähler metric on X with holonomy $\mathrm{SU}(4)$. Proposition 13.1.4 gives a torsion-free Spin(7)-structure (Ω, g) on X, and (X, Ω, g) is an ALE Spin(7)-manifold asymptotic to $\mathbb{R}^8/G$. Part (c) follows in the same way, using Theorems 9.3.3 and 9.3.4. □

The theorem gives us many examples of ALE and QALE Spin(7)-manifolds. We can make more examples by dividing the manifolds given by Theorem 13.1.10 by a freely acting finite group, as in Example 11.2.5 for the G_2 case. Examples of this will be given in §15.1.

It is a natural question whether there exist ALE or QALE Spin(7)-manifolds (X, Ω, g) with holonomy Spin(7), rather than some proper subgroup. The author has constructed examples of QALE 8-manifolds with holonomy Spin(7), using the method briefly described at the end of §11.2.1 for the G_2 case. We will not discuss this here, as it would take us too far afield. However, the author has so far been unable to find any examples of ALE 8-manifolds with holonomy Spin(7).

13.2 Orbifolds T^8/Γ, R-data, and resolutions of T^8/Γ

We now generalize the material of §11.3 and §11.4 to the Spin(7) case, defining notation to describe the singularities of orbifolds T^8/Γ with flat Spin(7)-structures (Ω_0, g_0), the idea of R-data, and the resolution (M, π) of T^8/Γ associated to a set of R-data for T^8/Γ. Sections 13.3–13.7 will prove that there exist torsion-free Spin(7)-structures on M. As we have met these ideas already in Chapter 11, we will be fairly brief.

13.2.1 *Orbifolds T^8/Γ and R-data*

Let (Ω_0, g_0) be the flat Spin(7)-structure on $\mathbb{R}^8$ given in Definition 10.5.1. Let Λ be a lattice in $\mathbb{R}^8$, and set $T^8 = \mathbb{R}^8/\Lambda$. Then (Ω_0, g_0) pushes down to T^8. Suppose Γ is a finite group of automorphisms of T^8 preserving (Ω_0, g_0). Then T^8/Γ is an orbifold. If A is a subgroup of Γ and F a subset of T^8, define the fixed point set Fix(A), the centralizer $C(F)$ and the normalizer $N(F)$ as in §11.3.

Definition 13.2.1 Define a finite set $\mathcal{L}$ of subsets of T^8 by

$$\mathcal{L} = \{F : F \text{ is a connected component of Fix}(A), \text{ where } A \text{ is a subgroup of } \Gamma\}.$$

Let I be an *indexing set* for $\mathcal{L}$, so that we may write $\mathcal{L} = \{F_i : i \in I\}$. Since $T^8 =$ Fix$(\{1\})$ is connected, we see that $T^8 \in \mathcal{L}$. Let the index of T^8 be 0, so that $F_0 = T^8$. Define a *partial order* $\succeq$ on I by $i \succeq j$ if $F_i \subseteq F_j$. Then $i \succeq 0$ for all $i \in I$. We shall also write $i \succ j$ if $i \succeq j$ and $i \neq j$.

Define $A_i = C(F_i)$ for each $i \in I$. Then A_i is a subgroup of Γ with $A_0 = \{1\}$, and F_i is a connected component of Fix(A_i). Define V_i to be the vector subspace of $\mathbb{R}^8$ fixed by A_i, and let $n_i = \dim V_i$ and $\Lambda_i = V_i \cap \Lambda$. Then Λ_i is a lattice in V_i, and F_i is isomorphic to V_i/Λ_i. Thus F_i is a torus T^{n_i} in T^8, and each tangent space $T_p F_i$ is naturally isomorphic to V_i.

For each $i \in I$, define W_i to be the perpendicular subspace to V_i in $\mathbb{R}^8$, using g_0. Then $\mathbb{R}^8 = V_i \oplus W_i$, and A_i acts on W_i, with $\mathbb{R}^8/A_i \cong V_i \times W_i/A_i$. Define B_i to be $N(F_i)/C(F_i)$. Then B_i acts naturally on F_i, V_i and W_i/A_i. For each $\gamma \in \Gamma$ and $i \in I$, let $\gamma \cdot i$ be the unique element of I such that $F_{\gamma \cdot i} = \gamma F_i$. This defines an action of Γ on I, which also satisfies $V_{\gamma \cdot i} = \gamma V_i$, $W_{\gamma \cdot i} = \gamma W_i$, and $A_{\gamma \cdot i} = \gamma A_i \gamma^{-1}$.

For each $x \in T^8$ there is a unique $i \in I$ with $x \in F_i$ and $C(\{x\}) = A_i$. Then T^8/Γ is locally isomorphic to $\mathbb{R}^8/A_i = V_i \times W_i/A_i$ near $x\Gamma$. Thus $x\Gamma$ is a singular point if and only if $i \neq 0$. That is, $x\Gamma$ is a singular point of T^8/Γ if and only if $x \in F_i$ for some $i \in I$ with $i \neq 0$, so the *singular set* S of T^8/Γ is

$$S = \bigcup_{i \in I \setminus \{0\}} F_i \Big/ \Gamma.$$

Here is the definition of R-data, short for *Resolution data*.

Definition 13.2.2 We define a *set of R-data* for T^8/Γ to be
 (a) a real local product resolution (Y_i, π_i) of W_i/A_i for each $i \in I$;
 (b) a torsion-free Spin(7)-structure (Ω_i, g_i) on $V_i \times Y_i$ for each $i \in I$; and
 (c) a diffeomorphism $\chi_{\gamma,i} : Y_i \to Y_{\gamma \cdot i}$ for each $i \in I$ and $\gamma \in \Gamma$.

These must satisfy the conditions:
 (i) $(V_i \times Y_i, \Omega_i, g_i)$ is both a *product* Spin(7)-*manifold* and a *QALE* Spin(7)-*manifold* asymptotic to $\mathbb{R}^8/A_i$ in the sense of Definitions 13.1.6 and 13.1.8, with projection $\mathrm{id} \times \pi_i : V_i \times Y_i \to V_i \times W_i/A_i \cong \mathbb{R}^8/A_i$;
 (ii) If $i \succeq j$ then $(V_i \times Y_i, \Omega_i, g_i)$ is asymptotic to $(V_j \times Y_j, \Omega_j, g_j)$ in the sense of Definition 13.1.8; and
 (iii) For each $i \in I$ and $\gamma \in \Gamma$, the following diagram commutes:

$$\begin{array}{ccc} Y_i & \xrightarrow{\chi_{\gamma,i}} & Y_{\gamma\cdot i} \\ \downarrow{\pi_i} & & \downarrow{\pi_{\gamma\cdot i}} \\ W_i/A_i & \xrightarrow{\gamma} & W_{\gamma\cdot i}/A_{\gamma\cdot i}, \end{array}$$

and the map $\gamma \times \chi_{\gamma,i} : V_i \times Y_i \to V_{\gamma \cdot i} \times Y_{\gamma \cdot i}$ satisfies $(\gamma \times \chi_{\gamma,i})^*(\Omega_{\gamma\cdot i}) = \Omega_i$ and $(\gamma \times \chi_{\gamma,i})^*(g_{\gamma\cdot i}) = g_i$.

13.2.2 Resolutions of T^8/Γ

Let T^8/Γ and (Ω_0, g_0) be as above, and suppose we are given a set of R-data (Y_i, π_i), (Ω_i, g_i), $\chi_{\gamma,i}$ for T^8/Γ, as in Definition 13.2.2. We shall use this to construct a resolution (M, π) of T^8/Γ. First we define tubular neighbourhoods O_i, P_i of F_i in T^8, as in §11.4.2.

Definition 13.2.3 Choose constants $\zeta_0, \zeta_1, \zeta_2, \zeta_4, \zeta_8 > 0$. For each $i \in I$ we have $n_i = \dim V_i$, so $n_i = 0, 1, 2, 4$ or 8 by Proposition 13.1.5. Define $O_i \subset P_i \subset T^8$ by

$$O_i = \{x \in T^8 : d(x, F_i) \leq \zeta_{n_i}\}, \quad P_i = \{x \in T^8 : d(x, F_i) < 2\zeta_{n_i}\},$$

where $d(\,,\,)$ is the distance in T^8 induced by g_0. The constants ζ_j must satisfy:
 (i) There are natural isomorphisms

$$O_i \cong F_i \times \overline{B}_i(\zeta_{n_i}) \quad \text{and} \quad P_i \cong F_i \times B_i(2\zeta_{n_i}),$$

where $\overline{B}_i(\zeta_{n_i})$ is the closed ball of radius ζ_{n_i} and $B_i(2\zeta_{n_i})$ the open ball of radius $2\zeta_{n_i}$ about 0 in W_i; and
 (ii) For each $i, j \in I$ with $i \not\succeq j$ and $j \not\succeq i$, we have

$$P_i \cap P_j \subseteq \bigcup_{\substack{k \in I: \\ k \succ i, k \succ j}} O_k.$$

When $F_i \cap F_j = \emptyset$ there are no $k \in I$ with $k \succ i, k \succ j$, so $P_i \cap P_j = \emptyset$.

We can always choose constants ζ_i to satisfy (i) and (ii). In the next result we use the O_i, P_i to define a special open cover $\{Q_i : i \in I\}$ for T^8. The proof follows that of Proposition 11.4.3.

Proposition 13.2.4 *For each $i \in I$, define $Q_i \subset T^8$ by*

$$Q_i = P_i \setminus \bigcup_{j \in I : j \succ i} O_j.$$

Then Q_i is open, $T^8 = \bigcup_{i \in I} Q_i$, and if $Q_i \cap Q_j \neq \emptyset$ then $i \succeq j$ or $j \succeq i$.

The group A_i acts on Q_i, and Q_i/A_i is naturally isomorphic to an open subset in $F_i \times W_i/A_i$. Also, B_i acts freely on Q_i/A_i, and the image of Q_i in T^8/Γ is isomorphic to $Q_i/N(F_i) = (Q_i/A_i)/B_i$.

This shows that the Q_i form an open cover for T^8, with simple rules for when the overlaps $Q_i \cap Q_j$ can be nonempty. Similarly, T^8/Γ is covered by open sets $(Q_i/A_i)/B_i$. As the B_i-action on Q_i/A_i is free, the singularities of $(Q_i/A_i)/B_i$ are just product singularities modelled on $V_i \times W_i/A_i$. We have found an open cover for T^8/Γ in which both the singularities of each open set, and the overlaps between open sets, are simple and easily understood.

Definition 13.2.5 Regard Q_i/A_i as an open subset of $F_i \times W_i/A_i$, as in Proposition 13.2.4. For each $i \in I$, define M_i to be the inverse image of Q_i/A_i under the map $\mathrm{id} \times \pi_i : F_i \times Y_i \to F_i \times W_i/A_i$. Then M_i is an open subset of $F_i \times Y_i$.

Define an *equivalence relation* '$\sim$' on the disjoint union $\coprod_{i \in I} M_i$ as follows. Let $j, k \in I$ and $x \in M_j$, $y \in M_k$. We say that $x \sim y$ if either

(a) there exists $\gamma \in \Gamma$ such that $i = \gamma \cdot k$ satisfies $i \succeq j$, $x \in M_j \cap (P_i \times Y_j)$ and $\mathrm{id} \times \psi'_j(x) = \gamma \times \chi_{\gamma,k}(y)$, where $\gamma \times \chi_{\gamma,k} : F_k \times Y_k \to F_i \times Y_i$, or

(b) there exists $\gamma \in \Gamma$ such that $i = \gamma \cdot j$ satisfies $i \succeq k$, $y \in M_k \cap (P_i \times Y_k)$ and $\gamma \times \chi_{\gamma,j}(x) = \mathrm{id} \times \psi'_k(y)$, where $\gamma \times \chi_{\gamma,j} : F_j \times Y_j \to F_i \times Y_i$.

Define the *resolution* M of T^8/Γ to be $\coprod_{i \in I} M_i / \sim$. It is easy to show that if $x, y \in M_i$ then $x \sim y$ if and only if $x = b \cdot y$ for some $b \in B_i$. Thus, the image of M_i in M is M_i/B_i, and M is covered by patches M_i/B_i for $i \in I$. But B_i acts freely on M_i, and so M_i/B_i is a manifold, and therefore M is a *compact 8-manifold*.

If $x \in M_i$ then $\mathrm{id} \times \pi_i(x) \in Q_i/A_i \subset T^8/A_i$. Let $\phi_i : T^8/A_i \to T^8/\Gamma$ be the natural projection, and let $[x]$ be the equivalence class of x under $\sim$. Define the *projection* $\pi : M \to T^8/\Gamma$ by $\pi([x]) = \phi_i(\mathrm{id} \times \pi_i(x))$. Then π is well-defined, continuous and surjective, and $\pi : M \setminus \pi^{-1}(S) \to T^8/\Gamma \setminus S$ is a diffeomorphism, where S is the singular set of T^8/Γ.

Finally we define special partitions of unity on T^8, T^8/Γ and M.

Definition 13.2.6 Choose a *partition of unity* $\{\eta'_i : i \in I\}$ on T^8 such that:

(i) $\eta'_i : T^8 \to [0, 1]$ is smooth, and zero outside Q_i for each $i \in I$;

(ii) $\eta'_i = \eta'_{\gamma \cdot i} \circ \gamma$ for all $i \in I$ and $\gamma \in \Gamma$, and $\sum_{i \in I} \eta'_i \equiv 1$; and

CONSTRUCTION OF VARIOUS EXTERIOR FORMS 349

(iii) For each $i \in I \setminus \{0\}$ we can identify T^8 with $F_i \times W_i$ near F_i. In a small neighbourhood of F_i, each function η'_j should be the pull-back to $F_i \times W_i$ of a smooth function on F_i. That is, each η'_j should be constant in the W_i directions near F_i in T^8.

We can always find a set of functions η'_i satisfying (i)–(iii). Let I/Γ be the set of orbits of Γ in I, and for each $i \in I$ define $[i] \in I/\Gamma$ by $[i] = \{\gamma \cdot i : \gamma \in \Gamma\}$. For each $[i] \in I/\Gamma$, define $\eta'_{[i]} = \sum_{j \in [i]} \eta'_j$. Then $\eta'_{[i]} : T^8 \to [0, 1]$ is Γ-invariant by part (ii), and so pushes down to T^8/Γ. Thus we have a set of smooth functions $\eta'_{[i]} : T^8/\Gamma \to [0,1]$ for $[i] \in I/\Gamma$, such that $\eta'_{[i]}$ is zero outside $(Q_i/A_i)/B_i$ and $\sum_{[i] \in I/\Gamma} \eta'_{[i]} \equiv 1$.

For each $[i] \in I/\Gamma$ define $\eta_{[i]} : M \to [0, 1]$ to be the pull-back $\pi^*(\eta'_{[i]})$ of $\eta'_{[i]}$ to M. Then parts (i)–(iii) above give

(a) $\eta_{[i]} : M \to [0, 1]$ is smooth, and zero outside M_i/B_i for each $[i] \in I/\Gamma$; and

(b) $\sum_{[i] \in I/\Gamma} \eta_{[i]} \equiv 1$.

To prove that $\eta_{[i]}$ is smooth in (a) we have to use parts (i) and (iii) above. In general the pull-back $\pi^*(f)$ of a smooth function f on T^8/Γ will be continuous on M, but not necessarily differentiable. But part (iii) ensures that $\pi^*(\eta'_{[i]})$ is smooth near the inverse image of every singular point.

13.3 Construction of various exterior forms

Next we define and discuss several families of exterior forms on the $V_i \times Y_i$ and on T^8/Γ, which will be needed in §13.4 and §13.5. We begin in §13.3.1 by defining 3-forms $\sigma_{i,j}$ as in §11.5, which roughly speaking satisfy $d\sigma_{i,j} = \Omega_i - \Omega_j$, where $i \succ j$ and (Ω_i, g_i), (Ω_j, g_j) are the Spin(7)-structures on $V_i \times Y_i$ and $V_j \times Y_j$. We will use the $\sigma_{i,j}$ to join together the Ω_i to get a closed 4-form on M.

In §13.3.2 and §13.3.3 we construct forms $\tau_i, \upsilon_i, \Xi, \Phi, \Psi$ and Υ which have no analogue in Chapter 11. Here is why we will need them. By Proposition 13.1.5, if $i \in I \setminus \{0\}$ then $n_i = \dim V_i$ is 0, 1, 2 or 4. Each $i \in I \setminus \{0\}$ contributes a piece $F_i \Gamma/\Gamma$ of dimension n_i to the singular set S of T^8/Γ. It will turn out later that the 'error' we make in defining a Spin(7)-structure (Ω^t, g^t) on the resolution of $F_i \Gamma/\Gamma$ is roughly of size $O(t^{8-n_i})$, where $t \in (0, 1]$ is a small parameter. Thus the largest errors come from the pieces $F_i \Gamma/\Gamma$ with $n_i = 4$.

For reasons to be explained in §13.5.2, we need to make special arrangements to reduce the size of the error in resolving the $F_i \Gamma/\Gamma$ with $n_i = 4$ from $O(t^4)$ to $O(t^{4+\epsilon})$, for some $\epsilon > 0$. To do this we extract the leading term in $\Omega_i - \Omega_0$, which will be responsible for the $O(t^4)$ error, and write it as $d\tau_i$, where τ_i is a special 3-form on $\mathbb{R}^8 \setminus V_i$. Then we use the τ_i to construct forms Ξ, Φ, Ψ and Υ on T^8/Γ, which will be used in §13.4 and §13.5 to cancel out the $O(t^4)$ part of the 'error' we make in resolving the $F_i \Gamma/\Gamma$.

13.3.1 The 3-forms $\sigma_{i,j}$

Let T^8/Γ be equipped with a flat Spin(7)-structure (Ω_0, g_0) and a set of R-data (Y_i, π_i), (Ω_i, g_i) and $\chi_{\gamma \cdot i}$. Suppose $i, j \in I$ with $i \succ j$. Then by Proposition 13.1.9 there is a 3-form $\sigma_{i,j}$ on $V_j \times Y_j \setminus V_i \times U'_j$, such that

$$(\mathrm{id} \times \psi'_j)^*(\Omega_i) - \Omega_j = d\sigma_{i,j} \quad \text{and}$$

$$\nabla^l \sigma_{i,j} = \sum_{k \in I': j \not\succeq k} O\big(\mu_{j,k}^{d_k} (v'_j)^{-1-l}\big) \tag{13.6}$$

on $V_j \times Y_j \setminus V_i \times U'_j$, for all $l \geq 0$, where $I' = \{m \in I : i \succeq m\}$. We choose the $\sigma_{i,j}$ to be invariant under the natural actions of Γ and V_i.

Now suppose $i, j, k \in I$ and $i \succ j \succ k$. Write $V'_k = V_k \cap W_i$ and $V''_k = V_k \cap W_j$, and let $\psi'_k : V'_k \times Y_k \to Y_i$ and $\psi''_k : V''_k \times Y_k \to Y_j$ be the natural maps. Then $(\mathrm{id} \times \psi''_k)^*(\sigma_{i,j})$, $\sigma_{i,k}$ and $\sigma_{j,k}$ are 3-forms on $V_k \times Y_k \setminus V_i \times U'_k$ satisfying

$$d\big((\mathrm{id} \times \psi''_k)^*(\sigma_{i,j})\big) = (\mathrm{id} \times \psi'_k)^*(\Omega_i) - (\mathrm{id} \times \psi''_k)^*(\Omega_j),$$

$$d\sigma_{j,k} = (\mathrm{id} \times \psi''_k)^*(\Omega_j) - \Omega_k \quad \text{and} \quad d\sigma_{i,k} = (\mathrm{id} \times \psi'_k)^*(\Omega_i) - \Omega_k,$$

$$\text{and so} \quad d\big((\mathrm{id} \times \psi''_k)^*(\sigma_{i,j}) + \sigma_{j,k} - \sigma_{i,k}\big) = 0.$$

Thus $(\mathrm{id} \times \psi''_k)^*(\sigma_{i,j}) + \sigma_{j,k} - \sigma_{i,k}$ is a closed 3-form. But $\sigma_{i,j}$, $\sigma_{j,k}$ and $\sigma_{i,k}$ are themselves only defined up to addition of closed 3-forms. We shall assume that the $\sigma_{i,j}$ have been chosen such that

$$(\mathrm{id} \times \psi''_k)^*(\sigma_{i,j}) + \sigma_{j,k} - \sigma_{i,k} = 0 \tag{13.7}$$

whenever $i \succ j \succ k$. It can be shown that this is possible.

13.3.2 The 3-forms τ_i and υ_i

Suppose that $i \in I$ with $n_i = 4$. Then $\dim V_i = \dim W_i = 4$, and so Y_i is an ALE manifold of dimension 4 with holonomy $SU(2)$, asymptotic to W_i/A_i. Now from §7.2 and Chapter 8 we know a great deal about such 4-manifolds Y_i, their metrics and asymptotic behaviour. We shall use this to describe more precisely how the Spin(7)-structure (Ω_i, g_i) on $V_i \times Y_i$ approaches the flat Spin(7)-structure (Ω_0, g_0) on $\mathbb{R}^8$.

First consider what §13.3.1 says when $n_i = 4$. Then the only $j \in I$ with $i \succ j$ is $j = 0$, since if $i \succ j$ then $\dim F_j > \dim F_i$, which forces $\dim F_j = 8$ by Proposition 13.1.5, so that $F_j = T^8$. Thus $I' = \{0, i\}$, and (13.6) becomes

$$(\mathrm{id} \times \psi'_0)^*(\Omega_i) - \Omega_0 = d\sigma_{i,0} \quad \text{and} \quad \nabla^l \sigma_{i,0} = O\big((v'_0)^{-3-l}\big)$$

on $\mathbb{R}^8 \setminus V_i \times U'_0$, for all $l \geq 0$, where U'_0 is the ball radius of $R_i > 0$ about 0 in W_i. Here $\mu_{0,i}$ is the distance to V_i in $\mathbb{R}^8$, and $v'_0 = \mu_{0,i} + 1$.

Now $\sigma_{i,0}$ is only defined up to the addition of closed 3-forms. It can be shown that we can choose $\sigma_{i,0}$ and a 3-form τ_i on $\mathbb{R}^8 \setminus V_i$ with the following properties:

(i) τ_i is invariant under three kinds of transformation of $\mathbb{R}^8 \setminus V_i$: firstly *translations* in V_i, secondly *dilations* of $\mathbb{R}^8$, and thirdly the natural action of $SU(2)$ on W_i, regarding $\mathbb{R}^8 \setminus V_i$ as $V_i \times (W_i \setminus \{0\})$. Here for each $t > 0$ the *dilation* $t : \mathbb{R}^8 \to \mathbb{R}^8$ maps v to tv in the obvious way. As τ_i is a 3-form, the invariance of τ_i under V_i translations and dilations implies that

$$\nabla^l \tau_i = O(\mu_{0,i}^{-3-l}) \quad \text{on } \mathbb{R}^8 \setminus V_i \text{ for all } l \geq 0. \tag{13.8}$$

Since the group generated by V_i translations, dilations and SU(2) rotations acts transitively on $\mathbb{R}^8 \setminus V_i$, we see that τ_i is determined by its value at a single point in $\mathbb{R}^8 \setminus V_i$.

(ii) $d\tau_i$ is an *anti-self-dual* 4-form on $\mathbb{R}^8 \setminus V_i$.
(iii) $\nabla^l(\sigma_{i,0} - \tau_i) = O\big((v_0')^{-4-l}\big)$ on $\mathbb{R}^8 \setminus V_i \times U_0'$ for all $l \geq 0$.

We can write τ_i explicitly as $\tau_i = \sum_{k=1}^3 \omega_k \wedge \lambda_k$, where the ω_k are constant self-dual 2-forms on V_i, and the λ_k are certain special 1-forms on $W_i \setminus \{0\}$. If we identify W_i with $\mathbb{H}$ with complex structures I, J, K, then each λ_k is a linear combination of the three 1-forms $I d(r^{-2})$, $J d(r^{-2})$ and $K d(r^{-2})$ on $\mathbb{H} \setminus \{0\}$. These have the property that $d\lambda_k$ is anti-self-dual on $W_i \setminus \{0\}$.

Since ω_k is self-dual on V_i and $d\lambda_k$ anti-self-dual on $W_i \setminus \{0\}$, it follows that $d\tau_i = \sum_{k=1}^3 \omega_k \wedge d\lambda_k$ is anti-self-dual on $V_i \times (W_i \setminus \{0\})$, which proves (ii). Basically, these conditions mean that τ_i is the highest-order $O\big((v_0')^{-3}\big)$ term in the asymptotic expansion of $\sigma_{i,0}$, and the remainder $\sigma_{i,0} - \tau_i$ decays with order at least $O\big((v_0')^{-4}\big)$. In our next result, for each $i \in I \setminus \{0\}$ we write $\sigma_{i,0}$ as a sum of τ_j and a remainder υ_i, which we estimate.

Proposition 13.3.1 *The 3-forms $\sigma_{i,j}$ may be chosen such that for each $i \in I \setminus \{0\}$ there exists a 3-form υ_i on $\mathbb{R}^8 \setminus V_i \times U_0'$ satisfying*

$$\sigma_{i,0} = \upsilon_i + \sum_{\substack{j \in I: \\ i \succeq j, \, n_j = 4}} \tau_j \quad \text{and} \quad \nabla^l \upsilon_i = O\big((v_0')^{-4-l}\big) \quad \text{for all } l \geq 0, \tag{13.9}$$

where $v_0'(x) = 1 + \min\{d(x, V_j) : j \in I \setminus \{0\}, i \succeq j\}$ on $\mathbb{R}^8 \setminus V_i \times U_0'$.

Proof We consider three cases. Firstly, suppose $n_i = 4$. Then the first equation of (13.9) becomes $\sigma_{i,0} = \upsilon_i + \tau_i$, so we define $\upsilon_i = \sigma_{i,0} - \tau_i$ and the second equation of (13.9) follows from part (iii) above.

Secondly, suppose that there are no $j \in I$ with $i \succeq j$ and $n_j = 4$. Then the first equation of (13.9) becomes $\sigma_{i,0} = \upsilon_i$. Thus (13.6) gives

$$\nabla^l \upsilon_i = \sum_{k \in I \setminus \{0\}: i \succeq k} O\big(\mu_{0,k}^{d_k}(v_0')^{-1-l}\big) \leq \sum_{k \in I \setminus \{0\}: i \succeq k} O\big((v_0')^{d_k - 1 - l}\big).$$

As $i \succeq k$ we have $n_k \neq 4$, and $n_k \neq 8$ as $k \neq 0$, so that $n_k = 0, 1$ or 2. Since $d_k = n_k - 6$ this gives $d_k \leq -4$. Therefore $\nabla^l \upsilon_i = O\big((v_0')^{-5-l}\big)$, which implies the second equation of (13.9).

Thirdly, consider the remaining cases, in which $n_i < 4$ and there is at least one $j \in I$ with $i \succeq j$ and $n_j = 4$. In this case the dominant terms in $\sigma_{i,0}$ are those $\sigma_{j,0}$ with $i \succeq j$ and $n_j = 4$, and in fact, following (13.6), it can be proved that we may choose $\sigma_{i,0}$ such that

$$\nabla^l \sigma_{i,0} - \sum_{\substack{j \in I: \\ i \succeq j, \, n_j = 4}} \nabla^l \sigma_{j,0} = \sum_{\substack{k \in I: \\ i \succeq k, \, n_k < 4}} O\big(\mu_{j,k}^{d_k}(v_j')^{-1-l}\big).$$

Every term on the right hand side of this equation is $O\big((v_0')^{-5-l}\big)$, as in the second case above. Substituting $\sigma_{j,0} = \tau_j + v_j$ gives

$$\nabla^l \sigma_{i,0} - \sum_{\substack{j \in I: \\ i \geq j,\, n_j = 4}} \nabla^l \tau_j = O\big((v_0')^{-5-l}\big) + \sum_{\substack{j \in I: \\ i \geq j,\, n_j = 4}} \nabla^l v_j.$$

Defining v_i by the first equation of (13.9), the left hand side of this equation is $\nabla^l v_i$. Thus, using the first case above to estimate $\nabla^l v_j$ we get the second equation of (13.9). □

13.3.3 *The forms* Ξ, Φ, Ψ *and* Υ

Let $i, j \in I$ with $i \geq j$ and $n_j = 4$. Then $Q_i \subset T^8$ may be identified with an open subset of $F_i \times W_i$, and $W_i \cong V_j' \times W_j$ where $V_j' = W_i \cap V_j$, so that Q_i is an open subset of $(F_i \times V_j') \times W_j$. But $F_i \times V_j' \cong V_j/\Lambda_i$, and so Q_i is isomorphic to an open subset of $(V_j/\Lambda_i) \times W_j$. Now τ_j is a 3-form on $V_j \times (W_j \setminus \{0\})$ and invariant under V_j translations, and thus τ_j pushes down to $Q_i \setminus F_j$.

Let η_i' be as in Definition 13.2.6, and define a 4-form Ξ on T^8 by

$$\Xi = \sum_{\substack{i,j \in I: \\ i \geq j,\, n_j = 4}} d\eta_i' \wedge \tau_j = \sum_{\substack{j \in I: \\ n_j = 4}} d\bigg(\sum_{\substack{i \in I: \\ i \geq j}} \eta_i'\bigg) \wedge \tau_j. \tag{13.10}$$

Since $d\eta_i'$ is zero outside Q_i and τ_j is defined in Q_i except at F_j, we see that $d\eta_i' \wedge \tau_j$ is smooth and well-defined on T^8 except perhaps at F_j. But as $\sum_{i \in I: i \geq j} \eta_i' \equiv 1$ near F_j, from the second part of (13.10) we see that the terms in τ_j cancel near F_j, and thus Ξ is well-defined and smooth on all of T^8. Also Ξ is clearly Γ-invariant, and so it pushes down to T^8/Γ.

By Hodge theory on T^8/Γ we can show that there exists $\Phi \in C^\infty(\Lambda_-^4 T^8/\Gamma)$ with $d\Phi = d\Xi$, and that Φ is unique up to the addition of an element of $\mathcal{H}_-^4$ on T^8/Γ. Thus $\Phi - \Xi$ is a closed 4-form on T^8/Γ, and defines a de Rham cohomology class $[\Phi - \Xi] \in H^4(T^8/\Gamma, \mathbb{R})$. By adding an element of $\mathcal{H}_-^4$ to Φ if necessary we can ensure that $[\Phi - \Xi]$ lies in $H_+^4(T^8/\Gamma, \mathbb{R})$, and this defines Φ uniquely. Thus, there exists $\Psi \in \mathcal{H}_+^4$ with $[\Psi] = [\Phi - \Xi]$. Then $\Phi - \Xi - \Psi$ is an exact 4-form on T^8/Γ, so there exists a smooth 3-form Υ on T^8/Γ with $d\Upsilon = \Phi - \Xi - \Psi$.

13.4 Introducing a real parameter $t \in (0, 1]$

We now generalize the material of §13.2 and §13.3 to include a real parameter $t \in (0, 1]$, in a similar way to the G_2 case in §11.5. Throughout this section let T^8/Γ be equipped with a flat Spin(7)-structure (Ω_0, g_0), and let (Y_i, π_i), (Ω_i, g_i) and $\chi_{\gamma, i}$ be a set of R-data for T^8/Γ, as in Definition 13.2.2.

13.4.1 Spin(7)-*structures* (Ω_i^t, g_i^t) *and resolutions* (M^t, g^t)

We begin by rescaling the (Y_i, π_i), (Ω_i, g_i) and $\chi_{\gamma, i}$ by a factor $t \in (0, 1]$ to get a new set of R-data (Y_i^t, π_i^t), (Ω_i^t, g_i^t) and $\chi_{\gamma, i}^t$.

Definition 13.4.1 Let (Y_i, π_i), (Ω_i, g_i) and $\chi_{\gamma,i}$ be a set of R-data for T^8/Γ. Fix $t \in (0, 1]$. For each $i \in I$, define a resolution (Y_i^t, π_i^t) of W_i/A_i by $Y_i^t = Y_i$ and $\pi_i^t = t\pi_i$, and a Spin(7)-structure (Ω_i^t, g_i^t) on $V_i \times Y_i^t$ by

$$\Omega_i^t = t^4(t^{-1} \times \mathrm{id})^*(\Omega_i) \quad \text{and} \quad g_i^t = t^2(t^{-1} \times \mathrm{id})^*(g_i), \tag{13.11}$$

where $t^{-1} \times \mathrm{id} : V_i \times Y_i^t \to V_i \times Y_i$ maps (v, y) to $(t^{-1}v, y)$ in the obvious way. Then (Ω_i^t, g_i^t) is a product Spin(7)-structure on $V_i \times Y_i^t$, and $(V_i \times Y_i^t, \Omega_i^t, g_i^t)$ is a QALE Spin(7)-manifold asymptotic to $\mathbb{R}^8/A_i$. For each $i \in I$ and $\gamma \in \Gamma$, define $\chi_{\gamma,i}^t : Y_i^t \to Y_{\gamma,i}^t$ by $\chi_{\gamma,i}^t = \chi_{\gamma,i}$.

It is easy to see that the (Y_i^t, π_i^t), (Ω_i^t, g_i^t) and $\chi_{\gamma,i}^t$ form a set of R-data for T^8/Γ. Therefore, following §13.2.2 we can define a resolution (M^t, π^t) of T^8/Γ using the R-data (Y_i^t, π_i^t), (Ω_i^t, g_i^t) and $\chi_{\gamma,i}^t$ instead of (Y_i, π_i), (Ω_i, g_i) and $\chi_{\gamma,i}$, but leaving O_i, P_i and Q_i unchanged and independent of t. Since the resolutions (M^t, π^t) of T^8/Γ form a smooth connected family, they are all diffeomorphic to the same compact 8-manifold M.

Definition 13.4.2 For each $t \in (0, 1]$, define a function $\rho_t : M^t \to (0, t^{-1}]$ by $\rho_t(z) = [t + d(\pi^t(z), S)]^{-1}$. We will use ρ_t to estimate forms on M^t.

Let $\zeta_0, \ldots, \zeta_8$ be as in Definition 13.2.3, and define subsets D^t, E^t in M^t for each $t \in (0, 1]$ by

$$D^t = \{x \in M^t : d(\pi^t(x), F_i\Gamma/\Gamma) > t^{1/3}\zeta_{n_i} \text{ for all } i \in I \setminus \{0\}\},$$
$$E^t = \{x \in M^t : d(\pi^t(x), F_i\Gamma/\Gamma) > 2t^{1/3}\zeta_{n_i} \text{ for all } i \in I \setminus \{0\}\},$$

where $d(\,,\,)$ is the distance in T^8/Γ. Then $E^t \subset D^t$ and neither D^t nor E^t intersect $(\pi^t)^*(S)$, so that D^t and E^t can also be regarded as subsets of T^8/Γ. Thus, the Spin(7)-structure (Ω_0, g_0) on T^8/Γ restricts to D^t and E^t. Observe also that $M_0^t/\Gamma \subseteq D^t$, with equality when $t = 1$.

It helps to think of D^t as an enlarged version of the subset M_0^t/Γ in M^t. We will often use D^t instead of M_0^t/Γ.

13.4.2 The forms $\sigma_{i,j}^t$, τ_i, υ_i^t and Ψ^t

Next we generalize the forms defined in §13.3 to include the parameter $t \in (0, 1]$. Suppose $i, j \in I$ with $i \succ j$. Define a 3-form $\sigma_{i,j}^t$ on $V_j \times Y_j^t \setminus V_i \times U_j^{t'}$ by $\sigma_{i,j}^t = t^4(t^{-1} \times \mathrm{id})^*(\sigma_{i,j})$, where $t^{-1} \times \mathrm{id} : V_i \times Y_i^t \to V_i \times Y_i$ maps (v, y) to $(t^{-1}v, y)$ in the obvious way. Keeping careful track of the powers of t, we find that (13.6) gives

$$(\mathrm{id} \times \psi_j^{t'})^*(\Omega_i^t) - \Omega_j^t = d\sigma_{i,j}^t \quad \text{and}$$
$$\nabla^l \sigma_{i,j}^t = \sum_{k \in I' : j \not\preceq k} t^{\dim W_k} O\big((\mu_{j,k}^t)^{d_k} (\nu_j^{t'})^{-1-l}\big) \tag{13.12}$$

on $V_j \times Y_j^t \setminus V_i \times U_j^{t'}$, for all $l \geq 0$.

As in §11.5, since the $\sigma_{i,j}^t$ are invariant under the natural actions of Γ and V_i, we see that $\sigma_{i,j}^t$ pushes down to the intersection of M_i^t/B_i and M_j^t/B_j in M^t. Thus, when $i \succ j$ we have Spin(7)-structures (Ω_i^t, g_i^t) on M_i^t/B_i and (Ω_j^t, g_j^t) on M_j^t/B_j, and a 3-form $\sigma_{i,j}^t$ on $(M_i^t/B_i) \cap (M_j^t/B_j)$ satisfying

$$\Omega_i^t - \Omega_j^t = d\sigma_{i,j}^t \quad \text{on } (M_i^t/B_i) \cap (M_j^t/B_j). \tag{13.13}$$

Suppose $i, j, k \in I$ and $i \succ j \succ k$. Then (13.7) implies that

$$\sigma_{i,j}^t + \sigma_{j,k}^t - \sigma_{i,k}^t = 0 \quad \text{on } (M_i^t/B_i) \cap (M_j^t/B_j) \cap (M_k^t/B_k). \tag{13.14}$$

Similarly we define υ_i^t on $\mathbb{R}^8 \setminus V_i \times U_0^{t''}$ by $\upsilon_i^t = t^4(t^{-1} \times \text{id})^*(\upsilon_i)$ for each $i \in I \setminus \{0\}$, and then (13.9) gives

$$\sigma_{i,0}^t = \upsilon_i^t + t^4 \sum_{\substack{j \in I: \\ i \succeq j, n_j = 4}} \tau_j \quad \text{and} \quad \nabla^l \upsilon_i^t = t^5 O\big((\upsilon_0^{t''})^{-4-l}\big) \tag{13.15}$$

on $\mathbb{R}^8 \setminus V_i \times U_0^{t''}$, for all $l \geq 0$. We can push $\sigma_{i,0}$, τ_j and υ_i^t down to $(M_i^t/B_i) \cap D^t$ in M^t. Here are estimates of all these forms on M^t.

Proposition 13.4.3 *Whenever $i, j \in I$ with $i \succ j$ we have*

$$|\nabla^l \sigma_{i,j}^t| = t^4 O(\rho_t^{1+l}) \quad \text{in } (M_i^t/B_i) \cap (M_j^t/B_j) \tag{13.16}$$

for all $l \geq 0$, and whenever $i \in I \setminus \{0\}$ we have

$$|\nabla^l \sigma_{i,0}^t| = t^4 O(\rho_t^{3+l}) \quad \text{in } (M_i^t/B_i) \cap D^t \tag{13.17}$$

for all $l \geq 0$. For each $j \in I$ with $n_j = 4$ there is a smooth 3-form τ_j which is defined on $(M_i^t/B_i) \cap D^t$ for any $i \in I$ with $i \succeq j$, and satisfies $|\nabla^l \tau_j| = O(\rho_t^{3+l})$. For each $i \in I \setminus \{0\}$ there is a 3-form υ_i^t on $(M_i^t/B_i) \cap D^t$ satisfying

$$\sigma_{i,0}^t = \upsilon_i^t + t^4 \sum_{\substack{j \in I: \\ i \succeq j, n_j = 4}} \tau_j \quad \text{and} \quad |\nabla^l \upsilon_i^t| = t^5 O(\rho_t^{4+l}) \tag{13.18}$$

in $(M_i^t/B_i) \cap D^t$ for all $l \geq 0$. Here ∇ and $|\,.\,|$ are defined using the metric g_j^t in (13.16) and g_0^t in (13.17) and (13.18).

Proof Clearly (13.8) gives $|\nabla^l \tau_j| = O(\rho_t^{3+l})$, and (13.15) implies (13.18). To prove (13.16), suppose $i, j \in I$ with $i \succ j$. Then $(v_j^{t'})^{-1} \leq \rho_t$ in M_j^t/B_j. If $k \in I'$ and $j \not\preceq k$ then $\zeta_{n_k} < \mu_{j,k}^t < 2\zeta_{n_i}$ in $(M_i^t/B_i) \cap (M_j^t/B_j)$, so $\mu_{j,k}^t = O(1)$. Hence, in $(M_i^t/B_i) \cap (M_j^t/B_j)$ we may replace $t^{\dim W_k} O\big((\mu_{j,k}^t)^{d_k}(v_j^{t'})^{-1-l}\big)$ in (13.12) by $t^{\dim W_k} O(\rho_t^{1+l})$.

But $\dim W_k$ is 4, 6, 7 or 8 by Proposition 13.1.5, so $\dim W_k \geq 4$. Thus (13.12) implies (13.16). The proof of (13.17) is the same, except that $\mu_{0,k}^t = O(1)$ does not hold in $(M_i^t/B_i) \cap D^t$. Instead we show that $t^{\dim W_k}(\mu_{0,k}^t)^{d_k} = t^4 O(\rho_t^2)$, which gives the extra factor ρ_t^2 in (13.17). $\square$

We shall need to lift the constant 4-form Ψ on T^8/Γ of §13.3.3 to a closed 4-form Ψ^t on all of M^t. The obvious candidate for Ψ^t is $(\pi^t)^*(\Psi)$, but this may be singular on $(\pi^t)^{-1}(S)$. So instead we take $\Psi^t = \Psi$ on D^t, and then extend Ψ^t smoothly over the rest of M^t.

Proposition 13.4.4 *There exists a closed 4-form Ψ^t on M^t such that $\Psi^t = \Psi$ on D^t, and $[\Psi^t] = (\pi^t)^*([\Psi])$ in $H^4(M^t, \mathbb{R})$, and $|\Psi^t| = O(1)$ and $|\nabla \Psi^t| = O(\rho_t)$ on M^t. Here ∇ and $|\,.\,|$ are defined using the metric g_i^t on M_i^t/B_i for any $i \in I$.*

We leave the proof as an exercise. It can be done by following the method used to write down φ^t and υ^t in §11.5.

13.5 Spin(7)-structures on M with small torsion

We shall now define a 1-parameter family of 4-forms ξ^t on M, for $t \in (0, 1]$, by gluing together the 4-forms Ω_i^t on the patches M_i^t/B_i using a partition of unity. The $\sigma_{i,j}^t$ are used to make ξ^t *closed*. Then in §13.5.1 we will use ξ^t to construct a Spin(7)-structure (Ω^t, g^t) on M with small torsion.

Definition 13.5.1 Let $\eta'_{[i]} : T^8/\Gamma \to [0, 1]$ be as in §13.2.2 for $[i] \in I/\Gamma$, and define $\eta_{[i]}^t : M^t \to [0, 1]$ by $\eta_{[i]}^t = (\pi^t)^*(\eta'_{[i]})$. Then as in parts (a) and (b) of Definition 13.2.6 we find that the $\eta_{[i]}^t$ form a smooth partition of unity on M^t. It is easy to show that

$$|\nabla^l \eta_{[i]}^t| = O(1) \quad \text{on } M_j^t/B_j \text{ for } l = 0, 1, 2 \text{ and all } j \in I, \tag{13.19}$$

where $|\,.\,|$ and ∇ are defined using the metric g_j^t. Let D^t, E^t be as in Definition 13.4.2, and choose a smooth function $f^t : M^t \to [0, 1]$ with $f^t \equiv 0$ outside D^t and $f^t \equiv 1$ in E^t. We can arrange that

$$|\nabla^l f^t| = O(t^{-l/3}) \quad \text{on } D^t \setminus E^t \text{ for } l = 0, 1, 2, \tag{13.20}$$

where $|\,.\,|$ and ∇ are defined using g_0^t. Define a 4-form ξ^t on M^t by

$$\xi^t = \Omega_i^t + \sum_{\substack{[j] \in I/\Gamma: \\ j > i}} d(\eta_{[j]}^t \sigma_{j,i}^t) - \sum_{\substack{[j] \in I/\Gamma: \\ i > j}} d(\eta_{[j]}^t \sigma_{i,j}^t) + t^4 d(f^t \Upsilon) + t^4 \Psi^t \tag{13.21}$$

on M_i^t/B_i, for each $i \in I$. Here we take $\eta_{[j]}^t \sigma_{j,i}^t \equiv 0$ and $\eta_{[j]}^t \sigma_{i,j}^t \equiv 0$ outside M_j^t/B_j and $f^t \Upsilon \equiv 0$ outside D^t in M^t.

By the reasoning used to prove (11.23) and (11.24) in §11.5, we find that

$$\begin{aligned}\xi^t = \Omega_i^t &+ \sum_{\substack{[j] \in I/\Gamma: \\ j > i}} \eta_{[j]}^t (\Omega_j^t - \Omega_i^t) - \sum_{\substack{[j] \in I/\Gamma: \\ i > j}} \eta_{[j]}^t (\Omega_i^t - \Omega_j^t) \\ &+ \sum_{\substack{[j] \in I/\Gamma: \\ j > i}} d\eta_{[j]}^t \wedge \sigma_{j,i}^t - \sum_{\substack{[j] \in I/\Gamma: \\ i > j}} d\eta_{[j]}^t \wedge \sigma_{i,j}^t + t^4 d(f^t \Upsilon) + t^4 \Psi^t\end{aligned} \tag{13.22}$$

in M_i^t/B_i, and thus we get the alternative expression for ξ^t in M_i^t/B_i

$$\xi^t = \sum_{[j]\in I/\Gamma} \eta_{[j]}^t \Omega_j^t + \sum_{\substack{[j]\in I/\Gamma: \\ j>i}} d\eta_{[j]}^t \wedge \sigma_{j,i}^t - \sum_{\substack{[j]\in I/\Gamma: \\ i>j}} d\eta_{[j]}^t \wedge \sigma_{i,j}^t + t^4 d(f^t \Upsilon) + t^4 \Psi^t.$$

The proof of the next result follows the first part of Proposition 11.5.4.

Proposition 13.5.2 *This 4-form ξ^t on M^t is well-defined, smooth and closed.*

In our next three results we estimate $\xi^t - \Omega_i^t$ on various regions in M^t.

Lemma 13.5.3 *Let ϵ_1, ϵ_2 be as in Proposition 10.5.9, and $i, j \in I$. Suppose $|\Omega_j^t - \Omega_i^t| \leq \epsilon_1$ on $(M_i^t/B_i) \cap (M_j^t/B_j)$. Then $|\pi_{27}(\Omega_j^t - \Omega_i^t)| \leq \epsilon_2 |\Omega_j^t - \Omega_i^t|^2$ on $(M_i^t/B_i) \cap (M_j^t/B_j)$. Here $|.|, \pi_{27}$ are defined using the Spin(7)-structure (Ω_i^t, g_i^t).*

Proof Apply Proposition 10.5.9 with $\Omega = \Omega_i^t$, $\phi = \Omega_j^t - \Omega_i^t$ and $\psi = 0$. Since $\Theta(\Omega_j^t) = \Omega_j^t$ as $\Omega_j^t \in C^\infty(\mathcal{A}M^t)$, eqn (10.20) reduces to

$$\pi_{27}(\Omega_j^t - \Omega_i^t) = -F(\Omega_j^t - \Omega_i^t).$$

The inequality in the lemma then follows from (10.21). □

Proposition 13.5.4 *For small $t \in (0, 1]$ and each $i \in I$ we have*

$$|\xi^t - \Omega_i^t| = t^4 O(\rho_t^2), \quad |\nabla(\xi^t - \Omega_i^t)| = t^4 O(\rho_t^3)$$
$$\text{and} \quad |\pi_{27}(\xi^t - \Omega_i^t)| = t^4 O(\rho_t) \tag{13.23}$$

on M_i^t/B_i. Here $\nabla, |.|$ and π_{27} are defined using the Spin(7)-structure (Ω_i^t, g_i^t).

Proof As we are in M_i^t/B_i and the terms $d(\eta_{[j]}^t \sigma_{j,i}^t)$ and $d(\eta_{[j]}^t \sigma_{i,j}^t)$ in (13.21) are zero outside M_j^t/B_j, eqn (13.16) shows that

$$|\nabla^l d(\eta_{[j]}^t \sigma_{j,i}^t)| = t^4 O(\rho_t^{2+l}) \quad \text{and} \quad |\nabla^l d(\eta_{[j]}^t \sigma_{i,j}^t)| = t^4 O(\rho_t^{2+l}).$$

Also one can show from (13.20) and Proposition 13.4.4 that

$$\begin{aligned} t^4 d(f^t \Upsilon) &= t^4 O(\rho_t), & \nabla(t^4 d(f^t \Upsilon)) &= t^4 O(\rho_t^2), \\ t^4 \Psi^t &= t^4 O(1) & \text{and} \quad \nabla(t^4 \Psi^t) &= t^4 O(\rho_t). \end{aligned} \tag{13.24}$$

Thus the first two equations of (13.23) follow from (13.21).

As $d\sigma_{i,j}^t = \Omega_i^t - \Omega_j^t$, eqn (13.16) gives $|\Omega_i^t - \Omega_j^t| = t^4 O(\rho_t^2) \leq t^2 O(1)$ on $(M_i^t/B_i) \cap M_j^t/B_j$. Thus $|\Omega_i^t - \Omega_j^t| \leq \epsilon_1$ when $t \in (0, 1]$ is small, and Lemma 13.5.3 shows that $|\pi_{27}(\eta_{[j]}^t(\Omega_i^t - \Omega_j^t))| = t^8 O(\rho_t^4)$ in $(M_i^t/B_i) \cap (M_j^t/B_j)$. Also (13.16) and (13.19) give $|d\eta_{[j]}^t \wedge \sigma_{i,j}^t| = t^4 O(\rho_t)$ in $(M_i^t/B_i) \cap (M_j^t/B_j)$. Combining these facts with (13.22) and (13.24) gives the third equation of (13.23). □

Proposition 13.5.5 *For small* $t \in (0, 1]$ *we have*

$$|\xi^t - \Omega_0| = t^4\, O(\rho_t^4), \quad |\nabla(\xi^t - \Omega_0)| = t^4\, O(\rho_t^5)$$
$$\text{and} \quad |\pi_{27}(\xi^t - \Omega_0)| = t^5\, O(\rho_t^4) \tag{13.25}$$

on E^t. *Here* ∇, $|.|$ *and* π_{27} *are defined using the* Spin(7)*-structure* (Ω_0, g_0).

Proof Putting $i = 0$ in (13.22) and recalling that $f^t = 1$, $\Psi^t = \Psi$ in E^t gives

$$\xi^t - \Omega_0^t = \sum_{\substack{[i]\in I/\Gamma: \\ i\neq 0}} \eta_{[i]}^t (\Omega_i^t - \Omega_0^t) + \sum_{\substack{[i]\in I/\Gamma: \\ i\neq 0}} d\eta_{[i]}^t \wedge \sigma_{i,0}^t + t^4 (d\Upsilon + \Psi).$$

Putting $\Omega_0^t = \Omega_0$ and using (13.18) to substitute for $\sigma_{i,0}^t$, we get

$$\xi^t - \Omega_0 = \sum_{\substack{[i]\in I/\Gamma: \\ i\neq 0}} \eta_{[i]}^t (\Omega_i^t - \Omega_0^t) + \sum_{\substack{[i]\in I/\Gamma: \\ i\neq 0}} d\eta_{[i]}^t \wedge \upsilon_i^t + t^4 \sum_{\substack{[i]\in I/\Gamma,\, j\in I: \\ i\geq j,\, n_j = 4}} d\eta_{[i]}^t \wedge \tau_j + t^4 (d\Upsilon + \Psi).$$

But identifying E^t with an open subset of T^8/Γ, the third term on the right hand side of this equation is $t^4\,\Xi$, where Ξ is defined in (13.10). Thus

$$\xi^t - \Omega_0 = \sum_{\substack{[i]\in I/\Gamma: \\ i\neq 0}} \eta_{[i]}^t (\Omega_i^t - \Omega_0^t) + \sum_{\substack{[i]\in I/\Gamma: \\ i\neq 0}} d\eta_{[i]}^t \wedge \upsilon_i^t + t^4 (\Xi + d\Upsilon + \Psi)$$
$$= \sum_{\substack{[i]\in I/\Gamma: \\ i\neq 0}} \eta_{[i]}^t (\Omega_i^t - \Omega_0^t) + \sum_{\substack{[i]\in I/\Gamma: \\ i\neq 0}} d\eta_{[i]}^t \wedge \upsilon_i^t + t^4 \Phi \tag{13.26}$$

on E^t, since $\Xi + d\Upsilon + \Psi = \Phi$ from §13.3.3. Equations (13.17) and (13.18) give

$$\begin{aligned} |\eta_{[i]}^t (\Omega_i^t - \Omega_0^t)| &= t^4\, O(\rho_t^4), & |\nabla(\eta_{[i]}^t (\Omega_i^t - \Omega_0^t))| &= t^4\, O(\rho_t^5), \\ |d\eta_{[i]}^t \wedge \upsilon_i^t| &= t^5\, O(\rho_t^4) & \text{and} \quad |\nabla(d\eta_{[i]}^t \upsilon_i^t)| &= t^5\, O(\rho_t^5). \end{aligned} \tag{13.27}$$

Also $|t^4\Phi| = t^4\, O(1)$ and $|\nabla(t^4\Phi)| = t^4\, O(1)$ as Φ is smooth and independent of t. Combining these estimates with (13.26) gives the first two equations of (13.25).

To finish the proof we use (13.27) and Lemma 13.5.3 as in the proof of Proposition 13.5.4 to show that $|\pi_{27}(\eta_{[i]}^t(\Omega_i^t - \Omega_0^t))| = t^8\, O(\rho_t^8)$. From §13.3.3 the 4-form Φ is anti-self-dual, and thus $\pi_{27}(\Phi) = 0$. These equations, (13.26) and the second line of (13.27) together prove the third equation of (13.25). $\square$

The third estimate $|\pi_{27}(\xi^t - \Omega_0)| = t^5\, O(\rho_t^4)$ in this proposition is important because it shows that $\pi_{27}(\xi^t - \Omega_0)$ is $O(t^5)$ on most of M^t. That is, $\pi_{27}(\xi^t - \Omega_0)$ is *smaller than* $O(t^4)$. We have achieved this by using the forms Ξ, Φ, Ψ and Υ to cancel out the highest-order $O(t^4)$ contributions to $\pi_{27}(\xi^t - \Omega_0)$. We will use this estimate to show that a certain norm of the torsion of the Spin(7)-structure (Ω^t, g^t) is also smaller than $O(t^4)$.

13.5.1 The definition of (Ω^t, g^t) and ϕ^t

Here is the definition of the Spin(7)-structure (Ω^t, g^t) and the 4-form ϕ^t.

Definition 13.5.6 Let ϵ_1 be the constant of Proposition 10.5.9. As $\rho_t \leqslant t^{-1}$ on M^t, eqn (13.23) shows that $|\xi^t - \Omega_i^t| = O(t^2)$ on M_i^t/B_i for all $i \in I$, and thus $|\xi^t - \Omega_i^t| \leqslant \epsilon_1$ on M_i^t/B_i when t is sufficiently small. Also $\rho^t = O(t^{-1/3})$ on E^t, so (13.25) shows that $|\xi^t - \Omega_0| = O(t^{8/3})$ on E^t. Thus $|\xi^t - \Omega_0| \leqslant \epsilon_1$ on E^t when t is sufficiently small. Choose $\epsilon \in (0, 1]$ such that $|\xi^t - \Omega_i^t| \leqslant \epsilon_1$ on M_i^t/B_i for all $i \in I$ and $|\xi^t - \Omega_0| \leqslant \epsilon_1$ on E^t when $t \in (0, \epsilon]$.

Then Proposition 10.5.9 shows that ξ^t lies in $\mathcal{T}M^t$ on M_i^t/B_i when $t \in (0, \epsilon]$. As this holds for all $i \in I$, we see that $\xi^t \in C^\infty(\mathcal{T}M^t)$. Therefore $\Theta(\xi^t)$ lies in $C^\infty(\mathcal{A}M^t)$ by Definition 10.5.8. Define $\Omega^t = \Theta(\xi^t)$, and let (Ω^t, g^t) be the induced Spin(7)-structure on M. Define a 4-form ϕ^t by $\phi^t = \xi^t - \Omega^t$. Then $\xi^t = \Omega^t + \phi^t$, and so $d\Omega^t + d\phi^t = 0$ as ξ^t is closed.

This 4-form ϕ^t has the following important properties.

Proposition 13.5.7 *For each $t \in (0, \epsilon]$ the 4-form ϕ^t satisfies $|\phi^t| \leqslant Ct^4 \rho_t$ and $|\nabla \phi^t| \leqslant C' t^4 \rho_t^3$ on all of M^t, and $|\phi^t| \leqslant C'' t^5 \rho_t^4$ on the subset E^t of M^t. Here ∇ and $|.|$ are taken with respect to g^t, and $C, C', C'' > 0$ are independent of t.*

Proof We apply Proposition 10.5.10 in M_i^t with $\Omega = \Omega_i^t$, $\phi = \xi^t - \Omega_i^t$, $\Omega' = \Omega^t$ and $\phi' = \phi^t$. Substituting (13.23) into (10.23) gives

$$|\phi^t| \leqslant \epsilon_4 \left\{ t^4 O(\rho_t) + \left(t^4 O(\rho_t^2)\right)^2 \right\} = t^4 O(\rho_t) \quad \text{and} \quad |\nabla \phi^t| = t^4 O(\rho_t^3)$$

on M_i^t/B_i, for each $i \in I$. Thus there exist $C, C' > 0$ independent of t such that $|\phi^t| \leqslant Ct^4 \rho_t$ and $|\nabla \phi^t| \leqslant C' t^4 \rho_t^3$ on M^t.

Similarly, we apply Proposition 10.5.10 in E^t with $\Omega = \Omega_0$, $\phi = \xi^t - \Omega_0$, $\Omega' = \Omega^t$ and $\phi' = \phi^t$. Substituting (13.25) into (10.23) gives

$$|\phi^t| \leqslant \epsilon_4 \left\{ t^5 O(\rho_t^4) + \left(t^4 O(\rho_t^4)\right)^2 \right\} = t^5 O(\rho_t^4) \quad \text{on } E^t.$$

Thus there exists $C'' > 0$ independent of t such that $|\phi^t| \leqslant C'' t^5 \rho_t^4$ on E^t. □

We can now now prove the main result of this section, the analogue of Theorem 11.5.7 in the G_2 case. In it we bound, by powers of t, various norms of the 4-form ϕ^t and analytic properties of the metric g^t defined above. The things we estimate are chosen to meet the needs of the analytic proofs in §13.6 and §13.7.

Theorem 13.5.8 *Let T^8/Γ be an orbifold of T^8 with a flat Spin(7)-structure (Ω_0, g_0). Suppose we are given a set of R-data for T^8/Γ, and (M, π) is the corresponding resolution. Then we can write down the following data on M explicitly:*

- *Constants $\epsilon \in (0, 1]$ and $\lambda, \mu, \nu > 0$,*
- *A Spin(7)-structure (Ω^t, g^t) on M for each $t \in (0, \epsilon]$, and*
- *A smooth 4-form ϕ^t on M with $d\Omega^t + d\phi^t = 0$ for each $t \in (0, \epsilon]$.*

These satisfy the three conditions
  (i) $\|\phi^t\|_{L^2} \leqslant \lambda t^{13/3}$ and $\|d\phi^t\|_{L^{10}} \leqslant \lambda t^{7/5}$;
  (ii) *the injectivity radius* $\delta(g^t)$ *satisfies* $\delta(g^t) \geqslant \mu t$; *and*
  (iii) *the Riemann curvature* $R(g^t)$ *satisfies* $\|R(g^t)\|_{C^0} \leqslant \nu t^{-2}$.
Here and all norms are calculated using the metric g^t.

Proof The Spin(7)-structure (Ω^t, g^t), 4-form ϕ^t and constant $\epsilon > 0$ are given in Definition 13.5.6, which also shows that $d\Omega^t + d\phi^t = 0$. It remains only to find constants $\lambda, \mu, \nu > 0$ such that (i)–(iii) hold.

From Proposition 13.5.7 we see that

$$\|\phi^t\|_{L^2} \leqslant Ct^4 \|\rho_t|_{M^t \setminus E^t}\|_{L^2} + C''t^5 \|\rho_t^4|_{E^t}\|_{L^2} \text{ and } \|d\phi^t\|_{L^{10}} \leqslant C't^4 \|\rho_t^3\|_{L^{10}}.$$

Using the method of the proof of Theorem 11.5.7 we can show that

$$\|\rho_t|_{M^t \setminus E^t}\|_{L^2} = O(t^{1/3}), \quad \|\rho_t^4|_{E^t}\|_{L^2} = O(t^{-2/3}) \text{ and } \|\rho_t^3\|_{L^{10}} = O(t^{-13/5}).$$

Hence there exists $\lambda > 0$ such that part (i) holds.

Parts (ii) and (iii) are elementary. The metric g_i^t is made by scaling g_i by a factor t. Thus $\delta(g_i^t) = t\delta(g_i)$ and $\|R(g_i^t)\|_{C^0} = t^{-2}\|R(g_i)\|_{C^0}$. We make g^t by gluing together the g_i^t on the patches M_i^t/B_i. It is clear that for small t, the dominant contributions to $\delta(g^t)$ and $\|R(g^t)\|_{C^0}$ come from $\delta(g_i^t)$ and $\|R(g_i^t)\|_{C^0}$ for some i, and these are proportional to t and t^{-2}. This proves (ii) and (iii) for some $\mu, \nu > 0$, and the theorem is complete. □

13.5.2 *Discussion of Theorem* 13.5.8

The 8-manifold M is a kind of 'blow-up' of the orbifold T^8/Γ in which each point s in the singular set S of T^8/Γ is replaced by a compact subspace $\pi^{-1}(s)$ of M, often a submanifold of M, or a union of several submanifolds. In the 1-parameter family of Spin(7)-structures (Ω^t, g^t) that we have constructed on M, we should think of t as a measure of the *diameter* of $\pi^{-1}(s)$ for each singular point s in T^8/Γ.

For example, if T^8/Γ has a singular T^4 locally isomorphic to $T^4 \times \mathbb{C}^2/\{\pm 1\}$, then M is locally diffeomorphic to $T^4 \times Y$, where Y is the blow-up of $\mathbb{C}^2/\{\pm 1\}$ at 0. Thus, each s in T^4 is replaced by a copy of $\mathbb{CP}^1 = \mathcal{S}^2$ in M, and T^4 is replaced by $T^4 \times \mathcal{S}^2$. The restriction of g^t to $T^4 \times \mathcal{S}^2$ is the product of a fixed metric on T^4, and the round metric on $\mathcal{S}^2$ with diameter πt.

The Spin(7)-structure (Ω^t, g^t) on M tends to the singular Spin(7)-structure (Ω_0, g_0) on T^8/Γ as $t \to 0$. Thus g^t approaches a singular metric as $t \to 0$, so for small t we expect g^t to be 'badly behaved' near $\pi^{-1}(S)$. Parts (ii) and (iii) of Theorem 13.5.8 measure this. The injectivity radius $\delta(g^t)$ is $O(t)$ and the Riemannian curvature $R(g^t)$ is $O(t^{-2})$ near $\pi^{-1}(S)$, so g^t has *short injectivity radius* and *large curvature* for small t. These both make analysis proofs on (M, g^t) more difficult.

Now (Ω^t, g^t) satisfies $d\Omega^t + d\phi^t = 0$. As the *torsion* $\nabla\Omega^t$ depends only on $d\Omega^t$ we see that $\nabla\Omega^t$ is determined by $d\phi^t$. Thus we may interpret ϕ^t as a *first integral of the torsion* of (Ω^t, g^t), since $\nabla\Omega^t$ depends on the first derivative of ϕ^t. We will need to estimate ϕ^t as well as $d\phi^t$ because of a trick in §13.7 involving integration by parts. Therefore part (i) of Theorem 13.5.8 gives estimates on the torsion of (Ω^t, g^t), and its first integral.

There are some important differences between the G_2 case of Chapter 11 and the Spin(7) case above, to do with the power of t in the inequality $\|\phi^t\|_{L^2} \leqslant \lambda t^{13/3}$ of part (i). To explain these differences, first recall the corresponding powers of t in Chapter 11. Part (i) of Theorem 11.5.7 gave the estimate $\|\psi\|_{L^2} \leqslant \lambda t^4$, where ψ is a first integral of the torsion of the G_2-structure (φ^t, g^t). Here the 4 in the power t^4 should be thought of as the codimension of the singularities of T^7/Γ.

For the proof of Theorem G2 in §11.8 to work, we needed $t^{-7/2}\|\psi\|_{L^2}$ to be sufficiently small. Here the 7 in $-7/2$ is $\dim T^7/\Gamma$, and the 2 comes from the L^2 norm. Thus, the proof in Chapter 11 succeeded because $4 > 7/2$, that is, the codimension of the singularities of T^7/Γ (four) is more than half the dimension of T^7/Γ (seven).

However, when we move up from 7 to 8 dimensions, there is a problem. If we had chosen to define the Spin(7)-structure (Ω^t, g^t) on M by following the method of §11.5.3 in the most obvious way, then we would have ended up with the estimate $\|\phi^t\|_{L^2} \leqslant \lambda t^4$ in part (i) of Theorem 13.5.8, where the power 4 is the codimension of the singularities of T^8/Γ.

In the proof of Theorem S2 in §13.7 we will need to assume that $t^{-8/2}\|\phi^t\|_{L^2}$ is small, where the 8 in $-8/2$ is $\dim T^8/\Gamma$ and the 2 comes from the L^2 norm. But the estimate $\|\phi^t\|_{L^2} \leqslant \lambda t^4$ only implies that $t^{-8/2}\|\phi^t\|_{L^2} \leqslant \lambda$. This does not force $t^{-8/2}\|\phi^t\|_{L^2}$ to be small, as λ may be quite big. So the simple method used in Chapter 11 does not quite work in the Spin(7) case.

To solve this problem we have adopted a more complicated definition of the Spin(7)-structure (Ω^t, g^t), for which we can prove the estimate $\|\phi^t\|_{L^2} \leqslant \lambda t^{13/3}$ in part (i) above. The proofs in §13.7 then work because $13/3 > 4$, and so $t^{-8/2}\|\phi^t\|_{L^2}$ is small when t is small. The basic idea is that we define Ω^t so as to cancel out the largest $O(t^4)$ contributions to the torsion, which come from the codimension 4 singularities. The sole purpose of the forms $\tau_i, \upsilon_i, \Xi, \Phi, \Psi, \Upsilon, \upsilon_i^t$ and Ψ^t is to achieve this.

13.6 An existence result for compact Spin(7)-manifolds

We shall now prove an existence theorem for torsion-free Spin(7)-structures on the resolutions M of T^8/Γ defined in §13.2. The proof depends upon the following two results, similar to Theorems G1 and G2 of §11.6.

Theorem S1 *Let μ, ν and t be positive constants, and suppose (M, g) is a complete Riemannian 8-manifold, whose injectivity radius $\delta(g)$ and Riemann curvature $R(g)$ satisfy $\delta(g) \geqslant \mu t$ and $\|R(g)\|_{C^0} \leqslant \nu t^{-2}$. Then there exist $C_1, C_2 > 0$ depending only on μ and ν, such that if $\chi \in L_1^{10}(\Lambda_-^4 T^*M) \cap L^2(\Lambda_-^4 T^*M)$ then*

$$\|\nabla \chi\|_{L^{10}} \leqslant C_1\big(\|d\chi\|_{L^{10}} + t^{-21/5}\|\chi\|_{L^2}\big) \tag{13.28}$$

$$\text{and} \quad \|\chi\|_{C^0} \leqslant C_2\big(t^{1/5}\|\nabla \chi\|_{L^{10}} + t^{-4}\|\chi\|_{L^2}\big). \tag{13.29}$$

The proof of this theorem is almost identical to the proof of Theorem G1 in §11.7, so we will omit it. Our next result will be proved in §13.7.

Theorem S2 *Let λ, C_1 and C_2 be positive constants. Then there exist positive constants κ, K such that whenever $0 < t \leqslant \kappa$, the following is true.*

Let M be a compact 8-manifold, and (Ω, g) a Spin(7)-structure on M. Suppose that ϕ is a smooth 4-form on M with $\mathrm{d}\Omega + \mathrm{d}\phi = 0$, and

(i) $\|\phi\|_{L^2} \leqslant \lambda t^{13/3}$ and $\|\mathrm{d}\phi\|_{L^{10}} \leqslant \lambda t^{7/5}$,

(ii) *if* $\chi \in L_1^{10}(\Lambda_-^4 T^*M)$ *then* $\|\nabla\chi\|_{L^{10}} \leqslant C_1\bigl(\|\mathrm{d}\chi\|_{L^{10}} + t^{-21/5}\|\chi\|_{L^2}\bigr)$, *and*

(iii) *if* $\chi \in L_1^{10}(\Lambda_-^4 T^*M)$ *then* $\|\chi\|_{C^0} \leqslant C_2\bigl(t^{1/5}\|\nabla\chi\|_{L^{10}} + t^{-4}\|\chi\|_{L^2}\bigr)$.

Let ϵ_1 and F be as in Proposition 10.5.9. Then there exists $\eta \in C^\infty(\Lambda_-^4 T^*M)$ satisfying $\|\eta\|_{C^0} \leqslant Kt^{1/3} \leqslant \epsilon_1$ and $\mathrm{d}\eta = \mathrm{d}\phi + \mathrm{d}F(\eta)$.

Assuming Theorems S1 and S2, we prove an existence theorem for torsion-free Spin(7)-structures on a compact 8-manifold M equipped with a Spin(7)-structure (Ω, g) with *small torsion*, in a suitable sense.

Theorem 13.6.1 *Let λ, μ, ν be positive constants. Then there exist positive constants κ, K' such that whenever $0 < t \leqslant \kappa$, the following is true.*

Let M be a compact 8-manifold, and (Ω, g) a Spin(7)-structure on M. Suppose that ϕ is a smooth 4-form on M with $\mathrm{d}\Omega + \mathrm{d}\phi = 0$, and

(i) $\|\phi\|_{L^2} \leqslant \lambda t^{13/3}$ and $\|\mathrm{d}\phi\|_{L^{10}} \leqslant \lambda t^{7/5}$,

(ii) *the injectivity radius $\delta(g)$ satisfies $\delta(g) \geqslant \mu t$, and*

(iii) *the Riemann curvature $R(g)$ satisfies $\|R(g)\|_{C^0} \leqslant \nu t^{-2}$.*

Then there exists a smooth, torsion-free Spin(7)-structure $(\tilde\Omega, \tilde g)$ on M with $\|\tilde\Omega - \Omega\|_{C^0} \leqslant K't^{1/3}$.

Proof By Theorem S1 and parts (ii), (iii) above there exist $C_1, C_2 > 0$ depending only on μ, ν, such that if $\chi \in L_1^{10}(\Lambda_-^4 T^*M)$ then (13.28) and (13.29) hold. This proves parts (ii) and (iii) of Theorem S2, and part (i) of Theorem S2 agrees with that of Theorem 13.6.1. Thus Theorem S2 gives constants κ, K depending only on λ, C_1, C_2, and hence on λ, μ, ν, such that if $0 < t \leqslant \kappa$ then there exists $\eta \in C^\infty(\Lambda_-^4 T^*M)$ satisfying $\|\eta\|_{C^0} \leqslant Kt^{1/3} \leqslant \epsilon_1$ and $\mathrm{d}\eta = \mathrm{d}\phi + \mathrm{d}F(\eta)$.

Now $\Omega + \eta \in C^\infty(\mathcal{T}M)$ by Proposition 10.5.9, as $|\eta| \leqslant \epsilon_1$. Let $\tilde\Omega = \Theta(\Omega + \eta)$. Then $\tilde\Omega \in C^\infty(\mathcal{A}M)$ by Definition 10.5.8, so $\tilde\Omega$ extends to a Spin(7)-structure $(\tilde\Omega, \tilde g)$ on M. Equation (10.20) shows that $\tilde\Omega = \Theta(\Omega + \eta) = \Omega + \eta - F(\eta)$, as η lies in $\Lambda_-^4 T^*M = \Lambda_{35}^4$, and so $\pi_1(\eta) + \pi_7(\eta) + \pi_{35}(\eta) = \eta$. Therefore

$$\mathrm{d}\tilde\Omega = \mathrm{d}\Omega + \mathrm{d}\eta - \mathrm{d}F(\eta) = -\mathrm{d}\phi + \mathrm{d}\eta - \mathrm{d}F(\eta) = 0,$$

using $\mathrm{d}\Omega + \mathrm{d}\phi = 0$ and $\mathrm{d}\eta = \mathrm{d}\phi + \mathrm{d}F(\eta)$. So $\mathrm{d}\tilde\Omega = 0$, and $(\tilde\Omega, \tilde g)$ is torsion-free by Proposition 10.5.3.

Finally, as $\tilde\Omega - \Omega = \eta - F(\eta)$ we have

$$|\tilde\Omega - \Omega| \leqslant |\eta| + |F(\eta)| \leqslant Kt^{1/3} + \epsilon_2(Kt^{1/3})^2,$$

estimating $F(\eta)$ with (10.21) and using $|\eta| \leqslant Kt^{1/3}$. Putting $K' = K + \epsilon_2 K^2 \kappa^{1/3}$ we have $|\tilde\Omega - \Omega| \leqslant K't^{1/3}$, as $t \leqslant \kappa$, and so $\|\tilde\Omega - \Omega\|_{C^0} \leqslant K't^{1/3}$, and the proof is complete. $\square$

Combining this with Theorem 13.5.8, we get the main result of the chapter:

Theorem 13.6.2 *Suppose T^8/Γ has a flat Spin(7)-structure (Ω_0, g_0) and a set of R-data (Y_i, π_i), (Ω_i, g_i) and $\chi_{\gamma,i}$, and let M be the resolution of T^8/Γ defined in §13.2. Then M is a compact 8-manifold, and there exist torsion-free Spin(7)-structures on M. In particular, if M is simply-connected and its Betti numbers satisfy $b^3 + b_+^4 = b^2 + 2b_-^4 + 25$, then M admits metrics with holonomy Spin(7).*

Proof By definition M is a compact 8-manifold. Let $\epsilon \in (0, 1]$ and $\lambda, \mu, \nu > 0$ be the constants given by Theorem 13.5.8. Then Theorem 13.6.1 gives constants $\kappa, K' > 0$ depending on λ, μ, ν. Choose $t > 0$ with $t \leq \epsilon$ and $t \leq \kappa$, and let (Ω, g) be the Spin(7)-structure (Ω^t, g^t) of Theorem 13.5.8, and ϕ the 4-form ϕ^t.

Theorem 13.5.8 shows that $d\Omega + d\phi = 0$, and (i)–(iii) of Theorem 13.5.8 give (i)–(iii) of Theorem 13.6.1. Therefore Theorem 13.6.1 shows that there exists a torsion-free Spin(7)-structure $(\tilde{\Omega}, \tilde{g})$ on M. If M is simply-connected and $b^3 + b_+^4 = b^2 + 2b_-^4 + 25$, then Theorem 10.6.8 proves that M admits metrics with holonomy Spin(7). □

13.7 The proof of Theorem S2

We shall now prove Theorem S2 of §13.6.

Theorem S2 *Let λ, C_1 and C_2 be positive constants. Then there exist positive constants κ, K such that whenever $0 < t \leq \kappa$, the following is true.*

Let M be a compact 8-manifold, and (Ω, g) a Spin(7)-structure on M. Suppose that ϕ is a smooth 4-form on M with $d\Omega + d\phi = 0$, and

(i) $\|\phi\|_{L^2} \leq \lambda t^{13/3}$ and $\|d\phi\|_{L^{10}} \leq \lambda t^{7/5}$,

(ii) *if* $\chi \in L_1^{10}(\Lambda_-^4 T^*M)$ *then* $\|\nabla \chi\|_{L^{10}} \leq C_1 \bigl(\|d\chi\|_{L^{10}} + t^{-21/5}\|\chi\|_{L^2}\bigr)$, *and*

(iii) *if* $\chi \in L_1^{10}(\Lambda_-^4 T^*M)$ *then* $\|\chi\|_{C^0} \leq C_2\bigl(t^{1/5}\|\nabla\chi\|_{L^{10}} + t^{-4}\|\chi\|_{L^2}\bigr)$.

Let ϵ_1 and F be as in Proposition 10.5.9. Then there exists $\eta \in C^\infty(\Lambda_-^4 T^*M)$ satisfying $\|\eta\|_{C^0} \leq Kt^{1/3} \leq \epsilon_1$ and $d\eta = d\phi + dF(\eta)$.

Proof We begin with the following result, similar to Proposition 11.8.1.

Proposition 13.7.1 *In the situation above, there exist positive constants κ, C_3 and K depending only on λ, C_1, C_2, such that if $t \leq \kappa$ then there exists a sequence $\{\eta_j\}_{j=0}^\infty$ in $C^\infty(\Lambda_-^4 T^*M)$ with $\eta_0 = 0$ satisfying the equation*

$$d\eta_j = d\phi + dF(\eta_{j-1}) \tag{13.30}$$

for each $j > 0$, and the inequalities

(a) $\|\eta_j\|_{L^2} \leq 4\lambda t^{13/3}$,

(b) $\|\nabla \eta_j\|_{L^{10}} \leq C_3 t^{2/15}$,

(c) $\|\eta_j\|_{C^0} \leq Kt^{1/3} \leq \epsilon_1$ *and*

(d) $\|\eta_j - \eta_{j-1}\|_{L^2} \leq 4\lambda 2^{-j} t^{13/3}$,

(e) $\|\nabla(\eta_j - \eta_{j-1})\|_{L^{10}} \leq C_3 2^{-j} t^{2/15}$,

(f) $\|\eta_j - \eta_{j-1}\|_{C^0} \leq K 2^{-j} t^{1/3}$.

Proof The proof is by induction in j, and will follow from the next two lemmas. At a number of points in the proof we shall need t to be smaller than some positive constant defined in terms of λ, C_1, C_2 and ϵ_1, ϵ_2, ϵ_3. As a shorthand we shall simply say that this holds since $t \leqslant \kappa$, and suppose without remark that κ has been chosen such that the relevant restriction holds.

Lemma 13.7.2 *Suppose by induction that $\eta_0, \ldots, \eta_k$ exist and satisfy* (13.30) *and parts* (a), (c) *and* (d) *of Proposition* 13.7.1 *for $j \leqslant k$. Then there exists a unique $\eta_{k+1} \in C^\infty(\Lambda_-^4 T^*M)$ satisfying* (13.30) *and parts* (a), (d) *for $j = k+1$, and such that $\eta_{k+1} - \phi - F(\eta_k)$ is L^2-orthogonal to $\mathcal{H}_-^4$.*

Proof As (M, g) is a compact, oriented Riemannian 8-manifold, it is easy to show using Theorem 1.1.3 that if $\chi \in C^\infty(\Lambda^5 T^*M)$ is exact then there exists $\xi \in C^\infty(\Lambda_-^4 T^*M)$ with $d\xi = \chi$. This ξ is unique up to the addition of an element of $\mathcal{H}_-^4$, so we can define ξ uniquely by specifying its L^2 inner product with $\mathcal{H}_-^4$. Note that $F(\eta_k)$ is well-defined as $|\eta_k| \leqslant \epsilon_1$ by part (c) of Proposition 13.7.1. Thus $\chi = d\phi + dF(\eta_k)$ is an exact 5-form, and there exists η_{k+1} satisfying (13.30) for $j = k+1$. We define η_{k+1} uniquely by requiring that $\eta_{k+1} - \phi - F(\eta_k)$ be L^2-orthogonal to $\mathcal{H}_-^4$.

It remains to show that parts (a) and (d) hold for $j = k+1$. We shall deal with the cases $k = 0$ and $k > 0$ separately. First suppose $k = 0$. Then $F(\eta_0) = 0$ as $\eta_0 = 0$, and $d\eta_1 = d\phi$ by (13.30). So $\eta_1 - \phi$ is a closed 4-form, and defines a cohomology class $[\eta_1 - \phi]$ in $H^4(M, \mathbb{R})$. But $\eta_1 - \phi$ is L^2-orthogonal to $\mathcal{H}_-^4$ by definition of η_1, so $[\eta_1 - \phi] \in H_+^4(M, \mathbb{R})$, and $[\eta_1 - \phi] \cup [\eta_1 - \phi] \geqslant 0$. Writing this as an integral over M and using the fact that $\eta_1 \in \Lambda_-^4 T^*M$ we get

$$\int_M \left\{ |\phi_+|^2 - |\eta_1 - \phi_-|^2 \right\} dV_g \geqslant 0,$$

where $\phi_\pm$ are the components of ϕ in $\Lambda_\pm^4 T^*M$. Thus $\|\eta_1 - \phi_-\|_{L^2} \leqslant \|\phi_+\|_{L^2}$, giving $\|\eta_1\|_{L^2} \leqslant 2\|\phi\|_{L^2} \leqslant 2\lambda t^{13/3}$ by (i) of Theorem S2. So (a), (d) hold for $j = 1$.

Next suppose $k > 0$. Define $\psi = \eta_{k+1} - \eta_k - F(\eta_k) + F(\eta_{k-1})$. Taking the difference of (13.30) for $j = k+1, k$ shows that $d\psi = 0$, and the definitions of η_k, η_{k+1} show that ψ is L^2-orthogonal to $\mathcal{H}_-^4$. So by the argument above $[\psi] \in H_+^4(M, \mathbb{R})$ and $[\psi] \cup [\psi] \geqslant 0$. Writing this as an integral over M and using the fact that $\eta_k, \eta_{k+1} \in \Lambda_-^4 T^*M$ we get

$$\int_M \left\{ |F(\eta_k)_+ - F(\eta_{k-1})_+|^2 - |\eta_{k+1} - \eta_k - F(\eta_k)_- + F(\eta_{k-1})_-|^2 \right\} dV_g \geqslant 0,$$

Thus $\|\eta_{k+1} - \eta_k\|_{L^2} \leqslant 2\|F(\eta_k) - F(\eta_{k-1})\|_{L^2}$. Applying (10.21) shows that

$$\|\eta_{k+1} - \eta_k\|_{L^2} \leqslant 2\epsilon_2 \bigl(\|\eta_k\|_{C^0} + \|\eta_{k-1}\|_{C^0} \bigr) \|\eta_k - \eta_{k-1}\|_{L^2}.$$

By part (c) of Proposition 13.7.1 for $j = k-1, k$ and since $t \leqslant \kappa$, we have $2\epsilon_2 \bigl(\|\eta_k\|_{C^0} + \|\eta_{k-1}\|_{C^0} \bigr) \leqslant \frac{1}{2}$. Therefore part (d) for $j = k+1$ follows from the above equation and part (d) for $j = k$, as we have to prove. Then part (a) for $j = k+1$ follows by induction from part (d) for $j = 1, \ldots, k+1$, and the lemma is complete. $\square$

For the next lemma, define $C_3 = 6\lambda C_1$ and $K = C_2(C_3 + 4\lambda)$.

Lemma 13.7.3 *Parts* (b), (c), (e) *and* (f) *of Proposition* 13.7.1 *hold for* $j = 1$. *Suppose by induction that* (13.30) *and parts* (a)–(f) *hold for* $j \leqslant k$, *and part* (d) *and* (13.30) *hold for* $j = k+1$. *Then parts* (b), (c), (e) *and* (f) *hold for* $j = k+1$.

Proof By (13.30) we have $d\eta_1 = d\phi$. Applying parts (i) and (ii) of Theorem S2 and part (d) of Proposition 13.7.1 for $j = 1$, we have

$$\|\nabla \eta_1\|_{L^{10}} \leqslant C_1(\lambda t^{7/5} + 2\lambda t^{2/15}).$$

Since $C_3 = 6\lambda C_1$ and $t \leqslant \kappa$, parts (b) and (e) of Proposition 13.7.1 hold for $j = 1$. Using parts (iii) of Theorem S2 and (d) and (e) of Proposition 13.7.1 for $j = 1$ we find that $\|\eta_1\|_{C^0} \leqslant \frac{1}{2}C_2(C_3 + 4\lambda)t^{1/3} = \frac{1}{2}Kt^{1/3}$. Thus parts (c) and (f) of Proposition 13.7.1 hold for $j = 1$, as we have to prove.

Now suppose that (13.30) and parts (a)–(f) hold for $j \leqslant k$, and part (d) and (13.30) hold for $j = k+1$. Taking the difference of (13.30) for $j = k+1, k$ we find that $d(\eta_{k+1} - \eta_k) = d\big(F(\eta_k) - F(\eta_{k-1})\big)$. Thus part (ii) of Theorem S2 and part (d) of Proposition 13.7.1 for $j = k+1$ show that

$$\|\nabla(\eta_{k+1} - \eta_k)\|_{L^{10}} \leqslant C_1\Big(\|d(F(\eta_k) - F(\eta_{k-1}))\|_{L^{10}} + 4\lambda 2^{-k-1}t^{2/15}\Big). \tag{13.31}$$

Equation (10.22) of Proposition 10.5.9 shows that

$$\|d(F(\eta_k) - F(\eta_{k-1}))\|_{L^{10}} \leqslant \epsilon_3 \Big\{ \|\eta_k - \eta_{k-1}\|_{C^0}\big(\|\nabla \eta_k\|_{L^{10}} + \|\nabla \eta_{k-1}\|_{L^{10}}\big)$$
$$+ \big(\|d\phi\|_{L^{10}} \|\eta_k - \eta_{k-1}\|_{C^0} + \|\nabla(\eta_k - \eta_{k-1})\|_{L^{10}}\big)\big(\|\eta_k\|_{C^0} + \|\eta_{k-1}\|_{C^0}\big) \Big\}.$$

Substituting parts (i) of Theorem S2 and (b), (c), (e) and (f) of Proposition 13.7.1 for $j = k-1, k$ into this equation gives $\|d(F(\eta_k) - F(\eta_{k-1}))\|_{L^{10}} \leqslant 2\lambda 2^{-k-1}t^{2/15}$, as $t \leqslant \kappa$. Thus (13.31) shows that $\|\nabla(\eta_{k+1} - \eta_k)\|_{L^{10}} \leqslant C_3 2^{-k-1}t^{2/15}$, which proves part (e) of Proposition 13.7.1 for $j = k+1$. Part (f) for $j = k+1$ then follows immediately from parts (d) and (e) for $j = k+1$ and part (iii) of Theorem S2. Parts (b) and (c) for $j = k+1$ follow by induction from parts (e) and (f) for $j = 1, \ldots, k+1$, and the lemma is finished. $\square$

Lemmas 13.7.2 and 13.7.3 establish the first step and the inductive step of our induction in j, and thus the sequence $\{\eta_j\}_{j=0}^{\infty}$ with $\eta_0 = 0$ exists in $C^{\infty}(\Lambda_-^4 T^*M)$ and satisfies (13.30) and parts (a)–(f) of Proposition 13.7.1. This completes the proof of Proposition 13.7.1. $\square$

Now parts (e) and (f) of Proposition 13.7.1 show that $\{\eta_j\}_{j=0}^{\infty}$ is convergent in $L_1^{10}(\Lambda_-^4 T^*M)$, by comparison with a geometric series. Let $\eta \in L_1^{10}(\Lambda_-^4 T^*M)$ be the limit of the sequence. Taking the limit in (13.30) and part (c) shows that $d\eta = d\phi + dF(\eta)$ and $\|\eta\|_{C^0} \leqslant Kt^{1/3} \leqslant \epsilon_1$, as we have to prove.

Thus, to complete the proof of Theorem S2 it remains only to show that η is smooth. We do this by writing η as the solution of an elliptic equation, and using elliptic regularity and the 'bootstrap method'. As η lies in $\Lambda^4_- T^* M$ we have $d^* \eta = *d\eta$ by (1.1), and so adding the equation $d\eta = d\phi + dF(\eta)$ and its Hodge star gives

$$(d + d^*)\eta = d\phi + *d\phi + dF(\eta) + *dF(\eta),$$

where η and both sides of the equation are regarded as sections of the vector bundle $V = \bigoplus_{i=0}^{8} \Lambda^i T^* M$ over M.

We rewrite this equation as

$$(d + d^*)\eta + G(\eta, \nabla\eta) = H(\eta, d\phi), \tag{13.32}$$

where G, H are smooth functions of their arguments, and $G(x, y)$ is linear in y with $G(0, y) = 0$. Define a first-order linear partial differential operator. $P : L^{10}_1(V) \to L^{10}(V)$ by $P(\zeta) = (d + d^*)\zeta + G(\eta, \nabla\zeta)$. Since $d + d^*$ is elliptic and ellipticity is an open condition, P is elliptic provided the perturbation $G(\eta, \nabla\zeta)$ is small enough. As $G(0, y) = 0$, this is true if η is small enough in C^0. But as $\|\eta\|_{C^0} \leqslant K t^{1/3}$ and $t \leqslant \kappa$, by making κ smaller if necessary we can assume that P is elliptic.

Thus $P(\eta) = H(\eta, d\phi)$ by (13.32), where P is elliptic. Now $\eta \in C^{0,1/5}(V)$ by Theorem 1.2.1, as $\eta \in L^{10}_1(V)$. Suppose $\eta \in C^{k,1/5}(V)$ for some $k \geqslant 0$. Then $H(\eta, d\phi)$ and the coefficients of P are both $C^{k,1/5}$, and so $\eta \in C^{k+1,1/5}(V)$ by Theorem 1.4.2, an elliptic regularity result. Hence by induction $\eta \in C^{k,1/5}(V)$ for all k, so η is smooth. This finishes the proof of Theorem S2. □

14

EXAMPLES OF COMPACT 8-MANIFOLDS WITH HOLONOMY Spin(7)

We now use the results of Chapter 13 to construct examples of compact 8-manifolds with holonomy Spin(7), in a similar way to the G_2 case of Chapter 12. Our main aim is to show that many such manifolds exist, and to calculate their Betti numbers (b^2, b^3, b^4).

We begin in §14.1 by discussing how to calculate the Betti numbers (b^2, b^3, b^4) of a resolution M of T^8/Γ, and to show that M has holonomy Spin(7). Then §14.2 gives a simple example of an orbifold T^8/Γ and its resolution M with holonomy Spin(7).

In §14.3 we construct examples of *Cayley 4-folds* Y in the 8-manifold M of §14.2. We find that Y can have a number of topological types, including T^4 and $K3$. Sections 14.4–14.6 study three more complicated orbifolds T^8/Γ and their resolutions. We find that one orbifold T^8/Γ can admit a large number of topologically distinct resolutions (M, π) with holonomy Spin(7). The examples of §14.4 and §14.5 both admit at least $2^{16} = 65\,536$ different resolutions.

Distinct resolutions may be diffeomorphic as 8-manifolds, so we have not found this many distinct 8-manifolds with holonomy Spin(7). But we do show that these examples yield manifolds with at least 54 and 98 different sets of Betti numbers (b^2, b^3, b^4) respectively. In total, the examples of this chapter yield 181 different sets of Betti numbers of compact 8-manifolds with holonomy Spin(7), which are listed in Tables 14.1, 14.2 and 14.3.

Two 8-manifolds with holonomy Spin(7) and the same Betti numbers can often be distinguished using the cup products on their cohomologies — examples of this are given in [120, §3.4] — so our constructions probably give many more than 181 topologically distinct 8-manifolds.

We could give lots more examples using the same techniques. However, we have chosen to close this chapter after studying only four orbifolds T^8/Γ, and move on to look at another way of constructing compact 8-manifolds with holonomy Spin(7), which will be explained in Chapter 15.

Sections 14.1–14.2 are based on the author's paper [120], with some alterations. The rest of the chapter is original research by the author, published here for the first time.

14.1 Calculating topological invariants of resolutions of T^8/Γ

Let Γ be a finite group of automorphisms of T^8 preserving a flat Spin(7)-structure (Ω_0, g_0) on T^8, and suppose (Y_i, π_i), (Ω_i, g_i) and $\chi_{\gamma,i}$ is a set of R-data for T^8/Γ, as in §13.2.1. Then Definition 13.2.5 gives a compact 8-manifold M, the *resolution* of T^8/Γ, and Theorem 13.6.2 proves that there exist torsion-free Spin(7)-structures on M.

TOPOLOGICAL INVARIANTS OF RESOLUTIONS OF T^8/Γ

In the rest of the chapter we will give many examples of such compact 8-manifolds M. For each we will want to know the most basic topological invariants — the fundamental group $\pi_1(M)$ and the Betti numbers b^2, b^3, b^4 and $b^4_\pm$. By Theorem 10.6.8, the torsion-free Spin(7)-structures (Ω, g) on M have holonomy Spin(7) if and only if M is simply-connected, and

$$b^3 + b^4_+ = b^2 + 2b^4_- + 25. \tag{14.1}$$

Thus, one approach would be to calculate the fundamental group and Betti numbers of each example M, and use them to prove that g has holonomy Spin(7), which is what we want to know. We explained in §11.1 how to find the fundamental group and Betti numbers of a resolution of T^7/Γ in the G_2 case, and the same methods work for T^8/Γ.

However, finding $\pi_1(M)$ is a rather tedious calculation, and is best avoided. Instead we can prove directly from the construction that g has holonomy Spin(7), using the theorem below, and then deduce from Theorem 10.6.8 that M is simply-connected and satisfies (14.1). One could just calculate b^2, b^3 and b^4, and then find $b^4_\pm$ using (14.1). But we prefer to calculate $b^4_\pm$ as well, and then use (14.1) to check that we have made no mistakes.

Here is a way of finding $\text{Hol}^0(g)$.

Theorem 14.1.1 *Suppose T^8/Γ has a flat Spin(7)-structure (Ω_0, g_0), and a set of R-data (Y_i, π_i), (Ω_i, g_i) and $\chi_{\gamma,i}$. Let M be the resolution of T^8/Γ given in Definition 13.2.5, and let (Ω, g) be a torsion-free Spin(7)-structure on M, which exists by Theorem 13.6.2.*

Regard Spin(7) as a group of automorphisms of $\mathbb{R}^8$, where $T^8 = \mathbb{R}^8/\Lambda$ for some lattice Λ in $\mathbb{R}^8$. Then $(V_i \times Y_i, \Omega_i, g_i)$ is a QALE Spin(7)-manifold asymptotic to $\mathbb{R}^8/A_i$, for each $i \in I \setminus \{0\}$. We can identify $\text{Hol}^0(g_i)$ with a subgroup of Spin(7) in a natural way. Then $\text{Hol}^0(g)$ is the subgroup of Spin(7) generated by the subgroups $\text{Hol}^0(g_i)$, for all $i \in I \setminus \{0\}$.

One can also show that $\text{Hol}(g)$ is the subgroup of Spin(7) generated by the subgroups $\text{Hol}(g_i)$ for all $i \in I \setminus \{0\}$, together with $\text{Hol}(g_0)$, where g_0 is the flat metric on T^8/Γ.

To prove the theorem we use a limiting argument as $t \to 0$ in the construction of Chapter 13. For small t, let $(\tilde{\Omega}^t, \tilde{g}^t)$ be the torsion-free Spin(7)-structure on M near to (Ω^t, g^t), constructed using Theorem 13.6.1. Then $(\tilde{\Omega}^t, \tilde{g}^t)$ is close to (Ω^t_i, g^t_i) on the portion of M modelled on $V_i \times Y^t_i$. Rescaling distances by a factor t^{-1}, we see that $(t^{-4}\tilde{\Omega}^t, t^{-2}\tilde{g}^t)$ is close to the Spin(7)-structure (Ω_i, g_i) on $V_i \times Y_i$.

But $\text{Hol}^0(t^{-2}\tilde{g}^t) = \text{Hol}^0(\tilde{g}^t) = \text{Hol}^0(g)$. So, taking a limit as $t \to 0$, we can show that $\text{Hol}^0(g_i) \subseteq \text{Hol}^0(g)$ for all $i \in I \setminus \{0\}$. Thus $\text{Hol}^0(g)$ contains the subgroup of Spin(7) generated by the subgroups $\text{Hol}^0(g_i)$, for all $i \in I \setminus \{0\}$. We then argue that this is the whole of $\text{Hol}^0(g)$ using Theorem 2.4.3. The details are left to the reader.

We explained in §12.1.2 how to calculate the Betti numbers $b^k(M)$ in the G_2 case, and exactly the same ideas can be used to work out b^2, b^3 and b^4 here. To calculate $b^4_\pm$ we use the same basic method: we first find $b^4_\pm(T^8/\Gamma)$, which are the dimensions of the

Γ-invariant subspaces of $\Lambda^4_\pm(\mathbb{R}^8)^*$, and then add on contributions from the resolution of the singular set S of T^8/Γ.

However, it requires some geometric insight to determine how much a resolution adds to b^4_+ and b^4_-. Here are some useful rules for doing this.

- Each connected component of S can be treated separately.
- If S has a component T^k locally modelled on $T^k \times (\mathbb{R}^{8-k}/G)$ for $k > 0$, then the resolution $T^k \times Y$ of these singularities makes equal contributions to b^4_+ and b^4_-. We prove this by showing that as T^k has an orientation-reversing automorphism for $k > 0$, so $T^k \times Y$ has an orientation-reversing automorphism, and $b^4_+(T^k \times Y) = b^4_-(T^k \times Y)$.
- More generally, it is in some sense true that all the singular points in T^8/Γ modelled on $\mathbb{R}^k \times (\mathbb{R}^{8-k}/G)$ for $k > 0$ contribute equally to b^4_+ and b^4_-, and only the resolution of points in T^8/Γ modelled on $\mathbb{R}^8/G$ with $\text{Fix}(G) = \{0\}$ can change the signature $b^4_+ - b^4_-$ of M.
- By Theorem 10.7.1, the dimension of the moduli space of holonomy Spin(7) metrics on M is $1 + b^4_-(M)$. Using this we find that the contribution of the resolution to b^4_- is the dimension of the set of possible choices of R-data.
- The resolution of any component of S adds at least 1 to b^4_-. Hence, if the resolution adds only 1 to b^4, then it fixes b^4_+ and adds 1 to b^4_-.
- If Y is a crepant resolution of an *isolated* singularity $\mathbb{C}^4/G$, we find that $b^4_-(Y) = b^2(Y)$ and $b^4_+(Y) = b^4(Y) - b^2(Y)$.

14.2 A simple example

We shall now explain in detail the simplest example of a compact 8-manifold with holonomy Spin(7) that the author knows, based on [120, Ex. 1]. Let $(x_1, \ldots, x_8)$ be coordinates on $\mathbb{R}^8$, and let (Ω_0, g_0) be the flat Spin(7)-structure on $\mathbb{R}^7$ given in Definition 10.5.1. Then

$$\Omega_0 = dx_{1234} + dx_{1256} + dx_{1278} + dx_{1357} - dx_{1368}$$
$$- dx_{1458} - dx_{1467} - dx_{2358} - dx_{2367} - dx_{2457}$$
$$+ dx_{2468} + dx_{3456} + dx_{3478} + dx_{5678}$$

by (10.19), where $dx_{ijkl} = dx_i \wedge dx_j \wedge dx_k \wedge dx_l$. Let $\mathbb{Z}^8$ act on $\mathbb{R}^8$ by translations in the obvious way, and let $T^8 = \mathbb{R}^8/\mathbb{Z}^8$. Then (Ω_0, g_0) descends to T^8. We shall regard $(x_1, \ldots, x_8)$ as coordinates on T^8, where each x_j lies in $\mathbb{R}/\mathbb{Z}$ rather than $\mathbb{R}$.

Let α, β, γ and δ be the involutions of T^8 defined by

$$\alpha : (x_1, \ldots, x_8) \mapsto (-x_1, -x_2, -x_3, -x_4, x_5, x_6, x_7, x_8), \tag{14.2}$$

$$\beta : (x_1, \ldots, x_8) \mapsto (x_1, x_2, x_3, x_4, -x_5, -x_6, -x_7, -x_8), \tag{14.3}$$

$$\gamma : (x_1, \ldots, x_8) \mapsto (\tfrac{1}{2} - x_1, \tfrac{1}{2} - x_2, x_3, x_4, \tfrac{1}{2} - x_5, \tfrac{1}{2} - x_6, x_7, x_8), \tag{14.4}$$

$$\delta : (x_1, \ldots, x_8) \mapsto (-x_1, x_2, \tfrac{1}{2} - x_3, x_4, -x_5, x_6, \tfrac{1}{2} - x_7, x_8). \tag{14.5}$$

By inspection, α, β, γ and δ preserve Ω_0, because of the careful choice of exactly which signs to change. It is easy to see that $\alpha^2 = \beta^2 = \gamma^2 = \delta^2 = 1$, and that $\alpha, \beta, \gamma, \delta$ all commute. Define Γ to be the group $\langle \alpha, \beta, \gamma, \delta \rangle$. Then $\Gamma \cong \mathbb{Z}_2^4$ is a group of automorphisms of T^8 preserving the flat Spin(7)-structure (Ω_0, g_0).

In the next two lemmas, we shall describe the singular set S of T^8/Γ. They are proved in the same way as Lemmas 12.2.1 and 12.2.2.

Lemma 14.2.1 *The fixed points of α, β, γ and δ are each 16 copies of T^4 in T^8, and the fixed points of $\alpha\beta$ are 256 points in T^8. These are the only nonidentity elements of Γ with fixed points in T^8. Also*

(i) *β acts trivially and $\langle \gamma, \delta \rangle$ acts freely on the set of 16 α T^4,*
(ii) *α acts trivially and $\langle \gamma, \delta \rangle$ acts freely on the set of 16 β T^4,*
(iii) *$\langle \alpha, \beta, \delta \rangle$ acts freely on the set of 16 γ T^4, and*
(iv) *$\langle \alpha, \beta, \gamma \rangle$ acts freely on the set of 16 δ T^4.*

Lemma 14.2.2 *The singular set S of T^8/Γ is the union of*

(i) *4 $T^4/\{\pm 1\}$ from the α fixed points,*
(ii) *4 $T^4/\{\pm 1\}$ from the β fixed points,*
(iii) *2 T^4 from the γ fixed points, and*
(iv) *2 T^4 from the δ fixed points.*

The union is not disjoint, as the sets (i) *and* (ii) *intersect in 64 points.*

As in §12.1.2, the Betti numbers $b^k(T^8/\Gamma)$ are the dimensions of the Γ-invariant subspaces of $\Lambda^k(\mathbb{R}^8)^*$. Calculation shows that there are no nonzero Γ-invariant 1, 2 or 3-forms, and that the Γ-invariant 4-forms are

$$\langle dx_{1234}, dx_{1256}, dx_{1278}, dx_{1357}, dx_{1368}, dx_{1458}, dx_{1467},$$
$$dx_{2358}, dx_{2367}, dx_{2457}, dx_{2468}, dx_{3456}, dx_{3478}, dx_{5678} \rangle.$$

Therefore the Betti numbers of T^8/Γ are $b^1 = b^2 = b^3 = 0$ and $b^4 = 14$. We can also easily see that $b_+^4 = b_-^4 = 7$.

Next we shall write down a set of *R-data* for T^8/Γ, as in §13.2. In the notation of §13.2.1, the elements of $\mathcal{L}$ are T^8, the four lots of 16 T^4 fixed by α, β, γ and δ, and the 256 points fixed by $\alpha\beta$. Let I be an indexing set for $\mathcal{L}$, so that $\mathcal{L} = \{F_i : i \in I\}$. Then $0 \in I$ with $F_0 = T^8$, and if $i \in I \setminus \{0\}$ then F_i is either one of the 64 T^4 fixed by α, β, γ or δ, or one of the 256 points fixed by $\alpha\beta$.

By Definition 13.2.2, a set of R-data for T^8/Γ is a choice of resolution (Y_i, π_i) for W_i/A_i and a torsion-free product Spin(7)-structure (Ω_i, g_i) on $V_i \times Y_i$ for each i, together with isomorphisms $\chi_{\gamma,i} : Y_i \to Y_{\gamma \cdot i}$ for each $\gamma \in \Gamma$ and $i \in I$, satisfying certain conditions.

When $F_i = T^4$ we find that $V_i \cong \mathbb{R}^4$, $W_i \cong \mathbb{C}^2$ and $A_i \cong \mathbb{Z}_2$, where A_i acts on W_i as multiplication by ± 1, so that $W_i/A_i \cong \mathbb{C}^2/\{\pm 1\}$. The conditions on (Y_i, π_i) and (Ω_i, g_i) imply that Y_i must be an ALE 4-manifold with holonomy SU(2), asymptotic to

$\mathbb{C}^2/\{\pm 1\}$. From §7.2, this implies that Y_i is one of the 3-dimensional family of *Eguchi–Hanson spaces*, described in Example 7.2.2. As a 4-manifold, Y_i is diffeomorphic to the blow-up of $\mathbb{C}^2/\{\pm 1\}$ at 0.

Choose Eguchi–Hanson spaces (Y_i, π_i) and corresponding Spin(7)-structures (Ω_i, g_i) on $V_i \times Y_i$ for all of the 64 $i \in I$ with $F_i = T^4$. Now these 64 choices are not independent, as we also need isomorphisms $\chi_{\epsilon,i} : Y_i \to Y_{\epsilon \cdot i}$ for all $i \in I$ and $\epsilon \in \Gamma$. Basically, we get to choose one Eguchi–Hanson space for each of the 4 α $T^4/\{\pm 1\}$, the 4 β $T^4/\{\pm 1\}$, the 2 γ T^4 and the 2 δ T^4 in T^8/Γ, so there are only 12 independent choices.

Let F_i be one of the 16 T^4 fixed by α, and F_j one of the 16 T^4 fixed by β. Then $F_i \cap F_j$ is a single point, one of the 256 points fixed by $\alpha\beta$. Every fixed point of $\alpha\beta$ can be written uniquely as $F_i \cap F_j$ for some i, j. Let $k \in I$ be the index of this point, so that $F_i \cap F_j = F_k$. Then we find that $V_k = \{0\}$, $W_k = \mathbb{R}^8$ and $A_k \cong \mathbb{Z}_2^2$, with $W_k/A_k \cong \mathbb{R}^4/\{\pm 1\} \times \mathbb{R}^4/\{\pm 1\}$.

In fact it is natural to write $W_k \cong W_i \oplus W_j$ and $W_k/A_k \cong W_i/A_i \times W_j/A_j$. This suggests that the resolution (Y_k, π_k) of W_k/A_k that we choose should satisfy $Y_k = Y_i \times Y_j$. As Y_i and Y_j carry metrics with holonomy SU(2), the product $Y_i \times Y_j$ has holonomy SU(2) × SU(2), and so we can define a Spin(7)-structure (Ω_k, g_k) on $Y_k = Y_i \times Y_j$ by Proposition 13.1.4.

It can be shown using part (ii) of Definition 13.2.2 that this is the only possible choice for Y_k and (Ω_k, g_k), once Y_i, Y_j and $(\Omega_i, g_i), (\Omega_j, g_j)$ are fixed. Thus the choice of R-data for the 64 T^4 in $\mathcal{L}$ determines the choice of R-data for the 256 points in $\mathcal{L}$. If we choose the R-data in this way, it satisfies all the conditions of Definition 13.2.2.

Thus we can choose a set of R-data $(Y_i, \pi_i), (\Omega_i, g_i)$ and $\chi_{\gamma,i}$ for T^8/Γ satisfying the required conditions. So by the results of Chapter 13, T^8/Γ has a resolution M which carries torsion-free Spin(7)-structures, and we prove:

Theorem 14.2.3 *There exists a set of R-data for the orbifold T^8/Γ defined above. Let M be the corresponding resolution of T^8/Γ, as in Definition 13.2.5. Then M has Betti numbers*

$$b^2 = 12, \quad b^3 = 16, \quad b^4 = 150, \quad b^4_+ = 107 \quad and \quad b^4_- = 43.$$

There exist metrics with holonomy Spin(7) *on M, which form a smooth family of dimension 44.*

Proof We constructed the R-data for T^8/Γ above. Theorem 13.6.2 proves that there exist torsion-free Spin(7)-structures (Ω, g) on M, and Theorem 14.1.1 shows that $\mathrm{Hol}^0(g)$ is the subgroup of Spin(7) generated by the subgroups $\mathrm{Hol}^0(g_i)$ for $i \in I \setminus \{0\}$.

If $F_i = T^4$ then $\mathrm{Hol}^0(g_i) \cong \mathrm{SU}(2)$, since Y_i is an ALE space with holonomy SU(2). However, the T^4 fixed by α, β, γ and δ yield four *different* SU(2) subgroups of Spin(7), and it is not difficult to show that these four SU(2) subgroups generate the whole of Spin(7). Hence $\mathrm{Hol}^0(g) = \mathrm{Hol}(g) = \mathrm{Spin}(7)$, as we want.

Next we calculate the Betti numbers of M. From above, T^8/Γ has Betti numbers $b^1 = b^2 = b^3 = 0$ and $b^4_+ = b^4_- = 7$. It can be shown that resolving each of the 4 α

and 4 β $T^4/\{\pm 1\}$ in T^8/Γ fixes b^3, adds 1 to b^2 and 3 to each of $b_\pm^4$. Each of the 2 γ and 2 δ T^4 in T^8/Γ adds 1 to b^2, 4 to b^3 and 3 to $b_\pm^4$.

In addition, resolving each of the 64 points in the intersection of the 4 α $T^4/\{\pm 1\}$ and the 4 β $T^4/\{\pm 1\}$ fixes b^2, b^3 and b_-^4 and adds 1 to b_+^4. Adding up all the contributions we get the Betti numbers in the theorem. Finally, Theorem 10.7.1 shows that the moduli space of holonomy Spin(7) metrics on M is a smooth manifold of dimension $1 + b_-^4 = 1 + 43 = 44$. □

Observe that the Betti numbers in the theorem satisfy (14.1), as they should. We can use also the method of §12.1.1 to show that M is simply-connected, and then Theorem 10.6.8 gives an alternative proof that g has holonomy Spin(7).

A number of variations on this example are given in [120, §3], related to it as the example of §12.3 is related to that of §12.2 in the G_2 case. We will not explain them, but instead move on to more elaborate cases.

14.3 Examples of Cayley submanifolds

Calibrated geometry was discussed in general terms in §3.7, and for the exceptional holonomy groups in §10.8. We saw in §10.8.3 that there is an interesting class of calibrated submanifolds in Spin(7)-manifolds (M, Ω, g), called *Cayley 4-folds*. We also showed in Proposition 10.8.6 that the fixed point set of an isometric involution $\sigma : M \to M$ with $\sigma^*(\Omega) = \Omega$ is automatically Cayley, provided it has the right dimension.

We now use this idea to find examples of Cayley submanifolds in the compact 8-manifold M with holonomy Spin(7) constructed in §14.2, following the example of §12.6 in the G_2 case. Throughout this section, let T^8, Γ, M and all other notation be as in §14.2.

Example 14.3.1 Define an isometry $\sigma : T^8 \to T^8$ by

$$\sigma : (x_1, \ldots, x_8) \mapsto (-x_1, \tfrac{1}{2} - x_2, x_3, x_4, x_5, x_6, -x_7, \tfrac{1}{2} - x_8).$$

Then σ commutes with Γ and $\sigma^*(\Omega_0) = \Omega_0$. The fixed points of σ in T^8 are 16 T^4, the fixed points of $\sigma\alpha\gamma$ are 256 points, and $\sigma\epsilon$ acts freely on T^8 for all $\epsilon \neq 1, \alpha\gamma$ in Γ. Also Γ acts freely on the 16 σ T^4 and the 256 $\sigma\alpha\gamma$ points.

Thus the fixed points of σ in T^8/Γ are the disjoint union of T^4 and 16 points, none of which intersect the singular set of T^8/Γ. When we resolve T^8/Γ in a σ-equivariant way to get M with holonomy Spin(7), the fixed points of σ in M are again the disjoint union of T^4 and 16 points. Proposition 10.8.6 shows that this T^4 is a Cayley 4-fold in M.

Example 14.3.2 Define an isometry $\sigma : T^8 \to T^8$ by

$$\sigma : (x_1, \ldots, x_8) \mapsto (\tfrac{1}{2} - x_1, -x_2, x_3, x_4, \tfrac{1}{2} - x_5, -x_6, x_7, x_8).$$

Then σ commutes with Γ and $\sigma^*(\Omega_0) = \Omega_0$. The fixed points of σ in T^8 are 16 T^4, the fixed points of $\sigma\alpha\beta\gamma$ in T^8 are 256 points, and $\sigma\epsilon$ has no fixed points in T^8 for all $\epsilon \neq 1, \alpha\beta\gamma$ in Γ. So far this is very like the previous example.

However, in this case $\langle \alpha, \beta, \gamma \rangle$ acts freely on the set of 16 σ T^4, but $\alpha\beta\delta$ fixes each T^4. Thus the fixed points of σ in T^8/Γ are the disjoint union of 2 copies of $T^4/\mathbb{Z}_2$ and 16 points, none of which intersect the singular set of T^8/Γ. Here the generator $\alpha\beta\delta$ of $\mathbb{Z}_2$ acts freely on each T^4 by

$$\alpha\beta\delta : (x_3, x_4, x_7, x_8) \mapsto (\tfrac{1}{2} + x_3, -x_4, \tfrac{1}{2} + x_7, -x_8).$$

So $T^4/\mathbb{Z}_2$ is a nonsingular 4-manifold. When we resolve T^8/Γ in a σ-equivariant way to get M with holonomy Spin(7), the fixed points of σ in M are again the disjoint union of 2 $T^4/\mathbb{Z}_2$ and 16 points. Proposition 10.8.6 shows that each $T^4/\mathbb{Z}_2$ is a Cayley 4-fold in M.

Example 14.3.3 Define an isometry $\sigma : T^8 \to T^8$ by

$$\sigma : (x_1, \ldots, x_8) \mapsto (-x_1, -x_2, -x_3, \tfrac{1}{2} - x_4, x_5, x_6, x_7, x_8).$$

Then σ commutes with Γ and $\sigma^*(\Omega_0) = \Omega_0$. The fixed points of σ in T^8 are 16 T^4, the fixed points of $\sigma\beta$ in T^8 are 256 points (which are contained in the 16 σ T^4), and $\sigma\epsilon$ has no fixed points in T^8 for all $\epsilon \neq 1$, β in Γ. The group $\langle \alpha, \gamma, \delta \rangle$ acts freely on the 16 T^4 fixed by σ, and β fixes each T^4, converting it into $T^4/\{\pm 1\}$.

Thus the fixed points of σ in T^8/Γ are the disjoint union of 2 copies of $T^4/\{\pm 1\}$. When we resolve T^8/Γ in a σ-equivariant way to get M with holonomy Spin(7), the fixed points of σ in M are 2 copies of $K3$ in M, as $K3$ is the resolution of $T^4/\{\pm 1\}$. Proposition 10.8.6 shows that each $K3$ is a Cayley 4-fold in M.

Example 14.3.4 Here is a more complex example. Define $\sigma : T^8 \to T^8$ by

$$\sigma : (x_1, \ldots, x_8) \mapsto (-x_1, x_2, -x_3, x_4, -x_5, x_6, -x_7, x_8).$$

Then $\sigma^*(\Omega_0) = \Omega_0$, and σ, $\sigma\alpha$, $\sigma\beta$ and $\sigma\alpha\beta$ each fix 16 T^4 in T^8, and $\sigma\alpha\beta\delta$ fixes 256 points, contained in the 16 $\sigma\alpha\beta$ T^4. The other elements of $\sigma\Gamma$ act freely. The sets of 16 T^4 from σ, $\sigma\alpha$, $\sigma\beta$ and $\sigma\alpha\beta$ all intersect, and are fixed by $\langle \alpha, \beta \rangle$.

Also $\langle \gamma, \delta \rangle$ acts freely on the sets of 16 σ, $\sigma\alpha$ and $\sigma\beta$ T^4, but γ acts freely and δ trivially on the 16 $\sigma\alpha\beta$ T^4. The fixed points of σ in T^8/Γ are the union of 3 sets of 4 $T^2/\{\pm 1\} \times T^2/\{\pm 1\}$ from the σ, $\sigma\alpha$ and $\sigma\beta$ fixed points in T^8, and 8 $T^4/\mathbb{Z}_2^3$ from the $\sigma\alpha\beta$ fixed points in T^8. This union is not disjoint.

In our previous examples there was only one σ-equivariant way to resolve T^8/Γ, but in this case there are topological choices. The fixed points of α and β in T^8 contribute two sets of 4 $T^4/\{\pm 1\}$ to the singular set of T^8/Γ. The resolution of each of these 8 $T^4/\{\pm 1\}$ introduces a class $[S^2]$ in $H_2(M, \mathbb{R})$. To define the action of σ on M we must choose $\sigma_* = \pm 1$ for each of these 8 homology classes, in a similar way to parts (a) and (b) of §12.3. The choices are independent, so there are $2^8 = 256$ possibilities.

Having chosen a σ-action on M, we can construct σ-invariant metrics with holonomy Spin(7) on M. The fixed points of σ in M form a nonsingular 4-manifold, not necessarily connected, and each component is a Cayley 4-fold by Proposition 10.8.6. The topology of these 4-manifolds depends on the action of σ on M, and is in general rather complicated. In the case when $\sigma_* = 1$ for all 8 classes, the fixed points of σ turn out to be the disjoint union of 12 $\mathcal{S}^2 \times \mathcal{S}^2$, and 8 copies of a simply-connected 4-manifold N with $b_+^2(N) = 1$ and $b_-^2(N) = 5$.

There are many other suitable involutions of M yielding Cayley 4-folds with a wide variety of topologies. We can also use the same technique to find Cayley 4-folds in other compact 8-manifolds with holonomy Spin(7).

14.4 A more complicated example

We now consider an orbifold T^8/Γ very similar to that in §14.2, but with a more complicated singular set that must be resolved using QALE manifolds with holonomy SU(3). It turns out that this leads to a striking increase both in the complexity of resolving T^8/Γ, and in the number of topologically distinct 8-manifolds that result.

Let T^8 and (Ω_0, g_0) be as in §14.2, and define $\alpha, \beta, \gamma, \delta : T^8 \to T^8$ by

$$\alpha : (x_1, \ldots, x_8) \mapsto (-x_1, -x_2, -x_3, -x_4, x_5, x_6, x_7, x_8), \tag{14.6}$$

$$\beta : (x_1, \ldots, x_8) \mapsto (x_1, x_2, x_3, x_4, -x_5, -x_6, -x_7, -x_8), \tag{14.7}$$

$$\gamma : (x_1, \ldots, x_8) \mapsto (\tfrac{1}{2} - x_1, \tfrac{1}{2} - x_2, x_3, x_4, -x_5, -x_6, x_7, x_8), \tag{14.8}$$

$$\delta : (x_1, \ldots, x_8) \mapsto (-x_1, x_2, -x_3, x_4, \tfrac{1}{2} - x_5, x_6, \tfrac{1}{2} - x_7, x_8). \tag{14.9}$$

These differ from (14.2)–(14.5) only in where $\tfrac{1}{2}$ appears. As before $\Gamma = \langle \alpha, \beta, \gamma, \delta \rangle$ is a group isomorphic to $\mathbb{Z}_2^4$, which acts on T^8 preserving (Ω_0, g_0). Here are the analogues of Lemmas 14.2.1 and 14.2.2 for this group Γ.

Lemma 14.4.1 *The fixed points of $\alpha, \beta, \gamma, \delta, \alpha\delta$ and $\beta\gamma$ are each 16 copies of T^4 in T^8, and the fixed points of $\alpha\beta$ are 256 points in T^8. These are the only nonidentity elements of Γ that have fixed points on T^8. Also*

(i) *β and δ act trivially and γ acts freely on the set of 16 α T^4,*

(ii) *α and γ act trivially and δ acts freely on the set of 16 β T^4,*

(iii) *β acts trivially and $\langle \alpha, \delta \rangle$ acts freely on the set of 16 γ T^4,*

(iv) *α acts trivially and $\langle \beta, \gamma \rangle$ acts freely on the set of 16 δ T^4,*

(v) *α acts trivially and $\langle \beta, \gamma \rangle$ acts freely on the set of 16 $\alpha\delta$ T^4, and*

(vi) *β acts trivially and $\langle \alpha, \delta \rangle$ acts freely on the set of 16 $\beta\gamma$ T^4.*

Lemma 14.4.2 *The singular set S of T^8/Γ is the union of*

(i) *8 $T^4/\langle \beta, \delta \rangle$ from the α fixed points,*

(ii) *8 $T^4/\langle \alpha, \gamma \rangle$ from the β fixed points,*

(iii) *4 $T^4/\langle \beta \rangle$ from the γ fixed points,*

(iv) *4 $T^4/\langle \alpha \rangle$ from the δ fixed points,*

(v) *4 $T^4/\langle \alpha \rangle$ from the $\alpha\delta$ fixed points, and*

(vi) *4 $T^4/\langle \beta \rangle$ from the $\beta\gamma$ fixed points.*

The union is not disjoint, as the sets (i), (ii) intersect in 64 points, the sets (i), (iv), (v) intersect in 16 T^2, and the sets (ii), (iii), (vi) intersect in 16 T^2.

Here the fixed points of $\langle \alpha, \beta \rangle$ in T^8 are $\{(x_1, \ldots, x_8) \in T^8 : x_j \in \{\mathbb{Z}, \frac{1}{2} + \mathbb{Z}\}\}$, which is 256 points. Their image in T^8/Γ is 64 points, the intersection of the sets (i) and (ii), and the singularities are locally modelled on $\mathbb{C}^2/\{\pm 1\} \times \mathbb{C}^2/\{\pm 1\}$.

The fixed points of $\langle \alpha, \delta \rangle$ in T^8 are

$$\{(x_1, \ldots, x_8) \in T^8 : x_1, x_2, x_3, x_4 \in \{\mathbb{Z}, \tfrac{1}{2} + \mathbb{Z}\}, \quad x_5, x_7 \in \{\tfrac{1}{4} + \mathbb{Z}, \tfrac{3}{4} + \mathbb{Z}\}\},$$

which is 64 T^2. Their image in T^8/Γ is 16 T^2, the intersection of the sets (i), (iv) and (v), and each T^2 is locally modelled on $T^2 \times \mathbb{C}^3/\mathbb{Z}_2^2$, where the generators α, δ of $\mathbb{Z}_2^2$ act on $\mathbb{C}^3$ by

$$\alpha : (z_1, z_2, z_3) \mapsto (-z_1, -z_2, z_3) \quad \text{and} \quad \delta : (z_1, z_2, z_3) \mapsto (z_1, -z_2, -z_3).$$

Similarly, the fixed points of $\langle \beta, \gamma \rangle$ in T^8 are

$$\{(x_1, \ldots, x_8) \in T^8 : x_1, x_2 \in \{\tfrac{1}{4} + \mathbb{Z}, \tfrac{3}{4} + \mathbb{Z}\}, \quad x_5, x_6, x_7, x_8 \in \{\mathbb{Z}, \tfrac{1}{2} + \mathbb{Z}\}\},$$

which is 64 T^2. Their image in T^8/Γ is 16 T^2, the intersection of the sets (ii), (iii) and (vi), and each T^2 is locally modelled on $T^2 \times \mathbb{C}^3/\mathbb{Z}_2^2$.

We wish to choose a set of R-data for T^8/Γ, as in §13.2. Now the set $\mathcal{L}$ is the union of T^8, the 6 lots of 16 T^4 fixed by $\alpha, \beta, \gamma, \delta, \alpha\delta$ and $\beta\gamma$, the 256 points fixed by $\langle \alpha, \beta \rangle$, and the 2 sets of 64 T^2 fixed by $\langle \alpha, \delta \rangle$ and $\langle \beta, \gamma \rangle$. So $\mathcal{L}$ has 481 elements.

Effectively, a set of R-data for T^8/Γ is a choice of

(a) An ALE 4-manifold Y_i with holonomy SU(2) asymptotic to $\mathbb{C}^2/\{\pm 1\}$ for each of the 96 T^4 in $\mathcal{L}$,

(b) A product $Y_i \times Y_j$ for each of the 256 points in $\mathcal{L}$, where Y_i, Y_j are both ALE 4-manifolds with holonomy SU(2) asymptotic to $\mathbb{C}^2/\{\pm 1\}$, and

(c) A QALE 6-manifold Y_k with holonomy SU(3) asymptotic to $\mathbb{C}^3/\mathbb{Z}_2^2$ for each of the 128 T^2 in $\mathcal{L}$.

These choices are not independent, because the maps $\chi_{\gamma,i}$ induce a lot of isomorphisms between the Y_i, and also part (ii) of Definition 13.2.2 relates the Y_i in (a) to limits of the $Y_i \times Y_j$ in (b) and the Y_k in (c).

As in §14.2, in (a) we must choose Y_i to be an *Eguchi–Hanson space*, one of a 3-dimensional family of ALE 4-manifolds with holonomy SU(2) diffeomorphic to the blow-up of $\mathbb{C}^2/\{\pm 1\}$ at 0. Also, as in §14.2, the choice of the $Y_i \times Y_j$ in (b) is entirely determined by the choice of the Y_i for the α and β T^4 in (a).

However, the possible choices in (c) are much more interesting. We have already studied this singularity $\mathbb{C}^3/\mathbb{Z}_2^2$ in Examples 9.3.6 and 9.9.9. We found that $\mathbb{C}^3/\mathbb{Z}_2^2$ has a number of topologically distinct resolutions and deformations, labelled (i)–(ix) in Example 9.9.9, which all carry QALE Kähler metrics with holonomy SU(3).

Our basic plan is that for each of the 32 T^2 in T^8/Γ coming from the fixed points of $\langle \alpha, \delta \rangle$ or $\langle \beta, \gamma \rangle$ we will choose one of cases (i)–(ix) in Example 9.9.9 to be the corresponding resolution Y_i in the R-data for T^8/Γ. This gives us a huge number of choices for the R-data of T^8/Γ.

But these choices are also subject to many consistency conditions, which rule out almost all the possibilities, but still leave a large number. Then for each of the allowed possibilities for the R-data, we want to calculate the Betti numbers of the associated resolution M of T^8/Γ. This is a task too big to attempt without the aid of a computer.

So instead we will simplify the problem by splitting it up into two stages, and limiting the kinds of resolution we consider. In the first stage we resolve $T^8/\langle\alpha,\beta\rangle$ to get $K3 \times K3$, such that the action of $\langle\gamma,\delta\rangle$ on $T^8/\langle\alpha,\beta\rangle$ lifts to $K3 \times K3$. Then $(K3 \times K3)/\langle\gamma,\delta\rangle$ is an orbifold. In the second stage we resolve $(K3 \times K3)/\langle\gamma,\delta\rangle$ to get M.

Note that this idea of first resolving the $\langle\alpha,\beta\rangle$ singularities and then resolving the $\langle\gamma,\delta\rangle$ singularities is just a way of simplifying the choice of R-data for T^8/Γ, and understanding the topology of the resulting resolution M. We do the actual construction of metrics with holonomy Spin(7) on M in one stage, not two.

14.4.1 Resolving $T^8/\langle\alpha,\beta\rangle$ in a $\langle\gamma,\delta\rangle$-equivariant way

Write $T^8 = T^4 \times T^4$, where (x_1, x_2, x_3, x_4) are coordinates on the first T^4, and (x_5, x_6, x_7, x_8) coordinates on the second. Then α acts only on the first T^4, and β only on the second, and $T^8/\langle\alpha,\beta\rangle = (T^4/\{\pm 1\}) \times (T^4/\{\pm 1\})$. But the natural resolution of $T^4/\{\pm 1\}$, using the Eguchi–Hanson space at each of the 16 singular points, is the $K3$ surface, as in Example 7.3.14.

Thus the natural resolution of $T^8/\langle\alpha,\beta\rangle$ is $K3 \times K3$. We shall write this as $K3_\alpha \times K3_\beta$ to distinguish the 2 copies of $K3$, since $K3_\alpha$ resolves the α singularities, and $K3_\beta$ resolves the β singularities. Now $\langle\gamma,\delta\rangle \cong \mathbb{Z}_2^2$ acts on $T^8/\langle\alpha,\beta\rangle$, and we wish to lift this $\langle\gamma,\delta\rangle$-action up to the resolution $K3_\alpha \times K3_\beta$. There are topological choices in doing this.

Consider first the action of $\langle\gamma,\delta\rangle$ on the first $T^4/\{\pm 1\}$ and its lift to $K3_\alpha$. The fixed points of γ in $T^4/\{\pm 1\}$ are $2\, T^2$, and γ acts freely on the 16 singular points in $T^4/\{\pm 1\}$. Thus there is only one way to lift the action of γ up to $K3_\alpha$, and the fixed points of γ in $K3_\alpha$ are 2 copies of T^2.

However, the fixed points of δ in $T^4/\{\pm 1\}$ are the union of 4 $T^2/\{\pm 1\}$ from the δ fixed points in T^4 and 4 $T^2/\{\pm 1\}$ from the $\alpha\delta$ fixed points in T^4, and these 2 sets of 4 $T^2/\{\pm 1\}$ intersect in the 16 singular points of $T^4/\{\pm 1\}$. In particular, δ fixes every singular point in the first $T^4/\{\pm 1\}$.

Resolving each singular point in $T^4/\{\pm 1\}$ introduces a copy of $\mathcal{S}^2$ in $K3_\alpha$, with homology class $[\mathcal{S}^2] \in H_2(K3_\alpha, \mathbb{R})$. As in parts (a) and (b) of §12.3, there are two topologically distinct ways to lift the action of δ up to $K3_\alpha$ near $\mathcal{S}^2$, with either (a) $\delta_*([\mathcal{S}^2]) = [\mathcal{S}^2]$ or (b) $\delta_*([\mathcal{S}^2]) = -[\mathcal{S}^2]$. As a shorthand we write $\delta_* = 1$ in case (a), and $\delta_* = -1$ in case (b).

Thus, we define the action of δ on $K3_\alpha$ by choosing $\delta_* = \pm 1$ for each of the 16 singular points in $T^4/\{\pm 1\}$. But γ, δ commute, and γ swaps the 16 points in pairs, so δ_* must have the same value for both points in each pair swapped by γ. Hence, to determine the action of $\langle\gamma,\delta\rangle$ on $K3_\alpha$ we must choose $\delta_* = \pm 1$ for each of the 8 pairs of singular points of the first $T^4/\{\pm 1\}$ swapped by γ, giving $2^8 = 256$ possible choices.

Now consider the action of $\langle\gamma,\delta\rangle$ on the second $T^4/\{\pm 1\}$, and its lift to $K3_\beta$. This is very similar, but the rôles of γ and δ are reversed. Here the fixed points of δ in $T^4/\{\pm 1\}$ and $K3_\beta$ are 2 T^2 and the δ action lifts uniquely. The fixed points of γ in $T^4/\{\pm 1\}$ are 2 sets of 4 $T^2/\{\pm 1\}$, meeting in the 16 singular points. To lift the action of γ to $K3_\beta$, we must choose $\gamma_* = \pm 1$ for each of the 8 pairs of singular points in the second $T^4/\{\pm 1\}$ swapped by δ. Again, there are $2^8 = 256$ possible choices.

Thus, there are $2^{16} = 65\,536$ topologically distinct choices for the action of $\langle \gamma, \delta \rangle$ on $K3_\alpha \times K3_\beta$. Choose one of these possible actions. Let $\delta_* = 1$ for k out of the 8 pairs of singular points of the first $T^4/\{\pm 1\}$ swapped by γ, so that $k = 0, 1, \ldots, 8$. Let Σ_α be the subset of $K3_\alpha$ fixed by δ. Then Σ_α is a compact, oriented 2-manifold.

Similarly, let $\gamma_* = 1$ for l out of the 8 pairs of singular points of the second $T^4/\{\pm 1\}$ swapped by δ, so that $l = 0, 1, \ldots, 8$. Let Σ_β be the subset of $K3_\beta$ fixed by γ, a compact, oriented 2-manifold. Using this notation, we prove:

Proposition 14.4.3 *The Betti numbers of* $(K3_\alpha \times K3_\beta)/\langle \gamma, \delta \rangle$ *are*

$$b^2 = k+l, \quad b^3 = 0, \quad b^4 = 142 - 2k - 2l, \quad b^4_+ = 103 - k - l \text{ and } b^4_- = 39 - k - l.$$

The singular set of $(K3_\alpha \times K3_\beta)/\langle \gamma, \delta \rangle$ *is the disjoint union of* $\Sigma_\alpha \times T^2$ *and* $T^2 \times \Sigma_\beta$, *where each singular point is locally modelled on* $\mathbb{R}^4 \times \mathbb{R}^4/\{\pm 1\}$.

Proof The cohomology of $(K3_\alpha \times K3_\beta)/\langle \gamma, \delta \rangle$ is the $\langle \gamma, \delta \rangle$-invariant part of the cohomology of $K3_\alpha \times K3_\beta$. Write $H^2(K3_\alpha, \mathbb{R}) = U_\alpha \oplus V_\alpha$, where $U_\alpha \cong \mathbb{R}^6$ is the lift of $H^2(T^4/\{\pm 1\}, \mathbb{R})$, and $V_\alpha \cong \mathbb{R}^{16}$ the extra cohomology introduced by the resolution of the singularities. Similarly, write $H^2(K3_\beta, \mathbb{R}) = U_\beta \oplus V_\beta$.

Then $H^2(K3_\alpha \times K3_\beta, \mathbb{R}) \cong U_\alpha \oplus V_\alpha \oplus U_\beta \oplus V_\beta$. Calculation shows that the $\langle \gamma, \delta \rangle$-invariant subspaces of $U_\alpha, V_\alpha, U_\beta$ and V_β are $0, \mathbb{R}^k, 0$ and $\mathbb{R}^l$ respectively. Thus the $\langle \gamma, \delta \rangle$-invariant subspace of $H^2(K3_\alpha \times K3_\beta, \mathbb{R})$ is $\mathbb{R}^k \oplus \mathbb{R}^l$. Also $H^3(K3_\alpha \times K3_\beta, \mathbb{R}) = 0$. Hence $(K3_\alpha \times K3_\beta)/\langle \gamma, \delta \rangle$ has $b^2 = k + l$ and $b^3 = 0$.

In the same way, $H^4(K3_\alpha \times K3_\beta, \mathbb{R})$ is given by

$$U_\alpha \otimes U_\beta \;\oplus\; U_\alpha \otimes V_\beta \;\oplus\; V_\alpha \otimes U_\beta \;\oplus\; V_\alpha \otimes V_\beta \;\oplus\; \mathbb{R}^2. \tag{14.10}$$

Calculation shows that the $\langle \gamma, \delta \rangle$-invariant subspaces of these five vector spaces have dimensions 12, $32 - 2l$, $32 - 2k$, 64 and 2 respectively. Adding these up shows that $(K3_\alpha \times K3_\beta)/\langle \gamma, \delta \rangle$ has $b^4 = 142 - 2k - 2l$. Furthermore, the $\mathbb{R}^{64}$ from $V_\alpha \otimes V_\beta$ contributes only to b^4_+, but the other four vector spaces contribute equally to b^4_+ and b^4_-. So we get $b^4_+ = 103 - k - l$ and $b^4_- = 39 - k - l$.

The fixed points of γ in $K3_\alpha$ are 2 T^2, and in $K3_\beta$ are Σ_β. So the fixed points of γ in $K3_\alpha \times K3_\beta$ are 2 copies of $T^2 \times \Sigma_\beta$. Now δ swaps the 2 T^2 in $K3_\alpha$ and takes Σ_β to itself, so δ swaps the 2 $T^2 \times \Sigma_\beta$ in $K3_\alpha \times K3_\beta$. Therefore the fixed points of γ contribute $T^2 \times \Sigma_\beta$ to the singular set of $(K3_\alpha \times K3_\beta)/\langle \gamma, \delta \rangle$.

Similarly, the fixed points of δ contribute $\Sigma_\alpha \times T^2$ to the singular set. Also $\gamma\delta$ has no fixed points in $K3_\alpha$ or $K3_\beta$. Thus the fixed points of γ and δ do not intersect, and the singular set of $(K3_\alpha \times K3_\beta)/\langle \gamma, \delta \rangle$ is the disjoint union of $\Sigma_\alpha \times T^2$ and $T^2 \times \Sigma_\beta$. These are both compact 4-manifolds, and it is easy to see that every singular point is locally modelled on $\mathbb{R}^4 \times \mathbb{R}^4/\{\pm 1\}$. □

14.4.2 A family of 8-manifolds resolving T^8/Γ

Next we must explain how to resolve the singularities of $(K3_\alpha \times K3_\beta)/\langle\gamma,\delta\rangle$ to get a nonsingular 8-manifold M. A similar problem was considered in §12.5.3, where we resolved $(T^3 \times K3)/\langle\beta,\gamma\rangle$ to get a compact 7-manifold, and we use an analogous approach here. We show that the singularities of $(K3_\alpha \times K3_\beta)/\langle\gamma,\delta\rangle$ can locally be regarded as singularities of a *complex* orbifold, and then resolve them by blowing up.

Let J be the unique complex structure on T^8 for which

$$(z_1, z_2, z_3, z_4) = (x_1 + ix_2, x_3 + ix_4, x_5 + ix_6, x_7 + ix_8)$$

are complex coordinates, where $z_j \in \mathbb{C}/(\mathbb{Z} + i\mathbb{Z})$. Then α, β and γ are holomorphic with respect to J, and δ is antiholomorphic. So $T^8/\langle\alpha,\beta\rangle$ is a complex orbifold. We can regard $K3_\alpha \times K3_\beta$ as a *complex* resolution or smoothing of $T^8/\langle\alpha,\beta\rangle$.

Hence $K3_\alpha \times K3_\beta$ also has a complex structure J, with respect to which γ is holomorphic and δ antiholomorphic. So $(K3_\alpha \times K3_\beta)/\langle\gamma\rangle$ is a complex orbifold, with complex structure J and singular set $2\, T^2 \times \Sigma_\beta$. The action of δ on $(K3_\alpha \times K3_\beta)/\langle\gamma\rangle$ takes J to $-J$ and swaps over the 2 copies of $T^2 \times \Sigma_\beta$.

Thus we can identify a neighbourhood of $T^2 \times \Sigma_\beta$ in $(K3_\alpha \times K3_\beta)/\langle\gamma,\delta\rangle$ with a neighbourhood of one of the copies of $T^2 \times \Sigma_\beta$ in $(K3_\alpha \times K3_\beta)/\langle\gamma\rangle$. So we can define a complex structure J near $T^2 \times \Sigma_\beta$ in $(K3_\alpha \times K3_\beta)/\langle\gamma,\delta\rangle$. This J does not extend to the whole of $(K3_\alpha \times K3_\beta)/\langle\gamma,\delta\rangle$, as δ is antiholomorphic.

Therefore the singularities $T^2 \times \Sigma_\beta$ in $(K3_\alpha \times K3_\beta)/\langle\gamma,\delta\rangle$ can be regarded as *complex* singularities, locally modelled on $\mathbb{C}^2 \times \mathbb{C}^2/\{\pm 1\}$. To resolve them we blow up along $T^2 \times \Sigma_\beta$, using J. This is a crepant resolution, as we need.

To resolve the δ singularities we define a complex structure J' on T^8 using

$$(w_1, w_2, w_3, w_4) = (x_1 + ix_3, x_2 - ix_4, x_5 + ix_7, x_6 - ix_8),$$

where $w_j \in \mathbb{C}/(\mathbb{Z} + i\mathbb{Z})$. Then α, β and δ are holomorphic with respect to J', and γ is antiholomorphic. As above, we define a complex structure J' near $\Sigma_\alpha \times T^2$ in $(K3_\alpha \times K3_\beta)/\langle\gamma,\delta\rangle$, and then resolve by blowing up using J'. In this way we resolve $(K3_\alpha \times K3_\beta)/\langle\gamma,\delta\rangle$ to get a nonsingular, compact 8-manifold M.

The crucial point in this whole exercise is that there exists a set of R-data for T^8/Γ, which yields this 8-manifold M as its resolution. One can rephrase the argument above in terms of R-data, and use it to determine the topological type and complex structure of each of the ALE and QALE manifolds in the R-data, and of the actions of the $\chi_{\gamma,i}$. We then use the results of Chapters 8 and 9 to obtain the metrics with holonomy SU(2) and SU(3) that are required.

Theorem 14.4.4 *There exists a set of R-data for T^8/Γ giving the compact 8-manifold M described above as its resolution. There exist metrics with holonomy $\mathrm{Spin}(7)$ on M. Let k, l be as in §14.4.1, and define $a = b^0(\Sigma_\alpha)$ and $b = b^0(\Sigma_\beta)$. Then M is simply-connected, with Betti numbers*

$$b^2 = k+l+a+b, \quad b^3 = 32-4k-4l+4a+4b, \quad b^4 = 206-10k-10l+6a+6b,$$
$$b^4_+ = 135 - 5k - 5l + 3a + 3b \quad \text{and} \quad b^4_- = 71 - 5k - 5l + 3a + 3b.$$

Proof We constructed the R-data for T^8/Γ above. Theorem 13.6.2 proves that there exist torsion-free Spin(7)-structures (Ω, g) on M, and as in the proof of Theorem 14.2.3, we can use Theorem 14.1.1 to show that $\text{Hol}(g) = \text{Spin}(7)$. Hence M is simply-connected, by Theorem 10.6.8.

We shall calculate the Betti numbers of M. As the blow up of $\Sigma_\alpha \times T^2$ and $T^2 \times \Sigma_\beta$ replaces each point by $\mathcal{S}^2$, it can be shown that

$$b^2(M) = b^2\big((K3_\alpha \times K3_\beta)/\langle \gamma, \delta \rangle\big) + b^0(\Sigma_\alpha \times T^2) + b^0(T^2 \times \Sigma_\beta),$$
$$b^3(M) = b^3\big((K3_\alpha \times K3_\beta)/\langle \gamma, \delta \rangle\big) + b^1(\Sigma_\alpha \times T^2) + b^1(T^2 \times \Sigma_\beta),$$
$$b^4_+(M) = b^4_+\big((K3_\alpha \times K3_\beta)/\langle \gamma, \delta \rangle\big) + b^2_-(\Sigma_\alpha \times T^2) + b^2_-(T^2 \times \Sigma_\beta),$$
$$b^4_-(M) = b^4_-\big((K3_\alpha \times K3_\beta)/\langle \gamma, \delta \rangle\big) + b^2_+(\Sigma_\alpha \times T^2) + b^2_+(T^2 \times \Sigma_\beta).$$

To understand these, note that the singularities of $(K3_\alpha \times K3_\beta)/\langle \gamma, \delta \rangle$ are modelled on $\mathbb{C}^2 \times \mathbb{C}^2/\{\pm 1\}$, and the resolution M is modelled on $\mathbb{C}^2 \times Y$, where Y is the blow up of $\mathbb{C}^2/\{\pm 1\}$ at 0. But the exceptional divisor $\mathbb{CP}^1$ in Y has self-intersection -2. Therefore a 2-homology class in $T^2 \times \Sigma_\beta$ or $\Sigma_\alpha \times T^2$ with positive (negative) self-intersection lifts to a 4-homology class in M with negative (positive) self-intersection. So we add $b^2_\mp(\Sigma_\alpha \times T^2) + b^2_\mp(T^2 \times \Sigma_\beta)$ to $b^4_\pm(M)$.

But we have $b^0(\Sigma_\alpha \times T^2) = b^0(\Sigma_\alpha)$, $b^1(\Sigma_\alpha \times T^2) = 2b^0(\Sigma_\alpha) + b^1(\Sigma_\alpha)$ and $b^2_+(\Sigma_\alpha \times T^2) = b^2_-(\Sigma_\alpha \times T^2) = b^0(\Sigma_\alpha) + b^1(\Sigma_\alpha)$, and similar equations for the Betti numbers of $T^2 \times \Sigma_\beta$. Combining these with the Betti numbers of M above, and the Betti numbers of $(K3_\alpha \times K3_\beta)/\langle \gamma, \delta \rangle$ in Proposition 14.4.3, we get:

$$b^2(M) = k + l + b^0(\Sigma_\alpha) + b^0(\Sigma_\beta), \tag{14.11}$$
$$b^3(M) = 2b^0(\Sigma_\alpha) + b^1(\Sigma_\alpha) + 2b^0(\Sigma_\beta) + b^1(\Sigma_\beta), \tag{14.12}$$
$$b^4_+(M) = 103 - k - l + b^0(\Sigma_\alpha) + b^1(\Sigma_\alpha) + b^0(\Sigma_\beta) + b^1(\Sigma_\beta), \tag{14.13}$$
$$b^4_-(M) = 39 - k - l + b^0(\Sigma_\alpha) + b^1(\Sigma_\alpha) + b^0(\Sigma_\beta) + b^1(\Sigma_\beta). \tag{14.14}$$

Now $b^0(\Sigma_\alpha) = a$ and $b^0(\Sigma_\beta) = b$, and so $b^2(\Sigma_\alpha) = a$ and $b^2(\Sigma_\beta) = b$ by Poincaré duality. But as in (12.13) one can show that $\chi(\Sigma_\alpha) = 4k - 16$ and $\chi(\Sigma_\beta) = 4l - 16$. Thus we find that $b^1(\Sigma_\alpha) = 16 + 2a - 4k$ and $b^1(\Sigma_\beta) = 16 + 2b - 4l$. Substituting these values into (14.11)–(14.14) gives the Betti numbers in the theorem, and the proof is complete. $\square$

Note that the Betti numbers in the theorem satisfy (14.1), as they should. Also we have $b^4_+ - b^4_- = 64$, whatever the values of k, l, a and b. These two equations imply that $2b^2 - 2b^3 + b^4 = 142$.

Here is why $b^4_+ - b^4_- = 64$. In §14.1 we explained that the only contributions to $b^4_+ - b^4_-$ come from the resolution of points in T^8/Γ modelled on $\mathbb{R}^8/G$ with $\text{Fix}(G) = 0$. In this case, we have 64 such points, each modelled on $\mathbb{R}^4/\{\pm 1\} \times \mathbb{R}^4/\{\pm 1\}$. They are all resolved in a standard way, which adds 1 to $b^4_+ - b^4_-$. Since we have no topological choices in resolving these 64 points, we find that $b^4_+ - b^4_-$ is independent of any choices made in constructing M.

14.4.3 *Calculating the possible Betti numbers*

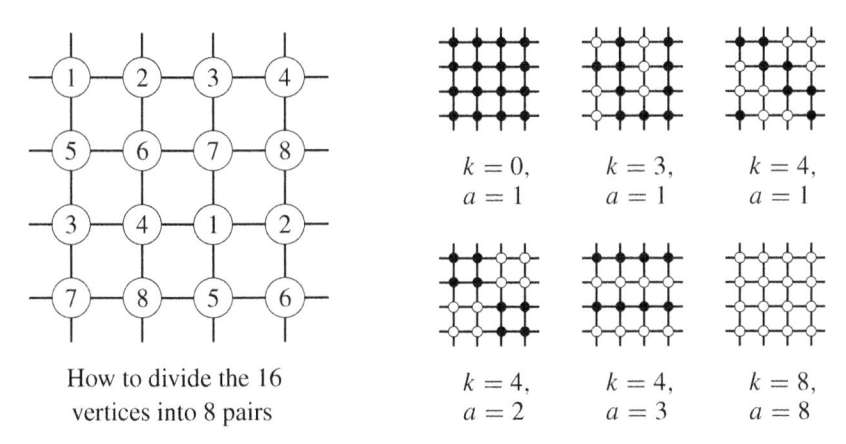

In §14.4.1 we resolved $T^8/\langle\alpha,\beta\rangle$ to get $K3_\alpha \times K3_\beta$, and found that there are $2^{16} = 65\,536$ topologically distinct ways of lifting the action of $\langle\gamma,\delta\rangle$ from $T^8/\langle\alpha,\beta\rangle$ to $K3_\alpha \times K3_\beta$. Then in §14.4.2 we showed that every one of these 65 536 possibilities leads to a resolution M of T^8/Γ with holonomy Spin(7), and evaluated the Betti numbers of M in terms of k, l, a and b.

Figure 14.1. Diagrammatic notation for finding k and a

Our goal now is to calculate and write down all sets of Betti numbers (b^2, b^3, b^4) realized by these 65 536 resolutions M. By Theorem 14.4.4, it is enough to work out the possible values of k, l, a and b. However, k and a depend only on the action of δ on $K3_\alpha$, and l and b only on the action of γ on $K3_\beta$. Thus the pairs (k, a) and (l, b) are independent. Furthermore, the possible values of (l, b) are the same as those of (k, a), so we only need to do the calculation once.

Recall that δ acts on the first $T^4/\{\pm 1\}$ fixing the 16 singular points. These 16 points are swapped in 8 pairs by γ. The lift of δ to $K3_\alpha$ acts with $\delta_* = 1$ on k out of the 8 pairs of 'S^2's resolving the 8 pairs of singular points, and $\delta_* = -1$ on the remaining $8 - k$ pairs. The fixed points of δ in $K3_\alpha$ are a compact 2-manifold Σ_α, and $a = b^0(\Sigma_\alpha)$, that is, a is the number of connected components of Σ_α. We want to calculate all the possible values of the pair (k, a).

Now we met an almost identical problem in §12.5, and in §12.5.2 we developed a diagrammatic notation for doing such calculations, which we can use again here. The notation involved drawing a 4×4 grid, and putting either 'o' or '•' at each vertex. In this case the 16 vertices are divided into 8 pairs, and for each pair we must either put 'o' at both vertices, or '•' at both vertices. We use 'o' for k pairs of vertices, and '•' for the remaining $8 - k$ pairs.

Figure 14.1 shows how the 16 vertices should be divided into pairs, and gives six examples of diagrams with different values of k and a. Section 12.5.2 explains how to find a for each diagram. The author used this notation to determine the values of the

Table 14.1 *Betti numbers* (b^2, b^3, b^4) *of the compact 8-manifolds with holonomy* Spin(7) *from the construction of* §14.4

(2, 40, 218)	(3, 36, 208)	(4, 32, 198)	(5, 28, 188)	(6, 24, 178)	(7, 20, 168)
(7, 28, 184)	(8, 16, 158)	(8, 24, 174)	(8, 32, 190)	(9, 12, 148)	(9, 20, 164)
(9, 28, 180)	(10, 8, 138)	(10, 16, 154)	(10, 24, 170)	(11, 12, 144)	(11, 20, 160)
(11, 28, 176)	(12, 8, 134)	(12, 16, 150)	(12, 24, 166)	(12, 32, 182)	(13, 12, 140)
(13, 20, 156)	(13, 28, 172)	(14, 8, 130)	(14, 16, 146)	(14, 24, 162)	(14, 32, 178)
(15, 12, 136)	(15, 20, 152)	(15, 28, 168)	(16, 16, 142)	(16, 24, 158)	(17, 12, 132)
(17, 20, 148)	(17, 36, 180)	(18, 16, 138)	(18, 24, 154)	(18, 32, 170)	(19, 20, 144)
(19, 28, 160)	(20, 16, 134)	(20, 24, 150)	(21, 20, 140)	(22, 24, 146)	(23, 20, 136)
(23, 28, 152)	(24, 24, 142)	(26, 24, 138)	(27, 28, 144)	(29, 28, 140)	(32, 32, 142)

pair (k, a), and found the thirteen possibilitities (0,1), (1,1), (2,1), (3,1), (4,1), (4,2), (4,3), (5,2), (5,3), (6,4), (6,5), (7,6) and (8,8). Since we can pick the pairs (k, a) and (l, b) independently from this list, this gives $13^2 = 169$ possibilities for the quadruple (k, a, l, b).

Thus, substituting these 169 possibilities into Theorem 14.4.4 gives us a list of Betti numbers of compact 8-manifolds with holonomy Spin(7) arising from this construction. In fact, as exchanging the pairs (k, a) and (l, b) gives the same Betti numbers we only have to consider $\binom{14}{2} = 91$ possibilities. Not all these possibilities yield distinct Betti numbers, and in fact we end up with only 54 sets of Betti numbers (b^2, b^3, b^4), which are listed in Table 14.1. They all satisfy $2b^2 - 2b^3 + b^4 = 142$, as in §14.4.2.

The construction above works in a very similar way to that of compact G_2-manifolds in §12.5. There we first resolved $T^7/\langle\alpha\rangle$ to get $T^3 \times K3$, and then resolved $(T^3 \times K3)/\langle\beta, \gamma\rangle$ to get M^7, whereas here we first resolve $T^8/\langle\alpha, \beta\rangle$ to get $K3_\alpha \times K3_\beta$, and then resolve $(K3_\alpha \times K3_\beta)/\langle\gamma, \delta\rangle$ to get M^8.

However, in §12.5 we found two different ways to resolve $(T^3 \times K3)/\langle\beta, \gamma\rangle$, described as 'crepant resolution' and 'smoothing', but above we used only the 'crepant resolution' method to resolve $(K3_\alpha \times K3_\beta)/\langle\gamma, \delta\rangle$. In fact we can also use the 'smoothing' method to resolve $(K3_\alpha \times K3_\beta)/\langle\gamma, \delta\rangle$, and so get more compact 8-manifolds with holonomy Spin(7) resolving T^8/Γ. But the calculations involved are rather complex, so we will not explain them here.

14.5 Another example with $\Gamma = \mathbb{Z}_2^4$

We now consider another orbifold T^8/Γ similar to those in §14.2 and §14.4. Let T^8 and (Ω_0, g_0) be as in §14.2, and define $\alpha, \beta, \gamma, \delta : T^8 \to T^8$ by

$$\alpha : (x_1, \ldots, x_8) \mapsto (-x_1, -x_2, -x_3, -x_4, x_5, x_6, x_7, x_8), \tag{14.15}$$

$$\beta : (x_1, \ldots, x_8) \mapsto (x_1, x_2, x_3, x_4, -x_5, -x_6, -x_7, -x_8), \tag{14.16}$$

$$\gamma : (x_1, \ldots, x_8) \mapsto (-x_1, -x_2, x_3, x_4, -x_5, -x_6, x_7, x_8), \tag{14.17}$$

$$\delta : (x_1, \ldots, x_8) \mapsto (\tfrac{1}{2} - x_1, x_2, \tfrac{1}{2} - x_3, x_4, \tfrac{1}{2} - x_5, x_6, \tfrac{1}{2} - x_7, x_8). \tag{14.18}$$

These differ from (14.2)–(14.5) and (14.6)–(14.9) only in where $\frac{1}{2}$ appears. As before $\Gamma = \langle \alpha, \beta, \gamma, \delta \rangle$ is a group isomorphic to $\mathbb{Z}_2^4$, which acts on T^8 preserving (Ω_0, g_0). Here are the analogues of Lemmas 14.4.1 and 14.4.2 for this group Γ.

Lemma 14.5.1 *The fixed points of* $\alpha, \beta, \gamma, \alpha\gamma, \beta\gamma, \alpha\beta\gamma$ *and* δ *are each* 16 *copies of* T^4 *in* T^8, *and the fixed points of* $\alpha\beta$ *are* 256 *points in* T^8. *These are the only nonidentity elements of* Γ *that have fixed points on* T^8. *Also* $\langle \alpha, \beta, \gamma \rangle$ *acts trivially and* δ *acts freely on each of the sets of* 16 T^4 *fixed by* $\alpha, \beta, \gamma, \alpha\gamma, \beta\gamma$ *and* $\alpha\beta\gamma$, *and* $\langle \alpha, \beta, \gamma \rangle$ *acts freely on the set of* 16 δ T^4.

Lemma 14.5.2 *The singular set S of* T^8/Γ *is the union of*

(i) 8 $T^2/\{\pm 1\} \times T^2/\{\pm 1\}$ *from the* α *fixed points,*
(ii) 8 $T^2/\{\pm 1\} \times T^2/\{\pm 1\}$ *from the* β *fixed points,*
(iii) 8 $T^2/\{\pm 1\} \times T^2/\{\pm 1\}$ *from the* γ *fixed points,*
(iv) 8 $T^2/\{\pm 1\} \times T^2/\{\pm 1\}$ *from the* $\alpha\gamma$ *fixed points,*
(v) 8 $T^2/\{\pm 1\} \times T^2/\{\pm 1\}$ *from the* $\beta\gamma$ *fixed points,*
(vi) 8 $T^2/\{\pm 1\} \times T^2/\{\pm 1\}$ *from the* $\alpha\beta\gamma$ *fixed points, and*
(vii) 2 T^4 *from the* δ *fixed points.*

The union is not disjoint, as sets (i)–(vi) *intersect in* 128 *points, sets* (i), (iii) *and* (iv) *intersect in* 32 $T^2/\{\pm 1\}$, *sets* (i), (v) *and* (vi) *intersect in* 32 $T^2/\{\pm 1\}$, *sets* (ii), (iii) *and* (v) *intersect in* 32 $T^2/\{\pm 1\}$, *and sets* (ii), (iv) *and* (vi) *intersect in* 32 $T^2/\{\pm 1\}$.

The fixed points of $\langle \alpha, \beta, \gamma \rangle$ in T^8 are $\{(x_1, \ldots, x_8) \in T^8 : x_j \in \{\mathbb{Z}, \frac{1}{2} + \mathbb{Z}\}\}$, which is 256 points. Their image in T^8/Γ is 128 points, the intersection of the sets (i)–(vi), and the singularities are locally modelled on 0 in $\mathbb{C}^4/\mathbb{Z}_2^3$, where the generators α, β, γ of $\mathbb{Z}_2^3$ act by

$$\alpha : (z_1, \ldots, z_4) \mapsto (-z_1, -z_2, z_3, z_4), \quad \beta : (z_1, \ldots, z_4) \mapsto (z_1, z_2, -z_3, -z_4)$$
$$\text{and} \quad \gamma : (z_1, \ldots, z_4) \mapsto (-z_1, z_2, -z_3, z_4).$$

This singularity $\mathbb{C}^4/\mathbb{Z}_2^3$ was considered in Example 9.3.7, where we showed it admits 48 topologically distinct crepant resolutions carrying QALE metrics with holonomy SU(4). Also, by deforming $\mathbb{C}^4/\mathbb{Z}_2^3$ and taking a crepant resolution of any remaining singularities, the author can count at least 14 other topologically distinct smoothings of $\mathbb{C}^4/\mathbb{Z}_2^3$ admitting QALE metrics with holonomy SU(4), following Example 9.9.9.

We wish to choose a set of R-data for T^8/Γ, as in §13.2. There are a very large number of topologically distinct ways in which we can do this — in particular, for each of 128 points we must choose one of at least $48 + 14 = 62$ desingularizations. These choices are not independent, but there is enough freedom to create a huge number of possibilities.

So, as in §14.4, we shall simplify the problem by splitting it up into two stages, and limiting the kinds of resolution we consider. We first resolve $T^8/\langle \alpha, \beta \rangle$ in a $\langle \gamma, \delta \rangle$-equivariant way to get $K3_\alpha \times K3_\beta$, and then resolve $(K3_\alpha \times K3_\beta)/\langle \gamma, \delta \rangle$ to get M. This is just a way of simplifying the choice of R-data for T^8/Γ, and understanding the topology of M. We do the actual construction of metrics with holonomy Spin(7) on M in one stage, not two.

14.5.1 Resolving $T^8/\langle \alpha, \beta \rangle$ in a $\langle \gamma, \delta \rangle$-equivariant way

As in §14.4.1 we write $T^8 = T^4 \times T^4$, so that $T^8/\langle \alpha, \beta \rangle = (T^4/\{\pm 1\}) \times (T^4/\{\pm 1\})$, and we resolve $T^8/\langle \alpha, \beta \rangle$ to get $K3_\alpha \times K3_\beta$. Again, $\langle \gamma, \delta \rangle \cong \mathbb{Z}_2^2$ acts on $T^8/\langle \alpha, \beta \rangle$, and we wish to lift this $\langle \gamma, \delta \rangle$-action up to $K3_\alpha \times K3_\beta$. However, in this case γ fixes the 16 singular points of both copies of $T^4/\{\pm 1\}$, and δ acts freely on both sets of 16 singular points. This is different to the case of §14.4.

Thus δ acts uniquely on $K3_\alpha$ and $K3_\beta$, but to define the action of γ on $K3_\alpha$ and $K3_\beta$ we must choose $\gamma_* = \pm 1$ for each of the 16 singular points in both copies of $T^4/\{\pm 1\}$. As δ swaps each set of 16 points in pairs, and we want γ, δ to commute, γ_* must take the same value for both points in each pair swapped by δ. So, choose $\gamma_* = \pm 1$ for each of the 8 pairs of singular points swapped by δ in both copies of $T^4/\{\pm 1\}$. There are $2^{16} = 65\,536$ possible choices. This defines an action of $\langle \gamma, \delta \rangle$ on $K3_\alpha \times K3_\beta$.

Let $\gamma_* = 1$ for k out of the 8 pairs of singular points of the first $T^4/\{\pm 1\}$ swapped by δ, and $\gamma_* = 1$ for l out of the 8 pairs of singular points of the second $T^4/\{\pm 1\}$ swapped by δ, so that $k, l = 0, 1, \ldots, 8$. Let Σ_α be the subset of $K3_\alpha$ fixed by γ, and Σ_β the subset of $K3_\beta$ fixed by γ. Then Σ_α and Σ_β are compact, oriented 2-manifolds. Here is the analogue of Proposition 14.4.3.

Proposition 14.5.3 *The Betti numbers of* $(K3_\alpha \times K3_\beta)/\langle \gamma, \delta \rangle$ *are*

$$b^2 = k+l, \quad b^3 = 0, \quad b^4_+ = 167 - 17k - 17l + 4kl \quad \text{and} \quad b^4_- = 39 - k - l.$$

The singular set of $(K3_\alpha \times K3_\beta)/\langle \gamma, \delta \rangle$ *is the disjoint union of* $(\Sigma_\alpha \times \Sigma_\beta)/\mathbb{Z}_2$ *and* 2 T^4, *where each singular point is locally modelled on* $\mathbb{R}^4 \times \mathbb{R}^4/\{\pm 1\}$.

Proof Let $U_\alpha, V_\alpha, U_\beta$ and V_β be as in the proof of Proposition 14.4.3, which we follow closely. Then the $\langle \gamma, \delta \rangle$-invariant subspaces of $U_\alpha, V_\alpha, U_\beta, V_\beta$ are $0, \mathbb{R}^k, 0, \mathbb{R}^l$ respectively. So $(K3_\alpha \times K3_\beta)/\langle \gamma, \delta \rangle$ has $b^2 = k + l$ and $b^3 = 0$, as in §14.4.

However, in the splitting (14.10) of $H^4(K3_\alpha \times K3_\beta, \mathbb{R})$ into 5 vector spaces, this time the $\langle \gamma, \delta \rangle$-invariant subspaces have dimensions $12, 32 - 2l, 32 - 2k, 128 - 16k - 16l + 4kl$ and 2 respectively. The difference is that the $\langle \gamma, \delta \rangle$-invariant subspace of $V_\alpha \otimes V_\beta$ had dimension 64 in §14.4, but $128 - 16k - 16l + 4kl$ here. We then get the values for $b^4_\pm$ above by the argument in Proposition 14.4.3.

The fixed points of γ in $K3_\alpha$ and $K3_\beta$ are Σ_α and Σ_β, so the fixed points in $K3_\alpha \times K3_\beta$ are $\Sigma_\alpha \times \Sigma_\beta$. Now δ acts freely on Σ_α and Σ_β, and hence on $\Sigma_\alpha \times \Sigma_\beta$. So the fixed points of γ contribute $(\Sigma_\alpha \times \Sigma_\beta)/\langle \delta \rangle$ to the singular set of $(K3_\alpha \times K3_\beta)/\langle \gamma, \delta \rangle$. Similarly, the fixed points of δ in $K3_\alpha$ and $K3_\alpha$ are both 2 T^2, and γ swaps the 2 copies in each case.

Thus the fixed points of δ in $K3_\alpha \times K3_\beta$ are 4 $T^2 \times T^2$ (that is, 4 T^4) which descend to 2 T^4 in the singular set of $(K3_\alpha \times K3_\beta)/\langle \gamma, \delta \rangle$. Also $\gamma \delta$ acts freely on $K3_\alpha \times K3_\beta$. Therefore the singular set of $(K3_\alpha \times K3_\beta)/\langle \gamma, \delta \rangle$ is the disjoint union of $(\Sigma_\alpha \times \Sigma_\beta)/\mathbb{Z}_2$ and 2 T^4. $\square$

14.5.2 A family of 8-manifolds resolving T^8/Γ

Next we shall resolve the singularities of $(K3_\alpha \times K3_\beta)/\langle\gamma,\delta\rangle$ to get a nonsingular 8-manifold M. We follow the method of §14.4.2, which was to show that each component of the singular set can locally be regarded as the singularities of a *complex* orbifold, and then resolve by blowing up using this complex structure.

The singular set of $(K3_\alpha \times K3_\beta)/\langle\gamma,\delta\rangle$ is the disjoint union of $(\Sigma_\alpha \times \Sigma_\beta)/\mathbb{Z}_2$ and $2\,T^4$. Resolving the $2\,T^4$ is straightforward: using the complex structure J' of §14.4.2 we identify $(K3_\alpha \times K3_\beta)/\langle\gamma,\delta\rangle$ with $T^4 \times \mathbb{C}^2/\{\pm 1\}$ near each T^4, and then blow up along $T^4 \times 0$ in the obvious way.

However, resolving the $(\Sigma_\alpha \times \Sigma_\beta)/\mathbb{Z}_2$ singularities is not quite as easy, because when Σ_α and Σ_β are connected there is no suitable complex structure defined near $(\Sigma_\alpha \times \Sigma_\beta)/\mathbb{Z}_2$. So instead we consider $(K3_\alpha \times K3_\beta)/\langle\gamma\rangle$, which has singular set $\Sigma_\alpha \times \Sigma_\beta$. The complex structure J of §14.4.2 is defined on $(K3_\alpha \times K3_\beta)/\langle\gamma\rangle$, so we can resolve by blowing up along $\Sigma_\alpha \times \Sigma_\beta$ using J, to get N say.

The action of δ on $(K3_\alpha \times K3_\beta)/\langle\gamma\rangle$ satisfies $\delta_*(J) = -J$, and lifts to a free antiholomorphic involution of the blow-up N. Thus, $N/\langle\delta\rangle$ provides a model for how to resolve the $(\Sigma_\alpha \times \Sigma_\beta)/\langle\delta\rangle$ singularities in $(K3_\alpha \times K3_\beta)/\langle\gamma,\delta\rangle$. In this way we resolve $(K3_\alpha \times K3_\beta)/\langle\gamma,\delta\rangle$ to get a nonsingular, compact 8-manifold M. As in §14.4.2, there exists a set of R-data for T^8/Γ, which yields M as its resolution.

Let k,l be as in §14.5.1, and define $a = b^0(\Sigma_\alpha)$ and $c = b^0(\Sigma_\beta)$. Then a is the number of connected components of Σ_α. Now δ takes Σ_α to itself, with $\delta^2 = 1$. So δ may fix some connected components of Σ_α, and swap other components in pairs. Let b be the number of pairs of connected components of Σ_α swapped by δ. Then $2b \leqslant a$, and δ fixes $a - 2b$ components of Σ_α. Similarly, let d be the number of pairs of connected components of Σ_β swapped by δ, so that $2d \leqslant c$. Using this notation, here is the analogue of Theorem 14.4.4.

Theorem 14.5.4 *Let M, k, l, a, b, c and d be as above. Then M is simply-connected, with Betti numbers*

$$b^2 = 2 + k + l + bc + ad - 2bd, \qquad b^3 = 8 + 8a + 8c - 2al - 2kc + 2ac,$$
$$b^4 = 346 - 50k - 50l + 12kl + 16a + 16c - 4al - 4kc + 4ac - 2bc - 2ad + 4bd,$$
$$b^4_+ = 237 - 33k - 33l + 8kl + 8a + 8c - 2al - 2kc + 2ac - bc - ad + 2bd,$$
$$b^4_- = 109 - 17k - 17l + 4kl + 8a + 8c - 2al - 2kc + 2ac - bc - ad + 2bd.$$

There exists a set of R-data for T^8/Γ giving this compact 8-manifold M as its resolution, and there exist metrics with holonomy Spin(7) *on M.*

Proof We calculate the Betti numbers of M by starting with the Betti numbers of $(K3_\alpha \times K3_\beta)/\langle\gamma,\delta\rangle$, which are given in Proposition 14.5.3, and adding on contributions from the resolution of each component of the singular set. The $2\,T^4$ components are easy — resolving each one adds 1 to b^2, 4 to b^3 and 3 to b^4_+ and b^4_-, as in Theorem 14.2.3. However, the contributions from the $(\Sigma_\alpha \times \Sigma_\beta)/\mathbb{Z}_2$ component require a little more thought.

In resolving the $\Sigma_\alpha \times \Sigma_\beta$ in $(K3_\alpha \times K3_\beta)/\langle\gamma\rangle$ to get N, we cut out $\Sigma_\alpha \times \Sigma_\beta$ and replace it with a $\mathbb{CP}^1$-bundle over $\Sigma_\alpha \times \Sigma_\beta$. We thus expect that

$$H^k(N, \mathbb{R}) \cong H^k\big((K3_\alpha \times K3_\beta)/\langle\gamma\rangle, \mathbb{R}\big) \oplus H^2(\mathbb{CP}^1, \mathbb{R}) \otimes H^{k-2}(\Sigma_\alpha \times \Sigma_\beta, \mathbb{R}).$$

When we resolve the $(\Sigma_\alpha \times \Sigma_\beta)/\langle\delta\rangle$ in $(K3_\alpha \times K3_\beta)/\langle\gamma, \delta\rangle$, the contribution to $H^k(M, \mathbb{R})$ is the δ-invariant subspace of $H^2(\mathbb{CP}^1, \mathbb{R}) \otimes H^{k-2}(\Sigma_\alpha \times \Sigma_\beta, \mathbb{R})$. But δ is antiholomorphic on $\mathbb{CP}^1$, and so $\delta_* = -1$ on $H^2(\mathbb{CP}^1, \mathbb{R}) \cong \mathbb{R}$. Hence, the contribution to $H^k(M, \mathbb{R})$ is the subspace of $H^{k-2}(\Sigma_\alpha \times \Sigma_\beta, \mathbb{R})$ on which $\delta_* = -1$.

Now the subspace of $H^{k-2}(\Sigma_\alpha \times \Sigma_\beta, \mathbb{R})$ on which $\delta_* = 1$ is $H^{k-2}\big((\Sigma_\alpha \times \Sigma_\beta)/\mathbb{Z}_2, \mathbb{R}\big)$. Thus, the contribution to $b^k(M)$ from resolving $(\Sigma_\alpha \times \Sigma_\beta)/\mathbb{Z}_2$ is $b^{k-2}(\Sigma_\alpha \times \Sigma_\beta) - b^{k-2}\big((\Sigma_\alpha \times \Sigma_\beta)/\mathbb{Z}_2\big)$. Including the contributions from the resolution of the $2\,T^4$, we get

$$b^2(M) = b^2\big((K3_\alpha \times K3_\beta)/\langle\gamma, \delta\rangle\big) + 2 + b^0(\Sigma_\alpha \times \Sigma_\beta) - b^0\big((\Sigma_\alpha \times \Sigma_\beta)/\mathbb{Z}_2\big),$$
$$b^3(M) = b^3\big((K3_\alpha \times K3_\beta)/\langle\gamma, \delta\rangle\big) + 8 + b^1(\Sigma_\alpha \times \Sigma_\beta) - b^1\big((\Sigma_\alpha \times \Sigma_\beta)/\mathbb{Z}_2\big),$$
$$b^4(M)_+ = b^4_+\big((K3_\alpha \times K3_\beta)/\langle\gamma, \delta\rangle\big) + 6 + b^2_-(\Sigma_\alpha \times \Sigma_\beta) - b^2_-\big((\Sigma_\alpha \times \Sigma_\beta)/\mathbb{Z}_2\big),$$
$$b^4(M)_- = b^4_-\big((K3_\alpha \times K3_\beta)/\langle\gamma, \delta\rangle\big) + 6 + b^2_+(\Sigma_\alpha \times \Sigma_\beta) - b^2_+\big((\Sigma_\alpha \times \Sigma_\beta)/\mathbb{Z}_2\big).$$

Here the relation between $b^4_\pm$ and $b^2_\mp$ is as in the proof of Theorem 14.4.4.

We have $b^0(\Sigma_\alpha) = b^2(\Sigma_\alpha) = a$ and $b^1(\Sigma_\alpha) = 16 + 2a - 4k$ as in Theorem 14.4.4, and similar formulae for $b^k(\Sigma_\beta)$, so by the Künneth formula we get

$$b^0(\Sigma_\alpha \times \Sigma_\beta) = ac, \qquad b^1(\Sigma_\alpha \times \Sigma_\beta) = 16a + 16c - 4al - 4kc + 4ac$$
$$\text{and}\quad b^2_\pm(\Sigma_\alpha \times \Sigma_\beta) = 128 - 32k - 32l + 8kl + 16a + 16c - 4al - 4kc + 3ac.$$

Considering the action of δ on $H^*(\Sigma_\alpha \times \Sigma_\beta, \mathbb{R})$, we find that

$$b^0\big((\Sigma_\alpha \times \Sigma_\beta)/\mathbb{Z}_2\big) = ac - bc - ad + 2bd,$$
$$b^1\big((\Sigma_\alpha \times \Sigma_\beta)/\mathbb{Z}_2\big) = 8a + 8c - 2al - 2kc + 2ac \quad\text{and}$$
$$b^2_\pm\big((\Sigma_\alpha \times \Sigma_\beta)/\mathbb{Z}_2\big) = 64 - 16k - 16l + 4kl + 8a + 8c - 2al - 2kc + ac + bc + ad - 2bd.$$

Combining the last 3 sets of equations with the values of $b^k\big((K3_\alpha \times K3_\beta)/\langle\gamma, \delta\rangle\big)$ in Proposition 14.5.3, we get the Betti numbers of M given in the theorem. The rest of the proof follows that of Theorem 14.4.4. $\square$

Note that the Betti numbers in the theorem satisfy (14.1), as they should. But this time $b^4_+ - b^4_- = 128 - 16k - 16l + 4kl$, so that $b^4_+ - b^4_-$ is not independent of the choice of resolution, as it was in §14.4. This is because of the topological choices involved in the resolution of the 128 points referred to in Lemma 14.5.2.

14.5.3 Calculating the possible Betti numbers

In §14.5.1 we resolved $T^8/\langle\alpha,\beta\rangle$ to get $K3_\alpha \times K3_\beta$, and found $2^{16} = 65\,536$ topologically distinct ways of lifting the action of $\langle\gamma,\delta\rangle$ from $T^8/\langle\alpha,\beta\rangle$ to $K3_\alpha \times K3_\beta$. Then in §14.5.2 we showed that every one of these 65 536 possibilities leads to a resolution M of T^8/Γ with holonomy Spin(7), and evaluated the Betti numbers of M in terms of k, l, a, b, c and d.

Now we wish to calculate and write down all sets of Betti numbers (b^2, b^3, b^4) realized by these 65 536 resolutions M. By Theorem 14.5.4, it is enough to work out the possible values of k, l, a, b, c and d. However, k, a and b depend only on the action of δ on $K3_\alpha$, and l, c and d only on the action of δ on $K3_\beta$. Thus the triples (k, a, b) and (l, c, d) are independent. Furthermore, the possible values of (l, c, d) are the same as those of (k, a, b).

In §14.4.3 we explained how to calculate the possibilities for the pair (k, a), using a diagrammatic notation illustrated in Fig. 14.1. We can easily use these diagrams to find b, the number of pairs of the a components of Σ_α swapped by δ. In this way the author has determined the following fourteen possibilities for the triple (k, a, b):

(0, 1, 0), (1, 1, 0), (2, 1, 0), (3, 1, 0), (4, 1, 0), (4, 2, 0), (4, 2, 1),
(4, 3, 1), (5, 2, 1), (5, 3, 1), (6, 4, 2), (6, 5, 2), (7, 6, 3), (8, 8, 4).

This gives $14^2 = 196$ possibilities for the sextuple (k, a, b, l, c, d).

Thus, substituting these 196 possibilities into Theorem 14.5.4 gives us a list of Betti numbers of compact 8-manifolds with holonomy Spin(7) arising from this construction. In fact, as exchanging the triples (k, a, b) and (l, c, d) gives the same Betti numbers we only have to consider $\binom{15}{2} = 105$ possibilities. Not all these possibilities yield distinct Betti numbers, and in fact we end up with 98 sets of Betti numbers (b^2, b^3, b^4), which are listed in Table 14.2.

Table 14.2 *Betti numbers (b^2, b^3, b^4) of the compact 8-manifolds with holonomy* Spin(7) *from the construction of §14.5*

(2,26,382)	(3,24,328)	(4,22,274)	(4,22,286)	(5,20,220)	(5,20,244)	(6,18,166)
(6,18,202)	(6,18,214)	(6,28,186)	(7,16,160)	(7,16,184)	(7,24,176)	(7,28,184)
(7,38,204)	(8,14,154)	(8,14,166)	(8,20,166)	(8,24,174)	(8,26,130)	(8,32,190)
(8,36,150)	(9,12,148)	(9,16,156)	(9,20,164)	(9,22,132)	(9,26,176)	(9,30,148)
(10,10,142)	(10,12,146)	(10,16,154)	(10,18,134)	(10,20,162)	(10,24,146)	(10,44,114)
(10,54,134)	(11,12,144)	(11,14,136)	(11,14,148)	(11,18,144)	(11,36,120)	(11,44,136)
(12,10,138)	(12,12,142)	(12,16,150)	(12,20,158)	(12,28,126)	(12,34,138)	(12,62,98)
(13,12,140)	(13,16,148)	(13,20,132)	(13,20,156)	(13,24,140)	(13,50,108)	(14,8,142)
(14,12,138)	(14,14,142)	(14,16,146)	(14,26,166)	(14,38,118)	(14,80,82)	(15,10,144)
(15,20,152)	(15,26,128)	(15,64,96)	(16,14,138)	(16,14,150)	(16,16,142)	(16,20,150)
(16,48,110)	(17,8,148)	(17,20,148)	(17,32,124)	(18,10,150)	(18,16,138)	(18,20,146)
(19,12,152)	(19,20,144)	(19,26,156)	(20,8,154)	(20,16,158)	(22,8,162)	(22,24,146)
(22,26,150)	(23,8,160)	(23,14,160)	(24,12,166)	(26,18,174)	(26,32,154)	(27,8,176)
(27,16,168)	(30,14,182)	(32,8,190)	(34,8,198)	(36,16,198)	(41,8,220)	(50,8,250)

14.6 An example with Γ nonabelian

We now consider a new orbifold T^8/Γ with Γ nonabelian. Let T^8 and (Ω_0, g_0) be as in §14.2, and define $\alpha, \beta, \gamma, \delta : T^8 \to T^8$ by

$$\alpha : (x_1, \ldots, x_8) \mapsto (-x_1, -x_2, -x_3, -x_4, x_5, x_6, x_7, x_8), \quad (14.19)$$

$$\beta : (x_1, \ldots, x_8) \mapsto (x_1, x_2, x_3, x_4, -x_5, -x_6, -x_7, -x_8), \quad (14.20)$$

$$\gamma : (x_1, \ldots, x_8) \mapsto (-x_2, x_1, -x_4, x_3, -x_6, x_5, -x_8, x_7), \quad (14.21)$$

$$\delta : (x_1, \ldots, x_8) \mapsto (x_3, -x_4, -x_1, x_2, x_7, -x_8, -x_5, x_6). \quad (14.22)$$

Then $\alpha, \beta, \gamma, \delta$ preserve (Ω_0, g_0). We have $\alpha^2 = \beta^2 = \gamma^4 = \delta^4 = 1$, and $\gamma^2 = \delta^2 = \alpha\beta$. Also α, β commute with all of $\alpha, \beta, \gamma, \delta$, and $\delta\gamma = \alpha\beta\gamma\delta$. Thus $\Gamma = \langle \alpha, \beta, \gamma, \delta \rangle$ is a nonabelian group of order 16, which acts on T^8 preserving (Ω_0, g_0). Here are the analogues of Lemmas 14.4.1 and 14.4.2 for this group Γ.

Lemma 14.6.1 *The fixed points of α and β are each 16 copies of T^4 in T^8 and the fixed points of $\alpha\beta$ are 256 points in T^8. The fixed points of $\gamma, \alpha\gamma, \beta\gamma$ and $\alpha\beta\gamma$ are the same set of 16 points. Similarly, the fixed points of $\delta, \alpha\delta, \beta\delta$ and $\alpha\beta\delta$ are a set of 16 points, and the fixed points of $\gamma\delta, \alpha\gamma\delta, \beta\gamma\delta$ and $\alpha\beta\gamma\delta$ are a set of 16 points.*

Of the 16 α T^4, two are fixed by $\langle \alpha, \beta, \gamma, \delta \rangle$, two are fixed by $\langle \alpha, \beta, \gamma \rangle$ and swapped by δ, two are fixed by $\langle \alpha, \beta, \delta \rangle$ and swapped by γ, two are fixed by $\langle \alpha, \beta, \gamma\delta \rangle$ and swapped by γ, and eight are fixed by $\langle \alpha, \beta \rangle$ and swapped in fours by $\langle \gamma, \delta \rangle$. The same applies to the 16 β T^4.

Lemma 14.6.2 *The singular set S of T^8/Γ is the union of*

(i) $2\ T^4/\langle \gamma, \delta \rangle$ from the α fixed points,
(ii) $3\ T^4/\mathbb{Z}_4$ from the α fixed points,
(iii) $2\ T^4/\mathbb{Z}_2$ from the α fixed points,
(iv) $2\ T^4/\langle \gamma, \delta \rangle$ from the β fixed points,
(v) $3\ T^4/\mathbb{Z}_4$ from the β fixed points, and
(vi) $2\ T^4/\mathbb{Z}_2$ from the β fixed points.

The union is not disjoint, as each of (i)–(iii) intersects each of (iv)–(vi) in finitely many points.

Now consider what the singular points of T^8/Γ look like. The four points $(0, \ldots, 0)$, $(0, 0, 0, 0, \frac{1}{2}, \frac{1}{2}, \frac{1}{2}, \frac{1}{2})$, $(\frac{1}{2}, \frac{1}{2}, \frac{1}{2}, \frac{1}{2}, 0, 0, 0, 0,)$ and $(\frac{1}{2}, \ldots, \frac{1}{2})$ in T^8 are fixed by the whole of Γ. Thus the corresponding four points in T^8/Γ are modelled on $\mathbb{C}^4/\langle \alpha, \beta, \gamma, \delta \rangle$, where $\alpha, \beta, \gamma, \delta$ act on $\mathbb{C}^4$ by

$$\alpha : (z_1, \ldots, z_4) \mapsto (-z_1, -z_2, z_3, z_4), \quad \beta : (z_1, \ldots, z_4) \mapsto (z_1, z_2, -z_3, -z_4),$$

$$\gamma : (z_1, \ldots, z_4) \mapsto (iz_1, iz_2, iz_3, iz_4), \quad \delta : (z_1, \ldots, z_4) \mapsto (\bar{z}_2, -\bar{z}_1, \bar{z}_4, -\bar{z}_3).$$

Note that as δ does not act holomorphically here, $\mathbb{C}^4/\langle \alpha, \beta, \gamma, \delta \rangle$ is *not* a complex singularity. So, how shall we find a suitable QALE Spin(7)-manifold asymptotic to $\mathbb{C}^4/\langle \alpha, \beta, \gamma, \delta \rangle$ to form part of the R-data for T^8/Γ? Here is a way. The quotient

$\mathbb{C}^4/\langle\alpha,\beta,\gamma\rangle$ is a complex singularity, and admits a unique crepant resolution X which can be described using toric geometry. The action of δ on $\mathbb{C}^4/\langle\alpha,\beta,\gamma\rangle$ lifts to a *free* antiholomorphic involution on X.

Thus, $X/\langle\delta\rangle$ is a nonsingular 8-manifold. By the results of Chapter 9 there exist QALE metrics with holonomy SU(4) on X, which are all δ-invariant, and so push down to metrics with holonomy $\mathbb{Z}_2 \ltimes \mathrm{SU}(4)$ on $X/\langle\delta\rangle$. Extending these metrics to torsion-free Spin(7)-structures in the usual way, we make $X/\langle\delta\rangle$ into a QALE Spin(7)-manifold asymptotic to $\mathbb{C}^4/\langle\alpha,\beta,\gamma,\delta\rangle$.

We have shown that although the orbifold groups of these four points in T^8/Γ are not contained in any SU(4) in Spin(7), we can still use the results of Chapter 9 to construct suitable QALE Spin(7)-manifolds X. In fact there are 9 topologically distinct ways of doing this, which have $\pi_1(X) = \mathbb{Z}_2$ in 3 cases (as above), and $\pi_1(X) = \mathbb{Z}_2^2$ in the other 6 cases.

Apart from these 4 points, T^8/Γ also has 18 points modelled on $\mathbb{C}^4/\langle\alpha,\beta,\gamma\rangle$, where α, β, γ act on $\mathbb{C}^4$ as above, and 54 points modelled on $\mathbb{C}^4/\langle\alpha,\beta\rangle$, that is, on $\mathbb{C}^2/\{\pm 1\} \times \mathbb{C}^2/\{\pm 1\}$. It is easy to find suitable R-data for these points. Every other singular point of T^8/Γ is locally modelled on $\mathbb{R}^4 \times \mathbb{R}^4/\{\pm 1\}$.

So, every singular point of T^8/Γ is of a kind that we know how to resolve with a QALE Spin(7)-manifold. We wish to put these QALE Spin(7)-manifolds together in a consistent way to get a set of R-data for T^8/Γ, and hence to define a resolution M of T^8/Γ with holonomy Spin(7). There are topological choices in doing this, which lead to many different 8-manifolds M.

To simplify the calculations we shall divide the resolution into two stages, as in §14.4 and §14.5. We first resolve $T^8/\langle\alpha,\beta\rangle$ to get $K3_\alpha \times K3_\beta$, in a $\langle\gamma,\delta\rangle$-invariant way, and then we resolve $(K3_\alpha \times K3_\beta)/\langle\gamma,\delta\rangle$ to get M. The author believes that this procedure gives all possible Spin(7) resolutions of T^8/Γ.

Note that the actions of γ and δ on $T^8/\langle\alpha,\beta\rangle$ satisfy $\gamma^2 = \delta^2 = 1$ and $\gamma\delta = \delta\gamma$, and so the group $\langle\gamma,\delta\rangle$ acting on $K3_\alpha \times K3_\beta$ is isomorphic to $\mathbb{Z}_2^2$, although the subgroup $\langle\gamma,\delta\rangle \subset \Gamma$ is nonabelian of order 8. The group acting on $K3_\alpha \times K3_\beta$ is really the quotient $\Gamma/\langle\alpha,\beta\rangle$, but we write it as $\langle\gamma,\delta\rangle$ for brevity of notation.

14.6.1 *Resolving $T^8/\langle\alpha,\beta\rangle$ in a $\langle\gamma,\delta\rangle$-equivariant way*

As in §14.4.1 and §14.5.1 we have $T^8/\langle\alpha,\beta\rangle = (T^4/\{\pm 1\}) \times (T^4/\{\pm 1\})$, and we resolve it to get $K3_\alpha \times K3_\beta$. We wish to lift the action of $\langle\gamma,\delta\rangle \cong \mathbb{Z}_2^2$ from $T^8/\langle\alpha,\beta\rangle$ up to $K3_\alpha \times K3_\beta$. First consider the action of $\langle\gamma,\delta\rangle$ on $T^4/\{\pm 1\}$.

The points $(0,0,0,0)$ and $(\frac{1}{2},\frac{1}{2},\frac{1}{2},\frac{1}{2})$ are fixed by $\langle\gamma,\delta\rangle$. Also, the points $(0,0,\frac{1}{2},\frac{1}{2})$ and $(\frac{1}{2},\frac{1}{2},0,0)$ are fixed by γ and swapped by δ, the points $(0,\frac{1}{2},0,\frac{1}{2})$ and $(\frac{1}{2},0,\frac{1}{2},0)$ are fixed by δ and swapped by γ, and the points $(0,\frac{1}{2},\frac{1}{2},0)$ and $(\frac{1}{2},0,0,\frac{1}{2})$ are fixed by $\gamma\delta$ and swapped by γ.

To define γ and δ near the resolution of $(0,\ldots,0)$ and $(\frac{1}{2},\ldots,\frac{1}{2})$, we must choose $\gamma_* = \pm 1$ and $\delta_* = \pm 1$ on the 2 homology classes $[\mathcal{S}^2]$ in $H_2(K3_\alpha, \mathbb{R})$ resolving these two points. But these choices are not independent. A careful investigation shows that at each point, we have three possibilities:

(a) $\gamma_* = 1, \delta_* = -1$ and $(\gamma\delta)_* = -1$,
(b) $\gamma_* = -1, \delta_* = 1$ and $(\gamma\delta)_* = -1$, and
(c) $\gamma_* = -1, \delta_* = -1$ and $(\gamma\delta)_* = 1$.

These are similar to (a)–(c) in §12.5.1, and are proved in the same way.

As $(0, 0, \frac{1}{2}, \frac{1}{2})$ and $(\frac{1}{2}, \frac{1}{2}, 0, 0)$ are fixed by γ we choose $\gamma_* = \pm 1$ to define the action of γ near the resolution of each. As δ swaps the two points, the values of γ_* must be the same for γ, δ to commute on $K3_\alpha$. Similarly, we choose $\delta_* = \pm 1$ for the pair $(0, \frac{1}{2}, 0, \frac{1}{2})$ and $(\frac{1}{2}, 0, \frac{1}{2}, 0)$, and $(\gamma\delta)_* = \pm 1$ for the pair $(0, \frac{1}{2}, \frac{1}{2}, 0)$ and $(\frac{1}{2}, 0, 0, \frac{1}{2})$.

Thus, to lift the action of $\langle \gamma, \delta \rangle$ to $K3_\alpha$ we have 3 choices for each of the 2 points in $T^4/\{\pm 1\}$ fixed by $\langle \gamma, \delta \rangle$, and 2 choices for each of 3 pairs of points fixed by γ, δ and $\gamma\delta$, so there are a total of $3^2 2^3 = 72$ topologically distinct ways that $\langle \gamma, \delta \rangle$ can act on $K3_\alpha$. The situation for $K3_\beta$ is exactly the same, so we also have 72 choices for the action on $K3_\beta$. So there are $72^2 = 5184$ topologically distinct ways to resolve $T^8/\langle \alpha, \beta \rangle$ to get $K3_\alpha \times K3_\beta$ in a $\langle \gamma, \delta \rangle$-equivariant way.

Choose one of these actions of $\langle \gamma, \delta \rangle$ on $K3_\alpha \times K3_\beta$, and define integers a_j, b_j, c_j and d_j for $j = 1, 2, 3$ as follows. Of the resolutions of the two points $(0, 0, 0, 0)$ and $(\frac{1}{2}, \frac{1}{2}, \frac{1}{2}, \frac{1}{2})$ in $K3_\alpha$, let $\gamma_* = 1$ on a_1 of them, $\delta_* = 1$ on a_2 of them, and $(\gamma\delta)_* = 1$ on a_3 of them. Then $a_j = 0, 1$ or 2 for $j = 1, 2, 3$. Moreover, since for each point exactly one of γ_*, δ_* and $(\gamma\delta)_*$ is 1 by (a)–(c) above, we have $a_1 + a_2 + a_3 = 2$.

Let $b_1 = 1$ if $\gamma_* = 1$ for the resolution in $K3_\alpha$ of the pair $(0, 0, \frac{1}{2}, \frac{1}{2})$ and $(\frac{1}{2}, \frac{1}{2}, 0, 0)$, and $b_1 = 0$ if $\gamma_* = -1$. Let $b_2 = 1$ if $\delta_* = 1$ in $K3_\alpha$ for the pair $(0, \frac{1}{2}, 0, \frac{1}{2})$ and $(\frac{1}{2}, 0, \frac{1}{2}, 0)$, and $b_2 = 0$ if $\delta_* = -1$. Let $b_3 = 1$ if $(\gamma\delta)_* = 1$ in $K3_\alpha$ for the pair $(0, \frac{1}{2}, \frac{1}{2}, 0)$ and $(\frac{1}{2}, 0, 0, \frac{1}{2})$, and $b_3 = 0$ if $(\gamma\delta)_* = -1$.

Similarly, of the resolutions of $(0, 0, 0, 0)$ and $(\frac{1}{2}, \frac{1}{2}, \frac{1}{2}, \frac{1}{2})$ in $K3_\beta$, let $\gamma_* = 1$ on $c_1, \delta_* = 1$ on c_2, and $(\gamma\delta)_* = 1$ on c_3. Then $c_j = 0, 1$ or 2 for $j = 1, 2, 3$, and $c_1 + c_2 + c_3 = 2$. Let $d_1 = 1$ if $\gamma_* = 1$ in $K3_\beta$ for the pair $(0, 0, \frac{1}{2}, \frac{1}{2})$ and $(\frac{1}{2}, \frac{1}{2}, 0, 0)$, and $d_1 = 0$ if $\gamma_* = -1$. Let $d_2 = 1$ if $\delta_* = 1$ in $K3_\beta$ for the pair $(0, \frac{1}{2}, 0, \frac{1}{2})$ and $(\frac{1}{2}, 0, \frac{1}{2}, 0)$, and $d_2 = 0$ if $\delta_* = -1$. Let $d_3 = 1$ if $(\gamma\delta)_* = 1$ in $K3_\beta$ for the pair $(0, \frac{1}{2}, \frac{1}{2}, 0)$ and $(\frac{1}{2}, 0, 0, \frac{1}{2})$, and $d_3 = 0$ if $(\gamma\delta)_* = -1$.

Then the a_j and b_j describe the action of $\langle \gamma, \delta \rangle$ on $K3_\alpha$, and the c_j and d_j describe the action of $\langle \gamma, \delta \rangle$ on $K3_\beta$. Also, a_j, b_j, c_j and d_j encode the values of γ_* when $j = 1$, of δ_* when $j = 2$, and of $(\gamma\delta)_*$ when $j = 3$. Using this notation, here is the analogue of Proposition 14.4.3.

Proposition 14.6.3 *The Betti numbers of* $(K3_\alpha \times K3_\beta)/\langle \gamma, \delta \rangle$ *are*

$$b^2 = 10 + \sum_{j=1}^{3}(b_j + d_j), \qquad b^3 = 0,$$

$$b^4_+ = 93 - \sum_{j=1}^{3}(b_j + d_j) + \sum_{j=1}^{3}(2a_j d_j + 2b_j c_j + 4b_j d_j),$$

$$b^4_- = 29 - \sum_{j=1}^{3}(b_j + d_j).$$

The singular set of $(K3_\alpha \times K3_\beta)/\langle\gamma,\delta\rangle$ is the disjoint union of $\sum_{j=1}^3 a_j c_j$ copies of $(\mathcal{S}^2 \times \mathcal{S}^2)/\mathbb{Z}_2$ and $\sum_{j=1}^3 (a_j d_j + b_j c_j + 2b_j d_j)$ copies of $\mathcal{S}^2 \times \mathcal{S}^2$, where each singular point is modelled on $\mathbb{R}^4 \times \mathbb{R}^4/\{\pm 1\}$.

Proof The Betti numbers of $(K3_\alpha \times K3_\beta)/\langle\gamma,\delta\rangle$ are determined as in Propositions 14.4.3 and 14.5.3, and we leave them as an exercise. Here is how to find the singular set of $(K3_\alpha \times K3_\beta)/\langle\gamma,\delta\rangle$. Consider first the action of γ on $T^4/\{\pm 1\}$, and its lift to $K3_\alpha$. Now γ fixes 4 points in $T^4/\{\pm 1\}$, which are singular, and are resolved with a copy of $\mathcal{S}^2$ in $K3_\alpha$.

For each of these 4 points we have chosen $\gamma_* = \pm 1$. It turns out that if $\gamma_* = 1$ then γ fixes the whole $\mathcal{S}^2$ in $K3_\alpha$, but if $\gamma_* = -1$ then γ is the antipodal map on $\mathcal{S}^2$, with no fixed points. But $\gamma_* = 1$ for $a_1 + 2b_1$ of the 4 points. Thus the fixed points of γ in $K3_\alpha$ are $a_1 + 2b_1$ $\mathcal{S}^2$. Similarly, γ fixes $c_1 + 2d_1$ $\mathcal{S}^2$ in $K3_\beta$. So the fixed points of γ in $K3_\alpha \times K3_\beta$ are $(a_1 + 2b_1)(c_1 + 2d_1)$ copies of $\mathcal{S}^2 \times \mathcal{S}^2$.

The action of δ on the $a_1 + 2b_1$ $\mathcal{S}^2$ fixed by γ in $K3_\alpha$ is the antipodal map on a_1 $\mathcal{S}^2$, and swaps the remaining $2b_1$ in pairs. Similarly, δ is antipodal on c_1 $\mathcal{S}^2$ in $K3_\beta$, and swaps $2d_1$ in pairs. Hence, of the $(a_1 + 2b_1)(c_1 + 2d_1)$ $\mathcal{S}^2 \times \mathcal{S}^2$ in $K3_\alpha \times K3_\beta$, δ fixes $a_1 c_1$ of them, and swaps $2a_1 d_1 + 2b_1 c_1 + 4b_1 d_1$ in pairs.

Thus the fixed set of γ contributes $a_1 c_1$ $(\mathcal{S}^2 \times \mathcal{S}^2)/\mathbb{Z}_2$ and $a_1 d_1 + b_1 c_1 + 2b_1 d_1$ $\mathcal{S}^2 \times \mathcal{S}^2$ to the singular set of $(K3_\alpha \times K3_\beta)/\langle\gamma,\delta\rangle$. In the same way, the fixed sets of δ and $\gamma\delta$ give $a_2 c_2$ and $a_3 c_3$ copies of $(\mathcal{S}^2 \times \mathcal{S}^2)/\mathbb{Z}_2$, and $a_2 d_2 + b_2 c_2 + 2b_2 d_2$ and $a_3 d_3 + b_3 c_3 + 2b_3 d_3$ copies of $\mathcal{S}^2 \times \mathcal{S}^2$ respectively. Clearly, each singular point is modelled on $\mathbb{R}^4 \times \mathbb{R}^4/\{\pm 1\}$. □

14.6.2 A family of 8-manifolds resolving T^8/Γ

Next we shall resolve the singularities of $(K3_\alpha \times K3_\beta)/\langle\gamma,\delta\rangle$ to get a nonsingular 8-manifold M, using the ideas of §14.4.2 and §14.5.2. To resolve each copy of $\mathcal{S}^2 \times \mathcal{S}^2$ in the singular set, we can define a complex structure J near $\mathcal{S}^2 \times \mathcal{S}^2$ in a natural way, and resolve by blowing up, as in §14.4.2.

To resolve each $(\mathcal{S}^2 \times \mathcal{S}^2)/\mathbb{Z}_2$, we apply the method of §14.5.2. Suppose for instance that this $(\mathcal{S}^2 \times \mathcal{S}^2)/\mathbb{Z}_2$ comes from the γ fixed points. Then near the corresponding singular $\mathcal{S}^2 \times \mathcal{S}^2$ in $(K3_\alpha \times K3_\beta)/\langle\gamma\rangle$ there is a complex structure J, with $\delta_*(J) = -J$. We resolve the $\mathcal{S}^2 \times \mathcal{S}^2$ in $(K3_\alpha \times K3_\beta)/\langle\gamma\rangle$ by blowing up to get N, and δ lifts to N, where it is free near the resolution of $\mathcal{S}^2 \times \mathcal{S}^2$. So $N/\langle\delta\rangle$ gives a local model for how to resolve this $(\mathcal{S}^2 \times \mathcal{S}^2)/\mathbb{Z}_2$.

Here is the analogue of Theorem 14.5.4 for this resolution M.

Theorem 14.6.4 *This compact 8-manifold M has Betti numbers*

$$b^2 = 10 + \sum_{j=1}^3 (b_j + d_j + a_j d_j + c_j b_j + 2b_j d_j), \qquad b^3 = 0,$$

$$b^4 = 122 - \sum_{j=1}^3 (2b_j + 2d_j) + \sum_{j=1}^3 (3a_j c_j + 4a_j d_j + 4b_j c_j + 8b_j d_j),$$

$$b^4_+ = 93 - \sum_{j=1}^3 (b_j + d_j) + \sum_{j=1}^3 (2a_j c_j + 3a_j d_j + 3b_j c_j + 6b_j d_j),$$

$$b^4_- = 29 - \sum_{j=1}^3 (b_j + d_j) + \sum_{j=1}^3 (a_j c_j + a_j d_j + b_j c_j + 2b_j d_j).$$

If $(a_j + b_j)(c_j + d_j) > 0$ for at least two j in $\{1, 2, 3\}$ then $\pi_1(M) = \{1\}$. If $(a_j + b_j)(c_j + d_j) = 0$ for exactly two j in $\{1, 2, 3\}$ then $\pi_1(M) = \mathbb{Z}_2$. If $(a_j + b_j)(c_j + d_j) = 0$ for all $j = 1, 2, 3$ then $\pi_1(M) = \mathbb{Z}_2^2$.

There exists a set of R-data for T^8/Γ giving this compact 8-manifold M as its resolution. There exist torsion-free Spin(7)-structures (Ω, g) on M, with $\text{Hol}(g) = \text{Spin}(7)$ if $\pi_1(M) = \{1\}$, $\text{Hol}(g) = \mathbb{Z}_2 \ltimes \text{SU}(4)$ if $\pi_1(M) = \mathbb{Z}_2$, and $\text{Hol}(g) = \mathbb{Z}_2^2 \ltimes \text{SU}(2)^2$ if $\pi_1(M) = \mathbb{Z}_2^2$.

Proof Resolving each $\mathcal{S}^2 \times \mathcal{S}^2$ in the singular set of $(K3_\alpha \times K3_\beta)/\langle\gamma, \delta\rangle$ fixes b^3 and adds 1 to b^2, b_+^4 and b_-^4. Resolving each $(\mathcal{S}^2 \times \mathcal{S}^2)/\mathbb{Z}_2$ fixes b^2 and b^3 and adds 1 to b_+^4 and b_-^4. Adding these contributions to the Betti numbers of $(K3_\alpha \times K3_\beta)/\langle\gamma, \delta\rangle$ in Proposition 14.6.3, we get the Betti numbers of M above.

The resolution of $(K3_\alpha \times K3_\beta)/\langle\gamma, \delta\rangle$ does not change the fundamental group, so that $\pi_1(M) \cong \pi_1\big((K3_\alpha \times K3_\beta)/\langle\gamma, \delta\rangle\big)$. Thus there are three cases. If at least two of γ, δ and $\gamma\delta$ have fixed points in $K3_\alpha \times K3_\beta$ then M is simply-connected. If exactly one of γ, δ and $\gamma\delta$ has fixed points then $\pi_1(M) = \mathbb{Z}_2$, and if none of γ, δ or $\gamma\delta$ has fixed points then $\pi_1(M) = \mathbb{Z}_2^2$.

But γ has fixed points in $K3_\alpha$ when $a_1 + b_1 > 0$, and fixed points in $K3_\beta$ when $c_1 + d_1 > 0$, so γ has fixed points in $K3_\alpha \times K3_\beta$ if and only if $(a_1 + b_1)(c_1 + d_1) > 0$. Similarly, $\delta, \gamma\delta$ have fixed points in $K3_\alpha \times K3_\beta$ if and only if $(a_j + b_j)(c_j + d_j) > 0$ for $j = 2, 3$ respectively. This gives the values of $\pi_1(M)$ in the theorem.

The existence of suitable R-data for T^8/Γ follows as in Theorem 14.4.4, and Theorem 13.6.2 then shows that there exist torsion-free Spin(7)-structures (Ω, g) on M. It remains only to identify the holonomy group $\text{Hol}(g)$. From the Betti numbers of M above we see that $\hat{A}(M) = 1$. But $\text{Hol}(g)$ can be deduced from $\hat{A}(M)$ and $\pi_1(M)$, as in Theorem 10.6.1 in the simply-connected case. □

Table 14.3 Betti numbers (b^2, b^3, b^4) of the compact 8-manifolds with holonomy Spin(7) from the construction of §14.6

(10,0,128) (11,0,126) (12,0,127) (12,0,130) (13,0,125) (13,0,128) (14,0,126)
(14,0,129) (14,0,132) (14,0,138) (15,0,124) (15,0,127) (15,0,130) (15,0,133)
(15,0,136) (16,0,128) (16,0,131) (16,0,134) (16,0,140) (17,0,132) (17,0,135)
(17,0,138) (17,0,144) (18,0,130) (18,0,133) (18,0,136) (18,0,139) (18,0,142)
(19,0,134) (19,0,137) (19,0,140) (19,0,146) (20,0,138) (20,0,141) (20,0,144)
(21,0,136) (21,0,142) (21,0,145) (21,0,148) (22,0,140) (22,0,143) (22,0,146)
(22,0,152) (22,0,158) (23,0,144) (23,0,147) (23,0,150) (23,0,156) (26,0,150)
(26,0,153) (26,0,156) (26,0,162)

Now we shall calculate the possible Betti numbers of compact 8-manifolds M with holonomy Spin(7) arising from this construction. We do this by substituting all consistent possibilities for the a_j, c_j, b_j and d_j into the formulae of Theorem 14.6.4. Recall that a_j and c_j are 0, 1 or 2 and satisfy $a_1 + a_2 + a_3 = c_1 + c_2 + c_3 = 2$, and b_j and d_j are 0 or 1.

Theorem 14.6.4 shows that M has holonomy Spin(7) if $\pi_1(M) = 1$, which happens if $(a_j + b_j)(c_j + d_j) > 0$ for at least two j in $\{1, 2, 3\}$. By symmetry in $j = 1, 2, 3$ we may as well assume that this holds for $j = 1$ and 2, so that we impose the conditions

$$a_1 + b_1 > 0, \quad c_1 + d_1 > 0, \quad a_2 + b_2 > 0 \quad \text{and} \quad c_2 + d_2 > 0.$$

Writing down all possible values of the $a_j, \ldots, d_j$ satisfying these conditions and substituting them into the Betti number formulae in Theorem 14.6.4, we obtain the set of all Betti numbers (b^2, b^3, b^4) of compact 8-manifolds M with holonomy Spin(7) arising from this construction. There are 52 sets of Betti numbers, which are listed in Table 14.3.

15

A SECOND CONSTRUCTION OF COMPACT 8-MANIFOLDS WITH HOLONOMY Spin(7)

In this chapter we will describe a second method for constructing compact 8-manifolds with holonomy Spin(7), in which one starts not with a torus T^8 but with a *Calabi–Yau 4-orbifold* Y with isolated singular points $p_1, \ldots, p_k$. We use algebraic geometry to find a number of suitable complex orbifolds Y, which in the simplest cases are hypersurfaces in *weighted projective spaces* $\mathbb{CP}^5_{a_0, \ldots, a_5}$.

Then, instead of a finite group Γ, we suppose we have an antiholomorphic, isometric involution $\sigma : Y \to Y$, whose only fixed points are $p_1, \ldots, p_k$. This involution does not preserve the SU(4)-structure on Y, but it does preserve the induced Spin(7)-structure. We think of σ as breaking the structure group of Y from SU(4) down to Spin(7). Define $Z = Y/\langle \sigma \rangle$. Then Z is an orbifold with isolated singular points $p_1, \ldots, p_k$, and the Calabi–Yau structure on Y induces a torsion-free Spin(7)-structure on Z.

If the singularities of Z are of a suitable kind, we can resolve them to get a compact 8-manifold M with holonomy Spin(7). To perform the resolution we need to find *ALE Spin(7)-manifolds* corresponding to the singularities of Z, as in Definition 13.1.7. Our construction then yields more examples of compact 8-manifolds M with holonomy Spin(7).

We calculate the Betti numbers $b^k(M)$ in each case. They turn out to be rather different to the Betti numbers we found in Chapter 13. In particular, in this new construction the middle Betti number b^4 tends to be rather large, as big as 11 662 in one example.

For simplicity we will assume that all the singularities of Y are modelled on $\mathbb{C}^4/\mathbb{Z}_4$, where the generator α of $\mathbb{Z}_4$ acts by

$$\alpha : (z_1, \ldots, z_4) \mapsto (iz_1, iz_2, iz_3, iz_4).$$

However, there are also other kinds of isolated singularities in Y and Z that could be incorporated in the construction.

We begin in §15.1 by giving some examples of ALE Spin(7)-manifolds. Section 15.2 then proves our main result, that given a Calabi–Yau 4-orbifold Y and an antiholomorphic involution $\sigma : Y \to Y$ satisfying certain conditions, we can construct a compact 8-manifold M with holonomy Spin(7).

We explain in §15.3 how to use the construction in practice, and ways of computing the Betti numbers of the resulting 8-manifolds M. Sections 15.4–15.7 apply the construction to generate new examples of compact 8-manifolds with holonomy Spin(7), and we finish in §15.8 with a discussion of our results and some research problems.

The material of this chapter is also published in [128].

15.1 ALE Spin(7)-manifolds

We construct some examples of ALE Spin(7)-manifolds, in the sense of Definition 13.1.7, which will be needed in §15.2.

15.1.1 *An example of an ALE* Spin(7)-*manifold*

Let $\mathbb{R}^8$ have coordinates $(x_1, \ldots, x_8)$ and Spin(7)-structure (Ω_0, g_0), as in Definition 10.5.1. Use the complex coordinates

$$(z_1, z_2, z_3, z_4) = (x_1 + ix_2, x_3 + ix_4, x_5 + ix_6, x_7 + ix_8)$$

to identify $\mathbb{R}^8$ with $\mathbb{C}^4$. Then $g_0 = |dz_1|^2 + \cdots + |dz_4|^2$, and $\Omega_0 = \frac{1}{2}\omega_0 \wedge \omega_0 + \operatorname{Re} \theta_0$, where ω_0 is the Kähler form of g_0 and $\theta_0 = dz_1 \wedge \cdots \wedge dz_4$ the complex volume form on $\mathbb{C}^4$.

Define $\alpha, \beta : \mathbb{C}^4 \to \mathbb{C}^4$ by

$$\begin{aligned} \alpha &: (z_1, \ldots, z_4) \mapsto (iz_1, iz_2, iz_3, iz_4), \\ \beta &: (z_1, \ldots, z_4) \mapsto (\bar{z}_2, -\bar{z}_1, \bar{z}_4, -\bar{z}_3). \end{aligned} \qquad (15.1)$$

Then $\alpha \in \mathrm{SU}(4) \subset \mathrm{Spin}(7)$ and $\beta \in \mathrm{Spin}(7)$, and α, β satisfy $\alpha^4 = \beta^4 = 1, \alpha^2 = \beta^2$ and $\alpha\beta = \beta\alpha^3$. Let $G = \langle \alpha, \beta \rangle$. Then G is a finite nonabelian subgroup of Spin(7) of order 8 which acts freely on $\mathbb{R}^8 \setminus \{0\}$.

Now $\mathbb{C}^4/\langle\alpha\rangle$ is a complex singularity, as $\alpha \in \mathrm{SU}(4)$. Let (Y_1, π_1) be the blow-up of $\mathbb{C}^4/\langle\alpha\rangle$ at 0. Then Y_1 is the unique crepant resolution of $\mathbb{C}^4/\langle\alpha\rangle$. The action of β on $\mathbb{C}^4/\langle\alpha\rangle$ lifts to a free antiholomorphic map $\beta : Y_1 \to Y_1$ with $\beta^2 = 1$. Define $X_1 = Y_1/\langle\beta\rangle$. Then X_1 is a nonsingular 8-manifold, and the projection $\pi_1 : Y_1 \to \mathbb{C}^4/\langle\alpha\rangle$ pushes down to $\pi_1 : X_1 \to \mathbb{R}^8/G$.

From Chapter 8 there exist ALE Kähler metrics g_1 on Y_1 with holonomy SU(4), which were in fact constructed explicitly in Example 8.2.5. Each such g_1 is invariant under the action of β on Y_1. Let ω_1 be the Kähler form of g_1, and $\theta_1 = \pi_1^*(\theta_0)$ the holomorphic volume form on Y_1. Then Proposition 13.1.4 defines a torsion-free Spin(7)-structure (Ω_1, g_1) on Y_1 with $\Omega_1 = \frac{1}{2}\omega_1 \wedge \omega_1 + \operatorname{Re} \theta_1$.

As $\beta^*(\omega_1) = -\omega_1$ and $\beta^*(\theta_1) = \bar{\theta}_1$, we see that β preserves (Ω_1, g_1). Thus (Ω_1, g_1) pushes down to a torsion-free Spin(7)-structure (Ω_1, g_1) on X_1. Then (X_1, Ω_1, g_1) is an *ALE* Spin(7)-*manifold* asymptotic to $\mathbb{R}^8/G$. The Betti numbers of X_1 are $b^1 = b^2 = b^3 = 0$ and $b^4 = 1$, and $\pi_1(X_1) = \mathbb{Z}_2$.

15.1.2 *A second ALE* Spin(7)-*manifold asymptotic to* $\mathbb{R}^8/G$

Define new complex coordinates $(w_1, \ldots, w_4)$ on $\mathbb{R}^8$ by

$$(w_1, w_2, w_3, w_4) = (-x_1 + ix_3, x_2 + ix_4, -x_5 + ix_7, x_6 + ix_8).$$

Then $g_0 = |dw_1|^2 + \cdots + |dw_4|^2$ and $\Omega_0 = \frac{1}{2}\omega_0' \wedge \omega_0' + \operatorname{Re} \theta_0'$, where ω_0' is the Kähler form of g_0 with respect to the complex structure induced by the w_j, and $\theta_0' = dw_1 \wedge \cdots \wedge dw_4$ is the complex volume form on $\mathbb{C}^4$.

As the action of SU(4) on $\mathbb{R}^8 = \mathbb{C}^4$ induced by the w_j preserves g_0, ω_0' and θ_0', it preserves (Ω_0, g_0). Thus the action of SU(4) on $\mathbb{R}^8$ compatible with the coordinates w_j is a subgroup of Spin(7). Note that this is a *different* SU(4) subgroup of Spin(7) to that considered above, induced by the z_j. In the coordinates w_j, we find that α, β act by

$$\alpha : (w_1, \ldots, w_4) \mapsto (\bar{w}_2, -\bar{w}_1, \bar{w}_4, -\bar{w}_3),$$
$$\beta : (w_1, \ldots, w_4) \mapsto (iw_1, iw_2, iw_3, iw_4).$$
(15.2)

Observe that (15.1) and (15.2) are the same, except that the rôles of α, β are reversed. Therefore we can use the ideas above again.

Let Y_2 be the crepant resolution of $\mathbb{C}^4/\langle\beta\rangle$. The action of α on $\mathbb{C}^4/\langle\beta\rangle$ lifts to a free antiholomorphic involution of Y_2. Let $X_2 = Y_2/\langle\alpha\rangle$. Then X_2 is nonsingular, and as above there exists a torsion-free Spin(7)-structure (Ω_2, g_2) on X_2, making (X_2, Ω_2, g_2) into an ALE Spin(7)-manifold asymptotic to $\mathbb{R}^8/G$.

Now (X_1, Ω_1, g_1) and (X_2, Ω_2, g_2) are clearly isomorphic as Spin(7)-manifolds, but they should be regarded as *topologically distinct* ALE manifolds, because the isomorphism between them acts nontrivially on $\mathbb{R}^8/G$. Thus, we have found two topologically distinct ALE Spin(7)-manifolds $(X_1, \Omega_1, g_1), (X_2, \Omega_2, g_2)$ asymptotic to the same singularity $\mathbb{R}^8/G$.

15.1.3 *Other examples of ALE Spin(7)-manifolds*

We can use the ideas above to construct other ALE Spin(7)-manifolds too. Here we very briefly describe two infinite families of ALE Spin(7)-manifolds X_1^n, X_2^n for $n = 1, 3, 5, \ldots$. For simplicity they will not be used in the rest of the chapter, although they easily could be.

Identify $\mathbb{R}^8$ and $\mathbb{C}^4$ as in §15.1.1. Let $n \geqslant 1$ be an odd integer, and define $\alpha, \beta, \gamma : \mathbb{C}^4 \to \mathbb{C}^4$ by

$$\alpha : (z_1, \ldots, z_4) \mapsto (e^{2\pi i/n} z_1, e^{-2\pi i/n} z_2, e^{2\pi i/n} z_3, e^{-2\pi i/n} z_4),$$
$$\beta : (z_1, \ldots, z_4) \mapsto (iz_1, iz_2, iz_3, iz_4),$$
$$\gamma : (z_1, \ldots, z_4) \mapsto (\bar{z}_2, -\bar{z}_1, \bar{z}_4, -\bar{z}_3).$$

Then $\alpha, \beta \in$ SU(4) and $\gamma \in$ Spin(7), and $G^n = \langle \alpha, \beta, \gamma \rangle$ is a finite nonabelian subgroup of Spin(7) of order $8n$ which acts freely on $\mathbb{R}^8 \setminus \{0\}$. Note that G^1 coincides with the group G of §15.1.1–§15.1.2.

We can construct a family of ALE Spin(7)-manifolds asymptotic to $\mathbb{R}^8/G^n$ as follows. The complex singularity $\mathbb{C}^4/\langle\alpha, \beta\rangle$ has a unique crepant resolution Y_1^n, which can be described explicitly using toric geometry. The action of γ on $\mathbb{C}^4/\langle\alpha, \beta\rangle$ lifts to a free antiholomorphic involution $\gamma : Y_1^n \to Y_1^n$, so that $X_1^n = Y_1^n/\langle\gamma\rangle$ is a nonsingular 8-manifold with a projection $\pi_1^n : X_1^n \to \mathbb{R}^8/G^n$.

By the results of Chapter 8, there exist ALE Kähler metrics g_1^n on Y_1^n with holonomy SU(4). We can choose g_1^n to be γ-invariant, and then the induced Spin(7)-structure (Ω_1^n, g_1^n) on Y_1^n is also γ-invariant, and pushes down to X_1^n, making $(X_1^n, \Omega_1^n, g_1^n)$ into an ALE Spin(7)-manifold asymptotic to $\mathbb{R}^8/G^n$. Using the idea of §15.1.2, we can also define a second ALE Spin(7)-manifold $(X_2^n, \Omega_2^n, g_2^n)$ asymptotic to $\mathbb{R}^8/G^n$.

15.2 Proof of the construction

Starting with a Calabi–Yau 4-orbifold Y with isolated singularities of a certain kind, and an antiholomorphic involution σ on Y, we will now construct a compact 8-manifold M by resolving $Z = Y/\langle\sigma\rangle$, and prove that there exist torsion-free Spin(7)-structures $(\tilde{\Omega}, \tilde{g})$ on M, which have holonomy Spin(7) if M is simply-connected. We use many of the ideas and results of Chapter 13 along the way.

15.2.1 A class of Spin(7)-orbifolds Z made from Calabi–Yau 4-orbifolds

We set out below the ingredients in our construction, and the assumptions they must satisfy.

Condition 15.2.1 Let (Y, J) be a compact complex 4-orbifold with $c_1(Y) = 0$, admitting Kähler metrics. Let σ be an antiholomorphic involution on Y. That is, $\sigma : Y \to Y$ is a diffeomorphism satisfying $\sigma^2 = \mathrm{id}$ and $\sigma^*(J) = -J$. Define $\alpha : \mathbb{C}^4 \to \mathbb{C}^4$ by

$$\alpha : (z_1, z_2, z_3, z_4) \longmapsto (iz_1, iz_2, iz_3, iz_4). \tag{15.3}$$

Then $\alpha^4 = 1$, so that $\langle\alpha\rangle \cong \mathbb{Z}_4$, and $\mathbb{C}^4/\langle\alpha\rangle$ has an isolated singular point at 0.

We require that the singular set of Y should be k isolated points $p_1, \ldots, p_k$ for some $k \geq 1$, each modelled on $\mathbb{C}^4/\langle\alpha\rangle$, and that the fixed set of σ in Y is exactly $\{p_1, \ldots, p_k\}$. We also suppose that $Y \setminus \{p_1, \ldots, p_k\}$ is simply-connected, and $h^{2,0}(Y) = 0$.

In the rest of the section we assume that Condition 15.2.1 holds.

Proposition 15.2.2 *There exists a Kähler metric g_Y on Y with holonomy* SU(4) *and Kähler form ω_Y, and a holomorphic volume form θ_Y on Y, which satisfy*

$$\sigma^*(g_Y) = g_Y, \quad \sigma^*(\omega_Y) = -\omega_Y \quad \text{and} \quad \sigma^*(\theta_Y) = \bar{\theta}_Y. \tag{15.4}$$

Define a 4-form Ω_Y on Y by $\Omega_Y = \frac{1}{2}\omega_Y \wedge \omega_Y + \mathrm{Re}(\theta_Y)$. Then (Ω_Y, g_Y) is a torsion-free Spin(7)*-structure on Y, which is invariant under σ.*

Proof We know Y admits Kähler metrics by Condition 15.2.1. Let g' be a Kähler metric on Y. Then $\sigma^*(g')$ is also a Kähler metric on Y, and so $g'' = g' + \sigma^*(g')$ is a σ-invariant Kähler metric on Y. Let κ be the Kähler class of g''. Then κ is σ-invariant, regarded as an equivalence class of Kähler metrics on Y.

As $c_1(Y) = 0$, by Theorem 6.5.6 there is a unique Ricci-flat Kähler metric g_Y in the Kähler class κ, with Kähler form ω_Y. Since κ is σ-invariant, this g_Y is σ-invariant by uniqueness of g_Y, and so the first two equations of (15.4) hold.

Proposition 6.1.1 shows that $\mathrm{Hol}^0(g_Y) \subseteq \mathrm{SU}(4)$. It follows from Theorem 3.4.1 that $\mathrm{Hol}^0(g_Y)$ is one of $\{1\}$, SU(2), SU(3), Sp(2) or SU(4). As $Y \setminus \{p_1, \ldots, p_k\}$ is simply-connected we can rule out $\{1\}$, SU(2) and SU(3), and we also have $\mathrm{Hol}(g_Y) = \mathrm{Hol}^0(g_Y)$. But Y cannot have holonomy Sp(2) either, as $h^{2,0}(Y) = 0$ by assumption. The only remaining possibility is that $\mathrm{Hol}(g_Y) = \mathrm{SU}(4)$.

Therefore Y has a holomorphic volume form θ_Y as in §6.1. Clearly $\sigma^*(\theta_Y)$ is proportional to $\bar{\theta}_Y$, and by multiplying θ_Y by $e^{i\phi}$ we can arrange that $\sigma^*(\theta_Y) = \bar{\theta}_Y$. Thus the third equation of (15.4) holds. Finally, (Ω_Y, g_Y) is a torsion-free Spin(7)-structure by Proposition 13.1.4, and it is σ-invariant by (15.4). □

In our next result, if Y is an orbifold and $p \in Y$ an orbifold point modelled on $\mathbb{R}^n/G$, then we say that the *tangent space* $T_p Y$ to Y at p is $\mathbb{R}^n/G$, in the obvious way. The proof is really only linear algebra, and we leave it as an exercise.

Proposition 15.2.3 *For each $j = 1, \ldots, k$ we can identify the tangent space $T_{p_j} Y$ to Y at p_j with $\mathbb{C}^4/\langle \alpha \rangle$ so that g_Y is identified with $|dz_1|^2 + \cdots + |dz_4|^2$ at p_j, and θ_Y is identified with $dz_1 \wedge \cdots \wedge dz_4$ at p_j, and $d\sigma : T_{p_j} Y \to T_{p_j} Y$ is identified with the map $\beta : \mathbb{C}^4/\langle \alpha \rangle \to \mathbb{C}^4/\langle \alpha \rangle$ given by*

$$\beta : (z_1, \ldots, z_4)\langle \alpha \rangle \longmapsto (\bar{z}_2, -\bar{z}_1, \bar{z}_4, -\bar{z}_3)\langle \alpha \rangle. \tag{15.5}$$

Now §15.1.1 defined a finite group $G = \langle \alpha, \beta \rangle$ acting on $\mathbb{R}^8$, and the definitions (15.3) and (15.5) of α and β above coincide with (15.1) in §15.1.1. Thus the singularities of $Z = Y/\langle \sigma \rangle$ are all modelled on $\mathbb{R}^8/G$, and we easily prove:

Corollary 15.2.4 *Define $Z = Y/\langle \sigma \rangle$. Then Z is a compact, real 8-dimensional orbifold. The Spin(7)-structure (Ω_Y, g_Y) on Y pushes down to give a torsion-free Spin(7)-structure (Ω_Z, g_Z) on Z. The singularities of Z are k points $p_1, \ldots, p_k$. For each $j = 1, \ldots, k$ there is an isomorphism $\iota_j : \mathbb{R}^8/G \to T_{p_j} Z$ which identifies the Spin(7)-structures (Ω_0, g_0) on $\mathbb{R}^8/G$ and (Ω_Z, g_Z) on $T_{p_j} Z$. Here G and (Ω_0, g_0) are defined in §15.1.1.*

15.2.2 Desingularizing Z to get a compact 8-manifold M

So far we have constructed a Spin(7)-orbifold (Z, Ω_Z, g_Z) with finitely many singular points $p_1, \ldots, p_k$, each modelled on the singularity $\mathbb{R}^8/G$ of §15.1.1. But sections 15.1.1 and 15.1.2 found two ALE Spin(7)-manifolds X_1 and X_2 asymptotic to $\mathbb{R}^8/G$. We shall now resolve each singular point p_j in Z using either X_1 or X_2 to get a compact 8-manifold M, in a similar way to the resolution of T^8/Γ in §13.2.2. To save time we include the parameter $t \in (0, 1]$ from the beginning, as in §13.4.

Definition 15.2.5 For each j let ι_j be as in Corollary 15.2.4, and let $\exp_{p_j} : T_{p_j} Z \to Z$ be the *exponential map*, which is well-defined as Z is complete. Then $\exp_{p_j} \circ \iota_j$ maps $\mathbb{R}^8/G$ to Z. Choose $\zeta > 0$ small, and let $B_{2\zeta}(\mathbb{R}^8/G)$ be the open ball of radius 2ζ about 0 in $\mathbb{R}^8/G$. Define $U_j \subset Z$ by $U_j = \exp_{p_j} \circ \iota_j\bigl(B_{2\zeta}(\mathbb{R}^8/G)\bigr)$, and $\psi_j : B_{2\zeta}(\mathbb{R}^8/G) \to U_j$ by $\psi_j = \exp_{p_j} \circ \iota_j$. Let $\zeta > 0$ be chosen small enough that U_j is open in Z and $\psi_j : B_{2\zeta}(\mathbb{R}^8/G) \to U_j$ is a diffeomorphism for $1 \leq j \leq k$, and that $U_i \cap U_j = \emptyset$ when $i \neq j$.

Proposition 15.2.6 *There is a smooth 3-form σ_j on $B_{2\zeta}(\mathbb{R}^8/G)$ for $1 \leq j \leq k$ and a constant $C_1 > 0$, such that $\psi_j^*(\Omega_Z) - \Omega_0 = d\sigma_j$ and $|\nabla^l \sigma_j| \leq C_1 r^{3-l}$ on $B_{2\zeta}(\mathbb{R}^8/G)$, for $l = 0, 1, 2$. Here $|.|$ and ∇ are defined using the metric g_0 on $B_{2\zeta}(\mathbb{R}^8/G)$, and $r : B_{2\zeta}(\mathbb{R}^8/G) \to [0, 2\zeta)$ is the radius function.*

Proof The derivative of $\exp_{p_j}$ at 0 is the identity map on $T_{p_j} Z$. Thus the derivative of ψ_j at 0 is $\iota_j : \mathbb{R}^8/G \to T_{p_j} Z$, and so $\psi_j^*(\Omega_Z)|_0 = \iota_j^*(\Omega_Z) = \Omega_0|_0$, since ι_j identifies

Ω_0 and Ω_Z. Therefore $\psi_j^*(\Omega_Z) = \Omega_0$ at 0 in $B_{2\zeta}(\mathbb{R}^8/G)$. As $\psi_j^*(\Omega_Z) - \Omega_0$ is a 4-form on a subset of $\mathbb{R}^8/G$, we can pull it back to $\mathbb{R}^8$, and regard $\psi_j^*(\Omega_Z) - \Omega_0$ as a 4-form on the ball $B_{2\zeta}(\mathbb{R}^8)$ of radius 2ζ in $\mathbb{R}^8$.

Then $\psi_j^*(\Omega_Z) - \Omega_0$ is a smooth G-invariant 4-form on $B_{2\zeta}(\mathbb{R}^8)$ which vanishes at 0. But G contains $-1 : \mathbb{R}^8 \to \mathbb{R}^8$, and any 4-form invariant under this map -1 has zero first derivative at 0. Hence $\psi_j^*(\Omega_Z) - \Omega_0$ vanishes to first order at 0 in $B_{2\zeta}(\mathbb{R}^8)$, and so by Taylor's Theorem we can show that $|\psi_j^*(\Omega_Z) - \Omega_0| = O(r^2)$ and $|\nabla \psi_j^*(\Omega_Z)| = O(r)$ on $B_{2\zeta}(\mathbb{R}^8)$.

Now Ω_Z and Ω_0 are closed, so that $\psi_j^*(\Omega_Z) - \Omega_0$ is closed, and as $B_{2\zeta}(\mathbb{R}^8/G)$ is contractible we can write $\psi_j^*(\Omega_Z) - \Omega_0 = d\sigma_j$ for some smooth 3-form σ_j on $B_{2\zeta}(\mathbb{R}^8/G)$. Since $\psi_j^*(\Omega_Z) - \Omega_0$ vanishes to first order at 0 we can easily arrange that σ_j vanishes to second order at 0, and therefore $|\nabla^l \sigma_j| = O(r^{3-l})$ for $l = 0, 1, 2$, using Taylor's Theorem as above. Thus there exists $C_1 > 0$ such that $|\nabla^l \sigma_j| \leqslant C_1 r^{3-l}$ on $B_{2\zeta}(\mathbb{R}^8/G)$, for $l = 0, 1, 2$ and $j = 1, \ldots, k$. □

Definition 15.2.7 Let the ALE Spin(7)-manifolds (X_n, Ω_n, g_n) and projections $\pi_n : X_n \to \mathbb{R}^8/G$ be as in §15.1.1–§15.1.2 for $n = 1, 2$. For each $t \in (0, 1]$ and $n = 1, 2$ let $X_n^t = X_n$, define a Spin(7)-structure (Ω_n^t, g_n^t) on X_n^t by $\Omega_n^t = t^4 \Omega_n$ and $g_n^t = t^2 g_n$, and define $\pi_n^t : X_n^t \to \mathbb{R}^8/G$ by $\pi_n^t = t\pi_n$. Then $(X_n^t, \Omega_n^t, g_n^t)$ is an ALE Spin(7)-manifold asymptotic to $\mathbb{R}^8/G$.

Following the proof of (13.12) we can show that there exists $C_2 > 0$ and a smooth 3-form τ_n^t on $\mathbb{R}^8/G \setminus B_{t\zeta}(\mathbb{R}^8/G)$, satisfying

$$(\pi_n^t)_*(\Omega_n^t) = \Omega_0 + d\tau_n^t \quad \text{and} \quad |\nabla^l \tau_n^t| \leqslant C_2 t^8 r^{-7-l} \quad \text{for } l = 0, 1, 2 \tag{15.6}$$

on $\mathbb{R}^8/G \setminus B_{t\zeta}(\mathbb{R}^8/G)$, where $|\,.\,|$ and ∇ are defined using the metric g_0.

For $j = 1, \ldots, k$, choose n_j to be 1 or 2. There are 2^k ways of defining the n_j. These choices are the analogue of a set of R-data for T^8/Γ, as in §13.2. We shall resolve each singular point p_j in Z using $X_{n_j}^t$ to get a 1-parameter family of resolutions (M^t, π^t) of Z, loosely following Definition 13.2.5.

Definition 15.2.8 For each $j = 1, \ldots, k$, define open subsets M_0^t in Z and M_j^t in $X_{n_j}^t$ for $1 \leqslant j \leqslant k$ by

$$M_0^t = Z \setminus \bigcup_{j=1}^k \psi_j\left(\overline{B}_{t^{4/5}\zeta}(\mathbb{R}^8/G)\right) \quad \text{and} \quad M_j^t = (\pi_{n_j}^t)^{-1}\left(B_{2t^{4/5}\zeta}(\mathbb{R}^8/G)\right).$$

That is, M_0^t is the complement in Z of the closed balls of radius $t^{4/5}\zeta$ about p_j for $1 \leqslant j \leqslant k$, and M_j^t is the inverse image of $B_{2t^{4/5}\zeta}(\mathbb{R}^8/G)$ in $X_{n_j}^t$.

Define an equivalence relation '$\sim$' on the disjoint union $\coprod_{j=0}^k M_j^t$ by $x \sim y$ if either

(a) $x = y$,

(b) $x \in M_j^t$ and $y \in U_j \cap M_0^t$ and $\psi_j \circ \pi_{n_j}^t(x) = y$, for some $j = 1, \ldots, k$, or

(c) $y \in M_j^t$ and $x \in U_j \cap M_0^t$ and $\psi_j \circ \pi_{n_j}^t(y) = x$, for some $j = 1, \ldots, k$.

Define the *resolution* M^t of Z to be $\coprod_{j=0}^k M_j^t / \sim$. It is easy to see that M^t is a compact 8-manifold. Define a projection $\pi^t : M^t \to Z$ by $\pi^t([x]) = x$ when $x \in M_0^t$, and $\pi^t([x]) = \psi_j \circ \pi_{n_j}^t(x)$ when $x \in M_j^t$ for some $j = 1, \ldots, k$, where $[x]$ is the equivalence class of x under $\sim$. Then π^t is well-defined, continuous and surjective, and $\pi^t : M^t \setminus \bigcup_{j=1}^k (\pi^t)^{-1}(p_j) \to Z \setminus \{p_1, \ldots, p_k\}$ is a diffeomorphism.

Since the resolutions (M^t, π^t) of Z form a connected family for $t \in (0, 1]$, they are all diffeomorphic to the same compact 8-manifold M. We can regard M_j^t as an open subset of M^t for $j = 0, \ldots, k$, and then the M_j^t form an *open cover* of M^t. If $1 \leq i, j \leq k$ and $i \neq j$ then $M_i^t \cap M_j^t = \emptyset$. The overlap $M_0^t \cap M_j^t$ is naturally isomorphic to an *annulus* in $\mathbb{R}^8/G$, with inner radius $t^{4/5}\zeta$ and outer radius $2t^{4/5}\zeta$. The reason for including the factors $t^{4/5}$ will be explained shortly.

We now calculate the fundamental group of M^t.

Proposition 15.2.9 *If $n_j = 1$ for $j = 1, \ldots, k$ then $\pi_1(M^t) \cong \mathbb{Z}_2$. Otherwise, M^t is simply-connected.*

Proof Since $Y \setminus \{p_1, \ldots, p_k\}$ is simply-connected by Condition 15.2.1 and σ acts freely on $Y \setminus \{p_1, \ldots, p_k\}$, we see that the fundamental group of $Z \setminus \{p_1, \ldots, p_k\}$ is $\mathbb{Z}_2$. The natural inclusion of $Z \setminus \{p_1, \ldots, p_k\}$ in M^t induces a homomorphism from $\pi_1(Z \setminus \{p_1, \ldots, p_k\})$ to $\pi_1(M^t)$, which is easily shown to be surjective. Also, as $X_{n_j}^t$ is X_1 or X_2 we have $\pi_1(X_{n_j}^t) \cong \mathbb{Z}_2$.

Therefore, $\pi_1(M_t)$ is $\mathbb{Z}_2$ if the generator of $\pi_1(Z \setminus \{p_1, \ldots, p_k\})$ projects to the nonzero element of $\pi_1(X_{n_j}^t)$ for all $1 \leq j \leq k$, and $\pi_1(M^t)$ is trivial otherwise. But calculation shows that the generator of $\pi_1(Z \setminus \{p_1, \ldots, p_k\})$ is nonzero in $\pi_1(X_{n_j}^t)$ if and only if $n_j = 1$. □

This shows that of the 2^k possible ways of choosing the n_j, one gives $\pi_1(M^t) = \mathbb{Z}_2$, and the remaining $2^k - 1$ all give simply-connected M^t.

15.2.3 A Spin(7)-*structure* (Ω^t, g^t) *on M^t with small torsion*

Each open subset M_j^t in M^t carries a torsion-free Spin(7)-structure, (Ω_Z, g_Z) for $j = 0$ and $(\Omega_{n_j}^t, g_{n_j}^t)$ for $1 \leq j \leq k$. We shall join these Spin(7)-structures together with a partition of unity to get a Spin(7)-structure (Ω^t, g^t) on M^t and estimate its torsion, as in §13.5.

Definition 15.2.10 Let $\eta : [0, \infty) \to [0, 1]$ be a smooth function with $\eta(x) = 0$ for $x \leq \zeta$ and $\eta(x) = 1$ for $x \geq 2\zeta$. Define a 4-form ξ^t on M^t by $\xi^t = \Omega_Z$ in $M_0^t \setminus \bigcup_{j=1}^k M_j^t$, and $\xi^t = \Omega_{n_j}^t$ in $M_j^t \setminus M_0^t$ for $1 \leq j \leq k$, and

$$\xi^t = \Omega_0 + d\big(\eta(t^{-4/5}r)\sigma_j\big) + d\big((1 - \eta(t^{-4/5}r))\tau_{n_j}^t\big) \quad \text{in } M_0^t \cap M_j^t \tag{15.7}$$

for $1 \leq j \leq k$, where we identify $M_0^t \cap M_j^t$ with an annulus in $\mathbb{R}^8/G$ in the natural way. Since $\Omega_Z = \Omega_0 + d\sigma_j$ and $\Omega_{n_j}^t = \Omega_0 + d\tau_{n_j}^t$ in $M_0^t \cap M_j^t$ it follows that ξ^t is smooth, and as Ω_Z, $\Omega_{n_j}^t$ and Ω_0 are closed, ξ^t is closed.

Lemma 15.2.11 *There exists $C_3 > 0$ such that for each $j = 1, \ldots, k$ and $t \in (0, 1]$, this 4-form ξ^t satisfies*

$$|\xi^t - \Omega_0| \leq C_3 t^{8/5} \quad \text{and} \quad |\nabla(\xi^t - \Omega_0)| \leq C_3 t^{4/5} \tag{15.8}$$

in $M_0^t \cap M_j^t$, where $|\,.\,|$ and ∇ are defined using the metric g_0.

Proof Expanding (15.7) we find that

$$\xi^t - \Omega_0 = \eta(t^{-4/5}r)\mathrm{d}\sigma_j + (1 - \eta(t^{-4/5}r))\mathrm{d}\tau_{n_j}^t + t^{-4/5}\eta'(t^{-4/5}r)\mathrm{d}r \wedge (\sigma_j - \tau_{n_j}^t)$$

in $M_0^t \cap M_j^t$. Since $t^{4/5}\zeta \leq r \leq 2t^{4/5}\zeta$, Proposition 15.2.6 and (15.6) show that

$$|\sigma_j| \leq 8C_1\zeta^3 t^{12/5}, \quad |\mathrm{d}\sigma_j| \leq 4C_1\zeta^2 t^{8/5}, \quad |\nabla\mathrm{d}\sigma_j| \leq 2C_1\zeta t^{4/5},$$
$$|\tau_{n_j}^t| \leq C_2\zeta^{-7} t^{12/5}, \quad |\mathrm{d}\tau_{n_j}^t| \leq C_2\zeta^{-8} t^{8/5} \quad \text{and} \quad |\nabla\mathrm{d}\tau_{n_j}^t| \leq C_2\zeta^{-9} t^{4/5}.$$

Combining these with the previous equation and using the facts that $|\mathrm{d}r| = 1$ and η' is bounded independently of t, we soon prove (15.8). $\square$

We can now explain why we chose the power $t^{4/5}$ in Definition 15.2.8. Suppose we had defined M^t and ξ^t using t^α in place of $t^{4/5}$, for some $\alpha \in [0, 1]$. Then in the calculation above the σ_j and $\tau_{n_j}^t$ terms would contribute $O(t^{2\alpha})$ and $O(t^{8-8\alpha})$ to $\xi^t - \Omega_0$ respectively, and so $\xi^t - \Omega_0$ would be $O(t^{2\alpha}) + O(t^{8-8\alpha})$. This is smallest when $2\alpha = 8 - 8\alpha$, that is, when $\alpha = 4/5$. So the power $t^{4/5}$ minimizes the size of $\xi^t - \Omega_0$.

Now we can define the Spin(7)-structures (Ω^t, g^t) on M^t.

Definition 15.2.12 Let ϵ_1 be as in Proposition 10.5.9, and choose $\epsilon \in (0, 1]$ such that $C_3\epsilon^{8/5} \leq \epsilon_1$. Suppose $t \in (0, \epsilon]$. Then $|\xi^t - \Omega_0| \leq C_3 t^{8/5} \leq \epsilon_1$ in $M_0^t \cap M_j^t$ for $1 \leq j \leq k$ by (15.8), and so ξ^t lies in $\mathcal{T}M^t$ on $M_0^t \cap M_j^t$ by Proposition 10.5.9. But ξ^t is Ω_Z or $\Omega_{n_j}^t$ outside the overlaps $M_0^t \cap M_j^t$, and thus $\xi^t \in C^\infty(\mathcal{T}M^t)$. For each $t \in (0, \epsilon]$ define $\Omega^t = \Theta(\xi^t)$, where Θ is given in Definition 10.5.8. Then $\Omega^t \in C^\infty(\mathcal{A}M^t)$, and so Ω^t extends to a Spin(7)-structure (Ω^t, g^t) on M^t. Define a 4-form ϕ^t on M^t by $\phi^t = \xi^t - \Omega^t$. Then $\mathrm{d}\Omega^t + \mathrm{d}\phi^t = 0$, as $\mathrm{d}\xi^t = 0$ on M^t.

15.2.4 *Existence of torsion-free* Spin(7)*-structures on* M

Next we shall show that (Ω^t, g^t) can be deformed to a torsion-free Spin(7)-structure on M when t is small. We begin with an analogue of Theorem 13.5.8.

Proposition 15.2.13 *In the situation above, there exist constants $\lambda, \mu, \nu > 0$ such that for all $t \in (0, \epsilon]$ we have*

(i) $\|\phi^t\|_{L^2} \leq \lambda t^{24/5}$ *and* $\|\mathrm{d}\phi^t\|_{L^{10}} \leq \lambda t^{36/25}$;
(ii) *the injectivity radius $\delta(g^t)$ satisfies $\delta(g^t) \geq \mu t$; and*
(iii) *the Riemann curvature $R(g^t)$ satisfies $\|R(g^t)\|_{C^0} \leq \nu t^{-2}$.*

Here all norms are calculated using the metric g^t on M^t.

Proof Outside the overlaps $M_0^t \cap M_j^t$ for $1 \leqslant j \leqslant k$ we either have $\xi^t = \Omega^t = \Omega_Z$ or $\xi^t = \Omega^t = \Omega_{n_j}^t$. In both cases $\phi^t = \xi^t - \Omega^t = 0$, and so ϕ^t is zero outside the $M_0^t \cap M_j^t$. In $M_0^t \cap M_j^t$ we apply Proposition 10.5.9 with $\Omega = \Omega_0$, $\phi = \xi^t - \Omega_0$, $\Omega' = \Omega^t$ and $\phi' = \phi^t$. The proposition shows that

$$\big|\phi^t\big|_{g^t} \leqslant \epsilon_4 \big(|\pi_{27}(\xi^t - \Omega_0)|_{g_0} + |\xi^t - \Omega_0|_{g_0}^2\big) \text{ and } \big|\nabla^{g^t}\phi^t\big|_{g^t} \leqslant \epsilon_5 \big|\nabla^{g_0}(\xi^t - \Omega_0)\big|_{g_0}.$$

Combining this with (15.8) gives

$$\big|\phi^t\big|_{g^t} \leqslant \epsilon_4 \big(C_3 t^{8/5} + C_3^2 t^{16/5}\big) \text{ and } \big|d\phi^t\big|_{g^t} \leqslant \big|\nabla^{g^t}\phi^t\big|_{g^t} \leqslant \epsilon_5 C_3 t^{4/5}.$$

Thus $|\phi^t| \leqslant C_4 t^{8/5}$ and $|\nabla \phi^t| \leqslant C_4 t^{4/5}$ in $M_0^t \cap M_j^t$ for some $C_4 > 0$ independent of t. Now each $M_0^t \cap M_j^t$ is an annulus in $\mathbb{R}^8/G$ with inner radius $t^{4/5}\zeta$ and outer radius $2t^{4/5}\zeta$, and the metric g^t on $M_0^t \cap M_j^t$ is close to the flat metric g_0 on $\mathbb{R}^8/G$. Therefore we can find $C_5 > 0$ independent of t such that $\sum_{j=1}^k \text{vol}(M_0^t \cap M_j^t) \leqslant C_5 t^{32/5}$. Hence

$$\int_{M^t} |\phi^t|^2 dV \leqslant (C_4 t^{8/5})^2 C_5 t^{32/5} \text{ and } \int_{M^t} |d\phi^t|^{10} dV \leqslant (C_4 t^{4/5})^{10} C_5 t^{32/5}.$$

Taking roots gives part (i) of the proposition, with $\lambda = C_4 \max(C_5^{1/2}, C_5^{1/10})$. Parts (ii) and (iii) are proved as in Theorem 13.5.8. □

Finally we can prove our main result, the analogue of Theorem 13.6.2.

Theorem 15.2.14 *Suppose Condition* 15.2.1 *holds, and let M be the compact 8-manifold defined in Definition* 15.2.8. *Then there exist torsion-free* Spin(7)-*structures $(\tilde{\Omega}, \tilde{g})$ on M. If $\pi_1(M) = \{1\}$ then $\text{Hol}(\tilde{g}) = \text{Spin}(7)$, and if $\pi_1(M) = \mathbb{Z}_2$ then $\text{Hol}(\tilde{g}) = \mathbb{Z}_2 \ltimes \text{SU}(4)$.*

Proof Let λ, μ, ν be as in Proposition 15.2.13. Then Theorem 13.6.1 gives a constant $\kappa > 0$. Choose $t > 0$ with $t \leqslant \epsilon \leqslant 1$ and $t \leqslant \kappa$. Let (Ω, g) be the Spin(7)-structure (Ω^t, g^t) on $M = M^t$, and ϕ the 4-form ϕ^t. Then $d\Omega + d\phi = 0$ by Definition 15.2.12, and parts (i)–(iii) of Proposition 15.2.13 imply (i)–(iii) of Theorem 13.6.1, as $t \leqslant 1$.

Therefore all the hypotheses of Theorem 13.6.1 hold, and the theorem shows that there exists a torsion-free Spin(7)-structure $(\tilde{\Omega}, \tilde{g})$ on M. It remains to identify the holonomy group $\text{Hol}(\tilde{g})$ of $\tilde{g}$. We can regard the Spin(7)-orbifold (Z, Ω_Z, g_Z) as the limit as $t \to 0$ of the Spin(7)-manifolds $(M, \tilde{\Omega}, \tilde{g})$. Because of this, it is not difficult to see that $\text{Hol}(g_Z) \subseteq \text{Hol}(\tilde{g})$.

Now $\text{Hol}(g_Z) = \mathbb{Z}_2 \ltimes \text{SU}(4)$, and thus $\mathbb{Z}_2 \ltimes \text{SU}(4) \subseteq \text{Hol}(\tilde{g}) \subseteq \text{Spin}(7)$. If $\pi_1(M) = \{1\}$ then $\text{Hol}(\tilde{g})$ is connected. But the only connected Lie subgroup of Spin(7) containing $\mathbb{Z}_2 \ltimes \text{SU}(4)$ is Spin(7), so $\text{Hol}(\tilde{g}) = \text{Spin}(7)$. On the other hand, if $\pi_1(M) = \mathbb{Z}_2$ then $\text{Hol}(\tilde{g}) \neq \text{Spin}(7)$ by Proposition 10.6.2. This forces $\text{Hol}^0(\tilde{g}) = \text{SU}(4)$, and it is then easy to show that $\text{Hol}(\tilde{g}) = \mathbb{Z}_2 \ltimes \text{SU}(4)$. □

Since by Proposition 15.2.9 we can always choose the n_j so that M is simply-connected, we can arrange for $\tilde{g}$ to have holonomy Spin(7). When $\pi_1(M) = \mathbb{Z}_2$, the complex orbifold Y has a crepant resolution $\tilde{Y}$, which admits Kähler metrics $\tilde{g}$ with holonomy SU(4), making it into a Calabi–Yau manifold. The action of σ on Y lifts to a *free* action of σ on $\tilde{Y}$, and so $M = \tilde{Y}/\langle\sigma\rangle$ is a compact 8-manifold. If we choose $\tilde{g}$ to be σ-invariant then it pushes down to M, and has holonomy $\mathbb{Z}_2 \ltimes \mathrm{SU}(4)$.

15.3 How to apply the construction

To use the construction of §15.2 we need a compact complex 4-orbifold Y with $c_1(Y) = 0$ admitting Kähler metrics, and an antiholomorphic involution σ on Y, such that $Z = Y/\langle\sigma\rangle$ has finitely many singular points $p_1, \ldots, p_k$, all modelled on the singularity $\mathbb{R}^8/G$ of §15.1.1. We now describe the methods we shall use to find such Y and σ, and how to calculate the Betti numbers of the resulting 8-manifold M.

15.3.1 *Ways of constructing Calabi–Yau 4-orbifolds Y*

Recall that in §6.7.2 we discussed a construction of Calabi–Yau manifolds by Candelas et al. [55]. In it one starts with a hypersurface Y of degree d in a *weighted projective space* $\mathbb{CP}^m_{a_0,\ldots,a_m}$, a natural class of compact complex orbifolds given in Definition 6.5.4. If Y is defined by a *transverse* polynomial then it is a *complex orbifold*, with singularities at the intersection of Y and the singular set of $\mathbb{CP}^m_{a_0,\ldots,a_m}$. Furthermore, if $d = a_0 + \cdots + a_m$ then $c_1(Y) = 0$, so that Y is a *Calabi–Yau orbifold*, of dimension $m-1$.

We shall use this idea, and generalizations of it, to find the Calabi–Yau 4-orbifolds Y that we need for our construction of compact 8-manifolds with holonomy Spin(7). Here is the simplest class of orbifolds Y that we will consider, which are hypersurfaces in $\mathbb{CP}^5_{a_0,\ldots,a_5}$.

Example 15.3.1 Let $a_0, \ldots, a_5$ be positive integers with highest common factor $\mathrm{hcf}(a_0, \ldots, a_5) = 1$, and let $d = a_0 + \cdots + a_5$. Usually we order the a_j with $a_0 \leqslant a_1 \leqslant \cdots \leqslant a_5$. Suppose that a_j divides d for $j = 0, \ldots, 5$, and define $k_j = d/a_j$. Define a hypersurface Y in $\mathbb{CP}^5_{a_0,\ldots,a_5}$ by

$$Y = \{[z_0, \ldots, z_5] \in \mathbb{CP}^5_{a_0,\ldots,a_5} : z_0^{k_0} + \cdots + z_5^{k_5} = 0\}.$$

Since $a_j k_j = d$, we see that $z_0^{k_0} + \cdots + z_5^{k_5}$ is a *weighted homogeneous polynomial of degree d* in the sense of §6.7, and it is also *transverse*.

Therefore Y is a complex orbifold, with singularities only at the intersection of Y with the singular set of $\mathbb{CP}^5_{a_0,\ldots,a_5}$. Since the degree d of Y satisfies $d = a_0 + \cdots + a_5$, we have $c_1(Y) = 0$ as in §6.7. Also Y admits Kähler metrics, as $\mathbb{CP}^5_{a_0,\ldots,a_5}$ is Kähler. So Y is a compact complex orbifold with $c_1(Y) = 0$, admitting Kähler metrics.

Now to apply the construction of §15.2 the singular points of Y must satisfy Condition 15.2.1. Thus we need to be able to find and describe the singular set S of Y. But S is the intersection of Y with the singular set of $\mathbb{CP}^5_{a_0,\ldots,a_5}$. Let $[z_0, \ldots, z_5] \in Y$, and let n be the highest common factor of those a_i for which $z_i \neq 0$. Then from Definition 6.5.4, if $n = 1$ then $[z_0, \ldots, z_5]$ is a nonsingular point of $\mathbb{CP}^5_{a_0,\ldots,a_5}$, and so of Y, and if $n > 1$ then $[z_0, \ldots, z_5]$ is a singular point, with orbifold group $\mathbb{Z}_n$.

Suppose $I \subset \{0, \ldots, 5\}$ with $|I| = l \geqslant 2$ and $\mathrm{hcf}(a_i : i \in I) = n > 1$. Define

$$S_I = \{[z_0, \ldots, z_5] \in Y : z_i \neq 0 \text{ if and only if } i \in I\}.$$

Then S_I is a nonempty subset of Y of dimension $l - 2$, and each point of S_I is singular in Y with orbifold group $\mathbb{Z}_n$. Every singular point in Y lies in some S_I. The conditions on the a_i above imply that $l = 2, 3$ or 4, so that S_I has dimension 0,1 or 2. In particular, the condition for Y to have singularities only of dimension 0 is easily stated:

Lemma 15.3.2 *The orbifold Y of Example* 15.3.1 *has isolated singularities if and only if any three of $a_0, \ldots, a_5$ have highest common factor* 1.

The requirement that Y satisfy Condition 15.2.1 is a strong restriction on $a_0, \ldots, a_5$, which admits only a few solutions. However, we can get many more suitable orbifolds Y by generalizing our construction a bit. Here are four ways in which we can do this.

- **Defining Y by a different polynomial.** We could define Y using some more general transverse weighted homogenous polynomial of degree d in $z_0, \ldots, z_5$, instead of $z_0^{k_0} + \cdots + z_5^{k_5}$. The requirement that a_j divides d for $j = 0, \ldots, 5$ is then replaced by some other condition on the a_j and d.

- **Dividing by a finite group.** Let W be a Calabi–Yau hypersurface in $\mathbb{CP}^5_{a_0,\ldots,a_5}$, as in §15.3.1, and let G be a finite group acting on W preserving its Calabi–Yau structure. Then $Y = W/G$ is a Calabi–Yau orbifold.

- **Partial crepant resolutions.** Let W be a Calabi–Yau hypersurface in $\mathbb{CP}^5_{a_0,\ldots,a_5}$ which has some singularities of the kind we want, together with other singularities that we don't want. We let Y be a partial crepant resolution of W, which resolves the singularities that we don't want, leaving those that we do.

- **Complete intersections in $\mathbb{CP}^m_{a_0,\ldots,a_5}$.** Rather than a hypersurface in $\mathbb{CP}^5_{a_0,\ldots,a_5}$, we take Y to be a *complete intersection* in $\mathbb{CP}^m_{a_0,\ldots,a_m}$ for $m > 5$. Complete intersections were discussed in §6.7.

We can also use combinations of these four techniques — for instance, we can take Y to be a partial crepant resolution of W/G, where W is a hypersurface in $\mathbb{CP}^5_{a_0,\ldots,a_5}$, and G a finite group acting on W.

15.3.2 *Antiholomorphic involutions $\sigma : Y \to Y$*

Suppose we have chosen an orbifold Y as above, with isolated singular points $p_1, \ldots, p_k$. The next ingredient in our construction is an antiholomorphic involution $\sigma : Y \to Y$, which should fix only $p_1, \ldots, p_k$. For example, suppose Y is a hypersurface in $\mathbb{CP}^5_{a_0,\ldots,a_5}$. Then to find σ we would look for an antiholomorphic involution $\sigma : \mathbb{CP}^5_{a_0,\ldots,a_5} \to \mathbb{CP}^5_{a_0,\ldots,a_5}$ with $\sigma(Y) = Y$, and restrict σ to Y.

The most obvious such σ maps $[z_0, \ldots, z_5] \mapsto [\bar{z}_0, \ldots, \bar{z}_5]$. But this will not do, as its fixed points are not isolated in Y. To get isolated fixed points we need to try something more subtle. Here is an example of the kind of thing we mean.

Example 15.3.3 In the situation of Example 15.3.1, suppose $a_0, \ldots, a_3$ are odd and a_4, a_5 even with $a_0 = a_1$, $a_2 = a_3$ and $a_4 = a_5$. Define $\sigma : \mathbb{CP}^5_{a_0,\ldots,a_5} \to \mathbb{CP}^5_{a_0,\ldots,a_5}$ by

$$\sigma : [z_0, \ldots, z_5] \mapsto [\bar{z}_1, -\bar{z}_0, \bar{z}_3, -\bar{z}_2, \bar{z}_5, \bar{z}_4].$$

As σ swaps the pairs z_0, z_1 and z_2, z_3 and z_4, z_5, we need $a_0 = a_1$, $a_2 = a_3$ and $a_4 = a_5$ for σ to be well-defined. Clearly σ is antiholomorphic, and $\sigma(Y) = Y$.

Now σ^2 acts by

$$\sigma^2 : [z_0, \ldots, z_5] \mapsto [-z_0, -z_1, -z_2, -z_3, z_4, z_5].$$

But putting $u = -1$ in (6.4) gives $[-z_0, -z_1, -z_2, -z_3, z_4, z_5] = [z_0, \ldots, z_5]$, as $a_0, \ldots, a_3$ are odd and a_4, a_5 even. Thus $\sigma^2 = 1$, and $\sigma : Y \to Y$ is an antiholomorphic involution.

It is not difficult to show that the fixed points of σ in $\mathbb{CP}^5_{a_0,\ldots,a_5}$ are

$$\{[0, 0, 0, 0, 1, e^{i\theta}] \in \mathbb{CP}^5_{a_0,\ldots,a_5} : \theta \in [0, 2\pi)\}.$$

Now $[0, 0, 0, 0, 1, e^{i\theta}]$ lies in Y if $1 + e^{k_5 i\theta} = 0$. The solutions of this equation are hcf(k_4, k_5) distinct points in Y.

Observe the trick we have used here: if $a_j = a_{j+1}$ then we can choose σ to act on the coordinates z_j, z_{j+1} by $(z_j, z_{j+1}) \mapsto (\bar{z}_{j+1}, -\bar{z}_j)$. All the fixed points of σ will then satisfy $z_j = z_{j+1} = 0$. By doing this with two pairs of coordinates, say z_0, z_1 and z_2, z_3, the fixed points of σ satisfy $z_0 = z_1 = z_2 = z_3 = 0$. Thus they will be of complex codimension 4 in Y, and will be *isolated*, as we want.

This trick can also be adapted to more general situations, in which Y is a quotient by a finite group, or a partial crepant resolution, and so on. Note that as σ^2 maps $(z_j, z_{j+1}) \mapsto (-z_j, -z_{j+1})$, care must be taken to ensure that $\sigma^2 = 1$.

15.3.3 *Calculating the Euler characteristic of Y*

To determine the Betti numbers of the 8-manifold M that we construct, we will need to know the *Euler characteristic* of Y. Now there are two different notions of the Euler characteristic of an orbifold, defined by Satake [190, §3.3]. The version we are interested in is the *ordinary Euler characteristic* $\chi(Y)$, which is an integer and satisfies $\chi(Y) = \sum_{j=0}^{2n}(-1)^j b^j(Y)$. There is also the *orbifold Euler characteristic* $\chi_V(Y)$, which is a rational number that crops up naturally in problems involving characteristic classes.

In the following example we explain an elementary and fairly crude method for finding $\chi(Y)$ in the case that Y is a hypersurface in $\mathbb{CP}^m_{a_0,\ldots,a_m}$, of the kind considered in Example 15.3.1. It is also possible to calculate $\chi_V(Y)$ using Chern classes and get $\chi(Y)$ by adding on contributions from the singular set (see for instance Hosono et al. [106, §2]), but we will not discuss this.

Example 15.3.4 Let $a_0, \ldots, a_m, k_0, \ldots, k_m$ and d be positive integers with $a_j k_j = d$ for $j = 0, \ldots, m$. For each $j = 0, \ldots, m$, define $Y_j \subset \mathbb{CP}^j_{a_0,\ldots,a_j}$ by

$$Y_j = \{[z_0, \ldots, z_j] \in \mathbb{CP}^j_{a_0,\ldots,a_j} : z_0^{k_0} + \cdots + z_j^{k_j} = 0\},$$

and define $\pi_j : Y_j \to \mathbb{CP}^{j-1}_{a_0,\ldots,a_{j-1}}$ by $\pi_j : [z_0, \ldots, z_j] \mapsto [z_0, \ldots, z_{j-1}]$.

Suppose for simplicity that a_i divides a_j for $0 \leqslant i < j \leqslant m$. Then for each j, π_j is a k_j-fold branched cover of $\mathbb{CP}^{j-1}_{a_0,\ldots,a_{j-1}}$, branched over Y_{j-1}. That is, if $p \in \mathbb{CP}^{j-1}_{a_0,\ldots,a_{j-1}}$ then $\pi_j^{-1}(p)$ is one point when $p \in Y_{j-1}$ and k_j points when $p \notin Y_{j-1}$. It follows that

$$\begin{aligned} \chi(Y_j) &= k_j \cdot \chi(\mathbb{CP}^{j-1}_{a_0,\ldots,a_{j-1}}) + (1 - k_j)\chi(Y_{j-1}) \\ &= k_j j + (1 - k_j)\chi(Y_{j-1}), \end{aligned} \quad (15.9)$$

since $\chi(\mathbb{CP}^{j-1}_{a_0,\ldots,a_{j-1}}) = j$. This equation gives $\chi(Y_j)$ in terms of $\chi(Y_{j-1})$. Hence by induction we can write $\chi(Y_m)$ in terms of $\chi(Y_0)$. But $Y_0 = \emptyset$ so that $\chi(Y_0) = 0$, and thus we determine $\chi(Y_m)$.

If a_i does not divide a_j for some $0 \leqslant i < j \leqslant m$, then π_j is also branched over other parts of $\mathbb{CP}^{j-1}_{a_0,\ldots,a_{j-1}}$. Let $p = [z_0, \ldots, z_{j-1}]$ be in $\mathbb{CP}^{j-1}_{a_0,\ldots,a_{j-1}} \setminus Y_{j-1}$, and let I be the set of i in $\{0, \ldots, j-1\}$ for which $z_i \neq 0$. Define $l = \mathrm{hcf}(a_i : i \in I)$ and $m = \mathrm{hcf}(l, a_j)$. Then it turns out that $\pi_j^{-1}(p)$ is $k_j m/l$ points in Y_j. Clearly $k_j m/l = k_j$ if $l = m$, that is, if l divides a_j.

Thus π_j is also branched over subsets of $\mathbb{CP}^{j-1}_{a_0,\ldots,a_{j-1}} \setminus Y_{j-1}$ corresponding to subsets $I \subseteq \{0, \ldots, j-1\}$ for which $l = \mathrm{hcf}(a_i : i \in I)$ does not divide a_j. To calculate $\chi(Y_j)$ in this case we must modify (15.9) by adding in contributions from each such I. We will explain this when we meet it in examples later.

15.3.4 *How to find topological invariants of Y, Z and M*

Here is a procedure for calculating the fundamental group and Betti numbers of Y, Z and M. The most difficult part is finding the Euler characteristic $\chi(Y)$, which we have already explained above.

(a) Calculate $\pi_1(Y)$, $H^2(Y, \mathbb{C})$ and $H^3(Y, \mathbb{C})$ explicitly. This can be done using the *Lefschetz Hyperplane Theorem*, Theorem 4.10.4. If Y is a hypersurface in $\mathbb{CP}^5_{a_0,\ldots,a_5}$ then $\pi_1(Y) = \{1\}$, $H^2(Y, \mathbb{C}) = \mathbb{C}$ and $H^3(Y, \mathbb{C}) = 0$. Verify that $\pi_1(Y \setminus \{p_1, \ldots, p_k\}) = \{1\}$ and $h^{2,0}(Y) = 0$, as in Condition 15.2.1.

(b) Compute the Euler characteristic $\chi(Y)$ of Y, as in §15.3.3.

(c) Calculate $H^2(Z, \mathbb{C})$ and $H^3(Z, \mathbb{C})$ explicitly, from $H^2(Y, \mathbb{C})$ and $H^3(Y, \mathbb{C})$. Note that $H^j(Z, \mathbb{C})$ is the σ-invariant part of $H^j(Y, \mathbb{C})$. Since σ swaps $H^{p,q}(Y)$ and $H^{q,p}(Y)$, it follows that $b^3(Z) = \frac{1}{2}b^3(Y)$.

(d) Compute the Euler characteristic $\chi(Z)$ of Z. If σ fixes k points in Y then this is given by $\chi(Z) = \frac{1}{2}(\chi(Y) + k)$.

(e) From (c) we know $b^2(Z)$ and $b^3(Z)$, and $b^1(Z) = 0$ as $\pi_1(Z)$ is finite. Thus we can calculate $b^4(Z)$ using the formula $b^4(Z) = \chi(Z) - 2 - 2b^2(Z) + 2b^3(Z)$.

(f) Now M was constructed in §15.2 by gluing $X_{n_1}, \ldots, X_{n_k}$ into Z, where $n_j = 1$ or 2 and X_1, X_2 are defined in §15.1.1–§15.1.2. It is easy to show that the Betti numbers of X_1 and X_2 are $b^1 = b^2 = b^3 = 0$ and $b^4 = 1$. Therefore the Betti numbers $b^j(M)$ satisfy

$$b^j(M) = b^j(Z) \text{ for } j = 1, 2, 3, \text{ and } b^4(M) = b^4(Z) + k. \tag{15.10}$$

Also, Proposition 15.2.9 gives $\pi_1(M)$.

(g) As M has metrics with holonomy Spin(7) or $\mathbb{Z}_2 \ltimes \text{SU}(4)$ by Theorem 15.2.14, we know that $\hat{A}(M) = 1$. Thus (10.24) gives

$$b^2(M) - b^3(M) - b^4_+(M) + 2b^4_-(M) + 25 = 0.$$

So we can calculate $b^4_\pm(M)$ using the equations

$$\begin{aligned} b^4_+(M) &= \tfrac{1}{3}\bigl(b^2(M) - b^3(M) + 2b^4(M) + 25\bigr) \\ \text{and} \quad b^4_-(M) &= \tfrac{1}{3}\bigl(-b^2(M) + b^3(M) + b^4(M) - 25\bigr). \end{aligned} \tag{15.11}$$

15.3.5 *A way of checking the answers*

If you make a mistake at some stage in these calculations, which is quite easy to do, then you are likely not to notice unless your values for $b^4_\pm(M)$ are not integers. Thus it is desirable to have some method for checking the answers. Here is a way of doing this. All of our examples have been checked for consistency in this way and others, but for brevity we will leave out the calculations.

Suppose we can compute the Hodge number $h^{3,1}(Y)$, using complex geometry. Then we can compute $b^4_-(Z)$ using the formula

$$b^4_-(Z) = h^{3,1}(Y) + b^2(Y) - b^2(Z) - 1.$$

But since X_1 and X_2 have $b^4_- = 1$, as in (15.10) we have $b^4_-(M) = b^4_-(Z) + k$. This gives an independent way of finding $b^4_-(M)$, which can be compared with your answer in part (g) above.

Now there is a complicated method for computing $h^{3,1}(Y)$ involving spectral sequences, and also a much simpler method called the 'polynomial deformation method' which *does not always give the right answer*. Both are discussed by Green and Hübsch [88]. Here is a sketch of the polynomial deformation method.

Suppose for simplicity that Y is a hypersurface of degree d in $\mathbb{CP}^5_{a_0,\ldots,a_5}$. As Y is a Calabi–Yau orbifold, $h^{3,1}(Y)$ is the dimension of the moduli space of complex structures on Y. We assume (this is *not* necessarily true) that every small deformation of Y is also a hypersurface of degree d in $\mathbb{CP}^5_{a_0,\ldots,a_5}$, and that two nearby isomorphic hypersurfaces Y, Y' of degree d are related by an automorphism of $\mathbb{CP}^5_{a_0,\ldots,a_5}$.

If these assumptions hold, then $h^{3,1}(Y) = m - n$, where m is the dimension of the space of hypersurfaces of degree d in $\mathbb{CP}^5_{a_0,\ldots,a_5}$, and n is the dimension of the automorphism group of $\mathbb{CP}^5_{a_0,\ldots,a_5}$. Both m and n are readily computed from $a_0, \ldots, a_5$ and d.

15.4 A simple example

Let Y be the hypersurface of degree 12 in $\mathbb{CP}^5_{1,1,1,1,4,4}$ given by

$$Y = \{[z_0, \ldots, z_5] \in \mathbb{CP}^5_{1,1,1,1,4,4} : z_0^{12} + z_1^{12} + z_2^{12} + z_3^{12} + z_4^3 + z_5^3 = 0\}.$$

Then $c_1(Y) = 0$, as $12 = 1 + 1 + 1 + 1 + 4 + 4$, and Y is Kähler as $\mathbb{CP}^5_{1,\ldots,4}$ is Kähler. Calculation shows that Y has three singular points $p_1 = [0, 0, 0, 0, 1, -1]$, $p_2 = [0, 0, 0, 0, 1, e^{\pi i/3}]$ and $p_3 = [0, 0, 0, 0, 1, e^{-\pi i/3}]$, satisfying Condition 15.2.1.

We use the method of §15.3.3 to calculate the Euler characteristic $\chi(Y)$.

Proposition 15.4.1 *The orbifold Y defined above has $\chi(Y) = 4887$.*

Proof Define Y_j and π_j as in §15.3.3, where $Y_5 = Y$. Then Y_1 is the set of 12 points $[z_0, z_1]$ in $\mathbb{CP}^1$ with $z_0^{12} + z_1^{12} = 0$, and so $\chi(Y_1) = 12$. Now $\pi_2 : Y_2 \to \mathbb{CP}^1$ is a 12-fold branched cover branched over Y_1, so by (15.9) we have

$$\chi(Y_2) = 12\chi(\mathbb{CP}^1) - 11\chi(Y_1) = 12 \cdot 2 - 11 \cdot 12 = -108.$$

Similarly, $\pi_3 : Y_3 \to \mathbb{CP}^2$ is a 12-fold branched cover branched over Y_2, so that

$$\chi(Y_3) = 12\chi(\mathbb{CP}^2) - 11\chi(Y_2) = 12 \cdot 2 - 11 \cdot (-108) = 1224.$$

And $\pi_4 : Y_4 \to \mathbb{CP}^3$ is a 3-fold branched cover of $\mathbb{CP}^3$ branched over Y_3, giving

$$\chi(Y_4) = 3\chi(\mathbb{CP}^3) - 2\chi(Y_3) = 3 \cdot 4 - 2 \cdot 1224 = -2436.$$

Finally, $\pi_5 : Y \to \mathbb{CP}^4_{1,1,1,1,4}$ is a 3-fold branched cover of $\mathbb{CP}^4_{1,1,1,1,4}$ branched over Y_4, and so

$$\chi(Y) = 3\chi(\mathbb{CP}^4_{1,1,1,1,4}) - 2\chi(Y_4) = 3 \cdot 5 - 2 \cdot (-2436) = 4887,$$

as we want. □

Proposition 15.4.2 *The Betti numbers of Y are*

$$b^0(Y) = 1, \ b^1(Y) = 0, \ b^2(Y) = 1, \ b^3(Y) = 0 \ \text{and} \ b^4(Y) = 4883.$$

Also $Y \setminus \{p_1, p_2, p_3\}$ is simply-connected and $h^{2,0}(Y) = 0$.

Proof The Lefschetz Hyperplane Theorem for orbifolds, Theorem 4.10.4, shows that $H^k(Y, \mathbb{C}) \cong H^k(\mathbb{CP}^5_{1,\ldots,4}, \mathbb{C})$ for $0 \leq k \leq 3$. Since $b^k(\mathbb{CP}^5_{1,\ldots,4})$ is 1 for k even with $0 \leq k \leq 10$ and 0 otherwise, this shows that $b^0(Y) = b^2(Y) = 1$ and $b^1(Y) = b^3(Y) = 0$, and so $b^4(Y) = 4883$ as $\chi(Y) = 4887$.

Theorem 4.10.4 also gives $\pi_1(Y) \cong \pi_1(\mathbb{CP}^5_{1,\ldots,4})$, so Y is simply-connected. As the nonsingular set of $\mathbb{CP}^5_{1,\ldots,4}$ is simply-connected, we can strengthen this to show that $Y \setminus \{p_1, p_2, p_3\}$ is simply-connected. The isomorphism $H^k(Y, \mathbb{C}) \cong H^k(\mathbb{CP}^5_{1,\ldots,4}, \mathbb{C})$ above identifies $H^{p,q}(Y)$ with $H^{p,q}(\mathbb{CP}^5_{1,\ldots,4})$, and so $h^{p,q}(Y) = h^{p,q}(\mathbb{CP}^5_{1,\ldots,4})$ for $p + q \leq 3$. Hence $h^{2,0}(Y) = 0$. □

Now define a map $\sigma : Y \to Y$ by

$$\sigma : [z_0, \ldots, z_5] \longmapsto [\bar{z}_1, -\bar{z}_0, \bar{z}_3, -\bar{z}_2, \bar{z}_5, \bar{z}_4].$$

As in Example 15.3.3, we find that σ is an antiholomorphic involution of Y, and that the fixed points of σ are exactly p_1, p_2, p_3. Thus Condition 15.2.1 holds for Y and σ. So we can apply the construction of §15.2, and resolve the orbifold $Z = Y/\langle\sigma\rangle$ to get a compact 8-manifold M. Choosing $n_j = 2$ for at least one $j = 1, 2, 3$, Proposition 15.2.9 shows that M is simply-connected, and Theorem 15.2.14 shows that M admits metrics with holonomy Spin(7).

Theorem 15.4.3 *This compact* 8*-manifold M has Betti numbers*

$$b^0 = 1, \quad b^1 = b^2 = b^3 = 0, \quad b^4 = 2446, \quad b^4_+ = 1639 \text{ and } b^4_- = 807.$$

There exist metrics with holonomy Spin(7) *on M, which form a smooth family of dimension* 808.

Proof We first calculate the Betti numbers of Z. As σ fixes 3 points in Y, by properties of the Euler characteristic we find that $\chi(Z) = \frac{1}{2}(\chi(Y) + 3)$. But $\chi(Y) = 4887$ by Proposition 15.4.1, so $\chi(Z) = 2445$. As $H^k(Z, \mathbb{C})$ is the σ-invariant part of $H^k(Y, \mathbb{C})$ we see from Proposition 15.4.2 that $b^0(Z) = 1$ and $b^1(Z) = b^3(Z) = 0$. Also $H^2(Y, \mathbb{C})$ is generated by $[\omega_Y]$ and $\sigma^*(\omega_Y) = -\omega_Y$, so σ acts as -1 on $H^2(Y, \mathbb{C})$, and $H^2(Z, \mathbb{C}) = 0$.

Thus $b^0(Z) = 1$, $b^1(Z) = b^2(Z) = b^3(Z) = 0$ and $\chi(Z) = 2445$, giving $b^4(Z) = 2443$. Equation (15.10) then gives the Betti numbers of M, and (15.11) gives $b^4_\pm$. Theorem 15.2.14 shows that there exist torsion-free Spin(7)-structures $(\tilde{\Omega}, \tilde{g})$ on M, with $\text{Hol}(\tilde{g}) = \text{Spin}(7)$ as M is simply-connected. By Theorem 10.7.1 the moduli space of metrics on M with holonomy Spin(7) is a smooth manifold of dimension $1 + b^4_-(M) = 808$. □

15.4.1 *A variation on this example*

Here is a variation on the above, using the idea of *partial crepant resolution* mentioned in §15.3.1. Let Y be as above, but define $\sigma' : Y \to Y$ by

$$\sigma' : [z_0, \ldots, z_5] \longmapsto [\bar{z}_1, -\bar{z}_0, \bar{z}_3, -\bar{z}_2, \bar{z}_4, \bar{z}_5].$$

Then σ' is an antiholomorphic involution of Y, which fixes the singular point $p_1 = [0, 0, 0, 0, 1, -1]$ in Y, and no other points. In particular, σ' swaps over the other two singular points p_2, p_3.

Thus Y and σ' do not satisfy Condition 15.2.1, because the fixed set of σ' is not the same as the singular set $\{p_1, p_2, p_3\}$ of Y. To rectify this we resolve the singular points p_2, p_3. Let Y' be the blow-up of Y at p_2 and p_3. This is a partial crepant resolution of Y, and so is also a Calabi–Yau orbifold.

Then Y' has just the one singular point p_1. The action of σ' on Y lifts to Y', with sole fixed point p_1. Thus Condition 15.2.1 holds for Y' and σ'. Therefore we can apply the construction of §15.2 to Y' and σ', so that $Z' = Y'/\langle\sigma'\rangle$ is a compact Spin(7)-orbifold with one singular point p_1 modelled on $\mathbb{R}^8/G$. Choosing $n_1 = 2$ we get a resolution M' of Z', which is a compact, simply-connected 8-manifold admitting metrics with holonomy Spin(7).

We shall calculate the topological invariants of Y' and M'.

Proposition 15.4.4 *The Betti numbers of Y' are*

$$b^0 = 1, \quad b^1 = 0, \quad b^2 = 3, \quad b^3 = 0 \text{ and } b^4 = 4885, \text{ so that } \chi(Y') = 4893.$$

Also, $Y' \setminus \{p_1\}$ is simply-connected and $h^{2,0}(Y') = 0$.

Proof By definition Y' is the blow-up of Y at p_2, p_3. Each blow-up fixes b^1 and b^3 and adds 1 to b^2 and b^4. So the Betti numbers of Y' follow from Proposition 15.4.2. As $Y \setminus \{p_1, p_2, p_3\}$ is simply-connected and $h^{2,0}(Y) = 0$, we see that $Y' \setminus \{p_1\}$ is simply-connected and $h^{2,0}(Y') = 0$. □

Here is the analogue of Theorem 15.4.3:

Theorem 15.4.5 *This compact 8-manifold M' has Betti numbers*

$$b^0 = 1, \quad b^1 = 0, \quad b^2 = 1, \quad b^3 = 0, \quad b^4 = 2444, \quad b^4_+ = 1638 \text{ and } b^4_- = 806.$$

There exist metrics with holonomy Spin(7) on M', which form a smooth family of dimension 807.

Proof As σ fixes 1 point in Y' we have $\chi(Z') = \frac{1}{2}(\chi(Y') + 1)$, so $\chi(Z') = 2447$ by the previous proposition. Since $H^k(Z', \mathbb{C})$ is the σ-invariant part of $H^k(Y', \mathbb{C})$ we have $b^0(Z') = 1$ and $b^1(Z') = b^3(Z') = 0$. Now $b^2(Y') = 3$, and $H^2(Y', \mathbb{C})$ is generated by $[\omega_{Y'}]$ and the cohomology classes dual to the two exceptional divisors $\mathbb{CP}^3$ introduced by blowing up p_2 and p_3. But σ' swaps p_2 and p_3, so σ'_* swaps the corresponding classes in $H^2(Y', \mathbb{C})$, and $\sigma'_*(\omega_{Y'}) = -\omega_{Y'}$ by definition. Therefore $H^2(Y', \mathbb{C}) \cong \mathbb{C} \oplus \mathbb{C}^2$, where σ'_* acts as 1 on $\mathbb{C}$ and -1 on $\mathbb{C}^2$. Hence $H^2(Z', \mathbb{C}) \cong \mathbb{C}$, and $b^2(Z') = 1$.

Thus $b^0(Z') = b^2(Z') = 1$, $b^1(Z') = b^3(Z') = 0$ and $\chi(Z') = 2447$, giving $b^4(Z') = 2443$. Equation (15.10) then gives the Betti numbers of M, and (15.11) gives $b^4_\pm$. Theorem 15.2.14 shows that there exist torsion-free Spin(7)-structures $(\tilde{\Omega}, \tilde{g})$ on M, with $\text{Hol}(\tilde{g}) = \text{Spin}(7)$ as M is simply-connected. By Theorem 10.7.1 the moduli space of metrics on M with holonomy Spin(7) is a smooth manifold of dimension $1 + b^4_-(M) = 807$. □

Observe that the Betti numbers of M and M' in Theorems 15.4.3 and 15.4.5 are very similar. It is an interesting question whether one can regard M and M' as two different resolutions of some singular Spin(7)-manifold M_0, not necessarily an orbifold. We leave this as a research exercise for the reader; the answer is not as simple as it looks.

15.5 Examples from hypersurfaces in $\mathbb{CP}^5_{a_0,\ldots,a_5}$

Here are three more examples based on hypersurfaces in $\mathbb{CP}^5_{a_0,\ldots,a_5}$.

15.5.1 A hypersurface of degree 16 in $\mathbb{CP}^5_{1,1,1,1,4,8}$

Let Y be the hypersurface of degree 16 in $\mathbb{CP}^5_{1,1,1,1,4,8}$ given by

$$Y = \{[z_0, \ldots, z_5] \in \mathbb{CP}^5_{1,1,1,1,4,8} : z_0^{16} + z_1^{16} + z_2^{16} + z_3^{16} + z_4^4 + z_5^2 = 0\}.$$

Then $c_1(Y) = 0$. We find that Y has two singular points $p_1 = [0, 0, 0, 0, 1, i]$ and $p_2 = [0, 0, 0, 0, 1, -i]$, both satisfying Condition 15.2.1.

Following Propositions 15.4.1 and 15.4.2, we find that $\chi(Y) = 9498$, and

Proposition 15.5.1 *The Betti numbers of Y are*

$$b^0 = 1, \quad b^1 = 0, \quad b^2 = 1, \quad b^3 = 0 \quad \text{and} \quad b^4 = 9494.$$

Also $Y \setminus \{p_1, p_2\}$ is simply-connected and $h^{2,0}(Y) = 0$.

Define an antiholomorphic involution $\sigma : Y \to Y$ by

$$\sigma : [z_0, \ldots, z_5] \longmapsto [\bar{z}_1, -\bar{z}_0, \bar{z}_3, -\bar{z}_2, \bar{z}_4, -\bar{z}_5].$$

The fixed points of σ are exactly the singular points p_1, p_2 of Y. Thus Condition 15.2.1 holds for Y and σ, and we can apply the construction of §15.2. Resolving $Z = Y/\langle\sigma\rangle$ gives a compact 8-manifold M. We choose at least one of n_1, n_2 to be 2, so that M is simply-connected. Then as in Theorem 15.4.3, we get:

Theorem 15.5.2 *This compact 8-manifold M has Betti numbers*

$$b^0 = 1, \quad b^1 = b^2 = b^3 = 0, \quad b^4 = 4750, \quad b^4_+ = 3175 \quad \text{and} \quad b^4_- = 1575.$$

There exist metrics with holonomy Spin(7) *on M, which form a smooth family of dimension* 1576.

15.5.2 A hypersurface of degree 24 in $\mathbb{CP}^5_{1,1,1,1,8,12}$

Let Y be the hypersurface of degree 24 in $\mathbb{CP}^5_{1,1,1,1,8,12}$ given by

$$Y = \{[z_0, \ldots, z_5] \in \mathbb{CP}^5_{1,1,1,1,8,12} : z_0^{24} + z_1^{24} + z_2^{24} + z_3^{24} + z_4^3 + z_5^2 = 0\}.$$

Then $c_1(Y) = 0$. We find that Y has one singular point $p_1 = [0, 0, 0, 0, -1, 1]$, which satisfies Condition 15.2.1.

Following Proposition 15.4.1, we find that $\chi(Y) = 23\,325$. Care is needed to get the right answer here. Define $\pi_5 : Y \to \mathbb{CP}^4_{1,1,1,1,8}$ by $\pi_5 : [z_0, \ldots, z_5] \mapsto [z_0, \ldots, z_4]$, and $Y_4 \subset \mathbb{CP}^4_{1,1,1,1,8}$ by

$$Y_4 = \{[z_0, \ldots, z_4] \in \mathbb{CP}^4_{1,1,1,1,8} : z_0^{24} + z_1^{24} + z_2^{24} + z_3^{24} + z_4^3 = 0\}.$$

Then π_5 is a double cover of $\mathbb{CP}^4_{1,1,1,1,8}$ branched over Y_4 *and the point* $[0, 0, 0, 0, 1]$ in $\mathbb{CP}^4_{1,1,1,1,8}$. Hence we get

$$\chi(Y) = 2\chi(\mathbb{CP}^4_{1,1,1,1,8}) - \chi(Y_4) - \chi([0, 0, 0, 0, 1]) = 9 - \chi(Y_4).$$

If we had not observed that π_5 is also branched over $[0, 0, 0, 0, 1]$ then we would have got $\chi(Y) = 23\,326$, which is incorrect.

As in Proposition 15.4.2, we show:

Proposition 15.5.3 *The Betti numbers of Y are*

$$b^0 = 1, \quad b^1 = 0, \quad b^2 = 1, \quad b^3 = 0 \quad \text{and} \quad b^4 = 23\,231.$$

Also $Y \setminus \{p_1\}$ is simply-connected and $h^{2,0}(Y) = 0$.

Define an antiholomorphic involution $\sigma : Y \to Y$ by

$$\sigma : [z_0, \ldots, z_5] \longmapsto [\bar{z}_1, -\bar{z}_0, \bar{z}_3, -\bar{z}_2, \bar{z}_4, \bar{z}_5].$$

The fixed points of σ are exactly the singular point p_1 of Y. Thus Condition 15.2.1 holds for Y and σ, and choosing the simply-connected resolution M of $Z = Y/\langle\sigma\rangle$, in the usual way we get:

Theorem 15.5.4 *This compact 8-manifold M has Betti numbers*

$$b^0 = 1, \quad b^1 = b^2 = b^3 = 0, \quad b^4 = 11\,662, \quad b^4_+ = 7783 \quad \text{and} \quad b^4_- = 3879.$$

There exist metrics with holonomy Spin(7) *on M, which form a smooth family of dimension 3880.*

This is the example with the largest value of b^4 known to the author.

15.5.3 A hypersurface of degree 40 in $\mathbb{CP}^5_{1,1,5,5,8,20}$

Here is a more complicated example, in which the hypersurface in $\mathbb{CP}^5_{a_0,\ldots,a_5}$ has other singularities which must first be resolved. Let W be the hypersurface of degree 40 in $\mathbb{CP}^5_{1,1,5,5,8,20}$ given by

$$W = \{[z_0, \ldots, z_5] \in \mathbb{CP}^5_{1,1,5,5,8,20} : z_0^{40} + z_1^{40} + z_2^8 + z_3^8 + z_4^5 + z_5^2 = 0\}.$$

Then $c_1(W) = 0$. The singularities of W are the disjoint union of the single point $p_1 = [0, 0, 0, 0, -1, 1]$ and the nonsingular curve Σ of genus 3 given by

$$\Sigma = \{[0, 0, z_2, z_3, 0, z_5] \in \mathbb{CP}^5_{1,1,5,5,8,20} : z_2^8 + z_3^8 + z_5^2 = 0\}.$$

The singularity at p_1 satisfies Condition 15.2.1. The singularity at each point of Σ is modelled on $\mathbb{C} \times \mathbb{C}^3/\mathbb{Z}_5$, where the generator β of $\mathbb{Z}_5$ acts on $\mathbb{C}^3$ by

$$\beta : (z_0, z_1, z_4) \mapsto (e^{2\pi i/5}z_0, e^{2\pi i/5}z_1, e^{-4\pi i/5}z_4).$$

Now the singularity $\mathbb{C}^3/\mathbb{Z}_5$ normal to Σ in W has a unique crepant resolution X, which can be described using toric geometry. Let Y be the partial crepant resolution of W which resolves the singularities at Σ using X, but leaves the singular point p_1 unchanged.

Proposition 15.5.5 *The Betti numbers of Y are*

$$b^0 = 1, \quad b^1 = 0, \quad b^2 = 3, \quad b^3 = 12, \quad \text{and} \quad b^4 = 7453.$$

Also $Y \setminus \{p_1\}$ is simply-connected and $h^{2,0}(Y) = 0$.

Proof Calculating the Betti numbers of W in the usual way gives

$$b^0(W) = 1, \quad b^1(W) = 0, \quad b^2(W) = 1, \quad b^3(W) = 0, \quad b^4(W) = 7449. \tag{15.12}$$

As W is modelled on $\mathbb{C} \times \mathbb{C}^3/\mathbb{Z}_5$ at each point of Σ, the resolution Y is modelled on $\mathbb{C} \times X$. Since $b^2(X) = b^4(X) = 2$, the Betti numbers of Y satisfy

$$b^k(Y) = b^k(W) + 2b^{k-2}(\Sigma) + 2b^{k-4}(\Sigma).$$

But Σ has genus 3, and so its Betti numbers are $b^0(\Sigma) = b^2(\Sigma) = 1$ and $b^1(\Sigma) = 6$. Combining this with (15.12) gives the Betti numbers of Y. The last part follows as in Proposition 15.4.2. □

Define $\sigma : W \to W$ by

$$\sigma : [z_0, \ldots, z_5] \mapsto [\bar{z}_1, -\bar{z}_0, \bar{z}_3, -\bar{z}_2, \bar{z}_4, \bar{z}_5].$$

The only fixed point of σ is p_1. Moreover, σ lifts to the resolution Y of W, and $\sigma : Y \to Y$ is an antiholomorphic involution which fixes only p_1 in Y. Thus Condition 15.2.1 holds for Y and σ, and we can apply the construction of §15.2, and resolve $Z = Y/\langle\sigma\rangle$ to get a simply-connected 8-manifold M. Proceeding in the usual way, the end result is

Theorem 15.5.6 *This compact 8-manifold M has Betti numbers*

$$b^0 = 1, \quad b^1 = b^2 = 0, \quad b^3 = 6, \quad b^4 = 3730, \quad b^4_+ = 2493 \text{ and } b^4_- = 1237.$$

There exist metrics with holonomy $\text{Spin}(7)$ *on M, which form a smooth family of dimension* 1238.

Note that $b^3 > 0$ in this example; this is because the resolution of the singular curve Σ contributes $H^1(\Sigma, \mathbb{C}) \otimes H^2(X, \mathbb{C}) = \mathbb{C}^6 \otimes \mathbb{C}^2 = \mathbb{C}^{12}$ to $H^3(Y, \mathbb{C})$. Half of this $\mathbb{C}^{12}$ is σ-invariant, and so pushes down to $H^3(Z, \mathbb{C})$ and lifts to $H^3(M, \mathbb{C})$.

15.6 A hypersurface of degree 8 in $\mathbb{CP}^5_{1,1,1,1,2,2}$ over $\mathbb{Z}_2$

Let W be the hypersurface of degree 8 in $\mathbb{CP}^5_{1,1,1,1,2,2}$ given by

$$W = \{[z_0, \ldots, z_5] \in \mathbb{CP}^5_{1,1,1,1,2,2} : z_0^8 + z_1^8 + z_2^8 + z_3^8 + z_4^4 + z_5^4 = 0\}.$$

Then $c_1(W) = 0$. We find that W has four singular points $p_1, \ldots, p_4$ modelled on $\mathbb{C}^4/\{\pm 1\}$, given by

$$[0, 0, 0, 0, 1, e^{\pi i/4}], \quad [0, 0, 0, 0, 1, e^{3\pi i/4}], \quad [0, 0, 0, 0, 1, e^{5\pi i/4}], \quad [0, 0, 0, 0, 1, e^{7\pi i/4}].$$

Define $\beta : W \to W$ by

$$\beta : [z_0, \ldots, z_5] \mapsto [iz_0, iz_1, iz_2, iz_3, z_4, z_5].$$

Then $\beta^2 = 1$, as $[z_0, \ldots, z_5] = [-z_0, -z_1, -z_2, -z_3, z_4, z_5]$ in $\mathbb{CP}^5_{1,1,1,1,2,2}$. The fixed set of β is the four points $p_1, \ldots, p_4$ together with the compact complex surface S in W, given by

$$S = \{[z_0, z_1, z_2, z_3, 0, 0] \in \mathbb{CP}^5_{1,1,1,1,2,2} : z_0^8 + z_1^8 + z_2^8 + z_3^8 = 0\}.$$

Thus $W/\langle\beta\rangle$ is a compact complex orbifold. Its singular set is the disjoint union of $p_1, \ldots, p_4$ and S. Each singular point p_j is modelled on $\mathbb{C}^4/\mathbb{Z}_4$, where the generator α of $\mathbb{Z}_4$ acts on $\mathbb{C}^4$ by (15.3). Each singular point in S is locally modelled on $\mathbb{C}^2 \times \mathbb{C}^2/\{\pm 1\}$.

Let Y be the blow-up of $W/\langle\beta\rangle$ along S. Because the singularities normal to S are modelled on $\mathbb{C}^2/\{\pm 1\}$, this is a partial crepant resolution. So Y is a compact complex orbifold with isolated singular points $p_1, \ldots, p_4$, modelled on $\mathbb{C}^4/\langle\alpha\rangle$. Now $c_1(W) = 0$, so $c_1(W/\langle\beta\rangle) = 0$, and as Y is a partial crepant resolution of $W/\langle\beta\rangle$ we see that $c_1(Y) = 0$.

Proposition 15.6.1 *The Betti numbers of Y are*

$$b^0 = 1, \quad b^1 = 0, \quad b^2 = 2, \quad b^3 = 0 \quad \text{and} \quad b^4 = 1806.$$

Also $Y \setminus \{p_1, \ldots, p_4\}$ is simply-connected and $h^{2,0}(Y) = 0$.

Proof As in Proposition 15.4.1, we find $\chi(W) = 2708$ and $\chi(S) = 304$. Thus

$$\chi(W/\langle\beta\rangle) = \tfrac{1}{2}\bigl(\chi(W) + \chi(4 \text{ points}) + \chi(S)\bigr) = \tfrac{1}{2}(2708 + 4 + 304) = 1508.$$

By the usual Lefschetz Hyperplane Theorem argument we find that W has $b^0 = b^2 = 1$ and $b^1 = b^3 = 0$, and it soon follows that $W/\langle\beta\rangle$ also has $b^0 = b^2 = 1$ and $b^1 = b^3 = 0$. Since $\chi(W/\langle\beta\rangle) = 1508$ we see that $b^4(W/\langle\beta\rangle) = 1504$.

Now Y is the blow-up of $W/\langle\beta\rangle$ along S, so that each point of S is replaced by a copy of $\mathbb{CP}^1$. It can be shown that the Betti numbers of Y satisfy

$$b^k(Y) = b^k(W/\langle\beta\rangle) + b^{k-2}(S). \tag{15.13}$$

But S can be thought of as an octic in $\mathbb{CP}^3$, and by the usual method we find that the Betti numbers of S are $b^0 = 1, b^1 = 0, b^2 = 302, b^3 = 0$ and $b^4 = 1$. Combining these with (15.13) and the Betti numbers of $W/\langle\beta\rangle$ above gives the Betti numbers of Y. The last part follows as usual. $\square$

Define an antiholomorphic involution $\sigma : W \to W$ by

$$\sigma : [z_0, \ldots, z_5] \longmapsto [\bar{z}_1, -\bar{z}_0, \bar{z}_3, -\bar{z}_2, \bar{z}_5, \bar{z}_4].$$

The fixed points of σ are exactly the singular points $p_1, \ldots, p_4$ of W. Also σ commutes with β, and acts freely on S. Hence σ pushes down to an antiholomorphic involution of $W/\langle\beta\rangle$, and lifts to the blow-up Y, to give an antiholomorphic involution $\sigma : Y \to Y$ with fixed points $p_1, \ldots, p_4$.

Thus Condition 15.2.1 holds for Y and σ, and in the usual way we choose a simply-connected resolution M of $Z = Y/\langle\sigma\rangle$ satisfying:

Theorem 15.6.2 *This compact 8-manifold M has Betti numbers*

$$b^0 = 1, \ b^1 = b^2 = b^3 = 0, \ b^4 = 910, \ b^4_+ = 615 \text{ and } b^4_- = 295.$$

There exist metrics with holonomy Spin(7) *on M, which form a smooth family of dimension* 296.

15.6.1 *A variation on this example*

We shall use the idea of §15.4.1 to make a second 8-manifold M' from the orbifold Y above. Let W and Y be as in §15.6.1, but define $\sigma' : W \to W$ by

$$\sigma' : [z_0, \ldots, z_5] \longmapsto [\bar{z}_1, -\bar{z}_0, \bar{z}_3, -\bar{z}_2, \bar{z}_4, i\bar{z}_5].$$

Then σ' pushes down to $W/\langle\beta\rangle$ and lifts to Y as above. However, this time σ' fixes the singular points $p_1 = [0, 0, 0, 0, 1, e^{\pi i/4}]$ and $p_2 = [0, 0, 0, 0, 1, e^{5\pi i/4}]$ in Y, but it swaps round $p_3 = [0, 0, 0, 0, 1, e^{3\pi i/4}]$ and $p_4 = [0, 0, 0, 0, 1, e^{7\pi i/4}]$.

Thus, Condition 15.2.1 does not hold for Y and σ', as the fixed set $\{p_1, p_2\}$ of σ' does not coincide with the singular set $\{p_1, \ldots, p_4\}$ of Y. So let Y' be the blow-up of Y at p_3 and p_4. Then Y' is a partial crepant resolution of Y, as the singularities at p_3, p_4 are modelled on $\mathbb{C}^4/\mathbb{Z}_4$. The singularities of Y' are p_1, p_2, and σ' lifts to an antiholomorphic involution of Y' fixing only p_1 and p_2.

We find the Betti numbers of Y' by adding contributions to those of Y, as in §15.4.1. Applying the construction of §15.2 to Y' and σ' gives a simply-connected 8-manifold M', such that

Theorem 15.6.3 *This compact 8-manifold M' has Betti numbers*

$$b^0 = 1, \ b^1 = 0, \ b^2 = 1, \ b^3 = 0, \ b^4 = 908, \ b^4_+ = 614 \text{ and } b^4_- = 294.$$

There exist metrics with holonomy Spin(7) *on M', which form a smooth family of dimension* 295.

15.7 Examples from complete intersections in $\mathbb{CP}^6_{a_0,\ldots,a_6}$

Here are some examples based on the intersection of 2 hypersurfaces in $\mathbb{CP}^6_{a_0,\ldots,a_6}$.

15.7.1 *The intersection of two octics in* $\mathbb{CP}^6_{1,1,1,1,4,4,4}$

Let Y be the complete intersection of two octics in $\mathbb{CP}^6_{1,1,1,1,4,4,4}$ given by

$$Y = \{[z_0, \ldots, z_5] \in \mathbb{CP}^6_{1,1,1,1,4,4,4} : z_0^8 + z_1^8 + 2iz_2^8 - 2iz_3^8 + z_4^2 - z_5^2 = 0,$$
$$2iz_0^8 - 2iz_1^8 + z_2^8 + z_3^8 + z_4^2 - z_6^2 = 0\}.$$

Then $c_1(Y) = 0$. We find that Y has four singular points $p_1 = [0, 0, 0, 0, 1, 1, 1]$, $p_2 = [0, 0, 0, 0, 1, -1, -1]$, $p_3 = [0, 0, 0, 0, 1, 1, -1]$ and $p_4 = [0, 0, 0, 0, 1, -1, 1]$, satisfying Condition 15.2.1.

By modifying the method of Proposition 15.4.1 we can show that $\chi(Y) = 2580$, and applying Theorem 4.10.4 twice we find that $b^k(Y) = b^k(\mathbb{CP}^6_{1,\ldots,4})$ for $0 \leq k \leq 3$. Thus we prove:

Proposition 15.7.1 *The Betti numbers of Y are*

$$b^0 = 1, \quad b^1 = 0, \quad b^2 = 1, \quad b^3 = 0 \quad \text{and} \quad b^4 = 2576,$$

Also $Y \setminus \{p_1, \ldots, p_4\}$ *is simply-connected and* $h^{2,0}(Y) = 0$.

Define an antiholomorphic involution $\sigma : Y \to Y$ by

$$\sigma : [z_0, \ldots, z_6] \longmapsto [\bar{z}_1, -\bar{z}_0, \bar{z}_3, -\bar{z}_2, \bar{z}_4, \bar{z}_5, \bar{z}_6].$$

The fixed points of σ are exactly the singular points $p_1, \ldots, p_4$ of Y, and Condition 15.2.1 holds for Y and σ. Proceeding in the usual way, we set $Z = Y/\langle\sigma\rangle$ and resolve Z to get a simply-connected 8-manifold M, which satisfies:

Theorem 15.7.2 *This compact 8-manifold M has Betti numbers*

$$b^0 = 1, \quad b^1 = b^2 = b^3 = 0, \quad b^4 = 1294, \quad b^4_+ = 871 \text{ and } b^4_- = 423.$$

There exist metrics with holonomy Spin(7) *on M, which form a smooth family of dimension 424.*

15.7.2 A variation on this example

Now let Y be as in §15.7.1, but define $\sigma' : Y \to Y$ by

$$\sigma' : [z_0, \ldots, z_6] \longmapsto [\bar{z}_3, -\bar{z}_2, \bar{z}_1, -\bar{z}_0, \bar{z}_4, \bar{z}_6, \bar{z}_5].$$

Then σ' is an antiholomorphic involution, with fixed points p_1 and p_2, which swaps round p_3 and p_4. Following the method of §15.4.1, define Y' to be the blow-up of Y at p_3 and p_4. Then Y' is a Calabi–Yau orbifold, σ' lifts to Y', and Condition 15.2.1 holds for Y' and σ'.

As usual we set $Z' = Y'/\langle\sigma'\rangle$ and resolve Z' to get a simply-connected 8-manifold M', such that

Theorem 15.7.3 *This compact 8-manifold M' has Betti numbers*

$$b^0 = 1, \quad b^1 = 0, \quad b^2 = 1, \quad b^3 = 0, \quad b^4 = 1292, \quad b^4_+ = 870 \text{ and } b^4_- = 422.$$

There exist metrics with holonomy Spin(7) *on M', which form a smooth family of dimension 423.*

15.7.3 The intersection of two 12-tics in $\mathbb{CP}^6_{3,3,3,3,4,4,4}$

Let $P(z_4, z_5, z_6)$ and $Q(z_4, z_5, z_6)$ be generic homogeneous cubic polynomials with real coefficients, and define W to be the complete intersection of two 12-tics in $\mathbb{CP}^6_{3,3,3,3,4,4,4}$ given by

$$W = \big\{[z_0, \ldots, z_5] \in \mathbb{CP}^6_{3,3,3,3,4,4,4} : z_0^4 + z_1^4 + z_2^4 + z_3^4 + P(z_4, z_5, z_6) = 0,$$
$$iz_0^4 - iz_1^4 + 2iz_2^4 - 2iz_3^4 + Q(z_4, z_5, z_6) = 0\big\}.$$

Then $c_1(W) = 0$. As P and Q are generic, the singular set of W is the disjoint union of the 9 points $p_1, \ldots, p_9$ given by

$$\{[0, 0, 0, 0, z_4, z_5, z_6] \in \mathbb{CP}^6_{3,3,3,3,4,4,4} : P(z_4, z_5, z_6) = Q(z_4, z_5, z_6) = 0\},$$

and the curve Σ of genus 33 given by

$$\Sigma = \{[z_0, z_1, z_2, z_3, 0, 0, 0] \in \mathbb{CP}^6_{3,3,3,3,4,4,4} : z_0^4 + z_1^4 + z_2^4 + z_3^4 = 0,$$
$$iz_0^4 - iz_1^4 + 2iz_2^4 - 2iz_3^4 = 0\}.$$

Each point p_j satisfies Condition 15.2.1, and each point of Σ is modelled on $\mathbb{C} \times \mathbb{C}^3/\mathbb{Z}_3$, where the action of $\mathbb{Z}_3$ on $\mathbb{C}^3$ is generated by

$$\beta : (z_4, z_5, z_6) \mapsto (e^{2\pi i/3}z_4, e^{2\pi i/3}z_5, e^{2\pi i/3}z_6).$$

Define an antiholomorphic involution $\sigma : W \to W$ by

$$\sigma : [z_0, \ldots, z_6] \longmapsto [\bar{z}_1, -\bar{z}_0, \bar{z}_3, -\bar{z}_2, \bar{z}_4, \bar{z}_5, \bar{z}_6].$$

Then the fixed points of σ are some subset of $\{p_1, \ldots, p_9\}$. Exactly which subset depends on the choice of P and Q, but σ must fix an odd number of the p_j, as the remaining p_j are swapped in pairs.

So let σ fix $2k + 1$ of the p_j, for some $k = 0, \ldots, 4$, and number the p_j such that σ fixes $p_1, \ldots, p_{2k+1}$ and swaps $p_{2k+2}, \ldots, p_9$ in pairs. Define Y_k to be the blow-up of W along Σ and at the points $p_{2k+2}, \ldots, p_9$. Then Y_k is a partial crepant resolution of W. Thus Y_k is a Calabi–Yau orbifold, with singular points $p_1, \ldots, p_{2k+1}$. Also σ lifts to Y_k to give an antiholomorphic involution $\sigma : Y_k \to Y_k$ with fixed points $p_1, \ldots, p_{2k+1}$.

It can be shown that we can choose P and Q so that k takes any value in $\{0, 1, 2, 3, 4\}$. For example, if $P = z_4^3 - z_5^3$ and $Q = z_4^3 - z_6^3$ then σ fixes only $p_1 = [0, 0, 0, 0, 1, 1, 1]$, so that $k = 0$, but if $P = z_4^2 z_5 - z_5^3$ and $Q = z_4^2 z_6 - z_6^3$ then σ fixes the 9 points $[0, 0, 0, 0, 1, z_5, z_6]$ for $z_5, z_6 \in \{1, 0, -1\}$, and $k = 4$.

Combining the methods used to prove Propositions 15.5.5 and 15.7.1, we get

Proposition 15.7.4 *The Betti numbers of Y_k are $b^0 = 1$, $b^1 = 0$, $b^2 = 10 - 2k$, $b^3 = 66$, $b^4 = 395 - 2k$, $b^4_+ = 262$ and $b^4_- = 133 - 2k$. Also $Y_k \setminus \{p_1, \ldots, p_{2k+1}\}$ is simply-connected, and $h^{2,0}(Y_k) = 0$.*

In the usual way we resolve $Z_k = Y_k/\langle \sigma \rangle$ to get M_k, which satisfies

Theorem 15.7.5 *For each $k = 0, \ldots, 4$ there is a compact 8-manifold M_k with Betti numbers $b^0 = 1$, $b^1 = 0$, $b^2 = 4 - k$, $b^3 = 33$, $b^4 = 200 + 2k$, $b^4_+ = 132 + k$ and $b^4_- = 68 + k$. There exist metrics with holonomy* Spin(7) *on M_k, which form a smooth family of dimension $69 + k$.*

These examples have the largest value of b^3 and the smallest values of b^4 that the author has found using this construction.

Table 15.1 *Betti numbers* (b^2, b^3, b^4) *of the compact 8-manifolds with holonomy* Spin(7) *in this chapter*

(4, 33, 200)	(3, 33, 202)	(2, 33, 204)	(1, 33, 206)	(0, 33, 208)
(1, 0, 908)	(0, 0, 910)	(1, 0, 1292)	(0, 0, 1294)	(1, 0, 2444)
(0, 0, 2446)	(0, 6, 3730)	(0, 0, 4750)	(0, 0, 11 662)	

15.8 A discussion of our results, and questions for future research

In Table 15.1 we give the Betti numbers (b^2, b^3, b^4) of the compact 8-manifolds with holonomy Spin(7) that we constructed in §15.4–§15.7.

The examples of §15.4–§15.7 are by no means all the manifolds that can be produced using the methods of this chapter, but only a selection chosen for their simplicity and to illustrate certain techniques. Readers are invited to look for other examples themselves; the author would be particularly interested in examples which have especially large or small values of b^4.

We have also chosen to restrict our attention in §15.2–§15.7 to orbifolds Y all of whose singularities are modelled on $\mathbb{C}^4/\mathbb{Z}_4$, where the generator α of $\mathbb{Z}_4$ acts as in (15.3). This is not a necessary restriction, and there are other types of singularities for Y and Z for which the construction would work, such as the $\mathbb{R}^8/G^n$ considered in §15.1.3, and which occur in suitable orbifolds Y. However, the author has not found many such Y; the $\mathbb{C}^4/\mathbb{Z}_4$ singularities do seem to be the easiest to construct.

We finish with a few remarks about the present state of knowledge about compact manifolds with exceptional holonomy, and where the subject may go in future. Nearly all of Chapters 10–15 has a rather narrow focus: we have sought to construct examples of compact manifolds with holonomy G_2 and Spin(7), and to calculate their Betti numbers. The author has aimed to develop the theory and machinery far enough that such examples can be produced easily and in large numbers.

Our understanding of compact manifolds with exceptional holonomy is in an early stage of development, and is at present a little like stamp-collecting: we get hold of examples any way we can, prizing those that are unusual in some way, but we have no systematic understanding of the set of all such manifolds, or even of roughly how many of them exist.

Also, if we only know the Betti numbers of a compact manifold with exceptional holonomy then we do not really understand it, any more than one can understand a city by knowing its radius and the number of its inhabitants. Out in the middle of the moduli space, away from the degenerate, nearly singular examples that we are able to construct, these manifolds have a rich geometrical structure that we know almost nothing about, and have no ways to explore.

There are many unanswered questions about compact manifolds with exceptional holonomy; here are some the author thinks are interesting.

- Which compact 7- and 8-manifolds admit metrics with holonomy G_2 and Spin(7)? Are there finitely or infinitely many, up to diffeomorphism? What are their Betti numbers?

The author believes that there should be only finitely many of each, but quite a large number, in the tens of thousands at least.

- Is there an analogue of the 'Mirror Symmetry' of Calabi–Yau 3-folds for compact manifolds with exceptional holonomy? If so, does this lead to any predictions about the sets of Betti numbers realized by compact manifolds with holonomy G_2 and Spin(7)?

  Probably it is too much to expect that the sets of Betti numbers should display a reflection symmetry, as in the Calabi–Yau case. But perhaps one could predict some other regularities in the data on physical grounds.

- At present we have only a local understanding of the moduli space of holonomy G_2 or Spin(7) metrics on a fixed compact 7- or 8-manifold. It would be interesting to form a more global picture — to work towards a 'Torelli Theorem' for such manifolds. For instance, we could aim to describe the image of the moduli space in $H^3(M, \mathbb{R}) \times H^4(M, \mathbb{R})$ in the G_2 case, or in $H^4(M, \mathbb{R})$ in the Spin(7) case.

  At the boundary of the moduli space, the metrics must become singular in some way. Thus, we should try and understand what kind of singularities can develop in a family of compact G_2-manifolds or Spin(7)-manifolds. Having identified some kind of singularity, we should consider what features it produces in the moduli space.

  Orbifold singularities we know about, but what about other kinds? For example, the metrics of Bryant and Salamon [47] give local models for conical singularities in compact G_2- and Spin(7)-manifolds.

- Some singularities of G_2-manifolds occur when associative 3-folds or coassociative 4-folds collapse to zero volume. The image of the moduli space in $H^3(M, \mathbb{R}) \times H^4(M, \mathbb{R})$ then has boundary on a real hyperplane in $H^3(M, \mathbb{R})$ or $H^4(M, \mathbb{R})$. So in simple cases at least, the moduli space of holonomy G_2 metrics on M may have its boundary on a finite collection of hyperplanes in $H^3(M, \mathbb{R}) \times H^4(M, \mathbb{R})$. The same idea should work for Cayley 4-folds in Spin(7)-manifolds.

Another fertile source of problems are the compact associative 3-folds and coassociative 4-folds in compact G_2-manifolds, and compact Cayley 4-folds in compact Spin(7)-manifolds.

- Can a compact 7-manifold with holonomy G_2 be fibred by compact coassociative 4-folds, with generic fibre T^4 or $K3$, and some singular fibres? Can a compact 8-manifold with holonomy Spin(7) be fibred by compact Cayley 4-folds, with generic fibre $K3$, and some singular fibres?

- One could try to define invariants of compact G_2-manifolds by counting associative 3-folds N, or of compact Spin(7)-manifolds by counting isolated Cayley 4-folds N. In each case we probably need to count with some weight depending on the topology of N.

  Such invariants will only really be interesting if they are invariant under deformations of the underlying torsion-free G_2- or Spin(7)-structure, or at least transform

in a predictable way under such deformations. A similar problem was considered by the author in [129] for special Lagrangian 3-folds in Calabi–Yau 3-folds.
- To answer the previous two questions, it will be essential to have a good understanding of what kind of singularities can develop in familes of calibrated submanifolds in exceptional holonomy manifolds.

  In a *generic* situation, such singularities should develop only at finitely many isolated points, and should be modelled on singularities of noncompact calibrated submanifolds in $\mathbb{R}^7$ or $\mathbb{R}^8$, with suitable asymptotic behaviour. So the first steps in this programme should be to construct and study examples of singular calibrated submanifolds in $\mathbb{R}^7$ and $\mathbb{R}^8$.

The mathematics of String Theory, when it is full-grown, may well answer some of these questions.

REFERENCES

[1] B. S. Acharya. $N = 1$ heterotic/M-theory and Joyce manifolds. *Nuclear Physics*, B475:579–596, 1996. hep-th/9603033.

[2] B. S. Acharya. Dirichlet Joyce manifolds, discrete torsion and duality. *Nuclear Physics*, B492:591–606, 1997. hep-th/9611036.

[3] B. S. Acharya. On mirror symmetry for manifolds of exceptional holonomy. *Nuclear Physics*, B524:269–282, 1998. hep-th/9707186.

[4] B. S. Acharya, M. O'Loughlin, and B. Spence. Higher-dimensional analogues of Donaldson–Witten theory. *Nuclear Physics*, B503:657–674, 1997. hep-th/9705138.

[5] S. Agmon, A. Douglis, and L. Nirenberg. Estimates near the boundary for solutions of elliptic partial differential equations satisfying general boundary conditions II. *Communications on pure and applied mathematics*, 17:35–92, 1964.

[6] D. V. Alekseevsky. Riemannian spaces with exceptional holonomy. *Functional Analysis and its Applications*, 2:97–105, 1968.

[7] D. V. Alekseevsky. Classification of quaternionic spaces with a transitive solvable group of motions. *Mathematics of the USSR. Izvestija*, 9:297–339, 1975.

[8] W. Ambrose and I. M. Singer. A theorem on holonomy. *Transactions of the American Mathematical Society*, 75:428–443, 1953.

[9] P. S. Aspinwall, B. R. Greene, and D. A. Morrison. Multiple mirror manifolds and topology change in string theory. *Physics Letters*, B303:249–259, 1993.

[10] P. S. Aspinwall, B. R. Greene, and D. A. Morrison. Calabi–Yau moduli space, mirror manifolds and space-time topology change in string theory. *Nuclear Physics*, B416:414–480, 1994.

[11] P. S. Aspinwall, B. R. Greene, and D. A. Morrison. Space-time topology change and stringy geometry. *Journal of Mathematical Physics*, 25:5321–5337, 1994.

[12] P. S. Aspinwall, C. A. Lütken, and G. G. Ross. Construction and couplings of mirror manifolds. *Physics Letters*, B241:373–380, 1990.

[13] M. F. Atiyah and N. J. Hitchin. *The Geometry and Dynamics of Magnetic Monopoles*. Princeton University Press, Princeton, 1988.

[14] M. F. Atiyah and G. Segal. On the equivariant Euler characteristic. *Journal of Geometry and Physics*, 6:671–677, 1989.

[15] M. F. Atiyah and I. M. Singer. The index of elliptic operators. III. *Annals of Mathematics*, 87:546–604, 1968.

[16] T. Aubin. Métriques riemanniennes et courbure. *Journal of Differential Geometry*, 4:383–424, 1970.

[17] T. Aubin. *Nonlinear Analysis on Manifolds. Monge–Ampère equations*, volume 252 of *Grundlehren der mathematischen Wissenschaften*. Springer–Verlag, New York, 1982.

[18] S. Bando, A. Kasue, and H. Nakajima. On a construction of coordinates at infinity on manifolds with fast curvature decay and maximal volume growth. *Inventiones mathematicae*, 97:313–349, 1989.

[19] S. Bando and R. Kobayashi. Ricci-flat Kähler metrics on affine algebraic manifolds. In T. Sunada, editor, *Geometry and Analysis on Manifolds*, volume 1339 of *Lecture Notes in Mathematics*, pages 20–31. Springer–Verlag, 1988.

[20] S. Bando and R. Kobayashi. Ricci-flat Kähler metrics on affine algebraic manifolds. II. *Mathematische Annalen*, 287:175–180, 1990.

[21] W. Barth, C. Peters, and A. Van de Ven. *Compact Complex Surfaces*, volume 4 of *Ergebnisse der Mathematik und ihre Grenzgebiete*. Springer–Verlag, New York, 1984.

[22] R. Bartnik. The mass of an asymptotically flat manifold. *Communications on pure and applied mathematics*, 39:661–693, 1986.

[23] V. V. Batyrev. Variations of the mixed Hodge structure of affine hypersurfaces in algebraic tori. *Duke Mathematical Journal*, 69:349–409, 1993.

[24] V. V. Batyrev. Dual polyhedra and mirror symmetry for Calabi–Yau hypersurfaces in toric varieties. *Journal of Algebraic Geometry*, 3:493–535, 1994. alg-geom/9310003.

[25] V. V. Batyrev and D. I. Dais. Strong McKay correspondence, string-theoretic Hodge numbers and mirror symmetry. *Topology*, 35:901–929, 1996.

[26] L. Baulieu, H. Kanno, and I. M. Singer. Special Quantum Field Theories in eight and other dimensions. *Communications in Mathematical Physics*, 194:149–175, 1998. hep-th/9704167.

[27] A. Beauville. Some remarks on Kähler manifolds with $c_1 = 0$. In K. Ueno, editor, *Classification of Algebraic and Analytic Manifolds, 1982*, volume 39 of *Progress in Mathematics*, pages 1–26. Birkhäuser, 1983.

[28] A. Beauville. Variétés Kählériennes dont la première classe de Chern est nulle. *Journal of Differential Geometry*, 18:755–782, 1983.

[29] A. Beauville et al. Géometrie des surfaces $K3$: modules et périodes, Séminaire Palaiseau 1981–2. *Astérisque*, 126, 1985.

[30] K. Becker, M. Becker, D. R. Morrison, H. Ooguri, Y. Oz, and Z. Yin. Supersymmetric cycles in exceptional holonomy manifolds and Calabi–Yau 4-folds. *Nuclear Physics*, B480:225–238, 1996. hep-th/9608116.

[31] M. Berger. Sur les groupes d'holonomie homogène des variétés à connexion affines et des variétés riemanniennes. *Bulletin de la Société Mathématique de France*, 83:279–330, 1955.

[32] J. Bertin and D. G. Markushevich. Singularités quotients non abéliennes de dimension 3 et variétés de Calabi–Yau. *Mathematische Annalen*, 299:105–116, 1994.

[33] A. L. Besse. *Einstein Manifolds*. Springer–Verlag, New York, 1987.

[34] O. Biquard. Sur les équations de Nahm et la structure de Poisson des algèbres de Lie semi-simples complexes. *Mathematische Annalen*, 304:253–276, 1996.

[35] S. Bochner. Vector fields and Ricci curvature. *Bulletin of the American Mathematical Society*, 52:776–797, 1946.

[36] E. Bonan. Sur les variétés riemanniennes à groupe d'holonomie G_2 ou Spin(7). *Comptes Rendus de l'Académie des Sciences. Série A, Sciences Mathematiques*, 262:127–129, 1966.

[37] A. Borel. Some remarks about Lie groups transitive on spheres and tori. *Bulletin of the American Mathematical Society*, 55:580–587, 1949.

[38] R. Bott and L. W. Tu. *Differential forms in Algebraic Topology*, volume 82 of *Graduate Texts in Mathematics*. Springer–Verlag, New York, 1995.

[39] J. P. Bourguignon et al. Première classe de Chern et courbure de Ricci: preuve de la conjecture de Calabi, Séminaire Palaiseau 1978. *Astérisque*, 58, 1978.

[40] C. Boyer. A note on hyper-Hermitian four-manifolds. *Proceedings of the American Mathematical Society*, 102:157–164, 1988.

[41] R. Brown and A. Gray. Riemannian manifolds with holonomy group Spin(9). In S. Kobayashi et al., editors, *Differential Geometry (in honour of Kentaro Yano)*, pages 41–59, Tokyo, 1972. Kinokuniya.

[42] R. L. Bryant. Metrics with holonomy G_2 or Spin(7). In *Arbeitstagung Bonn 1984*, volume 1111 of *Lecture Notes in Mathematics*, pages 269–277. Springer–Verlag, 1985.

[43] R. L. Bryant. Metrics with exceptional holonomy. *Annals of Mathematics*, 126:525–576, 1987.

[44] R. L. Bryant. A survey of Riemannian metrics with special holonomy groups. In *Proceedings of the International Congress of Mathematicians, Berkeley, 1986*, volume 1, pages 505–514. American Mathematical Society, 1987.

[45] R. L. Bryant. Classical, exceptional, and exotic holonomies: a status report. In A. L. Besse, editor, *Actes de la Table Ronde de Géométrie Différentielle (Luminy, 1992)*, volume 1 of *Seminaires et Congres*, pages 93–165, Paris, 1996. Société Mathematique de France.

[46] R. L. Bryant and F. R. Harvey. Some remarks on the geometry of manifolds with exceptional holonomy. preprint, 1994.

[47] R. L. Bryant and S. M. Salamon. On the construction of some complete metrics with exceptional holonomy. *Duke Mathematical Journal*, 58:829–850, 1989.

[48] D. Burns and M. Rapoport. On the Torelli problem for Kählerian $K3$ surfaces. *Annales scientifiques de l'École Normale Superieure*, 8:235–274, 1975.

[49] E. Calabi. The space of Kähler metrics. In *Proceedings of the International Congress of Mathematicians, Amsterdam, 1954*, volume 2, pages 206–207. North–Holland, Amsterdam, 1956.

[50] E. Calabi. On Kähler manifolds with vanishing canonical class. In *Algebraic geometry and topology, a symposium in honour of S. Lefschetz*, pages 78–89. Princeton University Press, Princeton, 1957.

[51] E. Calabi. Métriques kählériennes et fibrés holomorphes. *Annales scientifiques de l'École Normale Superieure*, 12:269–294, 1979.

[52] P. Candelas, X. de la Ossa, and S. Katz. Mirror symmetry for Calabi–Yau hypersurfaces in weighted $\mathbb{P}_4$ and extensions of Landau–Ginzburg theory. *Nuclear Physics*, B450:267–290, 1995.

[53] P. Candelas, X. C. de la Ossa, P. S. Green, and L. Parkes. A pair of Calabi–Yau manifolds as an exactly soluble superconformal theory. In S.-T. Yau, editor, *Essays on Mirror Manifolds*, pages 31–95. International Press, Hong Kong, 1992.

[54] P. Candelas, G. T. Horowitz, A. Strominger, and E. Witten. Vacuum configurations for superstrings. *Nuclear Physics*, B258:46–74, 1985.

[55] P. Candelas, M. Lynker, and R. Schimmrigk. Calabi–Yau manifolds in weighted $\mathbb{P}_4^*$. *Nuclear Physics*, B341:383–402, 1990.

[56] M. Cantor. Elliptic operators and the decomposition of tensor fields. *Bulletin of the American Mathematical Society*, 5:235–262, 1981.

[57] A. Chaljub-Simon and Y. Choquet-Bruhat. Problèmes elliptiques du second ordre sur une variété euclidienne à l'infini. *Annales Faculté Sciences Toulouse*, 1:9–25, 1978.

[58] G. Cheeger and M. Gromoll. The splitting theorem for manifolds of nonnegative Ricci curvature. *Journal of Differential Geometry*, 6:119–128, 1971.

[59] Y. Choquet-Bruhat and D. Christodoulu. Elliptic systems in $H_{s,\delta}$ spaces on manifolds which are Euclidean at infinity. *Acta Mathematica*, 146:129–150, 1981.

[60] L. A. Cordero and M. Fernández. Some compact solvmanifolds of G_2 coassociative type. *Analele Universitatii Bucuresti Matematica*, 36:16–19, 1987.

[61] G. De Rham. Sur la réductibilité d'un espace de Riemann. *Commentarii Mathematici Helvetici*, 26:328–344, 1952.

[62] L. Dixon, J. Harvey, C. Vafa, and E. Witten. Strings on orbifolds I. *Nuclear Physics*, B261:678–686, 1985.

[63] L. Dixon, J. Harvey, C. Vafa, and E. Witten. Strings on orbifolds II. *Nuclear Physics*, B274:285–314, 1986.

[64] S. K. Donaldson and R. Friedman. Connected sums of self-dual manifolds and deformations of singular spaces. *Nonlinearity*, 2:197–239, 1989.

[65] S. K. Donaldson and P. B. Kronheimer. *The Geometry of Four-Manifolds*. OUP, Oxford, 1990.

[66] D. G. Ebin. The moduli space of Riemannian metrics. In *Global Analysis*, volume 15 of *Proceedings of Symposia in Pure Mathematics*, pages 11–40. American Mathematical Society, 1968.

[67] T. Eguchi and A. J. Hanson. Asymptotically flat solutions to Euclidean gravity. *Physics Letters*, 74B:249–251, 1978.

[68] M. Fernández. A classification of Riemannian manifolds with structure group Spin(7). *Annali di matematica pura ed applicata*, 143:101–122, 1986.

[69] M. Fernández. An example of a compact calibrated manifold associated with the exceptional Lie group G_2. *Journal of Differential Geometry*, 26:367–370, 1987.

[70] M. Fernández. A family of compact solvable G_2-calibrated manifolds. *Tohoku Mathematical Journal*, 39:287–289, 1987.

[71] M. Fernández and A. Gray. Riemannian manifolds with structure group G_2. *Annali di matematica pura ed applicata*, 132:19–45, 1982.

[72] M. Fernández and T. Iglesias. New examples of Riemannian manifolds with structure group G_2. *Rendiconti del Circolo Matematico di Palermo*, 35:276–290, 1986.

[73] J. M. Figueroa-O'Farrill. Extended superconformal algebras associated with manifolds of exceptional holonomy. *Physics Letters*, B392:77–84, 1997.

[74] R. Friedman. Simultaneous resolution of threefold double points. *Mathematische Annalen*, 274:671–689, 1986.

[75] R. Friedman. On threefolds with trivial canonical bundle. In *Complex Geometry and Lie Theory*, volume 53 of *Proceedings of Symposia in Pure Mathematics*, pages 103–134. American Mathematical Society, 1991.

[76] T. Friedrich, I. Kath, A. Moroianu, and U. Semmelmann. On nearly parallel G_2-structures. *Journal of Geometry and Physics*, 23:259–286, 1997.

[77] A. Fujiki. On primitively symplectic Kähler V-manifolds of dimension 4. In K. Ueno, editor, *Classification of Algebraic and Analytic Manifolds, 1982*, volume 39 of *Progress in Mathematics*, pages 71–250. Birkhäuser, 1983.

[78] W. Fulton. *Introduction to Toric Varieties*. Number 131 in Annals of Mathematics Studies. Princeton University Press, Princeton, 1993.

[79] K. Galicki. A generalization of the momentum mapping construction for quaternionic Kähler manifolds. *Communications in Mathematical Physics*, 108:117–138, 1987.

[80] K. Galicki and H. B. Lawson. Quaternionic reduction and quaternionic orbifolds. *Mathematische Annalen*, 282:1–21, 1988.

[81] G. W. Gibbons and S. W. Hawking. Gravitational multi-instantons. *Physics Letters*, 78B:430–432, 1978.

[82] G. W. Gibbons, D. N. Page, and C. N. Pope. Einstein metrics on S^3, $\mathbb{R}^3$ and $\mathbb{R}^4$ bundles. *Communications in Mathematical Physics*, 127:529–553, 1990.

[83] D. Gilbarg and N. S. Trudinger. *Elliptic partial differential equations of second order*, volume 224 of *Grundlehren der mathematischen Wissenschaften*. Springer–Verlag, Berlin, 1977.

[84] M. Goresky and R. MacPherson. *Stratified Morse Theory*, volume 14 of *Ergebnisse der Mathematik und ihrer Grenzgebiete*. Springer–Verlag, Berlin, 1988.

[85] M. B. Green and D. J. Gross, editors. *Unified String Theories*. World Scientific, Singapore, 1986.

[86] M. B. Green, J. H. Schwarz, and E. Witten. *Superstring Theory*, volume 1 and 2. Cambridge University Press, Cambridge, 1987.

[87] P. Green and T. Hübsch. Calabi–Yau manifolds as complete intersections in products of complex projective spaces. *Communications in Mathematical Physics*, 109:99–108, 1987.

[88] P. Green and T. Hübsch. Polynomial deformations and cohomology of Calabi–Yau manifolds. *Communications in Mathematical Physics*, 113:505–528, 1987.

[89] P. Green and T. Hübsch. Connecting moduli spaces of Calabi–Yau threefolds. *Communications in Mathematical Physics*, 119:431–441, 1988.

[90] B. Greene and S.-T. Yau, editors. *Mirror Symmetry II*, volume 1 of *A.M.S. studies in Advanced Mathematics*. International Press, 1997.

[91] B. R. Greene and M. R. Plesser. An introduction to mirror manifolds. In S.-T. Yau, editor, *Essays on Mirror Manifolds*, pages 1–27. International Press, Hong Kong, 1992.

[92] R. E. Greene and H. Wu. Lipschitz convergence of Riemannian manifolds. *Pacific Journal of Mathematics*, 131:119–141, 1988.

[93] P. Griffiths and J. Harris. *Principles of Algebraic Geometry*. Wiley, New York, 1978.

[94] H. A. Hamm. Lefschetz theorems for singular varieties. In *Singularities*, volume 40, part 1 of *Proceedings of Symposia in Pure Mathematics*, pages 547–557, Rhode Island, 1983. American Mathematical Society.

[95] J. Harris. *Algebraic Geometry, a first course*, volume 133 of *Graduate Texts in Mathematics*. Springer–Verlag, New York, 1992.

[96] R. Hartshorne. *Algebraic Geometry*, volume 52 of *Graduate Texts in Mathematics*. Springer–Verlag, New York, 1977.

[97] R. Harvey. *Spinors and Calibrations*, volume 9 of *Perspectives in Mathematics*. Academic Press, San Diego, 1990.

[98] R. Harvey and H. B. Lawson. A constellation of minimal varieties defined over G_2. In *Partial Differential Equations and Geometry*, volume 48 of *Lecture Notes in Pure and Applied Mathematics*, pages 167–187, 1979.

[99] R. Harvey and H. B. Lawson. Calibrated geometries. *Acta Mathematica*, 148:47–157, 1982.

[100] S. Helgason. *Differential Geometry and Symmetric Spaces*. Academic Press, New York, 1962.

[101] H. Hironaka. On resolution of singularities (characteristic zero). In *Proceedings of the International Congress of Mathematicians, Stockholm, 1962*, pages 507–521. Mittag–Leffler, 1963.

[102] F. Hirzebruch. Some examples of threefolds with trivial canonical bundle. In *Collected papers*, volume II, pages 757–770. Springer–Verlag, Berlin, 1987.

[103] F. Hirzebruch and T. Höfer. On the Euler number of an orbifold. *Mathematische Annalen*, 286:255–260, 1990.

[104] N. Hitchin. Polygons and gravitons. *Mathematical Proceedings of the Cambridge Philosophical Society*, 85:465–476, 1979.

[105] N. J. Hitchin, A. Karlhede, U. Lindström, and M. Roček. Hyperkähler metrics and supersymmetry. *Communications in Mathematical Physics*, 108:535–589, 1987.

[106] S. Hosono, A. Klemm, S. Theisen, and S.-T. Yau. Mirror symmetry, mirror map and applications to complete intersection Calabi–Yau spaces. *Nuclear Physics*, B433:501–554, 1995. hep-th/9406055.

[107] T. Hübsch. Calabi–Yau manifolds – motivations and constructions. *Communications in Mathematical Physics*, 108:291–318, 1987.

[108] T. Hübsch. *Calabi–Yau Manifolds, a bestiary for physicists*. World Scientific, Singapore, 1992.

[109] D. Huybrechts. Compact hyperkähler manifolds: basic results. *Inventiones mathematicae*, 135:63–113, 1999. alg-geom/9705025.

[110] D. Huybrechts. The Kähler cone of a compact hyperkähler manifold. math/9909109, 1999.

[111] S. Iitaka. *Algebraic Geometry, an introduction to birational geometry of algebraic varieties*, volume 76 of *Graduate Texts in Mathematics*. Springer-Verlag, New York, 1982.

[112] Y. Ito. Crepant resolution of trihedral singularities. *Proceedings of the Japan Academy. Series A, Mathematical Sciences*, 70:131–136, 1994.

[113] Y. Ito. Crepant resolution of trihedral singularities and the orbifold Euler characteristic. *International Journal of Mathematics*, 6:33–43, 1995.

[114] Y. Ito and M. Reid. The McKay correspondence for finite subgroups of $SL(3, \mathbb{C})$. In M. Andreatta et al., editors, *Higher Dimensional Complex Varieties (Trento, June 1994)*, pages 221–240. de Gruyter, 1996.

[115] J. Jost and H. Karcher. Geometrische methoden zur Gewinnung von a-priori-Schanker für harmonische Abbildungen. *Manuscripta Mathematica*, 40:27–77, 1982.

[116] D. D. Joyce. The hypercomplex quotient and the quaternionic quotient. *Mathematische Annalen*, 290:323–340, 1991.

[117] D. D. Joyce. Compact hypercomplex and quaternionic manifolds. *Journal of Differential Geometry*, 35:743–761, 1992.

[118] D. D. Joyce. Explicit construction of self-dual 4-manifolds. *Duke Mathematical Journal*, 77:519–552, 1995.

[119] D. D. Joyce. Manifolds with many complex structures. *Oxford Quarterly Journal of Mathematics*, 46:169–184, 1995.

[120] D. D. Joyce. Compact 8-manifolds with holonomy $Spin(7)$. *Inventiones mathematicae*, 123:507–552, 1996.

[121] D. D. Joyce. Compact Riemannian 7-manifolds with holonomy G_2. I. *Journal of Differential Geometry*, 43:291–328, 1996.

[122] D. D. Joyce. Compact Riemannian 7-manifolds with holonomy G_2. II. *Journal of Differential Geometry*, 43:329–375, 1996.

[123] D. D. Joyce. Compact manifolds with exceptional holonomy. In J. E. Andersen, J. Dupont, H. Pedersen, and A. Swann, editors, *Geometry and Physics*, volume 184 of *Lecture notes in pure and applied mathematics*, pages 245–252, New York, 1997. Marcel Dekker.

[124] D. D. Joyce. Compact manifolds with exceptional holonomy. In *Proceedings of the International Congress of Mathematicians, Berlin, 1998*, volume II, pages 361–370, University of Bielefeld, 1998. Documenta Mathematica.

[125] D. D. Joyce. Hypercomplex algebraic geometry. *Oxford Quarterly Journal of Mathematics*, 49:129–162, 1998.

[126] D. D. Joyce. On the topology of deformations of Calabi–Yau orbifolds. Duke server alg-geom/9806146, 1998.

[127] D. D. Joyce. Asymptotically Locally Euclidean metrics with holonomy $SU(m)$. To appear in Annals of Global Analysis and Geometry. math.AG/9905041, 1999.

[128] D. D. Joyce. A new construction of compact 8-manifolds with holonomy Spin(7). To appear in *Journal of Differential Geometry*. math.DG/9910002, 1999.

[129] D. D. Joyce. On counting special Lagrangian homology 3-spheres. hep-th/9907013, 1999.

[130] D. D. Joyce. Quasi-ALE metrics with holonomy $SU(m)$ and $Sp(m)$. To appear in *Annals of Global Analysis and Geometry*. math.AG/9905043, 2000.

[131] D. D. Joyce. Compact Riemannian manifolds with exceptional holonomy. In C. LeBrun and M. Wang, editors, *Essays on Einstein manifolds*, volume V of *Surveys in Differential Geometry*, pages 39–66. International Press, 2000.

[132] S. Katz. Rational curves on Calabi–Yau threefolds. In S.-T. Yau, editor, *Essays on Mirror Manifolds*, pages 168–180. International Press, Hong Kong, 1992.

[133] S. Kobayashi and K. Nomizu. *Foundations of Differential Geometry*, volume 1. Wiley, New York, 1963.

[134] S. Kobayashi and K. Nomizu. *Foundations of Differential Geometry*, volume 2. Wiley, New York, 1963.

[135] K. Kodaira. On the structure of compact complex analytic surfaces. I. *American Journal of Mathematics*, 86:751–798, 1964.

[136] K. Kodaira. *Complex Manifolds and Deformation of Complex Structures*, volume 283 of *Grundlehren der mathematischen Wissenschaften*. Springer–Verlag, New York, 1986.

[137] J. Kollar. The structure of algebraic threefolds: an introduction to Mori's program. *Bulletin of the American Mathematical Society*, 17:211–273, 1987.

[138] J. Kollar. Minimal models of algebraic threefolds: Mori's program. *Astérisque*, 177–178:303–326, 1989.

[139] A. Kovalev. Nahm's equations and complex adjoint orbits. *Oxford Quarterly Journal of Mathematics*, 47:41–58, 1996.

[140] P. B. Kronheimer. The construction of ALE spaces as hyperkähler quotients. *Journal of Differential Geometry*, 29:665–683, 1989.

[141] P. B. Kronheimer. A Torelli-type theorem for gravitational instantons. *Journal of Differential Geometry*, 29:685–697, 1989.

[142] P. B. Kronheimer. A hyper-kählerian structure on coadjoint orbits of a semisimple complex group. *Journal of the London Mathematical Society*, 42:193–208, 1990.

[143] P. B. Kronheimer. Instantons and the geometry of the nilpotent variety. *Journal of Differential Geometry*, 32:473–490, 1990.

[144] S. Lang. *Real Analysis*. Addison–Wesley, Reading, Massachusetts, second edition, 1983.

[145] H. B. Lawson and M. L. Michelson. *Spin Geometry*. Princeton University Press, Princeton, 1989.

[146] C. LeBrun. Anti-self-dual Hermitian metrics on blown up Hopf surfaces. *Mathematische Annalen*, 289:383–392, 1991.

[147] C. LeBrun. Explicit self-dual metrics on $\mathbb{CP}^2 \# \cdots \# \mathbb{CP}^2$. *Journal of Differential Geometry*, 34:223–253, 1991.

[148] C. LeBrun. Scalar-flat Kähler metrics on blown-up ruled surfaces. *Journal für die Reine und Angewandte Mathematik*, 420:161–177, 1991.

[149] C. LeBrun and S. Salamon. Strong rigidity of positive quaternion-Kähler manifolds. *Inventiones mathematicae*, 118:109–132, 1994.

[150] C. LeBrun and M. Singer. A Kummer-type construction of self-dual 4-manifolds. *Mathematische Annalen*, 300:165–180, 1994.

[151] J. M. Lee and T. H. Parker. The Yamabe problem. *Bulletin of the American Mathematical Society*, 17:37–91, 1987.

[152] C. Lewis. Spin(7) *instantons*. PhD thesis, Oxford University, 1998.

[153] A. Lichnerowicz. Spineurs harmoniques. *Comptes Rendus de l'Académie des Sciences. Série A, Sciences Mathematiques*, 257:7–9, 1963.

[154] R. B. Lockhart. Fredholm properties of a class of elliptic operators on non-compact manifolds. *Duke Mathematical Journal*, 48:289–312, 1981.

[155] E. Looijenga. A Torelli theorem for Kähler–Einstein $K3$ surfaces. In *Geometry Symposium, Utrecht, 1980*, volume 894 of *Lecture Notes in Mathematics*, pages 107–112. Springer–Verlag, 1982.

[156] D. Lüst and S. Theisen. *Lectures on String Theory*, volume 346 of *Lecture Notes in Physics*. Springer, New York, 1989.

[157] D. G. Markushevich, M. A. Olshanetsky, and A. M. Perelomov. Description of a class of superstring compactifications related to semi-simple Lie algebras. *Communications in Mathematical Physics*, 111:247–274, 1987.

[158] J. McKay. Graphs, singularities, and finite groups. *Proceedings of Symposia in Pure Mathematics*, 37:183–186, 1980.

[159] R. C. McLean. Deformations of calibrated submanifolds. *Communications in Analysis and Geometry*, 6:705–747, 1998.

[160] R. McOwen. The behaviour of the Laplacian on weighted Sobolev spaces. *Communications on pure and applied mathematics*, 32:783–795, 1979.

[161] S. Merkulov and L. Schwachhöfer. Classification of irreducible holonomies of torsion-free affine connections. Technical report, University of Leipzig, 1997.

[162] J. Milnor and J. Stasheff. *Characteristic classes.* Number 76 in Annals of mathematics studies. Princeton University Press, Princeton, 1974.

[163] D. Montgomery and H. Samelson. Transformation groups of spheres. *Annals of Mathematics*, 44:454–470, 1943.

[164] S. Mori. Birational classification of algebraic threefolds. In *Proceedings of the International Congress of Mathematicians, Kyoto, 1990*, volume 1, pages 235–248. Springer–Verlag, Tokyo, 1991.

[165] C. B. Morrey. *Multiple Integrals in the Calculus of Variations*, volume 130 of *Grundlehren der mathematischen Wissenschaften*. Springer–Verlag, Berlin, 1966.

[166] D. R. Morrison. Mirror symmetry and rational curves on quintic 3-folds: a guide for mathematicians. *Journal of the American Mathematical Society*, 6:223–247, 1993.

[167] D. R. Morrison and G. Stevens. Terminal quotient singularities in dimensions three and four. *Proceedings of the American Mathematical Society*, 90:15–20, 1984.

[168] L. Nirenberg and H. Walker. The null spaces of elliptic partial differential operators in $\mathbb{R}^n$. *Journal of Mathematical Analysis and Applications*, 42:271–301, 1973.

[169] T. Oda. *Convex Bodies and Algebraic Geometry.* Springer–Verlag, New York, 1988.

[170] K. O'Grady. Desingularized moduli spaces of sheaves on a $K3$, I. Duke server alg-geom/9708009, 1997.

[171] K. O'Grady. Desingularized moduli spaces of sheaves on a $K3$, II. Duke server alg-geom/9805099, 1998.

[172] D. N. Page. A physical picture of the K3 gravitational instanton. *Physics Letters*, 80B:55–57, 1978.

[173] G. Papadopoulos and P. Townsend. Compactification of $D = 11$ supergravity on spaces of exceptional holonomy. *Physics Letters*, B357:300–306, 1995.

[174] H. Pedersen and Y. S. Poon. Deformations of hypercomplex structures. *Journal für die Reine und Angewandte Mathematik*, 499:81–99, 1998.

[175] Y. S. Poon and S. M. Salamon. Quaternionic Kähler 8-manifolds with positive scalar curvature. *Journal of Differential Geometry*, 33:363–378, 1991.

[176] M. Reid. Minimal models of canonical 3-folds. In S. Iitaka, editor, *Algebraic Varieties and Analytic Varieties*, number 1 in *Advanced Studies in Pure Mathematics*, pages 131–180, Amsterdam, 1983. North–Holland.

[177] M. Reid. The moduli space of 3-folds with $K = 0$ may nevertheless be irreducible. *Mathematische Annalen*, 278:329–334, 1987.

[178] M. Reid. Young person's guide to canonical singularities. In *Algebraic Geometry, Bowdoin 1985*, number 46 in *Proceedings of Symposia in Pure Mathematics*, pages 345–416. American Mathematical Society, 1987.

[179] M. Reid. McKay correspondence. alg-geom 9702016, 1997.

[180] R. Reyes Carrión. A generalization of the notion of instanton. *Differential Geometry and its Applications*, 8:1–20, 1998.

[181] S.-S. Roan. On the generalization of Kummer surfaces. *Journal of Differential Geometry*, 30:523–537, 1989.

[182] S.-S. Roan. On $c_1 = 0$ resolution of quotient singularity. *International Journal of Mathematics*, 5:523–536, 1994.

[183] S.-S. Roan. Minimal resolution of Gorenstein orbifolds. *Topology*, 35:489–508, 1996.

[184] S. M. Salamon. Quaternionic Kähler manifolds. *Inventiones mathematicae*, 67:143–171, 1982.

[185] S. M. Salamon. Differential geometry of quaternionic manifolds. *Annales scientifiques de l'École Normale Superieure*, 19:31–55, 1986.

[186] S. M. Salamon. *Riemannian Geometry and Holonomy Groups*, volume 201 of *Pitman Research Notes in Mathematics*. Longman, Harlow, 1989.

[187] S. M. Salamon. On the cohomology of Kähler and hyper-Kähler manifolds. *Topology*, 35:137–155, 1996.

[188] S. M. Salamon. Quaternion-Kähler geometry. In C. LeBrun and M. Wang, editors, *Essays on Einstein Manifolds*, volume V of *Surveys in Differential Geometry*, pages 83–122. International Press, 2000.

[189] I. Satake. On a generalization of the notion of manifold. *Proceedings of the National Academy of Sciences of the U.S.A.*, 42:359–363, 1956.

[190] I. Satake. The Gauss–Bonnet Theorem for V-manifolds. *Journal of the Mathematical Society of Japan*, 9:464–492, 1957.

[191] M. Schlessinger. Rigidity of quotient singularities. *Inventiones mathematicae*, 14:17–26, 1971.

[192] S. L. Shatashvili and C. Vafa. Exceptional magic. *Nuclear Physics B Proceedings Supplement*, 41:345–356, 1995.

[193] S. L. Shatashvili and C. Vafa. Superstrings and manifolds of exceptional holonomy. *Selecta Mathematica*, 1:347–381, 1995.

[194] J. Simons. On the transitivity of holonomy systems. *Annals of Mathematics*, 76:213–234, 1962.

[195] Y. T. Siu. Every $K3$ surface is Kähler. *Inventiones mathematicae*, 73:139–150, 1983.

[196] P. Slodowy. *Simple Singularities and Simple Algebraic Groups*, volume 815 of *Lecture Notes in Mathematics*. Springer–Verlag, Berlin, 1980.

[197] P. Spindel, A. Sevrin, W. Troost, and A. Van Proeyen. Extended super-symmetric σ-models on group manifolds. *Nuclear Physics*, B308:662–698, 1988.

[198] A. Strominger and E. Witten. New manifolds for superstring compactification. *Communications in Mathematical Physics*, 101:341–361, 1987.

[199] C. H. Taubes. The existence of anti-self-dual conformal structures. *Journal of Differential Geometry*, 36:163–253, 1992.

[200] C. H. Taubes. *Metrics, Connections and Gluing Theorems*, volume 89 of *Regional Conference Series in Mathematics*. American Mathematical Society, Rhode Island, 1996.

[201] R. Thomas. *Gauge Theory on Calabi–Yau Manifolds*. PhD thesis, Oxford University, 1997.

[202] G. Tian. Smoothness of the universal deformation space of compact Calabi–Yau manifolds and its Peterson–Weil metric. In S.-T. Yau, editor, *Mathematical Aspects of String Theory*, volume 1 of *Advanced series in Mathematical Physics*, pages 629–646. World Scientific, 1987.

[203] G. Tian. Gauge theory and calibrated geometry, I. preprint, 1998.

[204] G. Tian and S.-T. Yau. Existence of Kähler–Einstein metrics on complete Kähler manifolds and their applications to algebraic geometry. In S.-T. Yau, editor, *Mathematical Aspects of String Theory*, volume 1 of *Advanced series in Mathematical Physics*, pages 574–628. World Scientific, 1987.

[205] G. Tian and S.-T. Yau. Three-dimensional algebraic manifolds with $c_1 = 0$ and $\chi = -6$. In S.-T. Yau, editor, *Mathematical Aspects of String Theory*, volume 1 of *Advanced series in Mathematical Physics*, pages 543–559. World Scientific, 1987.

[206] G. Tian and S.-T. Yau. Complete Kähler manifolds with zero Ricci curvature. I. *Journal of the American Mathematical Society*, 3:579–609, 1990.

[207] G. Tian and S.-T. Yau. Complete Kähler manifolds with zero Ricci curvature. II. *Inventiones mathematicae*, 106:27–60, 1991.

[208] A. N. Todorov. Applications of Kähler–Einstein–Calabi–Yau metrics to moduli of $K3$ surfaces. *Inventiones mathematicae*, 61:251–265, 1980.

[209] A. N. Todorov. The Weil–Petersson geometry of the moduli space of $SU(n \geq 3)$ (Calabi–Yau) manifolds I. *Communications in Mathematical Physics*, 126:325–346, 1989.

[210] P. Topiwala. A new proof of the existence of Kähler–Einstein metrics on $K3$. I. *Inventiones mathematicae*, 89:425–448, 1987.

[211] C. Vafa. Modular invariance and discrete torsion on orbifolds. *Nuclear Physics*, B273:592–606, 1986.

[212] C. Vafa. String vacua and orbifoldized LG models. *Modern Physics Letters*, A4:1169–1185, 1989.

[213] C. Vafa. Mirror transform and string theory. In S.-T. Yau, editor, *Geometry, Topology and Physics – for Raoul Bott*, pages 341–356, Boston, 1995. International Press.

[214] J. Varouchas. Kähler spaces and proper open morphisms. *Mathematische Annalen*, 283:13–52, 1989.

[215] M. Verbitsky. Cohomology of compact hyperkähler manifolds. alg-geom/9501001, 1995.

[216] M. Verbitsky. Cohomology of compact hyperkähler manifolds and its applications. *Geometry and Functional Analysis*, 6:601–612, 1996. alg-geom/9511009.

[217] M. Verbitsky. Hyperholomorphic sheaves and new examples of hyperkähler manifolds. alg-geom/9712012, 1997.

[218] M. Y. Wang. Parallel spinors and parallel forms. *Annals of Global Analysis and Geometry*, 7:59–68, 1989.

[219] F. W. Warner. *Foundations of Differentiable Manifolds and Lie Groups*. Scott, Foresman and co., Illinois, 1971.

[220] P. M. H. Wilson. Kähler classes on Calabi–Yau threefolds – an informal survey. In S.-T. Yau, editor, *Essays on Mirror Manifolds*, pages 265–278. International Press, Hong Kong, 1992.

[221] P. M. H. Wilson. The Kähler cone on Calabi–Yau threefolds. *Inventiones mathematicae*, 107:561–583, 1992.

[222] E. Witten. Physics and geometry. In *Proceedings of the International Congress of Mathematicians, Berkeley, 1986*, volume 1, pages 267–303. American Mathematical Society, 1987.

[223] J. Wolf. Complex homogeneous contact manifolds and quaternionic symmetric spaces. *Journal of Mathematics and Mechanics*, 14:1033–1047, 1965.

[224] H. Yamabe. On an arcwise connected subgroup of a Lie group. *Osaka Journal of Mathematics*, 2:13–14, 1950.

[225] S.-T. Yau. On Calabi's conjecture and some new results in algebraic geometry. *Proceedings of the National Academy of Sciences of the U.S.A.*, 74:1798–1799, 1977.

[226] S.-T. Yau. On the Ricci curvature of a compact Kähler manifold and the complex Monge–Ampère equations. I. *Communications on pure and applied mathematics*, 31:339–411, 1978.

[227] S.-T. Yau. Compact three dimensional Kähler manifolds with zero Ricci curvature. In W. A. Bardeen and A. R. White, editors, *Symposium on Anomalies, Geometry, Topology (Argonne, 1985)*, pages 395–406, Singapore, 1985. World Scientific.

[228] S.-T. Yau, editor. *Mathematical Aspects of String Theory*. World Scientific, Singapore, 1987.

[229] S.-T. Yau, editor. *Essays on Mirror Manifolds*. International Press, Hong Kong, 1992.

[230] E. Zaslow. Topological orbifold models and quantum cohomology rings. *Communications in Mathematical Physics*, 156:301–331, 1993.

INDEX

∗, see Hodge star

Â-genus, 67, 164, 259
age grading, 130–132, 200
ALE manifold, 173–174
   analysis on, 178–182
   asymptotic coordinates, 174
   Calabi conjecture for, 186–188
   Eguchi–Hanson space, 153–154, 160
   examples, 177–178, 393–394
   Hodge theory, 183–184
   holonomy Spin(7), 345–346
   hyperkähler, 148, 153–155
   Kähler, 174, 201
   radius function, 174, 175
   Ricci-flat Kähler, 175–178
   structure group Spin(7), 344, 393–394
   weighted Hölder space, 176, 179
   weighted Sobolev space, 178
Ambrose–Singer Holonomy Theorem, 32, 37
associative 3-fold, 70, 264–265, 417
   examples, 326–327
Asymptotically Locally Euclidean, see ALE manifold
Atiyah–Singer Index Theorem, 19, 67, 259

Berger's Theorem, 55
Betti number, 2
   refined, 62, 245, 260
Bianchi identities, 36, 43, 45, 59
blow-up, 91–92, 293
Bochner Theorem, 64, 124
bootstrap method, 116, 303, 364

C^k space, 5
$C^{k,\alpha}(M)$, see Hölder space
Calabi conjecture, 98–100
   for ALE manifolds, 186–188
   for QALE manifolds, 225, 229
Calabi–Yau manifold, 56, 70, 82, 98, 121–147, 304
   constructions, 138–142, 146
   definition, 123
   deformations, 142–144
   Hodge numbers, 125
Calabi–Yau orbifold, 135, 146, 395–396, 401–402
calibrated geometry, 68–70, 264–269
calibrated submanifold, 68

calibration, 68
canonical bundle, 94
Cayley 4-fold, 70, 267–268, 417
   examples, 371–373
Cayley numbers, see octonions
characteristic class, 94
   Â-genus, 67, 164, 259
   first Chern class, 94, 97, 98, 122
   first Pontryagin class, 246–247, 259, 261
Cheeger–Gromoll Theorem, 64
Chow's Theorem, 90, 95, 126
coassociative 4-fold, 70, 265–267, 417
   examples, 327–328
cohomology
   de Rham, 2, 182
   Dolbeault, 77
   Hodge numbers, 85
   of sheaves, 89, 93
   other cohomology theories, 2
   Poincaré duality, 3
   with compact support, 182
compact linear map, 6, 103, 179, 187, 226
complete intersection, 139, 402, 413
complex manifold, 72–75
   biholomorphism, 74
   holomorphic function, 72
   holomorphic map, 74
   rigid, 92, 132, 139
complex projective space, 73
complex structure, 72–73
   almost, 72
complex symplectic manifold, 162–164
   irreducible, 162
   marked, 163
   moduli space, 163
connection, 20–25
   Levi-Civita, 40, 42–43
   on principal bundle, 23
   on tangent bundle, 33–38
   on vector bundle, 22
   torsion-free, 35–38
continuity method, 104
crepant resolution, 126–132
   of $\mathbb{C}^m/G$, 175, 210
curvature
   and holonomy groups, 32–33
   in principal bundles, 24
   in vector bundles, 22–23
   of Kähler metrics, 81–82

Ricci, 44
Riemann, 43–44
scalar, 44

de Rham Theorem, 2
deformation, 92–93
  of $\mathbb{C}^m/G$, 132, 201, 238
  of Calabi–Yau manifold, 142–144
  smoothing, 92, 313, 318, 324
  universal, 93
  versal, 93
Dirac operator, 65–68, 259, 260
divisor, 96–97
  exceptional, 91, 127
  prime, 96
Dolbeault cohomology, 77
Douady space, 166, 215
double point, 127
  ordinary, 127
  rational, 129

Ebin's Slice Theorem, 252
Eguchi–Hanson space, 153–154, 160, 177, 212, 311, 313, 370, 374
elliptic equation
  nonlinear, 250
elliptic operator, 7–19
  definition, 9, 11
  existence of solutions, 16–19
  kernel finite-dimensional, 17
  L^p estimate, 14
  nonlinear, 10, 100
  regularity, 13–16, 297, 303, 364
  Schauder estimate, 14–15, 180
  symbol, 9, 11
exterior form, see form

Fano manifold, 168
flop, 128
form, 1–4
  complex symplectic, 150
  G-structure splitting, 59–60
  Hermitian, 78
  holomorphic volume, 122
  hyperkähler 2-form, 151
  Kähler, 79, 122
  of type (p,q), 77
  on Kähler manifolds, 82–84
Frobenius Theorem, 48

G-structure, 38
G_2, definition, 242
G_2 holonomy, 57–58, 242–254
  associative 3-fold, 70, 264–265, 326–327, 417
  coassociative 4-fold, 70, 265–267, 327–328, 417

holonomy subgroups, 245
metrics Ricci-flat, 244
parallel spinors, 245
research problems, 303–305, 416–418
G_2 holonomy, compact manifolds
  Betti numbers of, 314, 315, 323, 325, 326, 332–334, 339
  constructions, 294–295, 303–305
  examples, 309–326, 329–338
  finding $\pi_1(M)$, 307–308
  finding Betti numbers, 308–309
  moduli space of, 251–254
  topology of, 245–247, 306–309
  with $\pi_1(M) \neq \{1\}$, 314–315
G_2-manifold, 243
  product, 274
  QALE, 272–278
  with holonomy $SU(2)$, 271
  with holonomy $SU(3)$, 272
G_2-orbifolds T^7/Γ, 278–280
  examples, 312, 315, 316, 329, 334–337
  R-data, 280–281
  resolutions of T^7/Γ, 282–284
G_2-structure, 243
  function Θ, 243, 249
  nearly parallel, 269
  positive 3-form, 243
  positive 4-form, 243
  small torsion, 284, 286
  splitting of forms, 244, 245
  torsion, 243
  torsion-free, 243
Green's representation, 14, 181

$\mathbb{H}$, see quaternions
harmonic coordinates, 298
Hilbert scheme, 166, 215
Hodge numbers, 85
  of hyperkähler manifolds, 164
Hodge star, 3, 60
  on Kähler manifolds, 83
Hodge theory, 4, 61–63
  on ALE manifolds, 183–184
  on Kähler manifolds, 84–85
Hölder space, 5–6
  weighted, 176, 179, 221
Hölder's inequality, 4
holomorphic vector bundle, 77
holonomy algebra, 28, 30, 45
holonomy group
  and cohomology, 59–64
  and curvature, 32–33
  classification, 55–59
  constant tensors, 34–35
  definition, 26, 29, 45

holonomy group *cont.*
    exceptional, *see* G_2 holonomy, Spin(7) holonomy, etc.
    for principal bundles, 29–32
    for vector bundles, 25–29
    restricted, 28, 29, 45
    Ricci-flat, 58
    Riemannian, 44–45
hypercomplex algebraic geometry, 171, 215
hypercomplex manifold, 148, 167, 170
hyperkähler manifold, 57, 98, 148–167, 169
    ALE, 148, 153–155
    examples, 165–167
    Hodge numbers, 164
    moduli space, 165
    QALE, 211, 213–215, 239
    twistor space, 151–152
hyperkähler quotient, 154, 169
hyperkähler structure, 148, 151
hypersurface, 96
    Calabi–Yau, 139
    degree d, 139
    in toric variety, 142
    in weighted projective space, 140, 401

Implicit Mapping Theorem, 7, 253, 263
injectivity radius, 5, 221, 292–295, 298, 359–361, 399
instanton, 169, 269
interior estimate, 16
intrinsic torsion, 40
Inverse Mapping Theorem, 7, 118, 228

$K3$ surface, 155–162
    examples, 156
    marked, 157
    moduli space, 157, 161
Kodaira Embedding Theorem, 96, 125
Kondrakov Theorem, 7, 16, 103
Kummer construction, 137, 156–157, 160–161
Kuranishi family, 94, 143
Kähler chamber, 158
Kähler class, 80, 85, 175, 210
Kähler cone, 85, 158, 175, 210
Kähler form, 79, 122
Kähler manifold, 56
Kähler metric, 79
Kähler potential, 80
Kähler potential type, 215, 229

$L^p(M)$, *see* Lebesgue space
$L^p_k(M)$, *see* Sobolev space
Laplacian, 3, 7, 10, 14, 61
    on ALE manifolds, 180–182
    on Kähler manifolds, 83, 105, 198

    on QALE manifolds, 220–225
Lebesgue space, 4–5
Lefschetz Hyperplane Theorem, 97, 140, 404
Levi-Civita connection, 40, 42–43
line bundle, 94–96
    ample, 95, 125
    canonical, 94, 122, 126
    first Chern class, 94, 97, 98, 122
    over $\mathbb{CP}^m$, 94
    positive, 96
    very ample, 95

maximum principle, 16, 222
McKay correspondence, 129, 131, 154, 309
metric
    Fubini–Study, 53, 80
    Hermitian, 78
    hyperkähler, 151
    Kähler, 79
    reducible, 46–50
    symmetric, 50–55
mirror symmetry, 142, 145–146
moduli space, 63, 169
    of complex symplectic manifolds, 163
    of G_2-manifolds, 251–254
    of hyperkähler manifolds, 165
    of $K3$ surfaces, 157, 161
    of Spin(7)-manifolds, 261–264
Monge–Ampère equation, 100

Newlander–Nirenberg Theorem, 72
Nijenhuis tensor, 72
node, 127

$\mathbb{O}$, *see* octonions
octonions, 57
orbifold, 132–136
    Calabi–Yau, 135, 146, 395–396, 401–402
    complex, 133, 166
    crepant resolution, 136–139
    group, 133
    Kähler, 134
    of torus, 137, 278–280, 346–347
    point, 133
    real, 133

parallel transport, 26
period domain, 158, 163
    hyperkähler, 161, 165
period map, 143, 157, 163
    hyperkähler, 161, 165
Picard group, 94
Poincaré duality, 3
principal bundle, 20
    connection, 23

QALE manifold
 analysis on, 219–225
 Calabi conjecture for, 225, 229
 definition, 208, 236–238
 examples, 212–215, 239–241, 276–278
 generalized, 235–241
 holonomy G_2, 277–278
 holonomy Sp(m), 211, 213–215, 239
 holonomy Spin(7), 345–346
 holonomy SU(m), 211–213, 239
 Kähler, 208–210, 238
 Kähler potentials, 215
 radius functions, 219
 Ricci-flat Kähler, 210–215, 238–239
 structure group G_2, 272–278
 structure group Spin(7), 344–346
 weighted Hölder space, 221
 weighted Sobolev space, 220
Quasi-ALE, see QALE manifold
quaternionic Kähler manifold, 57, 148, 167–168
quaternionic manifold, 148, 168–170
quaternions, 57, 149

R-data, 280–281, 346
Reduction Theorem, 31
resolution, 91
 crepant, 126–132
 local product, 205–207
 of T^7/Γ, 282–284
 of T^8/Γ, 348
 of $\mathbb{C}^m/G$, 128–132
 real, 236, 344
 real local product, 236, 274, 344
 small, 127
Ricci curvature, 44, 81
 and 1-forms, 63–64
Ricci form, 82, 98, 123
Riemann curvature, 43–44
Riemannian holonomy, see holonomy group
Riemannian metric, see metric

scalar curvature, 44
Schauder estimate, 14–15
scheme, 90
Schlessinger Rigidity Theorem, 132
self-dual 4-manifold, 168–170
sheaf, 88–89
 cohomology, 89, 93
 invertible, 126
singularity
 canonical, 127
 crepant resolution, 126–132
 deformation, 92–93
 Kleinian, 129
 nonisolated, 204
 of variety, 90
 resolution, 91
 terminal, 127, 130, 200
 smoothing, 92, 313, 318, 324
Sobolev Embedding Theorem, 6, 297
Sobolev space, 4–5
 weighted, 178, 220
Sp(m) holonomy, see hyperkähler manifold
special Lagrangian submanifold, 70, 304, 418
spin geometry, 64–68
spin structure, 65
Spin(7), definition, 255
Spin(7) holonomy, 57–58, 254–264
 Cayley 4-fold, 70, 267–268, 371–373, 417
 holonomy subgroups, 256
 metrics Ricci-flat, 256
 parallel spinors, 256
 research problems, 416–418
Spin(7) holonomy, compact manifolds
 Betti numbers of, 370, 380, 385, 390, 416
 constructions, 360–362, 395–401
 examples, 368–371, 373–391, 406–415
 moduli space of, 261–264
 topology of, 259–261, 366–368, 404–405
Spin(7)-manifold, 255
 ALE, 344, 393–394
 product, 343
 QALE, 344–346
 with holonomy SU(2), 342
 with holonomy SU(3), etc., 343
Spin(7)-orbifolds T^8/Γ, 346–347
 examples, 369, 373, 381, 386
 R-data, 346
 resolutions, 348
Spin(7)-orbifolds $Y/\langle\sigma\rangle$, 395–396
 examples, 406–415
 resolutions, 398
Spin(7)-structure, 255
 admissible 4-form, 255
 function (·), 257–258
 small torsion, 355, 398
 splitting of forms, 256, 260
 torsion, 255
 torsion-free, 255
spinor, 65
 harmonic, 67
 parallel, 66, 245, 256
Stokes' Theorem, 2
String Theory, 144–146
SU(m) holonomy, see Calabi–Yau manifold
symmetric space, 50–55
 constant curvature, 54
 definition, 50

Torelli Theorem
　Global, 158, 164
　Local, 158, 163
　Weak, 158
toric geometry, 129
torsion, 35–36
　of G-structure, 40
twistor space, 151–152, 168, 169

variety, 86–90
　abstract, 89
　affine, 87
　algebraic, 74–75, 87
　analytic, 90
　blowing up, 91–92
　deformation, 92–93
　morphism, 88
　projective, 87
　rational map, 88
　resolution, 91
　singular, 90
　toric, 129
vector bundle, 20
　connection, 22
　holomorphic, 77
　line bundle, *see* line bundle

weighted projective space, 133, 140, 401
Weitzenbock formula, 61, 125
Weyl group, 154–155, 241
Wolf space, 168

Zariski topology, 86, 87